P9-DGC-201

Terrain Analysis

COMMUNITY DEVELOPMENT SERIES

RICHARD P. DOBER, AIP, Series Editor

Bechtel/Enclosing Behavior
Becker/Housing Messages
Bednar/Barrier-Free Environments
Bell, Randall, and Roeder/Urban Environments and Human Behavior: An Annotated Bibliography
Branch/Planning Urban Environment
Branch/Urban Planning Theory
Coates/Alternative Learning Environments
Conway/Human Response to Tall Buildings
Finsterbusch and Wolf/Methodology of Social Impact Assessment
Golany/Strategy for New Community Development in the United States
Hesselgren/Man's Perception of Man-Made Environment: An Architectural Theory
Hester/Neighborhood Space: User Needs and Design Responsibility
Koncelik/Designing the Open Nursing Home
Küller/Architectural Psychology: Proceedings of the Lund Conference
Lang, Burnette, Moleski, and Vachon/Designing for Human Behavior: Architecture and the Behavioral Sciences
LaPatra/Applying the Systems Approach to Urban Development
Lawton, Newcomer, and Byerts/Community Planning for an Aging Society: Designing Services and Facilities
Michelson/Behavioral Research Methods in Environmental Design
Moore and Golledge/Environmental Knowing: Theories, Research, and Methods
Perks and Robinson/Canadian Planning Practices
Plēsums/Townframe: Environments for Adaptive Housing
Pollowy/The Urban Nest
Popko/Transitions: A Photographic Documentary of Squatter Settlements
Preiser/Facility Programming: Methods and Applications
Procos/Mixed Land Use: From Revival to Innovation
Rondinelli/Planning Development Projects
Sanoff/Methods of Architectural Programming
Sanoff/Designing with Community Participation
Sleeman and Rockwell/Instructional Media and Technology: A Professional's Resource
Stearns and Montag/The Urban Ecosystem: A Holistic Approach
Unesco/Planning Buildings and Facilities for Higher Education
Way/Terrain Analysis: A Guide to Site Selection Using Aerial Photographic Interpretation, 2nd ed.
Zube, Brush, and Fabos/Landscape Assessment: Values, Perceptions, and Resources

Terrain Analysis

A Guide to Site Selection Using Aerial Photographic Interpretation

Second Edition

Douglas S. Way
Graduate School of Design
Harvard University

CDS/1

DOWDEN, HUTCHINSON & ROSS, INC.
Stroudsburg, Pennsylvania

To
Bobbie, Danielle, Nicole, Noelle, and Stewart

TERRAIN ANALYSIS

85 84 83 2 3 4 5

Library of Congress Cataloging in Publication Data

Way, Douglas S
Terrain analysis.
(Community development series ; v. 1)
Includes bibliographies and index.
1. Landforms. 2. Petrology. 3. Building sites.
4. Aerial photography in geomorphology. I. Title.
GB406.W39 1978 551.4′028 77-20240
ISBN 0-87933-318-9

Distributed worldwide by Van Nostrand Reinhold Company Inc.,
135 W. 50th Street, New York, NY 10020.

Contents

Series Editor's Foreword

It seems fitting and appropriate that the first book in the Community Development Series be concerned about land, the forms it takes and the capabilities it has for satisfying human needs and uses. These include—as they must—the conservation and preservation of natural resources, restoration and renewal, and the development of new uses on the land.

While not neglecting the full opportunities it is in the latter category that Douglas Way makes a very special contribution to the literature. Without exhortation about past errors or mysticism about the future he provides the tools for solving very basic land use problems. Fundamental information is systematically organized. Techniques are straightforwardly described. The reader can sense for himself when additional help is needed, and where to get it.

Unquestionably this is a professional reference work. But it comes from a landscape architect with a sensitive eye, one who has crossed the boundaries of several disciplines to write about environmental issues and consequences. For that reason many non-professional readers will also benefit from reading this book, for the crossing was worth the effort, and the author describes it well.

Richard P. Dober, AIP

Preface

When the first edition of *Terrain Analysis* was published over five years ago, the perceived primary market was engineering and land planning offices. However, both the author and publisher were surprised by the wide and rapid acceptance of the book by professionals and educators. Although not originally conceived as a text, *Terrain Analysis* has been used by many groups to aid in the teaching of the landform technique. This growing demand has prompted the creation of a second edition which has been reorganized, updated, and enlarged to facilitate greater professional and academic use. A major section, Chapter 1, on remote sensing that has been added defines and summarizes the "state of the art." Chapter 2 outlines the approach to terrain analysis and has a comprehensive bibliography, and Chapter 3 describes detailed techniques of soil mapping. A section on mass wasting has been enlarged and several case studies have been added to illustrate the application of the terrain analysis technique. In addition, several more appendixes have been included, thus increasing references to various sources and hard-to-find supporting data. It is hoped that the second edition will continue to provide useful information to those interested in the teaching and application of terrain analysis.

I had two goals in mind while researching and writing this book. The first was to establish a systematic approach to terrain analysis which planners and other professionals involved in analyzing and making decisions about land areas could follow in order to achieve an understanding of land surface and subsurface conditions relevant to their program needs. A second goal was to provide a useful reference for planners, architects, and other professionals or students who are not specialists in geology or geomorphology but who wish to learn an interdisciplinary approach to the issues confronted in their professions.

Terrain analysis as described in this book is based upon the identification and interpretation of landforms caused by geological processes. The definition of a landform, developed in Chapter 2 and used throughout this volume, is: A landform is a terrain feature formed by natural processes, which has a definable composition and range of characteristics that occur wherever that landform is found. Landforms reflect similar surface-subsurface conditions; therefore, by identifying the landforms of a site, a land planner can acquire an understanding of the physical conditions of the site and can begin to identify problem areas. The content of landform terrain analysis touches the fields of geology, engineering, geomorphology, and pedology, by applying information and techniques from these professions to the data needs and concerns of the land planner and indicating where special consultants—geologists, engineers, soil scientists, hydrologists, ecologists—are needed to supplement the planner's knowledge and to help in decision making. It is hoped that this book will bridge the gaps now existing between many of these professions.

The major emphasis of this volume is on utilizing aerial photographic analysis as a technique for the identification and interpretation of landforms. This technique is based upon the systematic study of visual elements easily observed from aerial photographs, including topographic shape, drainage pattern, gully characteristics, photographic color or tone, erosional features, landform boundaries, land use types and distribution, vegetation types and distribution, and any other special features that may be present. These visual, or pattern, elements occur in combinations that are unique to each landform. Thus by following this "overview" technique of aerial photographic interpretation based on recognition of the pattern elements, the user quickly becomes familiar with the range of physical conditions of a site and can identify problem areas for detailed analysis.

The material presented in this book covers portions of the fields of geology, geomorphology, pedology, engineering and planning, and no one person could possess detailed knowledge of all of them. Therefore, I had to rely upon previous research by many individuals in the different professions to provide much of the information presented here. Their contributions are acknowledged in the reference sections accompanying each chapter.

Size and cost limitations have required that much of the subject matter be generalized or summarized, so that only a basic context or range of conditions that can be expected are presented. It must be assumed that a reader familiar with a specific physiographic region will be able to generate any additional data required, either by a more detailed photographic analysis or by the use of the proper consultants.

The book can be divided into three sections. Chapters 1 through 3 contain background information; Chapters 4 through 11 contain reference material for each landform; and Chapter 12 presents several case studies.

Chapter 1 presents a brief history and current technology in the field of remote sensing. The process and techniques of remote sensing and photointerpretation are discussed along with a summary of relevant applications to terrain analysis. In addition an extensive bibliography has been prepared focusing on remote sensing–terrain applications.

Chapters 2 and 3 more specifically introduce the technique of terrain analysis. They explain the principles of landform interpretation and describe the process of terrain analysis employing aerial photography. Included are definitions of the visual pattern elements by which landforms are identified and detailed soil conditions are mapped.

Chapters 4 through 11 comprise the second section. Chapter 4 is an introductory one that examines the engineering and construction issues associated with site development. Physical site requirements and engineering solutions are discussed for sewage disposal utilizing septic tank leaching fields, solid waste disposal through the use of sanitary land fills, trenching, excavation and grading, dewatering, sources of construction materials, compaction of fills, mass wasting and landslide susceptibility, groundwater supply, pond or lake construction, foundations, and highway construction.

For purposes of analysis, the individual landforms have been grouped into six categories: sedimentary, igneous, metamorphic, glacial, eolian (windlaid), fluvial (waterlaid), and mass wasting. Each of Chapters 5 through 11 is concerned with the landforms of one such category. An introduction to each chapter gives the characteristics of that group of landforms and provides a map showing its distribution throughout the United States and the world. At the end of the introduction a photographic interpretation chart summarizes the pattern elements associated with that group of landforms.

Each landform chapter contains a number of subsections, one for each landform type associated with that grouping. For example, under Metamorphic Rocks (Chapter 7) there are three landform types—slate, schist, and gneiss. Each landform is identified in terms of:

(1) Its classifications, subclassifications, and characteristic pattern elements (shown in table form); (2) its soil characteristics and their classifications; (3) its engineering and planning capabilities; and (4) photographic or cartographic examples (shown as a table listing government agency sources).

Chapter 12 includes several case studies illustrating the landform technique applied to different types and scales of projects. These range from fifty to over 26,000 acres and include site studies for housing, ski areas, recreation communities, and sand and gravel excavation.

Appendixes A, B, C, D, and E provide further supporting physical resource information useful to those involved with terrain analysis issues. Appendix A describes different soil classification schemes and procedures for field identification. Appendix B outlines data sources and governmental agencies that can provide aerial photographs and remote sensing images. Appendix C lists by county the published soil surveys available at this time from the Soil Conservation Service. Appendix D lists by state the geological and surficial geological maps available from state or federal agencies. Appendix E lists other physical resource maps which are important for classroom or reference use.

At the end of each landform chapter introduction, an aerial photographic interpretation summary chart illustrates the pattern elements related to that particular grouping of landforms. When attempting to identify an unknown landform, it may be useful to refer to these charts before consulting the specific landform sections for more detail.

Cambridge, Mass. Douglas S. Way

Acknowledgments

It is impossible to give proper credit to all the people who have contributed in one way or another to the creation of this book. Much of my basic philosophy about terrain analysis and the landform approach is derived from that of Donald J. Belcher, the Director of the Center for Aerial Photographic Studies at Cornell University. Many of his ideas, consciously or unconsciously, appear throughout this book, and his contributions and suggestions were sincerely appreciated. Among the others who were especially helpful and provided many suggestions were Olin W. Mintzer, Ohio State University, and Ernest E. Hardy, Cornell University, not to mention the geologists, engineers, planners and landscape architects with whom I have had numerous discussions.

Photographs were graciously supplied by the U.S. Geological Survey, the Agricultural Stabilization Conservation Service and Aerial Data Reductions, Riverside, New Jersey. Linda Somers of the Agricultural Stabilization Conservation Service was very helpful in researching photo coverage and in obtaining high quality prints. Becky Rollinger of the EROS Data Center was very helpful in providing requested images.

Many useful comments concerning the manuscript were given by Mr. Richard P. Dober who reviewed it in its entirety. The author appreciates the suggestions of Mr. Roy E. Hunt, professional engineer and partner in Joseph S. Ward and Associates, whose comments were useful in refining the technical engineering content.

I would like to express special appreciation to Joan Taylor and Susan Holmes, who labored diligently over this vast manuscript and helped me to organize much of the basic thought and format, and to Jean Poehler who skilfully typed the manuscript. The excellent illustrations and block diagrams were done by Benjamin Johnson and André Rojas, who had an inherent feel for the character of the various landscapes presented, and their graphic crew of Robert Fraser, Thierry Sprecher, and Timothy Dreese. Finally, special thanks is expressed to my wife and family, for helpful suggestions, for proofreading, and for living with scattered papers and books for the last year.

Appreciation and thanks are also given to all those involved in the case studies briefly outlined in Chapter 12, including Sasaki Associates, David P. Sugarman, Mt. Washington Development Corporation, Quechee Lakes Corporation and Planning Associates Collaborative.

To all these individuals my sincere thanks, for without their special contributions the completion of this book would not have been possible.

Research for this book was funded, in part, by: The National Endowment for the Arts: Washington, D.C. and The Milton Fund, Harvard University, Cambridge, Massachusetts. Generous support was provided by the Department of Landscape Architecture, Harvard University.

Part I

Remote Sensing and Landform Interpretation

CHAPTER 1

REMOTE SENSING

INTRODUCTION

The American Society of Photogrammetry (1975) defines *remote sensing* as

> . . . the measurement or acquisition of information on some property of an object or phenomenon by a recording device that is not in physical or intimate contact with the object or phenomenon under study; e. g. the utilization at a distance (as from aircraft, spacecraft, or ship) of any device and its attendant display for gathering information pertinent to the environment, such as measurements of force fields, electromagnetic radiation, or acoustic energy. The technique employs such devices as the camera, lasers, and radio frequency receivers, radar systems, sonar, seismographs, gravimeters, magnetometers, and scintillation counters.

For the applications discussed in this book, remote sensing associations with the electromagnetic spectrum will be emphasized. Remote sensing is the process of indirectly gathering information by using aerial photographs or images coupled with a process of inferential interpretation. Its role is merely to help supply information concerning an object or place. Remote sensing is not meant to be a replacement process; rather, it is to be integrated with a series of techniques that most effectively achieve a desired data collection task. It has been shown that professionals using the tools involved can organize and analyze data more quickly and accurately; remote sensing thus creates more time for planning and design services (Way, 1977). The technology of remote sensing has become very sophisticated with the advent of the space program. The material presented in this chapter attempts to provide a general overview of process, hardware, and issues relevant to applied terrain analysis.

History

The history of remote sensing is a vast subject; therefore, emphasis in this discussion is on the development of aerial photography. Not until 1839 were the first photographs taken in France; shortly thereafter in 1840 the first applications to topographic mapping were conceived by the Directory of the Paris Observatory. Early balloon flights were initiated by Gaspard Felix Tournachon who in 1858 ascended over Paris for the purpose of creating a topographic map. By 1871 photo technology had advanced to the stage where immediate development of the film was no longer necessary after exposure so that just the camera and film could be taken aloft. This new technology also facilitated the design of lighter weight cameras, such as those patented by Julius Neubronner in 1903, which were light enough to be carried by pigeons. During the late 1800s in the United States G. R. Lawrence developed kites as aerial platforms able to carry cameras in excess of 1,000 pounds. The first photographs from an airplane were (appropriately) taken by Wilbur Wright in 1909 over Centrocelli, Italy.

These early attempts at aerial photography were more a novelty than a serious science until the later stages of World War I when the value of aerial photography to intelligence-gathering efforts was demonstrated. The Second World War provided the impetus for continued refinement of cameras, lenses, mounts, and reconnaissance techniques. Near the war's end no major invasion or troop movement was initiated without first interpreting and analyzing aerial photographs of the affected territory. During these war years aerial photography was also recognized for its potential value in mineralogic investigation and mapping. Petroleum geologists, governmental agencies, and various oil companies used photographs as standard tools for preliminary reconnaissance. Systematic techniques of terrain analysis were developed in the 1940s and 1950s as many different professions recognized the contributions offered by photographic interpretation. The advent of the space program in the 1960s had a tremendous impact on the development of remote sensing hardware and resultant applications. Today in most professions remote sensing plays a significant role in data generation.

Purpose and Process

Civil engineers, landscape architects, and land planners are involved in many types of professional projects at a variety of scales and degrees of complexity. Generally, there is a site area and a development program for placement and design. To facilitate the required decisions, knowledge of the physical site conditions is necessary. The role of remote sensing is to help supply this information.

For the purpose of defining specific data needs, a project can be divided into its component issues and/or subproblems. Within the constraints of time, budget, and technology, the analysis techniques to be used are identified. Next, the major data categories that are required by the analysis procedures are identified. This identification results in a comprehensive data listing for the entire project. The status of the existing data is compared to the data requirements to determine the remaining data needs. An evaluation is then performed to decide what portion of the data can be satisfied by remote sensing and other techniques and at what cost. If the costs outweigh the benefits, the analysis procedures must be evaluated for possible compromise. Accurately evaluating the costs of supplying data via different types of remote sensors requires a knowledge of the principles of operation of each.

There is no set rule as to the minimum site size where remote sensing techniques are appropriate. Although it is felt that their efficiency greatly increases with site areas larger than fifty acres, remote sensing analysis can offer important information on smaller sites. Many critical soil and geological patterns can be recognized only from the air and may be missed during conventional ground survey operations.

The process of remote sensing can be simplified to illustrate the key elements and variables involved (see Figure 1.1). The sun generates electromagnetic energy that radiates and illuminates the earth. Much of tnis energy does not reach the earth's ground surface, as portions are reflected or absorbed by the atmosphere. The energy that does reach material on the earth's surface is reflected, absorbed, and later emitted in a pattern related to each type of material. Remote sensors can "sense" these reflections or emissions whose signatures can later be recognized and correlated to the presence of the material. Conceptually, the process is simple; in reality, many factors and variables create complexity.

Electromagnetic Spectrum

All objects at a temperature above absolute zero radiate electromagnetic energy through the oscillations of their atomic and molecular components. Our sun emits energy ranging from short wavelengths—gamma rays, X-rays, ultraviolet light, and the visible spectrum—to longer wavelengths—near-infrared rays, thermal infrared rays, microwaves, and very high frequency radio waves (Figure 1.2). The sun has an average temperature of 300°K and provides a broad range of wavelengths, thereby facilitating detection of both reflected and emitted radiation. There are over 60 octaves in the electromagnetic spectrum, but our eyes are sensitive to only one. In comparison, our ears are sensitive to over nine sound octaves. The use of remote sensors enables us to expand our perceptual range; films and imagery see wavelengths for us and present them in a form we can perceive.

The wavelengths of the electromagnetic spectrum are expressed in units that describe their length. The smallest is an angstrom, 1/10,000,000th of a millimeter, followed by a millimicron or nanometer, 1/1,000,000th of a millimeter, then a micron (the most common unit), 1/1,000th of a millimeter. Microwave and radar systems make reference to millimeters, centimeters, and occasionally meters.

Atmospheric Attenuation

Most of the electromagnetic spectrum is attenuated by the earth's atmosphere in some way before the energy strikes the ground surface. Particles of haze, smoke, dust, water vapor, water droplets, and various gases scatter, absorb, and reflect major portions of the sun's energy. A spectral window exists where the atmosphere has little or no blocking effect (Figure 1.2). Consideration of local forecasted atmospheric conditions that will further attenuate spectral transmission-reflectance is critical before any remote sensing mission is planned.

Reflectance Characteristics

Each object has its own unique conditions which influence the reflection, absorption, and emission of different portions of the electromagnetic spectrum (Figure 1.3). In the design of a remote sensing system, these characteristics are important so that compatible sensor type, filters, and time of acquisition are used. However, to make the problem more difficult, the characteristic reflectances are not always constant. Changes of moisture in soils, growth stages of crops, and even wind blowing across a field of grain will affect spectral reflectance. It is therefore important to consider the effects of different conditions such as time of day, season, moisture, and so on will have on the objects and materials being sensed.

Sensors

A variety of sensing devices that are sensitive to the reflectance-absorption-emission characteristics of different materials around spectral windows have been developed. Initially, military applications accounted for the research and development of these instruments, but in recent years, broader civilian applications have been found.

Sensors can be classified into two main groups: active or passive. Active sensors record reflected or emitted radiation and include camera films, scanners, radiometers, and television-like systems. Passive devices provide their own illumination and record the created reflectance in a manner similar to taking a photograph with a flash attachment. Radar is the most common example, and since it provides its own illumination, its use is possible during night hours or times of heavy overcast.

Each sensor, whether it uses film or other detectors, is sensitive to a limited range of reflected-emitted energy. The instruments and filtering devices selected for a mission must therefore be appropriate for recording the relevant spectral data necessary for later interpretation.

Platforms

A remote sensing system can be placed at almost any altitude by using the appropriate "platform." If large-scale images (greater than 1:12,000) are desired, lower altitudes are re-

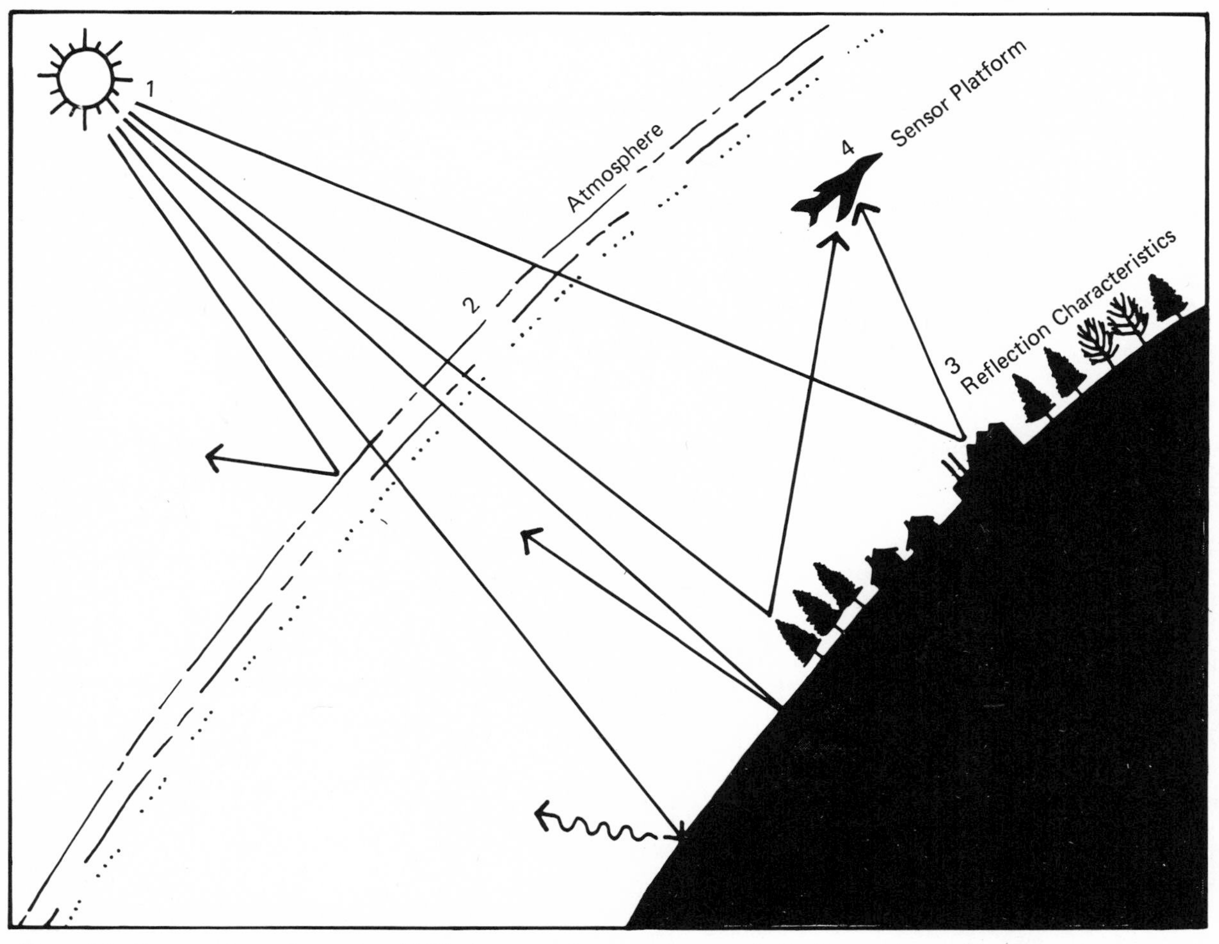

Figure 1.1. Process of remote sensing. (1) Electromagentic energy is generated by the sun, transmitted through space, and encounters the atmosphere. (2) Some radiation is reflected or absorbed while some penetrates and strikes the ground surface. (3) The exposed materials and objects reflect, absorb, and emit radiation in a manner characteristic of the composition of each material. (4) The reflected or emitted energy is recorded in a remote sensing system for real time or later display and interpretation.

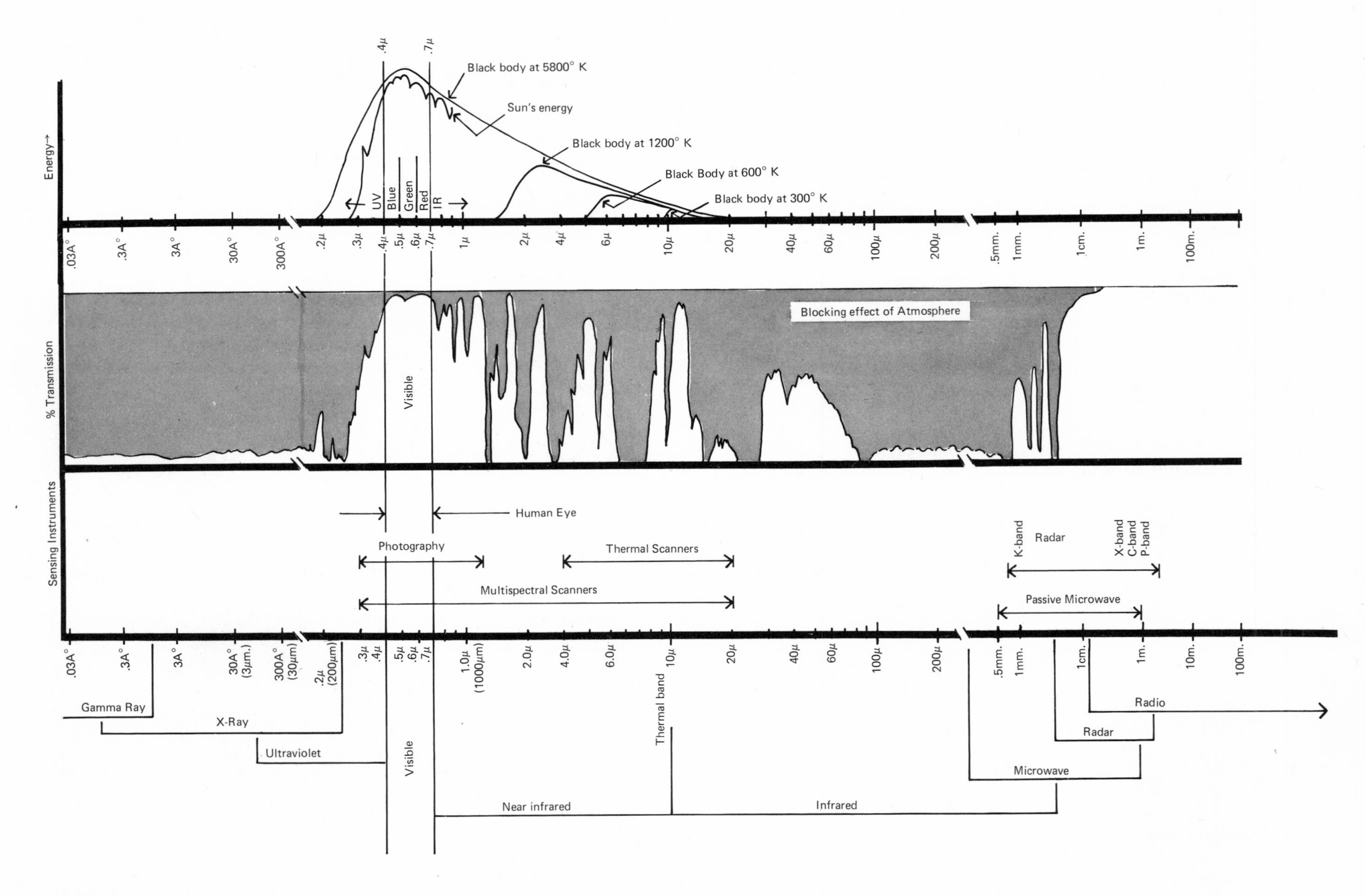

Figure 1.2. The electromagnetic spectrum illustrating atmospheric attenuation and general sensor categories.

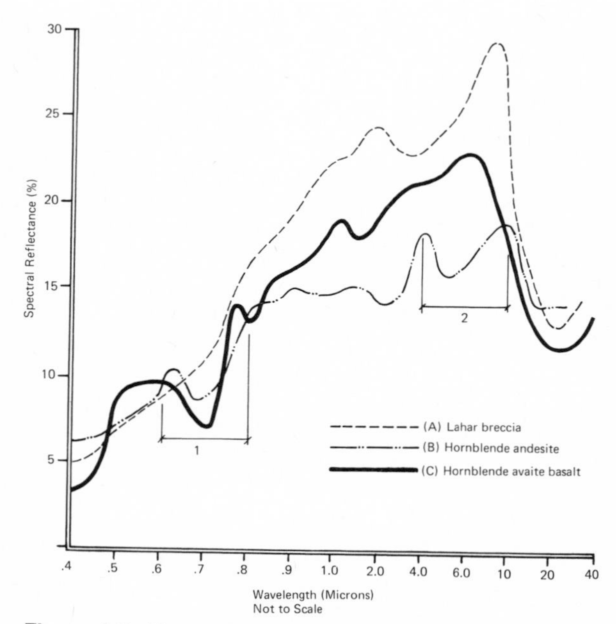

Figure 1.3. The spectral signatures of three materials shown for a limited portion of the spectrum. If a black-and-white photograph were taken of a narrow spectral segment near 1 all three materials would appear with a similar gray tone. Conversely, a photograph taken of segment 2 would show material A lightest; C, middle gray; and B, darkest. This procedure allows for obvious identification by gray scale. (Adapted from Mars, 1973.)

quired and are obtainable by using cranes, balloons, helicopters, or smaller fixed wing aircraft. Smaller scales (less than 1:60,000) can be obtained with jet aircraft, special high altitude reconnaisance aircraft, suborbital aircraft, and orbital satellites. Evaluation and selection of platforms includes, in addition to cost, consideration of weight, vibration, and accessibility.

AERIAL PHOTOGRAPHY

Camera, lens, and film systems that have been developed exclusively for aerial reconnaisance purposes operate in the 0.36 to 0.72 micron visual spectrum range. Some films are sensitive to the ultraviolet or near-infrared rays, up to 0.90 microns, and provide unusual reflectance data for analysis. High resolution and detail are possible with nonclassified commercially available systems; for example, studies by Raytheon have shown six-inch parking lot stripes on 1:120,000 scale photos.

Photo interpretation techniques have been well proven in over thirty years of development, testing, and varied applications. However, in order to take full advantage of the science, knowledge of the characteristics of the different film types and filtering modifications is essential. In addition, the factors that influence image quality and basics of organizing a photographic flight must be considered.

Photo Geometry Scale

Aerial photographs can be taken vertically or at some angle toward the horizon. At first, many photos were exposed obliquely since a larger area could be photographed in fewer frames and at lower altitudes. Today, most all aerial photos are taken vertically, especially if mapping applications are considered. Oblique photos are more appropriate for communication purposes where the perspective view aids perception of the scene.

The geometry of the aerial photograph reflects the altitude of the camera above the ground, the lens focal length, and the camera orientation. If the camera is vertically oriented, the photo scale is directly related to the lens focal length and altitude. If the altitude is increased, the ground area covered will increase; if the focal length is decreased, the ground area covered will decrease. Standard aerial lenses have focal lengths of either six, eight and one-fourth, or twelve inches. Figure 1.4 and Table 1.1 illustrate the interrelationships of scale, focal length, and altitude.

The relationship of camera focal length and altitude is used to determine the representative fraction or photo scale:

$$\text{RF}\left(\begin{matrix}\text{representative}\\ \text{fraction}\end{matrix}\right) = \frac{\text{Camera focal length (ft.)}}{\text{Altitude above ground (ft.)}}$$

If any two factors are known, the third can be calculated. For example, a camera focal length of six inches at an altitude of 12,000 feet:

$$\text{RF} = \frac{0.5\text{ ft.}}{12{,}000\text{ ft.}} = 1{:}24{,}000$$

(one unit of the photograph equals 24,000 units on the ground)

Conversely, a scale of 1:10,000 obtained with a six-inch lens would require an altitude of 5,000 feet.

Image Displacement

The scale on photo images will be distorted by two factors: changes of ground relief and parallax. The ground surface under the flight path is rarely flat, and many distortions are created as the topography changes (Figure 1.4). Hilltops closer to the camera will appear with a larger scale than valley bottoms. Even if the terrain is flat, the zone directly beneath the camera is slightly closer than the area near the edges of the photo, which results in minor scale distortion. Because of these problems, it is difficult to measure direct distances on photographs without establishing control measures and adopting photogrammetric procedures.

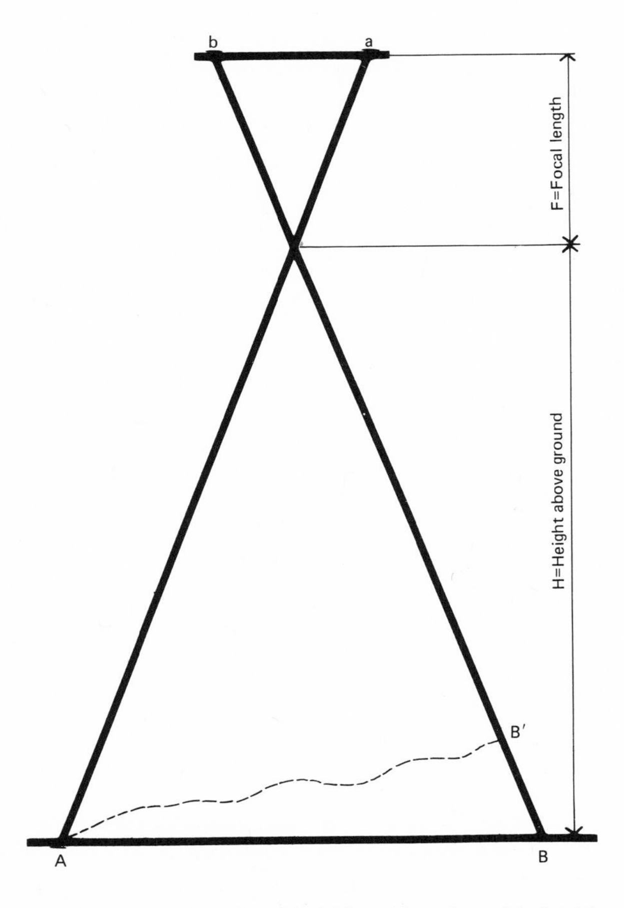

Figure 1.4. The relationship of focal length and height to resultant photographic scale. When rugged terrain is encountered (line A-B′), the varying height results in areas closest the lens having scales larger than areas more distant.

Table 1.1. Map scales and equivalents

Ratio scale	Feet per inch	Inches per 1,000 feet	Inches per mile	Meters per inch	Acres per square inch	Square inches per acre	Square miles per square inch
1: 500	41.667	24.00	126.72	12.700	0.0399	25.091	0.00006
1: 600	50.00	20.00	105.60	15.240	.0574	17.424	.00009
1: 1,000	83.333	12.00	63.36	25.400	.1594	6.273	.00025
1: 1,200	100.00	10.00	52.80	30.480	.2296	4.356	.00036
1: 1,500	125.00	8.00	42.24	38.100	.3587	2.788	.00056
1: 2,000	166.667	6.00	31.68	50.800	.6377	1.568	.00100
1: 2,400	200.00	5.00	26.40	60.960	.9183	1.089	.0014
1: 2,500	208.333	4.80	25.344	63.500	.9964	1.004	.0016
1: 3,000	250.00	4.00	21.12	76.200	1.4348	.697	.0022
1: 3,600	300.00	3.333	17.60	91.440	2.0661	.484	.0032
1: 4,000	333.333	3.00	15.84	101.600	2.5508	.392	.0040
1: 4,800	400.00	2.50	13.20	121.920	3.6731	.272	.0057
1: 5,000	416.667	2.40	12.672	127.000	3.9856	.251	.0062
1: 6,000	500.00	2.00	10.56	152.400	5.7392	.174	.0090
1: 7,000	583.333	1.714	9.051	177.800	7.8117	.128	.0122
1: 7,200	600.00	1.667	8.80	182.880	8.2645	.121	.0129
1: 7,920	660.00	1.515	8.00	201.168	10.00	.100	.0156
1: 8,000	666.667	1.500	7.92	203.200	10.203	.098	.0159
1: 8,400	700.00	1.429	7.543	213.360	11.249	.089	.0176
1: 9,000	750.00	1.333	7.041	228.600	12.913	.077	.0202
1: 9,600	800.00	1.250	6.60	243.840	14.692	.068	.0230
1: 10,000	833.333	1.200	6.336	254.000	15.942	.063	.0249
1: 10,800	900.00	1.111	5.867	274.321	18.595	.054	.0291
1: 12,000	1,000.00	1.0	5.280	304.801	22.957	.044	.0359
1: 13,200	1,100.00	.909	4.800	335.281	27.778	.036	.0434
1: 14,400	1,200.00	.833	4.400	365.761	33.058	.030	.0517
1: 15,000	1,250.00	.80	4.224	381.001	35.870	.028	.0560
1: 15,600	1,300.00	.769	4.062	396.241	38.797	.026	.0606
1: 15,840	1,320.00	.758	4.00	402.337	40.000	.025	.0625
1: 16,000	1,333.333	.750	3.96	406.400	40.812	.024	.0638
1: 16,800	1,400.00	.714	3.771	426.721	44.995	.022	.0703
1: 18,000	1,500.00	.667	3.52	457.201	51.653	.019	.0807
1: 19,200	1,600.00	.625	3.30	487.681	58.770	.017	.0918
1: 20,000	1,666.667	.60	3.168	508.002	63.769	.016	.0996
1: 20,400	1,700.00	.588	3.106	518.161	66.345	.015	.1037
1: 21,120	1,760.00	.568	3.00	536.449	71.111	.014	.1111
1: 21,600	1,800.00	.556	2.933	548.641	74.380	.013	.1162
1: 22,800	1,900.00	.526	2.779	579.121	82.874	.012	.1295
1: 24,000	2,000.00	.50	2.640	609.601	91.827	.011	.1435
1: 25,000	2,083.333	.480	2.534	635.001	99.639	.010	.1557
1: 31,680	2,640.00	.379	2.000	804.674	160.000	.006	.2500
1: 48,000	4,000.00	.250	1.320	1,219.202	367.309	.003	.5739
1: 62,500	5,208.333	.192	1.014	1,587.503	622.744	.0016	.9730
1: 63,360	5,280.00	.189	1.000	1,609.347	640.00	.0016	1.0000
1: 96,000	8,000.00	.125	.660	2,438.405	1,469.24	.0007	2.2957
1: 125,000	10,416.667	.096	.507	3,175.006	2,490.98	.0004	3.8922
1: 126,720	10,560.00	.095	.500	3,218.694	2,560.00	.0004	4.00
1: 250,000	20,833.333	.048	.253	6,350.012	9,963.907	.0001	15.5686
1: 253,440	21,120.00	.047	.250	6,437.389	10,244.202	.0001	16.00
1: 500,000	41,666.667	.024	.127	12,700.025	39,855.627	.000025	62.2744
1: 1,000,000	83,333.333	.012	.063	25,400.050	159,422.507	.0000062	249.0977

Image Quality

Image quality and the factors that influence it are paramount in the perception of details and terrain patterns in aerial photography. Image quality can be defined by resolution, granularity, acutance, and exposure density. Granularity and acutance are largely a function of the type of film used; resolution and exposure density are affected by: (1) physical factors such as illumination, reflectance, scattering, and sun angle; (2) camera characteristics including lens, image motion compensating devices and shutter; (3) the film-filter combinations, and (4) the film processing and printing.

Resolution

Resolution refers to the level of detail or smallest object that can be recognized on an aerial photograph. Most often it is expressed in lines per millimeter of a resolving power test chart that is placed on the ground and captured in the photograph for later analysis. The expression of ground resolution is more easily conceived of and defined as the smallest object that can be identified and distinguished from another adjacent object of the same size. Therefore, if two adjacent objects, each one foot in diameter, can be identified, the resulting resolution is stated as one foot. Linear features do not follow this rule. The contrast that may occur along a linear edge enables the boundary to be identified even though it may be caused by a thin cable, path, or road. For example, a one-half inch diameter electric cable (light tone) crossing a river (dark tone) would show clearly on most conventional aerial photographs (1:20,000–1:40,000).

Granularity

Granularity of photographic emulsions affects the resolution and the amount of detail that can be observed on photo images. It increases directly with the film speed; high speed films such as Super XX have a high ASA rating but are more granular than lower speed films. High-Definition Aerial Film 3404 has a resolving power and granularity value several magnitudes superior to Super XX 5425. To insure the highest level of detail, the slowest film that exposure permits should be used for aerial photographs.

Acutance

Acutance refers to the film's ability to record edge sharpness. It does not improve resolution but does have an effect on perceived detail. The property is illustrated by exposing a sharp knife edge on the film to produce a sharp edge contrast. Subsequent examination by a densitometer determines the rate of transition across the exposure boundary (Figure 1.5).

Exposure Density

Physical Factors of Illumination. Light scattering, sun angle, and spectral sensitivity have a direct bearing on photo exposure quality. Illumination or luminous flux falling on an area must be compatible for proper film exposure and maximization of contrasts. Reflectance from different materials varies widely and is greatly influenced by altitude. Large-scale photos (1:12,000) may have contrast ratios of 1,000:1, whereas the same area photographed at higher altitude (1:120,000) may be 5:1 (National Academy of Sciences, 1970). This reduction of contrast at higher altitudes can be detrimental to photointerpretation procedures if color tone is a critical identifier.

Light scattering is created by spectral energy encountering particles such as dust, smoke, water vapor, and various gases. The ultraviolet and blue ranges of the spectrum are most vulnerable; a yellow (minus blue) filter is used to minimize this effect. The particles that cause scattering occur below 30,000 feet; the problem still exists but does not increase at higher altitudes.

Sun angle influences the quantity and quality of the illumination. For example, Boston, Massachusetts receives approximately 11,000 foot candles (fc) illumination at apparent noon compared to less than 5,000 fc at the same time in December. Similarly, a dropoff occurs several hours before and after noon. To insure proper illumination and exposure of aerial films, photographs should be obtained within two hours of apparent noon. An exception should be made, however, if water penetration is desired;

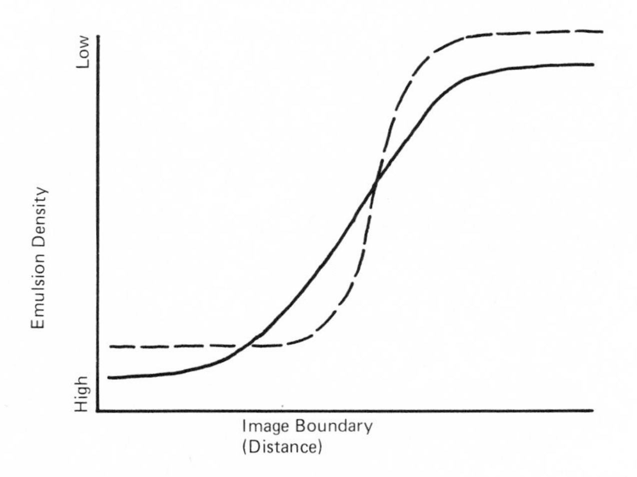

Figure 1.5. Acutance characteristic of film.

a high sun angle risks the appearance of the reflected sun image in the water.

Spectral balance is modified by the sun angle and the scattering effects previously mentioned. Low sun angles result in blue scattering due to the greater amount of atmosphere and particles encountered. Film exposed when the sun is at its highest angle will minimize atmospheric spectral attenuation.

Cameras and Lenses. Cameras and high-quality lenses have been developed for aerial sensing applications. Most cameras use nine and one-half inch roll film and incorporate vacuum plates, vibration dampening mounts, and in some instances image motion compensating devices. An excellent listing and documentation of various camera systems appears in the *Manual of Remote Sensing* (1975).

Film and Filters. The four major film types—black-and-white, black-and-white infrared, color, and color infrared—are available at a variety of speeds and types (positive and negative). Given the sensitivities of each, proper selection for a data acquisition problem will maximize efficiency. Compatible filters must be used during film exposure to ensure the inherent film-spectral relationships and to minimize any adverse atmospheric conditions. These combinations are more fully discussed in the section on film types.

Exposure and Printing. Even in the instances where all factors are optimized and image quality is seemingly guaranteed, photographs can be rendered useless if the most basic operation, film exposure, is completed irresponsibly. Fortunately such errors are rare, but all camera settings and exposure calculations should be double checked.

Sloppy printing can produce a worthless set of photos from otherwise excellent negatives. Contrasts can be lost by printing on low contrast paper, and color balance can be completely modified. When ordering prints for purposes of photointerpretation, contrast, paper weight, paper stability, print finish, and color balance should be specified.

Table 1.2. The mapping/contour interval/accuracy relationships commonly utilized in photogrammetric topographic mapping.*

Manuscript scale	Contour interval (C.I.)	Horizontal accuracy 90% 1/40 ft	10% 1/20 ft	Vertical accuracy 90% ½ C.I.	10% 1 C.I.
1″ = 30′	1′	0.75′	1.5′	0.5′	1′
1″ = 40′	1′	1.00′	2.0′	0.5′	1′
1″ = 50′	2′	1.25′	2.5′	1.0′	2′
1″ = 100′	2′	2.50′	5.0′	1.0′	2′
1″ = 200′	5′	5.00′	10.0′	2.5′	5′
1″ = 400′	10′	10.00′	20.0′	5.0′	10′

*Courtesy of Aerial Data Reduction Associates, Pennsauken, N.J. 08110.

†A map manuscript at 100 feet per inch has a 2-foot contour interval of which 90% must be within plus or minus 1 foot of its true elevation and 10% must be within 2 feet of its true elevation. Ninety percent of the point locations for the same map must be within 2½ feet of their true coordinates, and 10% must be within 5 feet.

Flying Format

Aerial photographs are typically oriented vertically. The goal is to produce a series of overlapping images that facilitate three dimensional stereo viewing and photogrammetric mapping (see Tables 1.2 and 1.3). The flight lines follow parallel strips in which each individual frame overlaps the next by about 60%. Parallel adjacent strips overlap by approximately 30% (Figure 1.6). The plane camera systems must be kept level, and pitch and yaw must be minimized to obtain correct overlap and minimum distortion. The orientation of the flight strips is a function of the shape of the area flown. The goal is to achieve the least expensive coverage format (i.e., minimizing turns and short strips).

Once printed, the photos are commonly assembled into an uncontrolled mosaic that acts as an index for illustrating specific photo cover-

Table 1.3. The contour/photography scale relationships commonly utilized in photogrammetric aerial mapping*

Direct manuscript (drawing) scale	Contour interval	Photography scale	Flight height†	Area covered by model (acres)
1″ = 30′	1′	1″ = 300′	1,800′	45
1″ = 40′	1′ or 2′	1″ = 400′	2,400′	80
1″ = 50′	1′ or 2′	1″ = 500′	3,000′	130
1″ = 100′	2′	1″ = 800′	4,800′	330
1″ = 200′	5′	1″ = 1,800′	10,800′	1,685
1″ = 400′	10′	1″ = 3,600′	21,600′	6,745

*Table based on high order stereoplotters. Courtesy of Aerial Data Reduction Associates, Pennsauken, N.J. 08110.

†Above average ground elevation.

age (Figure 1.7). In some instances the flight lines and photo frames are drawn on maps in lieu of photo indexes.

Film Types

There are over twenty aerial films, negative and positive, commercially available from U.S. manufacturers. These films offer different speeds, resolutions, and spectral sensitivities (Table 1.4). The following summary description places them into four groups: panchromatic black-and-white, black-and-white infrared, color, and color infrared. Each has its own particular advantages and disadvantages for recognition and interpretation of various materials depending upon the pattern of spectral reflectance and the film's ability to capture and portray the same.

The first step in any photointerpretation procedure concerns defining the specific categories to be indentified and determining the most effective film type, scale, associated resolution, and acquisition date. If selecting prints from many types of existing photographs, a comparative analysis should be completed by using samples of each to test their utility (Table 1.5).

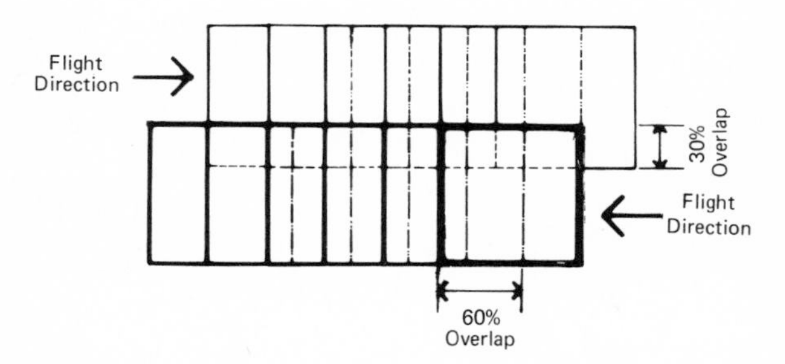

Figure 1.6. Format of the flying pattern in vertical aerial photography. Photographs are taken along a flight line with an image overlap of 60%. The next flight line has a side overlap of the previous flight line of 30%.

Black-and-White

Black-and-white panchromatic film has proved to be the most versatile for mapping and interpretation and is the most widely used. It was developed first and has since become the workhorse of all aerial photography. It has been used widely for data interpretation in the study of land use, vegetation, wildlife, geomorphology, soils, geology, hydrology, oceanography, and social demographic factors.

Black-and-white panchromatic films record the reflected visible spectrum from 0.4 to 0.7 microns (see Table 1.6) and present the image in gray tones. Speed and resolution range from excellent with High-Definition Aerial (AFS speed 8) to the grainier but faster Tri-X Aerographic (AFS speed 650). Normal exposure is through a 12 yellow filter (minus blue) to aid haze penetration. Heavier filters (15G dark yellow or 25A red) may be required if thicker haze is encountered. The addition of these filters influences the exposure, and adjustments must be made accordingly. For example, the 25A red filter has an associated exposure factor of 4.0.

The advantages of black-and-white photography lie in the versatility, long working history, and broad exposure range of the film. Compared to other film types, acquisition and processing costs are lowest. The exposure latitude coupled with the many filtering choices make it a most flexible film and one that can be used with a minimum of risk.

The major disadvantage of black-and-white

Figure 1.7 Photographic index.

film is the fact that it is black-and-white. Interpreters must learn to correlate gray tones to more familar colors. There is some evidence that on large interpretive projects, the mundane, unexciting gray tones induce fatigue and the associated factor of greater error. Management should anticipate this problem and seek ways of minimizing the possible boredom.

Many interpretive techniques are possible with black-and-white photographs; *The Manual of Remote Sensing* published by the American Society of Photogrammetry in 1975 provides a comprehensive summary. It is generally agreed by those in the photointerpretation profession that black-and-white panchromatic films offer the lowest risk, lowest cost of coverage, and the best general interpretive product.

Black-and-White Infrared

Black-and-white infrared film has special properties that are valuable assets to photointerpretation. The film records 0.4 to 0.9 microns, that is, the visible spectrum plus the nonvisible near-infrared. Even though the film is sensitive to thermal (heat) radiation on long exposure, the photo image obtained through conventional exposure is *not* a picture of heat. Normal exposure with a 25A red filter effectively removes the blue and green while retaining the red and near-infrared. For special purposes, exposure may be made through a 89B filter that eliminates all of the visible spectrum and provides a photograph completely illuminated by near-infrared. The resolution capability of black-and-white infrared is similar to the faster panchromatic films; the major difference lies in its ability to capture a portion of the reflected near-infrared. In some instances this ability provides a distinct interpretive advantage. For example, infrared photographs show most vegetation as light tones, thereby facilitating identification of plant life. Many timber companies and the U. S. Forest Service use black-and-white infrared photographs to aid in forest inventories and growth-yield projections.

Black-and-white infrared photos will show a high contrast between wet and dry areas because water absorbs much infrared energy and so appears dark. Thus, drainage patterns, swamps, and shorelines appear emphasized. These photos do not provide suitable images for stereoplotting topographic mapping, as the

Table 1.4. Aerial photographic films*

Film Name	Film Number	Nominal Base Thickness (μm)	Aerial Film Speed (AFS)	Aerial Exposure Index (AEI)	Developer or processes	Resolving Power, lines/mm target-object contrast		Diffuse RMS granularity[f]
						1000:1	1.6:1	
Plus-X Aerographic	2402	100	200	80	D-19	100	50	19
Tri-X Aerographic	2403	100	650[c]	250	D-19	80	20	33
Double-X Aerographic	2405	100	320[c]	125	DK-50	80	40	26
Panatomic-X Aerial	3400	60	64	20	D-19	160	63	16
Plus-X Aerial	3401	60	200	64	D-19	100	40	30
High-Definition Aerial	3414	60	8	2.5	D-19	630	250	9
	1414	40	8	2.5	D-19	630	250	9
Infrared Aerographic[a]	2424	100	200	100	D-19	80	32	33
Multispectral Infrared Aerial[a]	SO-289	100	100			200	80	12
Aerochrome Infrared[b]	2443	100	40	10	EA-5	63	32	17
	3443	60	40	10	EA-5	63	32	17
High Definition	SO-127	100	6[c]		EA-5	160	50	9
Aerochrome Infrared[b]	SO-131	60	6[c]		EA-5	160	50	9
	SO-130	40	6[c]		EA-5	160	50	9
Aerocolor Negative	2445	100	100[c]	32[e]	Aero-Neg. color proc.	80	40	13
Ektachrome MS Aerographic	2448	100	32[c]	6[e]	EA-4	80	40	12
Aerial Color	SO-242	60	6	2	ME-4 (modified)	200	100	11
	SO-255	40	6	2	ME-4 (modified)	200	100	11
Ektachrome EF Aerographic	SO-397	100	64[c]	12	EA-4	63	32	13
	SO-154	60	64[c]	12	EA-4	63	32	13
Water Penetration Aerial Color	SO-224	100	40[c]		EA-5	125	50	24
Anscochrome	D/200	100	90[d]	30	AR-1C	100	50	25
Anscochrome	D/500	100	230[d]	75	AR-1C	80	40	45

[a]Speeds are for no filter.

[b]Speeds are for a Wratten number 12 filter.

[c]Effective film speeds are relative speeds, found by comparison to films whose speeds are determined according to ANSI Standard PH2.34, 1969.

[d]At time of compilation the manufacturer had not released an AFS for this film. The number shown is an estimate determined by multiplying the AEI by a factor of about 3.

[e]Haze filters such as HF-3, HF-4, and HF-5 may be used without changing exposure.

[f]The RMS-Diffuse granularity value represents 1000 times the standard deviation in diffuse visual density produced by scanning a uniformly exposed and developed sample with an f/2.0 system of 48-μm scanning spot size.

high contrast in shadow areas makes visual penetration to the ground difficult. However, this high contrast shadowing can be an aid in some interpretations. The lighter toned appearance of most broad-leaved plants compared to conifers is a result of the higher level of contained shadows in the conifers, not a change in infrared reflectance.

Costs of this film are somewhat higher than those of panchromatic black-and-white due to its limited production, shorter shelf life, and lower flexibility of exposure. It is very effective for haze penetration when used with the 25A or 89B filters, but its poor resolution makes it undesirable for this purpose alone. Its ability to emphasize drainage patterns and wet areas makes it useful for landform terrain analysis, especially if photos are taken when vegetation is dormant. For general photographic interpretation purposes, black-and-white infrared should be used with panchromatic black-and-white, rather than as a substitute.

Color

Color film has been steadily improved to offer higher resolution, better color balance, faster speeds, and greater exposure latitude. It is available in positive or negative formats (the latter is compatible with making color or black-and-white prints or glass diapositives for mapping). The major difference between color and panchromatic black-and-white film is the added aspect of color. The human eye can distinguish over 100 times more color (combinations of hue, value, and chroma) than gray scales, that is, a factor of 20,000 to 200 (Evans, 1948). This factor is critical for some interpretive purposes: Mineralogic and geologic studies and soil mapping use color to distinguish major rock, mineral, and soil units, while detailed vegetation typing during peak fall coloration can be accomplished only with color. (See Figure 1.8 in color section.)

Color films are sensitive to 0.36 to 0.72 microns, and resolution is similar to that of mid-range panchromatic black-and-white films. Film speed is very high, the same as High Definition Aerial black-and-white. Normal exposure below 10,000 feet is made through either HF-1, HF-2, or HF-3 filters, which tend to minimize blueing effects. No significant reduction in exposure is required.

A special color film, Kodak SO-224, has been produced for optimal water penetration. It is especially sensitive at the 0.48 and 0.55 micron range where water penetration is high. (Infrared films would be very poor for water penetration as their normal filtering blocks this portion of the spectrum.)

In certain regions of the country, optimum mapping conditions for color photography may be difficult to find. In northern latitudes there may be only a few weeks between the time the snow melts and the time the deciduous trees bloom during which the sun is at proper angle. Furthermore, the visibility-haze conditions for proper color balance are more stringent than those for black-and-white photography since minus blue haze penetration filters cannot be used. Thus it may be impossible to obtain an

Table 1.5. Photographic type-scale-date evaluation*

			SOIL (UNIFIED)	SOIL DEPTH	DEPTH TO W.T.	PERC. RATE	EXISTING LAND USE	WATER	VEGETATION TYPE	VEGETATION HEIGHT	VEGETATION DENSITY	TOTAL RATING
1:4800	Spring	B&W	A	A	A	A	A	A	B	A	A	35
1:7200	Spring	B&W	A	A	A	A	A	A	B	A	A	35
	Fall	C	D	D	D	D	A	B	A	A	A	23
		CIR	D	D	D	D	A	A	A	A	A	24
1:12000	Spring	B&W	A	B	B	B	A	A	B	A	A	32
	Fall	C	D	D	D	D	A	B	A	A	A	23
1:20000	Spring	B&W	B	C	C	C	B	B	C	B	B	23
	Summer	B&W	D	D	D	D	B	C	C	B	B	17
	Fall	B&W	D	D	D	D	B	C	B	B	B	18
1:40000	Fall	B&W	D	D	D	D	B	C	C	C	C	15
1:60000-70000	Spring	B&W	C	C	C	C	B	D	D	D	C	16
	Summer	CIR	D	D	D	D	B	C	D	D	C	13
	Fall	B&W	D	D	D	D	B	D	D	D	C	12
		C	D	D	D	D	B	D	C	D	C	13
		CIR	D	D	D	D	B	C	C	D	C	14

*From Way, 1977.
Relative ratings: A is best, D worst. Total number of points determined by assigning 4 to A, 3 to B, 2 to C, and 1 to D.

Table 1.6. Film and filter combinations

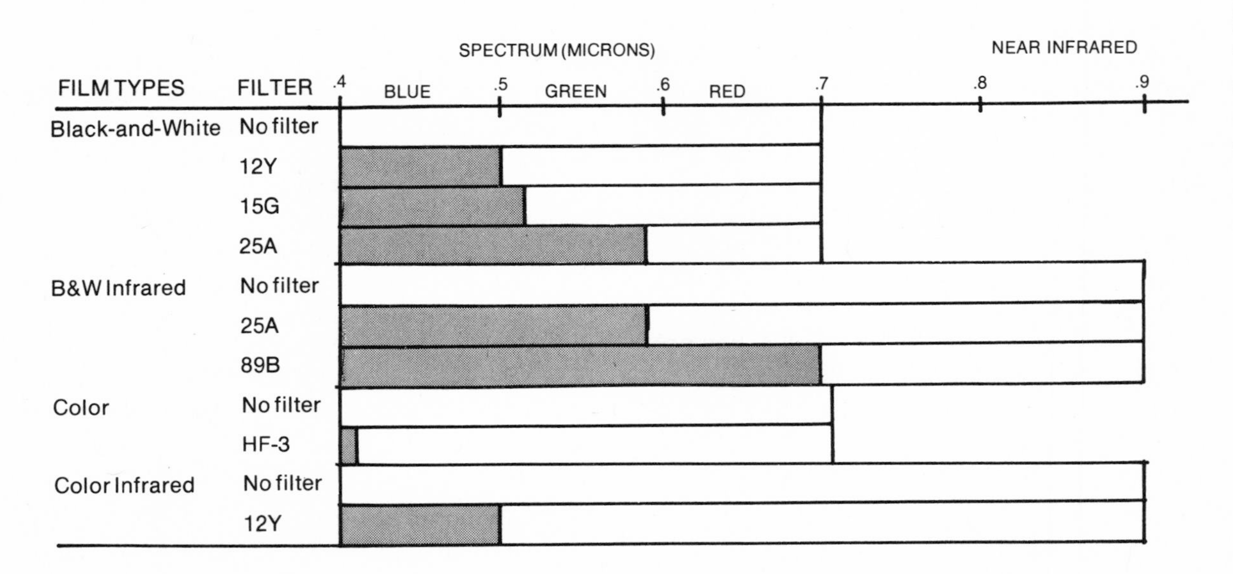

optimum flying day in the spring in northern latitudes; the flight would then have to be postponed for a year.

Many studies have suggested that color photography offers the best overall photographic sensor for general data identification and interpretation. However, some qualified interpreters have found that color does not consistently offer sufficient data-gathering potential over panchromatic black-and-white to justify the higher cost. Some studies have indicated that the additional chroma and tonal distinctions that can be made on color film may provide too much information and may confuse the interpretater with nonsignificant or noncorrelating color changes. Color film may be of value for general use by those not well-trained in interpretation techniques, since the color image is easier to relate to normal, on-the-ground visual appearance. In addition, the more interesting virtue of color seems to produce less fatigue in photo-interpreters and therefore fewer errors.

Color processing and printing is more expensive and takes longer than for black-and-white. More important, uniform color balance is difficult to maintain from roll to roll and during different print runs. If interpretations are to be based upon minor chroma and tonal differences, it would be difficult to establish a standard identifying key to apply to several rolls. It is anticipated, however, that future advances in color technology will minimize these difficulties.

Color Infrared

Color infrared photographs offer a wide tonal and hue advantage, similar to that of color photography, in addition to their sensitivity to the near-infrared, as discussed for black-and-white infrared film. Sensitivity ranges from 0.36 to 0.90 microns, but normal exposure through a 12 yellow filter removes most of the blue. The resolution and speed has been improved and is now similar to that of ordinary color film. The dye-coupling process of infrared film produces the false colors that appear on the final transparancies or prints. Table 1.7 outlines the relationships of color and color infrared film and shows how resultant colors are dye-coupled and produced.

The primary advantage of color infrared lies in its ability to capture reflectances in the near-infrared and in its presentation of false colors. Healthy vegetation, such as broad-leaved deciduous trees, is a high reflector of near-infrared energy and is recorded with a brilliant red or pink color (see Figure 1.8 in color section). Some studies have shown that diseased vegetation has a reduced ability to reflect infrared energy and therefore does not appear as

Table 1.7. Color and color infrared film characteristics*

Color Film				
Reflected Spectrum	Blue	Green	Red	
Dye Coupling	Yellow	Magenta	Cyan	
Resulting Color	Blue	Green	Red	
Color Infrared Film				
Reflected Spectrum	Blue	Green	Red	Infrared
Dye Coupling	Filtered	Yellow	Magenta	Cyan
Resulting Color		Blue	Green	Red

*Color film is sensitive to blue, green, and red reflected light. The respective dye coupling is with yellow, magenta, and cyan. The processing and printing using color addition yields blue, green, and red that copy the original visual appearance. Color infrared is sensitive to blue, green, red, and near-infrared. The blue is removed by filtering, and the remaining green, red, and near-infrared are paired with yellow, magenta, and cyan. By color subtraction the resulting colors of blue, green, and red are presented. Thus, a high reflector of near-infrared energy, such as healthy vegetation, will appear red.

brilliant red but perhaps as orange, yellow, or even white (Murtha, 1972). Due to this unique ability color infrared has been used to monitor vegetation and crop areas for the effects of blights, pests, and drought.

Color infrared is useful for making distinctions between broad groups of vegetation, such as deciduous and coniferous species. Vegetation density may be more easily determined with this type of film, especially in detecting areas of sparse vegetive cover over bare rock and soil. Thus sand dune regions and coastal areas can be observed to determine the amount of stabilizing vegetation present.

As in the case of black-and-white infrared film, color infrared emphasizes through color and tonal differences the distinctions between water-covered areas and dry land. Stream channels can be accurately located, shorelines plotted, high and low watermarks defined, soil moisture differences identified, water-vegetive boundaries plotted along marshes and bogs, and so on. These identifications can also be made, but perhaps with more difficulty, by using black-and-white panchromatic or black-and-white infrared film.

Color infrared film is slightly more expensive to contract than ordinary color film. The limited exposure latitude, short shelf life, and general unfamilarity contribute to this cost. While it has been determined that color infrared images, like those in ordinary color, are more interesting and exciting to interprete than black-and-white images, this benefit is somewhat countered by the confusion many novice interpreters face in attempting to analyze the complexities of false color.

For purposes of terrain analysis, some benefits may be obtained from using color infrared film, but its usefulness in this application remains debatable. Therefore, the additional flying and handling expenses should be carefully weighed. Color infrared has been proven most effective for vegetation surveys, but when it is used, it should be supplemented with coverage using color and black-and-white panchromatic films. Color and tonal shifts between and within each different film can thereby be studied in detail via a system of cross-checking and comparison.

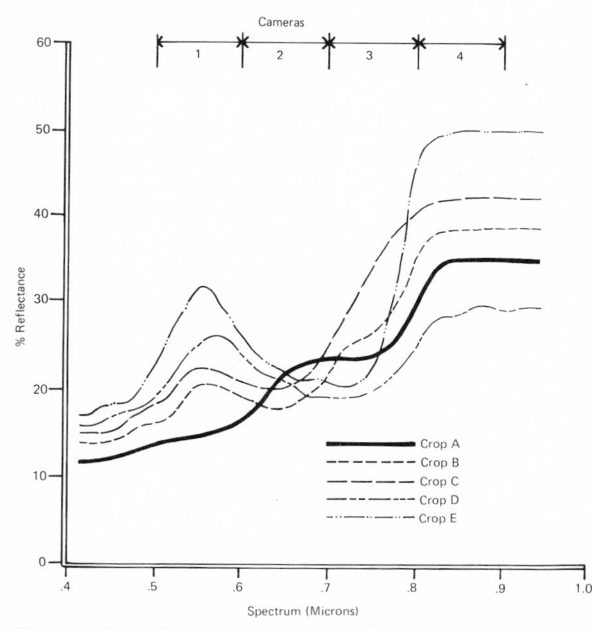

Figure 1.9. Multispectral recognition. The spectral signatures of five different crops are illustrated. A four-camera multispectral system is filtered to provide camera 1 with the spectral zone 0.5 to 0.6 microns; camera 2, 0.6 to 0.7 microns; camera 3, 0.7 to 0.8 microns; and camera 4, 0.8 to 0.9 microns. Crop types E, A, and D can easily be distinguished by gray tone on photos from camera 4, and B can be interpreted from C on photos from camera 3. Photos from cameras 1 and 2 would be used to check interpretations. Thus by making image comparisons from the four photo sets, the five crop types can be identified by gray scales. (The same process can be used for rock and soil types.)

Multispectral Photography

Multispectral photography involves the use of multiple cameras to make simultaneous exposures. Different filter-film combinations are employed to capture selected spectral segments. Subsequent image superimposition and color intensity mixing enables the photointerpreter to maximize the identification of spectral signatures of different materials (Figure 1.9).

The advantage of multispectral systems over a single type of photography or imagery lies in the analytic capability gained by being able to compare spectral responses that in combination may identify a material. These systems facilitate the use of computers for automatic processing where spectral responses can be analyzed and matched to previously defined spectral responses in order to identify combinations associated with types of vegetation, materials, and so on. As now used by some groups, this type of system can automatically inventory crop production, natural resources, and land use and cover, and if used over time, monitor changes.

Orthophotography

The many distortions inherent in aerial photographs make them unsuited for controlled map images unless they are taken over flat land. However, the preparation of orthophotographs can remove the effects of tilt, terrain relief, and lens aberrations while retaining a photoimage (American Society of Photogrammetry, 1966).

The preparation of orthophotographs involves placing glass photo positive plates in a double projection plotting instrument and projecting the stereoimage onto an opaque screen with a movable slit. The film below it can be moved in the z direction while the slit is moved in the x or y direction. This procedure compensates for the distortion effects of topographic relief as the photoimage is scanned. In effect, the focal length of the projection system is changed as the ground plane changes to correct the scale. The orthophoto can be used as a controlled base for contour plotting and allow for measurements to be made accurately. Large area resource inventories are most efficiently completed from orthophotos, especially when the information is computerized. Digital measurements can be taken directly from the orthophoto base as the data is being interpreted from the photoimage.

Photo and Image Interpretation

Principles of photointerpretation have been applied to aerial photographs since shortly after the discovery of the photographic process in 1839. Although originally directed toward military applications, the technique is now well established as a scientific tool. Several factors make aerial photography and imagery unique for data collection: the photos and images represent a record of conditions that existed during exposure; they present an overview of relatively large land areas, and the three-dimensional viewing aspect facilitates depth perception and interpretation.

Principles of Object Recognition

Aerial photographs and images are usually taken vertically to satisfy mapping requirements, but the earth's perspective is presented in an unfamiliar manner. The photointerpreter must work with equipment and employ techniques that are compatible with this perspective. Stereoscopes are used to provide three dimensional viewing, and various visual indicators are observed to aid in object recognition. These indicators include size, shape, pattern, tone and color, texture, and shadow. Many objects show unique combinations of these indicators that occur whenever similar objects are found. For example, a golf course exhibits tonal variations among the tees, greens, fairways, and sand traps along with the size, shape, and pattern relationships of these areas. Similar combinations of patterns are associated with all golf courses.

Size. Object size, a function of photograph scale, is one of the most important pattern indicators. It allows distinctions to be made between trees and shrubs, houses and gravestones, rural roads and superhighways.

Shape. Some objects have unique shapes that are critical to their recognition. Examples include airports, highway cloverleaf interchanges, and racetracks.

Pattern. Aerial remote sensing techniques provide an advantage over conventional ground data collection procedures due to the unique overview perspective obtained. Pattern characteristics are difficult to observe on the ground but can be seen easily from the air. The connective pattern or drainage is important for interpretation of rock types. Similarly, pipelines, highways, and railroads can be differentiated by their inherent pattern characteristics.

Tone. The gray tone of photographs and images may suggest different soil types, moisture conditions, crops, and manmade materials. On color photographs, hue and chroma may correlate to known conditions serving as an interpretive indicator.

Texture. Texture influences photographic gray tone, but alone, it can provide valuable interpretive data. Some crop and vegetation types with the same tone may be distinguished by texture. The density or texture of drainage patterns may help to suggest rock types and hydrologic conditions.

Shadow. In some instances shadow is a primary element in image recognition. Vertical smokestacks, firetowers, or oil derricks can not be seen easily but their shadows are readily recognizable. Many foresters observe tree shadows in order to identify branching patterns that can then be correlated with certain species.

Stereoviewing

Aerial photographs are normally taken in flight lines with individual frames consecutively overlapping by 60%. This procedure allows for viewing in three dimensions or stereo. Once obtained, the aerial photographs of a site should be examined and then placed in position for stereoviewing as follows. Orient the photographs so that all the numbers are either to the left or to the right of the observer. Then place the photographs so that the areas with the same image overlap. (This overlap area accounts for approximately 60% of both photographs.) Locate an object that appears on *both* photographs, such as a pond or building or a stand of vegetation, and separate the photographs so that there are approximately 2¼ inches between the objects (it is easiest if the object selected is near the edge of one of the photographs). Place the stereoglasses directly over the two objects. After the glasses have been corrected for proper eye separation, view the two objects through the glasses and adjust the photographs slightly to align the images so that when viewed, they become one image appear-

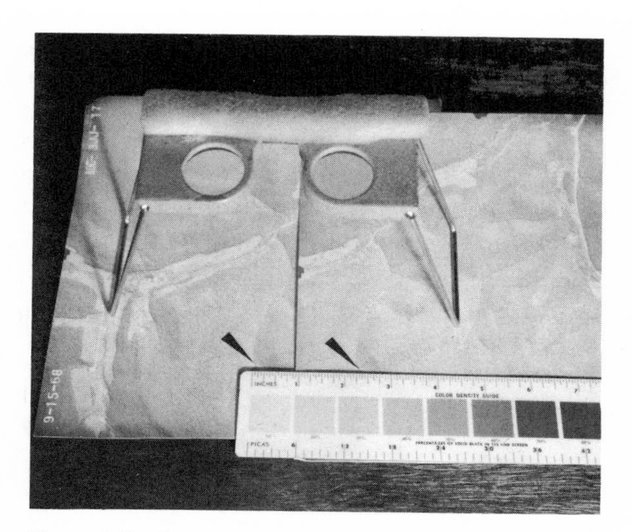

Figure 1.10. The photographs are arranged for stereoviewing by overlapping the images, separating them by approximately 2¼ inches, and centering the stereoviewer directly over the edge of the overlapping photograph. It may be useful to place a strip of foam rubber along the top edge of the stereoviewer for comfort during long periods of use.

ing in three dimensions (see Figure 1.10). If difficulty is encountered in obtaining a stereo-image, quickly close one eye and then the other alternately and see whether the two objects are close to being superimposed upon one another; then adjust the photographs to correct the alignment. Before attempting this procedure for the first time, the stereovision test (Figure 1.11) should be viewed.

Photointerpretation Equipment

Many companies manufacture equipment that may be useful to photointerpreters. Very little is necessary but access to the following may be useful: a good pocket stereoscope, mirror stereoscope, light table, acetate, engineers scale, various drafting instruments, china markers, dot grids and other measurement overlays, carbon tetrachloride for cleaning photos, and perhaps a parallax bar or wedge for measuring object height.

A stereoscope determines the amount of area observed and the quality of the perceived image. Selection and use of this instrument can have an important influence on photointerpretation. Most applications require at minimum a pocket type stereoscope with at least 2X magnification. This type of stereoscope is valuable for everyday office and field use because of its small size, low cost, and adaptability. When more precision is desired, a mirror or reflecting stereoscope can be used. Some have interchangeable optics that allow viewing of the entire 60% image overlap area at one time or high magnification on a selected area. Those with zoom lenses can accomplish these magnification changes without disturbing the photographs or the stereoscope. More sophisticated models have image transfer capabilities whereby one photo image can be overlaid on another photo or map for comparison and/or mapping. These devices are invaluable for large mapping projects in which data must be trans-

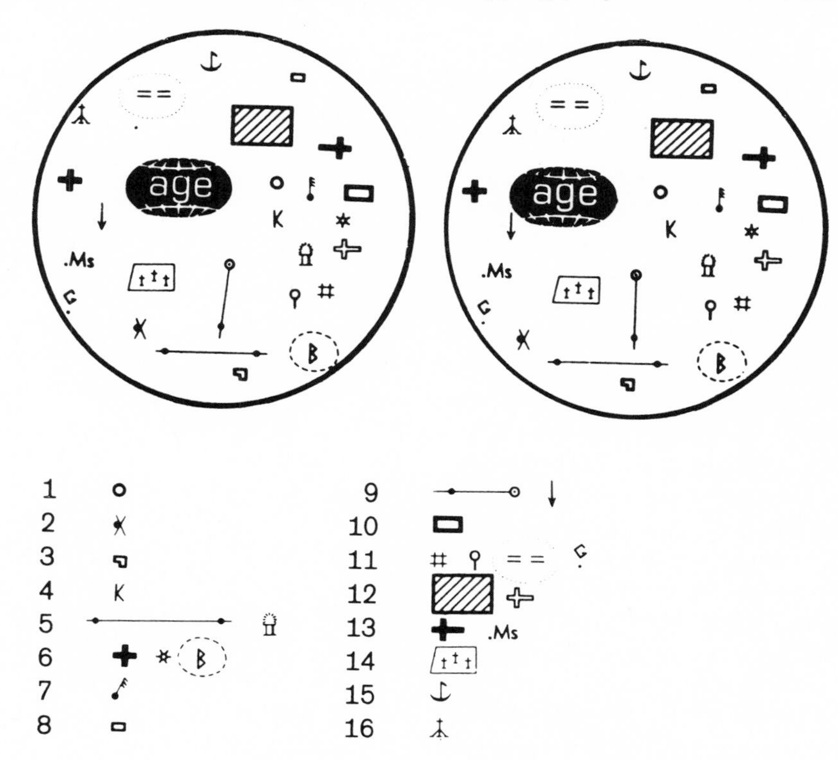

Figure 1.11. Stereovision test. Different amounts of parallax have been introduced into these symbols. Cover the legend and try to determine the relative heights of the objects in the stereogram; 1 is the highest and 16 the lowest. (Courtesy of Alan Gordon Enterprises, Inc.)

ferred from photographs or images to base maps of different scale. Expensive, sophisticated equipment, however, is not a prerequisite for accurate photointerpretation. While the equipment can contribute to the speed and efficiency of the interpretive operation, human judgment must still be made.

The various companies that manufacture and distribute photointerpretation equipment can be contacted for catalogs and price lists of their offerings. *Photogrammetric Engineering and Remote Sensing* carries advertising of many of these companies.

NONPHOTOGRAPHIC SENSORS

Camera sensing systems, while satisfying many applications and data-gathering needs, have several major shortcomings: the photographic output is not well suited for providing electronic input for computer analysis; the films are limited to the visual and near infrared spectrum; the films require reflected illumination for exposure; and the correct exposure is limited by haze, smoke, and clouds. Many nonphotographic sensors exist that do not have these limitations: They can provide imagery through haze and even clouds; they are sensitive to a spectral range from microwaves to ultraviolet light; their electronic output is directly compatible with transmission and further analysis; and some can be operated at night.

The different types of detectors and systems are quite numerous and complex in operation. *The Manual of Remote Sensing* by the American Society of Photogrammetry (1975) provides an excellent summary description of different sensors and their technical aspects. The following discussion concentrates on those devices which may have a distinct application to terrain analysis.

Scanners

Scanners are systems with various detectors that incorporate a scanning process. Those with photomultiplier detectors operate in the near-ultraviolet and visible spectrum, while those with extended sensitivities range into the far-infrared. Scanners offer an advantage over conventional photographic equipment in that in addition to their possessing extended range, they are able to use several channels simultaneously (American Society of Photogrammetry, 1975). These abilities further facilitate analysis potential through the use of multispectral techniques.

Scanners are part of a larger group of electro-optical sensors that detect segments of the electromagnetic spectrum and create an electronic signal from which a photograph-like image can be produced. The image collection process usually incorporates an optical system, hence the name "electro-optical." There are many types of these devices including thermal scanners, multispectral scanners, and television cameras.

Most scanners operate with a rotating mirror that scans the ground perpendicular to the platform (Figure 1.12). The mirror reflects the electromagnetic energy onto the detector whose output is variable electronic signals. These signals are amplified and displayed on a cathode-ray-tube (CRT) printer and recorded on film for visual interpretation. In addition, the signals are placed on computer tape that serves as the original record and is available for image enhancement programs and further analysis. The final image created is similar to a black-and-white photograph: Light tones infer high reflectance of the spectral band sensed; dark tones infer little reflection. An important objective of object recognition, similar to that of the photographic processes, is to maximize contrast between the object and its immediate background. Contrast will be influenced by the time of day imagery is acquired, especially in the far-infrared bands.

The resolution capabilities of scanners do not approach those of photographic systems but they are better than most of the other nonphotographic techniques. Scanners with rotating mirrors have a one-square-foot ground resolution from an altitude of 1,000 feet. That finer resolution is possible is illustrated by the LANDSAT satellite in which scan widths of 79 meters are sensed from an orbit of 500 nautical miles.

Thermal Infrared Scanners

Infrared or thermal imagery acquired by scanners has proved to be of great value for certain very specific data collection projects. The sensors detect reflected or emitted thermal infrared radiation and present a photograph of an electronic image that shows warmer objects as light tones and cold objects as dark tones (Figure 1.13). The gray tones of the image can be accurately measured by a densitometer; they indicate relative temperature differences that can be correlated with ground measurements to determine and map temperatures within plus or minus one-half degree Centigrade. However, the thermal image presented is a function of temperature conditions that exist at the surface of the ground, water, or object, and subsurface mixing in water and underground variances do not show unless they influence the surface temperature. For example, the detection and monitoring of hot water discharge from power plants is illustrated two dimensionally on thermal imagery; the underwater temperatures and mixing conditions are not shown.

There are many applications of thermal imagery in which temperature data is used directly or is used as an indicator for other conditions. The U. S. Forest Service uses thermal infrared scanners to detect and map forest fires and their hot spots through thick clouds of smoke or at night (American Society of Photogrammetry, 1968). Ecologists use this sensor to map thermal differences in water bodies caused by heat pollution, volcanic activity, and geothermal activity, as in Yellowstone Park. In arctic areas, snow bridges and crevasses can be identified.

Many experiments in terrain analysis have attempted to map underground voids but have met with mixed results. In Texas, test mapping of several sites in limestone materials has shown that some voids can be found, if they are close to the surface and large enough to influence surface temperature. Other relevant applications include rock unit differentiation, location of springs and seepage zones, mapping of soil depth to near surface bedrock, depth to water table, and identification and monitoring of underground mine fires. The use of this sensor can be justified for the acquisition of specific thermal data, but it is relatively expensive to use for general purposes.

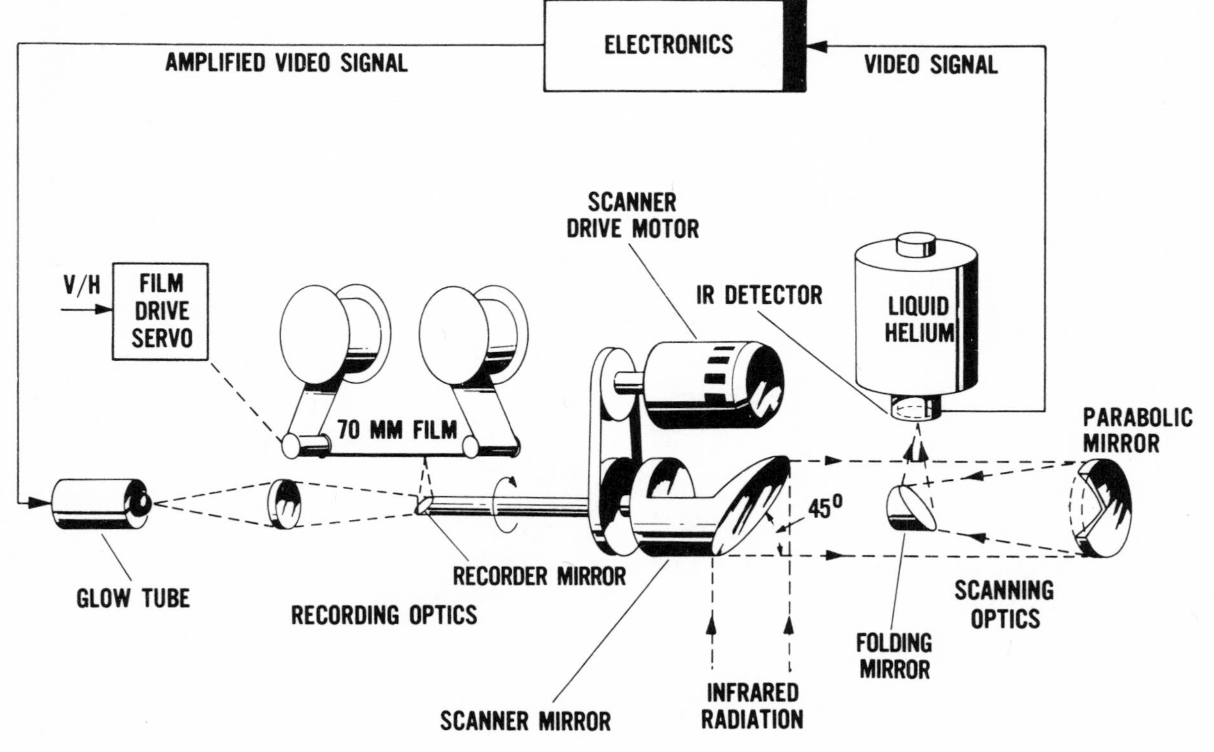

Figure 1.12. Typical infrared scanner. (Courtesy of Environmental Analysis Department, HRB/Singer, Inc.)

Microwave Sensors

Microwave remote sensors are unique in their "all weather" ability to measure through clouds or light rain and at night. The longer wavelengths used (1 millimeter to 1 meter), and their properties do not allow the fine resolution obtainable with infrared and visible spectrum sensors. Wavelength frequency, polarization, and viewing angle (some vertical and some oblique) control the response of microwave energy and therefore greatly influence the design of microwave systems. Passive microwave sensors use radiometers to measure emitted and reflected radiation from other sources; active systems provide their own illumination.

Active Microwave (RADAR)

The active systems that illuminate a scene through transmission of radio waves are classified as RADAR (Radio Dectection and Ranging). Since the properties of the transmitted energy are accurately known, wave travel time and velocity from objects can easily be calculated for image formation. Most radar systems operate with wavelengths from 0.5 centimeters to 1 meter; the shorter wavelengths are best for highest detail, and the longer are more effective for cloud, rain, and light vegetation penetration. Resolution is controlled by complex interactions of wavelength used, antenna length, and effects of synthetic aperture if used. If all factors are maximized, resolution of less than ten feet is easily obtainable.

Radar systems can be carried in aircraft or spacecraft, although in the latter difficulties related to the system's high weight and power needs may be encountered. In operation, the radio signals are transmitted and synchronized

Figure 1.13. Thermal imagery and analysis. A thermal plume can be recognized by its light tone (high temperature) in the Allegheny River, Pennsylvania. The annotated gray scale is produced by computer analysis which is necessary for the delineation of isothermal contours. (Courtesy of the Environmental Analysis Department, HRB-Singer, Inc.)

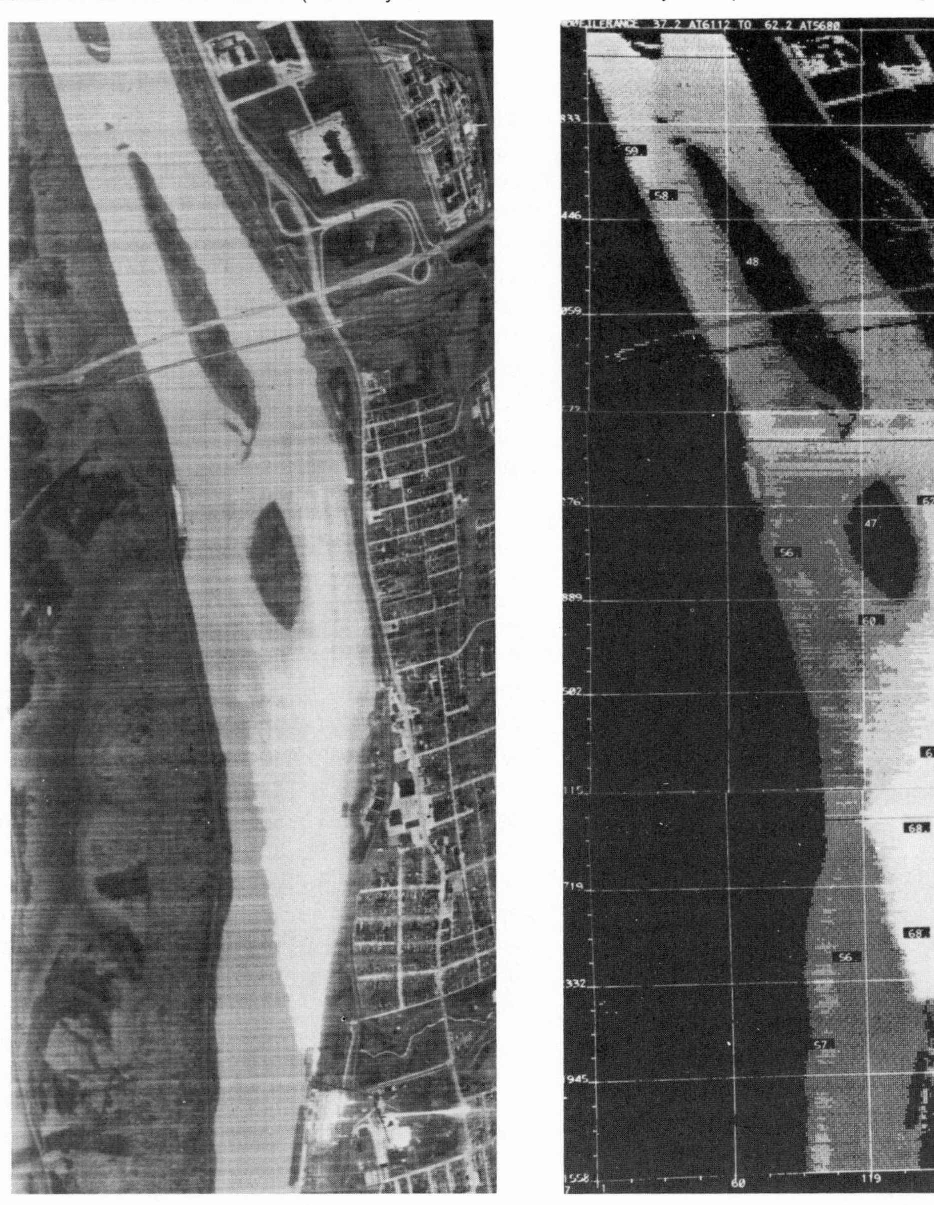

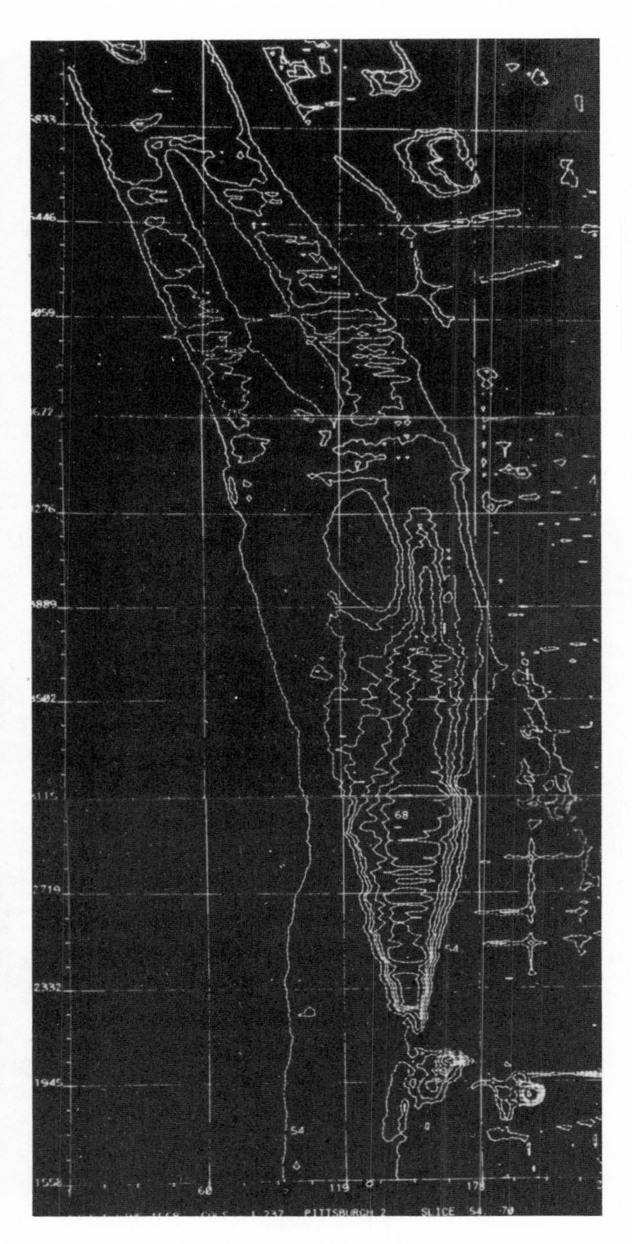

Thermal Image

Annotated Gray Scale (°F)

Isothermal Contours

for comparision of reflected waves that are first processed for range and intensity and then viewed on a CRT and recorded. After adjustment and enhancement procedures, final images are produced and photographed. The image is not acquired vertically beneath the plane, but rather obliquely to one or both sides at angles from 20 or 30 degrees to 75 or 80 degrees, thereby encompassing approximately 50 degrees of view (Figure 1.14). If both sides are sensed simultaneously, a mosaic can be constructed from all left or all right images and then labeled directionally (i.e., east looking or west looking). Radar acquisition systems operating in this manner are titled SLAR (Side-Looking Airborne Radar).

The most effective application of SLAR relates to its all weather capability. Many areas have resisted conventional mapping because they contain combinations of adverse conditions such as cloud cover, remoteness, and rugged terrain. Large areas such as these can be efficiently mapped and images interpreted in a manner similar to that used in black-and-white photography for geologic, geomorphologic,

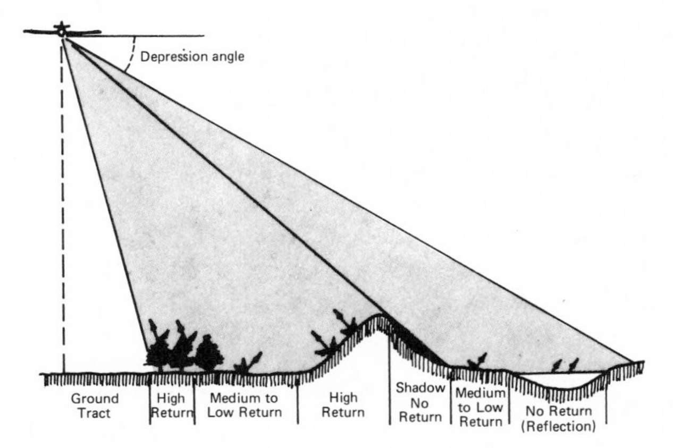

Figure 1.14. Basic geometry of Side Looking Airborne Radar (SLAR).

soils, mineralogy, and hydrologic data. Much of Panama and the Amazon River basin in Brazil have been mapped and interpreted in this manner.

Geologic mapping can be greatly aided by the use of radar over conventional methods (Wing, 1971). The slight vegetation penetration ability at longer wavelengths removes any masking effect, thereby facilitating better definition of structures, contacts, lithology, and topographic roughness (Figure 1.15). Soil moisture and wetness significantly affect radar reflectance, however, and resolution constraints make detailed soil mapping more efficient from conventional aerial photography.

Unfortunately, only large area surveys can justify the present high cost of radar surveys. However, much of the United States has been imaged for testing purposes; companies involved in radar sensing or the Department of the Interior should be contacted for coverage available.

Passive Microwave

Passive microwave sensors do not provide their own illumination but have sensitivities similar to short wavelength radar, 0.1 millimeter to 3 centimeters. The microwave energy radiated from objects is a function of reflection, transmission, emission, and temperature, all of which can be influenced and affected. For example, emitted radiation from underlying rocks can be transmitted through the soil and emitted along with the soil's own characteristics of emission/reflection. Several properties of a material determine the image appearance sensed by microwave radiometers. Object shape or mass, viewing angle, surface texture, and the material's absorption coefficient determine its microwave emittance, transmission, and reflectance. In addition, the object temperature is of primary importance.

The instrumentation of passive microwave systems includes a recovery antenna for a radiometer, an amplifying network, and a video recording and/or viewing system. Resolution is best at lowest wavelengths, but in this range it is adversely affected by atmospheric attenuation. Longer wavelengths are not adversely attenuated and achieve sensitivities of ½°K from aircraft and 2°K from space platforms.

The low resolution qualities of passive microwave sensors make terrestrial applications unattractive. However, the low operating cost and suitability for space platform use make them very attractive for oceanographic and meteorologic studies.

U. S. SPACE PROGRAMS

It has long been recognized that remote sensing from satellite platforms could provide a unique perspective for earth resources surveys. The early weather satellites of the late 1950s provided the first synoptic observations. These were followed by the "pretty pictures" of the manned programs of Mercury, Gemini, and Apollo, most of which were used only for promotional purposes. Not until the 1970s, during the ERTS, SKYLAB, and LANDSAT II missions, was the full potential of earth resources surveys from satellite platforms tested and evaluated.

The use of satellites for remote sensing platforms presents both some major advantages and difficulties over the use of aircraft platforms. The major advantage lies in the regular predictable coverage and in a higher viewing altitude that can capture a large area in one

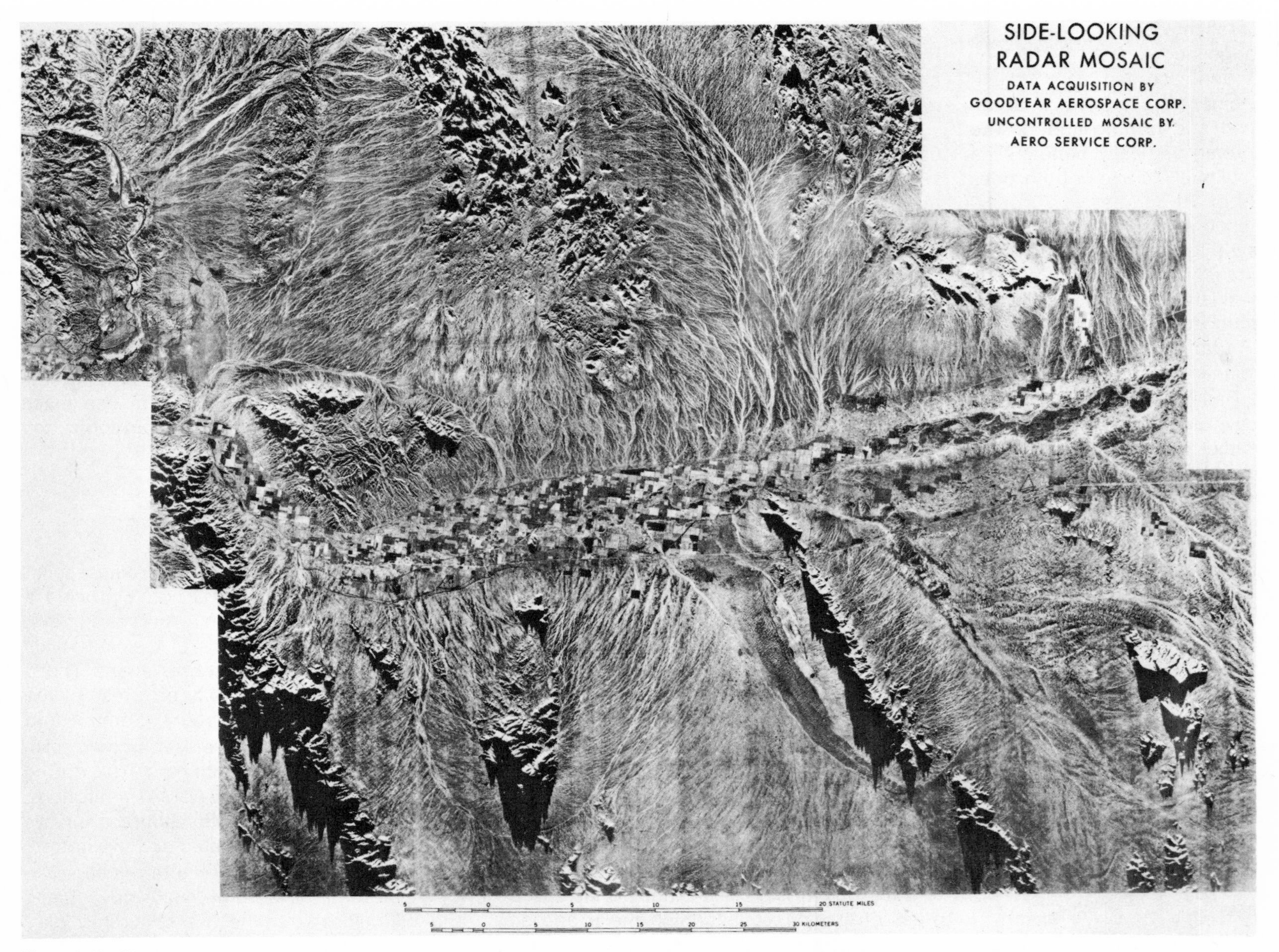

Figure 1.15. Side looking airborne radar imagery taken over the Phoenix, Arizona, region shows the ability of this imagery to portray surface topographic form. Several flight lines have been combined in this example to form a mosaic. Courtesy of Aero Service Corporation.

Figure 1.8. Color and color infrared photographs taken during October with peak foilage coloration facilitating vegetive identification. Plymouth County, Massachusetts, 1:7.200.

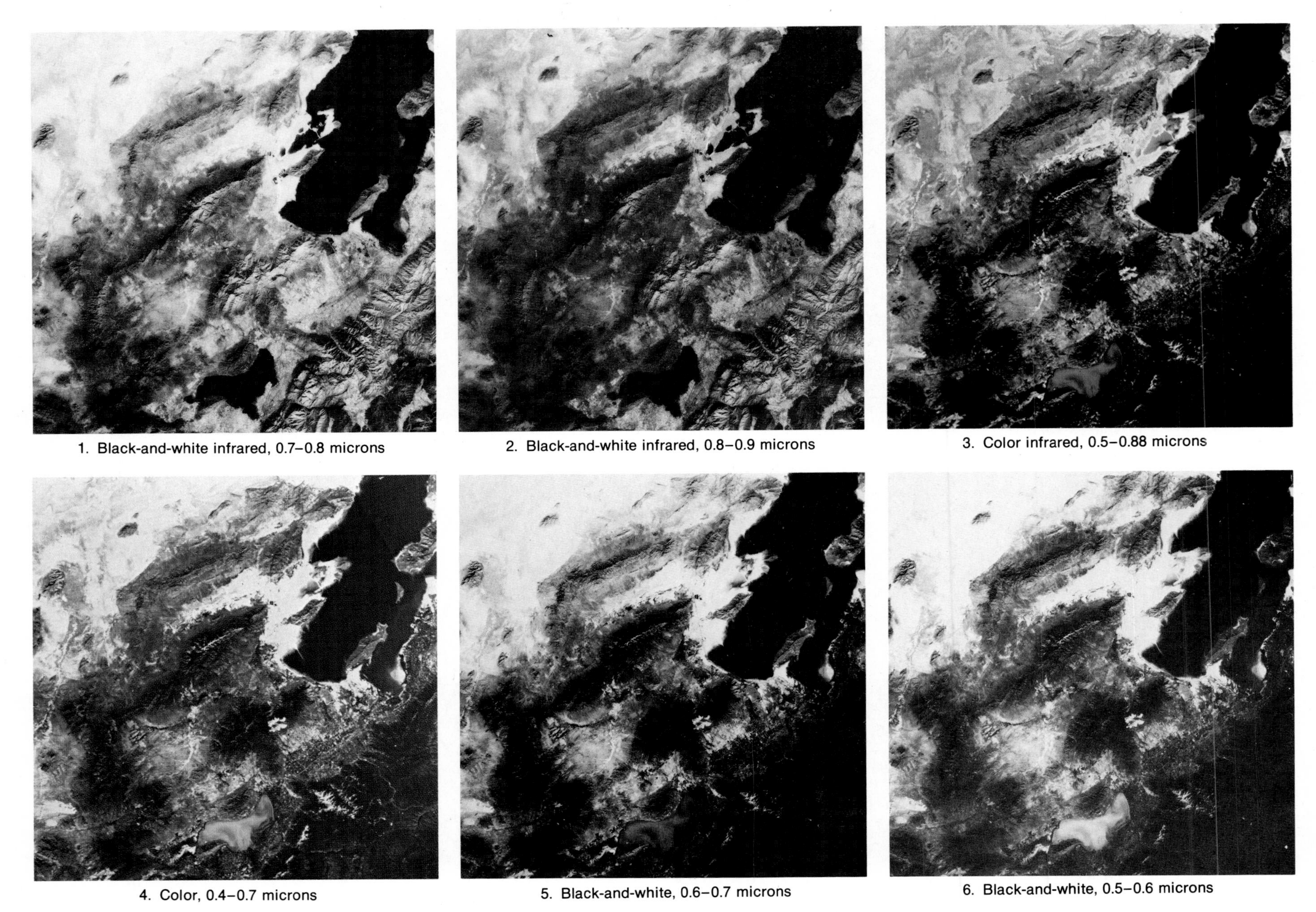

Figure 1.16. Skylab S190A Multispectral Camera. Exposures taken over Salt Lake City, Utah, illustrate the different film types and filtering effects obtained from this camera system.

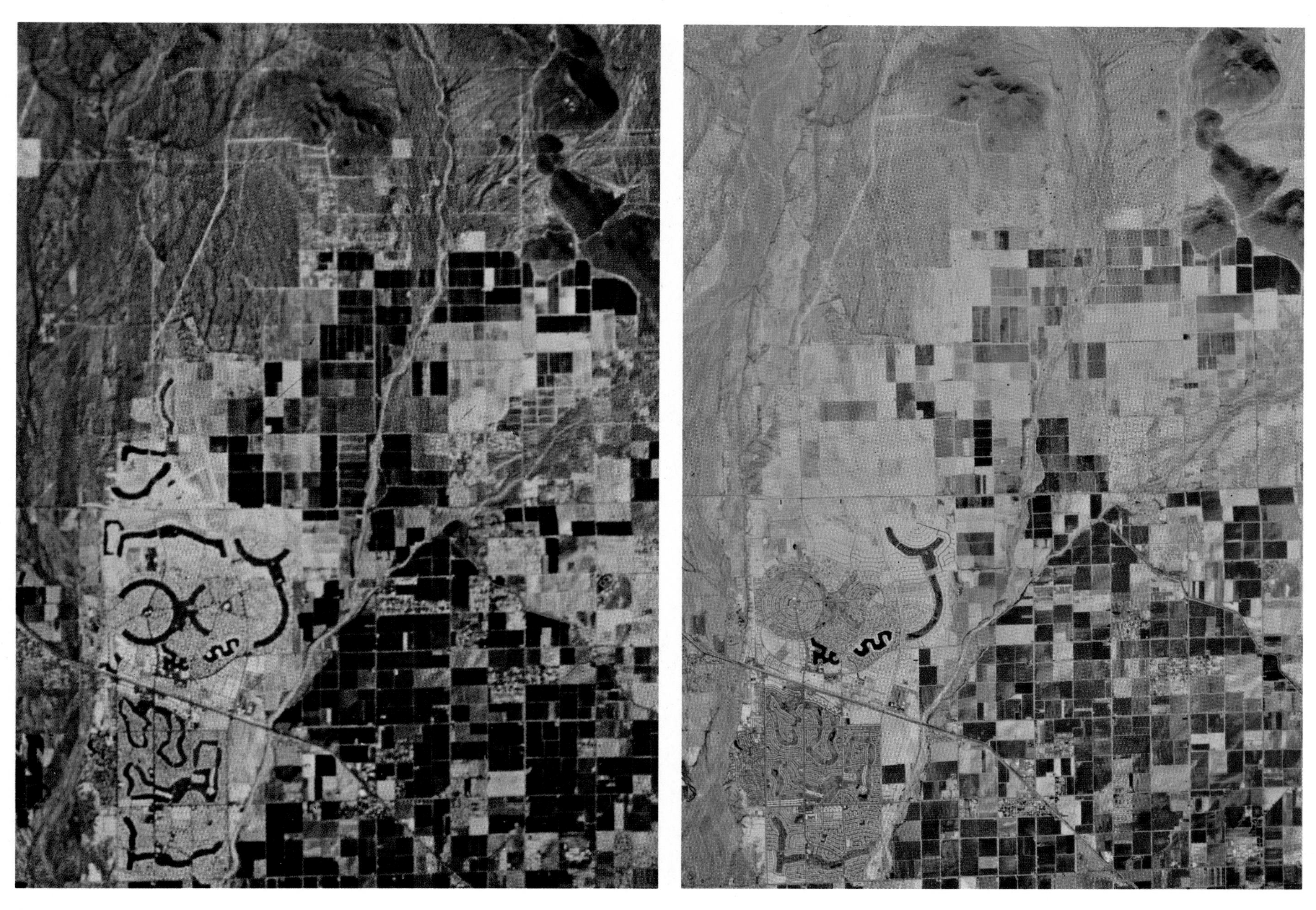

Figure 1.17. An enlarged section of a color exposure made through Skylab S190B high resolution Earth Terrain Camera is compared to a high altitude color infrared photograph. Note the high level of detail apparent on the Skylab enlargement (left) where individual houses can be identified. The area shown is Sun City, Arizona, near Phoenix. Scale, 1:125,000.

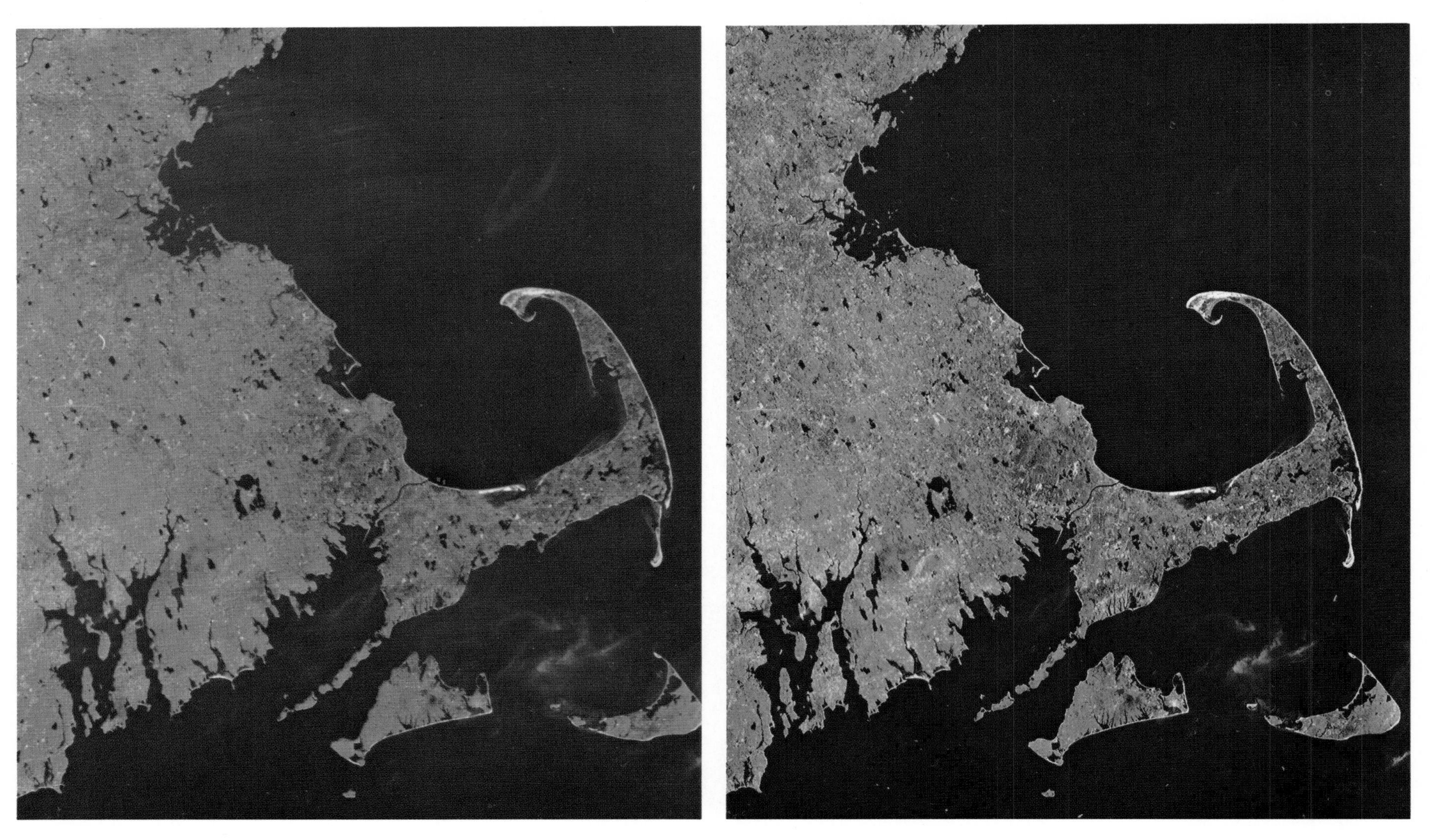

Figure 1.20. A sample of a conventional LANDSAT print (left) next to a computer digital enhanced version illustrates the increased detail available from this process. The computer program modifies the tonal contrast and enhances edges to effect greater apparent detail. The example, acquired July 6, 1976, shows Cape Cod and Boston, Massachusetts. Photographs prepared by the EROS Data Center, Sioux Falls.

frame. Once launched, a satellite can cover all portions of the earth's surface every X number of days if in polar orbit, or if located over the equator and orbiting at the same rate as the earth's rotation, can remain "stationary" over the ground surface (geosynchronous orbit). A high resolution geosynchronous satellite, Earth Observation Satellite (SEOS), for earth resources applications was proposed by the Department of the Interior and NASA in the early 1970s. SEOS would contain a minimum of three spectral bands operating at an altitude of 36,000 km and resolving picture elements of 20 to 50 meters. Unfortunately, this program has not yet been implemented.

Orbital satellites present a problem of imagery return; the image has to be either transmitted in electrical form or returned in capsules via reentry and recovery. Due to the expense and difficulties associated with the latter, electronic transmission is generally used.

Problems of vibration present in aircraft are eliminated in satellite platforms but maintenance flexibility is lost. Once a system is in orbit it is prohibitively difficult to service, which creates a use factor of high risk (Belcher et. al., 1967). The ERTS A satellite early in its career had electrical malfunctions that for a short time threatened its continued operation.

Early Unmanned Satellites

Most of the early satellites launched during the late 1950s and through the 1960s were test platforms for space remote sensing instrumentation. These included the Nimbus and Application Technology Satellite (ATS) programs directed toward meteorological applications. Early missions used instruments similar to television systems for cloud mapping and subsequent meteorological interpretations. Later versions have included a multitude of radiometers and spectrometers along with associated recording, transmitting, and power generation facilities. The results of these early missions have contributed to the concept and design of the Synchronous Meteorologic Satellite (SMS) launched in 1974. This satellite, part of the World Weather Watch program, is in stationary orbit, and it provides imagery of cloud cover, data on space environment, and a relay for other meteorological information.

Manned Spacecraft Missions

All of the manned orbital missions have been experimental and multipurpose and have not been directed exclusively to remote sensing. The resulting compromises of time, space, and weight have determined the type and amount of imagery-photography obtained. Later missions did place a greater emphasis on earth resources surveys, but early operations were primarily directed toward spacecraft system development and evaluation. Photographs from these missions can be used for large-scale terrain analysis, and their availability should be investigated as normal project procedure. Coverage and image quality can be quickly determined by the User Services Department of the EROS Data Center, Sioux Falls, South Dakota.

Mercury

The four orbital Mercury missions were directed toward systems testing but also included experiments in terrain photography. Cameras of 70 and 35 mm format were hand held, and a variety of films were exposed through the space vehicle windows.

Gemini

The ten Gemini flights provided additional earth photography by using mostly 70 mm cameras. Viewing was still done through windows except for short periods when the hatch was opened. Film types included color, color infrared, and some high resolution black-and-white.

Apollo

The Apollo program allowed for extended orbits, which created an opportunity to take many photographs, while the larger-sized spacecraft provided greater operational equipment flexibility. The program objective was lunar exploration, but many practical applications were shown for the acquired earth photography. Apollo 6, which was unmanned in earth orbit, obtained many photos; Apollo 9 provided for testing of multispectral equipment that was the prototype for the remote sensing ERTS program. Exposure of 70 mm cameras was accomplished either through windows or from mounts in the Scientific Instrument Module. More sophisticated sensing was directed at the lunar surface by using an 18-inch (610 mm) optical-bar panoramic camera with black-and-white high resolution film from which one-meter boulders critical to lunar landing stability could be identified. About 25% of the lunar surface was photographed and mapped from three missions.

Two books available from the Government Printing Office illustrate many of the photographs obtained from the Gemini and Apollo missions: *This Island Earth* (1970); and *Earth Photographs from Gemini III, IV, and V* (1967).

Skylab

Skylab provided for extended manned earth orbital flights that allowed for crew replacement, continued use of equipment, and extended experiments. It was placed in an orbit altitude of 235 nm (435 km) with a 53 degree inclination from the equator. This location provided a one orbit duration of 93 minutes and a repetitive ground tracking of five days. Mission one, SL-1, in June of 1973 included the launch and deployment of the Skylab module followed by three manned missions: mission two, SL-2 (May 25–June 22, 1973); mission three, SL-3 (July 28–September 25, 1973); and mission four, SL-4 (November 16–February 8, 1974). Over 35,000 frames of photography and 230,000 feet of earth resources magnetic tape was acquired.

The remote sensing package consisted of a 70 mm multispectral camera bank (S190A), a high resolution Earth Terrain Camera (S190B), an infrared spectrometer associated with a thirteen channel multispectral scanner, a microwave radiometer, scatterometer, an altimeter for enhancement of data interpretation, and an L-Band microwave radiometer for correction of water-vapor attenuations in the other microwave radiometers. The primary imagery for earth resources interpretation was to be supplied by the two camera systems, the S190A and S190B, and the multispectral scanner. Unfortunately, the multispectral scanner encountered operation difficulties and satisfactory imagery from it was not obtained. All of the earth's surface was not photographed. Consultation for coverage available should be directed to either the EROS Data Center or Chapter X and XI in *Skylab Earth Resources Data Catelog* (U. S. Government Printing Office, 1974).

Skylab provided for an indepth earth resources survey operation by using photographic and imagery instruments. Maintenance by crews meant that film and filters could be changed and returned to earth for processing. The 35,000 photographs acquired have contributed greatly to terrain, mineralogic, and land use studies, thereby demonstrating the value of manned remote sensing missions.

Multispectral Camera (S190A). The multispectral camera bank consisted of six high precision 70 mm cameras with different film-filter combinations that recorded, in total, the spectrum from 0.4 to 0.9 microns. The cameras were simultaneously exposed to capture a view of 88 nm (163 km) square with resolution as low as 100 feet (30 m) but averaging 150–250 feet (46–76 m). Table 1.8 outlines the station number for each of the S190A cameras, each's film-filter combination with resultant spectral sensitivity and ground resolution, as well as mission and roll numbers. The multispectral photographs provide an opportunity for multispectral analysis and recognition of spectral signatures of different crops and materials. (See Figure 1.16 in color section.) Several studies by Itek Corporation have illustrated crop typing and water quality analysis from sequential Skylab multispectral photography.

Earth Terrain Camera. The Earth Terrain Camera (S190B) has provided the highest quality photographic images of any manned

Table 1.8. Skylab S190A Multispectral Camera data

Sta	Filter	Filter bandpass, micrometer	Film type*	Estimated ground resolution††, feet (meters)	Mission & Roll no.		
					SL-2†	SL-3	SL-4
1	CC	0.7 – 0.8	EK 2424 (B&W infrared)	240 – 260 (73 – 79)	01‡,07,13	19,25,31, 37,43	49§,55,61, 67,73, A1,1B
2	DD	0.8 – 0.9	EK 2424 (B&W infrared)	240 – 260 (73 – 79)	02,08,14	20,26,32, 38,44	50§,56,62, 68,74,A2,2B
3	EE	0.5 – 0.88	EK 2443 (color infrared)	240 – 260 (73 – 79)	03,09,15	21,27,33, 39,45	51§,57,63, 69,75, A3,3B
4	FF	0.4 – 0.7	SO-356 (hi-resolution color)	130 – 150 (40 – 46)	04,10,16	22,28,34, 40,46	52§,58,64,70 76,A4,4B
5	BB	0.6 – 0.7	SO-022 (PANATOMIC-X B&W)	100 – 125 (30 – 38)	05,11,17	23,29,35, 41,47	53§,59,65, 71,77,A5,5B
6	AA	0.5 – 0.6	SO-022 (PANATOMIC-X B&W)	130 – 150 (40 – 46)	06,12,18	24,30,36, 42,48	54§,60,66, 72,78,A6,6B

* Eastman Kodak Company.
† SL-1 was the launch of Skylab without crew.
†† At low contrast.
‡ Note that all roll numbers are 2-digit numbers. Single-digit numbers were used in other cameras.
§ Without filter.

space mission. It contained an 18-inch (457 mm) f/4 lens that when used with high resolution films provided excellent detail (Figure 1.17 in color section). Several film types were exposed (Table 1.9) in a five-inch (127 mm) format, with each frame covering a 59 nm (109 km) square ground area. Sufficient detail is present on these photos to facilitate large- and medium-scale terrain analysis studies, and their availability should be investigated for any project area.

Multispectral Scanner. The multispectral scanner had operational difficulties, and imagery of acceptable quality was not obtained. The scanner was to provide imagery from thirteen channels ranging from 0.41 microns to thermal infrared at 12.50 microns. A scan width of 40 nm (74 km) was covered and processed on five-inch (127 mm) film for each band. Color and color infrared composites were to have been available.

Table 1.9. Skylab S190B Earth Terrain Camera data

Film type*	Wratten filter	Filter bandpass, micrometer	Estimated Ground resolution††, feet (meters)	Mission & Roll no.		
				SL-2	SL-3	SL-4
SO-242 (hi-resolution color)	none	0.4 – 0.7	70 (21)	81	83,84, 86,88	90,91, 92,94
EK 3414 (hi-definition B&W)	12†	0.5 – 0.7	55 (17)	82	85	89
EK 3443 (SL-2 & SL-3) (infrared color)	12	0.5 – 0.88	100 (30)	–	87	–
SO-131 (SL-4) (hi-resolution infrared color)	12	0.5 – 0.88	75 (23)	–	–	93

* Eastman Kodak Company.
¾ "Minus blue" filter.
†† At low contrast.

Space Shuttle

The Space Shuttle program defined in 1972 and expected to be operational in the 1980s promises to continue the exploration of earth resources through manned remote sensing operations. The concept of a reusable space vehicle minimizes much of the high cost and expendability associated with earlier space efforts. Specific payload instrumentation is flexible for each flight and will probably include multispectral photographic and scanning devices, thermal sensors, and camera systems of higher resolution than that of Skylab.

Erts–Landsat

The Earth Resources Technology Satellite (ERTS), launched July 23, 1972, was designed to test and evaluate the feasibility of mapping earth resources from unmanned space remote sensor platforms. After three years of operation, ERTS-I (now renamed LANDSAT I) exceeded original expectations, and the second ERTS satellite, LANDSAT II, was launched in November of 1975. Its instrumentation is nearly identical to that of the first ERTS-LANDSAT satellite (see Figure 1.18). The satellites are in a near-polar sun-synchronous orbit at an altitude of approximately 500 nm (900 to 950 km), which allows for full earth coverage every 18 days.

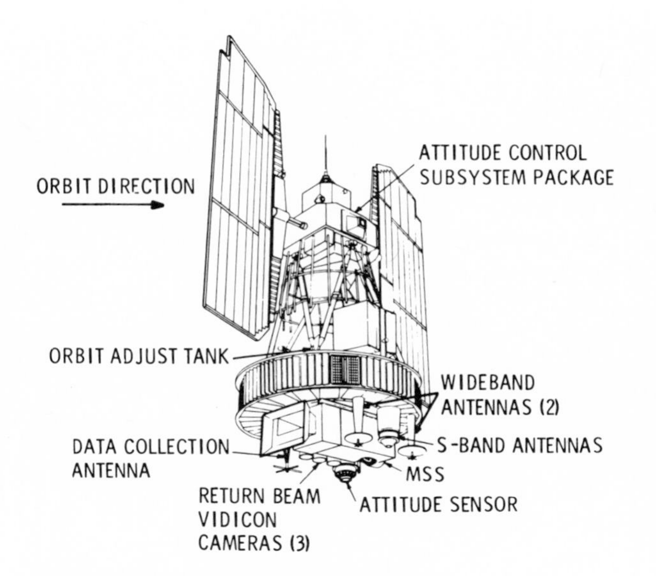

Figure 1.18. The ERTS–LANDSAT Satellite.

Both LANDSAT I and II have similar remote sensing capabilities: a data collection system (DCS) for receiving and relaying data from ground stations; a four channel multispectral scanner subsystem (MSS), and three return beam vidicon (RBV) cameras. The RBV system on LANDSAT I originally was intended to operate as the primary imagery supplier with the MSS to be used as a backup system. During the first week of operation a malfunction precluded further use of the RBV but did not seriously jeopardize the other instruments. A decision was made to use the MSS as the primary system; in retrospect, it far exceeded predicted quality expectations. From this experience, LANDSAT II was designed with MSS as the primary and RBV as the secondary system for image acquisition.

The thousands of high quality, cloud-free images provided by LANDSAT I and II have enabled many countries to undertake critical predevelopment surveys, identify potential energy resources, and monitor changes in the living environment. The high level of benefits from this experimental program should encourage a continued orbital unmanned sensing program.

Data Collection System (DCS). The Data Collection System (DCS) provides a data transfer function from remote ground Data Collection Platforms (DCP) to main ground receiving stations. Up to eight sensors can be included in a single DCP as designed by the user agency or researcher. The pertinent data is transmitted when the satellite is within range of both its ground station and the DCP. Data relayed to a ground receiving station is further relayed to the Ground Data Handling System (GDHS) at Greenbelt, Maryland. The information is then formatted, cataloged, and transferred to the various users.

The DCP records data continuously and transmits a 38 millisecond burst of data for all sensor channels every three minutes. LANDSAT receives these transmissions only when it is in range of a DCP simultaneously with a ground receiving station, which occurs anywhere in North America every twelve hours. At this time the data is transferred by real-time relay from the satellite to the ground station.

The potential of this type of data network is perhaps underrated. Earth scientists have always been plagued by difficulties of data collection, and this procedure can simply and quickly monitor hundreds of environmental situations. As a research experiment, the Army Crops of Engineers has many DCPs scattered over New England monitoring rainfall and streamflow. A high rainfall or building flood crest can quickly be identified, thereby facilitating early hydrologic management decisions.

Multispectral Scanner Subsystem (MSS). The Multispectral Scanner Subsystem (MSS) has been the imagery workhorse of the LANDSAT satellites. It operated without flaw on ERTS-I for over three years and surpassed preflight estimates of quality and resolution. The MSS is a four-band scanner sensitive to solar reflected energy from 0.5 through 1.1 microns. The bands are referenced as 4, 5, 6, and 7. Band 4 is sensitive from 0.5 to 0.6 microns (green); band 5, from 0.6 to 0.7 microns (red); band 6, from 0.7 to 0.8 microns (near-infrared); and band 7, from 0.8 to 1.1 microns (also near-infrared).

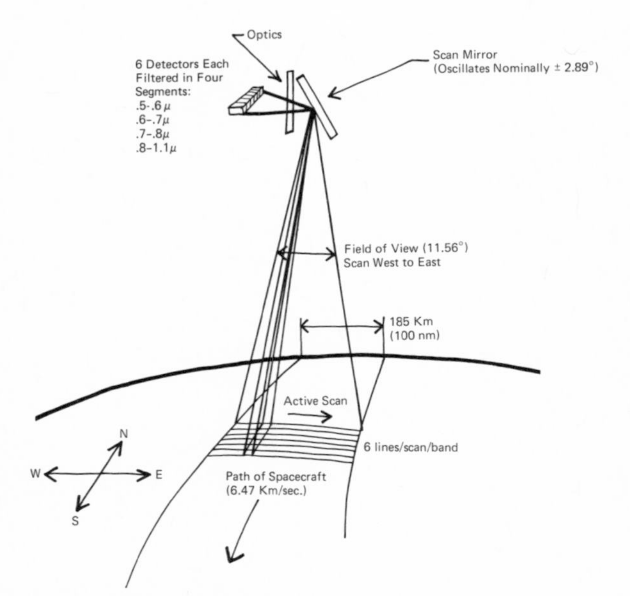

Figure 1.19. The LANDSAT MSS Scanner.

The scanner (Figure 1.19) operates with a scan mirror oscillating from west to east through an arc of 11.56 degrees and covering a ground area of 100 nm (185 km). One swing of the mirror includes six scan lines each 79 meters wide on the ground. The resulting smallest bit of electrical input (pixel) covers a ground area of 79 by 57 meters. As one scan of six lines is observed, the spacecraft advances exactly 474 meters at a speed of 6.47 km per second. Thus the next set of scanned lines aligns exactly with the previous set. The light from each set of six scanned lines is transmitted by optical fibers to four detectors, each of which produces an electronic signal. These signals are multiplexed into a serial digital stream and encoded to a ground station. Processing of signals into images and recording on film is done by the NASA Data Processing Facility (NDPF) at Goddard Spacecraft Center, with negative film copies sent to other image-processing facilities such as the EROS Data Center. The final images cover a 100 nm (185 km) by 100 nm area.

Resolution of MSS images is controlled primarily by the pixel size, quality of the copy negative used, and whether image enhancement procedures are employed. Examination of publically ordered images indicates a ground resolution of approximately 100 meters or 300 feet. However, under controlled conditions, as reported by the Earth Satellite Corporation in 1976, resolution by processing direct from computer-compatible tapes can approach 79 meters (see Figure 1.20 in color section).

Several MSS output products are available from the image-processing centers, including various black-and-white positive and negative

Table 1.10. Scale comparison of LANDSAT MSS, Skylab S190A Multispectral Camera, and Skylab S190B Earth Terrain Camera

ERTS – LANDSAT (MSS)		SKYLAB S190A (Multispectral)		SKYLAB S190B (Earth Terrain Camera)	
Scale	Image size in inches (cm)	Scale	Image size in inches (cm)	Scale	Image size in inches (cm)
1:3,369,000	2.2 × 2.2 (5.58 × 5.58)				
		1:2,850,000	2.25 × 2.25 (5.72 × 5.72)		
1:1,000,000	7.3 × 7.3 (18.5 × 18.5)	1:1,000,000	6.41 × 6.41 (16.29 × 16.29)		
				1:950,000	4.5 × 4.5 (11.43 × 11.43)
1:500,000	14.6 × 14.6 (37.1 × 37.1)	1:500,000	12.83 × 12.83 (32.5 × 32.5)	1:500,000	8.55 × 8.55 (21.72 × 21.72)
1:250,000	29.2 × 29.2 (74.2 × 74.2)	1:250,000	25.65 × 25.65 (65.15 × 65.15)	1:250,000	17.10 × 17.10 (43.43 × 43.43)
				1:125,000	34.2 × 34.2

films and prints of the four bands, color composites from three spectral bands, and computer-compatible tapes. (See Table 1.10 for a scale comparison of MSS imagery and that obtained by Skylab.) The MSS color composites simulate the appearance of color infrared photographs: Band 4 (green) is assigned yellow, band 5 (red) is assigned magenta, and band 6 or 7 (near-infrared) is assigned cyan. Projecting white light through the three overlaid color transparences creates the false color composite (see Figure 1.20).

Return Beam Vidicon (RBV). The Return Beam Vidicon (RBV) system covers the same ground area as the MSS scanner, but is usually not operated concurrently; it is saved as a backup system. It consists of three television-like cameras each filtered as follows: channel 1 is sensitive from 0.475 to 0.575 microns (green); channel 2, from 0.580 to 0.680 microns (red); and channel 3, from 0.690 to 0.830 microns (near-infrared). The quality is similar to, but not as good as that provided by the MSS system.

REFERENCES

Adams, W., L. Lepley, C. Warren, and S. Chang, "Coastal and Urban Surveys with IR," *Photogrammetric Engineering,* Vol. 36, No. 2, pp. 173–180, 1970.

American Society of Photogrammetry, *Manual of Photographic Interpretation,* Falls Church, Va., 1960.

American Society of Photogrammetry, *Manual of Photogrammetry,* Vols. I and II, Falls Church, Va., 1966.

American Society of Photogrammetry, *Manual of Color Aerial Photography,* Falls Church, Va., 1968.

American Society of Photogrammetry, "Select Papers on Remote Sensing," *Selected Proceedings of the Third Conference on Remote Sensing,* Ann Arbor, Mich., 1968.

American Society of Photogrammetry, *Manual of Remote Sensing,* Vols. 1 and 2, Falls Church, Va., 1975.

Anshutz, G., and A. H. Stallard, "An Overview of Site Evaluation," *Photogrammetric Engineering,* Vol. 33, No. 12, pp. 1381–1396, 1967.

Anson, A., "Color Aerial Photos in the Reconnaissance of Soils and Rocks," *Photogrammetric Engineering,* Vol. 36, No. 4, p. 343, 1970.

Ashley, M. D., and R. E. Roger, "Tree Heights and Upper Stem Diameters," *Photogrammetric Engineering,* Vol. 35, No. 2, pp. 136–146, 1969.

Avery, T. E., *Interpretation of Aerial Photographs,* 2nd ed., Burgess, Minneapolis, 1968.

Baker, R. D., "Aerial Photographs in the Forest," *Photogrammetric Engineering,* Vol. 33, No. 12, pp. 1373–1376, 1967.

Barr, D. J., and M. D. Hensey, "Industrial Site Study with Remote Sensing," *Photogrammetric Engineering,* Vol. 40, No. 2, 1974.

Barr, D. J., and R. D. Miles, "Techniques for Utilizing Side-Looking Airborne Radar (SLAR) Imagery in Regional Highway Planning," *Highway Research Board Special Report No. 102,* National Academy of Sciences, Washington, D.C., 1969.

Belcher, D., et al., *Potential Benefits to be Derived from Applications of Remote Sensing of Agricultural, Forests and Range Resources,* Center for Aerial Photographic Studies, Cornell University, Ithaca, N.Y., 1967.

Belcher, D. J., "The Engineering Significance of Landforms," *Highway Research Board Bulletin,* No. 13, 1948.

Bird, S.J.G., "Environmental Criteria for Recreationally Oriented Highway Planning," *Highway Research Board Report No. 452,* National Academy of Sciences, Washington, D.C., 1973.

Bock, P., and J. G. Barmby, "Survey Effectiveness of Spacecraft Remote Sensors," *Photogrammetric Engineering,* Vol. 35, No. 8, pp. 756–762, 1969.

Blythe, R., and E. Kurath, "Infrared and Water Vapor," *Photogrammetric Engineering,* Vol. 33, No. 7, pp. 772–777, 1967.

Bradie, R. A., "SLAR Imagery for Sea Ice Studies," *Photogrammetric Engineering,* Vol. 33, No. 7, pp. 763–766, 1967.

Chaves, J. R. and R. L. Schuster, "Use of Aerial Color Photography in Materials Survey," *Highway Research Board Report No. 63,* National Academy of Sciences, Washington, D.C., 1964.

Christian, C. S., J. N. Jennings, and C. R. Twidale, "Guidebook to Research Data for Arid Zone Development," *Arid Zone Research,* UNESCO, Paris, pp. 51–65, 1957.

Christian, C. S., "The Concept of Land Units and Land Systems," *Proceedings of the Ninth Pacific Science Congress,* Vol. 20, pp. 74–81, 1958.

Christian, C. S., "Aerial Surveys and Integrated Studies," *Proceedings of the Toulouse Conference, 1964,* UNESCO, 1968.

Ciesla, W. M., J. C. Bell, and J. W. Curun, "Color Photos and the Southern Pine Beetle," *Photogrammetric Engineering,* Vol. 33, No. 8, pp. 883–888, 1968.

Cochrane, G. R., "False-Color Film Fails in Practice," *Photogrammetric Engineering,* Vol. 34, No. 11, pp. 1142–1146, 1968.

Colvocoresses, A. P., "Erts-A Satellite Imagery," *Photogrammetric Engineering,* Vol. 36, No. 6, pp. 555–560, 1970.

Colwell, R. N., "Some Practical Applications of Multiband Spectral Reconnaissance," *American Scientist,* Vol. 49, No. 1, pp. 9–36, 1961.

Colwell, R. N., "Uses and Limitations of Multi-Spectral Remote Sensing," *Proceedings of the 4th Symposium on Remote Sensing of the Environment,* University of Michigan, Ann Arbor, 1966.

Crandall, C. J., "Radar Mapping in Panama," *Photogrammetric Engineering,* Vol. 35, No. 7, pp. 641–646, 1967.

Currey, D. T., "Identifying Flood Water Movement," *Remote Sensing of the Environment,* Vol. 6, No. 1, 1977.

Danko, J. A., Jr., "A New Concept in Orthophotography," *Photogrammetric Engineering,* Vol. 39 (November) 1973.

DeCaprio, G. R., and J. E. Wasielewski, "Radar, Image Processing and Interpreter Performance," *Photogrammetric Engineering,* Vol. 42, No. 8, 1976.

Dellwig, L. F. and C. Burchell, "Side-Look Radar: Its Uses and Limitations as a Reconnaissance Tool," *Highway Research Board Report No. 421,* National Academy of Sciences, Washington, D.C., 1972.

DeLoach, W. C., "Remote-Sensing Applications to Environmental Analysis," *Highway Research Board Report No. 452,* National Academy of Sciences, Washington, D.C., 1973.

Eastman Kodak Co., *Kodak Wratten Filters,* Eastman Kodak Co., Rochester, N.Y., 1965.

El-Ashry, M. R., and H. R. Wanless, "Shoreline Features and Their Changes," *Photogrammetric Engineering,* Vol. 33, No. 2, pp. 184–189, 1967.

Evans, R. M., *An Introduction to Color,* Wiley, New York, 1948.

Eyre, L. A., B. Adolphus, and M. Amiel, "Census Analysis and Population Studies," *Photogrammetric Engineering,* Vol. 36, No. 5, pp. 460–466, 1970.

Fischer, W. A., "Spectral Reflectance Measurements as a Basis for Film-Filter Selection for Photographic Differentiation of Rock Units," *Geologic Survey Research,* 1960.

Fischer, W. A., "Color Aerial Photography in Geologic Investigations," *Photogrammetric Engineering,* Vol. 28, No. 1, p. 133, 1962.

Fischer, W. A., "Examples of Remote Sensing Applications to Engineering," *Highway Research Board Special Report No. 102,* National Academy of Sciences, Washington, D.C., 1969.

Garofalo, D. and F. Wobber, "Solid Waste and Remote Sensing," *Photogrammetric Engineering,* Vol. 40, No. 2, 1974.

Gausman, H. W., W. A. Allen, R. Cardenas, and R. L. Bowen, "Color Photos, Cotton Leaves and Soil Salinity," *Photogrammetric Engineering,* Vol. 36, No. 5, pp. 454–459, 1970.

Gilbertson, B., T. G. Longshaw, and R. P. Viljoen, "Multispectral Aerial Photography as Exploration Tool," *Remote Sensing of the Environment,* Vol. 5, No. 2, 1976.

Hammac, J. C., "Landsat Goes to Sea," *Photogrammetric Engineering,* Vol. 43, (June) 1977.

Harris, W. D. and M. J. Umbach, "Underwater Mapping," *Photogrammetric Engineering,* Vol. 38, No. 8, 1972.

Heath, G. R., "Hot Spot Determination," *Photogrammetric Engineering,* Vol. 39, (November) 1973.

Helgenson, G. A., "Water Depth and Distance Penetration," *Photogrammetric Engineering,* Vol. 36, No. 2, pp. 164–172, 1970.

Henderson, F. M., "Radar for Small-Scale Land-Use Mapping," *Photogrammetric Engineering,* Vol. 41, No. 3, 1975.

Hills, G. A., "The Use of Aerial Photography in Mapping Soil Sites," *Forestry Chronicle,* Vol. 26, No. 1, pp. 4–37, 1950.

Hitchcock, H. C., T. L. Cox, F. P. Baxter, and C. W. Smart, "Soil and Land Cover Overlay Analysis," *Photogrammetric Engineering,* Vol. 41, (December) 1975.

Holmes, R. F., "Engineering Materials and Side-looking Radar," *Photogrammetric Engineering,* Vol. 33, No. 7, pp. 767–770, 1967.

Institute of Science and Technology, *Proceedings of the 3rd Symposium on Remote Sensing of the Environment,* University of Michigan, Ann Arbor, 1965.

International Remote Sensing Institute, "Remote Sensing," Vol. I and II, *1st Annual IRSI Symposium Proceedings,* Sacramento, Calif., 1969.

Kedar, E. Y., W. Paterson, and S. Hsu, "Earthquake-Risk Mapping," *Photogrammetric Engineering,* Vol. 39, (August) 1973.

Keene, D. F. and W. G. Pearcy, "High-Altitude Photos of the Oregon Coast," *Photogrammetric Engineering,* Vol. 39, (February) 1973.

Kiefer, R. W., "Landform Features in the United States," *Photogrammegic Engineering,* Vol. 33, No. 2, pp. 174–182, 1967.

Kristof, S. J. and A. L. Zachary, "Mapping Soil Features from Multispectral Scanner Data," *Photogrammetric Engineering,* Vol. 40, No. 12, 1974.

Kuhl, A. D., "Color and IR Photos for Soils," *Photogrammetric Engineering,* Vol. 36, No. 5, pp. 475–482, 1970.

Laprade, G. L., and E. S. Leonardo, "Elevations from Radar Imagery," *Photogrammetric Engineering,* Vol. 35, No. 4, pp. 366–371, 1969.

Latham, J. P., and R. E. Witmer, "Comparative Waveform Analysis of Multisensor Imagery," *Photogrammetric Engineering,* Vol. 33, No. 7, pp. 779–786, 1967.

Laver, D. T., "Multispectral Sensing of Forest Vegetation," *Photogrammetric Engineering,* Vol. 35, No. 4, pp. 346–354, 1969.

Leachtenauer, J. C., "Photo Interpretation Test Development," *Photogrammetric Engineering,* Vol. 39 (November) 1973.

Leamer, R. W., D. A. Weber, and C. L. Wiegand, "Pattern Recognition of Soils and Crops from Space," *Photogrammetric Engineering,* Vol. 41, No. 4, 1975.

Legault, R. R., and F. C. Polcyn, "Investigations of Multi-Spectral Image Interpretation," *Proceedings of the 3rd Symposium on Remote Sensing of the Environment,* University of Michigan, Ann Arbor, 1964.

Lent, J. D., and G. A. Thorley, "Some Observations on the Use of Multiband Spectral Reconnaissance for the Inventory of Wildland Resources," *Remote Sensing of Environment,* Vol. 1, No. 1, pp. 31–45, 1969.

Lepley, L. K., "Coastal Water Clarity from Space Photographs," *Photogrammetric Engineering,* Vol. 34, No. 7, pp. 667–673, 1968.

Letourneaux, P. J. "Improving Quality of Aerial Color Prints," *Photogrammetric Engineering,* Vol. 35, No. 2, pp. 147–152, 1969.

Lins, Jr., H. F., "Land-Use Mapping From Skylab S-190B Photography," *Photogrammetric Engineering,* Vol. 42, No. 3, 1976.

Lowe, D. S., and C. L. Wilson, "Multispectral Scanning Systems: Their Features and Limitations," *Highway Research Board Report No. 421,* National Academy of Sciences, Washington, D.C., 1972.

Lueder, D. R., *Aerial Photographic Interpretation: Principles and Application,* McGraw-Hill, New York, 1959.

Lyon, R. J. P., "The Multiband Approach to Geological Mapping from Orbiting Satellites: Is it Redundant or Vital?" *Remote Sensing of Environment,* Vol. 1, No. 4, pp. 237–244, 1970.

MacConnell, W. P., and P. Stoll, "Evaluating Recreational Resources of the Connecticut River," *Photogrammetric Engineering,* Vol. 35, No. 7, pp. 686–692, 1969.

Mairs, R. L., and D. K. Clark, "Remote Sensing of Estuarine Circulation," *Photogrammetric Engineering,* Vol. 39 (September) 1973.

Mars, R. W., "Applications of Remote-sensing Techniques to the Geology of the Bonanza Volcanic Center," *Remote Sensing Report 73-1,* Colorado School of Mines, Golden, 1973.

McCue, G. A., and J. Green, "Roughness of Simulated Planetary Terrain," *Photogrammetric Engineering,* Vol. 36, No. 3, pp. 273–279, 1970.

McDowell, D. Q., and M. R. Specht, "Spectral Reflectance Using Aerial Photographs," *Photogrammetric Engineering,* Vol. 40, No. 5, 1974.

McEwen, R. B., W. J. Kosco, and V. Carter, "Coastal Wetlands Mapping," *Photogrammetric Engineering,* Vol. 42, No. 2, 1976.

Malila, W. A., "Multispectral Techniques for Image Enhancement and Discrimination," *Photogrammetric Engineering,* Vol. 34, No. 6, pp. 566–575, 1968.

Merifield, P. M., J. Cronin, L. L. Foshee, S. J. Gawarecky, J. T. Neal, R. E. Stevenson, R. O. Stone, and R. S. Williams, "Satellite Imagery of the Earth," *Photogrammetric Engineering,* Vol. 35, No. 7, pp. 654–668, 1969.

Meyer, M. P., and L. Calpouzos, "Detection of Crop Diseases," *Photogrammetric Engineering,* Vol. 34, No. 6, pp. 554–557, 1968.

Meyer, M. P., and H. A. Maklin, "P.I. Techniques for Ektachrome IR Transparencies," *Photogrammetric Engineering,* Vol. 35, No. 11, pp. 1111–1114, 1969.

Mintzer, O. W., "Remote Sensing for Engineering of Terrain-Photographic Systems," Ohio State University, Columbus, unpublished report, 1968.

Moessner, K. E., "Comparative Usefulness of Three Parallax Measuring Instruments in the Measurement and Interpretation of Forest Stands," *Photogrammetric Engineering,* Vol. 27, No. 5, pp. 705–709, 1961.

Moore, R. K., "Heights from Simultaneous Radar and Infrared," *Photogrammetric Engineering,* Vol. 35, No. 7, pp. 649–651, 1969.

Morain, S. A., and D. S. Simonett, "K-Band Radar in Vegetation Mapping," *Photogrammetric Engineering,* Vol. 33, No. 7, pp. 730–740, 1967.

Murtha, P. A., *A Guide to Air Photo Interpretation of Forest Damage in Canada,* Canadian Forestry Service, Ottawa, 1972.

Musgrove, R. G., "Photometry for Interpretation," *Photogrammetric Engineering,* Vol. 35, No. 10, pp. 1015–1023, 1969.

Myers, U. I., and M. D. Heilman, "Thermal Infrared for Soil Temperature Studies," *Photogrammetric Engineering,* Vol. 35, No. 10, pp. 1024–1032, 1969.

National Academy of Sciences, *Remote Sensing: with Special Reference to Agriculture and Forestry,* Washington, D.C., 1970.

National Aeronautics and Space Administration, *This Island Earth,* Washington, D.C., 1970.

National Aeronautics and Space Administration, *Earth Photographs from Gemini III, IV, and V,* Washington, D.C., 1967.

Neubert, R. W., "Sick Trees" *Photogrammetric Engineering,* Vol. 35, No. 5, pp. 472–475, 1969.

Newton, A. R., "Pseudostereoscopy with Radar Imagery," *Photogrammetric Engineering,* Vol. 39 (October) 1973.

Noble, V. E., "Ocean Swell Measurements from Satellite Photographs," *Remote Sensing of Environment,* Vol. 1, No. 3, pp. 151–154, 1970.

Northrop, K. G., and E. W. Johnson, "Forest Cover Type Identification," *Photogrammetric Engineering,* Vol. 36, No. 5, pp. 483–490, 1970.

Nunnally, N. R., "Integrated Landscape Analysis with Radar Imagery," *Remote Sensing of Environment,* Vol. 1, No. 1, pp. 1–6, 1969.

Nunnally, N. R., and Witmer, R. E., "Remote Sensing for Land-Use Studies," *Photogrammetric Engineering,* Vol. 36, No. 5, pp. 449–453, 1970.

Olson, C. E., L. W. Tombaugh, and H. C. Davis, "Inventory of Recreation Sites," *Photogrammetric Engineering,* Vol. 35, No. 6, pp. 561–568, 1969.

Otterman, J. and Fraser, R. S., "Earth-Atmospheric System and Surface Reflectances in Arid Regions from LANSAT MSS Data," *Remote Sensing of the Environment,* Vol. 5, No. 4, 1976.

Parry, J. T., W. R. Cowan, and J. A. Heginbottom, "Soils Studies Using Color Photos," *Photogrammetric Engineering,* Vol. 35, No. 1, pp. 44–56, 1969.

Parry, J. T., W. R. Cowan, and J. A. Heginbottom, "Color for Coniferous Forest Species," *Photogrammetric Engineering,* Vol. 35, No. 7, pp. 669–678, 1969.

Pease, R. W., D. A. Nichols, "Energy Balance Maps from Remotely Sensed Imagery," *Photogrammetric Engineering,* Vol. 42, No. 11, 1976.

Pestrong, R., "Multiband Photos for a Tidal Marsh," *Photogrammatic Engineering,* Vol. 35, No. 5, pp. 453–470, 1969.

Philpotts, L. E., and V. R. Wallen, "IR Color for Crop Disease Identification," *Photogrammetric Engineering,* Vol. 35, No. 11, pp. 1116–1125, 1968.

Quiel, F., "Thermal/IR in Geology," *Photogrammetric Engineering,* Vol. 41, No. 3, 1975.

Raines, G. L., and K. Lee, "In Situ Rock Reflectance," *Photogrammetric Engineering,* Vol. 41, No. 2, 1975.

Reeves, F. B., "Mensuration: Color vs. Pan," *Photogrammetric Engineering,* Vol. 36, No. 3, pp. 239–244, 1970.

Reimold, R. J., J. L. Gallagher, and D. E. Thompson, "Remote Sensing of Tidal Marsh," *Photogrammetric Engineering,* Vol. 39, (May) 1973.

Reinheimer, C. J., C. L. Rudder, and J. L. Berrey, "Detection of Petroleum Spills," *Photogrammetric Engineering,* Vol. 39 (December) 1973.

Rib, H. T., "An Optimum Multisensor Approach for Detailed Engineering Soils Mapping," 2 Vols., *Joint Highway Research Project,* Report No. 22, Purdue University, Lafayette, Ind., 1966.

Rib, H. T., and R. D. Miles, "Automatic Interpretation of Terrain Features," *Photogrammetric Engineering,* Vol. 35, No. 2, pp. 153–164, 1969.

Richardson, A. J., A. H. Gerbermann, H. W. Gausman, and J. A. Cuellar, "Detection of Saline Soils with Skylab Multispectral Scanner Data," *Photogrammetric Engineering,* Vol. 42, No. 5, 1976.

Richter, D. M., "Sequential Urban Change," *Photogrammetric Engineering,* Vol. 35, No. 8, pp. 764–770, 1969.

Richter, D. M., "An Airphoto Index to Physical and Cultural Features in the Western United States," *Photogrammetric Engineering,* Vol. 33, No. 12, pp. 1402–1419, 1967.

Ritchie, J. C., F. R. Schiebe, and J. R. McHenry, "Remote Sensing of Suspended Sediments in Surface Waters," *Photogrammetric Engineering,* Vol. 42, No. 12, 1976.

Roberts, L. H., "Electron Viewer for Multiband Imagery," *Photogrammetric Engineering,* Vol. 39 (February) 1973.

Rohde, W. G., and C. E. Olson, "Detecting Tree Moisture Stress," *Photogrammetric Engineering,* Vol. 36, No. 6, pp. 561–566, 1970.

Rosenshein, J. S., C. R. Goodwin, and A. Jurado, "Bottom Configuration and Environment of Tampa Bay," *Photogrammetric Engineering,* Vol. 43 (June) 1977.

Ross, D. S., "Atmospheric Effects in Multispectral Photos," *Photogrammetric Engineering,* Vol. 39 (April) 1973.

Ross, D. S., "Simple Multispectral Photos and Color Viewing," *Photogrammetric Engineering,* Vol. 39 (June) 1975.

Sabins, F. E., Jr., "Flight Planning for Thermal IR," *Photogrammetric Engineering,* Vol. 39 (January) 1973.

Sabins, F. F., Jr., "Recording and Processing Thermal IR Imagery," *Photogrammetric Engineering,* Vol. 39 (August) 1973.

Sabins, F. F., "Infrared Imagery and Geologic Aspects," *Photogrammetric Engineering,* Vol. 33, No. 7, pp. 743–750, 1967.

Sayn-Wittgenstein, L., "Recognition of Tree Species on Air Photographs by Crown Characteristics," *Photogrammetric Engineering,* Vol. 27, No. 5, pp. 792–809, 1961.

Schepis, E. L., "Time-Lapse Remote Sensing in Agriculture," *Photogrammetric Engineering,* Vol. 34, No. 11, p. 1166, 1968.

Scherz, J. P., D. R. Graff, and W. C. Boyle, "Photographic Characteristics of Water Pollution," *Photogrammetric Engineering,* Vol. 35, No. 1, pp. 38–43, 1969.

Seher, J. S., and P. T. Tueller, "Color Aerial Photos for Marshland," *Photogrammetric Engineering,* Vol. 39 (May) 1973.

Siegal, B. S., and M. J. Abrams, "Geologic Mapping Using LANDSAT Data," *Photogrammetric Engineering,* Vol. 42, No. 3, 1976.

Siegal, B. S., and A.F.H. Goetz, "Effect of Vegetation on Rock and Soil Type Discrimination," *Photogrammetric Engineering,* Vol. 43, (February) 1977.

Silvestro, F. B., "Object Detection Enhancement," *Photogrammetric Engineering,* Vol. 35, No. 6, pp. 555–559, 1969.

Simons, J. H. "Some Applications of Side-Looking Airborne Radar," Paper written for Contract AN/APQ-56, Raytheon Company, Alexandria, Va.

Slater, P. N., and R. A. Schowengerdt, "Sensor Performance for Earth Resources," *Photogrammetric Engineering,* Vol. 39 (February) 1973.

Smith, W. L., ed., *Remote Sensing Applications for Mineral Exploration,* Dowden, Hutchinson, & Ross, Stroudsburg, Pa., 1977.

Specht, M. R., "IR and Pan Films," *Photogrammetric Engineering,* Vol. 36, No. 4, pp. 360–364, 1970.

Specht, M. R., D. Needler, and N. L. Fritz, "New Color Film for Water Penetration," *Photogrammetric Engineering,* Vol. 39 (April) 1973.

Stafford, D. B., and J. Langfelder, "Air Photo Survey of Coastal Erosion," *Photogrammetric Engineering,* Vol. 37, No. 6, 1971.

Steiner, D., "Time Dimension for Crop Surveys from Space," *Photogrammetric Engineering,* Vol. 36, No. 2, p. 187, 1970.

Steiner, D., and H. Haefner, "Grey Tone Distortion on Aerial Photographs—A Problem for the Quantified and Automated Photointerpreptation of Terrain Cover Types," Paper presented at the 30th Annual Meeting of the American Society of Photogrammetry, 1964.

Steinitz, C. S., T. Murray, D. Sinton, and D. Way, *A Comparative Evaluation of Resource Analysis Methods,* U.S. Army Corps of Engineers, Research Contract DACW33-68-CO152, Harvard University, Graduate School of Design, Cambridge, Mass., 1969.

Stephens, P. R., "Comparison of Color, Color Infrared, and Panachromatic Aerial Photography," *Photogrammetric Engineering,* Vol. 42, No. 12, 1976.

Stingelin, R. W., "Airborne Infrared Imagery and Its Limitations in Civil Engineering Practice," *Highway Research Board Report No. 421,* National Academy of Sciences, Washington, D.C., 1972.

Strandberg, C. H., *Aerial Discovery Manual,* John Wiley, New York, 1967.

Strandberg, C. H., "Photoarchaeology," *Photogrammetric Engineering,* Vol. 33, No. 10, pp. 1152–1157, 1970.

Strong, A. E., and I. S. Ruff, "Utilizing Satellite-Observed Solar Reflections from the Sea Surface as an Indicator of Surface Wind Speeds," *Remote Sensing of Environment,* Vol. 1, No. 3, pp. 181–185, 1970.

Struble, R. A., and O. W. Mintzer, "Combined Investigation Techniques for Procuring Highway Design Data," *Proceedings of the 18th Annual Highway Geology Symposium,* Purdue University, Lafayette, Ind., 1967.

Tuyahow, A. J., and R. K. Holz, "Remote Sensing of a Barrier Island," *Photogrammetric Engineering,* Vol. 39 (February) 1973.

Ulaby, F. T., and J. McNaughton, "Classification of Physiography from ERTS Imagery," *Photogrammetric Engineering,* Vol. 41, No. 8, 1975.

Veign, J. L., and F. B. Reeves, "A Case for Orthophoto Mapping," *Photogrammetric Engineering,* Vol. 39 (October) 1973.

Viksne, A., T. C. Liston, and C. D. Sapp, "SLR Reconnaissance of Panama," *Photogrammetric Engineering,* Vol. 36, No. 3, pp. 253–259, 1970.

Van Lopik, J. R., A. E. Pressman, and R. L. Ludhum, "Mapping Pollution with Infrared," *Photogrammetric Engineering,* Vol. 34, No. 6, pp. 561–564, 1968.

Way, D. S., "Aerial Photography for Residential Subdivision Site Selection," *Proceedings of the ASP 43rd Annual Meeting,* ASP, Washington, D.C., pp. 319–329, (March) 1977.

Welch, R., "Skylab S-190B ETC Photo Quality," *Photogrammetric Engineering,* Vol. 42, No. 8, 1976.

Wellar, B. S., "Remote Sensing and Urban Information Systems," *Photogrammetric Engineering,* Vol. 39 (October) 1973.

Whipple, J. M., "Survellance of Water Quality," *Photogrammetric Engineering,* Vol. 39, No. 2, 1973.

Whittlesey, J. H., "Tethered Balloon for Archaeological Photos," *Photogrammetric Engineering,* Vol. 36, No. 2, p. 181, 1970.

Williams, R. S., and T. R. Ory, "Infrared Imagery Mosaics for Geological Investigations," *Photogrammetric Engineering,* Vol. 33, No. 12, pp. 1377–1380, 1967.

Willow Run Laboratories, *Proceedings of the Fourth Symposium on Remote Sensing of the Environment,* Infrared Physics Laboratory, University of Michigan, Ann Arbor, 1966.

Wing, R. S., "Structural Analysis from Radar Imagery; Eastern Panamanian Isthmus," *Modern Geology,* Vol. 2, pp. 1–21, 1971.

Wise, D. U., "Radar Geology and Pseudo-Geology on an Appalachian Piedmont Cross Section," *Photogrammetric Engineering,* Vol. 33, No. 7, pp. 752–761, 1967.

Wittgenstein, M., "A Report on Application of Aerial Photography to Urban Land Use Inventory, Analysis and Planning," *Photogrammetric Engineering,* Vol. 22, 1956.

Wobber, F. J., C. E. Wier, T. Leshendok, and W. Beeman, "Coal Refuse Site Inventories," *Photogrammetric Engineering,* Vol. 41, No. 9, 1975.

Woober, F. J., and R. R. Anderson, "ERTS Data for Coastal Management," *Photogrammetric Engineering,* Vol. 39 (June) 1975.

Work, E. A., Jr., and D. S. Gilmer, "Utilization of Satellite Data for Inventorying Prairie Ponds and Lakes," *Photogrammetric Engineering,* Vol. 42, No. 5, 1976.

Yost, E., and S. Wenderoth, "Multispectral Color for Agriculture and Forestry," *Photogrammetric Engineering,* Vol. 37, No. 6, 1971.

Zaitz, C. E., "Resources and Cadastral Mapping of Panama," *Photogrammetric Engineering,* Vol. 35, No. 8, pp. 772–778, 1969.

CHAPTER 2

LANDFORMS AND AERIAL PHOTOGRAPHIC TERRAIN ANALYSIS

INTRODUCTION

It is generally agreed that any site development undertaken today must be compatible with both the possibilities and limitations of our natural environment and its resources. Predevelopment analyses must, therefore, take into account the problems presented by physical site factors as well as those posed by factors such as economics, politics, and sociology. This book presents to land planners a method of site analysis which is sensitive to the many interrelationships of surface and subsurface terrain conditions, upon which design concepts and forms can evolve. The physical site factors considered include: geology (rock type, fractures, faults, attitudes of beds), soils (type, textural composition, organic content, moisture content, depth to bedrock, depth to water table), water (drainage pattern density, flow capacity, groundwater flow, yield, quality), vegetation (association, type, height, density), and minerals (type, grade distribution).

A useful method of terrain analysis must accomplish two goals. It must provide the necessary information about physical site factors and at the same time take into account and be sensitive to the interactions of these factors. The earliest approaches to terrain analysis of this kind were developed in the late 1940s by professionals with a range of backgrounds in civil engineering and in the earth sciences (see, for example, Belcher et al., 1951). The methods they developed were based upon the identification of landforms. Donald Belcher, continuing the landform analysis concept in the early 1950s, developed many engineering applications that employed these techniques. Interest in these and other methods of terrain analysis (Steinitz et al., 1969) has recently increased due to the widespread concern with environmental issues.

Planners trying to solve environmental problems need an information-gathering technique that will provide very complete and accurate physical resource data. Unfortunately, important types of physical site data for most of the land area of the United States are either lacking or available only in a highly fragmented form. Even where the information does exist, it is often outdated, incomplete, or in a form not relevant to the application at hand. Land planners, therefore, have been commonly in the position of having to base their analyses and decisions on questionable existing data.

Recently, however, the technology of remote sensing has become much more highly developed, and the availability of aerial photographs and imagery has greatly increased. Satellite and high-altitude aircraft sensing systems now coming into use offer a further range of photographic and imagery materials from which a vast amount of data can be generated.

By using aerial photographic interpretation, planners, engineers, landscape architects, and other professionals engaged in site development can obtain a clear understanding of physical site conditions, for the unique overview position allows visual examination of large land areas in a way that is impossible during a usual field survey. This same technique can be used to identify those areas of a site which will require more detailed investigations, such as borings, seismic surveys, resistivity measures, percolation tests, or other types of field surveys.

AERIAL PHOTOGRAPHIC TERRAIN ANALYSIS

Aerial photographic terrain analysis is the systematic study of visual elements relating to the origin, morphologic history, and composition of distinct landscape units that appear in aerial photographs. Through the analysis of pattern elements visually apparent on an aerial photograph, the composition or parent material of a site is interpreted or inferred.

The visual pattern elements examined in photoanalysis include topographic form, drainage patterns, gully characteristics, erosional features, landform boundaries, color or photographic tone, land use pattern and distribution, vegetation pattern and distribution, and any other special features that may be present. The photoanalyst examines each element in three ways: separately, in relation to one another, and in relation to the entire pattern. Interpretation of the visual pattern elements is made by formulating a hypothesis which indicates the landforms and morphologic units of a particular site. The hypothesis is verified—and modified if necessary—by examining certain elements in more detail, consulting other sources of information, and making field checks. The entire process relies upon the fact that similar landforms under similar conditions have the same combinations of the same pattern elements. For example, beach ridges in Florida have the same visual appearance as beach ridges in China, and residual shale formations in North Dakota are similar to those found in California. These similarities are obviously true only in general terms, however, since microclimate, vegetation, moisture, and other conditions slightly modify and influence factors such as the composition of the residual soils.

Furthermore, in the approach presented in this book the earth's surface is described in terms of distinct, idealized categories of landforms, each with its own characteristics. In actuality, however, landforms are found in complex combinations and transitions, and understanding these complex situations requires knowing the basic forms. With experience and continued application of the techniques described here, analysis and interpretation of the more complex landforms will be facilitated.

Two Approaches

Many different groups and individuals have developed techniques of terrain analysis directed toward engineering applications. Having an overview of several of these methodologies is useful before learning the procedure of any one technique. Thus, two of the most widely practiced methods are presented here: The Australian "terrain unit" and the United States "landform" systems. The serious student of terrain analysis should research the associated references and become familiar with the details of these two and other approaches.

The Australian "Terrain Unit" Approach

In 1946 Australia established the Northern Australia Regional Survey (now the Division of Land Research) within the Council for Scientific and Industrial Research (now the Commonwealth Scientific and Industrial Research Organization or CSIRO) to undertake land reconnaisance surveys of large land areas of Australia. To achieve this mapping and analysis goal, a land classification system was needed which would generalize pertinent land resource information and group it into relevant categories. During the 1950s G. A. Stewart and C. S. Christain of the CSIRO developed a hierarchical classification of land systems, units, and sites that allowed for broad study and interpretation for a variety of use and management alternatives. Shortly thereafter, the Division of Soil Mechanics of the CSIRO began to modify the classification scheme for engineering problems and applied it to several test areas. At this same time other parallel terrain approaches were being developed in the United Kingdom and South Africa (Stewart, 1968). In 1965 a conference was held in Darwin at which representatives from all major countries reviewed and explored a more uniform international terrain classification system. The next year the Pattern-Unit-Component-Evaluation (PUCE) program evolved to serve as an encompassing system of classification directed toward engineering applications.

The PUCE scheme is based upon the unique definition of any area by its topographic slope, structure, lithology, soil, vegetation, and hydrologic characteristics. For each defined zone, engineering information is assembled to provide users with appropriate data for site evaluation. Four hierarchical levels are included: terrain province, terrain pattern, terrain unit, and terrain component. The four levels are illustrated in Figure 2.1, and their characteristics are defined in Table 2.1.

The U.S. "Landform" Approach

The use of aerial photographs in the development of terrain analysis for engineering applications was pioneered in the early 1940s by Donald J. Belcher, a civil engineer at Purdue University. His early studies suggested that soils and topography could be correlated to large soil unit areas and that they could be recognized on aerial photographs. He also demonstrated that soils developed from similar parent materials in similar climates and topographic positions would have associated definable engineering properties. Belcher (1943) summarized this concept as follows:

> Regardless of geographic distribution, soils developed from the similar parent materials under the same conditions of climate and relief are related and will have similar engineering properties which in comparable positions will present common construction problems.

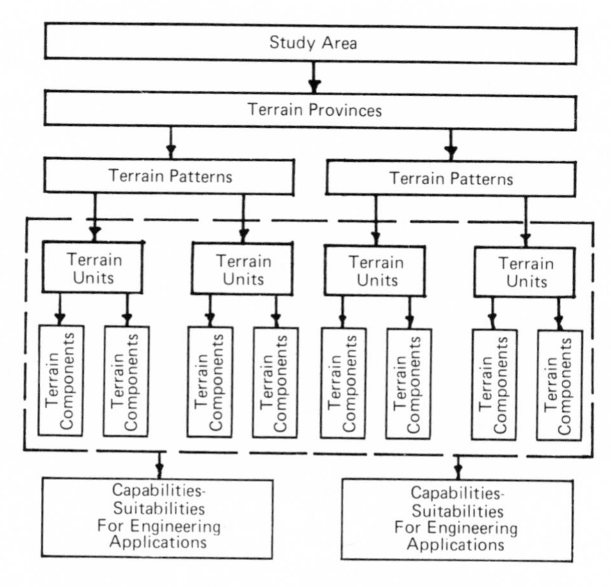

Figure 2.1. PUCE mapping units. The study area is first defined by geological differences, terrain provinces, which are mapped at a scale of 1:250,000. Terrain patterns include relief amplitudes and drainage basins mapped at the same scale, whereas terrain units include major soil and topographic areas and are mapped at 1:50,000. Terrain components, the most detailed distinction, are usually not mapped but would be similar to a detailed soil survey but would include associated vegetation.

Table 2.1. PUCE units and associated characteristics

Stage in terrain classification	Map scale	Terrain factors used for description	Terrain factors suitable for quantitative expression: Factors	Method and scale
Terrain province	1:250,000	Geology	Properties of geologic materials	Air photointerpretation or geological maps 1:1,000,000
Terrain pattern	1:250,000 plus block diagram	Geomorphology; basic char. of soil, rock, vegetation common among constituent terrain units; drainage pattern	Relief amplitudes, stream frequencies	Airphoto and/or ground study 1:10,000
Terrain unit	1:50,000	Physiographic unit; principal characteristics of soil, rock, and vegetation	Dimensions of physiographic unit (relief amplitude, length, width)	Airphoto and/or ground study 1:10,000
Terrain component	Usually not mapped but described	Physiographic component, lithology, soil type, vegetive association	Dimensions of physiographic component (relief amplitude length, width, slopes)	Measure on site
			Dimensions of vegetation (height, diameter, spacing)	Measure on site
			Dimensions of vegetation (height, diameter, spacing)	Measure on site
			Dimensions of surface obstacles including rock outcrops and termitaria	Measure on site
			Properties of earthen materials throughout profile (depth, particle size gradation, consistence strength, permeability, suction, mineralogy)	Measured in the field or through laboratory procedures
			Quantities of earthen materials	Measured or estimated on site

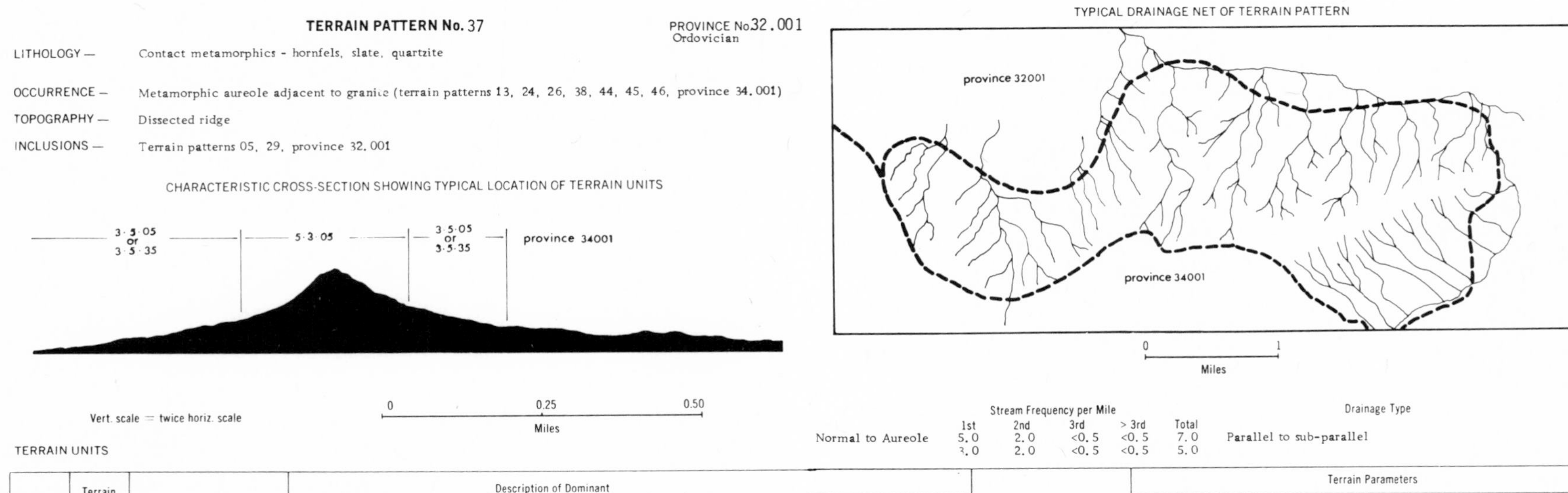

	Stream Frequency per Mile					Drainage Type
	1st	2nd	3rd	> 3rd	Total	
Normal to Aureole	5.0	2.0	<0.5	<0.5	7.0	Parallel to sub-parallel
	3.0	2.0	<0.5	<0.5	5.0	

TERRAIN UNITS

Number	Terrain Pattern Area (%)	Occurrence	Description of Dominant				Inclusions
			Topography	Soil	Land use	Vegetation	
3.5.05 or 3.5.35	70	Adjacent to and below dissected rigdes	Dissected slope interfluves: convex to 10° depressions: concave to 5°	Interfluves and upper slopes: rock outcrop, pockets of shallow brown mottled clayey silty gravel (GM-GC) slopes: and depressions: duplex grey-brown clayey silt to 6 in. over yellow-red medium to heavy-textured clay to 12 in. (ML/CL-CH;Dy) over decomposed rock	Unused or pasture	Mostly woodland, some areas cleared, grassland, sparse tussock grass, scattered trees, yellow box, grey box, occasional yellow gum	Terrain patterns 05, 29, province 32.001
5.3.05	30	Semi-continuous; sub-linear	Dissected ridge crests: undulating, convex to 20° slope: convex to 20°	Rock outcrop, pockets of brown mottled clayey silty gravel (decomposed rock) to 6 in. over rock	Unused or pasture	Mostly woodland, some areas cleared, grassland, sparse tussock grass, scattered trees, yellow box, grey box, occasional yellow gum	-
9.2.00	<1	Drainage	Minor stream channel Banks: convex to 60°, occasionally concave to vertical, floors: flat	Banks: duplex grey-brown clayey silt to 10 in. over yellow mottled medium to heavy-textured clay to 3ft (ML/CL-CH;Dy) over decomposed rock floors: fine gravelly, silty sand (GP-SM) over rock, areas of rock outcrop	Unused	Bare, yellow box, grey box, occasional yellow gum, river red gum fringing	-
9.4.05 or 9.4.35	<1	Drainage leading to minor stream channel	Gully slopes: concave to 20° sloping to 20°	Upper slopes: rock outcrop, pockets of shallow brown mottled clayey silty gravel (GM-GC) lower slopes: duplex dark grey-brown clayey silt to 4 in. over yellow-red medium to heavy-textured clay to 1ft (ML/CL-CH;Dy) over decomposed rock	Unused	Woodland, yellow box, grey box, occasional yellow gum, river red gum	-

Terrain Parameters			
Terrain Unit No.	Max. Local Relief Amplitude (ft)	Length of Terrain Unit	Width of Terrain Unit
3.5.05 or 3.5.35	100	Extensive	500 yards
5.3.05	200	Extensive	1000 yards
9.2.00	20	Extensive	20 yards
9.4.05 or 9.4.35	50	500 yards	50 yards

DIAGRAMMATIC REPRESENTATION OF TOPOGRAPHY AND ARRANGEMENT OF TERRAIN UNITS WITHIN TERRAIN PATTERN

9 4 05
3 5 05
5 3 05
soil
slate

Figure 2.2. Terrain pattern number 37 with its associated terrain units. (Grant, K., 1972, Terrain Classification for Engineering Purposes of the Melbourne Area, Victoria Division of Applied Geomechanics Technical Paper No. 11, CSIRO, Melbourne.)

Belcher has also contended that the landforms and associated properties recur whenever the landforms are found under similar conditions (Belcher, 1948). Belcher's studies in the early 1950s showed how detailed engineering soils data including soil type, texture, depth of regolith, soil moisture, and depth of groundwater could be extracted from aerial photographs. He has outlined how these characteristics could be used to estimate such soil properties as permeability, volume change, shear strength, Atterburg limits, and landslide characteristics (Belcher et al., 1952). This data could then be applied to engineering and construction issues, such as sources of construction materials, excavation and grading, trenching, and the construction of highways, foundations, and airfields.

The publication of Belcher's *Landform Reports* in 1952 (under a contract with the Office of Naval Research) established the first fully documented approach to landform analysis. Belcher became Director of the Center for Aerial Photographic Studies at Cornell and continued to make major contributions toward the evolution of applied engineering aerial photo interpretation. In addition, other civil engineers have learned and taught the technique, thereby producing further refinements and variations (Lueder, 1950 and 1959; Mintzer and Frost, 1952; Miles, 1962; Keifer, 1966; Rib 1966; Mollard, 1972).

Identifying and mapping landform units is the process of systematically analyzing the visual pattern elements present on aerial photographs and extracting land resource and terrain information by deductive processes. Typically, the results would include information on landform type and origin, soil and rock characteristics and properties, depth of soil over bedrock, hydrologic properties, depth to water table, groundwater data, and a sense of past, present, and future geomorphologic processes. The key to this process is the identification of landform units that provide the base for physical property and soil interpretations.

Landforms are defined in this approach as *land units that have resulted from constructional or destructional processes that when found under similar conditions (such as age, climate, weathering, erosion, attitude) will exhibit a definable range of visual and physical characteristics.* Accordingly, clay shale found under similar conditions in different places will have a predictable appearance on aerial photographs and a definable range of soil characteristics and properties (for example, such properties as the engineering indexes that allow for evaluation of relative construction performance). The use of the term *landform* in this context includes some land units of the third through fifth order (Enzmann, 1968).

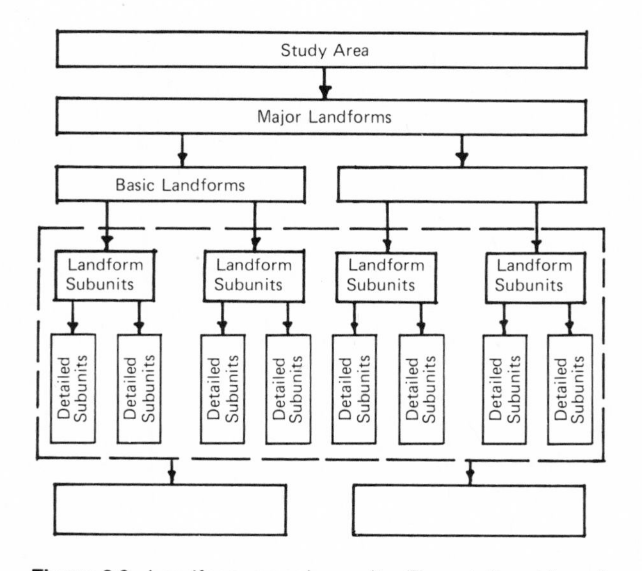

Figure 2.3. Landform mapping units. The relationships of major and secondary landform units is similar in hierarchy to that of the PUCE method. However, the process does not necessarily follow a unit hierarchy. The mapping proceeds toward, and may illustrate, the final most detailed units, but it does not show how they can be aggregated into larger categories.

In the identification of landform units, the visual elements portrayed on aerial photographs are examined: topographic relief, drainage pattern, photographic color and tone, erosion-gully analysis, and land use and cover characteristics. Certain patterns occurring either singly or in combination can be correlated with the actual presence of predicted conditions, and thus by analogy, unknown conditions can be identified and mapped (Belcher, 1942).

The procedure listed below is typically followed during landform identification and mapping (see Figure 2.3 and Table 2.2).

All existing information is gathered for the entire region, including the definition of climatic controls on weathering, processes of mass wasting and erosion, and previous erosional surfaces and zones of differential erosion.

The site region is observed and interpreted for the major landform units (similar to basic regional bedrock and surficial geology). See Figure 2.4

The photographs and available topographic maps are interpreted and homogeneous zones of texture and drainage are delineated.

Onsite investigations refine preliminary landform boundaries.

Larger-scaled photographs are used to delineate major soil boundaries which serve as landform subunits.

Table 2.2. Landform units and associated characteristics

Stage in terrain classification	Map scale	Terrain factors used for description	Terrain factors suitable for quantitative expression Factors	 Method and scale
Major landforms	1:250,000	Geology or major surficial geol. categories (i.e., glacial fluvial)	Properties of geologic materials	Air photointerpretation (1:70,000 or satellite) or geologic maps (1:1,000,000)
Landforms	1:60,000-1:10,000	Geomorphology/geology; basic characteristic of rock and soil (parent materials)	Stream pattern, density relief amplitudes	Air photointerpretation and ground surveys 1:60,000 - 1:20,000
Subunits (see text)	1:4,800-1:1,200	Soil types, texture, depth, stoniness, erosion, ground water, topographic slopes	Properties of earthen materials throughout profile (depth, particle size, gradation, wetness, strength, permeability, mineralogy, organic matter)	Air photointerpretation, ground surveys, and laboratory procedures. 1:12,000 - 1:4,800
			Quality and quantity of groundwater resources	Air photointerpretation and ground surveys
			Dimensions of topographic relief (amplitude, length, width, slope)	Air photointerpretation and ground surveys

Additional soil subunits are mapped, depending upon the detail and nature of data requirements. This phase includes onsite surveys and laboratory analyses.

Basic landform areas can vary in size from a clay shale region of hundreds of square kilometers to a glacial drumlin one-fourth of a square kilometer. Subsequent subunit delineation is based upon differences in soil characteristics and is very similar in logic to the breakdown used by the Soil Conservation Service (SCS) of the U. S. Department of Agriculture: The basic landforms (zones of similar parent material) correspond to major soil units; the next order of landform subunits corresponds to soil series which are generally alike in profile thickness and depth to seasonal high water table but are not alike in surface texture. These land form subunits are followed by soil types that have common surface texture and then soil phases that include demarcation of topographic slope, degree of erosion, and stoniness. Unlike the SCS classification, however, the detailed mapping of landforms is concerned with the entire soil profile to bedrock.

Comparison and Evaluation

The landform method is directed to site and project scale analysis in which the proposed development and associated engineering issues define the nature of the data collected and hence the particular subdivision of land units used. The process is used by many engineers for a variety of purposes; thus it is not surprising to find discrepancies in definitions of generic suborders. Conversely, the Australian PUCE method is the established approach of the Soil Mechanics section of the CSIRO and necessitates a more ordered generic classification that can be applied to any part of the country. Given the international character of its formation, it is also logical that this technique is adaptable and usable by many countries. The final engineering applications are made from the data available at the unit or component scale. The style of analysis is as follows: Given the nature and organization of the data at the unit or component level, how can it be analyzed for _____? This approach is sound *if* the data categories at the unit and component level have been previously well defined and determined to be relevant. The landform system is directed in its mapping process to provide a set of desired end-product categories related to the engineering problem at hand. Hence, the approach styles are quite different: The PUCE is more general and allows easier implementation and uniform interpretation of mapping units; the landform technique is more project scale and site specific.

The mapping units used in the two approaches are similar. The major landforms and the terrain provinces are both attempts to map geology. The basic landforms and terrain patterns map homogeneous zones of earthen materials (parent materials), but the PUCE system includes vegetation. The landform subunits and the terrain units/components may or may not be similar, depending upon how the subunit landforms are defined.

Both approaches facilitate and encourage the use of remote sensors (primarily air photointerpretation) for data generation and mapping.

Figure 2.4. Major and basic landforms mapped in a portion of Lycoming and Clinton Counties, Pennsylvania. (A) Shale residual soil. (B) Flood plain; 1 = alluvial flood plain; 2 = alluvial fan; 3 = colluvial materials associated with zone C. (C) Tilted interbeded sedimentary rocks, 1 = quartzitic sandstone; 2 = colluvial valley deposits—may be underlaid by shale. (D) Karstic limestone residual soil, 1 = fringing colluvium associated with materials in C and E; 2 = karstic limestone—note sinkholes and karstic drainage pattern. (E) Quartzitic sandstone residual soil.

This data is supplemented when necessary by ground surveys and laboratory analysis. The largest units, terrain provinces and major landforms, facilitate mapping from satellite imagery; the smaller units, terrain patterns and components and the basic landforms and their subunits, require aerial photographs with scales from 1:70,000 to 1:4,800.

The landform approach is used privately by engineers and land planners on a project-by-project basis; therefore, study areas have a wide variance in size from multistate regions through several acres. The PUCE method is directed toward large area mapping where little previous information exists. The terrain units and components are applicable to project scale design and can be applied to small sites. It is hypothesized that as the site becomes larger and more complex and the number of engineering/construction options increases, the benefits of using these approaches and aerial photographs will also increase.

Each technique encourages the collection of all previously available material before initiating the process. However, much of the information required to determine the mapping zones does not exist and must be interpreted from aerial photographs. The map unit zones are defined with two dimensional boundaries even though the natural ground and subsurface conditions do not inherently have this characteristic. Neither approach employs a transitional boundary delineation. The specific data collected is similar in both methods.

The two approaches analyzed here do a credible job of defining and describing the landscape in data categories relevant to site evaluation for engineering/construction issues. However, for accurate analysis and design, all relevant data must be considered. Do the methods satisfy the total requirements needed for the analysis of various engineering and construction issues? Have any important data needs been overlooked? Neither approach satisfactorily describes or analyzes the processes present on the land units. These processes in themselves may present more critical engineering issues than simply the nature of the underlying materials. An approach is needed that is similar to these but that expands its dimension to encompass both process and time (past and future).

Principles of Landform Interpretation

Terrain analysis as presented in this book is based on the definition of *landform* given above, and it uses aerial photographic interpretation for landform identification. Several essential facts, however, should be kept in mind in following the method of interpretation described here.

(1) The use of a key format for the identification of landforms is based upon the principles (a) that the same landform, regardless of its location, if found under the same approximate environmental conditions will exhibit similar identifying pattern elements, and (b) that each landform has a characteristic range of soil and rock composition.

(2) Changes in environmental conditions usually change the appearance and significance of pattern elements. Thus, as a result of different weathering-erosional processes and rates, the pattern characteristics of a landform in an arid climate differ in appearance and importance from the pattern characteristics of the same landform found in a humid climate. If the interpreter has a clear understanding of weathering processes and of the erosional characteristics of each landform, he should be able to recognize the pattern elements of a landform in different climates or in different stages of the erosion cycle. The landform identification keys presented in this volume distinguish, when they are significant, differences in the appearance of the pattern elements caused by arid and humid climates; minor interpolations can be made to determine how these patterns appear in semiarid and subhumid climates. Furthermore, most of the landform keys describe the characteristics associated with landforms in the mature stage of erosion. It should be noted that the same landform shows different visual and engineering characteristics in the later stages of erosion.

(3) The quality, date, scale, and type of aerial photograph significantly influence the visibility of the pattern elements of landforms. The most useful photographs for terrain analysis show exposed soils several days after a heavy rainfall. Also, in most regions, the best time for taking photographs is during the beginning of the growing season, for soil tones and microfeatures are not hidden by crop or tree cover at that time and therefore can be easily distinguished. Numerous articles on this subject have appeared in the magazine *Photogrammetric Engineering,* and they should be consulted for help in determining the best type of film, filter, scale, and date to emphasize a particular pattern element or combination of elements.

(4) By using aerial overviews, the photoanalyst can observe land patterns that cannot be perceived from ground surveys and can also identify and map the extent of visual patterns more easily. For instance, an area of heavily wooded, hummocky topography that may actually be a pitted outwash landform could be mistaken on the ground for a type of moraine.

(5) The use of aerial photographic interpretation does not mean the end of detailed, on-site field investigations. In many instances the services of geological, soil, or geophysical consultants are necessary in order to obtain detailed information not within the realm of photographic interpretation. The use of the photographic interpretation process is not meant to eliminate the need for these specialists but rather to identify how and where they can be used more effectively.

(6) No matter how detailed and scrupulous a photointerpretation may be, a field check is a prerequisite for final judgment and conclusions. As a rule, aerial photographic interpretation is never expected to stand alone, although as experience is gained in a range of recurring situations the amount of time needed for field checking can certainly be decreased. However, the aerial photograph analyst must always guard against overconfidence and inexperience.

(7) In mapping landform, soil, and rock conditions, finite boundaries must be drawn for conditions that do not actually have finite boundaries. Thus, a certain degree of error must be expected, even in the most detailed mapping. All interpretations should be accompanied by written qualifications, which state the levels of accuracy of the information mapped so that the client or reader is not misled. If certain areas cannot be shown within the limits of accuracy of the investigation, they should be indicated and, if necessary, additional field investigations requested to obtain the data.

(8) A strong background in the earth sciences (geology, geomorphology, botany, ecology, pedology, hydrology) seems to be the best preparation for becoming a competent photointerpreter. Professionals from these areas have consistently proved this point. Another well-equipped group is comprised of those, such as land planners and allied professionals, who have wide experience in mapping land resources, identifying them in the field, and interpreting their capabilities for land uses. It should be noted, though, that most of these individuals, even those having high aptitudes for photoanalytic and interpretive techniques, initially draw differing conclusions from their observations because they rely upon observations that have meaning in terms of their past experiences.

Landform: Definitions

The term *landform* is used in many land classification systems, by geologists, civil engineers, landscape architects, and planners alike. It is important to realize that there is no accepted standard definition of this term.

Geologists commonly use "landform" to describe the geomorphic characteristics of a region. However, their definition of the term includes formations that vary widely in size and that may or may not share common parent soil materials (for example, fault block, mountain, or peneplain, although called "landforms," do not have within each a definable composition). The geological use of "landform" thus seems to combine a description of surface expression with an inference of structure or attitude, and such a definition is not stringent enough for terrain analysis.

Engineers involved in terrain analysis have developed definitions and classifications of landforms that are more useful to land planners. Both of the following definitions stipulate that for a land unit to be classified as a landform, it must retain the same basic composition and characteristics whatever its geographic region. Donald Belcher, Director of the Center for Aerial Photographic Studies at Cornell University, uses the following definition of a landform: "The earth's features may be divided into landforms so that each form presents separate and distinct soil characteristics, topography, rock materials, and groundwater conditions. The recurrence of the landform, regardless of the location, implies a recurrence of the basic characteristics of that landform" (Belcher, 1948). Belcher's definition includes land features formed by transported and residual soils, as well as those formed by bedrock type and attitude.

Donald R. Lueder, who has also developed many photointerpretive techniques, has formulated a definition slightly different from Belcher's: "A unit landform may be defined as a terrain feature or terrain habit, usually of the third order, created by natural processes in such a way that it may be described and recognized in terms of typical features wherever it may occur, and which, when identified, provides dependable information concerning its own structure and either composition and texture or uniformity" (Lueder, 1959). Lueder defines landforms as terrain features of the third order. First-order forms include continents and ocean basins; mountain ranges are relief features of the second order; and third-order formations include valleys, basins, ridges, and cliffs. The main innovation in Lueder's definition is his reference to the scale of the formations.

Many planners and landscape architects think of landforms in a strictly visual sense, and thus from their viewpoint, landforms include valleys, mountains, cliffs, depressions, plains, and so on. The limitations of this definition are obvious; only a visual description is given and there is no reference to composition or size.

The definition of landform to be used in this book incorporates the definitions introduced by Belcher and Lueder, while enlarging on the descriptive definition of visual characteristics: *Landforms are natural terrain units (including geologic elements and transported or residual soils) that, where developed under similar conditions of climate, weathering, erosion, and mass wasting, will exhibit a predictable range of physical and visual characteristics. Therefore, soils developed from similar parent materials (under similar conditions) are related and have similar engineering properties.* Thus specific distinctions can be made among landform units, by which it is possible to describe unique topography, composition, or structure or to make visual distinctions relevant to planning issues and capabilities.

Influences of Climate, Weathering, and Erosion

The accurate identification of landforms and their physical site conditions depends upon an understanding of the complex of geological processes called *landmass denudation.* Landmass denudation is part of the larger cyclical movement by which the earth's surface is constantly worn down and renewed (see Figure 2.5).

During the process of landmass denudation, massive, exposed bedrock is first weathered into smaller units by exposure to air, water, and organisms. The loosened, weathered particles that result are then moved downslope, principally by gravity, a process defined as mass wasting. Eventually, the particles are picked up, transported from their place of origin, and deposited elsewhere by glacial, fluvial, or eolian

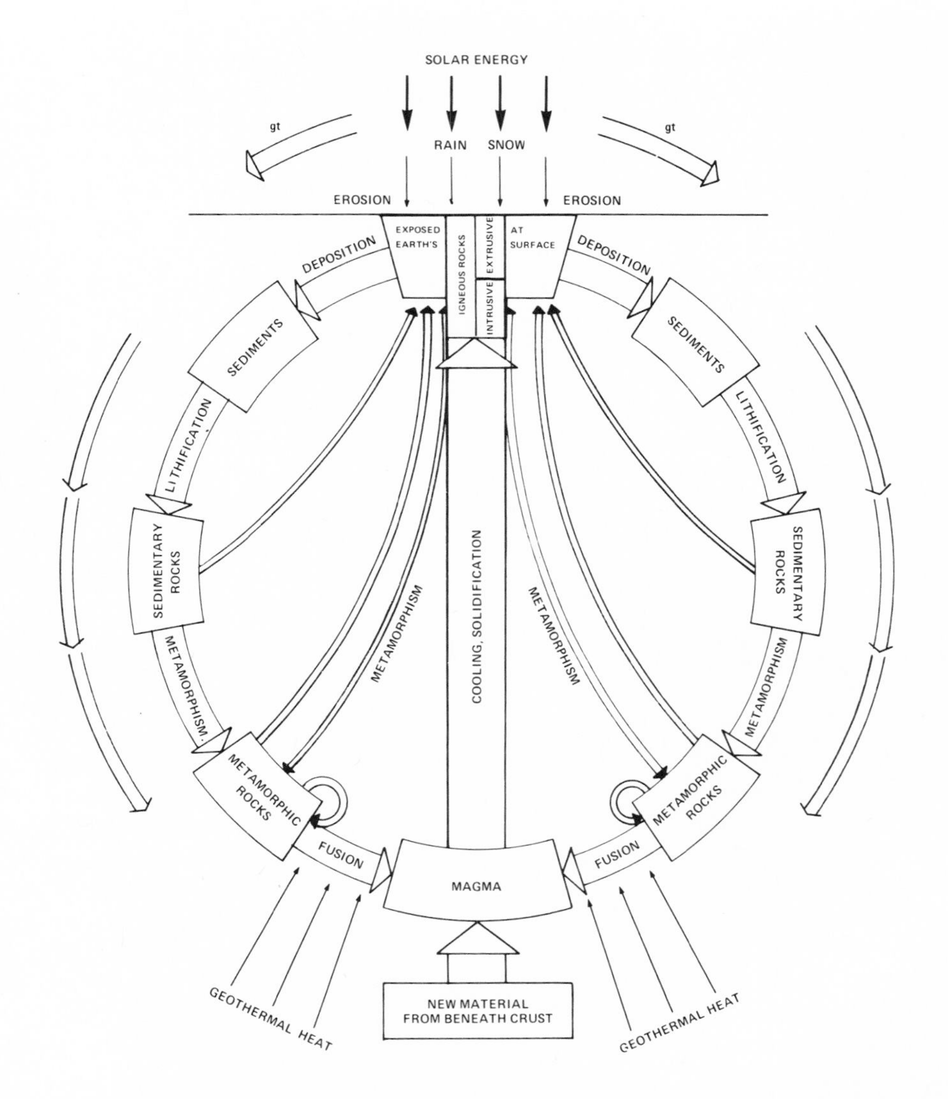

Figure 2.5. Rock cycle. Starting at the top of the diagram, one sees how snow, rain, and solar energy react physically and chemically with rocks (sedimentary, igneous, or metamorphic) to produce a soil debris that is then eroded and deposited. Sediments accumulating in deep strata are either re-exposed and eroded or undergo lithification and become sedimentary rocks. The sedimentary rocks are then either re-exposed and eroded or become metamorphic rocks through the applied heat and pressure of metamorphism. Metamorphic rocks are exposed and eroded or, as a result of additional heating (fusion) which destroys all traces of earlier rock composition, dissolve into the magma of the earth's interior. When magma cools and solidifies, it forms igneous rock. If magma cools beneath the surface, it is called intrusive igneous rock (granitic types). (After C. R. Longwell et al., *Physical Geology,* John Wiley, New York, 1969, p. 102.)

means. After deposition these particles may again become part of a rock mass, and the entire process can begin again.

This chapter reviews the influences of climate on the physical processes of weathering, mass wasting, and erosion that act to denude the earth's surface.

Climate

Temperature, humidity, rainfall, and frequency of climatic change all affect the processes of weathering and erosion, and in each climatic zone these factors have an effect upon the pattern signatures of the different landforms. For example, landforms in humid climates tend to have a subdued topography and deep soil profiles, since the dense rainfall facilitates both mechanical and chemical weathering and because the lush vegetative development stabilizes the disintegrated rock and adds organic material to the soil. In arid climates landforms are generally more rugged and without significant soil development over rock features. In these regions the lack of both precipitation and vegetation results in slower weathering, but when storms occur they tend to be severe, thereby causing rapid erosion of any materials that may have disintegrated.

The pattern elements that identify a landform are greatly affected by these climatic differences; therefore the landform keys in Chapters 5 through 11 show, whenever appropriate, the variations that can be expected between humid and arid climates. Minor interpolations may be necessary for other climatic zones such as subhumid or subarid.

It is difficult to incorporate the world's many different climatic conditions into a simple classification system; climatologists have tried many different ways of categorizing and mapping homogeneous climate zones. Maps can be prepared to show temperature and precipitation ranges, but to explain the world's climatic intricacies, other factors, such as the distribution of precipitation during the year and the interrelationships of temperature and precipitation, must be included. In 1918, Wladimir Köppen of the University of Graz, Austria, devised a useful worldwide classification scheme which combined fixed values (monthly and yearly) of temperature and precipitation with limits for known vegetative and soil boundaries. In this system a hierarchical listing of climatic zones is designated by a letter code.

The Köppen system, however, is too detailed for use in terrain analysis, where it is better to work with a system that aggregates small climatic subdivisions into larger, homogeneous zones. Therefore, the classification used in this book includes only the four major zones: arctic, arid, humid, and tropical (Figure 2.6). It has been found that these will provide the base needed for categorizing weathering processes and describing vegetative and soil zones.

The following descriptions of the four major climatic zones illustrate the strong influence of climate upon mechanical and chemical weathering processes, soil formation, and topographic appearance.

The four climatic zone subgroups—subtropical, subhumid, semiarid, and subarctic—are not included in the following descriptions, but their characteristics of temperature and precipitation are similar to their respective major climatic types. For example, a subhumid climate is similar to that of a humid climate, but the precipitation is less, typically 20 to 40 inches.

Arctic Climates. These typically cold, dry areas are found in northern Canada and Siberia and along the coast of Greenland. Precipitation is under 10 inches per year; the average January temperature is –20 to –30°F, and the average July temperature is 40°F. Because of the low temperatures, little moisture is available for the maintenance of plant life except from minor surface melting during the summer season. The predominant form of weathering is frost wedging, which in the uplands results in cliffs covered with rock talus. The lowland areas contain indicators of permafrost, such as polygonal frost wedges and beaded ponds along streams.

Arid Climates. Arid climates are widespread throughout the world; they cover approximately 25% of the world's land area. Arid climates occur in the west-southwestern United States, in a narrow strip along the west coast of South America, in most of central Asia, in the northern third of Africa, and in the central portion of Australia. These areas are characterized by less than 20 inches of annual rainfall. They can be warm or cold; thus, average January temperatures range from 20 to 90°F, and average July temperatures range from 60 to 90°F. Evaporation exceeds rainfall, causing groundwater, if present, to be drawn to the surface and evaporated. This evaporation results in the formation of calcium carbonate, gypsum, or alkali deposits in the upper horizons of the soil. Rocks containing many different types of minerals (polycrystalline) weather more rapidly than monomineralic rocks such as limestone. Limestone and sandstone rocks form upland cliffs and cap rocks in these regions, while polycrystalline rocks, such as granites and metamorphics, occupy the more weathered lowlands. Such vegetation as occurs is widely spaced and takes the form of shrubs and salt-tolerant bushes, with bunch grass being present in semiarid climates. Both mechanical and chemical weathering processes take place at a slow

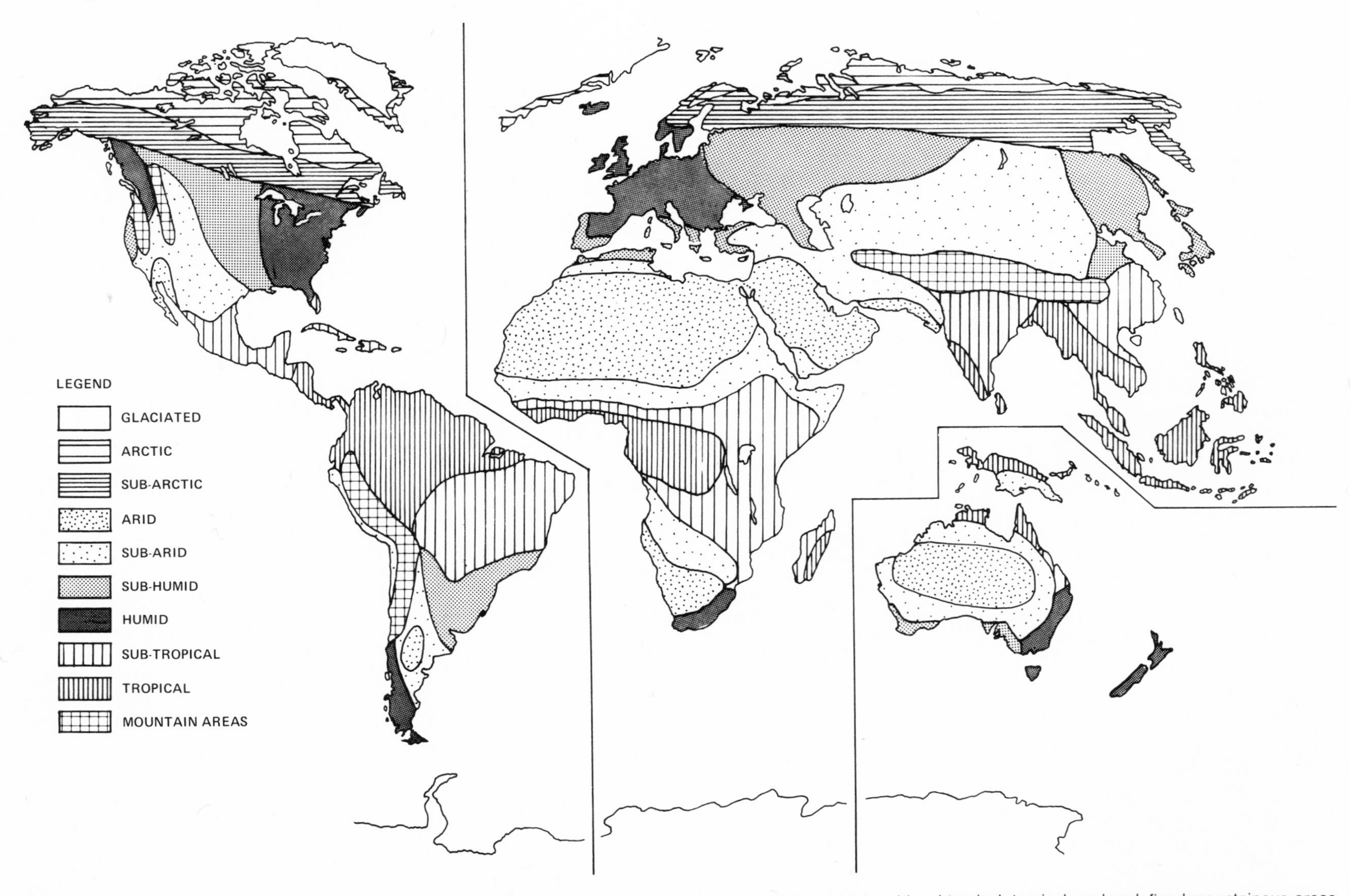

Figure 2.6. The basic climatic zones of the world, including glaciated, arctic, subarctic, arid, subarid, subhumid, humid, subtropical, tropical, and undefined mountainous areas.

rate because of the low rainfall. Chemical decomposition involving water dominates over mechanical disintegration, and yields a soil almost entirely free of organic matter or interconnected roots. This unconsolidated soil is extremely susceptible to wind erosion and forms windlaid landforms.

Humid Climates. Humid climates are characterized by rainfall that exceeds evaporation and by the accumulation of humus faster than microorganisms can consume it. Humid climates exist in the eastern half of the United States, southern South America, the western coast of Australia, most of New Zealand, the Japanese Islands, and in a belt across central Asia. Precipitation varies from 40 to 80 inches per year; average temperatures in January range from 20 to 60°F, and in July from 50 to 80°F. Both mechanical and chemical weathering take place, with frost wedging being a principal cause of deep rock disintegration. Soils are leached in the surface horizons and create a silica-rich residue with a high organic content. The soils develop deep profiles; broad topographic forms result, further shaped by soil creep and water erosion. Resistant bedrock occasionally outcrops along hillsides; soil depths are generally shallow on the ridges and deeper in the valleys.

Tropical Climates. Tropical climates are characterized by hot, rainy conditions combined with a short dry season. Tropical climates include most of South and Central America, southern Mexico, southern Florida, Cuba, the West Indies, the south central portion of Africa, southern India, and southeast Asia. Precipitation is generally over 60 to 80 inches per year; the average temperature range in both January and July is 70 to 90°F. Chemical weathering processes are dominant as a result of the heavy amounts of rainfall and the high temperatures. The residual soils within these regions contain high concentrations of aluminum and iron oxides which, in some instances, are mined as ores. Tropical soils such as laterite and bauxite are generally unsuited for agricultural production. Limestone rock regions in tropical climates present some of the world's most spectacular topography, for caves, needlelike spires, and steep conical hills develop from the rapid dissolving of calcium carbonate by carbonic acid.

Weathering

Before the pattern elements on aerial photographs can be interpreted and physical site conditions evaluated, understanding the interrelated processes of weathering is necessary. Although climate is the greatest single influence on weathering in significantly determining the topographic appearance of any given rock landform, many other related factors also determine the effectiveness of weathering or influence the rate at which a material weathers. These are: the geological type of the rock and its mineral composition and structure, humidity, temperature, rate of climatic change, vegetation, topographic slope, and exposure.

Weathering is the process by which rock is altered through exposure to air, water, and organisms; the process includes various chemical and mechanical actions. By definition, weathering takes place where oxygen can penetrate. Weathering actions occur near or at the surface of the earth and cause decomposition and disintegration of rock. (Transportation of the loosened particles, or erosion, is not a part of the weathering process; it is discussed later in this chapter.)

Most rock structures undergo both mechanical disintegration and chemical decomposition. Mechanical disintegration is the reduction of rock into smaller units by physical agents such as temperature, frost, thermal expansion-contraction, pressure release, root pressure, undermining, and abrasion produced by water, wind, and ice. In chemical decomposition, rocks and minerals are broken down into smaller units by chemical activities such as oxidation, carbonation, hydrolysis, hydration, base exchange, and chelation. Typically, complex compounds tend to break down into simpler, more stable ones. Chemical decomposition has its greatest effect in regions of warm temperature and high humidity, while mechanical disintegration, which is induced primarily by temperature changes, is most active in arid regions and on exposed mountains. The combination of decomposition and disintegration of rock surfaces provides the basic material for the formation of soils.

Mechanical Weathering Processes

Mechanical weathering, or disintegration, takes place as rocks are broken into smaller units by physical compression or splitting forces. There are four major processes of disintegration: (1) pressure release created when overburden material is removed and the rock is exposed to the surface; (2) growth of ice or salt crystals in cracks and pores of the rock; (3) expansion-contraction caused by rapid heating and cooling; and (4) splitting forces created by root pressure.

Pressure Release. Most rock structures are formed deep beneath the earth's surface under the tremendous pressure of overriding rock layers, sediments, or water. When these rocks are exposed by erosion and/or uplift, the rock layers expand. In measurements recorded at granite quarries at Stone Mountain, Georgia, expansion of 0.05% occurs along the length of freshly cut granite. (This increase amounts to ⅛ inch per 10 feet of length.) The effects of pressure release can be seen in the domelike appearance of certain granitic landforms or in the sheeting joints parallel to the surface which develop in flat-lying sedimentary rocks.

Pressure release is usually the first step in the process of rock weathering. The fractures and cracks that develop during this stage allow water, ice, roots, acids, and so on to penetrate the rock mass, which thus further decomposes and disintegrates the rock structure.

Crystal Growth. The growth of ice or salt crystals exerts forces great enough to exceed the tensile strength of rock. (Freezing water measured under laboratory conditions has been found to exert 2,100 tons of pressure per square foot.) The growth of ice crystals within cracks and fractures efficiently disintegrates rock by slowly flaking off chips. The most obvious effects of such ice crystal wedging can be found above the timberline in middle-high latitudes; here, angular blocks formed by the separation and shattering forces of ice crystals cover the ground surface. (Such shattered-rock fields are called *felsenmeers,* or boulder fields.)

Water-soluble salts carried in surface runoff are a major weathering force in arid climates. As the water evaporates, salts precipitate from the supersaturated solutions, and salt crystals form in rock fractures. As the crystals grow, they exert tremendous expansive forces and can thus disintegrate any rock into which the water has penetrated. In some instances, the salts react chemically with minerals contained in the rock itself, which further increases the rate of weathering.

Thermal Expansion-Contraction. There are many conflicting opinions concerning the ability of thermal heating and cooling to weather rock significantly. Laboratory experiments in which thermal limits found in nature were exceeded have not produced stresses approximating those required to initiate rock disintegration. However, the cumulative effects of a continuing process have not been fully investigated. This form of weathering, if it exists, is very slow, but such a process would provide an explanation as to how rocks in arid deserts in Australia become shattered when no other weathering process is apparent.

Root Action. Vegetative growth results in another important form of weathering. Rootlets, which can penetrate even the smallest cracks in rock and cause them to expand, allow water to enter the rock mass. When the vegetation dies, the remaining organic material still retains moisture, enabling frost action to be more efficient. The organic material also decomposes into organic acids which initiate chemical weathering. And, as residual soil material accumulates, larger species of vegetation can establish themselves, thus continuing the process.

Chemical Weathering Processes

Chemical weathering, or decomposition, is the result of changes in the chemical composition of rock; by these changes new minerals, more stable and better suited to the lower temperatures and pressures found at the earth's surface, are formed. Chemical changes involve several processes; the major ones are oxidation, hydration, carbonation, and desilication.

Oxidation. Oxidation is an exothermic (heat-evolving) reaction which produces new compounds of greater volume and lower density. This process is most effective in rocks in which compounds of iron and water are present; the resulting ferric compounds contain more oxygen than the original compounds. Slight oxidation of iron produces the mineral hematite, which is reddish in color. Exposed rock surfaces showing red or yellowish colors can indicate weathering by oxidation, but samples should be split and the natural internal colors examined in order to make an accurate identification of the rock.

Hydration. In hydration, rock minerals combine with water molecules. Rocks that have undergone hydration almost double in volume, appear dull and lusterless, and flake easily where exposed to the air. In expanding, the surface rock material flakes off, thus exposing new areas to be similarly weathered. Examples of the hydration process can be found in such tropical soils as laterites. These begin as kaolinite (a clay product of weathered orthoclase feldspar) and hydrate into rich aluminum or iron oxides which can be mined as ore. Limonite, a naturally occurring iron oxide, results from hydration and is used in the production of iron. Some clay minerals hydrate and dehydrate alternately, depending upon the amount of available moisture. Such soils present severe construction problems because of their tendency to expand after heavy rains and later, when dry, to contract.

Carbonation. Carbonation is the process in which carbonates are formed from reactions between carbonic acid and minerals. Carbonic acid occurs naturally as carbon dioxide; it combines with water in the atmosphere and is carried to the earth's surface as rain.

In limestone regions carbonic acid reacts with calcite to form calcium and bicarbonate ions in solution. The resulting calcium bicar-

Table 2.3. Relative resistance of rocks to weathering and erosion*

Rock type		Resistance	Physiographic forms
Igneous			
Fine-textured			
Dark (basic)	Basalt	Usually resistant	Escarpments and flows
Medium	Andesite	Usually resistant	Not widespread
Light (siliceous)	Rhyolite	Usually resistant	Bluffs and cliffs
Coarse-textured			
Dark (basic)	Gabbro	Usually very resistant	Escarpments and domes
Medium	Syenite	Usually resistant	Uplands
Light (siliceous)	Granite	Usually resistant except in arid regions	Domes and uplands
Sedimentary			
Fine-grained			
Loose	Clay	Weak but forms vertical walls	Badlands
Consolidated	Shale	Usually weak	Gentle slopes and lowlands
Limy-loose	Marl	Very weak	Low valleys
Limy-consolidated	Limestone	Weak in humid; resistant in arid	Karstland or table rocks
Coarse-grained			
Loose	Sand	Usually weak	Lowlands
Consolidated	Sandstone	Resistant if strong cementing agent	Cliffs, plateaus, caprocks
Very coarse			
Loose	Gravel	Moderately resistant	Often caps uplands
Consolidated	Conglomerate	Very resistant	Ridges and mountains
Metamorphic			
Sedimentary origin			
Shale	Slate	Weak	Lowlands
Limestone	Marble	Weak	Lowlands
Sandstone	Quarzite	Very resistant	Ridges, knobs, monadnocks
Igneous or sedimentary origin			
Banded	Gneiss	Usually very resistant	Uplands
Schistose	Schist	Usually resistant	Uplands and ridges

*After A. K. Lobeck, *Geomorphology*, McGraw-Hill, New York, p. 40, 1933.

bonate is about 30 times more soluble than calcium carbonate, which thus accounts for the rapid erosion of limestone landforms.

Desilication. One of the most common chemical weathering reactions is *desilication*, or the hydrolysis of feldspar minerals by carbonic acid. The products of a hydrolysis reaction with feldspar are a clay mineral, silica in solution, and a carbonate or bicarbonate solution of potassium, calcium, or sodium. Calcium plagioclase feldspar weathers most easily, followed by sodium plagioclase feldspar and potassium orthoclase feldspar.

Other Chemical Weathering Processes. Other chemical weathering reactions include base exchange and chelation. Base exchange involves cation transfer between a solution and a mineral solid, in which one cation replaces other cations. Chelation involves the complex assimilation of metallic cations into hydrocarbon molecules.

Susceptibility of Rock Types to Weathering

Depending upon their process of formation and their mineralogical content, different types of rock vary in their susceptibility to weathering by the processes just discussed (see Table 2.3). In general, igneous rocks are most susceptible to chemical weathering, for they originate deep within the earth and are not in harmony with an exposed environment. Also, the large, varied, crystalline structure of igneous rocks is more susceptible to weathering agents than the monocrystalline structure of, for instance, sedimentary rocks. Sedimentary rock consists of lithified, previously weathered residues of igneous, metamorphic, or sedimentary rocks and any weak minerals they contained were decomposed before the sedimentary rock was formed. Therefore, in sedimentary rocks the cementing agents that bond the particles of sand and clay together determine the relative weathering resistance of a particular rock. Quartzites, conglomerates, and sandstones are commonly cemented by a matrix of silica, which is a very resistant material, and rocks formed in this way tend to dominate their surrounding materials.

Erosion

The transporting of previously weathered earthen materials is called *erosion*. Some of these materials are moved downslope by mass wasting; others are picked up and directly transported by glacial, fluvial, or eolian processes. The interaction of climate, structure, material, vegetation, topography, and time influence the ultimate landscape appearance, and variation in one element will influence that appearance. The most significant factors, however, are climate, material, structure, and time. The first and last factors, climate and time, most directly influence the general regional landscape appearance. Material and structure are most significant in determining more detailed landscape differences and are discussed separately in the landform chapters of this book. The more gradual landscape evolution as controlled by climate and time is well illustrated by Davis (1899) in his erosion cycle concept, which while not technically accurate, provides a sequence that is readily conceptualized. Figures 2.7 and 2.8 illustrate these sequences as they may develop in humid and arid climates.

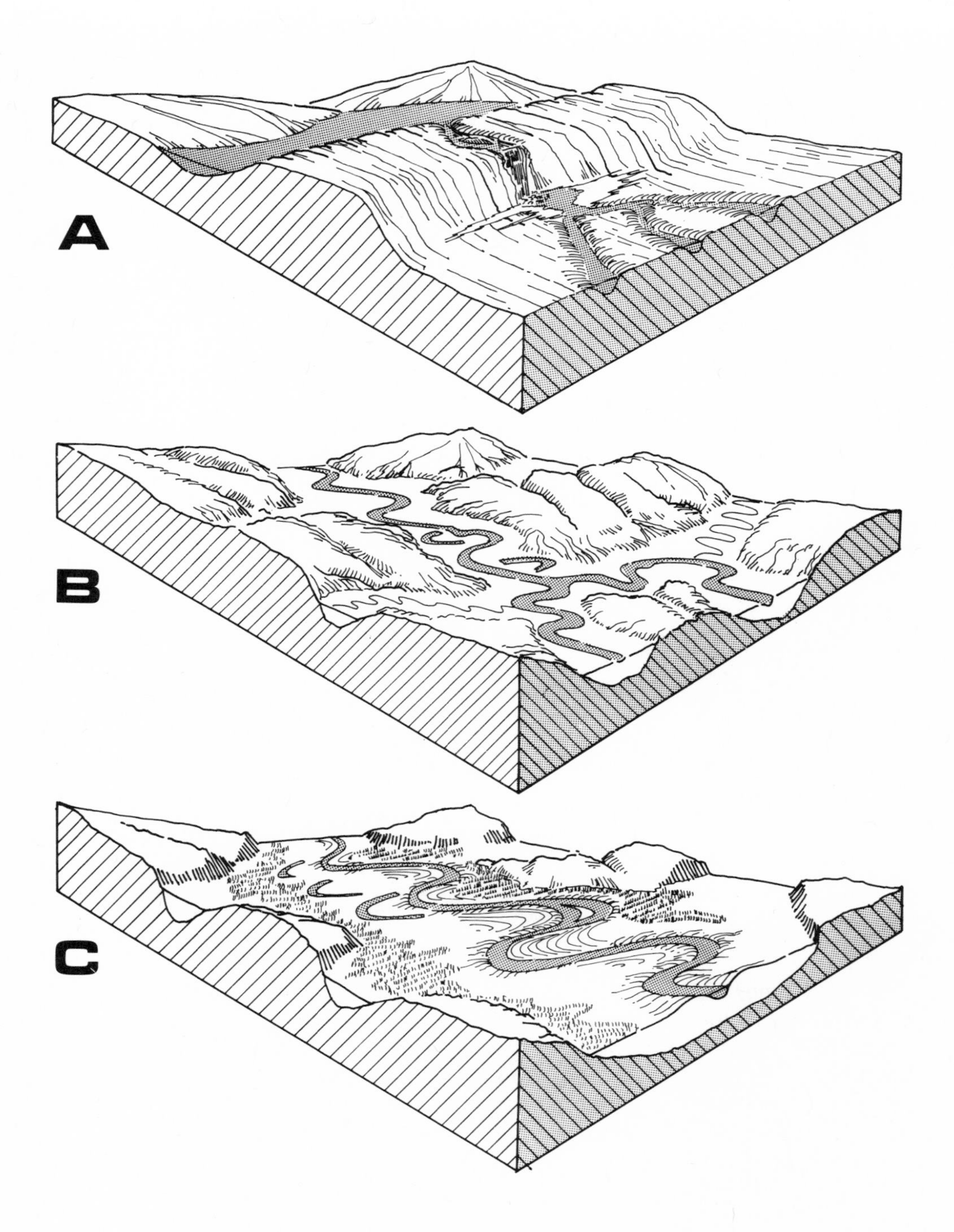

Figure 2.7. Three landscape topographic sequences which are the result of long-term erosion in a humid climate. Following the Davis (1899) model, rapid uplift is assumed, and the humid climate establishes a related soil and vegetation cover complex. The dense mat of plant growth shields the soil surface from direct contact with rain droplets, thus facilitating slow absorption by the soil. However, after the soil is saturated, overland flow starts, eroding the regolith and eventually transporting the weathered materials to the oceans. The youth stage of the sequence (A) is characterized by a land surface with broad, flat uplands and a small, narrow stream valley; maturity (B) appears as a highly dissected topography with narrow, ridged divides; and old age (C) develops an erosional surface with little relief except for occasional *monadnocks* (hills of residual rock which have a slightly higher degree of resistance to erosion). If there is subsequent uplift of the region after achieving plantation, the process of erosion will be reinstituted; this time, however, the stream system forms entrenched meanders.

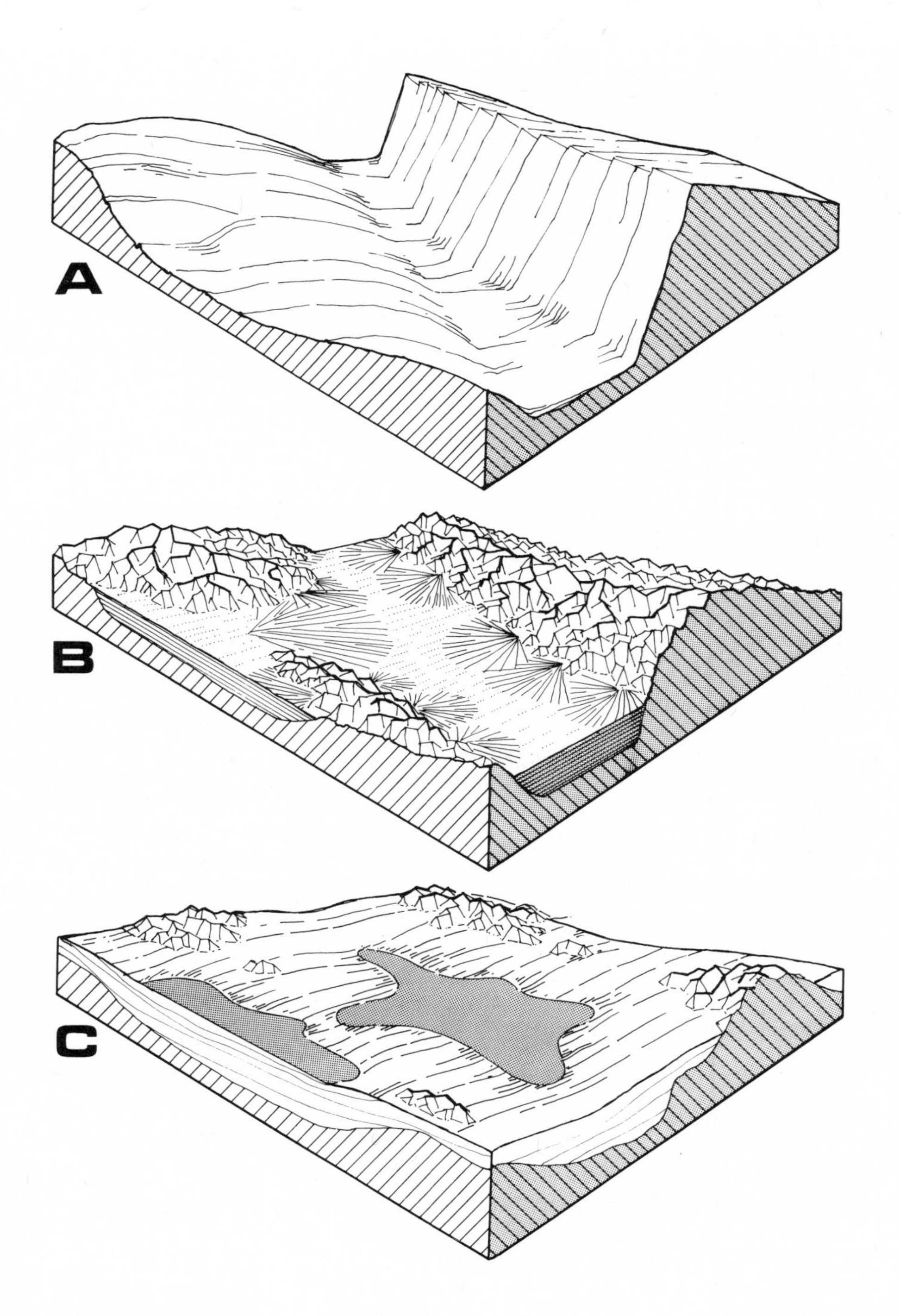

Figure 2.8. The erosion sequence in an arid climate following the Davis (1899) model, in which rapid uplift is inferred. Arid climates are characterized by evaporation that exceeds rainfall, and therefore the development of any significant vegetative cover is impossible. During the rainfalls that do occur, water droplets striking the bare soil surface splash loose the soil sediments which are easily transported either by overland flow or by streams. However, in contrast to humid climates, this flow rarely reaches the ocean; it is quickly absorbed in alluvial soils and deposits its sediment load upon the ground surface. Wind erosion is effective in removing many of the fine soil fractions, while coarse materials may accumulate and move in dunes. The youth stage of the erosion sequence (A) is characterized by a rugged, mountainous form typically having many fault depressions. The mature stage (B) has a highly dissected upland and valleys filled with alluvial debris, forming alluvial fans and valley fills. Large, central depressions form temporary lakes or playas. In the old stage of erosion (C), alluvial filling of the valleys is almost complete, with small residual rock forms projecting through. Relief is low, and large depressions may contain arid lakes or playas. (After A.N. Strahler, *Introduction to Physical Geography*, Wiley, New York, p. 312, 1965.)

The Process of Terrain Analysis Through the Use of Aerial Photographs

Aerial photographic analysis of a given site must be carried out very systematically, for carelessness at this stage of an investigation will affect the accuracy of all later results. Stereovision should always be used, and all other available sources of information, such as photomosaics, topographic maps, soil reports, groundwater reports, and so on, should be drawn upon. A correctly conducted analysis must incorporate the following steps:

(1) The site boundaries should be delineated on a map, preferably a U.S. Geological Survey quadrangle. Sources of stereophotographic coverage should be investigated. Attention should be given to the desired scale, type, date, and quality of photographs available with respect to the particular area and task of investigation. If photographs are not available, specifications for a photoflight should be formulated.

(2) The climatic zone should be defined, and other local conditions that may be reflected in the cultural or physical patterns on the photographs should be noted. Broad types of landform groups should be identified as well. For example, is the area glaciated, or is it in a typical sedimentary, igneous, or metamorphic rock region? This procedure eliminates many landforms from consideration and allows concentration on one or several major groups.

(3) Photomosaics or other map coverage of the immediately surrounding regions should be obtained to aid in the identification of the regional context or relevant bordering conditions.

(4) Reports and/or maps concerning the region in question should be obtained and reviewed; examples are Soil Conservation Service soil reports, well records, hydrological studies, and topographic, surficial geological, and bedrock geology maps. Existing material of this kind provides the analyst with clues by which to verify or modify his initial interpretations. (Unfortunately, substantial existing information is rarely available, thereby forcing a greater dependence upon field checking.)

(6) After preliminary examination of the photographs, boundaries of areas of water, land, or texture should be drawn on them by using a wax crayon on an acetate overlay. The simple distinctions at this stage are hilly, level, or a nontopographic classification of beach areas.

(7) The photographs should be examined and analyzed with respect to the basic pattern elements of topography, drainage patterns, gully characteristics, erosional features, landform boundary, color or photographic tone, distribution of land uses, vegetative distribution, and special features. Each pattern element should be examined separately in relation to each other element and in relation to the entire pattern.

(8) The pattern elements should be interpreted and a hypothesis formulated concerning the basic geomorphic structure and composition of the site in question. Consult the landform interpretation tables in Chapters 5–11 for the range of possible landforms and their associated pattern elements. The hypothesis should be further refined by reexamining the photographs for microfeatures and resolving any conflicts that may exist. If all the basic and micropattern elements interrelate and account for one another, the hypothesis concerning the basic parent material and associated landforms is supported.

(9) The interpretation should be field checked to verify the hypothesis and to add the specific information needed for detailed mapping of soils, depth to water table, depth to bedrock, and so on.

The interpretive process should be done initially on a region larger than the site area and then narrowed to include more detail. For example, drainage patterns may be missed if too small an area is used for beginning observations. The larger overview allows for a proper regional setting to be defined and major geologic and landform units identified. Subunits and additional detail can then be added within the correct context.

Landform Pattern Elements for Aerial Photographic Interpretation

Correctly interpreting and identifying a landform requires knowing and being able to define each of its characteristic pattern elements. The photoanalyst first investigates the photographic coverage of the entire site and then identifies the characteristics of the pattern elements. These include topography, type, and texture of the drainage pattern, photographic tone and/or color, gullies or other erosional features, vegetation and land use, and miscellaneous features such as fractures, outcrops, and geobotanical indicators. In the landform sections of this book (Chapters 5–11) the characteristic pattern elements of each landform are presented. In the following discussion the pattern elements and their subcategories are described.

Topography

In the photographic keys in the landform sections of this book, small cross-sectional drawings of idealized landform surface-subsurfaces show the forms of hills, the relationship of hilltops to depressions, and the relative steepness of slopes. The topography of a landform is verbally described by indicating its degree of dissection and continuity. Typical descriptions are: flat; flat table rocks; massive hills; soft, rounded hills; steep, rounded hills; karst topography; terracing; parallel ridges; saw-toothed ridges; bold, domelike hills; A-shaped hills; parallel laminations; undulating; snakelike ridges; conical hills; pitted plains. Other descriptions indicate the planimetric shape of a landform, such as fan-shaped, star-shaped, crescent-shaped, or drumlin-shaped.

Drainage

Drainage is studied according to its pattern type and its texture or density. It is probably the most important single identifier of landforms. Drainage pattern analysis can give a great deal of information concerning the parent rock and soil materials, since these influence how and to what extent the water runs off or drains from a landform surface. (This type of analysis can also be made by using topographic maps.)

Drainage patterns are classified as regional or local, depending upon the scale of image observed, but 1:40,000–1:60,000 stereopairs should give sufficient coverage for regional analysis. Photographs at scales larger than 1:40,000 should be supplemented by maps that allow study and identification of regional patterns, since clues to landform identification commonly occur at the regional rather than the local level.

Drainage Textures. Drainage patterns are classified by their *density of dissection,* or texture, and by their type of pattern form. Drainage texture is indicated by three categories, fine, medium, and coarse, as based upon its appearance in photographs of different scales (see Figure 2.9 and Table 2.4).

Fine-Textured. Fine-textured patterns typically indicate high levels of surface runoff, impervious bedrock, and soils of low permeability. Distances between drainage tributaries is short, and averages less than ¼ inch (6 mm) on 1:20,000 scaled photographs.

Medium-Textured. The spacing of first-order streams is less than fine textured but more than coarse textured. The amount of runoff is medium compared to that associated with fine and coarse textures. Soil textures are typically neither fine nor coarse but contain mixtures of materials.

Coarse-Textured. These textures generally indicate a more resistant bedrock which may be permeable and which forms coarse, permeable soils. First-order streams are typically over 2 inches (50 mm) apart and carry relatively little runoff.

Drainage Pattern Types. An examination of drainage pattern types considers the entire drainage pattern, the gullies or first areas of channelized flow, the tributaries, and the major channels which may themselves be depositing eroded materials and forming surficial, water-laid landforms. They are best observed on high altitude photographs and images of scales of approximately 1:60,000 or even 1:120,000. The different patterns can indicate specific rock types, soil materials, rock attitude and structure, and drainage conditions. The pattern categories to be discussed in this section include those that enable major inferences (see Figures 2.10 to 2.12).

Table 2.4 Relationships of photographic scale to drainage pattern

Photo scale	Coarse	Medium	Fine
1:20,000	> 2″	¼ - 2″	< ¼″
	> 50mm.	6 - 50mm.	< 6mm.
1:60,000	> 5/8″	3/32 - 5/8″	< 3/32″
	> 16mm.	2 - 16mm.	< 2mm.
1:120,000	> 5/16″	3/64″ - 5/16″	< 3/64″
	> 8mm.	1 - 8mm.	< 1mm.

Dendritic. The dendritic is the most common drainage pattern and is characterized by a treelike branching system in which the tributaries join the gently curving mainstream at acute angles. The occurrence of this drainage system indicates homogeneous, uniform soil and rock materials and is typified by landforms of soft sedimentary rocks, volcanic tuff, dissected deposits of thick glacial till, and old, dissected coastal plains.

Angulate. The angulate pattern is a variation of the dendritic or trellis system in which faults, fractures, or jointing systems have modified the classic form. Sharp, angular bends are common in the mainstream; tributaries demonstrate control by rock features. The type and direction of angulations may also indicate the specific rock type. For example, sandstone tends to develop parallel jointing patterns, while limestone develops joints that intercept one another at acute angles.

Rectangular. Rectangular patterns are also variations of a dendritic system. Here the tributaries join the mainstream at right angles and form rectangular shapes controlled by bedrock jointing, foliations, or fracturing. The stronger or harsher the pattern, the thinner the soil cover. These patterns are often formed in slate, schist, and gneiss, in resistive sandstone in arid climates, or in sandstone in humid climates if little soil profile has developed.

Annular. This type of pattern is developed on topographic forms usually similar to those associated with radial patterns, but in this case the bedrock joints or fracturing control the parallel

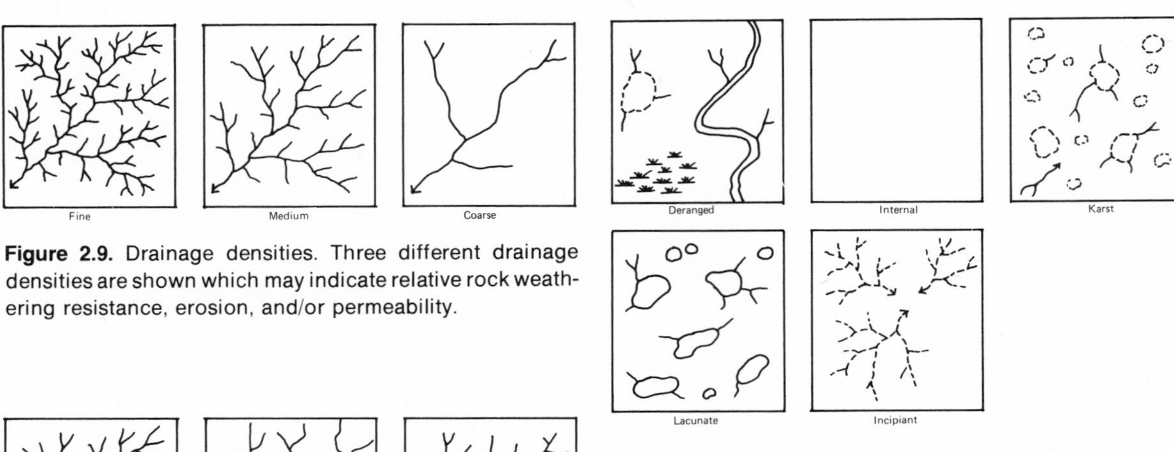

Figure 2.9. Drainage densities. Three different drainage densities are shown which may indicate relative rock weathering resistance, erosion, and/or permeability.

Figure 2.11. Drainage patterns (nonintegrated).

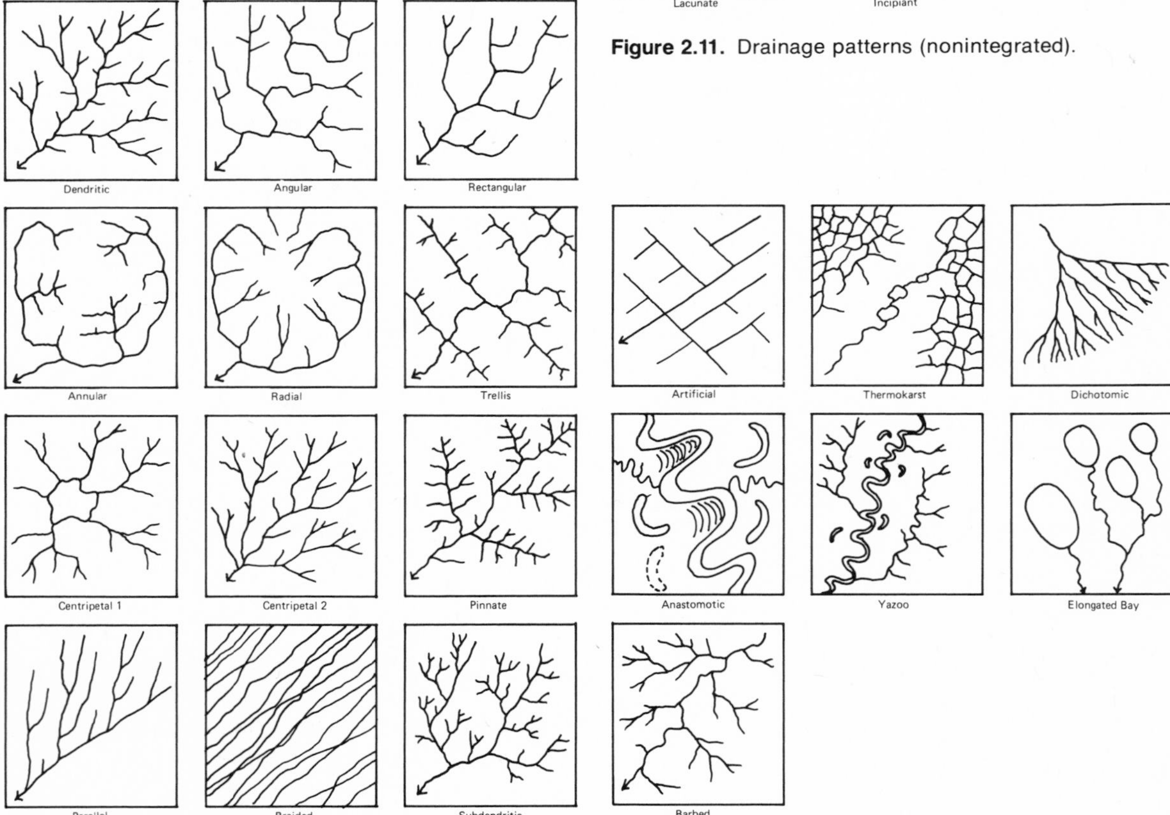

Figure 2.10. Drainage patterns (integrated).

Figure 2.12. Miscellaneous patterns.

tributaries. Granitic or sedimentary domes may develop this type of pattern.

Radial (Centrifugal). A circular network of almost parallel channels flowing away from a central high point characterizes this pattern. A major collector stream is usually found in a curvilinear alignment around the bottom of the elevated topographic feature. Volcanoes, isolated hills, and domelike landforms exhibit this type of drainage network.

Trellis. Trellis patterns are modified dendritic forms with parallel tributaries and short parallel gullies occurring at right angles. This pattern indicates a bedrock structure rather than a type of bedrock and usually indicates tilted, interbedded, sedimentary rocks in which the main, parallel channels follow the strike of the beds.

Centripetal. This pattern is a variation of the radial system in which the drainage is directed downward toward a central point. This form generally indicates a basin or sink depression, or the end of an eroded anticline or syncline.

Pinnate. Pinnate patterns are actually modified dendritic patterns that indicate a high silt content of the soil. They are typically found in loess or in fine-textured flood plains. Drainage follows a featherlike branching pattern in which tributaries intersect mainstreams at angles that are slightly acute upstream.

Parallel. Parallel drainage systems develop on homogeneous, gentle, uniformly sloping surfaces whose main collector streams may indicate a fault or fracture. Tributaries characteristically join the mainstream at approximately right angles. Such landforms as young coastal plains and large basalt flows are excellent regional examples of this drainage pattern.

Braided. Braided patterns are typically found spread across large alluvial plains in arid regions. The fluvial system is at times heavily loaded with bedload, which causes channel clogging and overflowing that results in the associated stream's seeking a more efficient alignment. This continuing channel shifting accounts for the braided appearance, and its presence is associated with coarse soil materials. Occasionally, a braided pattern can be found along the bottom of a much larger drainage channel.

Subdendritic. Many drainage patterns present complex combinations of several types. The subdendritic pattern as shown in Figure 2.10 has a primary pattern that is rectangular and somewhat parallel, but the first order streams are dendritic. This difference may indicate that the underlying rock influencing the second- and third-order drainage is different from that affecting the first-order drainage. For example, in flat beds of shale over sandstone, the first-order streams would be controlled by the shale, and the deeper, higher order channels contained in and controlled by the sandstone.

Deranged. These patterns represent nonintegrated drainage systems resulting from a relatively young landform having flat or undulating topographic surface and a high water table. Depressions contain swamps, bogs, marshes, ponds, or lakes. Regional streams may meander through the area but do not influence its drainage. These forms commonly occur on young, thick till plains, end moraines, and flood plains.

Internal. The lack of an integrated drainage system is significant in the identification of landforms and parent materials. It is usually associated with granular materials having high permeability or with porous rock materials. Landforms with this apparent lack of surface drainage include outwash terraces, alluvium, beach ridges, and sand dunes.

Dichotomic. This pattern occurs on alluvial fans and has many braided channels. Its shape and braided patterns suggest a constructional alluvial landform.

Anastomotic. Meander floodplains develop an anastomotic pattern which may contain oxbow lakes, meander cutoffs, and point bar deposits. The pattern is associated with those floodplains that have been subject to meander erosion-deposition.

Yazoo. A yazoo pattern is formed when a river is subject to over-the-bank flooding that creates a natural levee that blocks drainage across the floodplain into the river. The tributaries are then deflected downstream parallel to the main river until a break is found.

Elongated Bay. These parallel elliptical forms (Carolina Bays) are lowlands that drain into a parallel dendritic type pattern. This pattern is unique to the coastal plain area of the eastern United States.

Artificial. In humid climates in landforms of flat topography occupying low positions, artificial drainage structures may be built in an attempt to lower the elevation of the water table. Dead furrows and ditches in fields and along roads indicate this condition; these are commonly found in glacial lake beds and low, broad flood plains. Irrigation ditches in arid and semiarid climates designed to bring water into the fields should not be confused with drainage structures.

Thermokarst. These patterns develop in poorly drained, fine-grained sediments and in organic materials in regions of permafrost. Freezing causes many cracks to develop, and these form polygonal and hexagonal shapes; thawing causes slumping, settlement, and depressions. Streams crossing the area may connect rounded "button" depressions that have a beaded appearance. This type of drainage pattern, with its associated hexagons and beaded ponds, indicates the existence or previous presence of permafrost conditions.

Karst. The karst drainage pattern is associated with the surface-subsurface drainage network and results from solution weathering of limestone. A few streams may be found that end in sinkholes, many scattered sinkhole depressions, and some gullies that lead into drainage ravines or sinkholes.

Lacunate. This pattern is found on geologically young formations where an integrated runoff pattern has not yet formed. It consists of many depressions with some gullying and is found in conjunction with thick deposits of glacial till and clay lakebeds.

Incipient. This pattern shows the very beginnings of a drainage pattern typically found in thick areas of glacial till. The gullying system is weakly structured in a dendritic type pattern that is identifiable on aerial photographs by the alignment of the dark photographic tones.

Several drainage patterns are associated with rather specific, not commonly occurring conditions. When found, these patterns may indicate permafrost, certain types of floodplains, or man's intervention and attempts to improve drainage.

Barbed. These patterns occur within other drainage systems that have been modified through warping or topographic uplift. The resulting barbed or spurred appearance indicates a degree of tectonic disruption.

Tone

Tones on aerial photographs indicate surface and near-surface ground conditions, such as relative moisture and texture. The measurement of tone is relative, and since variation is found in different photographs taken under different conditions, tones should be compared only within sets of stereopairs. Photographs made during optimum moisture conditions, preferably several days after a heavy spring rain, exhibit good tonal range and thus facilitate detailed soil identification and mapping. Photographic tone can vary in shade, uniformity, and sharpness of boundary; these photographic qualities indicate conditions of soil composition and moisture.

Interpretation of photographic tone is difficult and requires some experience before accurate inferences can be made. Obstructions such as vegetation or cultivation often either mask tones or have tones of their own that can be confused with those of the natural soil conditions. All tonal characteristics described in this section refer to natural, exposed soil tones, not to those caused by crop or vegetative cover. With experience the interpreter will be able to identify tones across gridded agricultural fields and through light or scattered forest cover (Figure 2.13).

Tone Variation. Tonal variations are relative measures with no absolute scale. As a general rule, light tones indicate well-drained, coarse soils, and dark tones indicate poorly drained, fine soils. (See Figure 2.13.)

White. White or bright photographic tones indicate well-drained, coarse, dry soil materials such as exposed sand or gravel. In agricultural or developed areas, bare spots and/or borrow pits underlaid by these types of soil materials appear as very light tones.

Light Gray. These tones are generally associated with soils that contain mainly coarse textures mixed with some fines. These soils tend to be droughty, and they contain little associated organic matter.

Dull Gray. Soils exhibiting these tones generally consist of fine textures, and they have fairly good profile development and organic content. Poor drainage is inferred from the darker tone. However, some rock materials, such as basaltic lava, naturally exhibit dark tones; these should not be confused with the soil characteristics associated with this tone. Cross-checking with other factors helps to eliminate potential errors in interpretation.

Dark Gray or Black. Dark or black tones indicate poor internal drainage and/or a water table near the surface. Organic content is high, and fine textures predominate.

Tone Uniformity. The distribution of tones over a photograph indicates the relative uniformity or homogeneity of the soil and rock materials. Thus uniform tones represent little change in moisture or texture, whereas many different tones indicate differing soil conditions. (See Figure 2.14.)

Uniform. Uniform tones, typified by lake beds, continental alluvium, and flat, thickly bedded sedimentary rocks, indicate uniform soil texture and moisture conditions.

Mottled. Mottled tones indicate significant changes in soil moisture and/or texture within short distances; these changes result in the presence of many puffy light and dark tones. Typically, darker tones indicate depressions, and lighter tones indicate slightly drier areas. Till plains, coastal plains, limestone in humid climates, and infiltration basins in terraces, flood plains, or outwashes all typically exhibit mottled tones.

Banded. Banding occurs where there are linear-shaped differences in soil or rock texture, drainage, or moisture availability. In transported soil landforms banded tones indicate wet and dry areas and are associated with meander scrolls in flood plains, ancient out-

Figure 2.13. Types of photographic tone. (A) White; (B) light gray; (C) dull gray; (D) black. Note sharpness of boundary around A compared to zone B. This sharpness indicates a high change in moisture and texture conditions.

Figure 2.14. Types of tone uniformity. (A) Uniform. (B) Mottled. (C) Banded. (D) Scrabbled.

wash channels, ripple marks in lake beds, beach ridges and linear sand dunes, and so on. Interbedded sedimentary rocks or highly foliated rocks containing seep zones or areas of different moisture availability appear banded because of the distribution of vegetation; the dark bands are moister and therefore covered with vegetation. Finally, in cases in which rock is exposed, bandings may represent differences in the natural rock color.

Scrabbled. Scrabbled tones are common in arid regions where alkali deposits are found on the ground surface. The tonal pattern is very irregular, blotchy, and fine textured in appearance. Young volcanic lava flows sometimes also have this same tone.

Sharpness of Tone Boundary. The sharpness of the boundary between photographic tones aids in identifying soil textures as a factor of

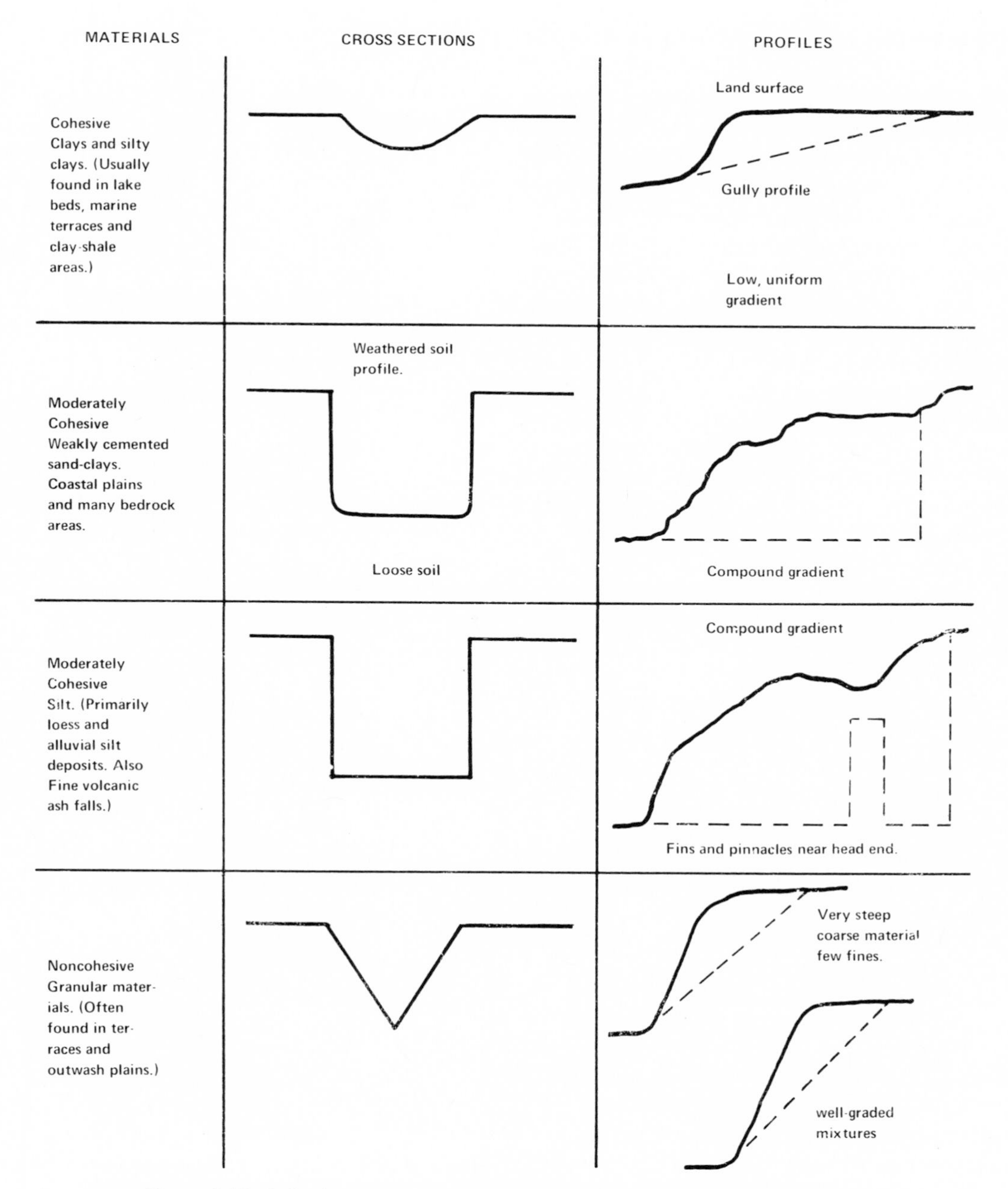

Figure 2.15. Gully characteristics: cross sections, profiles, and associated soils.

their differences in moisture retention ability (Figure 2.13).

Sharp, Distinct. Sharp, distinct boundaries indicate quick changes in moisture content and are related to coarse soils which do not have the ability to hold water. Rapid drainage in these soils causes low areas near the water table to be wet and higher areas to be dry.

Gradual, Fuzzy. Light and dark tones with fuzzy, indistinct, gradual boundaries indicate fine-textured soils and gradual changes in moisture content between high and low areas.

Gullies

Gullies are formed when sheet runoff collects in channelized flow and by eroding the bottom, forms the first-order drainage system. As the gullies erode through the surface soils, they adopt characteristic cross-sectional shapes which reflect the textural composition and cohesiveness of the surrounding soils. Identification and recognition of gully shapes aids in the interpretation and mapping of general soil materials. Figures 2.15 and 2.16 illustrate the typical cross-sections, profiles, and related materials associated with each type of gully.

The gullies occurring in certain landforms are described as having white fringes around their planimetric outline. These commonly occur in limestone, dolomite, or Illinoian glacial till and indicate the presence of relatively lighter, exposed subsurface profiles.

Vegetation and Land Use

The pattern of vegetation, as distributed across landforms, is very useful as an indicator of soil conditions. The presence or absence of vegetative cover helps in distinguishing the texture, permeability, and moisture retention

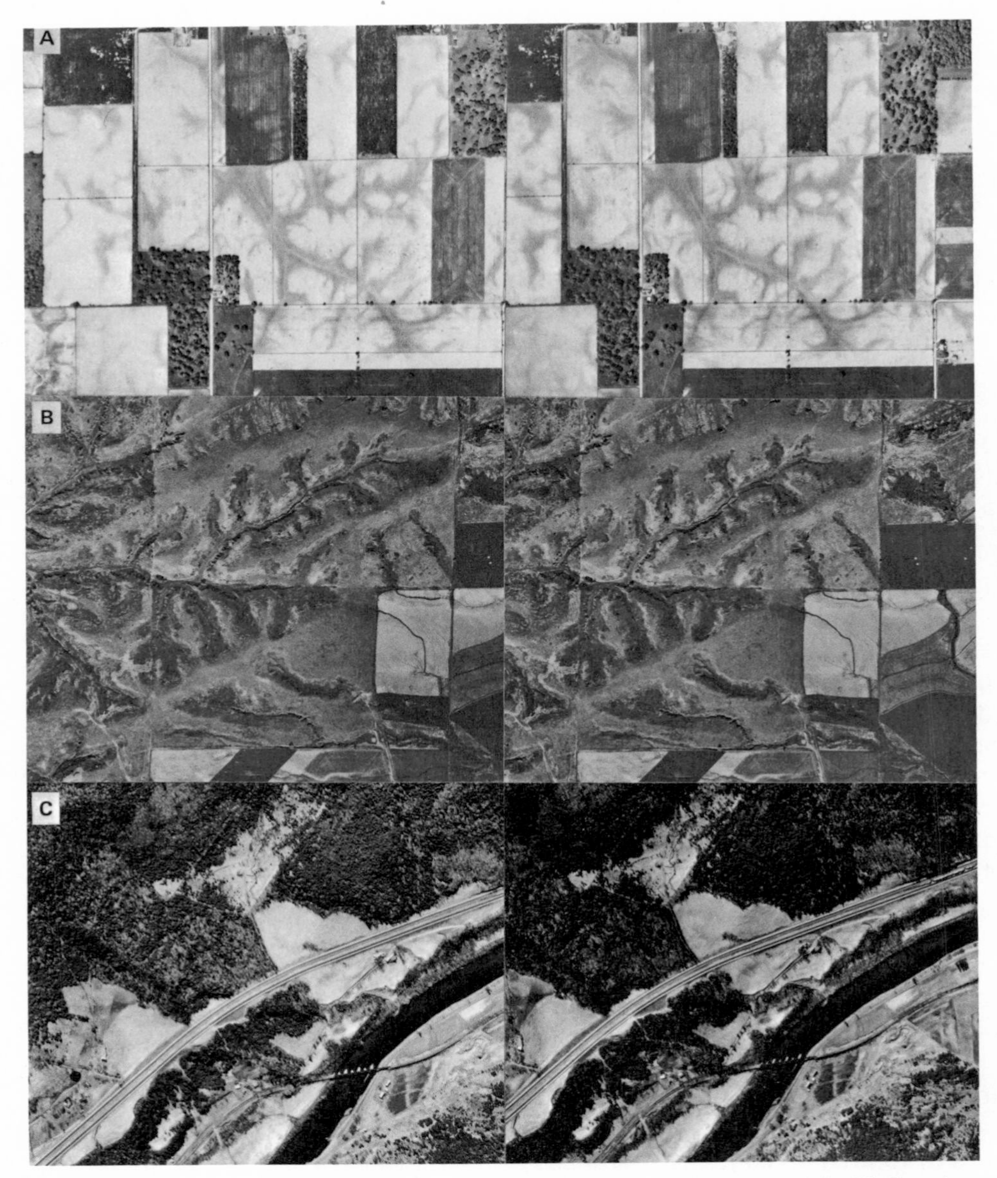

Figure 2.16. Typical gullies. (A) Gentle sag-and-swale gully indicating fine, cohesive materials. (B) Box-shaped gully indicating moderately cohesive silt. (C) V-shaped gullies indicating granular, noncohesive materials.

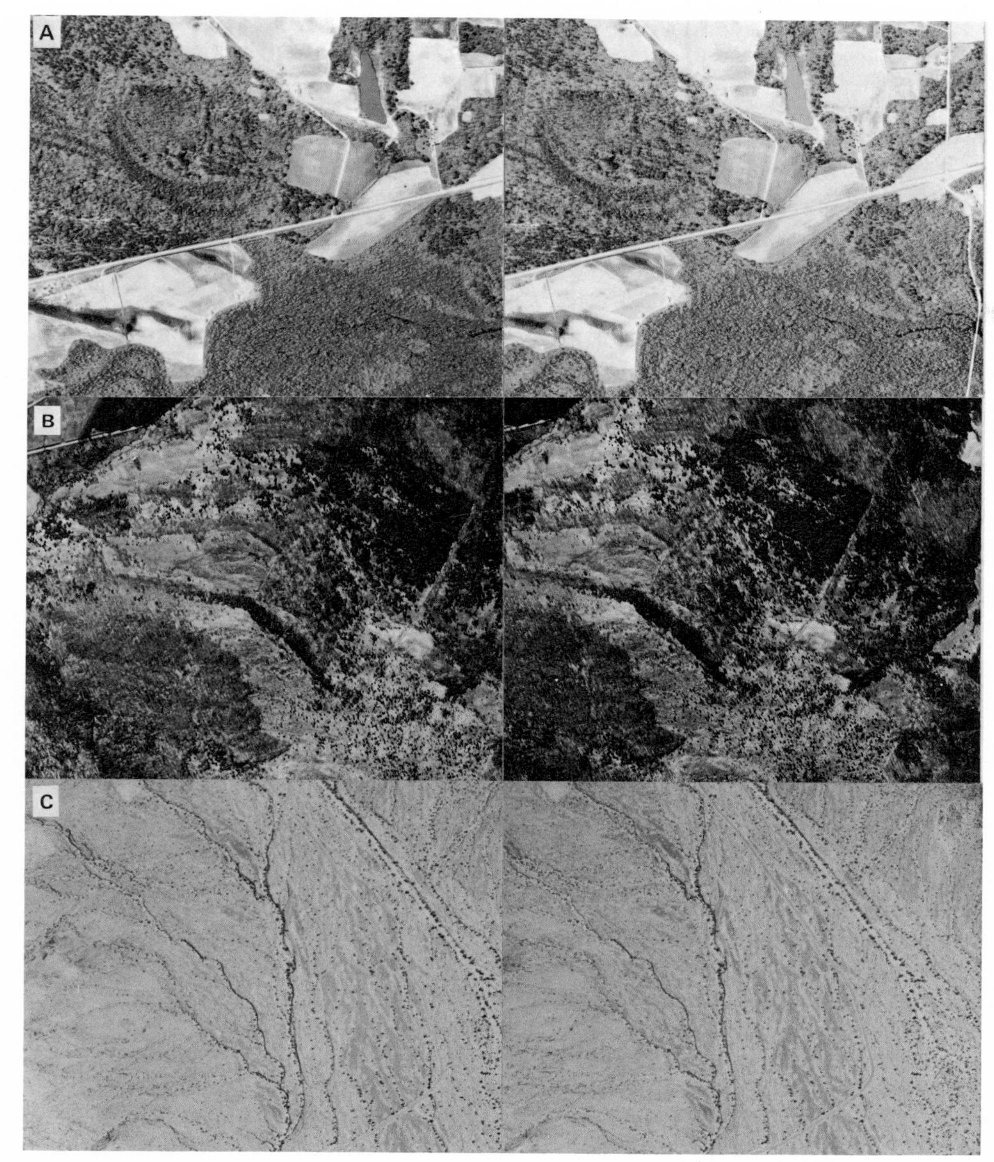

Figure 2.17. Vegetation as an indicator of physical site conditions. (A) Land left in vegetation indicates a wet condition too severe for agriculture. (B) Steep, rocky slopes are in vegetation cover, reflecting severe conditions not presently suitable for land use development. (C) Drainage courses in an arid climate emphasized by vegetation seeking greater moisture availability.

availability of soils, but local experience is necessary in making accurate and detailed correlations. Different species and vegetative associations indicate different conditions, but the same species can occur in different soil conditions in different climatic zones. For example, in New England red maple can be found in uplands and lowlands in a wide variety of soil conditions. The photographic charts presented in this volume indicate only the general presence or absence of vegetation and its mix and distribution with other land uses, such as agriculture. Generally, the occurrence of vegetation in agricultural areas indicates land that is not suitable for cultivation for a variety of reasons; such land may be poorly drained, steep, dry, rocky, and so on (Figure 2.17).

Land use patterns are valuable as indicators of different soil conditions, such as those caused by drainage. Most development tends to be located on the best, least expensive, least maintenance-prone sites available within an immediate region. In deep soil areas the higher, drier sites are the most valuable for structural locations, as shown by the usual location of town centers and even individual farm units. In rocky regions with thin soils, lowland sites with deeper soils adjacent to natural drainage courses and transportation systems are preferred. Railroad and major highway alignments give important clues to soil-rock conditions, since they typically follow the path of least construction cost, stay above 50- and 100-year flood plains, and, where cuts have been made, reveal the nature of the material of the subsurface.

REFERENCES

American Society of Photogrammetry, *Manual of Photo Interpretation,* Falls Church, Va., 1960.

American Society of Photogrammetry, *Manual of Color Aerial Photography,* Falls Church, Va., 1968.

American Society of Photogrammetry, *Manual of Remote Sensing,* Vols. 1 and 2, Falls Church, Va., 1975.

Aitchison, G. D., and K. Grant, "The PUCE Programme of Terrain Description, Evaluation, and Interpretation for Engineering Purposes," *Proceedings of the Fourth Regional Conference for Africa on Soil Mechanics and Foundation Engineering,* Cape Town, S. Africa, 1967.

Aitchison, G. D., and K. Grant, "Proposals for the Application of the PUCE Programme of Terrain Classification and Evaluation to Some Engineering Problems," *CSIRO Division of Soil Mechanics Research Paper No. 119,* Melbourne, 1968.

Avery, T. E., *Interpretation of Aerial Photographs,* 2nd ed., Burgess, Minneapolis, 1968.

Beckett, P.H.T., and R. Webster, "A Review of Studies on Terrain Evaluation by the Oxford-MEXE-Cambridge group, 1960–1969," *MEXE Report No. 1123,* London, 1970.

Belcher, D. J., "The Use of Soil Maps in Highway Engineering, Proceedings of the 28th American Road School," *Engineering Bulletin No. 26,* Engineering Extension Department, Purdue University, Layfayette, Ind., 1942.

Belcher, D. J., "The Engineering Significance of Soil Patterns," *Proceedings of the 23rd Annual Meeting,* Highway Research Board, Washington, D.C., 1943.

Belcher, D. J., "The Engineering Significance of Landforms," *Highway Research Board Bulletin,* No. 13, 1948.

Belcher, D. J., *Landform Reports,* Under contract with the Office of Naval Research, Cornell University, Ithaca, N.Y., 1952.

Berry, L., and B. P. Ruxton, "Notes on Weathering Zones and Soils on Granitic Rocks in Two Tropical Regions," *Journal of Soil Science,* Vol. 10, pp. 54–63, 1959.

Branch, M. C., *City Planning and Aerial Information,* Harvard University Press, Cambridge, Mass., 1971.

Bridges, E. M., and Doornkamp, J. C., "Morphological Mapping and the Study of Soil Patterns," *Geography,* Vol. 48, No. 2, p. 175–181, 1963.

Brink, A.B.A., T. C. Partridge, R. Webster, and A.A.B. Williams, "Land Classification and Data Storage for the Engineering Usage of Natural Materials," *Proceedings of the Symposium on Terrain Evaluation for Engineering, Australian Road Research Board 4th Conference,* Melbourne, 1968.

Bryan, K., "The Retreat of Slopes," *Annals of the Association of American Geographers,* Vol. 30, 1940.

Bryan, K., and C. C. Albritton, "Soil Phenomena as Evidences of Climatic Changes," *American Journal of Science,* Vol. 241, pp. 469–490, 1943.

Carlston, C. W., "Drainage Density and Streamflow," *U.S. Geological Survey, Professional Paper 422-C,* Washington, D.C., 1963.

Chapman, C. A., "Control of Jointing by Topography," *Journal of Geology,* Vol. 66, 1958.

Chikishev, A. G., ed., *Landscape Indicators,* Plenum, New York, 1973.

Chorley, R. J., ed., *Water, Earth and Man,* Methuen, London, 1969.

Christian, C. S., "Aerial Surveys and Integrated Studies," *Proceedings of the Toulouse Conference, 1964,* UNESCO, 1968.

Christian, C. S., "The Concept of Land Units and Land Systems," *Proceedings of the Ninth Pacific Science Congress,* Vol. 20, pp. 74–81, 1958.

Cooke, R. U., and A. Warren, *Geomorphology in Deserts,* University of California Press, Berkeley, 1973.

Davis, W. M., "The Geographical Cycle," *Geographical Journal,* Vol. 14, 1899.

Davis, W. M., "The Geographical Cycle in an Arid Climate," *Journal of Geology,* Vol. 13, 1905.

Davis, W. M., "Sheetfloods and Streamfloods," *Bulletin of the Geological Society of America,* Vol. 49, 1938.

Dedman, E. V., and J. L. Culver, "Airborne Microwave Radiometer Survey to Detect Subsurface Voids," *Highway Research Board Report No. 421,* National Academy of Sciences, Washington, D.C., 1972.

Donovan, D. T., *Stratigraphy,* Murby, London, 1966.

Enzman, R. D., "Geomorphology—Expanded Theory," in R. Fairbridge, ed., *The Encyclopedia of Geomorphology,* Reinhold, N.Y., p. 408, 1968.

Fairbridge, R. W., ed., *Encyclopedia of Geomorphology,* Reinhold, New York, 1968.

Fenneman, N. M., *Physiography of the Western United States,* McGraw-Hill, New York, 1931.

Fenneman, N. M., *Physiography of the Eastern United States,* McGraw-Hill, New York, 1938.

Flint, R. F., *Glacial and Pleistocene Geology,* Wiley, New York, 1957.

Frost, R. E., and O. W. Mintzer, "The Influence of Topographic Position in Airphoto Interpretation of Permafrost," *Highway Research Board Bulletin,* No. 28, 1950.

Goldich, S. S., "A Study in Rock Weathering," *Journal of Geology,* Vol. 46, pp. 17–58, 1938.

Grant, K., "A Terrain Evaluation System for Engineering, CSIRO, Australian Division of Soil Mechanics," *Technical Publication No. 2,* Melbourne, 1968.

Grant, K., "Terrain Classification for Engineering Purposes of the Melbourne Area, Victoria," *Division of Applied Geomechanics Technical Paper No. 11,* CSIRO, Melbourne, 1972.

Griffiths, J. F., *Applied Climatology: An Introduction,* Oxford University Press, London, 1966.

Hamblin, W. K., and J. D. Howard, *Physical Geology: Laboratory Manual,* 2nd ed., Burgess, Minneapolis, 1967.

Heath, G. R., "A Comparison of Two Basic Theories of Land Classification and Their Adaptability to Regional Photo-Interpretation Key Techniques," *Photogrammetric Engineering,* Vol. 22, pp. 144–168, 1956.

Horton, R. E., "Erosional Development of Streams and their Drainage Basins," *Bulletin of the Geological Society of America,* Vol. 56, 1945.

Howard, J. A., *Aerial Photo-Ecology,* American Elsevier, New York, 1970.

Hunter, G. T., et al., "Critical Terrain Analysis," *Photogrammetric Engineering,* Vol. 36, No. 9, pp. 939–952, 1970.

Keller, W. D., *The Principals of Chemical Weathering,* Lucas, Columbia, Mo., 1957.

Kendrew, W. G., *Climatology,* 2nd ed., Oxford University Press, Fair Lawn, N.J., 1957.

Kiefer, R. W., *Airphoto Interpretation for Soil Studies,* Engineering Experiment Station, University of Wisconsin, Madison, 1966.

Kiefer, R. W., "Terrain Analysis for Metropolitan Fringe Area Planning," *Journal of the Urban Planning and Development Division,* Proceedings of the American Society of Civil Engineers, UP4,93., 1967.

King, L. C., "Cannons of Landscape Evolution," *Bulletin of the Geological Society of America,* Vol. 64, 1953.

Klingebiel, A. A., and P. H. Montgomery, "Land Capability Classification," *S.C.S. Agriculture Handbook No. 210,* U.S. Department of Agriculture, Washington, D.C., 1061.

Köppen, W., *Grundriss der Klimakunde,* Walter De Gruyter, Berlin, 1931.

Lacate, D. S., "A Review of Land Type Classification and Mapping," *Land Economics,* Vol. 37, pp. 271–280, 1961.

Lo, C. P., and F. Y. Wong, "Micro-Scale Geomorphology Features," *Photogrammetric Engineering,* Vol. 39, (December) 1973.

Lobeck, A. K., *Geomorphology, An Introduction to the Study of Landscapes,* McGraw-Hill, New York, 1939.

Longwell, C. R., R. F. Flint, and J. E. Sanders, *Physical Geology,* Wiley, New York, 1969.

Lueder, D. R., "A System for Designating Map-Units on Engineering Soil Maps," *Highway Research Board Bulletin,* No. 28, 1950.

Lueder, D. R., "The Preparation of an Engineering Soil Map of New Jersey," *Symposium on Surface and Subsurface Reconnaisance,* American Society of Testing and Materials, Philadelphia, 1951.

Lueder, D. R., *Aerial Photographic Interpretation: Principles and Application,* McGraw-Hill, New York, 1959.

Mabbutt, J. A., "Review of Concepts of Land Classification," in G. A. Stewart, ed., *Land Evaluation,* Macmillan of Australia, Melbourne, 1968.

McHarg, I. L., *Design with Nature,* Natural History Press, Garden City, N.Y., 1969.

Melton, M. A., "An Analysis of the Relations Among Elements of Climate, Surface Properties, and Geomorphology," *Technical Report II,* Department of Geology, Columbia University, New York, 1957.

Melton, M. A., "Correlation Structure of Morphometric Properties of Drainage Systems and their Controlling Agents," *Journal of Geology,* Vol. 66, 1958.

Melton, M. A., "List of Sample Parameters of Quantitative Properties of Landforms," *Technical Report No. 16,* Office of Naval Research, Department of Geography, Columbia University, New York, 1958.

Miles, R. D., "A Concept of Landforms, Parent Materials, and Soils in Air Photo Interpretation Studies for Engineering Purposes," *Transactions of the Symposium on Photo Interpretation,* International Society of Photogrammetry, Delft, The Netherlands, 1962.

Mitchell, C., *Terrain Evaluation,* Longman, London, 1973.

Mintzer, O. W., and R. E. Frost, "How to Use Air Photo Maps for Material Surveys," *Highway Research Board Bulletin,* No. 62, 1952.

Mintzer, O. W., *A Comparative Study of Photography for Soils and Terrain Data,* Army Engineer Topographic Laboratories, Fort Belvoir, Va., 1968.

Mintzer, O. W., "Remote Sensing for Engineering of Terrain-Photographic Systems," Ohio State University, Columbus, O., unpublished report, 1968.

Mollard, J. D., *Landforms and Surface Materials of Canada: A Stereoscopic Airphoto Atlas and Glossary,* Mollard Associates, Regina, Saskatchewan, 1972.

Myers, B. J., "Rock Outcrops Beneath Trees," *Photogrammetric Engineering,* Vol. 41, No. 4, 1975.

Odum, E. P., *Fundamentals of Ecology,* 3rd ed., W. B. Saunders, Philadelphia, 1971.

Olgyay, V., *Design with Climate,* Princeton University Press, Princeton, N.J., 1963.

Parry, J. T., J. A. Heginbottom, and W. R. Cowan, "Terrain Analysis in Mobility Studies for Military Vehicles," *Land Evaluation,* ed. by G. A. Stewart, Macmillan of Australia, Melbourne, 1968.

Penck, A., "Climatic Features in the Land Surface," *American Journal of Science,* Series 4, Vol. 19, 1905.

Quiel, F., "Thermal /IR in Geology," *Photogrammetric Engineering,* Vol. 41, No. 3, 1975.

Raines, G. L., and K. Lee, "In Situ Rock Reflectance," *Photogrammetric Engineering,* Vol. 41, No. 2, 1975.

Ray, R. G., "Aerial Photographs in Geologic Interpretation and Mapping," *U.S. Geological Survey Professional Paper No. 373,* Washington, D.C., 1960.

Renwick, C. C., "Land Assessment for Regional Planning," in G. A. Stewart, ed., *Land Evaluation,* Macmillan of Australia, Melbourne, 1968.

Rib, H. T., "Utilization of Photo Interpretation in the Highway Field," *Highway Research Board Bulletin,* No. 109, 1966.

Saint Onge, D. A., "Geomorphic Maps," *Encyclopedia of Geomorphology,* ed. by R. Fairbridge, Reinhold, New York, 1968.

Shelton, J. S., *Geology Illustrated,* W. H. Freeman, San Francisco, 1966.

Siegal, B. S., and M. J. Abrams, "Geologic Mapping Using LANDSAT Data," *Photogrammetric Engineering,* Vol. 42, No. 3, 1976.

Smith, H.T.U., "Physical Effects of Pleistocene Climatic Changes in Non-glaciated Areas," *Bulletin of the Geological Society of America,* Vol. 60, 1949.

Smith, K. G., "Standards of Grading Texture of Erosional Topography," *American Journal of Science,* Vol. 648, pp. 655–668, 1950.

Smith, W. L., ed., *Remote Sensing Applications for Mineral Exploration,* Dowden, Hutchinson, & Ross, Stroudsburg, Pa., 1977.

Sparks, B. W., *Rocks and Relief,* St. Martin's Press, New York, 1971.

Spurr, S. H., *Photogrammetry and Photo-Interpretation with a Section on Applications to Forestry,* Ronald Press, New York, 1960.

Steinitz, C. S., T. Murray, D. Sinton, and D. Way, *A Comparative Evaluation of Resource Analysis Methods,* U.S. Army Corps of Engineers, Research Contract DACW33-68-C-0152, Harvard University, Graduate School of Design, Cambridge, Mass., 1969.

Steward, G. A., ed., *Land Evaluation, Papers of a CSIRO Symposium Organized in Cooperation with UNESCO, Canberra, August 1968,* Macmillan of Australia, Melbourne, 1968.

Strahler, A. N., and A. H. Strahler, *Environmental Geoscience,* Hamilton Publishing, Santa Barbara, Calif., 1973.

Strandberg, C. H., *Aerial Discovery Manual,* Wiley, New York, 1967.

Thomas, M. F., *Tropical Geomorphology,* Macmillan, London, 1974.

Thornbury, W. D., *Regional Geomorphology of the United States,* Wiley, New York, 1965.

Tricart, J., *Landforms of Humid Tropics, Forest, and Savannas,* St. Martin's Press, New York, 1972.

Twidale, C. R., *Structural Landforms,* M.I.T. Press, Cambridge, Mass., 1971.

Ulaby, F. T., and J. McNaughton, "Classification of Physiography from ERTS Imagery," *Photogrammetric Engineering,* Vol. 41, No. 8, 1975.

Verstappen, H., and R. A. Van Zuidam, "The ITC System of Geomorphological Surveying," *ITC Textbook of Photo-Interpretation,* Delft, The Netherlands, 1968.

Way, D. S., *Air Photo Interpretation for Land Planning,* Department of Landscape Architecture, Harvard University, Cambridge, Mass., 1968.

Way, D. S., "Aerial Photography for Residential Subdivision Site Selection," *Proceedings of the ASP 43rd Annual Meeting,* ASP, Washington, D.C., pp. 319–329, March 1977.

West, T. R., "Engineering Soil Mapping from Multispectral Imagery Using Automatic Classification Techniques," *Highway Research Board Report No. 421,* National Academy of Sciences, Washington, D.C., 1972.

Wood, W. F., and J. B. Snell, "A Quantitative System for Classifying Landforms, HQ Quartermaster Research and Engineering Command," *Technical Report No. EP-124,* Natick, Mass., 1960.

Wright, R. L., *A Geomorphological Approach to Land Classification,* PhD Thesis, University of Sheffield, 1967.

Yatsu, E., *Rock Control in Geomorphology,* Sozosha, Tokyo, 1966.

Zakrzewska, B., "Trends and Methods in Land Form Geography," *Annals of the Association of American Geographers,* Vol. 57, 1957.

CHAPTER 3

SOILS AND THEIR AERIAL PHOTOGRAPHIC INTERPRETATION

SOILS

When development activities are undertaken upon land surface, land planners must be concerned with and understand the properties of soils. Those who ignore or misinterpret the characteristics of soil cover or geological foundations may encounter structural failures or higher construction and maintenance costs. Furthermore, if development designs are to be in harmony with the site, the specifications for foundations, roadways, excavation, and so on must be based upon the engineering properties of the soils present.

Definitions of Soil

There are many different definitions of soil. Agronomists and pedologists define soil as the upper portion of the regolith, consisting of mineral matter originating from rock weathering, living and dead organic matter originating from biological processes, soil moisture containing both organic and mineral matter in colloidal state or solution, and soil air. It should be noted that all the elements included in this definition are those necessary to the growth of vegetation.

Geologists or engineers usually define soil more broadly to include all loose, unconsolidated material above bedrock which has been weathered from its original condition. This definition includes the pedologists' materials but adds to them broken rock, volcanic ash, alluvium, eolian sand, glacial material, or any other product of residual or transported weathered rock material. Land planners, concerned with determining the capabilities of land areas to support different types of land uses, tend to favor this definition because it includes all material to bedrock.

The solid matter of soil results from chemical and mechanical weathering. Weathered particles can be classified according to their content and whether or not they have been moved from their original place of weathering as follows: (1) *residual soils* are materials formed in place through weathering of hard or soft bedrock; (2) *transported soils* are materials transported from their places of origin by wind, water, glacial ice, or gravity and redeposited; and (3) *cumulose soils* are materials organic in nature and formed in place from the decomposition of peat and other organic material.

Table 3.1 lists these soil categories and their related landforms.

Soil Formation

Five major factors determine the rate of formation and the general type of a soil: the parent bedrock or transported material, climate, vegetation, slope, and time. All these factors are continuously at work in influencing soil composition. They are independent of one another, and a change in only one element can create a different soil type. For example, a forest fire burning a large area can create a change in the vegetative cover which within a short period of time will change the soil character.

The soil-forming processes create horizontal soil layers of differing textural composition and mineralogical content. A vertical section of these layers is called a soil profile, and the profile is usually described as having A, B, C, and sometimes D, horizons (Figure 3.1). The A horizon is the uppermost portion of the soil; it contains humus and organic matter and represents the zone of maximum removal or leaching of materials. The B horizon contains the highest accumulation of minerals leached from the A horizon. The C horizon consists of slightly altered or oxidized soil of the parent material. The

Table 3.1. General soil groups and landforms

Soil group	Origin	Landform*
Residual	Sedimentary rocks	Shale, Sandstone, Limestone, Interbedded
	Igneous rocks	Granite, Basalt
	Metamorphic rocks	Slate, Schist, Gneiss
Transported	Glacial	Till plain, End moraine, Drumlin
	Fluvial	Esker, Kame, Outwash
	Lacustrine	Lake bed, Beach ridge
	Eolian	Sand dune, Loess
	Fluvial	Flood plain, Delta, Alluvium
	Marine	Tidal flat, Coastal plain, Beach ridge
	Lacustrine	Lake bed
Cumulose	Organic	Peat bog, Organic deposits

* More specific landform definition and classification is presented within each landform chapter.

D horizon is any stratum beneath the three soil horizons which is not parent material but which affects the overlying soil.

Soil Composition

Every soil consists of various proportions of three states of matter—solids, water, and gas (air)—and the interrelationships among these elements determine many of the engineering characteristics of soils.

The size of the solid particles in the soil directly determines many of the soil's behavioral properties. Very general values relating to bearing and water-holding capacity and susceptibility to frost heave can be estimated if the percentages of the textural composition are known. Soils are divided into textural categories of coarse and fine; coarse soils consist of boulders, gravels, and sand; fine soils contain silts and clays. Soil classification systems have many different standards of textural definition, as illustrated in Table 3.2.

Some soil classification systems refer to soil textures as well or poorly graded. A well-graded soil contains a large assortment of particle sizes; the particles in poorly graded soils are all approximately the same size. Gradation is also a determinant of certain engineering behavioral characteristics of soils.

Primary soil structure is determined by the arrangement of the soil particles. The most common types of structure are homogeneous, varved, flocculent, blocky, and platy.

Fissures or cracks in the soil are not part of the soil structure but result from the secondary effects of mass wasting, desiccation, or displacement from overload.

Physical Properties of Soils

The physical properties of soil are of great importance to planners, architects, and engineers because of their concern with how and where structures should be placed upon the land's surface. The nature of these properties is dependent upon the grain size and shape of the soil particles, the amount and type of cementa-

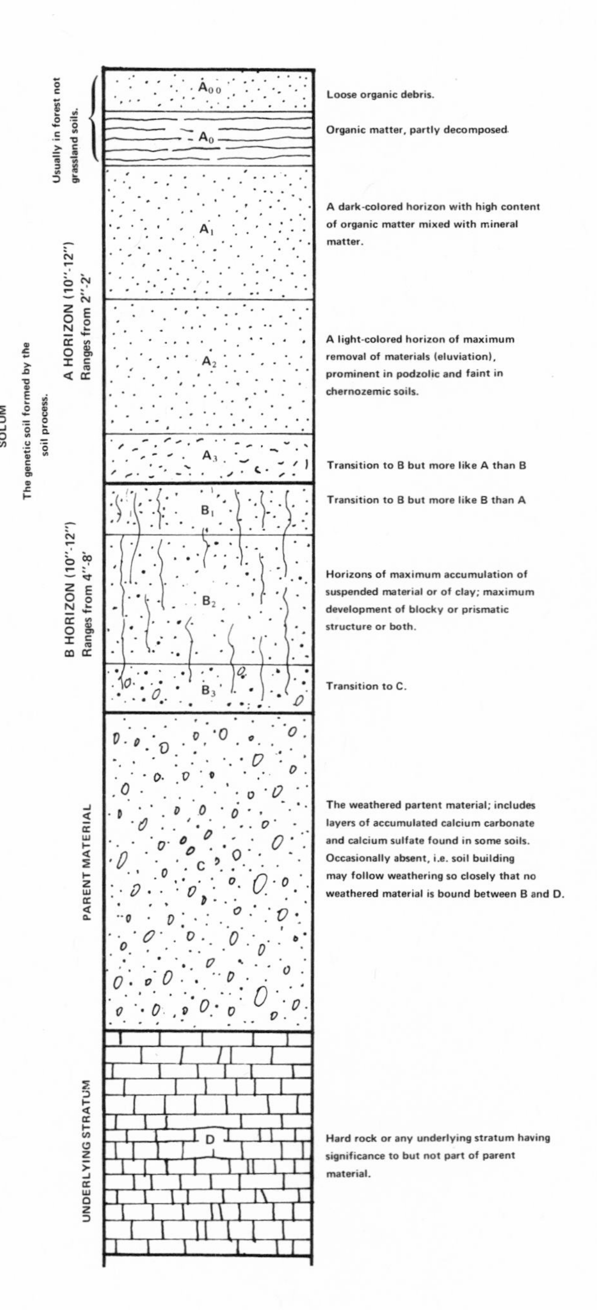

Figure 3.1. A hypothetical soil profile showing all major soil horizons. No one soil has all of these profiles. (After U.S. Department of Agriculture Yearbook, 1957.)

tion, and the stress history. Many factors are important in determining these properties: parent material, mineralogical composition, organic matter content, climate, age, method of transportation deposition, method and degree of compaction, soil texture, soil gradation, and soil structure. The properties usually measured include permeability, shear strength, and volume change.

Permeability refers to the ability of water to flow or pass through the soil. Soil texture primarily determines the relative permeability of soils, but permeability is also influenced by soil structure and the degree of compaction. The degree of permeability determines the general potential of soils for septic tank leaching fields and for operations such as dewatering of excavations.

Shear strength is measured by the shear stress that can be sustained without excessive deformation under specified conditions of rate of loading, confining pressure, and pore water pressure. For cohesionless soils, such as sands and inorganic silts, the distribution of grain sizes and the shape of the grains are the most significant determinants of shear strength. The component of shear strength that does not depend upon interparticle friction is known as cohesion. Cohesion is an attribute of the shearing strength of the cement or of the adsorbed water films that separate individual grains at their points of contact (Stokes and Varnes, 1955).

Volume change deformations can occur in soil masses both naturally and as a result of the application or removal of external loads. Deformations are caused by (1) a lowering of the groundwater table, which allows compression; (2) shrinking and swelling phenomena resulting from build-up or release of moisture in the soil's pore water; and (3) frost heave. Compression takes place as pore water is squeezed from the soil and the void ratio is decreased. It may take place naturally, if the groundwater table is lowered by the overlying weight of the soil, or it

Table 3.2. Relationships of textural descriptions to particle diameter size*

American Society for Testing and Materials

American Association of State Highway Officials soil classification

U.S. Department of Agriculture soil classification

Federal Aviation Agency soil classification

Unified soil classification (Corps of Engineers, Department of the Army, and Bureau of Reclamation)

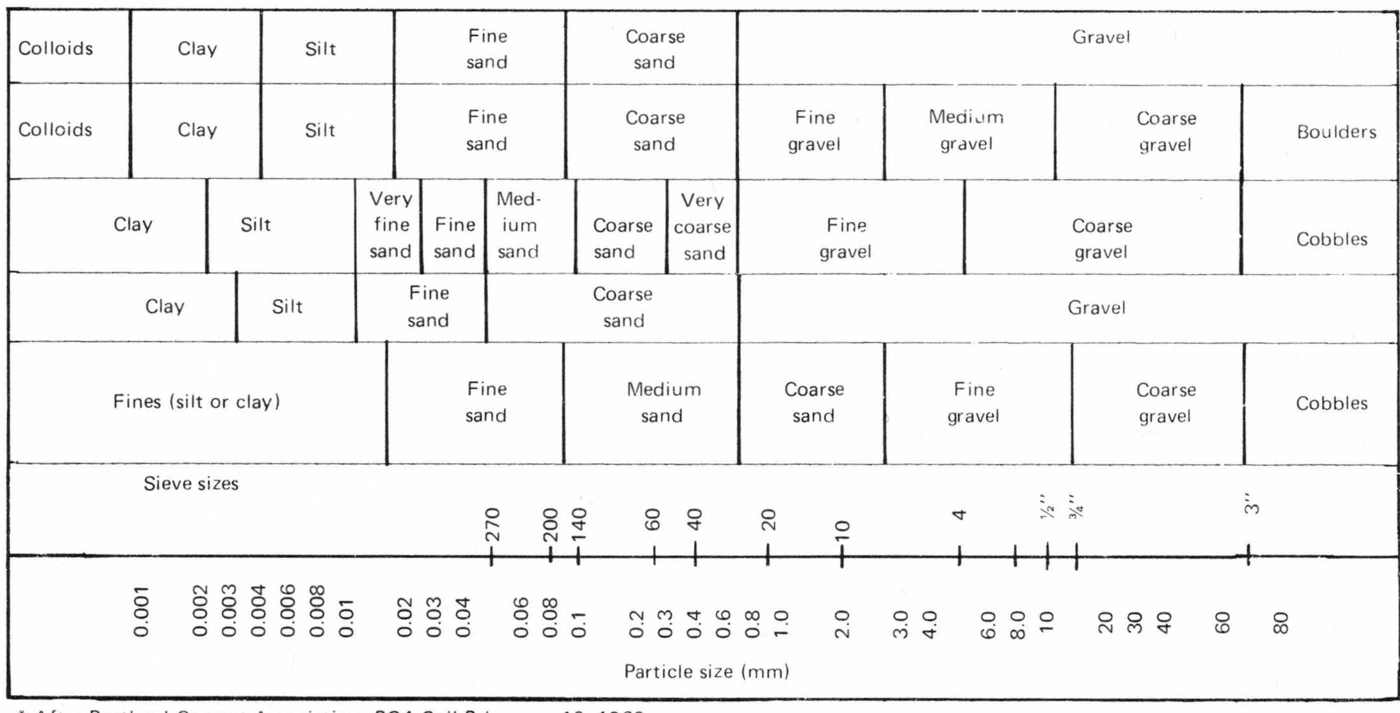

* After Portland Cement Association, *PCA Soil Primer*, p. 10, 1962.

can occur under an applied load. The rate of compression or consolidation depends upon the rate at which pore water escapes and hence upon the permeability of the soil (Stokes and Varnes, 1955).

In waterlaid, fine-grained soils, especially clay, volume changes are caused by the accumulation and release of capillary tensile stresses within the soil's pore water and by the degree of attractiveness of some clay minerals for water. As desiccation or drying occurs, the sediments shrink in volume.

The third type of volume change, deformation by frost heave, occurs when ice lenses form near the soil surface and are fed moisture from the water table by capillary action. Water freezing near the surface within the zone of frost penetration forms ice lenses, and the increased volume of the ice formation displaces the ground surface upward. In the spring the surface ice melts and saturates the upper soil layer with water that eventually percolates back to the water table. The old ice cavities are now filled with air, and if compacted by gravity or traffic, will form a depression on the surface. Granular soils usually do not present frost heave problems because of their capillary system's inability to carry water into the frost zone. To counteract frost heave in fine-grained soils susceptible to this condition either the soils must be covered with thick layers of coarse fill or the water table must be lowered.

Other physical properties of soils are plasticity and elasticity. *Plasticity* refers to the soil's ability to undergo permanent deformation without significant volume change, elastic rebound, or rupture. It is a colloidal property and is a function of mineralogy, grain shape and size, surface area, and absorption of water films. Plastic deformation takes place when stresses accumulate, thereby causing small slippages at grain-to-grain contact points between particles, and minute structural collapses occur. The Unified soil classification system, described later in this chapter, is based on the relative levels of plasticity of both organic and inorganic soils. *Elasticity* is a measure used primarily in highway engineering. It refers to a soil's ability to return to its original volume after the application and release of an applied load. The actual rebounding is generally controlled by the elasticity of individual mineral particles and the deformation of certain soil structures.

SOIL MAPPING FROM AERIAL PHOTOGRAPHS

Soil scientists, because of their training, find it very difficult to consider mapping soils in any way other than by observation of the soil profile. To them, interpretation and mapping from aerial photographs represent a high degree of subjective judgment, and they do not readily admit that since no two observers feel the same textures, observe the same colors, smell the same odors, or even visualize the same number of horizons, profile examination is also a matter of subjective judgment. Even soil maps force the delineation of finite soil boundaries on conditions that are not finite, and therefore pedological classification and mapping itself has an inherent and not always predictable degree of error.

It is, however, possible to use a combination of aerial photographic interpretation and field checks to identify and map soils. The method presented in this section and discussed in more detail for each landform is not intended to replace soil field mapping but rather to complement it. The ability to map soils accurately by using photographic interpretation depends upon the mapper's knowledge of the types of soils found in the region and on his interpretive experience, the quality, scale, and date of available photographs, the strategic location of field checks, and the degree of detail desired in the final map. If a combination of photographic interpretation and field investigation produces soil data that is accurate within a tolerable margin of error, then the method is justified in its economy and rapidity.

Much of the basic technique for mapping soils and parent materials from aerial photographs was developed by Belcher (1948). By means of interpretation of, or inference from, visible elements, such as drainage, erosion, soil color, topography, vegetation, land use, and microfeatures, Belcher was able to identify soil parent materials.

Interpretation of soils from aerial photographs begins with the identification of landforms having particular shapes and topographic positions that represent materials deposited by geomorphic processes. Identification of the basic landforms reduces and defines the range of soils likely to be found; for example, residual soils formed from sedimentary rock shale do not occur in windlaid loess deposits, and glacial outwash does not have the same composition as glacial till. The regional distribution and location of landforms gives additional clues to the soils that comprise them; glacial till in the Midwest, for example, contains many silt and clay particles and is predominately comprised of silty clay loams, silt loams, or clay loams; glacial till found over most of New England contains more sand, as a result of its origin in igneous rock regions, and is classified as sandy loam and fine sandy loam. Different climatic regions and differences in vegetation cover affect both the type and characteristics of the soils and their appearance on aerial photographs. Once the landforms are recognized and incorporated into a hypothesis concerning the formation of the surficial geology of the area being mapped, details on the photographs may offer clues to the presence of more specific soil types. Elements on the photographs, such as tonal change, or lack of it, type of tone, microdrainage features, including first-order streams and gullies, gully cross-sections, erosional features, slope, vegetative cover and land use are studied in order to formulate an interpretive hypothesis concerning the kinds and distribution of soil types. Locations for field checks from which to gather additional data and to verify the hypothesis can also be established from aerial photographs.

Photographic scales larger than 1:12,000 are preferred for detailed soil characteristic mapping while those from 1:4,800 to 1:7,200 are optimum (Way, 1977). At these scales, microtopography can be easily observed as can rock outcrops, large boulders, gullies, erosional features, and subtle tones.

Soil Texture

Mapping of soil by the texture of the parent material begins with the delineation of the basic landform units that define ranges of expected soil characteristics. Additional patterns such as erosional indicators, microdrainage, microtopography, and tones are interpreted in detail to refine further the expected range. For example, an aerial photograph of a site in the Midwest, where glacial tills are typically composed of silty clay loam, silt loam, clay, clay loam, silty clay, and loam, may show mottled tones with gradual, fuzzy boundaries and white fringing gullies. These patterns correlate with high silt content soils and would therefore suggest silty clay or silty clay loam. A low wet area on the same site would appear dark and drab and may show evidence of some artificial drainage, which would suggest poorly drained clay or silt clay soils. Its soil profile would be gray with mottling as a result of a high water table and imperfect drainage.

The subtle differences of tone, color, microtopography, and erosional features can provide valuable clues for a detailed determination of soil texture. In addition, background field knowledge of the area where interpretations are to be made is important and determines interpretive speed, confidence, and accuracy.

The Unified soil classification system can be used to provide soil textural distinctions and has the added advantage of providing useful engineering capability ratings (see Appendix A). It can be organized in a hierarchical pattern which facilitates mapping at both a general and a more detailed level (see Table 3.3).

In the first level, as on small-scale photographs (i.e., 1:60,000 and greater) where vast areas of land are covered, broad textural distinctions between soils can be used to indicate a predominance of gravel (G), sand (S), fines (F), or peat (Pt). When large land areas are surveyed at this scale in a general reconnaissance, these distinctions provide sufficient information to identify primary areas for more detailed investigation. In most cases textural distinctions are made from aerial photographs by relating the parent landform, climatic region, and photographic tone.

The second level of detail in the Unified system breaks the four first-level soil types into eight subcategories by making distinctions within each category between the distribution of textures and the liquid limit. Soil reconnaissance maps of medium scale (1:24,000 to 1:60,000) use this level of classification to describe generalized soil characteristics and as an

Table 3.3. Hierarchy of the Unified soil classification system

Third level	Second level	First level
GW (well-graded gravel)	CG (clean gravels)	G (gravel)
GP (poorly graded gravel)		
GM (gravel-silt mixtures)	FG (gravels and fines)	
GC (gravel-clay mixtures)		
SW (well-graded sand)	CS (clean sands)	S (sand)
SP (poorly graded sand)		
SM (sand-silt mixtures)	FS (sand and fines)	
SC (sand-clay mixtures)		
ML (silts: low plasticity)	FL (fines: silt and clay); liquid limit <50	F (fines: silts and clays)
CL (clays: low-medium plasticity)		
OL (organic silts: low plasticity)	FO (fines: organic silts and clay); high and low plasticity	
OH (organic clays: high plasticity)		
MH (silts-inorganic: low plasticity)	FH (fines: silt and clay); liquid limit >50	
CH (clays: high plasticity)		
Pt (peat-highly organic)	Pt (Peat—highly organic)	Pt (Peat)

interim step in identifying areas of high priority for more detailed mapping and field investigation. It has been found that it is possible by detailed interpretation of a site's landforms and by identification of parent materials from aerial photographs to provide most of the information necessary for mapping these eight categories.

The third level of detail describes the final 15 categories of the Unified system by noting differences in grading, organic content, and liquid limit. A greater degree of field checking is usually necessary in order to substantiate these categories, but it is important to use these classifications when land areas are being evaluated for construction-related land uses. Photos scaled 1:4,800 to 1:7,200 are best for this level of detail.

Soil Depth Over Bedrock

Mapping soil (regolith) depths to massive, unweathered bedrock from aerial photographs can be accurately accomplished in some locations. Most success occurs in situations where the rock is covered by less than ten feet of soil; the characteristics of the rock, topography, drainage, and tones are still apparent and suggest a rock-controlled topography. Areas of soil deeper than ten feet begin to mask these patterns effectively, and the pattern characteristics of the soils begin to predominate. Drainage patterns, however, do not follow this rule, and bedrock control may still be inferred in areas where the residual soils are greater than fifty feet thick. For example, the deep schist residual soils (40 to 60 feet) in the Columbia, Maryland, region have associated rectangular drainage patterns that indicate the presence of metamorphic rock.

Residual Soils

In humid climates several general rules can be followed for soil depth mapping. Rock outcrops and shallowest soils usually occur along the upper hillside slopes in a dissected topography. Hilltops, depending upon their size, will contain relatively deep soils, and lower hillside slopes will accumulate deeper deposits from slope processes of creep, slumping, and erosion. It should be noted that predicting the properties of these lower slope zones is most difficult due to the complex nature of the processes by which they were formed. Major valleys contain fluvial transported landforms that can be assumed to be deep (greater than twenty feet). Some field sampling is necessary to establish the basis of the interpretation, but correlations between soil depth zones and topography can be made, thereby simplifying the mapping operation.

Transported Soils

Soil depth mapping is more difficult in transported soil regions since the depositional process covers the underlying materials. Thus, the masking aspects are more effective compared to the same depth of residual soil. In addition in residual soil areas it is difficult to make accurate assumptions between topography and soil depth ranges. For example, in the glaciated area of New England, a large hill could consist of thinly veneered till over bedrock or deep till. In these mapping situations, knowledge of soil variances in the area becomes an important interpretive factor in pattern correlation and keying.

Where soil deposits are less than twenty feet thick, several general rules can be applied. The rock type should first be identified to establish meaning to microtopographical features by determining whether they are associated with the rock or with deeper soil deposits. Topographic features associated with the rock type can be assumed to be rock controlled and hence have thin (less than 10 feet) soil cover. All rock outcroppings and exposures should be mapped to further indicate the presence of rock controlled topography. The areas immediately adjacent to the rock outcrop should be carefully observed to determine how far the zone of very thin (1 to 3 feet) soil cover extends. These thin soils will show definite rock related photographic tones, microtopography, and have many outcrops. Any slight fracturing and joints, even if covered by 1 to 3 feet of soil cover, should still be apparent on the large-scaled photographs. Soils deeper than approximately 10 feet will not show these characteristics. The topographic surfaces and contours will tend to be less jagged and more uniform. The remaining soil zone, 3 to 10 feet, is the most difficult to map. It is best to delineate all other zones first, that is, the exposed zone, the 1 to 3 feet range, and the range greater than 10 feet. The remaining area is then the 3 to 10 feet zone. The location of the boundary between the greater than 10 feet zone and less than 10 feet zone is the most inaccurate part of the map and requires the greatest number of field samples to clarify its location. (The depth categories used in this discussion are arbitrary; such categories should be derived specifically for each project. See, for example, Figures 3.2 and 3.3.)

Depth to Water Table

The depth to the groundwater table is difficult to map and requires local knowledge of the expected ranges of conditions. As a further complication, the water table elevation may fluctuate widely within one season, thus necessitating further definition of what is average, seasonally high, or low. Well records and boring logs are extremely helpful in providing initial data and should be used when available.

Water Table Depth in Soils

Several general rules can be applied to mapping water table elevation in deep soils. First, the water table will express a regional gradient in the direction of its flow. This gradient is most easily observed by plotting different water elevations from well records or by the elevations of ponds where the water table encounters the surface (see Figure 3.2). When several points of water table elevation are known, the gradient between them can be calculated. If the soils are very granular and no recharge is introduced through infiltration, the gradient between points is a straight line. If infiltration recharge occurs, the gradient is bowed toward the surface. This bowing is greater when the soils are fine textured and thus slow infiltration.

Mapping of near-surface (less than 10 feet) seasonal high water table elevations in humid climates is not difficult from aerial photographs. The procedure requires previous mapping of landform, general soil, and soil depth since the water table depth may be affected by these factors. If a rock type is impervious, drainage will be slow, and water, controlled by the underlying rock topography, will pond within the soil profile. Porous rocks will allow more rapid internal draining. Actual mapping proceeds with the collection of well records indicating the well locations and water elevations on basemaps. Then, surface water and wet areas, which may represent contacts of the groundwater table, are delineated from the photos. The well records and the surface contacts allow for an initial hypothesis to be constructed regarding the water table subsurface elevations and direction of flow. The areas adjacent to the wet zones are carefully studied, and a boundary estimating the 1 to 3 foot category is drawn. This estimate is made by studying the topography, associated vegetation, and soil tones in the photos. The deep (greater than 10 foot) water table areas can be identified from the well records, landform and associated soil characteristics, and topographic elevation. The final category, 3 to 10 feet can be divided into several subunits such as 3 to 5 feet and 5 to 10 feet if desired. These additional divisions can be mapped on the basis of the topography and data concerning the textures of the soils and the effect on the water table surface. The boundary location be-

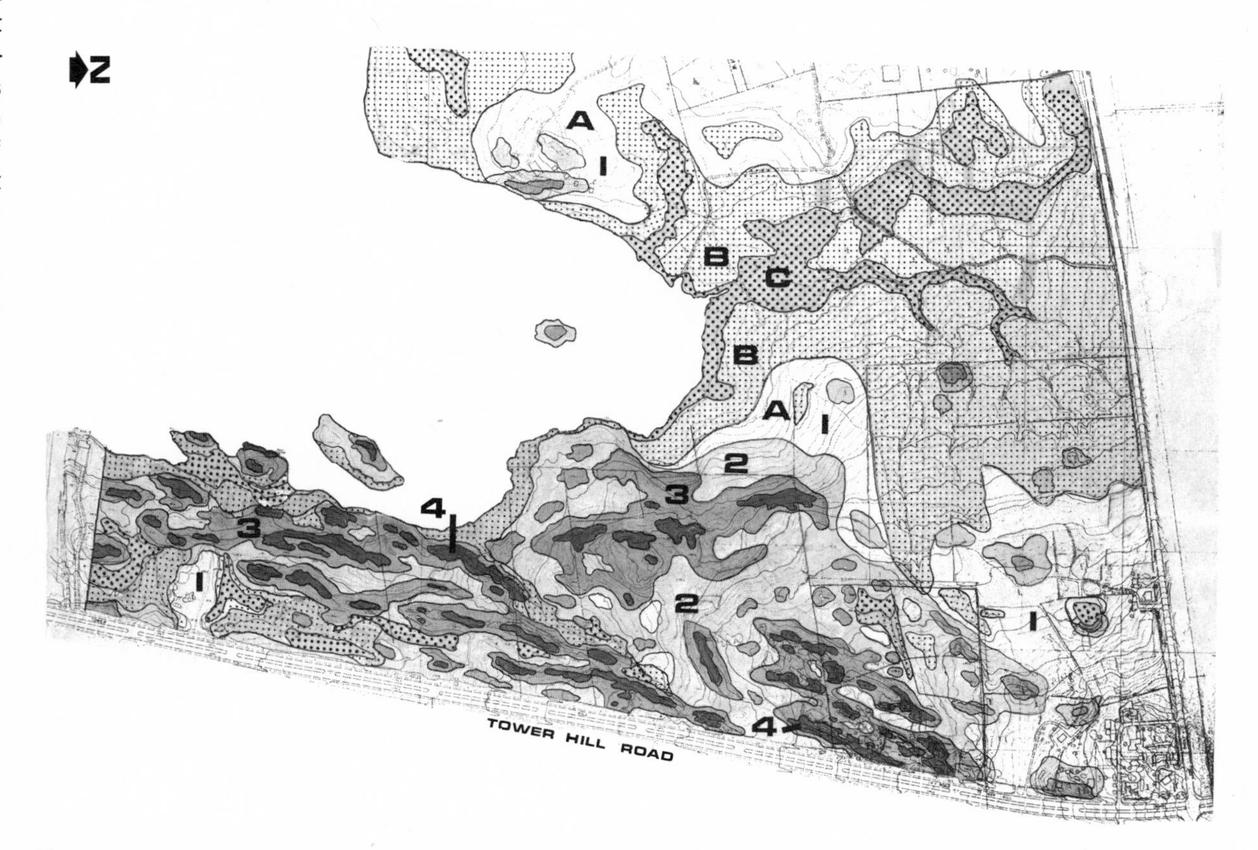

Figure 3.2. Depth to bedrock and depth to seasonal high water table illustrated for a site in southern Rhode Island (glacial tills over crystalline bedrock). Depth of soil over bedrock: (1) = greater than 7 feet; (2) = 3 to 7 feet; (3) = 1 to 3 feet; and (4) = exposed rock. Depth to seasonal high water table: (A) = greater than 5 feet; (B) = 1 to 5 feet; and (C) = 0 to 1 foot.

tween the greater than 5 feet zone and less than 5 feet zone is usually the most difficult to map accurately and will require the greatest field checking. These mapping operations will seem difficult at first until experience and knowledge of expected local conditions are acquired. (The depth categories used here are for discussion purposes; specific categories should be established for each separate project.)

Groundwater in Rocks

Mapping groundwater aquifers in rock requires a more technical set of interpretations that include the *underlying* rock structure. The rock types on the ground surface can easily be identified, and the associated drainage patterns can suggest which materials may be acting as subsurface aquifers. Dip and strike angles, joints, fractures, faults, and other anomalies must be considered in forming a hypothesis of the underground rock structure which would control subsurface groundwater flow. From this, aquifer zones can be followed through a region.

For local groundwater availability studies, the rock type, association with other rock types, their mineralogic composition, cracking (faults, fractures, and joints), spatial extent, and associated surface hydrology determine how much groundwater may be available from different locations. Porous sandstones and soluble limestones can provide high yields, whereas shales, which are relatively impervious, allow for very slow recharge. Shale layers on both sides of a sandstone can act as aquicludes, thereby creating a piping effect through the sandstone. If the water flow is under pressure, a well drilled in such a situation would be artesian if the water level is higher than the elevation of the local groundwater table. Impervious rock in some instances may provide high yields of water. Highly weathered and fractured granite allows rapid

Figure 3.3. A section of the Rhode Island site shown in Figure 3.2 presented in stereo. The arrows indicate exposed rock. Photographs taken April 21, 1972, 1:4,800. Aerial Date Reductions, Pennsauken, New Jersey.

infiltration through the many cracks and can deliver high yields of water resources.

In essence this type of analysis requires a thorough understanding of the regional geology and the surface hydrologic network. Their interaction can be hypothesized and documented by available well records and boring logs to provide reasonably accurate estimates of groundwater flow, recharge, and quality.

REFERENCES

Asphalt Institute, *Soil Manual for the Design of Asphalt Pavement Structures,* College Park, Md., 1969.

Belcher, D. J., "The Engineering Significance of Soil Patterns," *Photogrammetric Engineering,* No. 2, 1945.

Belcher, D. J., "The Engineering Significance of Landforms," *Highway Research Board Bulletin,* No. 13, 1948.

Buringh, P., and W. I. Liere, "Example of a Reconnaissance Soil Map Produced by the Pedological Analysis of Aerial Photographs Followed by the Study of Soils in the Field," *Proceedings, 5th International Congress of Soil Science,* Vol. 4, p. 338, 1961.

Casagrande, A., "Classification and Identification of Soils," *Transportation,* American Society of Civil Engineers, 1948.

Chorley, R. J., Ed., *Water, Earth and Man,* Methuen, London, 1969.

Curtis, L. F., "Soil Classification and Photo Interpretation," *Proceedings, Symposium on Photo Interpretation—Commission VII,* Delft, The Netherlands, pp. 153–158, 1963.

Crandell, F. J., "Ground Vibrations Due to Blasting and Its Effect upon Structures," *Journal of the Boston Society of Civil Engineers,* 1949.

Dedman, E. V., and J. L. Culver, "Airborne Microwave Radiometer Survey to Detect Subsurface Voids," *Highway Research Board Report No. 421,* National Academy of Sciences, Washington, D.C., 1972.

Federal Housing Administration, Architectural Standards Division, *Engineering Soil Classification for Residential Developments,* Washington, D.C., 1961.

Felt, E. J., "Soil Series Names as a Basis for Interpretative Soil Classification for Engineering Purposes," *Symposium on the Identification and Classification of Soils,* Special Technical Publication No. 113, American Society for Testing and Materials, Philadelphia.

Frost, R. E., "Factors Limiting the Use of Aerial Photographs for the Analysis of Soil and Terrain," *Photogrammetric Engineering,* Vol. 19, No. 3, pp. 427–437, 1953.

Highway Research Board, "Soil Exploration and Mapping," *Highway Research Board Bulletin,* No. 28, Washington, D.C., 1950.

Hitchcock, H. C., T. L. Cox, F. P. Baxter, and C. W. Smart, "Soil and Land Cover Overlay Analysis," *Photogrammetric Engineering,* Vol. 41, (December) 1975.

Holden, A., "Engineering Soil Mapping from Airphotos," *Photogrammetria,* Vol. 23, No. 6, pp. 185–199, 1968.

Jarvis, R. A., "The Use of Photo Interpretation for Detailed Soil Mapping," *Proceedings, Symposium on Photo Interpretation,* Delft, The Netherlands, p. 177, 1962.

Jenkins, D. S., et al., *The Origin, Distribution and Air Photo Interpretation of United States Soils,* Technical Development Report No. 52, U.S. Department of Commerce, Civil Aeronautics Administration, Washington, D.C., 1946.

Kellogg, C. E., *The Principals of Planning Based on Soil Surveys,* U.S. Department of Agriculture, Washington, D.C., December 1970.

Kiefer, R. W., *Airphoto Interpretation for Soil Studies,* Engineering Experiment Station, University of Wisconsin, Madison, December 1966.

Kiefer, R. W., "Sequential Aerial Photography and Imagery for Soil Studies," *Highway Research Board Report No. 421,* National Academy of Sciences, Washington, D.C., 1972.

Kristof, S. J., and A. L. Zachary, "Mapping Soil Features from Multispectral Scanner Data," *Photogrammetric Engineering,* Vol. 40, No. 12, 1974.

Leamer, R. W., D. A. Weber, and C. L. Wiegand, "Pattern Recognition of Soils and Crops from Space," *Photogrammetric Engineering,* Vol. 41, No. 4, 1975.

Lo, C. P., and F. Y. Wong, "Micro-Scale Geomorphology Features," *Photogrammetric Engineering,* Vol. 39 (December) 1973.

Lyman, A.K.B., "Compaction of Cohesionless Foundation Soils by Explosives," *Transportation,* American Society of Civil Engineers, Vol. 107, 1942.

Matalucci, R. V., and M. Abdel-Hady, "Surface and Subsurface Exploration by Infrared Surveys," *Highway Research Board Special Report No. 102,* National Academy of Sciences, Washington, D.C., 1969.

Mintzer, O. W., *A Comparative Study of Photography for Soils and Terrain Data,* Army Engineer Topographic Laboratories, Fort Belvoir, Va., 1968.

Taylor, D. W., *Fundamentals of Soil Mechanics,* Wiley, New York, 1948.

Terzaghi, K., and R. B. Peck, *Soil Mechanics in Engineering Practice,* Wiley, New Jery, 1967.

U.S. Department of Agriculture, *Aerial-Photo Interpretation in Classifying and Mapping Soils,* Handbook No. 294, Washington, D.C., 1966.

Morse, P. K., and T. H. Thornburn, "Reliability of Soil Maps," *Proceedings of the 5th International Conference of Soil Mechanics and Foundation Engineering,* pp. 259–262.

Myers, B. J., "Rock Outcrops Beneath Trees," *Photogrammetric Engineering,* Vol. 41, No. 4, 1975.

Pasto, J. K., "Soil Mapping by Stereoscopic Interpretation of Airphotos," *Papers and Proceedings of the Soil Science Society of America—Division V,* November 20, 1952.

Piech, K. P., and J. E. Walker, "Interpretation of Soils," *Photogrammetric Engineering,* Vol. 40, No. 2, 1974.

Poland, J. F., "Land Subsidence Due to Ground-Water Development," *American Society of Civil Engineers Journal, Irrigation and Drainage Division,* Vol. 84, 1958.

Pomerening, J. A., and M. G. Cline, "The Accuracy of Soil Maps Prepared by Various Methods that Use Aerial Photographic Interpretation," *Photogrammetric Engineering,* Vol. 19, No. 5, pp. 809–830, 1953.

Portland Cement Association, *PCA Soil Primer,* Chicago, 1962.

Rib, H. T., and R. D. Miles, "Multisensor Analysis for Soils Mapping," *Highway Research Board Special Report No. 102,* National Academy of Sciences, Washington, D.C., 1969.

Richardson, A. J., A. H. Gerbermann, H. W. Gausman, and J. A. Cuellar, "Detection of Saline Soils with Skylab Multispectral Scanner Data," *Photogrammetric Engineering,* Vol. 42, No. 5, 1976.

Siegal, B. S., and A.F.H. Goetz, "Effect of Vegetation on Rock and Soil Type Discrimination," *Photogrammetric Engineering,* Vol. 43 (February) 1977.

Soil Conservation Service, *Soil Taxonomy: Agricultural Handbook, No. 436,* U.S. Department Agriculture, Washington, D.C., 1975.

Stokes, W. L., and D. J. Varnes, *Glossary of Selected Geologic Terms,* Colorado Scientific Society, Denver, 1955.

Stone, R. O., and J. Dugundji, "A Study of Micro-relief, its Mapping, Classification, and Quantification by Means of Fourier Analysis," *Engineering Geology,* Vol. 1, No. 2, pp. 89–187, 1965.

U.S. Department of Agriculture, "Nonfarm Uses of Soil Surveys," *Soil Conservation,* Vol. 32, No. 1, Washington, D.C., 1967.

U.S. Department of Agriculture and U.S. Department of Housing and Urban Development, "Soil, Water and Suburbia," *Proceedings, National Conference on Soil, Water and Suburbia,* Washington, D.C., 1967.

U.S. Department of Agriculture, *Soil Classification: A Comprehensive System—7th Approximation,* Washington, D.C., 1960. (Supplement issued March 1967.)

U.S. Department of Agriculture, *Soil Survey Manual,* Handbook No. 18, Washington, D.C., 1951.

U.S. Department of the Interior, Bureau of Reclamation, *Earth Manual: A Guide to the Use of Soils as Foundations and as Construction Materials for Hydraulic Structures,* Washington, D.C., 1963.

Wagner, T. W., "Multispectral Remote Sensing of Soil Areas: A Kansas Study," *Highway Research Board Report No. 421,* National Academy of Sciences, Washington, D.C., 1972.

Way, D. S., "Aerial Photography for Residential Subdivision Site Selection," *Proceedings of the ASP 43rd Annual Meeting,* ASP, Washington, D.C., pp. 319–329, March 1977.

West, T. R., "Engineering Soil Mapping from Multispectral Imagery Using Automatic Classification Techniques," *Highway Research Board Report No. 421,* National Academy of Sciences, Washington, D.C., 1972.

Willis, E. A., "Discussion: A Study of Lateritic Soils," *Proceedings,* Highway Research Board, Washington, D.C., 1948.

Woods, K. B., R. D. Miles, and C. W. Lovell, Jr., "Origin, Formation, and Distribution of Soils in North America," in G. A. Leonards, ed., *Foundation Engineering,* McGraw-Hill, New York, Chapter 1, 1962.

Yong, R. N., and B. P. Warkentin, "Soil Freezing and Permafrost," in *Introduction to Soil Behavior,* Macmillan, New York, Chapter 12, 1966.

Part II

LANDFORMS AND ISSUES OF SITE DEVELOPMENT

CHAPTER 4

INTRODUCTION TO SITE DEVELOPMENT

The following seven chapters discuss the major landforms, their formative and interpretive characteristics, associated soil conditions, and issues of site development. As an introduction, this chapter describes a range of engineering and construction problems that should be considered early in the planning process before site development is initiated. Physical site requirements and capabilities are discussed for sewage disposal utilizing septic tank leaching fields, solid waste disposal through the use of sanitary land fills, trenching, excavation and grading, dewatering, sources of construction materials, compaction of fills, pond or lake construction, and foundation and highway construction. The discussions are not meant to be complete; rather their purpose is to review briefly the general requirements of these engineering construction categories. More detailed information can be obtained from many of the references listed at the end of the chapter.

SEWAGE DISPOSAL

Within the overall problem of sewage disposal, only on-site septic tank leaching fields are discussed in relation to the engineering requirements associated with each landform. A septic tank system operates by carrying sewage effluent out of a residential living unit to a septic tank unit where it is temporarily stored while bacteria work to decompose much of the solid matter. The liquid effluent slowly overflows the tank unit and is carried by pipes to a leaching field. The leaching field contains a series of segmented or perforated clay pipes, buried approximately 2 feet beneath the surface in rows 6 to 8 feet apart. The effluent slowly seeps through the perforations or open joints of the pipes into the soil, and as it filters through the soil the process of decomposition by aerobic bacteria continues. In a properly working system, this decomposition-filtering process is completed before any impervious substrata or the water table are encountered. However, if insufficiently treated sewage reaches the water table or ground surface, contamination of the surface-subsurface water resources will occur, thereby presenting a serious health hazard.

The capability of a landform to support this type of sewage system depends upon its characteristics of slope, depth to water table, depth to bedrock or other impervious stratum, and soil percolation rate. Most states have legal minimum standards which attempt to guarantee and protect the quality of the surface and groundwater resources when such systems are installed. For instance, a typical state standard specifies that a site have a slope of less than 12%, a seasonal high water table more than 4 feet below the ground surface, bedrock or other impervious material more than 4 feet beneath the trench bottom, soils of percolation rates greater than 1 inch per hour, distances to streams or steep embankments of at least 50 feet, and distances to wells of at least 100 feet. Unfortunately, however, no legal maximum percolation standards have been established for very granular soil conditions in which there is danger of the effluent draining too quickly through the subsurface and encountering the water table before it is sufficiently filtered and decomposed.

When septic tank systems are being considered, sanitary engineers should be consulted, for they are aware of any specific difficulties in installing septic systems in the various landforms and may be able to offer suggestions for other suitable onsite systems.

SOLID WASTE DISPOSAL

In many states recent legislation has forced municipalities to change their methods of solid waste disposal to a sanitary landfill system. In this type of operation, the wastes are deposited in thin layers in the disposal pit and are covered daily with soil material. If operated and sited correctly, this method of disposal is very effective and nonpolluting; there is virtually no problem of offensive odors or rats.

The ideal location for a sanitary landfill operation is a natural or manmade depression underlaid by an impervious stratum; such a situation prevents the leakage of leachate (dissolved contaminants from refuse) into the groundwater resource. Typically, the seasonal high groundwater table should be at least 10 feet beneath the bottom of the depression to be filled to allow any leachate that may inadvertently leak out to be adequately filtered. The soil materials covering each day's deposit of wastes should be adequate to provide an impervious layer, so that rainwater does not penetrate and form leachate.

The type of sanitary landfill method discussed here should not be confused with the methods employed by many existing dumps. Many of these disposal areas violate one or more of the site criteria outlined in this section and thereby become contaminants of hydrological resources. Sufficient depth to water table is the most commonly violated criterion; in fact many existing dumps are located in swamps and wetlands. Many of these sites were originally selected because the land was inexpensive; new wetland legislation in many states will help to prohibit this activity on such sites.

Each of the landforms in Chapters 5 through 10 is examined with respect to the site requirements for a sanitary landfill—depth to seasonal high water table, soil permeability, and suitability of cover materials. The siting constraints most often occurring within each landform are also identified.

TRENCHING

Trenching operations, as considered in this volume, refer to the shallow (less than 20 feet), temporary trench excavations used for the installation of subsurface linear utilities such as liquid-carrying pipelines (sewers, water), gaslines, and electrical cables.

Physical site conditions—soil composition, cohesion, water table depth, soil depth, stoniness, and rock type—influence the difficulties or costs encountered during trenching operations. If rock occurs within the soil profile, costs will increase, since rock is difficult to remove and will possibly require blasting and heavy equipment. Costs will also increase if the water table requires drainage or if organic soils are involved. Methods of bracing shallow trenches are basically standardized, depending upon the soil engineering properties and the depth of the trench. Deep trenches present a more complex problem, and detailed analysis is needed to determine the best procedures and bracing techniques.

Capabilities or relative difficulty cost factors of trenching operations are considered in each of the landform sections in the following chapters. Costs are expressed in terms of a factor or a percentage representing the relative per-unit cost existing under the conditions identified for each landform. All costs are compared to those for shallow trenching operations in which deep, dry, moderately cohesive materials are being removed. These conditions represent a cost unit of 1.0. It has been found that rock requiring blasting increases costs by a factor of 4 to 5; a high water table within 1 foot of the surface increases costs by 35 to 40%, and organic materials with a high water table increase costs by a factor of 3 or 4.

It should be realized that actual operational costs fluctuate from region to region and season to season and are dependent upon such factors as the total amount of material to be removed, the availability of the necessary equipment, manpower, and so on. Therefore, the relative costs given in the landform sections allow sites to be evaluated only in general terms.

EXCAVATION AND GRADING

Excavation and grading operations as considered here relate to the removal of earth materials to create either large, flat sites for land uses such as an airport, an industrial park, or a large, commercial shopping center, or pit excavations for the construction of foundation structures. The same physical site conditions described in the previous section influence the relative difficulties and costs incurred during removal and grading operations. Costs as presented in the landform sections are given relative values in relation to a base cost (1.0) of removing the same amount of deep, dry, moderately cohesive material. Thus rock requiring blasting increases costs by a factor of 4 to 5, a high water table within 1 foot of the surface increases costs by 35 to 40%, and organic materials with a high water table increase costs by a factor of 3 to 4. Increases in total excavation and grading per unit cost caused by changes in slope are not included in these estimates, but obviously the presence of higher, steeper slopes requires removal of more total material to obtain the desired grades when excavating. Such added constraints as drainage of high water tables, workability of materials in cold climates, and so on are discussed where appropriate. Again, the goal is to be able to estimate relative costs and to make a quick evaluation of the problems of different sites.

DEWATERING

During trenching or excavation procedures, the flow of water from the water table into the excavation must be decreased to an insignificant amount in order to allow a safe, efficient removal operation. Soil permeability is the most critical factor in determining the amount of water that enters a given size of excavation over time. The method of drainage chosen depends upon both the permeability of the soil and the size of the excavation. The following discussion summarizes the techniques used in soils of different permeabilities.

Soils of low permeability or nonuniform, dense soils can usually be suitably drained from open *sumps* (shallow pits at the bottom of open excavations) where the water can collect and be pumped out of the excavation. This practice is common in shallow excavations. The same technique can be used in silts and fine sands, although softening and sloughing of the slope base may occur together with the formation of boils (springs in the excavation bottom which release a mixture of soil and water). If boils form, the ground surface surrounding the excavation will subside. However, sheet piles driven some distance below the excavation grade can be used to decrease the potential formation of boils. The piles act to intercept part of the seepage and in effect lower the hydraulic gradient.

Uniform soils of medium permeability are more suitably drained by *well points* or filter wells surrounding the excavation. In this technique steel pipes (approximately 2 inches in diameter) with screened, perforated ends are placed 3 to 6 feet apart around the edge of an excavation at a depth lower than the base of the proposed excavation. A series of well points is connected by a header leading to a pump which acts by suction to remove water and lower the water table. Each set of well points can remove only an 18-foot depth of water; if the proposed foundation base is deeper, several sets of well points may be used. For deep excavations (greater than 50 feet) with braced, vertical sides, deep wells containing submerged electrical or turbine pumps can be used efficiently. If the permeability *decreases,* however, and deep well pumps are therefore not economical, jet-eductor pumps may be used but their per-unit discharge and efficiency are much lower. Soils of high permeability, such as loose sands and gravels, are easily drained by pumping from well points or deep wells.

Soils that are uniform but have low permeability are most efficiently stabilized by *vacuum pumps.* In this technique well points are jetted into the ground to create holes 10 to 12 inches in diameter which are immediately filled with sand that acts as a filter. The top of the hole is sealed and a vacuum pump is attached, thereby lowering the pressure within the hole. The atmospheric pressure acting on the surrounding water initiates flow into the hole, and the resulting water is then removed by the vacuum pump. As this process takes place, the shearing resistance of the soil greatly increases and allows vertical excavation walls to stand without support. For vacuum pumping to be efficient, however, the system must be in operation for a long period of time, perhaps up to several weeks.

Cohesionless or slightly cohesive fine silts and uniform silty soils of low permeability may be so soft that the vacuum method will not prevent bottom heaving. However, *electro-osmosis,* a method by which seepage pressures are created by electricity, may be efficient for stabilizing excavated slopes below the groundwater table. Other excavation alternatives, such as dredging or the use of compressed air, should also be considered.

CONSTRUCTION MATERIALS (GENERAL SUITABILITIES)

Construction materials, as discussed in the landform sections of this book, are evaluated relative to the general capability of the various landforms to supply them. Ratings are: not suitable, poor, fair, good, excellent.

Topsoil. Those considered most suitable are soils having textures of silty clay loams, silt loams, loams, and sandy clay loams. All of these should have a suitable supply of organic material.

Sand. The best sources of sand are those soils and parent materials that naturally contain poorly graded sand. Fine sands are not rated as high as coarse sands. Ratings are generally high for landforms that contain coarse, stratified deposits.

Gravel. Capabilities for sources of gravel are best when poorly graded gravels occur within the landform. Ratings are generally high where coarse, stratified deposits occur.

Aggregate. Aggregate is used mostly as an inert filler in concrete or macadam, but it is also used by itself for railroad ballast and filter beds. Coarse, stratified deposits are generally the best sources. Materials that are of the desired size but that have weak beddings or foliations are considered not suitable, since they wear and weather quickly.

Surfacing. These materials, used in the surfacing of secondary roads, are best when they contain a mixture of fine and coarse soils. This provides a surface with suitable binder that will not develop corrugated ruts (typical when binders—fines—are insufficient). However, an excess of fines creates surfaces that are slippery when wet and dusty when dry.

Borrow. Soil and parent materials are evaluated for their suitability as borrow for fills. The coarser soils are most desirable, since they are readily compacted and have high bearing capacities. Fine soils require compaction, but some may be difficult to compact and require compaction to occur under narrow limits of moisture content. Mixtures that are too dry are difficult to compact and create dust and erosion problems; those that are too wet become plastic, are difficult to compact, and are slippery. Problems of soil compaction are discussed in the next section.

Building Stone. Some landforms and unweathered parent materials are excellent sources of building stone or decorative materials. Sandstone, limestone, gneiss, marble, and slate are commercially quarried for use as building materials.

SOIL COMPACTION

When soils are excavated to be used as fill, many of their original characteristics are significantly modified; in particular, the average porosity, permeability, and compressibility all increase. Therefore, to avoid excessive settlement of the fill over a long period of time, it is now common practice to compact soils in layers as they are deposited for fills. The extent to which any one soil can be compacted under a given procedure is primarily dependent upon the water content of the soil. The greatest degree of compaction occurs when the soils have a value known as the optimum moisture content. The following discussion summarizes the common methods employed for the compaction of artificial fills and some of those used for the compaction of undisturbed, naturally occurring soil masses.

Cohesionless soils (sand and gravel) can be compacted by vibrating, watering, and rolling, or by various combinations of these. Vibrators are most commonly used but are more effective if used with heavy rollers. Watering produces compaction as a result of the seepage pressure of percolating water settling the unstable groups of soil particles; the resulting fill has characteristics similar to hydraulic fill. Compaction by watering has two major disadvantages: (1) the process requires a vast amount of water, and (2) the resulting compaction may be insufficient. For these reasons many engineers do not recommend this practice. Rollers alone are not sufficient for the compaction of granular soils unless the materials are saturated; but saturation is difficult both to accomplish and to maintain during operations because of the associated high permeability of the soils.

Moderately cohesive soils (sandy or silty) are best compacted by rollers. Vibrators and watering are inefficient, because even a slight cohesiveness prevents the particles from settling into stable positions. In order to accomplish the highest degree of compaction of moderately cohesive soils, the material should be deposited in thin layers at its optimum moisture content and compacted with either pneumatic-tired or sheepsfoot rollers. Pneumatic-tired rollers are used on slightly cohesive, mixed-grain, gravel-silt combinations, sandy soils, and nonplastic silty soils. Plastic silt and clay soils are typically compacted with sheepsfoot rollers.

Cohesive soils (clay and clay silts) are best compacted in thin layers by using sheepsfoot rollers while the water content is slightly greater than the plastic limit. If the clay is excessively wet, the rollers sink into the surface or the soil sticks in clumps to the roller; if it is too dry, the clay chunks do not readily compact, leaving many open spaces between chunks.

Naturally occurring soil strata and existing fills can be compacted to minimize future potential settlement. However, since the soil material exists as a mass, the compaction techniques for thin layers previously discussed are not applicable. The type of technique most efficient for a given situation depends primarily upon the soil cohesion and the distance to the water table.

Cohesionless natural soil masses of loose sand and gravel are best compacted by vibration using pile driving, vibroflotation, or dynamite. Pile driving is the most common technique; the resulting vibrations allow the sand grains to assume more stable configurations. In vibroflotation a vibrator is jetted into the soil and slowly withdrawn as it is vibrating. The combination of the vibration and the water jets acts to compact the surrounding soil material. This method is efficient only in clean sand, however. The vibrations caused by many scattered, subsurface dynamite explosions have proved successful in the compaction of loose sands.

Slightly cohesive natural soil masses can be compacted efficiently by using pile drivers above the water table. The compaction process here does not depend upon physical vibrations but upon a reduction in the void spaces by static pressure. Beneath the water table the effect of pile drivers decreases rapidly, since permeability decreases. Therefore, to obtain compaction at greater depths, it is necessary to use drains to lower the water table.

Very soft, compressible natural soil masses (clays, loose silts, and organic materials) can be compacted by surcharging. Here, the area to be compacted is covered with fill of sufficient weight to cause compaction within an allowable time limit. When using this process, drainage of the underlying materials must be insured by the use of sand drains and by placing a pervious drainage blanket under the fill.

In the landform sections of this volume, the associated problems and recommended processes of compaction are described for each landform. These discussions are found under the heading "Borrow" in the sections entitled "Sources of Construction Materials."

GROUNDWATER SUPPLY

The capability of a landform to supply groundwater for municipal needs depends upon many factors, but the permeability of the water-bearing strata is the primary variable. Very permeable rock and soil materials allow rapid recharge of water that is removed by pumping. Materials of low permeability, however, do not facilitate water flow, for here the ability of the soil or rock to retain water exceeds the forces applied by pumping operations. Although some rock types are practically impervious, they may contain fractures and joints in that act as seepage corridors and thus may supply moderate yields.

The groundwater table as referred to in this discussion is defined as the level at which permanent, saturated conditions are found; this condition may exist in either soil or rock. The seasonal high water table in soils may or may not reflect the elevation of the groundwater table; indeed, in some cases the seasonal high water table is the result of a *perched* condition; that is, where the saturated zone in the soil is separated from the main body of groundwater by unsaturated rock or soil.

The potential capabilities of each landform to supply sufficient groundwater flow for municipal use are evaluated in the landform chapters, and both desirable and constraining conditions are identified. It should be noted that detailed analysis and on-site test wells are always required early in the planning process in order to determine actual capabilities.

POND OR LAKE CONSTRUCTION

In a search for potential sites for agricultural ponds, large lakes, or even regional reservoirs, data supplied by aerial photographic interpretation and landform analysis can be used to identify the most attractive sites for detailed field investigations. Many factors determine the final feasibility of such sites, but the major factors that must be identified and described include: (1) the physical characteristics of the valley, including landforms, soils, area, and dimensions, (2) the occurrence of ancient landslides or potentially unstable zones, (3) the characteristics of the watershed area, including drainage network, vegetative cover, land uses, and potential sedimentation, (4) the available supply of construction materials to be used for dam construction or to provide a suitable bottom seal, and (5) the physical site characteristics that affect the location and design of a dam, a spillway, or other major associated structures.

Throughout the landform chapters the capabilities of each landform are discussed with regard to their suitability for siting small ponds, lakes, or large reservoirs. Evaluations are made that take into account (1) the suitabilities of associated soils as impervious bottom materials for preventing leakage, (2) problems of the placement of dam structures, and (3) possible situation hazards, with recommendations for their prevention by proper watershed management.

FOUNDATIONS

The *foundation* of a structure is the part that acts exclusively to transfer the load or weight onto natural ground. The type of foundation used for a specific building problem is determined from onsite borings and engineering analysis, but the foundation type is also chosen in relation to the proposed structural framing system. Several types of foundations can be used, depending upon the type, sequence, depth, and characteristics of the underlying soils and rock materials. These include footings, rafts, piles, and piers.

Spread foundations (either footings or rafts) are used to transfer loads when suitable bearing strata are near the surface of the ground. *Footings* are designed to support segments of a building separately; thus single columns are supported by individual footings, a group of columns by a combined footing, and walls by a continuous footing. The most important factor in footing design is the correct evaluation of the greatest pressure that can be applied to the underlying materials without causing failure or excessive settlement. This measure, the allowable bearing value, is determined by engineering tests and analysis. The depth of footing must be sufficient to insure that volume changes from wetting and drying or frost heave will not affect the footing. Typically, a depth of 4 feet is satisfactory except where frost penetration is extremely deep or where, in clay materials, deep volume changes occur.

Raft (mat) foundations are more economical than footings if the area to be covered by the footings exceeds half the total area coverage of the building. A raft foundation is simply a large footing, but through its design the distribution of pressure of the applied load to the soils is changed. In the case of footings, the pressures are transfered individually to the soil material beneath each footing, and each footing settles independently as though the others did not exist; if the soils were homogeneous, the settlement would be uniform, but since homogeneous soils do not normally occur, the settlement reflects variations in the compressibility of the underlying soils. In raft foundations, however, the pressure transfer acts as though the soils were more homogeneous, and the spread of the raft has the effect of compensating for weak zones. Settlement still occurs for the entire raft and extends to greater depths than it does with footings, but it is more consistent than that associated with footings.

In some instances, footing foundations are placed on natural rafts. These occur when firm, thick strata overlay weaker, more compressible strata. Settlement is similar to that of a raft foundation.

Pier and pile foundations are used when they are the least expensive method of load transfer. *Pier* foundations differ from footings primarily in their ratio of foundation depth to width of footing base. This value ranges from 0.25 to 1 for footings but is commonly greater than 5 for piers. Piers are typically of larger diameter than piles and may have an enlarged base to help disperse the applied load. Piers are installed by driving a steel casing into the ground, excavating its interior, and filling it with concrete. *Piles* are installed by their being driven into the ground with a pile driver or by filling previously driven steel casings with concrete. The specific type of pile that is best for a given problem is determined by economic considerations and the conditions imposed by the character of the job. Piles are categorized by the manner in which they function: Floating piles are used in fine-grained soils of low permeability where a firm substratum does not exist within a certain distance from the surface; friction piles are used in coarse-grained very permeable soils; and point-bearing piles are used where a firm underlying stratum is found beneath weak materials. When piles are driven, a large amount of the driving force may be expended in overcoming the skin friction of the soil along the sides of the pile. In some cases, even with point-bearing piles, it may become extremely difficult to penetrate materials such as soft clays to reach the firm substratum; in these cases piers may be more feasible.

HIGHWAY CONSTRUCTION

Much of the physical and cultural data required for preliminary highway corridor location can be gathered from aerial photograph and landform interpretation. Routes avoiding zones of obvious high construction costs can quickly be identified, as can alternative routes of least construction cost. The application of aerial photographic interpretation to the data needs of highway construction can continue through the design process of corridor selection, alignment, and centerline location.

There are many complex factors and issues that must be considered before the decision on a final best route for a given location can be made. The following physical and cultural data can be interpreted from aerial photographs to aid in the decision making: (1) variety, type, and nature of the landforms in relation to their load-carrying ability, volume of earthwork, character of material to be excavated, and stability of cut slopes; (2) location of water resources, including drainage network, flooding, depth to water table, springs, and seepage zones; (3) soils susceptible to volume change; (4) landslides, including ancient and potential high-risk zones; (5) potential sources of construction materials (base, subbase, aggregate); (6) distribution and amount of vegetation that indicates clearing and grubbing costs; and (7) costs of right-of-way with respect to land acquisition and the number of displaced persons.

Each landform section in the following chapters indicates site constraints which may be encountered during highway construction and emphasizes alignment difficulties, major earthwork operations, potential tunneling, or major bridging. Other concerns—such as landslides and sources of construction materials—are discussed separately under other headings in the landform sections.

REFERENCES

American Society of Photogrammetry, *Manual of Photo Interpretation,* Washington, D.C., 1960.

Belcher, D. J., "The Engineering Significance of Landforms," *Highway Research Board Bulletin,* No. 13, 1948.

Chaves, J. R., and R. L. Schuster, "Use of Aerial Color Photography in Materials Survey," *Highway Research Board Report No. 63,* National Academy of Sciences, Washington, D.C., 1964.

Coates, D. R., *Geomorphology and Engineering,* Dowden, Hutchinson & Ross, Stroudsburg, Pa., 1976.

Dallavia, L., *Estimating General Construction Costs,* 2nd ed., F. W. Dodge, New York, 1957.

DePina, E. J., and L. A. McMahon, eds., *Dodge Estimating Guide for Public Works Construction,* McGraw-Hill, New York, 1969.

Fischer, W. A., "Examples of Remote Sensing Applications to Engineering," *Highway Research Board Special Report No. 102,* National Academy of Sciences, Washington, D.C., 1969.

Garofalo, D., and F. Wobber, "Solid Waste and Remote Sensing," *Photogrammetric Engineering,* Vol. 40, No. 2, 1974.

Havers, J. A., and F. W. Stubbs, eds., *Handbook of Heavy Construction,* 2nd ed., McGraw-Hill, New York, 1971.

Highway Research Board, *Landslides and Engineering Practice,* Special Report Number 29, 1958.

Hodgson, J. H., *Earthquakes and Earth Structure,* Prentice-Hall, Englewood Cliffs, N.J., 1964.

Lee, F. T., "Engineering Geology," *Geotimes,* Vol. 20, No. 1, 1975.

Legget, R. F., *Geology and Engineering,* McGraw-Hill, New York, 1939.

Leonards, G. A., Ed., *Foundation Engineering,* McGraw-Hill, New York, 1962.

Lobeck, A. K., *Geomorphology, An Introduction to the Study of Landscapes,* McGraw-Hill, New York, 1939.

Lueder, D. R., *Aerial Photographic Interpretation: Principles and Application,* McGraw-Hill, New York, 1959.

Mansur, C. I., and R. I. Kaufman, "Dewatering," in G. A. Leondards, ed., *Foundation Engineering,* McGraw-Hill, New York, Chapter 3, 1962.

Parizek, R. R., "Impact of Highways on the Hydrogeologic Environment," in D. R. Coates, ed., *Environmental Geomorphology,* State University of New York, Binghamton, N.Y., 1971.

Schultz, J. R., and A. B. Cleaves, *Geology in Engineering,* Wiley, New York, 1955.

Sherard, J. L., R. J. Woodward, S. F. Gizienski, and W. A. Clevenger, *Earth and Earth-Rock Dams, Engineering Problems of Design and Construction,* Wiley, New York, 1967.

Stingelin, R. W., "Airborne Infrared Imagery and Its Limitations in Civil Engineering Practice," *Highway Research Board Report No. 421,* National Academy of Sciences, Washington, D.C., 1972.

Taber, S., "Freezing and Thawing of Soils as Factors in the Destruction of Road Pavements," *Public Roads,* Vol. 11, 1930.

Teng, W. C., *Foundation Design,* Prentice-Hall, Englewood Cliffs, N.J., 1962.

Tornley, J. H., *Foundation Design and Practice,* Columbia University Press, New York, 1951.

Todd, D. K., *Ground Water Hydrology,* Wiley, New York, 1959.

Tomlinson, M. J., *Foundation Design and Construction,* Wiley, New York, 1963.

Tschebotarioff, G. P., *Soil Mechanics, Foundations and Earth Structures, An Introduction to the Theory and Practice of Design and Construction,* McGraw-Hill, New York, 1951.

CHAPTER 5

SEDIMENTARY ROCKS

INTRODUCTION

Sedimentary rocks are formed by the deposition of sediments transported by streams, ocean or wave currents, ice, or wind. Most sediments are remnants of previously decomposed and disintegrated igneous, sedimentary, and metamorphic rocks, but some are derived from chemical reactions and organic sources. When a particular transporting agent can no longer carry their mass, sediments are deposited. Variations in the velocity of the transporting agent produce layers or beds whose particles vary in texture, and it is the presence of these beds that distinguishes sedimentary from igneous rocks, which tend to be massive and nonbedded. The bedding planes in sedimentary rock are originally laid down parallel to the earth's surface but later may be tilted to any angle by movements of the earth's crust.

Sedimentary rocks are of common occurrence, accounting for approximately 75% of the earth's exposed land surface. There are two categories of sedimentary rock. The first category includes rocks that are clastic or fragmental, such as shales, sandstones, and conglomerates, formed from particles originating from other rocks. The second category includes sedimentary rocks formed from chemical and biochemical (organic) sediments precipitated from solution (examples are calcium carbonate, and limy parts of organisms such as corals, algae, foraminifera, clams, and snails). Rocks formed from these sediments include limestones, gypsums, and salt. Large organic or swamp deposits upon lithification become coal and are commonly found interbedded with other sedimentary materials. Coal is not discussed separately in the landform sections because of its small areal occurrence.

Most sedimentary deposits originate underwater in oceans, brought there by stream and river systems. Sediments carried by streams comprise a variety of gravels, sands, silts, and clays; the resulting deposits vary in texture according to the distance from shore and water velocities at the time of deposition (Figure 5.1). Thus along coastal areas, where wave and current velocities are highest, gravels and coarse sands are deposited which may eventually become conglomerates. As the distance from the coast increases and the water becomes deeper, wave and current velocities decrease, depositing sands and fine sands

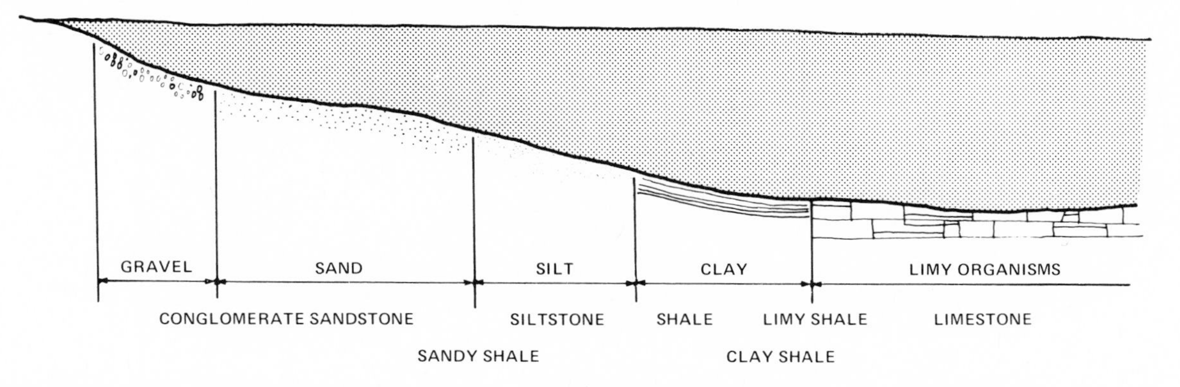

Figure 5.1. Sediments are deposited in relation to distance from the shoreline, water depth, and textural size of the sediment particles.

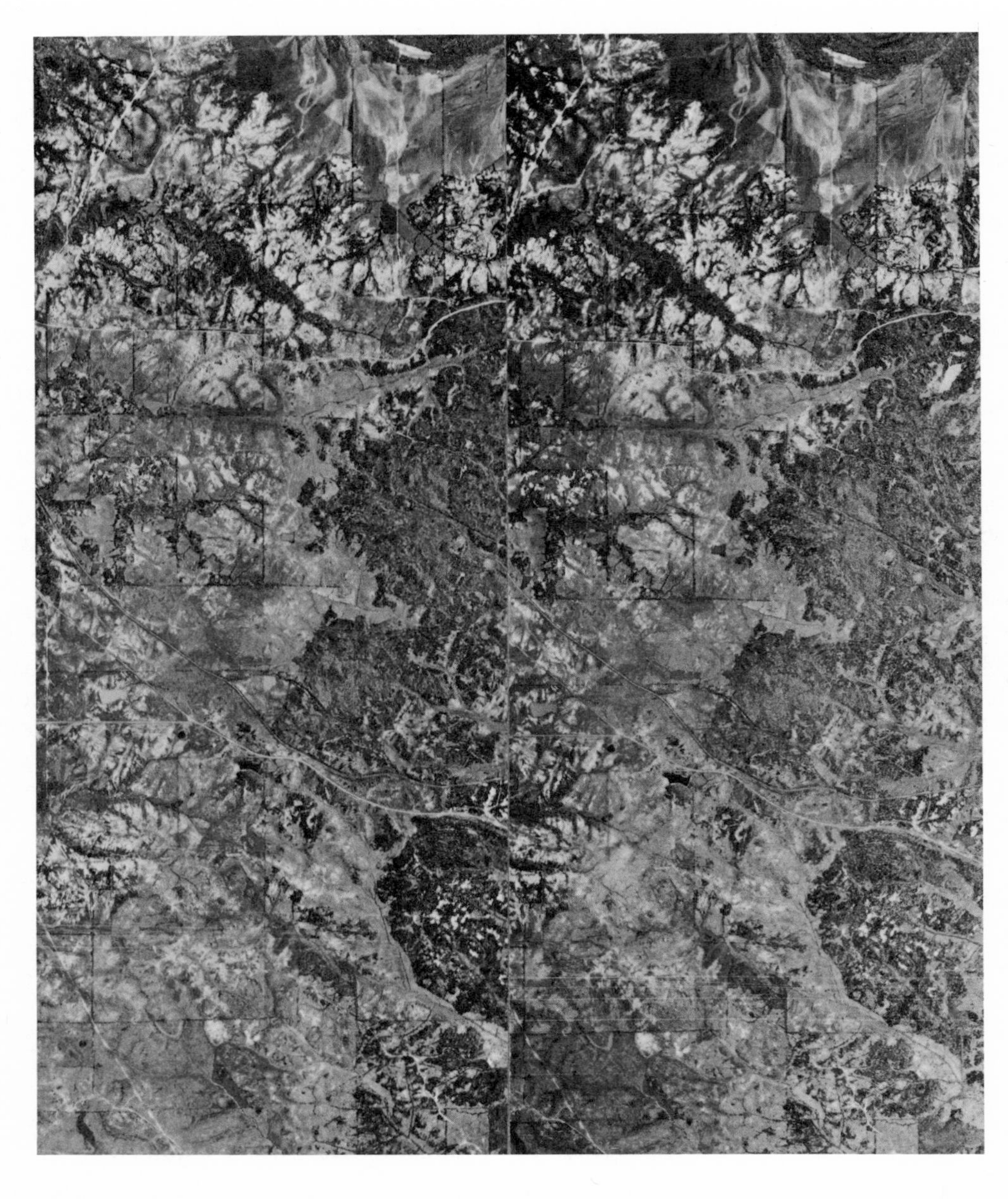

Figure 5.2. Chalk in Sumter County, Alabama, is similar in appearance to clay shale, having a fine, dendritic drainage patern. The light soil tones caused by the calcareous materials are easily distinguished. Photographs by the U.S. Geological Survey, Alabama 1-ABC, 1:69,000 (5750′/″), March 19, 1952.

which are potential sandstones. Sandstones account for approximately 15% of the rocks occurring at the earth's surface. Further from the shoreline wave and current velocities continue to decrease, and finer silts and clays are deposited in extensive layers of mud which upon lithification become shales. Shales are the most common of the sedimentary rocks found in the world and account for approximately 52% of the earth's exposed land surface. Finally, in the deeper, calmer reaches of the ocean, precipitation of calcium carbonate or the settling of the limy debris of marine life takes place. These materials become limestones and account for approximately 7% of the earth's exposed land surface.

Sedimentary rock is formed from sediment deposits when lithification, or consolidation of loose materials into solid rock, occurs. Lithification is caused either by compaction of sediments, by their own weight and by water pressure, or by cementation, in which dissolved minerals form chemical bonds between particles. Over time ocean bottom deposits may be uplifted and become part of the earth's exposed land mass. The drying process associated with exposure creates joints or cracks which are perpendicular to the original bedding planes and have no lateral movement.

A long process of deposition without significant changes in sea level results in thick sedimentary layers. Thinly bedded rocks occur as interbedded layers of varying texture, and are the result of fluctuations in sea level and/or seasonal changes from wet to dry. During dry seasons, that is, the lack of rainfall decreases runoff and erosion, allowing only the very fine sediments to be transported by low-velocity streams. The particles, being very fine, remain suspended and are carried to and deposited in calm, deep-water areas of the ocean. During wet seasons, however, runoffs are heavier, and greater amounts of bedload and sediments are carried by the higher velocities of the hydrological system. The thin beds of clay deposited previously are now covered with materials slightly coarser in texture. As this process continues, thin, interbedded layers form; when these layers are lithified into solid rock, zones of weakness exist between the beds of varying texture. This becomes evident when the material is exposed to weathering and erosion; flaking and disintegration, resulting from rapid moisture penetration along the bedding plains, quickly occur.

Regions of sedimentary rock are classified as they are bedded, in thick or thin layers. Thicker deposits also occur in interbedded sequences. Alternating deposits of shales and limestones are common, since they share the same conditions for deposition. Interbedded sandstones and shales are also common for the same reason. Sandstones and limestones, however, are not commonly found interbedded since their deposition areas are farthest apart.

In this section the sedimentary landforms that commonly occur upon the exposed surface of the earth are defined and described according to their categories—clastic or organic-chemical.

Clastic Rocks

Conglomerate

Conglomerates contain a variety of coarse materials, primarily consolidated gravels, and may be classified in a variety of ways: by texture, as pebble, cobble, and boulder conglomerates; by origin, as glacial, fluvial, estuarine, lacustrine, crush, and marine conglomerates; and by the nature of the material, as granite or chert conglomerates. Conglomerates contain a large

Figure 5.3. The salt dome in Iberia Parish, Louisiana, shows the rounded character typical of these forms which are common throughout the coastal plain region of eastern Texas, Louisiana, and Mississippi. Photographs by the Agricultural Stabilization Conservation Service, CEH-3DD-149, 150, 1:20,000 (1667′/″), December 9, 1963.

Table 5.1. Sedimentary rocks: summary chart

Landform	Topography	Drainage*	Tone	Gullies	Vegetation and land use
Sandstone					
Humid	Massive, steep slopes	Dendritic, C	Light	Few (V-shaped)	Forested
Arid	Flat table rocks	Angular dendritic, M to F	Light (banded)	Few to none	Barren
Shale					
Humid	Soft hills	Dendritic, M to F	Mottled to dull	Soft, U-shaped	Cultivated
Arid	Steep, rounded hills	Dendritic, F	Light (banded)	Steep-sided	Barren or badlands
Limestone					
Humid	Karst topography	Internal	Mottled	White-fringed	Cultivated
Arid	Flat table rocks	Angular dendritic, M to F	Light	Few to none	Barren
Tropical	Tropical karst	Internal	Uniform light	None	Barren or forested
Dolomite					
Humid†	Hill and valley	Angular dendritic, M	Light gray	Soft, U-shaped	Cultivated and forested
Coral					
Tropical‡	Terraced or reef	Internal	White to gray	None	Barren or forested
Flat, interbedded (thick bedded)					
Humid	Terraced hillsides	Dendritic, M to C	Subdued bands	Varies	Cultivated and forested
Arid	Terraced hillsides	Dendritic, M to F	Banded	Varies to few	Barren
Flat, interbedded (thin bedded)					
Humid	Uniform slopes	Dendritic, M	Medium gray	Soft, U-shaped	Cultivated and forested
Arid	Minor terraces	Dendritic, F	Faint, thin bands	Few to none	Barren
Tilted, interbedded					
Humid	Parallel ridges	Trellis, M	Faint banding	Varies	Forested and cultivated
Arid	Saw-toothed ridges	Trellis, F	Banded	Varies	Barren

*C, coarse; M, medium; F, fine.
†Characteristics of dolomite in arid climates appear similar to those listed for arid limestone.
‡For all practical purposes coral formations are found only in tropical climates.

proportion of sands and other fines which fill the spaces between the larger, usually rounded, coarser materials. Some conglomerates contain angular fragments, and these are generally called breccias; they result from landslides, talus material, or fragments carried from steep ravines during flash floods.

Sandstone

Sandstone consists of consolidated, cemented sand grains which are either rounded or angular. These rocks generally consist of quartz grains, but a siliceous composition is not always present. Sandstones have a range of textural grades extending from conglomerates to shales. The properties and strengths of the various sandstones depend upon the cementing agents that bond the particles together.

Siltstone

Siltstone is very similar to sandstone, since it too is composed of fine sands and/or silts. It is usually classified as a variety of shale, since it tends to be fissile, and is sometimes referred to as mudstone.

Shale

Shale is a general term describing lithified silts or clays which are fissile or tend to break down into thick sheets along and parallel to bedding planes. Shales comprise a wide variety of types; they are classified by composition and textural differences and include siltstones, sandy shales, clay shales, limy shales, and so on.

Pyroclastic Rocks

Pyroclastic rocks, which originate from volcanic explosions, are discussed more fully in the section on igneous rocks since they commonly occur in association with igneous rocks. The most common type is mentioned here.

Tuff

Tuff is formed from fine-grained volcanic ash. Some tuffs are deposited while at high temperatures thus causing the particles of ash to fuse together and form a hard rock known as welded tuff.

Organic and Chemical Rocks

Limestone

Limestones are sedimentary rocks that contain the mineral calcium carbonate in proportions significant enough to influence the weathering and overall characteristics of the rock. The content of calcium carbonate can range from 40% to as much as 98% or more. Dolomitic limestones consist predominantly of the mineral dolomite, which is calcium magnesium carbonate. Other limestones, known as coquina, consist of organic shell fragments cemented together. Limestones containing clay and sand impurities are referred to as lime shales, sandy limestones, or limy sands. Chalk is a very pure, white limestone (Figure 5.2).

Coal

Coal is an organic type of sedimentary rock consisting of carbonaceous materials formed through partial decomposition of vegetative debris from vast bog deposits. It is classified by its degree of decomposition, beginning with peat and grading through lignite to metamorphosed coal or anthracite. Coal is usually found with shale and sandstone.

Evaporites

Evaporites are chemical sediments (salts) formed in desert lakes or constricted ocean depressions where quick water evaporation has caused crystallization of the contained salts. Common evaporites include anhydrite (calcium sulfate), gypsum, and halite (rock salt). Gypsum is commonly associated in layers with limestone and shale, while halite is found with shale. Formations known as salt plugs or domes are found where a thick column of salt has penetrated surrounding sedimentary rock; it is believed that these plugs are formed as a result of the lower specific gravity of the salt in relation to the overlying rocks (Figure 5.3).

Distribution

Sedimentary rocks are found throughout the world and account for approximately 75% of the earth's exposed land surface. The sedimentary regions described in the following listing are the major areas that contain well-consolidated sedimentary rocks. Weakly consolidated sedimentary rocks, occurring as coastal plains

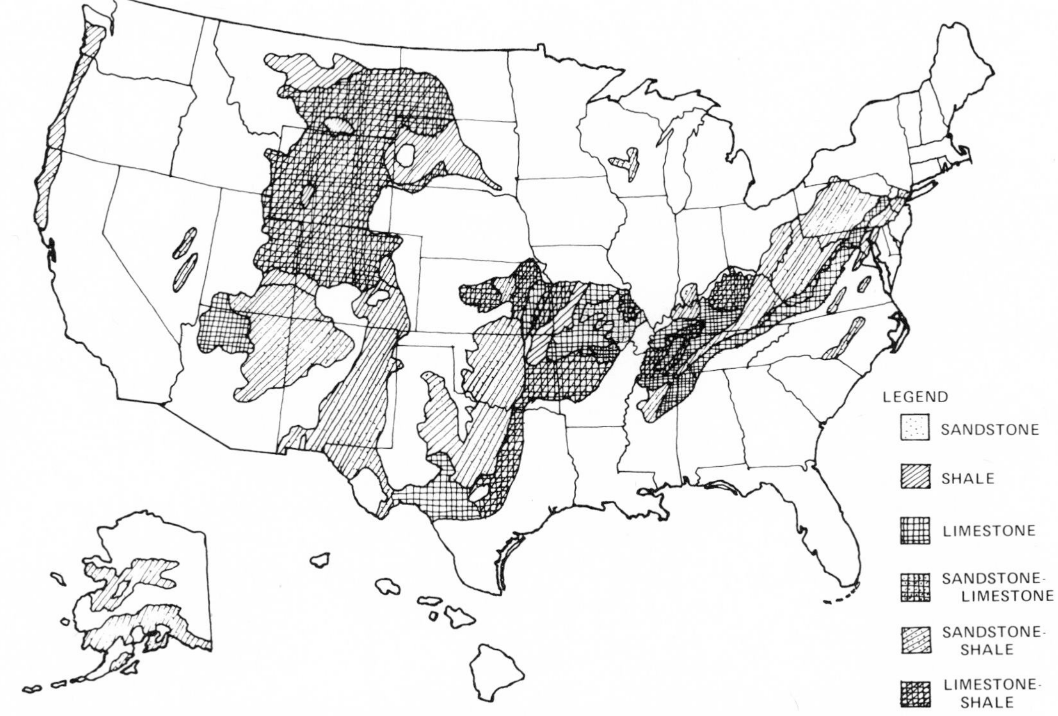

Figure 5.4. Distribution of well-consolidated, sedimentary rock parent materials in the United States. Note—glaciated areas are not shown. (After Belcher, D.J., and Associates, Inc., Ithaca, New York, "Origin and Distribution of United States Soils," 1946.)

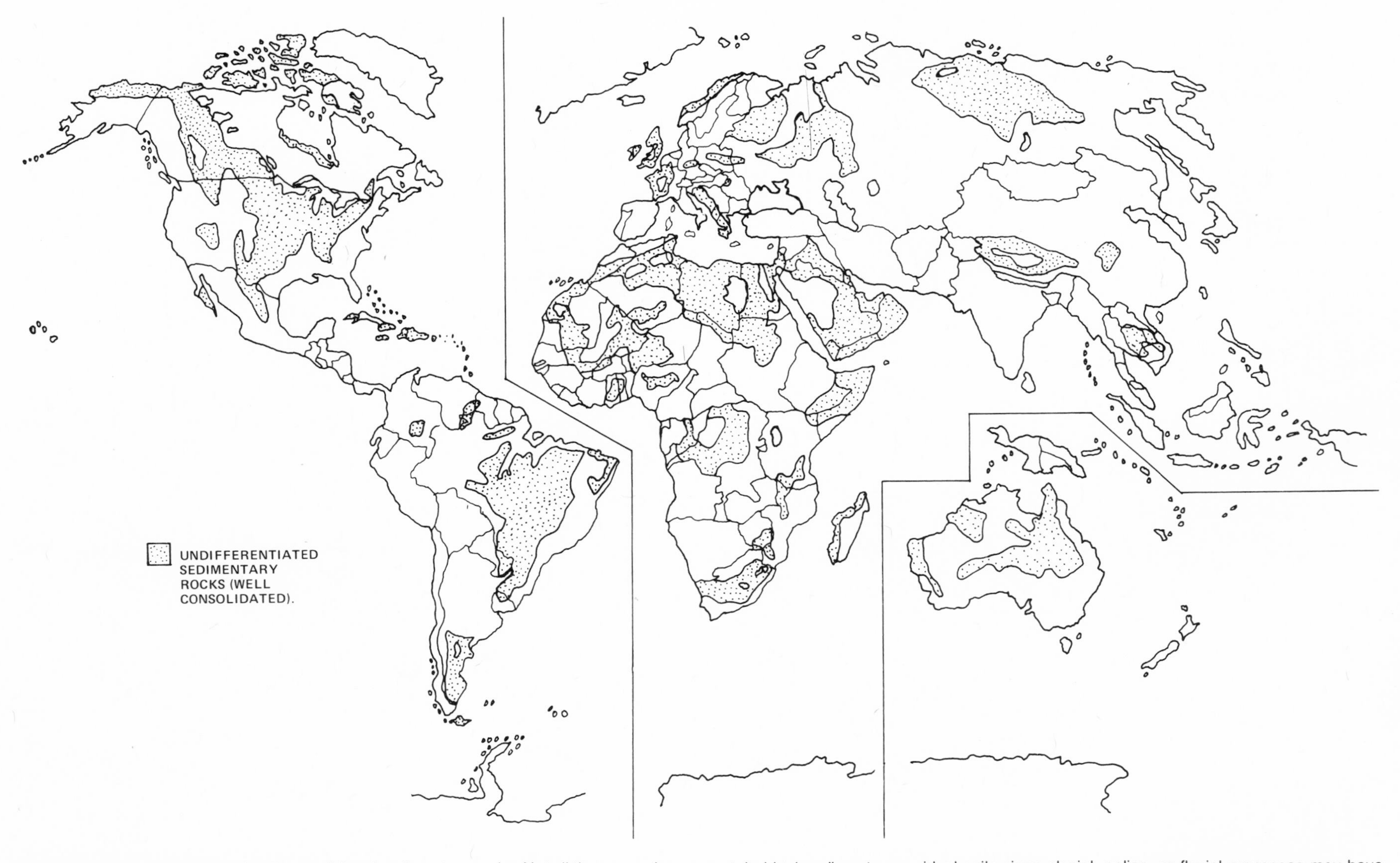

Figure 5.5. World distribution of well-consolidated sedimentary rocks. Not all the areas shown contain ideal sedimentary residual soils since glacial, eolian, or fluvial processes may have further modified them. (After Trewartha, G. T., Robinson, A. H., and Hammond, E. H., *Elements of Geography,* McGraw-Hill, New York, 1967)

and alluvium, are not included here. No distinctions between sedimentary rock types or bed attitudes are made for the world description, since most sedimentary occurrences are interbedded to a certain degree.

North America

United States. Figure 5.4 shows the distribution of sedimentary rocks in the United States. The regions shown include only those where residual soils have developed from weathered sedimentary rocks. Major portions of the United States contain sedimentary rocks of all types and attitudes, especially the areas adjacent to the Appalachian Mountains, the central plains, and scatterings throughout the Rocky Mountains.

Canada. A band of interbedded sedimentary rock extends across most of Saskatchewan and eastern Alberta, across the Northwest Territories just west of Great Bear Lake and into the Alaskan north slope.

South and Central America

Sedimentary rocks are found in southern Argentina and throughout parts of Brazil.

Africa

Sedimentary rocks are widespread across Africa with almost all countries having surface sedimentary occurrences and some residual soils.

Europe

Sedimentary rocks are found associated with most European mountain ranges. Southern England, Ireland, and France have major formations.

Asia

A large portion of western Russia and central Siberia consists of sedimentary rock, with scattered deposits occurring through Tibet. Smaller exposures can be found in Cambodia and Thailand.

Australasia

Sedimentary rocks of a well-consolidated nature are found in most of Queensland, parts of South Australia, and parts of Western Australia.

Pacific Region

Well-consolidated sedimentary rocks do not occur significantly in the Pacific Island chain, except for islands of coral formations and coquina.

Caribbean Region

Sedimentary deposits, mostly limestone, are found in parts of Cuba, Haiti, the Dominican Republic, and Puerto Rico.

REFERENCES

American Society of Photogrammetry, *Manual of Photo Interpretation,* American Society of Photogrammetry, Falls Church, Va., Chapter 4, 1960.

Barrell, J., "Marine and Terrestrial Conglomerates," *Bulletin of the Geological Society of America,* Vol. 36, 1925.

Blatt, H., G. Middleton, and R. Murray, *Origin of Sedimentary Rocks,* Prentice-Hall, Englewood Cliffs, N.J., 1972.

Bradley, W. C., "Large-scale Exfoliation in Massive Sandstones of the Colorado Plateau," *Bulletin of the Geological Society of America,* Vol. 74, 1963.

Chorley, R. J., ed., *Water, Earth and Man,* Methuen, London, 1969.

Fenneman, N. M., *Physiography of the Western United States,* McGraw-Hill, New York, 1931.

Fenneman, N. M., *Physiography of the Eastern United States,* McGraw-Hill, New York, 1938.

Folk, R. L., *Petrology of Sedimentary Rocks,* Hemphill's, Austin, Texas, 1968.

Garrels, R. M., and C. L. Christ, *Solutions, Minerals, and Equalibria,* Harper & Row, New York, 1965.

Gill, E. A., "Coal Exploration," *Photogrammetric Engineering,* Vol. 33, No. 2, pp. 157–161, 1967.

Grout, F. F., "The Clays and Shales of Minnesota," *U.S. Geological Survey Bulletin,* No. 678, 1919.

Hamblin, W. K., and J. D. Howard, *Physical Geology: Laboratory Manual,* 2nd ed., Burgess, Minneapolis, 1967.

Hsu, K. J., "Chemistry of Dolomite Formation," *Carbonate Rocks Development in Sedimentology,* Elsevier, Amsterdam, pp. 170–191, 1967.

Jennings, J. N., *Karst,* M.I.T. Press, Cambridge, Mass., 1970.

Lobeck, A. K., *Geomorphology,* McGraw-Hill, New York, 1939.

Lobeck, A. K., *Things Maps Don't Tell Us,* Macmillan, New York, 1956.

Longwell, C. R., R. F. Flint, and J. E. Sanders, *Physical Geology,* Wiley, New York, Chapters 16 and 17, 1968.

Pettijohn, F. J., "Classification of Sandstones," *Journal of Geology,* Vol. 62, 1954.

Pettijohn, F. J., *Sedimentary Rocks,* Harper, New York, 1957.

Reineck, H.-E., and I. B. Smith, *Depositional Sedimentary Environments,* Springer-Verlag, New York, 1975.

Rutland, R. W. R., "Graphic Determination of Slope and of Dip and Strike," *Photogrammetric Engineering,* Vol. 30, No. 2, pp. 178–184, 1969.

Shrock, R. R., "A Classification of Sedimentary Rocks," *Journal of Geology,* Vol. 56, 1948.

Shrock, R. R., *Sequence in Layered Rocks,* McGraw-Hill, New York, 1948.

Strahler, A. N., *Introduction to Physical Geography,* Wiley, New York, Chapters 25 and 26, 1965.

Sweeting, M. M., "The Karstlands of Jamaica," *Geological Journal,* Vol. 124, 1958.

Sweeting, M. M., "The Weathering of Limestones," in G. H. Dury, ed., *Essays in Geomorphology,* Heinemann, London, pp. 177–210, 1966.

Sweeting, M. M., et al., "Denudation in Limestone Regions," *Geographical Journal,* Vol. 131, 1965.

Tallman, S. F., "Sandstone Types: Their Abundance and Cementing Agents," *Journal of Geology,* Vol. 57, 1949.

Thornbury, W. D., *Regional Geomorphology of the United States,* Wiley, New York, 1965.

Trainer, F. W., and R. L. Ellison, "Fracture Traces in the Shenandoah Valley, Virginia," *Photogrammetric Engineering,* Vol. 33, No. 2, pp. 190–199, 1967.

Twenhofel, W. H., *Principles of Sedimentation,* McGraw-Hill, New York, 1950.

Sandstone

Introduction

Sandstone is a sedimentary rock consisting of consolidated, cemented sand grains which are either rounded or angular. It can occur interbedded with other sedimentary rocks or in layers as thick as 400 feet. The term *sandstone* implies a high quartz content, but this component is not always present. Sandstones are formed from water or windlaid sediments. Waterlaid sandstones were deposited in water adjacent to ocean shorelines, the deposits accumulating in flat beds; wind-deposited materials are generally thickly layered and exhibit cross-bedding. In cross-bedding the sequential bedding planes are deposited at various angles to the horizon. Such a rock clearly reflects its origin in layers of sand accumulated in ancient sand dune deposits.

Sandstones are classified according to textural compositions and cementing agents. The most coarse-textured, the conglomerates, grade into coarse sandstone; the finer sandstones grade into sandy shales and siltstone. Cementing agents function to form chemical bonds between the particles and are carried through the sandstone during lithification. The strength of the cementing agent determines the rock's relative resistance to weathering and many of its characteristics. Silica, calcium carbonate, clay, and iron oxides are the most common cementing agents. The strongest are silica and iron oxides (which have a variety of colors such as red, yellow, and brown). Calcium carbonate and clay cementing agents form weaker bonds, resulting in more rapid rates of weathering and erosion.

Sandstone, because of its process of formation, typically contains bedding planes and joints. The joints originate when the material is uplifted and dried and represent parting planes along which there is little if any movement in respect to the opposite walls. In humid climates initial drainage patterns are established along these joints, and even when a deep residual soil is formed the drainage pattern exhibits angularities controlled by the jointing system. This is most apparent in the more resistant sandstones. In arid climates the joint patterns can be readily observed on the surface of the ground; an aerial view shows many symmetrical, rectangular, and rhomboidal shapes.

Sandstone is an important rock: its high silica content makes it valuable for glass and for industrial chemicals, and it can also be used as a resistant building material.

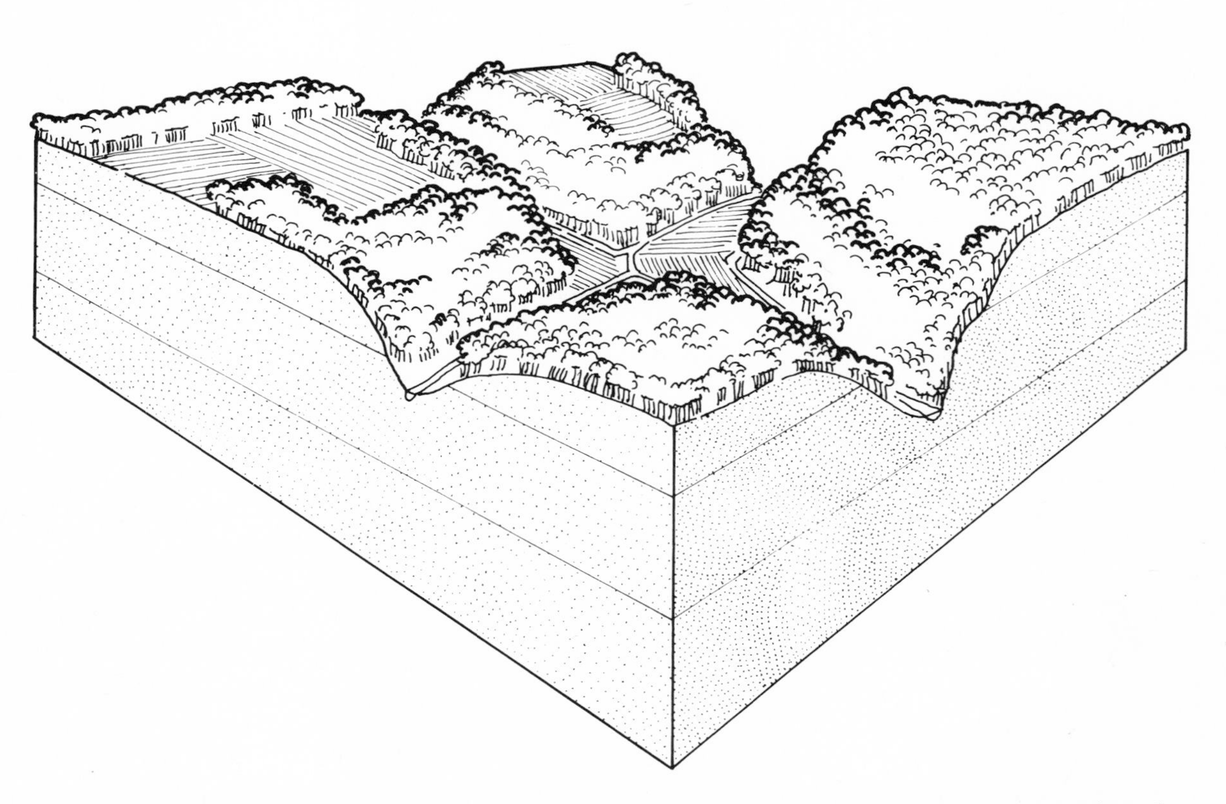

Figure 5.6. Sandstone: humid climate. Flat, thickly bedded sandstone regions in humid climates present an absorptive, closed visual environment. Most circulation systems, such as highways, occur in the major valleys, where enclosure is greatest, although if the topography is not well dissected, circulation systems are generally located along the ridgelines, thereby offering greater viewing capacity. Even on the ridges, however, views are limited, since all hilltops tend to occur at the same elevation. Most of the valleys, slopes, and ridges are covered with forest growth which contributes to the absorptive abilities of the landscape. Valleys that contain other landforms (such as waterlaid) may have development patterns; in these cases the steep, wooded slopes facing the valley help to maintain the rural visual character of the valley towns, and removal of the vegetation by development will have a severe visual impact.

Table 5.2. Sedimentary rocks: Sandstone (humid and arid)

Humid

Topography

Massive, steep slopes

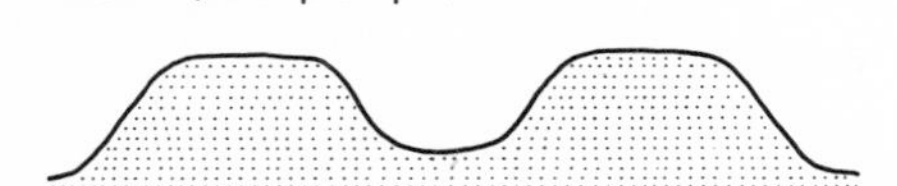

Sandstones tend to be relatively resistant to weathering because of the strength of their cementing agents. Therefore, in humid climates they produce a massive, bold topography with steep sideslopes. The residual soils tend to be very shallow along the ridgelines but thicker at lower elevations, owing to the accumulation of colluvium. Since sandstone is usually the most resistant sedimentary rock in humid climates, it tends to occur as an overlying cap rock. Where sandstones encounter other sedimentary rocks, there is generally a sharp boundary.

Drainage

Dendritic: Coarse

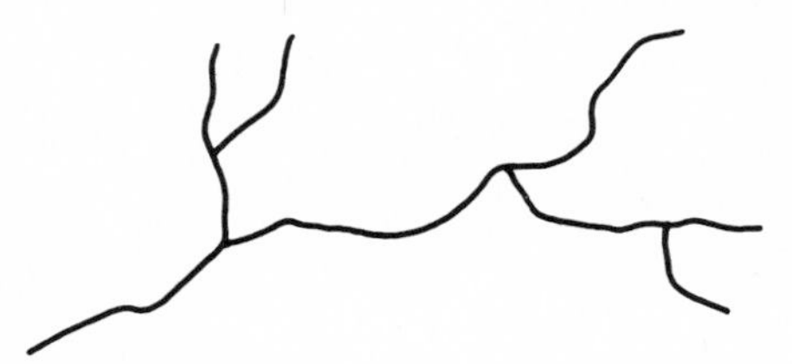

In humid climates the drainage pattern for sandstones is generally dendritic although, depending on the influence of the jointing pattern, it may also be somewhat angular or even rectangular. The drainage system texture is usually coarse but may also be medium, with minor tributaries joining the next higher stream order at right angles.

Tone

Light

The photographic tones are light because both the landform and the residual soils are well drained. Banding indicates bedding planes of other sedimentary deposits.

Gullies

Few, V-shaped

Few gullies are observed because of the porosity of the soil and rock, which decreases runoff. Any gullies present are V-shaped.

Vegetation and land use

Forested

Sandstone regions in humid climates do not develop residual soils of sufficient depth or nutrient value for intensive agricultural use. Because of the rugged topography and soil conditions, the land is covered with forests with little ground cover. The forest cover, if undisturbed, tends to be uniform, reflecting the uniformity of composition of the residual soils. Major changes in vegetative associations are found along hillsides, where there are different climatic conditions and orientation. Since major valleys in such regions have deeper soils and greater available moisture, they may be in agricultural use. Settlement patterns and circulation systems tend to follow the major valley

Arid

Topography

Flat table rocks

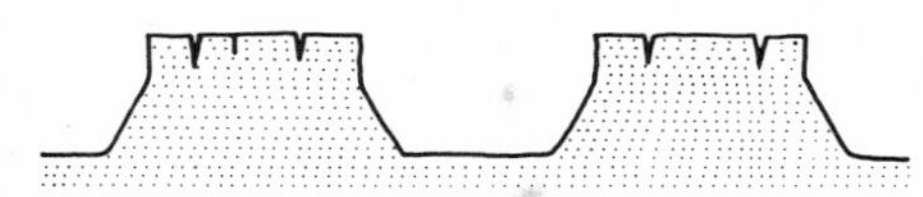

Sandstone deposits in arid climates are resistant to the forces of weathering and form dominant formations in the landscape. Fracture or joint patterns, having the appearance of rectangular blocks, can be readily observed since there is little soil cover. Cliffs may occur where sandstones overlie weaker sedimentary rocks, and wind erosion and blowing sand may modify these vertical elements, rounding and carving them into streamlined shapes. If the beds are flat, the hilltops tend to be of equal elevations.

Drainage

Angular dendritic

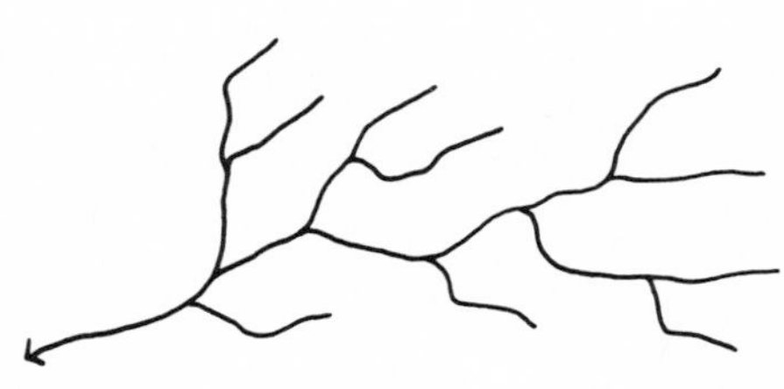

In arid climates the lack of residual soil cover allows the joints in sandstone to have maximum control over the drainage pattern. Thus, it is usually an angular dendritic or even rectangular pattern, medium to coarse in texture. Many streams are intermittent, so large areas of no apparent drainage may be observed.

Tone

Light, banded

Jointing and bedding patterns can be readily observed in arid climates, owing to the lack of significant soil cover. The overall tone is light, indicating the dryness of the rock and any residual soils.

Gullies

None

Gullies do not occur, owing to lack of significant soil cover and rainfall.

Vegetation and land use

Natural cover

Because of the lack of soil and the ruggedness of the topography, sandstone formations in arid climates generally do not have significant land use or vegetative patterns. In interbedded rock areas there may be some vegetation concentrated along hillside bedding planes associated with seepage zones. Highways and railroads transecting these regions select corridors along valley floors or plateaus.

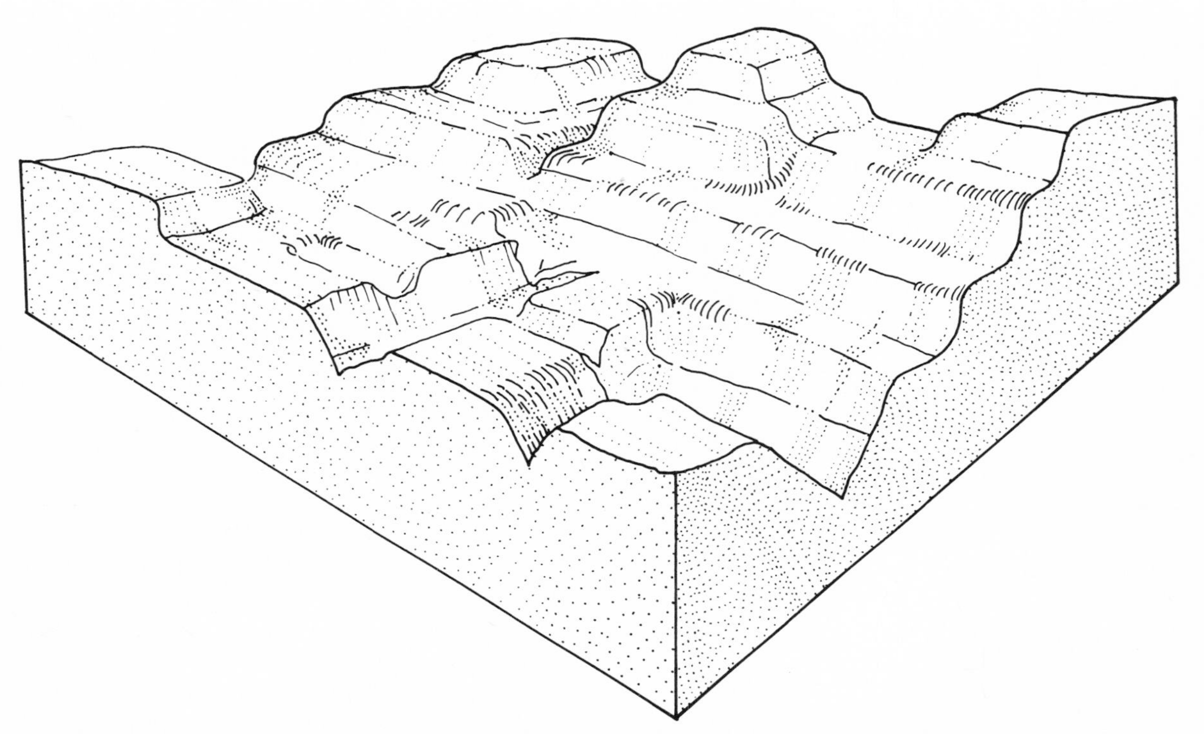

Figure 5.7. Sandstone: arid climate. In arid climates the landscape is rugged, and this is emphasized by the natural colors of the exposed sandstone. Little vegetation is present, thus making any structure placed upon a hilltop or plateau visible from great distances. The rugged topography offers diversity of spatial variation, creating regional and semi-regional views along with many enclosed spaces.

Interpretation of Pattern Elements

The pattern elements characteristic of sandstone are presented in Table 5.2. The table is divided into two sections to distinguish the pattern differences that occur between arid and humid climates; the pattern elements are those associated with massive sandstone deposits several hundred feet in thickness. Sandstone in humid climates is characterized by massive, rounded hills with steep sideslopes, a coarse dendritic drainage pattern, light tones, and forest cover. In arid climates these landforms are comprised of flat table rocks of equal elevation, an angular, coarse, dendritic drainage system, light tones with some banding, and no established traditional land uses.

Soil Characteristics

Sandstone residual soils form thin, relatively sterile profiles over fractured bedrock. Little soil development occurs in arid climates, since wind and water erosion remove weathered particles and deposit them in other landforms. In humid climates surface horizons are commonly sand loams; subsurface horizons are sandy loams or gravelly sand loams. Many rock fragments can be found throughout the profile, especially in the C horizon.

U.S. Department of Agriculture Classification

The U.S. Department of Agriculture typically classifies residual sandstone surface horizons as sandy loam, loam, loamy sand, gravelly sandy loam, and occasionally silt loam. The subsurface or C horizon in humid climates is classified as sandy loam, gravelly sandy loam, and loamy sands. Rock fragments are indicated in the soil series description as rocky, stony, or very stony.

Unified and AASHO Classifications

Surface horizons for sandstone residual soils are commonly designated SM, SM-SC, and occasionally ML. The parent material in the C horizon is most commonly SM, SM-SC, or GM, SC, GC, with ML being less common (Figure 5.11). The AASHO classifications most common for parent materials of sandstone are A-4, A-2-4, A-2, and A-1-b.

Water Table

In arid climates the water table, if present, occurs within the rock structure because of the lack of residual soil cover. In humid climates the seasonal high water table is usually not near the ground surface, because of the high porosity and permeability of the rock structure and the well-drained nature of the residual soil, but is found within the rock structure at depths dependent upon local conditions. These aquifers offer a very high potential yield capacity for individual and municipal supply, although care should be taken not to contaminate water entering them since the natural flow through the rock structure will disperse the pollutant throughout the aquifer. (See below, "Groundwater Supply.")

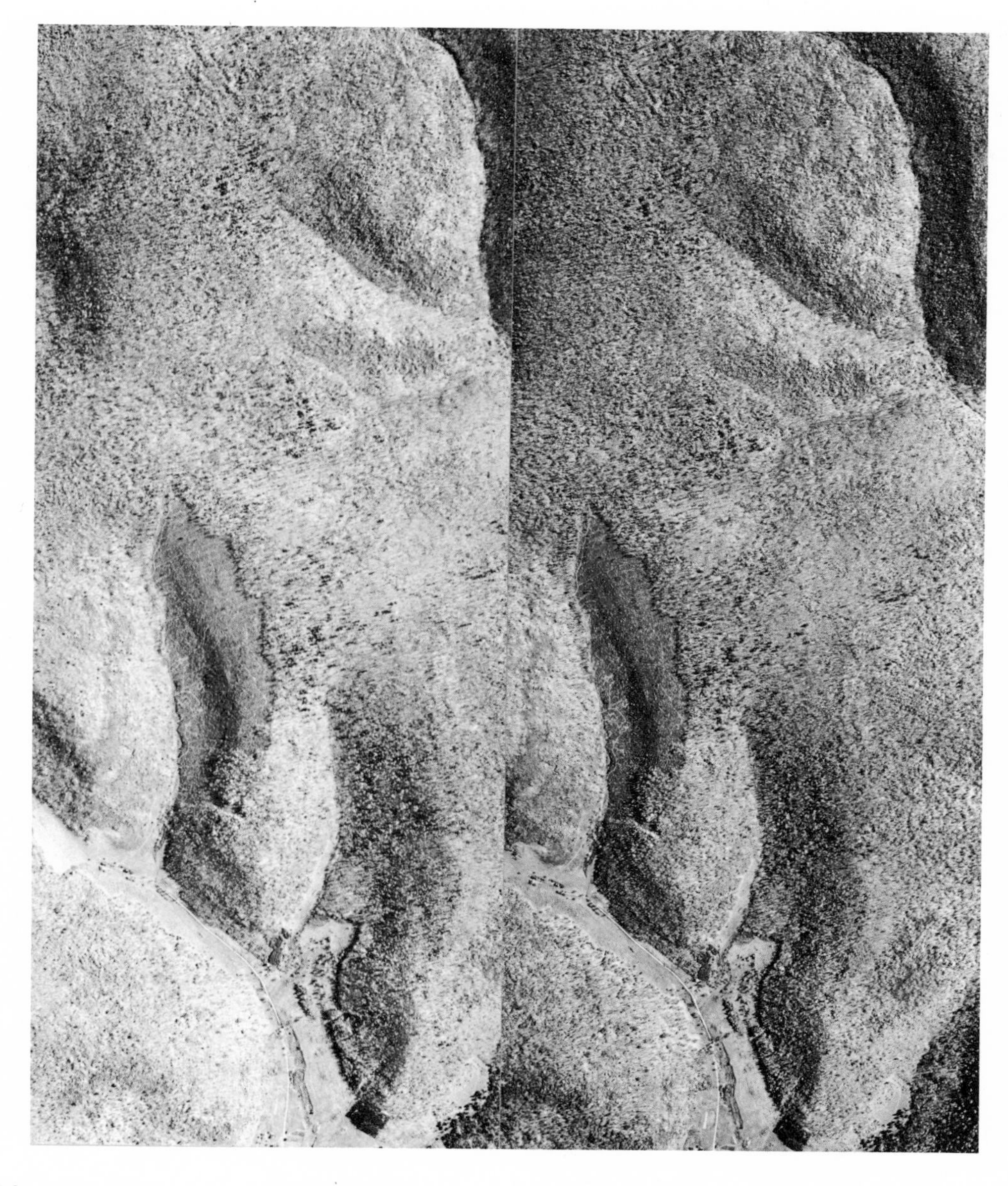

Figure 5.8. Sandstone in a humid climate in McKean County, Pennsylvania, is characterized by a rugged topography, coarse dendritic drainage, and heavy vegetation cover. Photograph by the Agricultural Stabilization Conservation Service, APL-116-20,21, 1:20,000 (1667'/"), October 22, 1940.

Drainage

Sandy-loam residual soils developed in humid climates provide good initial surface drainage which is augmented by the high internal drainage capacity of the bedrock. In thin soil areas the drainage may be excessive for agricultural production.

Soil Depth to Bedrock

Because of the resistant and homogeneous nature of bedrock, weathering processes can only slowly break down sandstone rock to form a residual soil. Therefore, relatively thin soil profiles are formed which even in humid climates are only 2 to 5 feet in thickness, grading into rock fragments and massive rock. In arid climates little soil cover is found since it is readily removed by erosive forces. The thickness of the residual soils found in different climatic areas is dependent upon the strengths of the cementing agents of the particular soils.

Issues of Site Development

Sewage Disposal

In both humid and arid climates, the use of septic tank leaching fields in sandstone regions is severely limited by the thin soil cover, rugged topography, and high permeability of the rock structure. Humid climates may have deeper soil deposits in the valleys, as a result of associated fluvial landforms, and thus may offer greater potential suitability. Any intense development in these regions should consider use of sewage treatment systems to guard against potential pollution of groundwater resources.

Solid Waste Disposal

Sites for sanitary landfill operations are difficult to locate in sandstone residual soil areas because

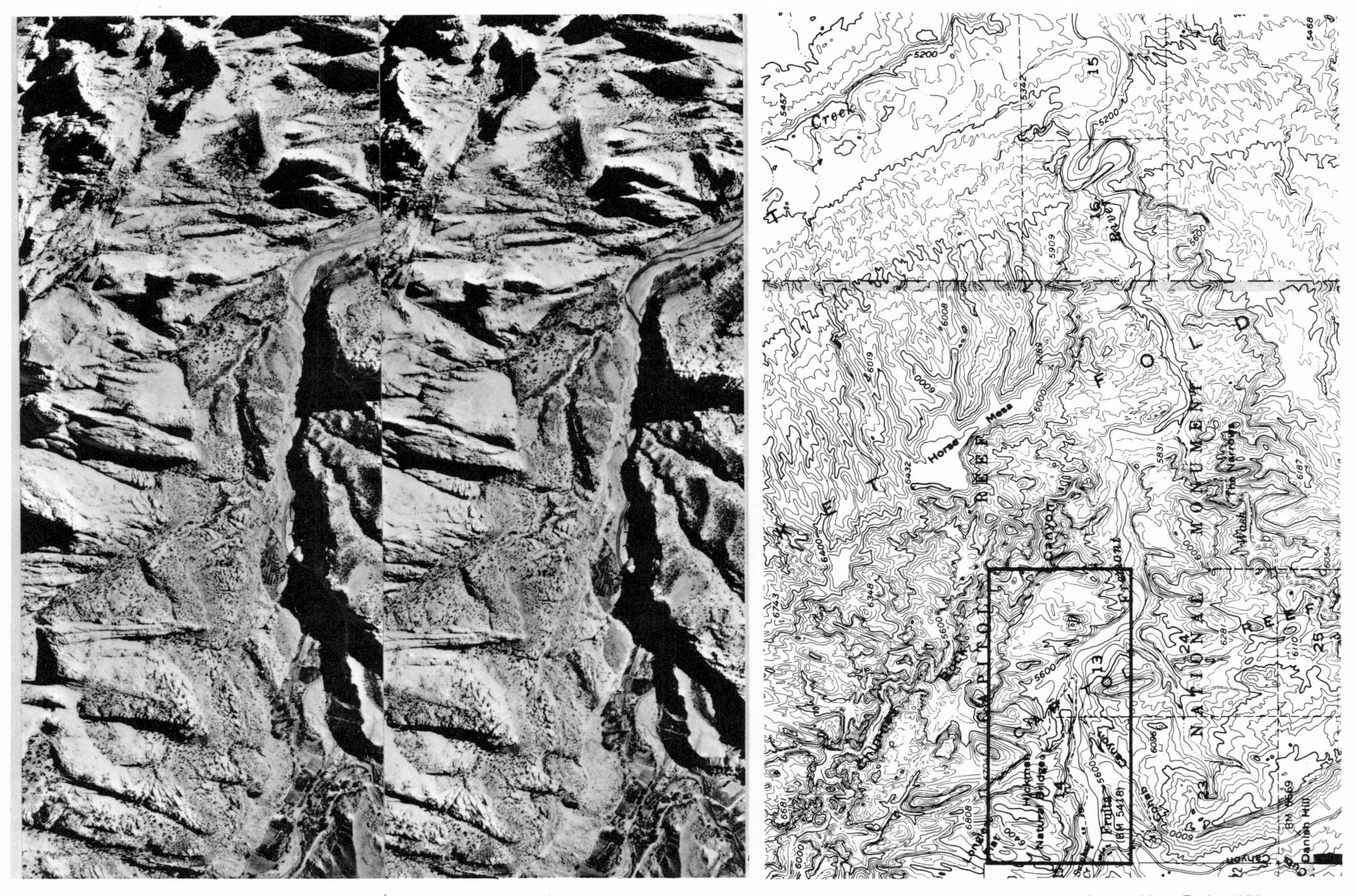

Figure 5.9. (A) Sandstone in an arid climate in Wayne County, Utah, is characterized by a rugged topography and the tendency of the jointing pattern to be parallel. Photographs by the Soil Conservation Service, DKT-7-21,22, 1:20,000 (1667′/″), October 29, 1950.

Figure 5.9. (B) U.S. Geological Survey Map: Fruita (15′).

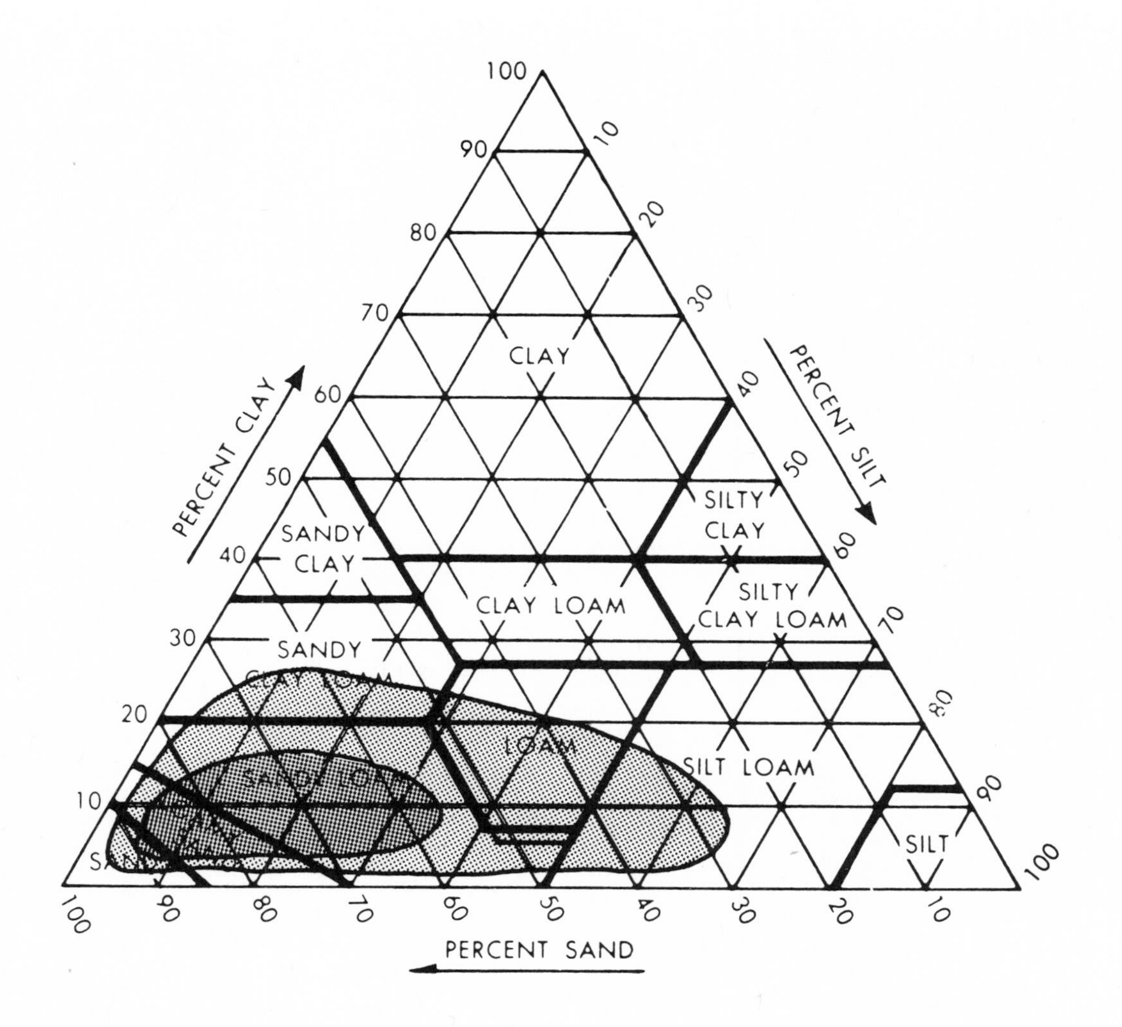

Figure 5.10. The U.S. Department of Agriculture soil textural triangle indicates sand, loamy sand, and sandy loam as the dominant materials formed from sandstone.

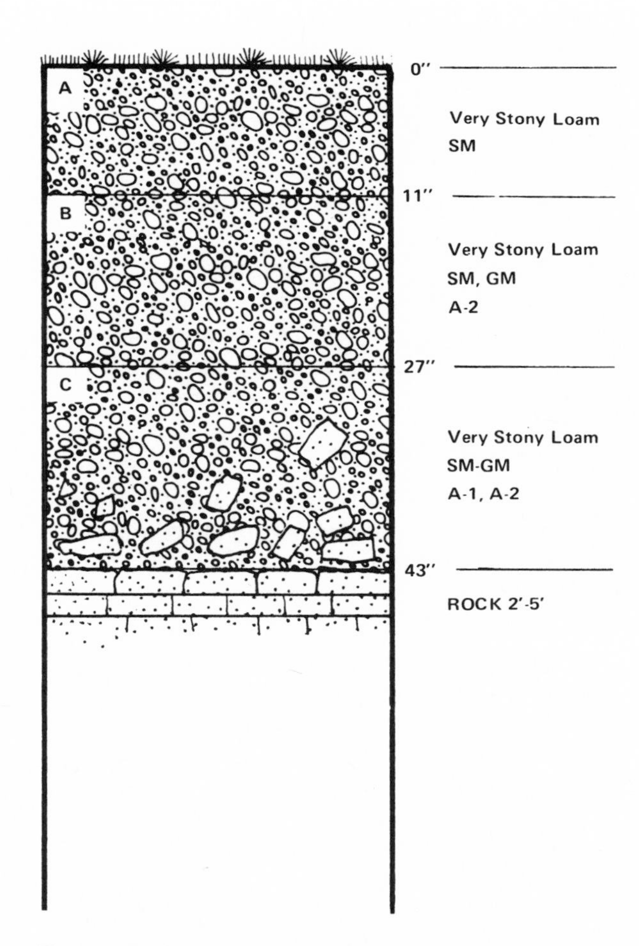

Figure 5.11. A typical soil profile developed from sandstone materials in a humid climate.

of the rugged topography, steep slopes, lack of soil cover and depth, and high permeability of the bedrock. Valleys may have deeper soil deposits developed upon other associated fluvial landforms, but care must still be taken not to pollute the groundwater system.

Trenching

Trenching costs are relatively high in sandstone regions because of the associated rugged topography, thin soils, and massive bedrock. In humid climates shallow trenches may not encounter severe rock conditions, since several feet of soil cover exist over fractured, partially weathered rock. Costs may be slightly higher than removing a dry, deep, moderately cohesive soil, owing to the occurrence of many boulders and rock fragments in the subsoil. The existing jointing pattern may provide natural trenches for the location of pipelines.

In arid climates trenching is more expensive, since little soil cover exists, and thus blasting is required, costing four to five times as much as removing equivalent amounts of deep, dry, moderately cohesive soil. The fractures and jointing patterns, and their alignment, may facilitate the placement of blasting charges. Normally, groundwater or seepage zones are not encountered in arid regions unless the water table is high or slight interbedding of materials has occurred.

Excavation and Grading

Most sandstone regions in humid climates are characterized by massive, rugged hills with steep slopes; thus creation of a large, flat site necessitates a great deal of excavation and grading. Natural sites suitable for large-area land uses, such as airports, are difficult to locate. The shallow soil depths to bedrock, especially along ridgelines, present the added expense of blasting and rock removal. Typical costs for rock removal and blasting are four to five times those for removing dry, deep, moderately cohesive soil materials. Rock presplitting is feasible, and cut surfaces can easily hold a vertical face in massive formations. In arid climates the topography may have plateaus large enough to facilitate large, flat-area land uses. Any grading or excavation that is necessary encounters the expense and difficulty of rock blasting and removal. For large operations in both climates, drilling equipment and power shovels are necessary. Seepage may occur if the water table or water-bearing strata are encountered.

Construction Materials (General Suitabilities)

Topsoil: Poor to Not Suitable. Sandy loam or loamy sand soils are rather sterile and droughty as sources of topsoil materials.

Sand: Fair to Poor. In humid climates availability of sand depends upon the specific composition of the rock and its cementing agent. Deposits are relatively thin. In arid climates sand deposits can be found as windlaid landforms whose materials originated from weathered sandstone.

Gravel: Not Suitable. Riverwash in valleys or other associated waterlaid landforms may serve as a potential source of materials.

Aggregate: Excellent if well cemented. **Fair to Poor** if weakly cemented.

Surfacing: Fair to Poor. If sufficient clays are present to act as binder, residual sandstone soils are satisfactory for surfacing materials. Without sufficient fines the lack of binder encourages rapid erosion and the appearance of corrugated surfaces.

Borrow: Good to Poor. The thin soils do not provide much material that can be removed, although spoilings from blasting and rock removal can provide a good source of coarse fills. The bases of steep slopes may contain alluvial materials of greater depth. Compaction of moderately cohesive residual sandstone soils for fills is best accomplished with pneumatic-tired rollers.

Building Stone: Excellent. Resistant sandstone is an important source of building stone.

Mass Wasting and Landslide Susceptibility

Unless underlaid by weaker materials, thickly bedded, flat-lying sandstone is very stable and unlikely to be a source of landslides. The residual soils are moved downslope by a combination of creep, frost action, and erosion.

Groundwater Supply

During the formation of sandstone, only a few of the pore spaces between sand grains are filled by cementing agents. Thus many open spaces are left which can be occupied by gaseous or liquid matter. Sandstones act as giant sponges, absorbing and holding vast amounts of groundwater; they also have a very high recharge capability. In many regions sandstones act as water aquifers even when the rock material does not occur on the actual ground surface.

The high permeability of sandstone and its resulting aquifer capabilities require that caution be exercised in the disposal of wastes and pollutants. Contaminants in the groundwater system will dilute themselves as they spread but will not disappear unless they decompose over time. Chemical pollutants are the most critical, since they do not decompose and it is impossible to clean a contaminated subsurface aquifer.

Sandstones, when interbedded with other, impervious sedimentary materials, can serve as water-carrying systems; seepage along with heavier vegetative growth can be expected along hillsides where water-carrying rocks are exposed.

Pond or Lake Construction

Creation of reservoirs, ponds, or lakes in sandstone regions is difficult because of the seepage likely to occur through the bedrock under a water feature or through the rock around a dam. Dam sites for large projects must be thoroughly investigated by competent engineers or geologists to determine potential problems. Small water features can be created if enough soil cover to prevent excessive seepage losses exists or is placed over the bedrock; also, there are commercial products available to aid in creating a more impervious layer along the bottom of such ponds or lakes. Additional problems occur in arid climates, since large quantities of water are lost by evaporation or by seepage of runoff into the groundwater system.

Table 5.3. Sandstone: aerial photographic and cartographic references

State	County	Photo Agency	Photograph No.	Scale	Date	Corresponding quadrangle and geologic maps from USGS	Corresponding soil reports from SCS
Arizona	Navajo	USGS*	Arizona 3-ABC	1667'/"	9/16/51	Agathla Peak (15'), Dinnehotso (15')	–
Arizona	Navajo	USGS*	Arizona 9-ABC	4500'/"	9/20/54	Flagstaff (1:250,000)	–
Colorado	Montezuma	USGS*	Colorado 7-AB	5000'/"	11/9/54	Mesa Verde National Park (1:31,250)	–
Pennsylvania	Forest	ASCS	APJ-61-58,59,60	1667'/"	6/7/39	Warren (1:250,000)	–
Pennsylvania	Forest	ASCS	APJ-4V-18,19	1667'/"	8/28/58	Marienville (15')	–
Pennsylvania	McKean	ASCS	APL-116-18,19,20,21	1667'/"	10/22/40	Warren (1:250,000)	–
Pennsylvania	McKean	ASCS	APL-IJJ-69,70	1667'/"	–	Warren (1:250,000)	–
Pennsylvania	Potter	ASCS	AQC-3JJ-16,17,18	1667'/"	9/15/68	Gaines (15'), Genesse (15')	1958
Utah	Wayne	USGS*	Utah 7-AB	1667'/"	10/29/50	Fruita (15')	–
Utah	Wayne	USGS*	Utah 8-AB	2640'/"	7/17/38	Notom (15')	–
Utah	Wayne	USGS*	Utah 9-AB	2640'/"	7/12/39	Factory Butte (15')	–
Utah	San Juan	USGS*	Utah 13-ABCD	1667'/"	9/21/51	Upheaval Dome (15')	Area 1962
Utah	Washington, Kane	USGS*	Utah 14-ABC	5280'/"	6/16/53	Zion National Park (1:31,250)	–
Virginia	Augusta	USGS*	Virginia 2-AB	833'/"	3/13/55	Parnassus (15') Prof. Paper No. 347	1937
Wisconsin	Iowa	ASCS	WT-3JJ-197,198	1667'/"	5/19/68	Spring Green (15')	1962
Wisconsin	Juneau	ASCS	BHT-4T-109,110	1667'/"	7/19/57	Madison (1:250,000)	1911†

*From *Selected Aerial Photographs of Geologic Features in the United States*, U.S. Geological Survey Prof. Paper No. 590
†Out of print.

Foundations

Since most sandstones are well-consolidated and well-cemented, they have high load-bearing capacities and provide suitable foundation platforms. Depending on the structural system being used, structures may either be loaded directly onto the rock, using piers, or may bear on the residual soils, using spread foundations. Because the residual soil cover is shallow, spread foundations, footings, and rafts are commonly used. Residential units typically include continuous footings with connecting slabs and basements. To eliminate excessive grading in residential developments along hillsides, foundations should be used that are sympathetic to the slope and rock conditions. Column footings, for instance, allow for changes in floor elevations by stepping the housing units along the slopes.

Highway Construction

The rugged topography, steep sideslopes, thin soils, and massive bedrock of sandstone formations cause many difficulties in highway corridor location. Major ridges, plateaus, or valleys can be followed to minimize grading, tunneling, and bridging, but road cuts necessitate the use of heavy equipment and blasting, and transecting valleys may require massive fills or major bridge structures. Most highways in these regions are curvilinear, in the attempt to minimize cuts by crossing the hillslopes diagonally across the contours.

Shale

Introduction

Shale is a general term for lithified muds, clays, and silts which are fissile and break along planes paralleling the original bedding planes. It is the most common sedimentary rock, occurring as 52% of the earth's exposed land surface. These rocks are formed from fine sediments deposited in deep ocean areas and compacted under pressure. Little cementation is found in shales, owing to the fineness of the particle size which inhibits penetration by cementing agents. Shale has very low permeability; in comparison, sandstone is approximately 100 times more permeable. Because of their low permeability and their compactness, most shale deposits are relatively impervious and have high levels of runoff. These characteristics greatly influence and control both weathering rates and the processes that act upon shale landforms. Shales are generally the most easily weathered rocks of any associated sedimentaries and therefore usually occupy a subdued or lowland position in the topography.

Unlike other sedimentary rocks, shales tend to have the same topographic appearance, regardless of the position or attitude of the bedding planes. Sandy shales, however, being slightly more resistant, may, if the beds are tilted, exhibit a saw-toothed topography, similar to weak sandstone. The section on interbedded sedimentary rocks deals with the influence of tilted beds on the pattern and engineering characteristics of shales.

Interpretation of Pattern Elements

A wide variety of shales are included under the sedimentary rock classification, including conglomerate shales, sandy shales, clay shales, and limy shales. The descriptions in the photographic analysis key, however, apply only to clay shales, for the following reasons: conglomerate shales are very rare and do not cover any significant areas of the earth's surface; the characteristics of sandy shales are similar to those of sandstones; and, finally, the characteristics of limy shales are similar to those of clay shales. For the purposes of terrain analysis, the key is divided into two parts, describing the pattern elements associated with clay shales in humid and arid climates.

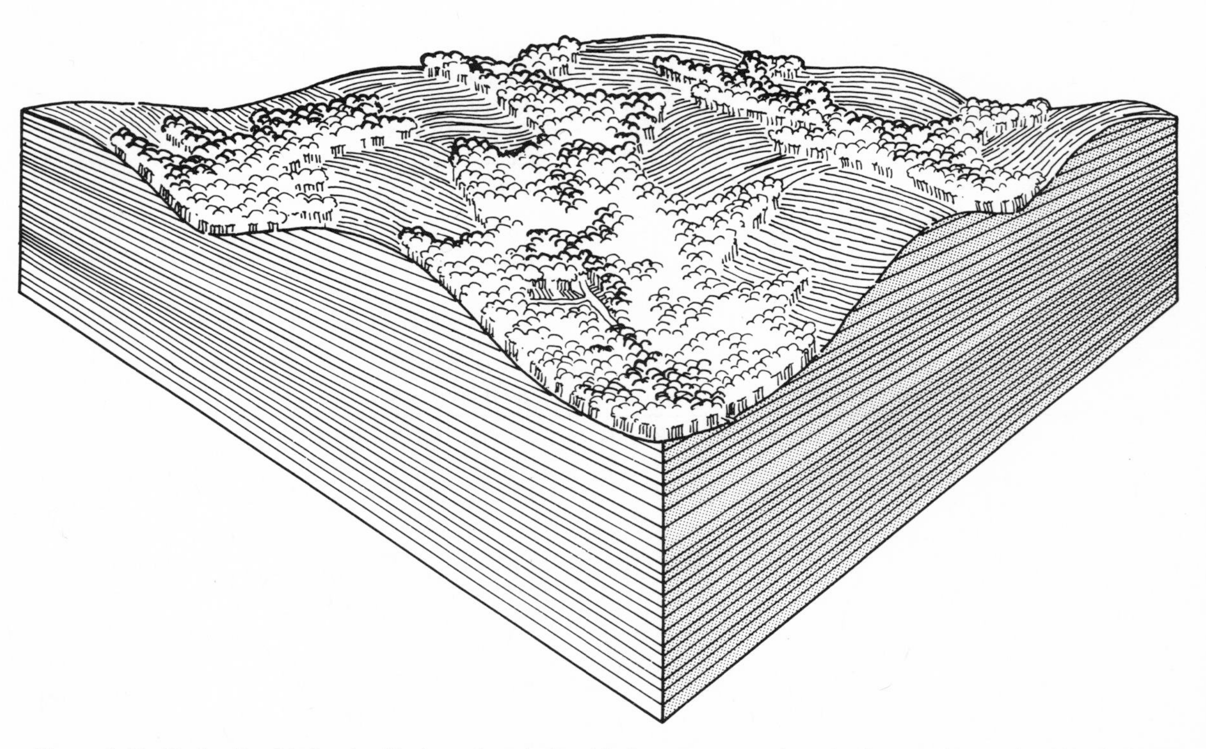

Figure 5.12. Shale: Humid climate. Shale regions in humid climates are moderately dissected by a dendritic drainage pattern. The gently rolling uplands are cultivated or in forest cover. Even if cultivated, the slopes facing the drainage system generally remain wooded, providing rich wildlife habitats. Vegetative associations vary according to changes in microclimate and orientation. Few swamps and natural water bodies are found because of the extensive drainage system. The residual soils are fairly homogeneous over large areas and do not significantly determine changes in vegetative associations, except in major valleys. The overall landscape pattern forms a rather treelike corridor system, reflecting the drainage pattern and the steep, uncultivated slopes.

Traveling through a shale landscape in a humid climate exposes the viewer to a rolling topography covered with farms, pasture land, and some crops, with many subtle changes in scene but few regional views. The smooth, undulating topography offers limited visual spatial change; that is, hilltops tend to occur at the same elevation, and, though major valleys offer some degree of visual closure, the valley walls are not usually steep enough to create tight spatial enclosure.

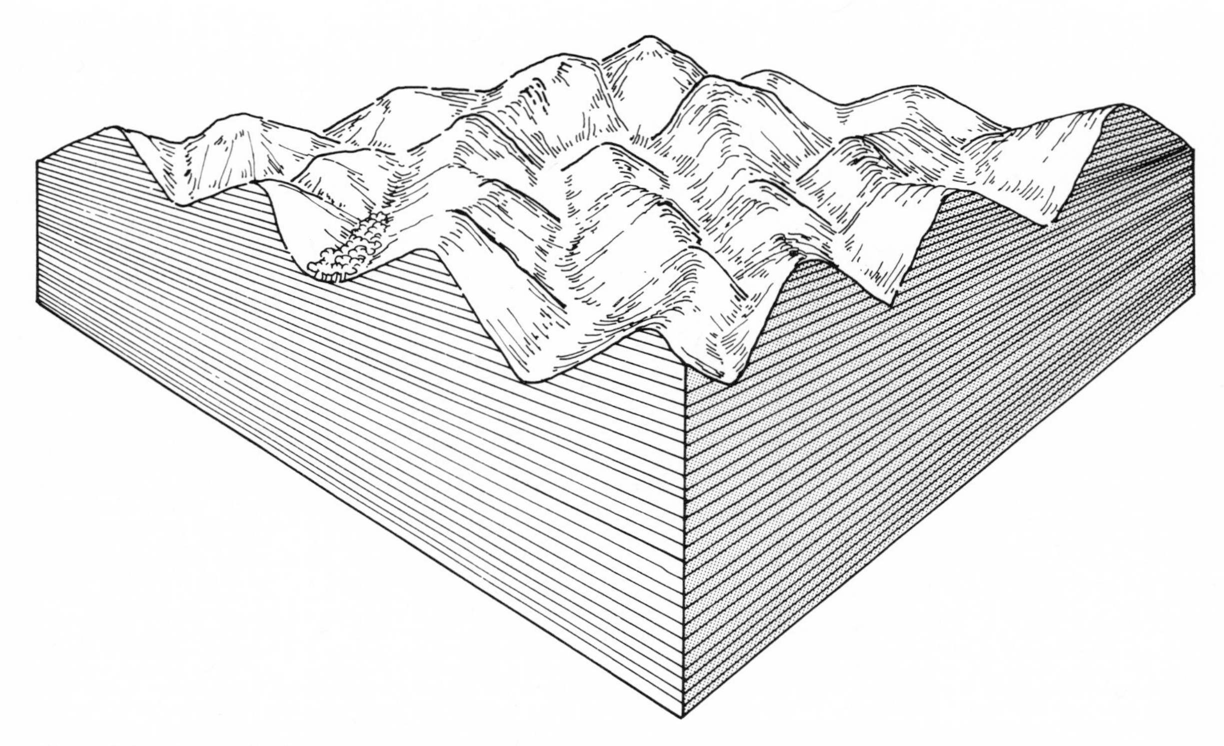

Figure 5.13. Shale: arid climate. Landscapes of shale regions in arid climates appear very rugged, since the topography is dissected to a high degree. Little vegetation is present, owing to the poor ability of the soil to retain moisture. Some scattered, scrubby growth is found, but little diversity of associations exists over large areas. Most of the drainage system is intermittent and contains no natural large bodies of water or swamps. The high degree of dissection of the topography and the exposed banding provide a picturesque landscape. Visually, arid shale regions contain tight areas of tight spatial enclosure created by the rugged topography, steep sideslopes, and small valleys. This type of landscape tends to be very absorptive, since most circulation and viewing occur in the valleys. Many of the hills are at equal elevations, although a few sometimes offer potential viewing capacity. Since the hilltops are rounded, however, it is difficult to travel along the ridgelines where regional views could be obtained. The lack of vegetative cover permits the natural color of the landscape to be observed.

Clay shales in humid climates are characterized by soft, rounded hills controlling a medium-to-fine dendritic drainage pattern; photographic tones are light gray, exhibiting some mottling. The vegetation and land use patterns are mixed, with the steeper valley slopes remaining in forest cover and the uplands in cultivation. The topography and the drainage pattern usually serve as the key identifiers.

In arid climates these forms develop a badland topography and a fine dendritic drainage pattern which is characteristic of arid clay shales. Some light bands illustrating bedding planes may be visible along hillsides which are covered with a vegetation of scattered scrub growth and grasses.

Soil Characteristics

Residual soils developed from shale landforms consist primarily of fine particles. Soils are thin to moderately deep to bedrock, and grade into rock fragments and then into weathered shale in the subsoil. In arid climates the soils tend to be highly saline, in amounts excessive for agricultural productivity, and very corrosive to pipelines.

U.S. Department of Agriculture Classification

The U.S. Department of Agriculture commonly classifies residual shale surface soils as clay or silty clay and the subsurface soils as shaly clay, platy clay, platy clay shale, silty clay, silt loam, loam, and silty clay loam.

Unified and AASHO Classifications

Residual shale surface soils are commonly described as CH, CL, or CL-CH, with the parent material being classified as ML, CL, MH, CH, CL-CH, and occasionally SC, CL-SC, or even GC to indicate platy or shaly materials (Figure 5.17). The AASHO classification system commonly classifies shale residual soils as A-6, A-4, or A-7. Occasionally, A-2 is used to describe soils containing many shaly or platy fragments.

Water Table

In arid climates a surface water table is not generally found in residual shale soils, since runoff is rapid and the moisture retention capability of the soil is very low. In humid climates soils have a much higher moisture retention capability. Long, steady rainfalls can saturate the residual soil, creating a temporary water table near the surface. Frost heave difficulties are encountered in frost regions.

Table 5.4. Sedimentary rocks: Shale (humid and arid)

Humid

Topography	*Drainage*	*Tone*	*Vegetation and land use*
Soft Hills	Dendritic: Medium to fine	Mottled, dull	Cultivated
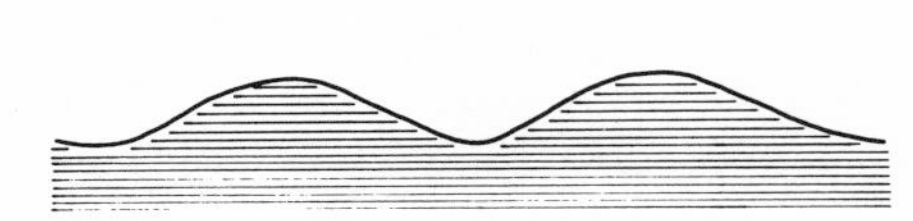 A smooth, sag-and-swale topography occurs, appearing as soft hills and mounds. Sharp breaks in slopes are neither common nor stable. The attitude of the bedding layers does not affect the appearance of the topography, and it is difficult to observe bedding planes, because of the deep soil profiles found in this climatic zone.	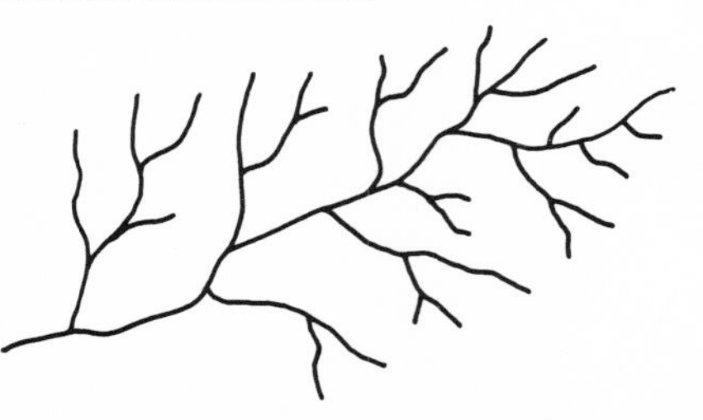 The soft materials exert no control over the drainage system, allowing a medium to fine dendritic pattern to develop freely. No angularity is found, and tributaries enter streams of the next order at acute angles.	The fine cohesive materials comprising the residual soils show as dull gray tones with some mottling. The mottling is caused by slight differences in moisture and organic content. *Gullies* Sag and swale Gentle sag-and-swale type gullies are common, reflecting the cohesive nature of the residual soils.	Humid climate shale regions are either intensely cultivated or heavily forested, depending upon the slopes and the depth of the residual soil cover. In cultivated areas the steeper sideslopes tend to be left in forest cover, thus emphasizing the dendritic pattern. Field patterns are generally square or rectangular, becoming irregular only if drainage courses are intersected. Highways are usually in gridded form unless major drainage systems are encountered, when they become curvilinear.

Arid

Topography	*Drainage*	*Tone*	*Vegetation and land use*
Steep sideslopes	Dendritic: Fine	Light and dull	Barren
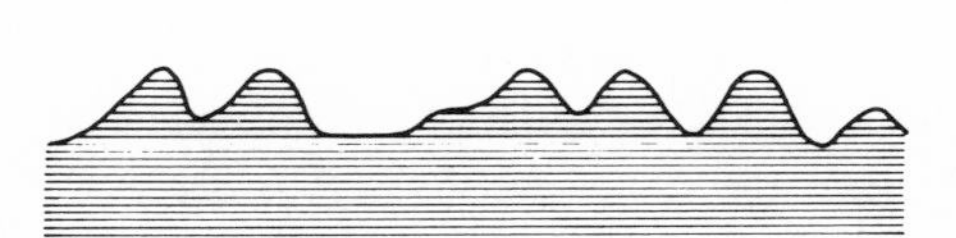 The topography is characterized by steep sideslopes and is highly dissected, reflecting the soft nature of the rock. Ridgelines are rounded, and faint bedding planes may be observed along the sides of hills.	 Shale regions show a very finely dissected pattern, dendritic in nature, reflecting rapid runoff and the impervious nature of the shale.	The tone is uniform, light and dull, with some banding visible along hillsides. Soil color and tonal differences reflect different mineralogical composition rather than textural differences. *Gullies* Steep-sided Sag-and-swale gullies are found, but with very steep sides. Silty shales develop the vertical gully walls characteristic of silty materials.	In highly dissected shale regions, the land is barren and covered with scattered shrubby grasses. In regions not so severely dissected, some farming, concerned with ranching or grain crops, may be carried on. Few highways exist except in those areas not highly dissected. Many of the highly dissected clay shale regions of the United States are known as "badlands" and do not have any traditional land uses.

Drainage

Soil internal drainage in shale regions is very slow because of the fine textural composition of the residual soils. Drainage of runoff depends upon the surface systems that carry the major part of the rainfall. Very little moisture is absorbed by the soils, especially in arid climates where rainfall is of short duration.

Soil Depth to Bedrock

The depth of soil cover over bedrock is rather thin, ranging from little or no cover in arid climates to 3 to 8 feet in humid climates. In humid climates

Figure 5.14. (A) Shale in a humid climate in Hunterdon County, New Jersey, is characterized by a medium-textured dendritic drainage pattern and a gentle, undulating topography. Photograph by the Agricultural Stabilization Conservation Service, CMY-2R-15,16, 1:20,000 (1667′/″), August 3, 1956.

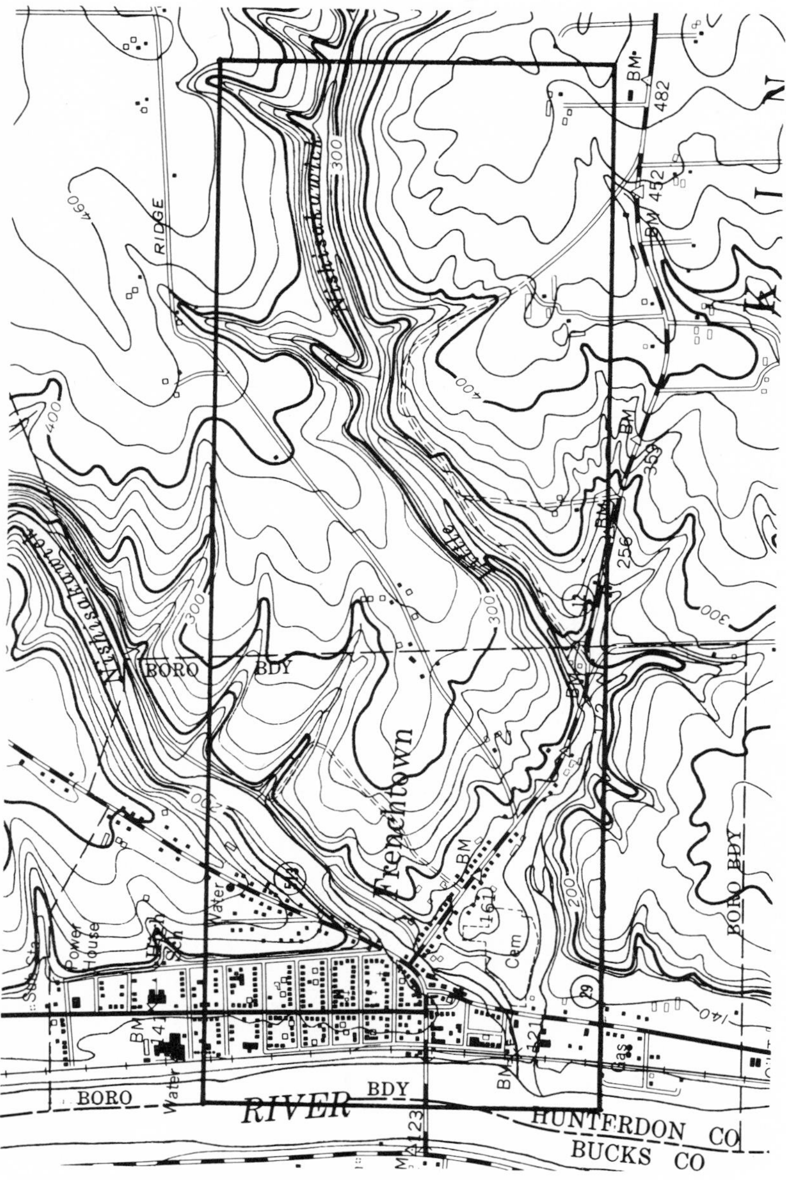

Figure 5.14. (B) U.S. Geological Survey Quadrangle: Frenchtown (7½′).

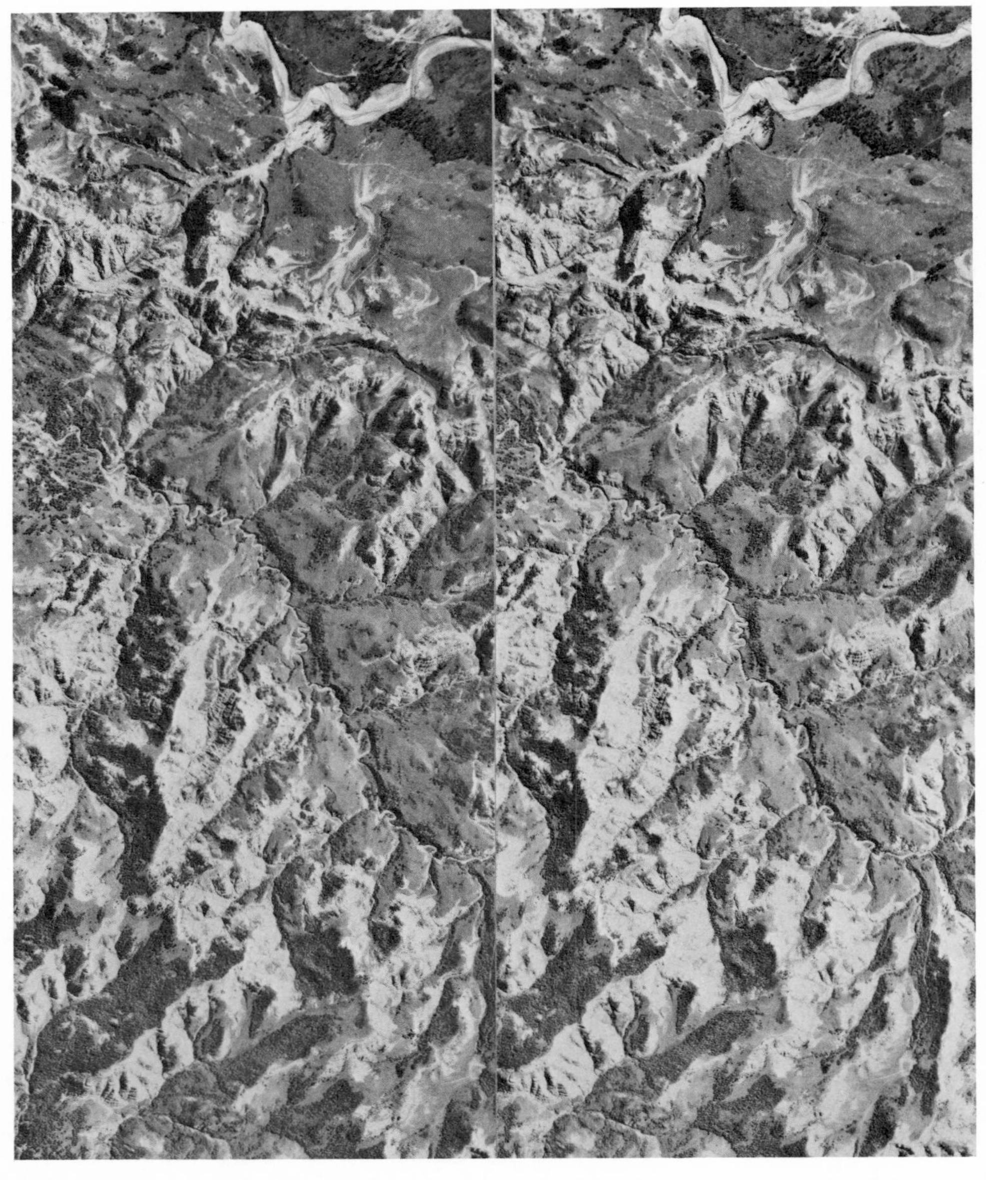

Figure 5.15. Shale in an arid climate in Dunn County, North Dakota, is characterized by a fine-textured dendritic drainage pattern and rounded hills. The light tones and gullies suggest limy shale. Photograph by the Agricultural Stabilization Conservation Service, bccp-3GG-18,19, 120,000 (1667'/"), September 6, 1966.

greater depths are found along the lower slopes of hills where materials have been transported downslope by creep, frost action, and erosion.

Issues of Site Development

Sewage Disposal

In shale regions in humid climates on-site sewage disposal systems, such as septic tank leaching fields, are not suitable because of the impervious nature of the residual soil and the high water table. Sewage leaching beds have been constructed in these areas, but in most instances these have not provided satisfactory systems. Sewage treatment is therefore needed for any intensive development, or the surface water system will become contaminated.

In arid climates the same problems exist in relation to sewage disposal, but they are more serious because of the thinner cover of residual soils. Many farmers utilize cesspool-type arrangements which require periodic emptying. Intensive development always requires a treatment system if the quality of the surface water system is to be maintained.

Solid Waste Disposal

Sites for sanitary landfill operations are difficult to locate in residual shale soil areas because of the thin soil cover, high water table, and shaly, platy subsoil. Large valleys in these regions may contain deeper deposits and may be suitable for landfill if the depth to water table is greater than 10 to 15 feet. However, in intensely developed areas incineration of solid wastes will probably be necessary if the quality of the hydrological resources is to be maintained.

Trenching

Trenching in shale regions, even through bedrock, is not significantly expensive or difficult

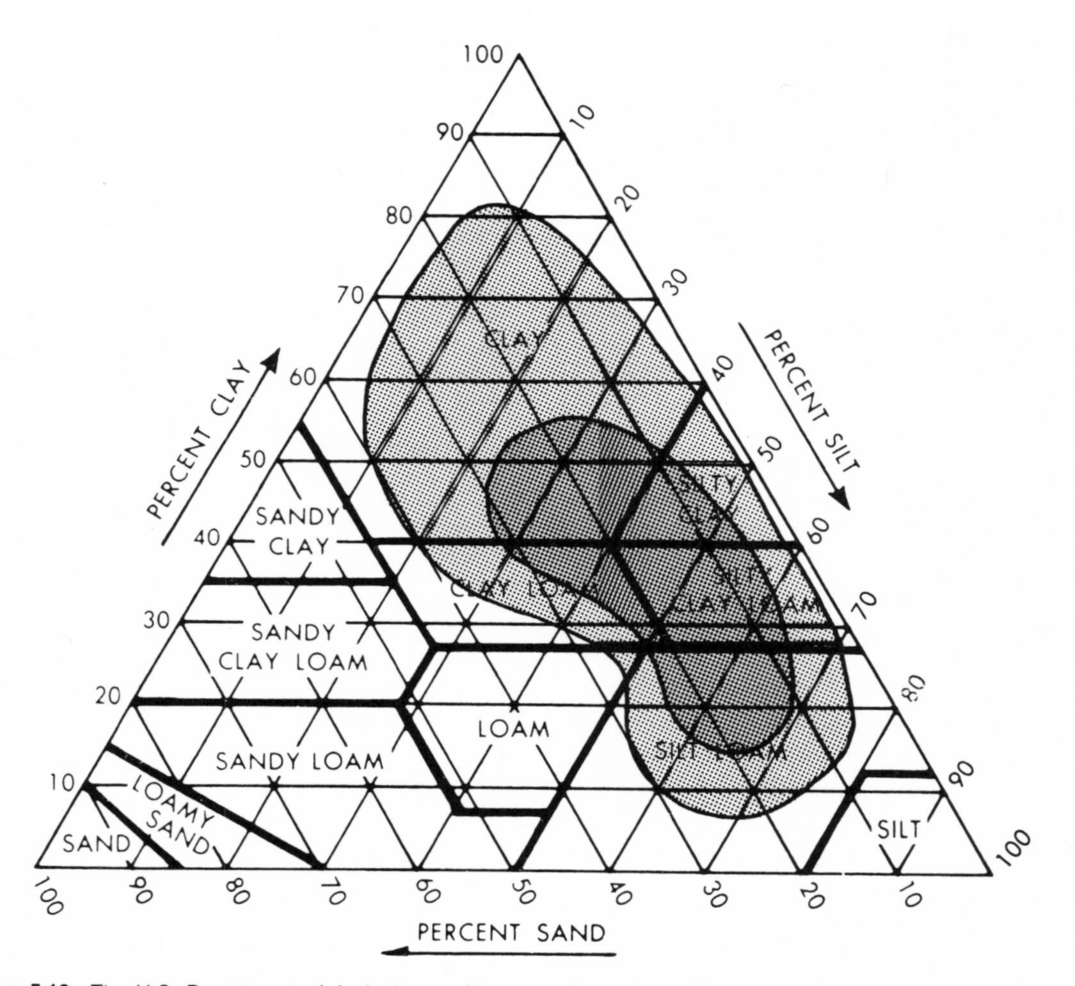

Figure 5.16. The U.S. Department of Agriculture soil texture triangle indicates clay, silty clay, silty clay loam, and silt loam as dominant materials for residual soils of clay shales.

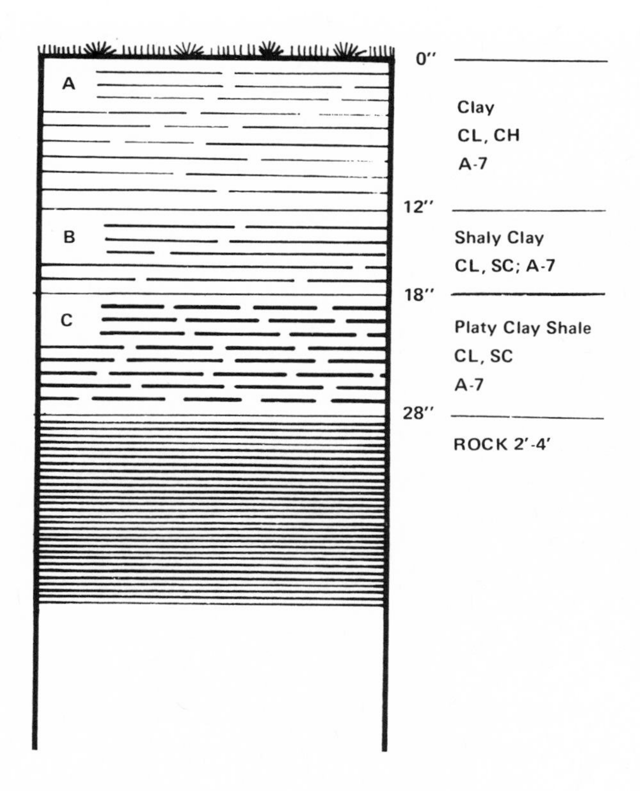

Figure 5.17. A typical residual soil profile of clay shale formed in a semiarid climate.

because of the softness of the rock. Costs per cubic yard for rock removal are 1.8 to 1.2 times greater than those for removing the same amount of deep, dry, moderately cohesive soil. Shale does not need blasting, and removal can be accomplished with power shovels without difficulty. In arid regions linear trenching alignments are difficult to locate because of the ruggedness of the topography and steep sideslopes. Seepage is minimal unless the shale is interbedded with other, more pervious materials or is highly fractured.

Excavation and Grading

Excavation and grading are not difficult in shale regions even though bedrock is encountered in most operations. Since the rock is soft, it can be removed by power shovels without blasting, and costs are therefore only slightly increased. Per-cubic-yard costs are approximately 1.5 to 1.2 times those encountered in removing a deep, dry, moderately cohesive soil. Seepage is generally not encountered unless the shale is interbedded with other, more pervious materials.

Construction Materials (General Suitabilities)

Topsoil: Generally Poor to Unsuitable. The thin soil cover, platy subsoil, and partly weathered rock provide a poor topsoil containing little organic matter.

Table 5.5. Shale: aerial photographic and cartographic references

State	County	Photo Agency	Photograph no.	Scale	Date	Corresponding quadrangle and geologic maps from USGS	Corresponding soil reports from SCS
Montana	Blaine	USGS*	Montana 8-ABC	1667'/"	8/22/56	Bird Rapids (7½')	–
Montana	Judith Basin	ASCS	NA-4FF-208-210	1667'/"	8/14/65	Great Falls (1:250,000)	1967
Montana	Wilbaux	ASCS	AZL-1LL-104-106	3334'/"	7/24/70	Glendive (1:250,000)	1958
New Jersey	Hunterdon	ASCS	Photographic Index	5280'/"	1956	Frenchtown (7½'), Stockton (7½'), Pittstown (7½')	–
New Jersey	Hunterdon	ASCS	CMY-2R-15-16	1667'/"	8/3/56	Frenchtown (7½')	–
New Jersey	Hunterdon	ASCS	CMY-2R-36-37	1667'/"	8/3/56	Pittstown (7½'), Stockton (7½')	–
New Mexico	San Juan	USGS*	New Mexico 1-ABCD	4500'/"	2/5/54	Ship Rock (15')	–
North Dakota	Billings	ASCS	CCN-5T-159,160,161	1667'/"	8/1/57	Watford City (1:250,000)	1944
North Dakota	Billings	ASCS	CCN-4T-161,162,163	1667'/"	9/23/57	Watford City (1:250,000)	1944
North Dakota	Dunn	ASCS	CCP-3GG-48-50	1667'/"	9/6/66	Watford City (1:250,000)	–
North Dakota	Dunn	ASCS	CCP-3GG-16-18	1667'/"	9/6/66	Watford City (1:250,000)	–
Pennsylvania	Columbia	ASCS	AQR-105-36-37	1667'/"	6/24/39	Williamsport (1:250,000)	1967

*From *Selected Aerial Photographs of Geologic Features in the United States,* U.S. Geological Survey Prof. Paper No. 590.

Sand: Not Suitable

Gravel: Not Suitable. Any shale fragments of this size wear and/or weather quickly into silts and clays.

Aggregate: Not Suitable. Most shales are too soft for use as aggregate materials.

Surfacing: Poor. The weakness of the material causes it to wear fast, but it is occasionally used for secondary roads.

Borrow: Good to Poor. For some purposes, such as highway road fill, excavated shale rock may make suitable fill material when compacted. Many residual soils are either not suitable or poor for fill, however, since they are difficult to work and have tendencies toward high volume change and high plasticity levels. Compaction for fills is accomplished under optimum moisture conditions by using sheepsfoot rollers. Silty, moderately cohesive soils may be compacted with pneumatic-tired rollers.

Building Stone: Not Suitable

Mass Wasting and Landslide Susceptibility

In thickly bedded clay shale regions, landslides are not common, and the major forms of mass wasting are soil creep and slumps. Slumps can be expected if the natural slopes are steepened by construction or erosion. However, if clay shales are interbedded with other, more pervious materials, landslides will be common under several conditions. For instance, tilted shales interbedded with limestones and/or sandstones are very unstable, and such situations should be investigated in detail by competent engineers or geologists before planning decisions are made.

Groundwater Supply

Similar to sandstone, shale has a high porosity; that is, many spaces between particles. Its permeability, however, is 1/100 that of sandstone, because the voids between the particles are not aligned so as to allow any flow of water through its structure. Therefore, very low water yields are found within shale materials. However, shale is always interbedded at least to some degree with other materials, and water-bearing veins of sandstone and/or limestone can usually be found. In thickly bedded clay shale areas, deep wells can usually locate water-bearing strata beneath the shale layer.

Pond or Lake Construction

Water features can be created in shale regions with little difficulty, since the rock structure and residual soils are impervious to water and the dendritic system provides many options for dam sites. Some compaction of the residual soils may be necessary if they are used as impervious bottom materials. However, siltation and future maintenance of the reservoir will be problems, since the runoff feeding such a pond contains easily eroded silts and clays. Care should be taken to protect the watershed from siltation by maintaining heavy vegetative cover adjacent to the first-order gully system (first tributaries of channelized flow).

Foundations

Residual soils formed from shale materials consist primarily of silts, clays, and shale fragments grading into shale bedrock. The foundations typically used on shale residual soils are footings and occasionally rafts, but piers are common for transferring greater loads to bedrock. Footings are most common for residential structures; continuous footings with basement slabs are typical. Tilted shales interbedded with thin layers of soft clay can fail upon overloading; early analysis and investigations by competent engineers and geologists are recommended in these situations. (See "Tilted Interbedded Sedimentary Rocks.")

Highway Construction

In humid climates major highways follow curvilinear alignments, reflecting the undulating character of the terrain. Minor highways may follow grid patterns on smaller scales. The granular construction materials needed may not be available within shale areas, and it may be necessary to transport these materials from other landforms. Grading operations in thick shales in humid climates are minimal, reflecting the undulating character of the topography.

In arid climates highway construction is more difficult, because of the highly dissected topography, and requires a great deal of excavation and grading. The fine drainage pattern creates many gullies and valleys that must be crossed, and these require culverts and bridges. Few highways are found in these regions because of the difficulties of constructing them, but when they occur they generally follow major valleys.

Limestone

Introduction

Limestones are generally classified as sedimentary rocks having calcium carbonate in proportions great enough to influence their weathering and overall characteristics. The calcium carbonate content can range from 40 to 98% or more. Certain limestones contain impurities, such as clay, silts, and sand. Limestones with a high clay content are called limy shales; those with a sand content are categorized as limy sands or even sandstones. Limestones are fairly abundant, being found on approximately 7% of the exposed land surface of the earth.

The landscape pattern and engineering characteristics of limestone are determined by the various processes by which it is formed. These processes include organic or biological deposition, inorganic or chemical precipitation, and chemical reactions which affect the molecular structure of the limestone. The most common process is that of deposition of biological debris, wherein small diatoms and sea animals die and are deposited on the ocean bottom by gravity. If the area of deposition is far from coastal areas, pure, thick layers or beds of limestone are formed. Such deposits usually cover large areas, and they account for most of the limestone found throughout the world. Another variety of organic limestone is formed from the corals and sea animals surrounding warm-water islands. If an island is uplifted or the sea level lowered, the coral deposits are exposed and become part of the land mass. Many islands in the Caribbean and the South Pacific consist of and are fringed by such old reef terraces and active coral growth.

Inorganic chemical precipitation accounts for the formation of impure limestone deposits. Freshwater rivers, carrying large amounts of dissolved calcium carbonate, combine with salt water, forming a precipitate of lime which is deposited on the ocean bottom. The deposits are not pure, since other materials (silt, sand, and clay) carried by coastal currents become mixed and deposited with the lime precipitate. These deposits are very similar to soft shales in their weathering and erosion processes and have similar patterns and engineering characteristics.

Other limestones, such as dolomitic and cherty limestones, have undergone chemical reactions so that they consist of compounds other than pure calcium carbonate. Thus dolomitic limestone is formed when magnesium (Mg^{++}) ions replace calcium (Ca^{++}) ions to form magnesium carbonate. This material is more resistant to weathering than other limestones and does not develop the karst topography associated with pure deposits of calcium carbonate limestones in humid or tropical climates. Cherts are concentrations, in the form of nodular masses or distinct beds, of a very dense, siliceous rock exhibiting a variety of colors and containing microcrystalline quartz and microfibrous chalcedonic quartz as well as other minor impurities. The origin of chert has never been satisfactorily explained, but it is assumed that it is formed by precipitation in seawater.

Limestones are of many different colors, depending on their mineralogical composition and the amount and type of impurities present. Most pure varieties are white or light gray; impure limestones are commonly darker in tone, being gray or even black and sometimes having streaks or blotches of yellows, reds, or browns resulting from the oxidation of contained and exposed iron minerals. Limestone can be easily identified in the field since it readily effervesces with such common acids as acetic acid (vinegar) or citric acid (lemon juice).

Weathering Characteristics

Limestone weathers by a chemical process known as carbonation which involves the formation of carbonates from the reaction of carbonic acid and calcite. The resulting calcium bicarbonate is approximately 30 times more soluble in water than the original calcium carbonate. Water seeping through jointing patterns in limestone bedrock dissolves the calcium bicarbonate and forms subterranean channels and solution cavities. Occasionally, a roof of one of these channels collapses, forming a depression or sinkhole upon the ground surface.

It has been estimated that 1 year of rainfall on 1 acre of land in the region of Mammoth Cave, Kentucky, dissolves 25 cubic feet of limestone rock and that continued weathering could lower the ground surface approximately 1 foot every 1000 years. Erosion through limestone strata ceases only if the water table or impervious shales or other rocks are encountered, or if the drainage of internal water is impeded by clogging by silts and clays. Photointerpreters concerned with terrain analysis should be familiar with the various levels of the erosion cycle associated with limestone (Figure 5.18).

Interpretation of Pattern Elements

The photographic interpretation key for limestones includes pure, thickly bedded deposits commonly found in humid, arid, and tropical climates; dolomitic and cherty limestones of humid climates; and coral formations found in tropical climates. The dolomitic and cherty limestone deposits found in arid climates have characteristics similar to those associated with pure limestone in arid climates so they are not discussed separately. Coral is discussed only in relation to tropical climates since it occurs significantly only

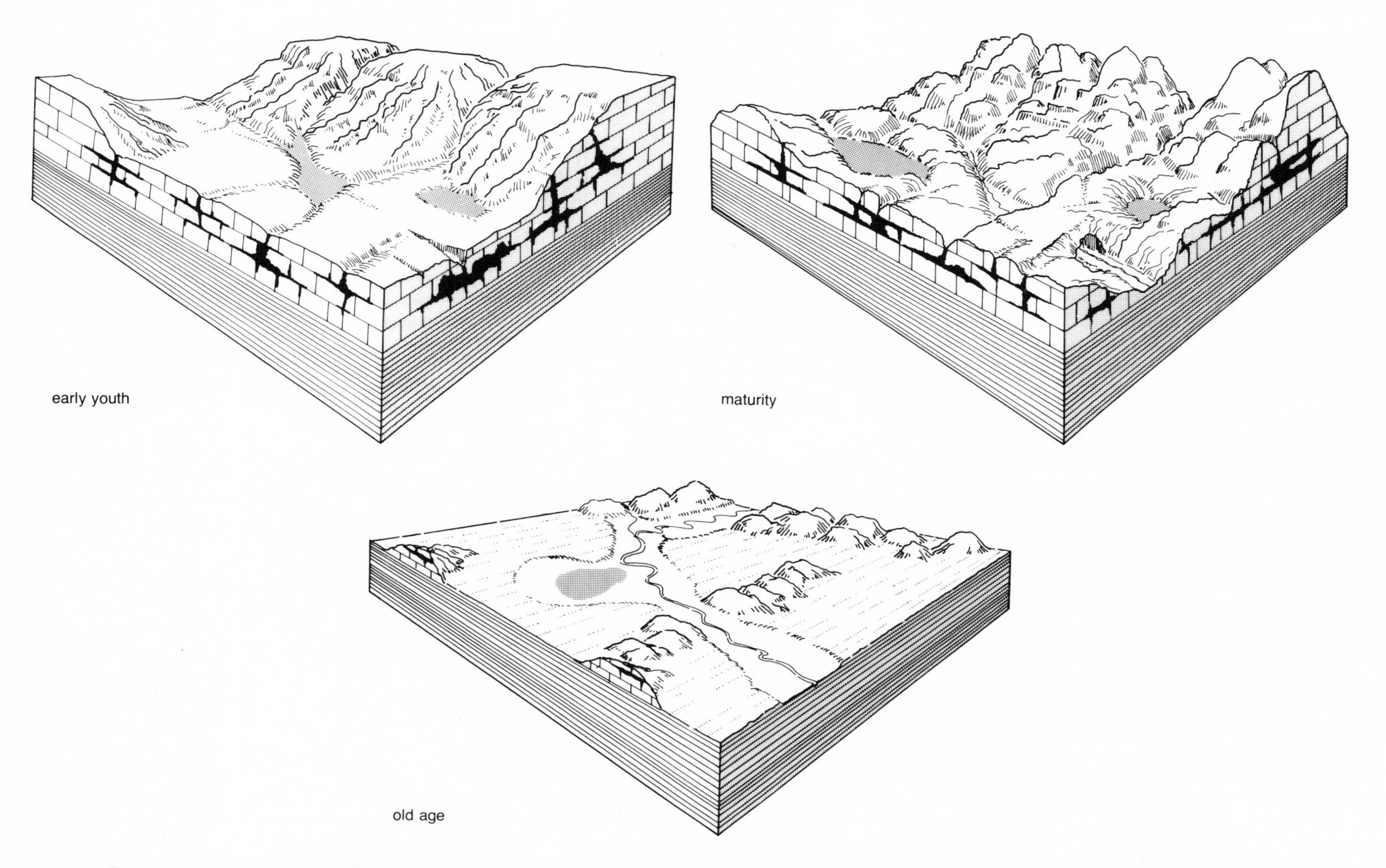

Figure 5.18. Erosion sequences of limestone. The three block diagrams illustrate the evolution of karst topography through the stages of increasing denudation.

in tropical, warm-water regions. Mature landscapes of thick limestone deposits in humid climates are easily recognized by their characteristic karst (sinkhole) topography and associated internal drainage. Limestone in arid climates, including dolomite, forms resistant table rocks with vertical faces; it is similar in appearance to sandstone, but the jointing pattern is more irregular. Tropical limestone develops very hummocky, haystack, karst topography which is characteristic only of this landform and therefore permits instant recognition. Dolomitic or cherty limestone in humid climates exhibits an angular dendritic pattern of moderate or coarse texture and white-fringed gullies; where chert occurs, white, residual chert blocks can be observed scattered across the surface. Coral is recognized by its association with water, its flat surface combined with jagged cliffs, and its lack of any evident surface drainage.

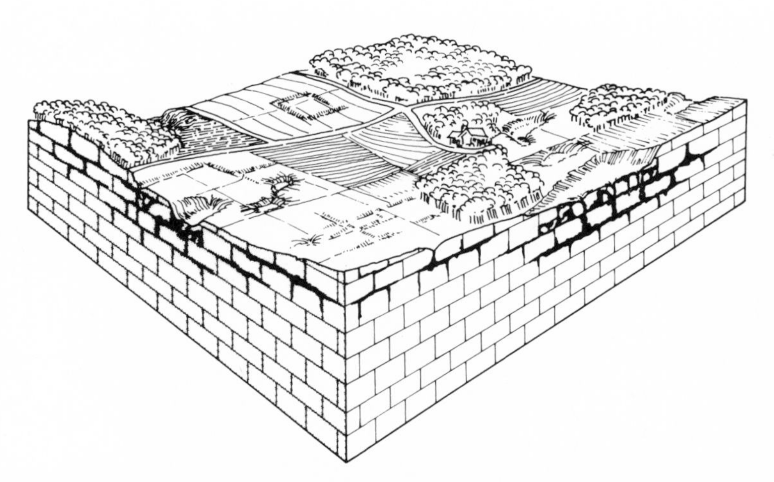

Figure 5.19. Limestone: humid climate. Limestone regions in humid climates develop an undulating topogaphy, spotted with scattered sinkholes. Most such areas are cleared and under cultivation, having irregular fields and small woodlots associated with farmsteads. The landscape is rather open and does not provide significant spatial variation. . Large swamps or surface water features are not common, owing to the high permeability of the residual soil and the bedrock. Some sinkholes may be ponded, and this may indicate the elevation of the groundwater table.

Soil Characteristics

As limestone materials weather, the rock decomposes and disintegrates into fine silt- and clay-sized particles. The residual soil commonly develops a structure that allows excellent drainage and permeability. If, however, this structure is destroyed, the soil is no longer well drained. Dolomitic limestone residual soils may contain more clays; chert fragments are present in cherty limestone soils. The residual soils from coral formations are thin and develop into red, silty clays which are well drained, having high liquid limits and plasticity indexes.

U.S. Department of Agriculture Classification

Surface horizons in limestone regions are typically classified by the U.S. Department of Agriculture as silt loam or silty clay loam. Parent materi-

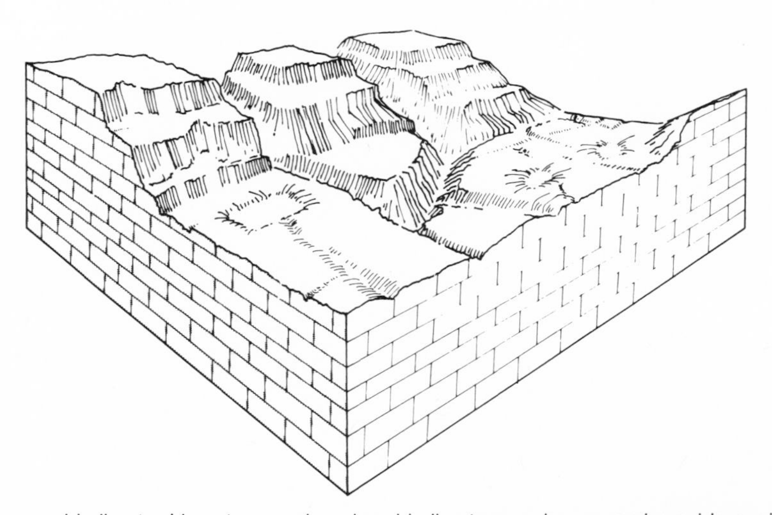

Figure 5.20. Limestone: arid climate. Limestone regions in arid climates are barren and provide a wide range of spatial variation and viewing capacity. The flat table rocks commonly associated with limestones in these areas provide viewing platforms of the surrounding region. If the beds are flat, other limestone buttes or plateaus may exist at the same elevations. Surface water features are not common, and these regions have little if any vegetative cover. Limestone table rocks are visually sensitive because of their elevated positions; the lowlands between buttes are visually more absorptive.

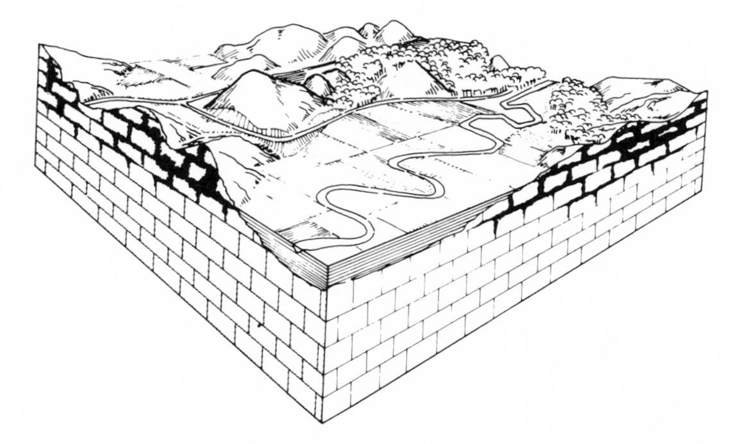

Figure 5.21. Limestone: tropical climate. Tropical limestone regions present one of the most visually absorptive natural landscapes. The great amount of vertical relief over short horizontal distances creates pits and peaks which effectively screen virtually any type of development. Hilltops bordering lowlands are visually more sensitive, but it is not probable that they would be considered for any type of construction or development. Swamps or standing-water features are not commonly found, owing to the high permeability of the residual soil and the bedrock.

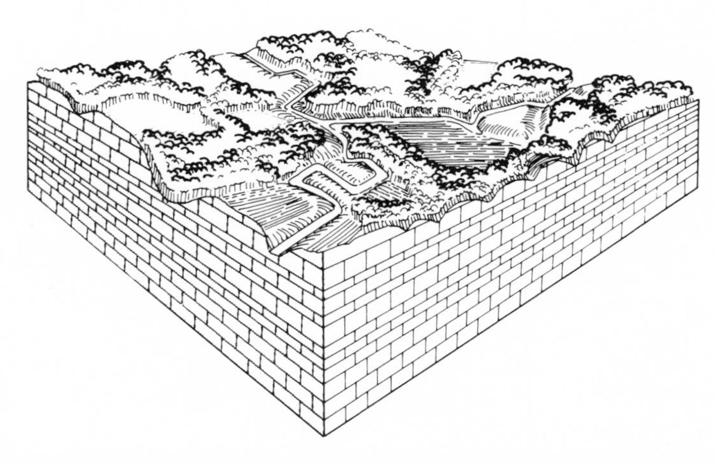

Figure 5.22. Dolomitic limestone: humid climate. Dolomitic and cherty limestone regions in humid climates are very similar in appearance to sandstone landscapes in humid climates. There are many hills and valleys resulting from the dendritic dissection of the land surface. Sideslopes bordering the drainage system are wooded, with irregular fields located in the uplands. The wooded corridors following the drainage system provide excellent cover for wildlife. The many hills and valleys provide for a wide range of spatial variation, but the hilltops generally offer more of a "good valley view" perspective rather than significant regional views. Dolomite in arid climates appears and has characteristics similar to arid limestone.

als, or the C horizon, in limestone profiles are generally classified as clay, silty clay, or silty clay loam. Many of the cherty and dolomitic limestone residual soils are described as having chert fragments or flaggy clay in the subsoil.

Unified and AASHO Classifications

Limestone surface residual soils are classified by the Unified system as ML-CL, ML, or CL. Those containing chert fragments may be classified as GM. The C horizon is commonly described as CL, CH, MH, ML, CH-MH, or occasionally ML-CL. In tropical climates the lower horizons are commonly CH or MH; lower horizons with many chert fragments are classified as GC or GM. The AASHO classification system commonly categorizes limestone and coral residual soils as A-6 or A-7.

Water Table

The depth to water table is extremely variable in limestone regions, even in humid and tropical climates. Depths are generally 12 to 20 or more feet. Ponded sinkholes may give some indication of the level of the water table. Corals are so porous that it is difficult for them to hold a significant water table above sea level.

Drainage

Residual limestone soils are very fine in textural composition, but the soil structure allows high permeability and rapid internal drainage except during the wet season of the year. The bedrock, especially in humid and tropical climates, contains many joints and solution cavities, thus increasing the drainage capability of the surface soil. Coral formations are honeycombed and spongelike and therefore drain very rapidly when elevated above sea level.

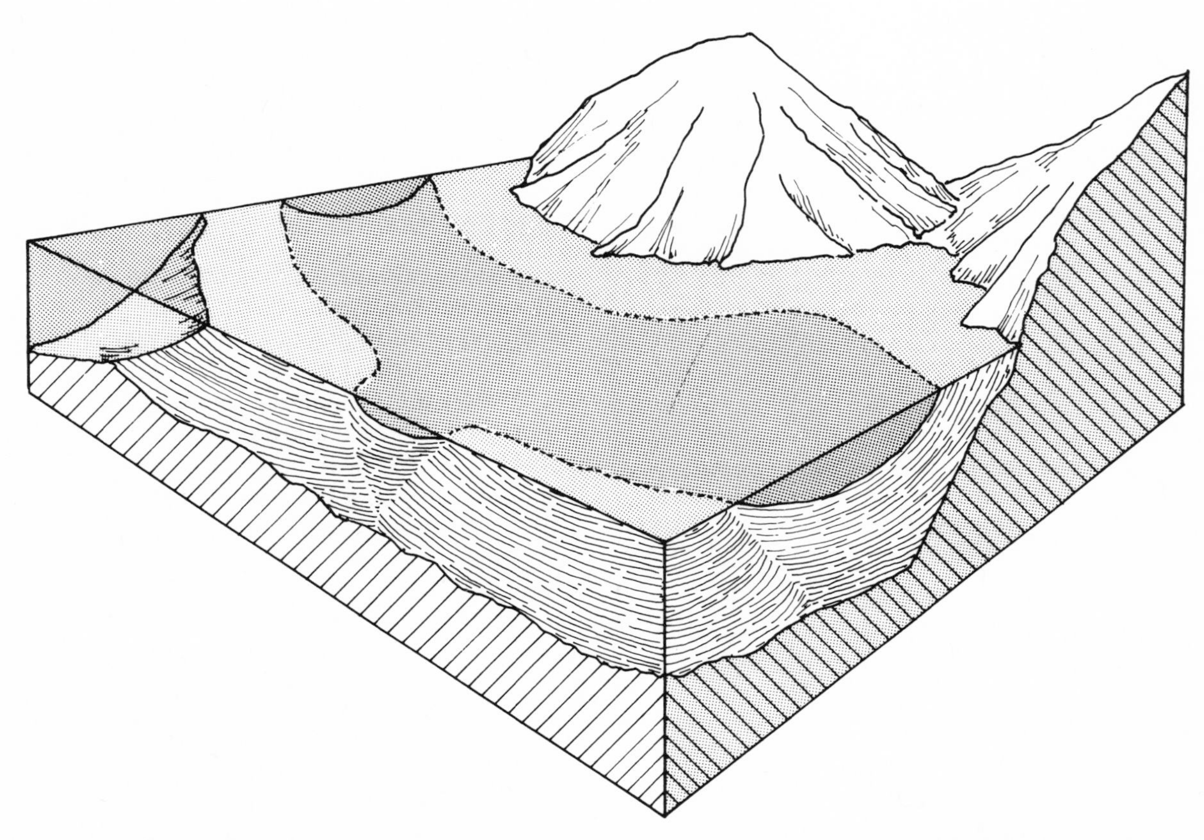

Figure 5.23. Coral. Coral formations are generally not large enough to determine a regional visual character. Elevated formations are flat, having jagged cliffs along the seaward edge; as a result of the porous nature of the material, no surface water features are found, and vegetation is shrubby even in tropical climates. Coral formations close to sea-level elevations develop covers of dense forest.

Soil Depth to Bedrock

Residual limestone soils in humid climates develop profiles of varying depth but are commonly 10 to 15 feet over weathered bedrock. Arid limestone residual soils, if present at all, are very thin since they are readily removed by wind and water erosion processes. Coral residual soils are shallow, several feet in depth, since such formations are relatively young.

Issues of Site Development

Sewage Disposal

In all limestone regions on-site sewage disposal systems are generally unsuitable because of the varying and unpredictable depths to water table and bedrock. Contamination of the subsurface groundwater system is easily caused by septic tank leaching fields. Dolomitic limestone regions without solution formations may offer sites suitable for individual on-site disposal systems, for these soils, unless disturbed, have suitable percolation rates. Treatment systems should be constructed for handling liquid wastes created by intensive development.

Solid Wastes

Sanitary landfill operations are difficult to locate in limestone regions because of the possibility of contaminating groundwater resources by leaching. The unpredictable depths to bedrock and water table mean there is a high risk of groundwater contamination unless the situation is investigated in detail. The fine-textured soils are suitable for covering solid wastes in a sanitary landfill operation, for they are impervious enough to inhibit infiltration of rainwater and to slow leaching.

Trenching

In humid limestone regions trenching operations encounter unpredictable bedrock conditions and depths which influence removal costs. In arid climates trenching operations are concerned with removal of large quantities of massive bedrock. Typical costs per cubic yard for removing massive limestone bedrock are approximately four to five times higher than for removing the same amount of deep, dry, moderately cohesive soil. Blasting is also necessary. Limestone residual soils can be removed with little difficulty, but they are difficult to work. Coral forms may be either soft and easily trenched by power equipment, or crystalline and require heavy equipment and blasting. In all limestone regions seepage zones may be encountered in deep trenching operations.

Excavation and Grading

Excavation and grading of large sites in humid limestone regions are difficult because of the vari-

Table 5.6. Sedimentary rocks: Limestone (humid and arid)

Humid

Topography	*Drainage*	*Tone*	*Vegetation and land use*
Karst	Internal	Mottled	Cultivated
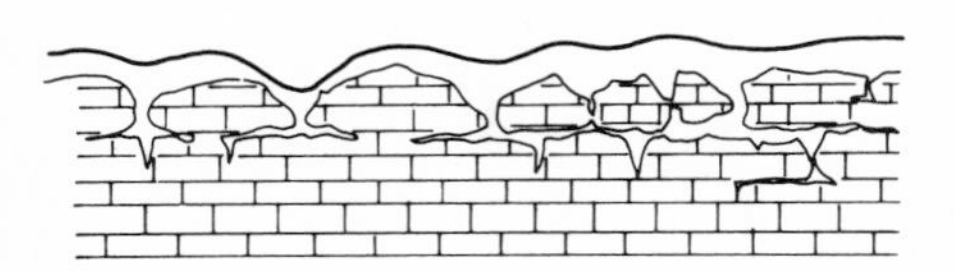	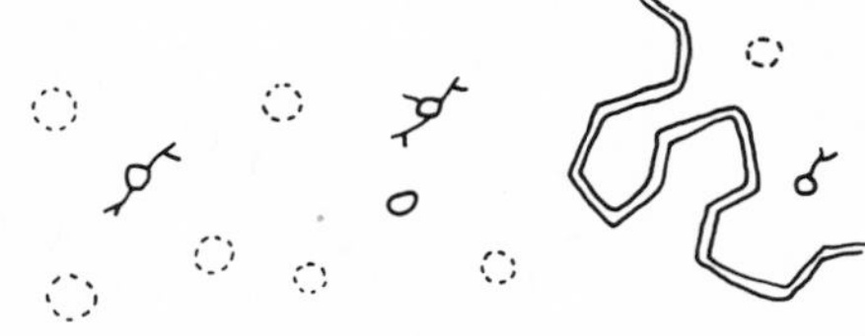	The general tone in limestone regions is a mottled gray. Sinkholes containing ponded water have darker tones and regular outlines.	Limestone regions in humid climates are cultivated and cleared of forest growth. Small woodlots may be associated with farmsteads. Field patterns are irregular, being controlled by the sinkholes and depressions of the karst topography. Highways are generally curvilinear and attempt to avoid crossing sinks.
		Gullies	
		White gullies	
Chemical weathering dissolves the rock along jointing and bedding planes, thus developing a collapsing surface of sinkholes and depressions known as "karst topography." The ground surface is undulating and forms indistinct transitional boundaries with other, associated sedimentary rocks. Sinkholes are rounded in flat-laying beds, elongated in tilted beds.	The solution cavities within the rock and the high permeability of the residual soil cause humid limestone regions to be drained internally, leaving little water to be collected in a surface water system. Major streams follow angular alignments of old jointing patterns. Typical sinkholes average 10 to 40 feet in depth and 50 to 500 feet in width.	Few are present, but short, white-fringed gullies occur around sinkholes. If present, they reflect a cohesive soil with a sag-and-swale section.	

Arid

Topography	*Drainage*	*Tone*	*Vegetation and land use*
Table Rocks	Angular dendritic: Medium to fine	Light	Barren
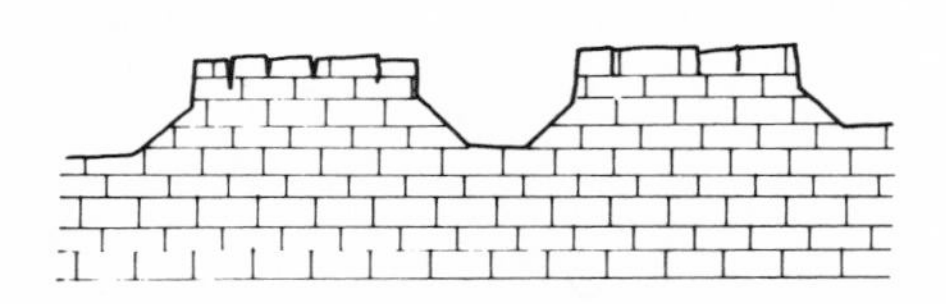	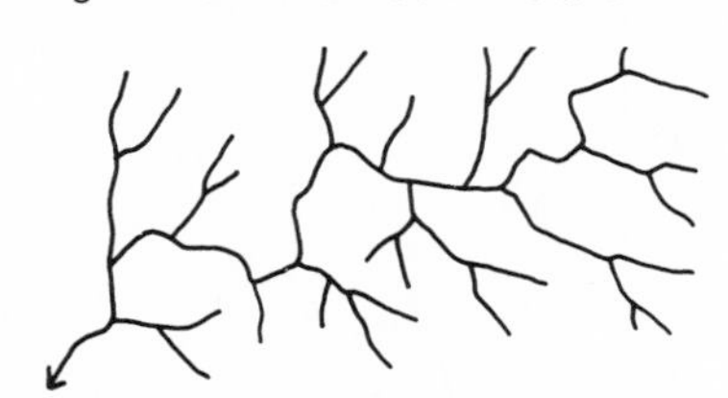	Limestone in arid regions shows as a uniform light tone.	The resistant nature of the bedrock, the thinness of the residual soil, and the arid climate are not favorable conditions for the establishment of agriculture or vegetative growth. Some shrubby vegetation may perhaps be observed in major valleys, but generally there is no form of traditional land use.
		Gullies	
		Few gullies	
Since little moisture is available for chemical weathering in arid climates, the limestones present erode very little. Pure, thick limestone deposits form cap or table rocks, developing none of the characteristics associated with karst topography.	The surface drainage system is well developed (karst topography does not exist in arid climates.) The pattern is very angular, following jointing alignments in the bedrock, and is medium to fine textured. All but major streams are intermittent.	Very little residual soil is present, for it is quickly eroded and removed by processes of wind and water.	

Figure 5.24. (A) Limestone in a humid climate in Hardin County, Kentucky, shows the typical development of karst topography. Photographs by the Agricultural Stabilization Conservation Service, ALN-7V-34,35, 1:20,000 (1667′/″), January 2, 1959.

Figure 5.24. (B) U.S. Geological Service Quadrangle: Vine Grove (7½′), Flaherty (7½′).

Table 5.7. Sedimentary rocks: Tropical limestone, dolomitic or cherty limestone (humid), and coral

Tropical limestone

Topography

Tropical karst

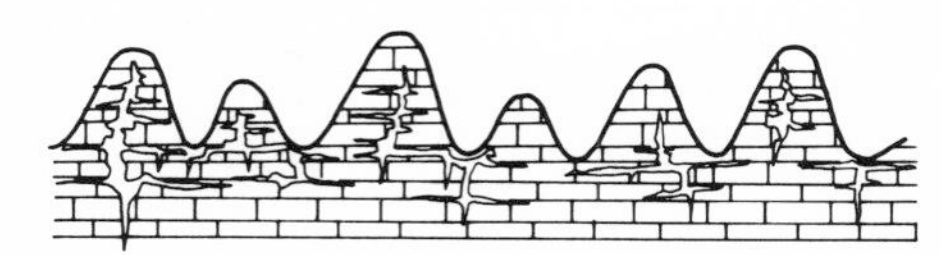

A very rugged, tropical karst topography is developed with conical hills and large depressions (up to 1000 feet deep and 1 mile in diameter).

Drainage

Internal

All drainage is internal through the highly permeable soil and bedrock.

Tone

Uniform light

The high permeability of the residual soil and bedrock, in combination with their natural color, present uniform light tones.

Gullies

None

Vegetation and land use

Barren or forested

The steep, concial hills can be either forested or barren. Lower hills may be cultivated.

Dolomitic or cherty limestone

Topography

Hill and valley

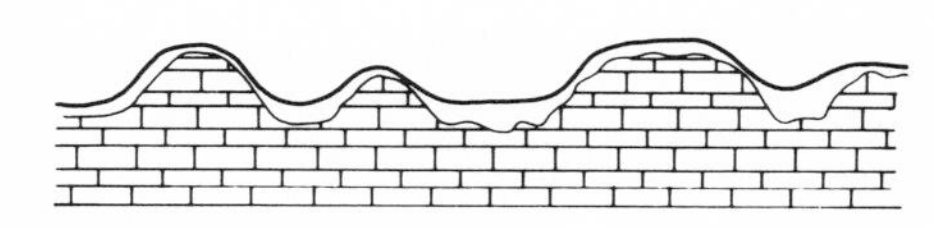

These areas are similar to sandstone in appearance, with little or few solution cavities or sinkholes. Nodules of chert can be observed in open fields.

Drainage

Angular dendritic: Medium

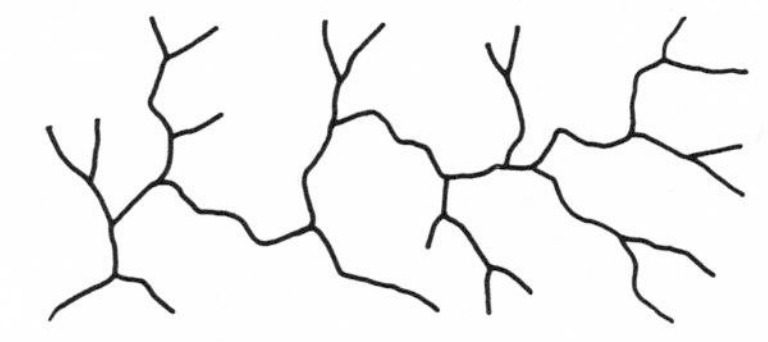

A dendritic pattern of medium texture is characteristic in humid climates, with angularity reflecting jointing control.

Tone

Light gray

Light tones with minor mottling occur in dolomitic and cherty areas.

Gullies

Sag and swale

Gullies of a sag-and-swale nature, indicating cohesive materials, are associated with these forms.

Vegetation and land use

Cultivated and wooded

These regions are cultivated, and the forms of its irregular fields are controlled by the dissection of the drainage system. The drainage system is usually wooded.

Coral

Topography

Terraced

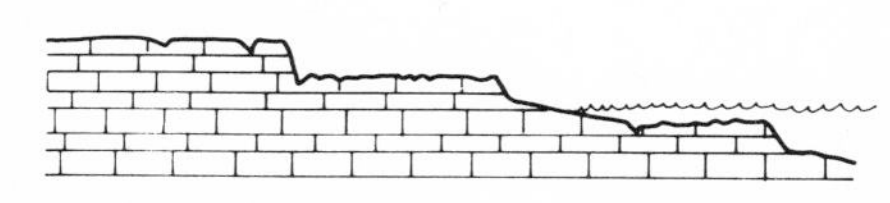

Elevated reefs appear flat, with a steep, jagged cliff along the seaward edge. Sea-level reefs either encircle (barrier type) or appear as a white, discontinuous fringe along the edge of an island.

Drainage

Internal

Coral is very porous and spongy and therefore does not develop either a surface drainage pattern or sinkhole topography.

Tone(Coral)

White or gray

Exposed coral at sea level is brilliant white; elevated coral develops a thin soil cover and is therefore more gray in tone.

Gullies

None

Vegetation and land use

Forested or barren

Because of their porous nature, elevated coral formations are very droughty and do not support much vegetation. Reefs at sea level are commonly covered with forests.

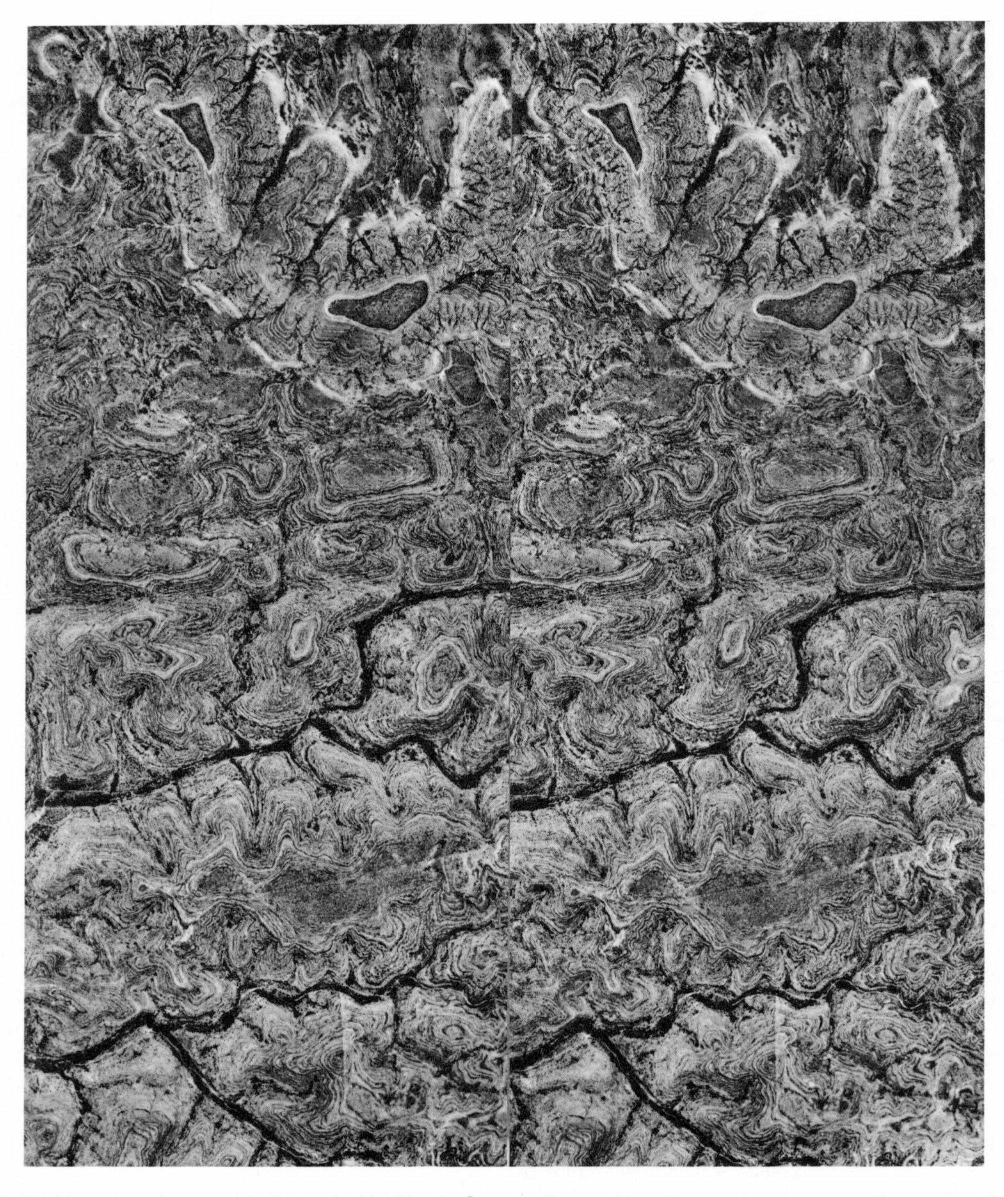

Figure 5.25. Limestone in an arid climate in Val Verde County, Texas. The thin bands are caused by slight differences in bedding. Photographs by the U.S. Geological Survey, Texas 5-AB, 1:46,000 (3833′/″), April 24, 1950.

Figure 5.26. Limestone in a tropical climate in Puerto Rico shows the typical domelike karst topography developed in these climates. Photographs by the U.S. Geological Survey, Puerto Rico 5-AB, 1:20,000 (1667′/″), February 8, 1963.

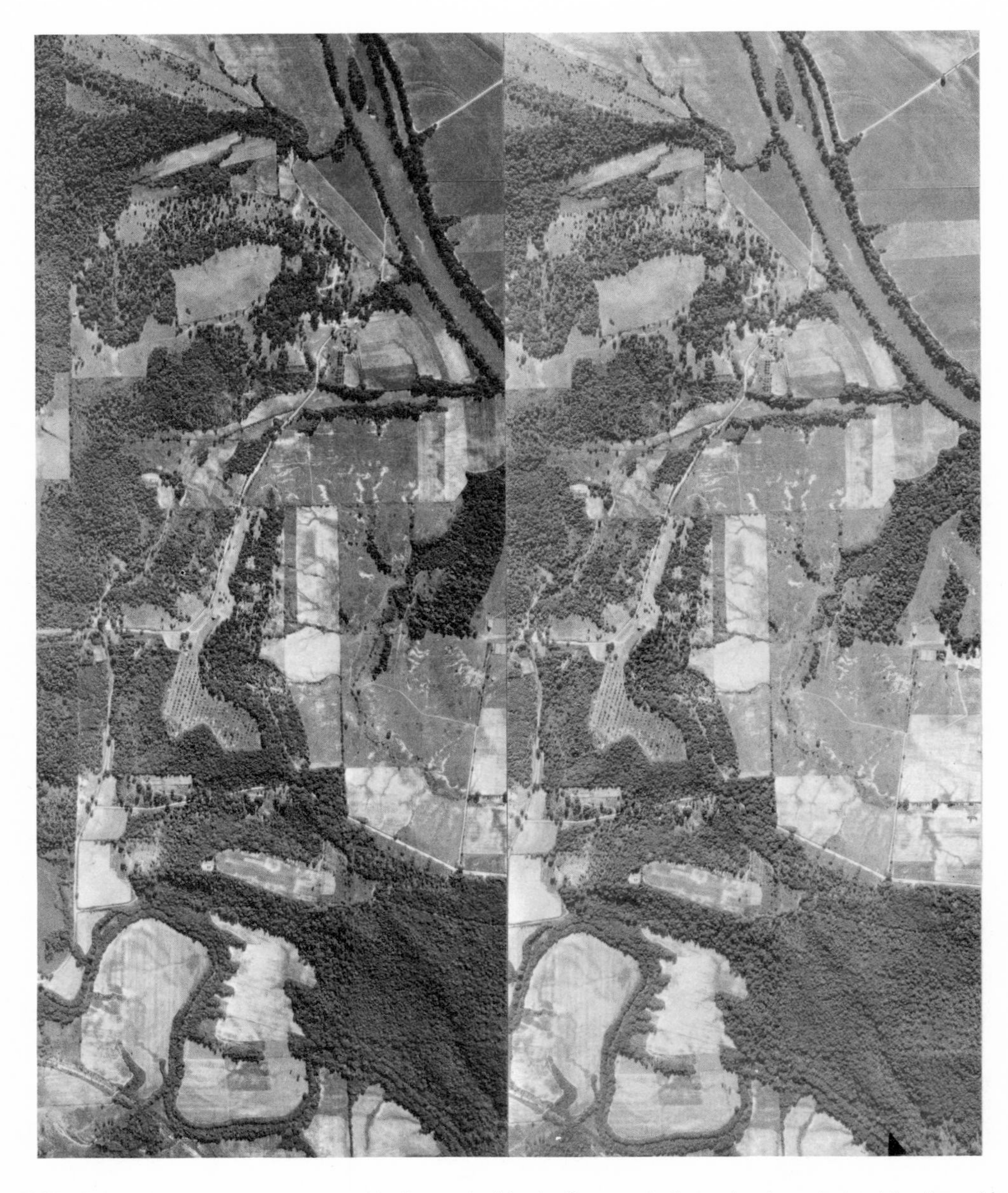

Figure 5.27. Dolomitic limestone in a humid climate in Martin County, Indiana, is characterized by the white gullies, by lack of development of solution (karst) topography, and by angular drainage channels. Photographs by the Agricultural Stabilization Conservation Service, CMH-3A-91,92 (1667′/″), August 1, 1940.

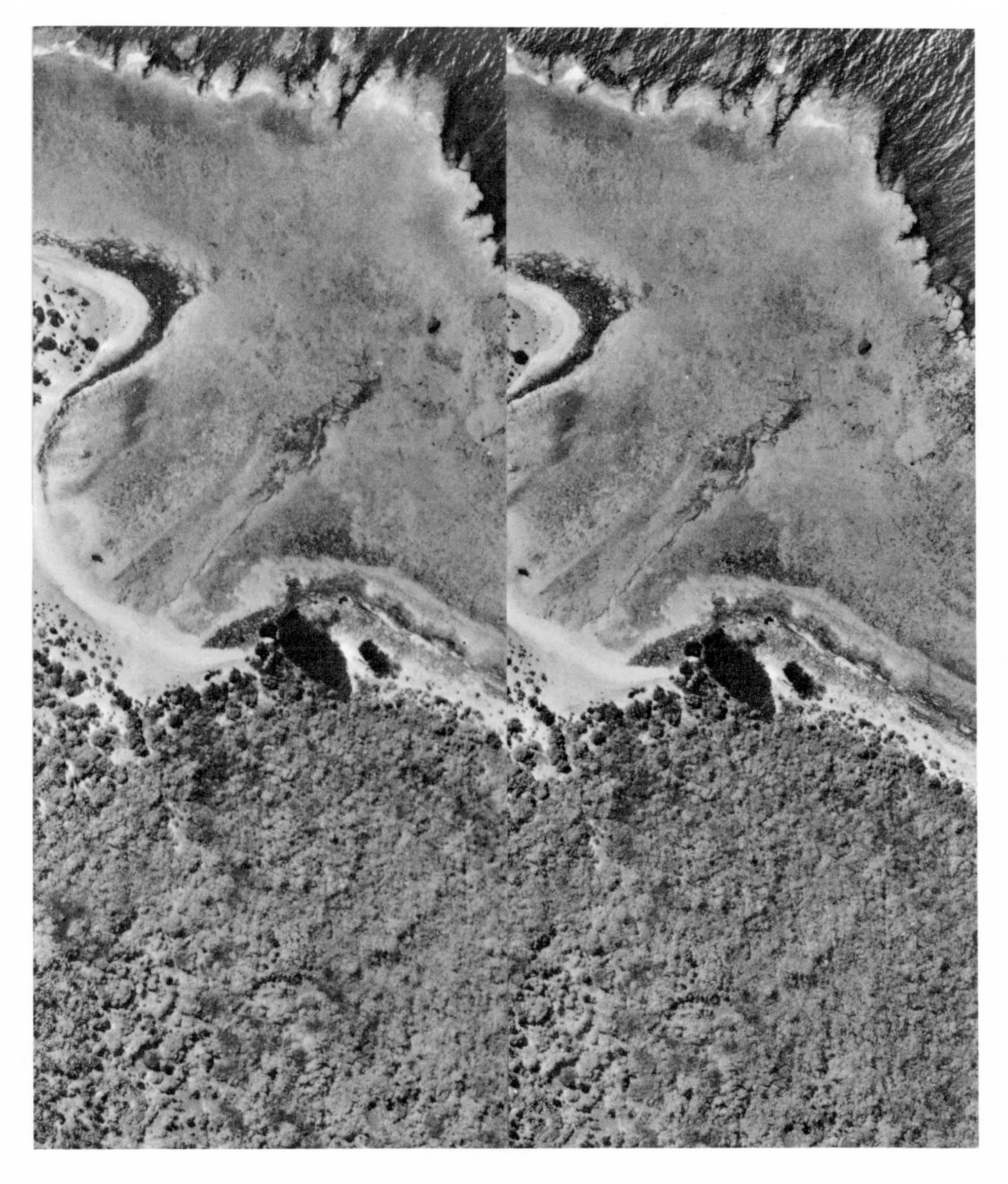

Figure 5.28. A fringing coral reef along the edge of Enirik Island at the southernmost point of Bikini Atoll shows damage from large storm waves, indicated by the comb-tooth pattern. Photographs by the U.S. Geological Survey, Bikini 2-AB, 1:5,000 (417′/″), date unknown.

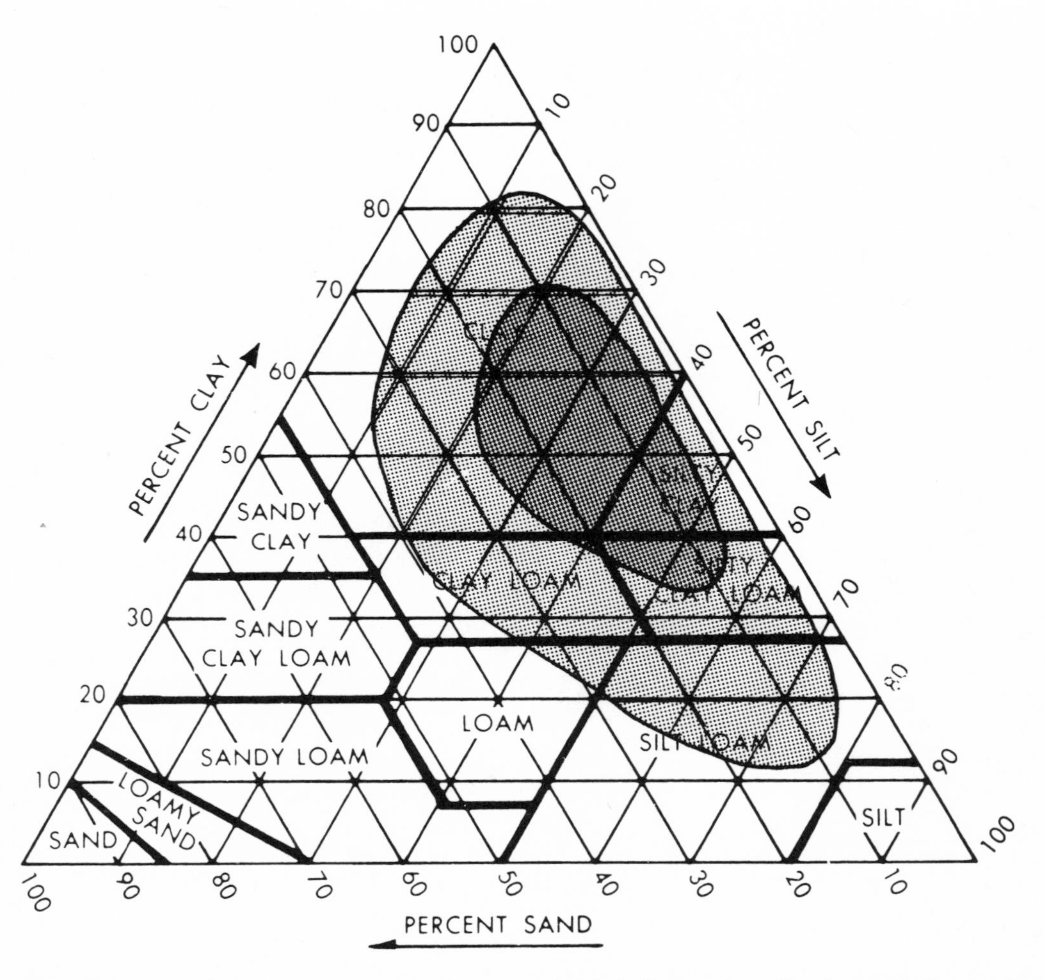

Figure 5.29. The U.S. Department of Agriculture soil texture triangle indicates clay, silty clay, and silty clay loam as dominant materials.

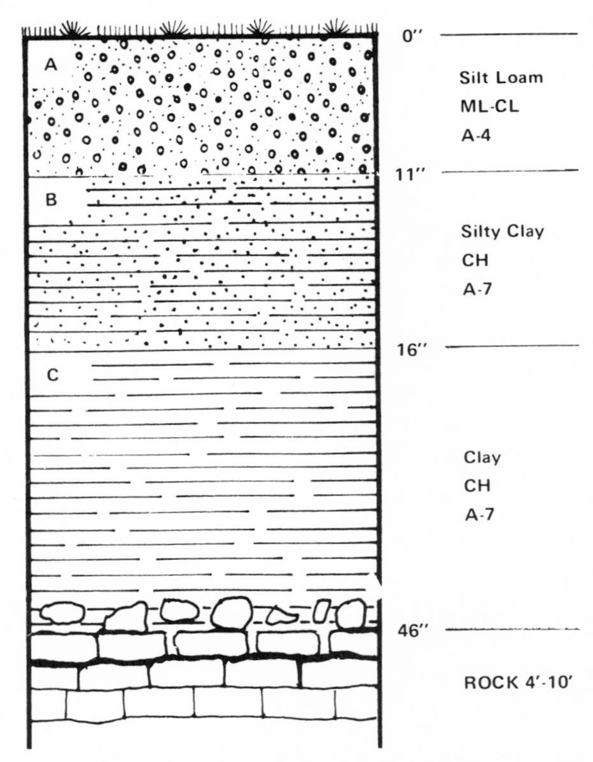

Figure 5.30. A typical limestone residual soil profile as found in a humid climate.

able depths to bedrock and water table. Typical per-cubic-yard costs of removing rock material are four to five times higher than for removing the same amount of deep, dry, moderately cohesive soil. Massive bedrock is encountered in deep cuts and requires blasting. Deep residual soils are hard to work, owing to their high plasticity and poor drainage characteristics when disturbed. Limestone regions in tropical areas are so rugged that large-scale excavation and grading is rarely feasible. The karst topography encountered in this climate presents one of nature's most rugged land surfaces. Coral formations are naturally flat along their upper surface and do not require any significant grading and excavation to create a flat site. Soft corals can be excavated by power scrapers or draglines; the crystalline variety requires the use of explosives.

Construction Materials (General Suitabilities)

Topsoil: Fair. These soils are generally highly plastic, especially in tropical regions or as residuum of coral formations. They are difficult to work.

Sand:Not Suitable

Gravel: Not Suitable. Some soil classification systems, such as the Unified, indicate gravel in

Table 5.8. Limestone: aerial photographic and cartographic references

State or Country	County	Photo Agency	Photograph no.	Scale	Date	Corresponding quadrangle and geologic maps from USGS	Corresponding soil reports from SCS
Arizona	Coconino	USGS*	Arizona 6-ABC	5000'/"	10/11/55	National Park (1:48,000) (1:48,000)	—
Bikini	Enirik Island	USGS†	Bikini 2-ABCD	417'/"	Unknown	Prof. Paper No. 260-A	—
Florida	Sarasota	ASCS	DEW-1T-101-104	1667'/"	2/23/57	Sarasota (7½'), Bee Ridge (7½')	1959 1946
Indiana	Martin	ASCS	CMH-3A-90,94	1667'/"	8/1/40	—	—
Kansas	Barton	ASCS	CHG-1FF-197-200	1667'/"	10/8/65	Ellinwood (7½')	—
Kansas	Clark	ASCS	CHE-2AA-199-201	1667'/"	6/19/60	Ashland (1:250,000)	—
Kentucky	Edmonson, Barrow	USGS*	Kentucky 4-AB	1200'/"	2/22/54	Survey Park (7½') GQ-183	—
Kentucky	Hardin	ASCS	ALN-4V-137-139	1667'/"	10/1/58	Flaherty (7½')	—
Kentucky	Hardin	ASCS	ALN-7V-34-35	1667'/"	1/2/59	Vine Grove (7½')	—
Kentucky	Hardin	ASCS	ALN-4V-123-124	1667'/"	10/1/58	Winchester (1:250,000)	—
Kentucky	Warren	USGS*	Kentucky 2-AB	1200'/"	2/12/54	Bristow (7½')	1904‡
Palau	Palau Island	USGS†	Palau 3-AB	3333'/"	1/10/48	—	—
Puerto Rico	Manati	USGS*	Puerto Rico 5-AB	1667'/"	2/8/63	Manti (7½'), Barceloneta (7½') I-334	1942
Puerto Rico	Utuado	USGS*	Puerto Rico 3-ABC	1667'/"	3/21/63	Bayaney (7½'), Utuado (7½') oil and gas preliminary map No. 83	1942
Texas	Ector, Crane	USGS*	Texas 3-AB	5280'/"	5/3/54	Pecos (1:250,000)	—
Texas	Val Verde	USGS*	Texas 5-AB	3833'/"	4/24/50	Bakers Crossing (15')	—
Virginia	Rockingham	ASCS	DJN-6T-72,73	1667'/"	10/13/57	Bridgewater (7½')	—
West Virginia	Wood	ASCS	ANG-35-115-116	1667'/"	9/23/39	—	1970
Wisconsin	Iowa	ASCS	WT-4CC-136-139	1667'/"	8/28/62	Montfort (7½')	1962

*From *Selected Aerial Photographs of Geologic Features in the United States*, U.S. Geological Survey Prof. Paper No. 590.
†From *Selected Aerial Photographs of Geologic Features outside the United States,* U.S. Geological Survey Prof. Paper No. 591.
‡Out of print.

the parent materials of limestone, such as GM or GC. These gravelly textures are usually caused by weathered rock or chert fragments in the subsoil.

Aggregate: Limestone—Excellent; Coral—Poor

Surfacing: Good to Excellent. Crushed limestone rock materials make good surfacing materials for secondary roads.

Borrow: Fair to Poor. Residual soils of limestone are rated poor for use as road fill since they are hard to work and plastic. However, suitable compaction of the plastic, clayey soil for fills can be accomplished under optimum moisture conditions by using sheepsfoot rollers.

Building Stone: Fair to Good. Limestone is used for building stone and tombstones, although its disadvantage is that it weathers greatly over time.

Mass Wasting and Landslide Susceptibility

Limestones are not susceptible to landslides except along the faces of vertical cliffs or where well-developed joints parallel hillslopes. If limestone is interbedded with impervious clay or shale, however, slides are common. The retention and accumulation of water in voids greatly increases the rock mass, thereby creating unstable conditions along slopes and escarpments.

Groundwater Supply

Suitable water supplies can usually be found in limestone regions, because of the many solution cavities and jointing patterns associated with this rock type. The water is usually very hard and may require minor treatment to soften it (that is, to remove the dissolved carbonates). In regions of karst topography, if the groundwater is contaminated, the pollutants will travel many miles underground; the pollutants tend not to be decomposed or absorbed within the subsurface water system, for they are not exposed to sunlight nor is the water filtered through granular stream bed materials.

Pond or Lake Construction

Construction of ponds and lakes in limestone regions involves difficulties related to the high porosity of the rock and its associated solution cavities. Dams are difficult to site, since the water can seep around the structure through subterranean channels which are enlarged over time. Limestone residual soils in their natural state are slightly permeable and require compaction to destroy the natural soil structure; this results in the formation of a suitable impervious seal. There are some maintenance problems in minimizing siltation of the water body because of the fine textures of the residual soil. Care should be taken to protect the vegetative cover within the watershed, especially along gullies where channelized drainage begins and along the shorelines of major tributaries.

Foundations

Many of the residual soils developed from limestone and associated rocks contain fine silt and clays subject to medium or high volume change as they become wet and are dried. Foundations must therefore be designed to transfer the bearing load to the rock structure or to soils of suitable bearing capacity not subject to volume change below the ground surface. Shallow foundations can be used, but early, detailed, subsurface exploration is necessary, especially where large loads are considered. If footings are proposed, test borings will be necessary under each footing base since future enlargement of solution cavities, if present, may facilitate excessive settlement and possibly failure. Footings and rafts can be used only if cavities are not found and other requirements are satisfied. Gravel beds can be used for raft foundations to provide a stable, well-drained base over the plastic, silty clay residual soils. Continuous wall footings with basement slabs are common for residential structures. Foundations used to transfer large loads may include piers or point-bearing piles, depending upon the specific site conditions. Large solution cavities may require filling with force-pumped concrete or other suitable materials, but this procedure is not only expensive but somewhat risky since some cavities may be inadvertently missed. Large-area land uses, such as shopping centers and airports, may require capping and sealing of sinkholes.

Highway Construction

Many problems are encountered in building highways across the different types of limestone in various climatic zones. In humid climates of karst topography, alignments are difficult, owing to the existence of sinkholes and depressions. Depths to bedrock are not easily predicted, making it hard to estimate accurately grading and excavation costs. Excavation of bedrock requires the use of explosives, although jointing patterns and subterranean channels may facilitate the placement of charges. Suitable materials for construction can usually be located with little difficulty.

In arid climates limestone does not develop significant solution topography, and it occupies table rock positions or broad plateaus. These sites are naturally flat and do not require much grading unless highly dissected.

Dolomitic and cherty limestone areas may have some solution cavities located within the bedrock structure, and the landscape is dissected; these factors cause alignment difficulties and necessitate greater quantities of cut and fill. Blasting is necessary for bedrock removal, and seepage zones may be encountered during excavation.

Tropical limestone regions present one of the most difficult landscapes for highway construction. The rapid topographic changes and great amounts of exposed rock demand vast grading operations for the simplest of highways. Tropical limestone areas are usually bypassed and tend to remain largely undeveloped. Only naturally dissected landforms approaching a peneplain contain enough flat land between the remaining conical hills or monadnocks so that highway grading is not difficult.

Coral regions have a relatively flat topography and do not require an extensive amount of grading. Coral is a good source of base course material since it is very porous and naturally well drained. Soft varieties of coral can be bulldozed and, after they are crushed, wet with salt water, and compacted, become recemented. These softer corals, however, also contain impurities which may cause the road surface to become slippery when wet unless it is paved or covered with other materials.

Interbedded Sedimentary Rocks: Flat-Lying

Introduction

The previous sections of this chapter have discussed each sedimentary rock type separately, for although landforms classified as shale, sandstone, or limestone may not consist entirely of those rocks, the individual rock types are present in amounts significant enough to control pattern elements and determine engineering capabilities. However, all sedimentary rocks are interbedded to various degrees, according to their formative processes. Interbedded sediments can result from changes in water level during deposition, or from an increase or decrease in the sediment load and type in water currents. A common cause of thin interbedding is wet season–dry season runoff changes in coastal river systems.

Under ideal conditions, sediments deposited underwater grade from coarse materials near the coastal regions where water velocities are high and depths are shallow, to fine materials further out in the ocean where velocities are low and depths are greater. It follows that coastal areas are associated with deposits of gravels, forming conglomerates; as depths increase, the gravels grade into sand deposits, forming sandstone, followed by sandy shales, then shales, clay shales, and limy shales, grading into thinly bedded limestones and finally into pure, thick deposits of limestone. Therefore, rock types commonly found interbedded were formed in close proximity under similar conditions; such combinations include conglomerates with sandstones, sandstones with shales, and shales with limestones.

Figure 5.31. Thickly bedded sedimentary rocks in humid climates produce a terraced, rugged topography with forested hillsides and cultivated uplands. Major valleys typically have villages and towns located in relation to the natural water system. The landscape tends to be visually absorptive, with the valley walls creating spatial enclosure and barriers. Viewed from the valleys, the most sensitive visual areas are upland rim edges where the slopes meet the hilltops. Thinly bedded sedimentary deposits in humid climates have much the same visual character, except that the hillsides are not terraced.

Interpretation of Pattern Elements

This section describes interbedded sedimentary rocks in which two or three rock types are involved and whose bedding attitudes are flat. For the purposes of the key, flat-lying interbedded sedimentary rocks are categorized as thick beds if they are more than 25 feet in thickness, and as thin beds if they are less than 25 feet in thickness. Thick beds have significant effects on topographic characteristics, resulting in terracing. Thin interbedded deposits, however, keep the characteristics of the predominant rock type.

Soil Characteristics

The residual soil formed in humid climates from interbedded sedimentary deposits is typically shallow and silty and contains many rock fragments within the subsoil. Greater accumulation of fragments can be expected along the lower edge of sandstone outcrops. Thick interbedded

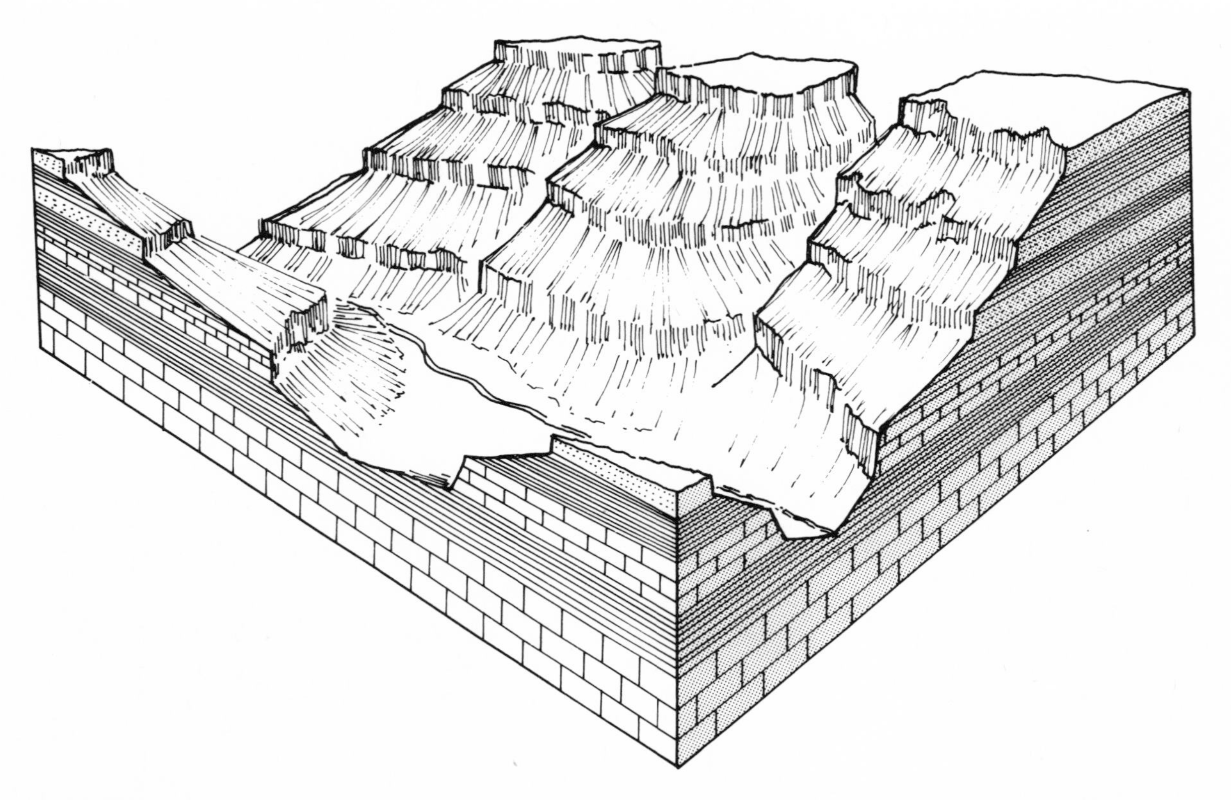

Figure 5.32. All interbedded sedimentary deposits in arid climates appear barren, since they are covered only lightly by scattered grasses and shrubs. Thickly bedded deposits determine the distribution of vegetative associations: for instance, in a shale-sandstone deposit the vegetation is most dense along the moister sandstone layers. (In such regions the sandstone appears darker than the shale and should not be confused with the natural tones of the shale materials.) Even thinly bedded deposits may show the tonal bandings caused by preferences of vegetation for hillside sites of slightly greater moisture. Other concentrations of vegetation may be found in drainage courses in valleys. Visually, interbedded rocks in arid climates are characterized by a very rugged, picturesque topography. Hilltops occur at the same elevations, offering viewing platforms for regional views. The deep valleys have a definite feeling of spatial closure. Many exciting rock features and forms are found in interbedded arid regions; sand grains carried by the wind abrade and round rock formations into natural sculpture, and there are many exposed outcroppings and massive cliffs.

sedimentary rocks develop residual soils common to those associated with each rock type. Sandstones interbedded with shales develop more sands; limestone-shales develop more clays.

In arid climates soil profiles are thin, and talus debris accumulates along lower slopes and alluvial deposits in the valleys.

U.S. Department of Agriculture Classification

The individual soil characteristics defined under the individual sedimentary rock sections should be consulted as specific rock types are identified. Thin interbedded residual soils have combined textures and are classified as silt loam, silty clay loam, silty clay, clay loam, and, occasionally, clay, sandy loam, sandy clay loam, and sandy clay (Figure 5.36). Rock fragments occur in the subsoil, and chert fragments are found scattered with residual soils of cherty limestone.

Unified and AASHO Classifications

Most profile descriptions include CL, ML, CH, MH, and ML-CL, with SC, GM, or SM occasionally occurring. Gravel categories are used to indicate the existence of partially weathered rock fragments, flaggy clay, shale, or chert nodules. The AASHO clasification system categorizes commonly occurring soils as A-6, A-7, A-4, and A-2-4. Coarse-textured, partially weathered rock is classified as A-2 and A-1.

Water Table

The depth to water table in residual soils is variable, depending upon the specific composition of the materials and the sequence of interbedding. Sandstone and limestone residual soils tend to be well drained; shale is not. Sandstone or limestone interbedded with shale are inhibited in their internal drainage, creating seepage zones along hillsides. Groundwater is found within these rock structures, sandstone being the best potential aquifer, followed by limestone. Shale is impervious and does not contain any significant water resources.

Drainage

The occurrence of shale, especially in thinly bedded situations, inhibits internal drainage and creates more surface runoff. If residual soils from limestone are disturbed, internal soil drainage will be decreased because the original soil then becomes almost impervious. Sandstones develop a coarser soil and have better drainage tendencies.

Table 5.9. Sedimentary rocks: Thick interbedded (humid and arid)

Humid

Topography	*Drainage*	*Tone*	*Vegetation and land use*
Terraced hillsides 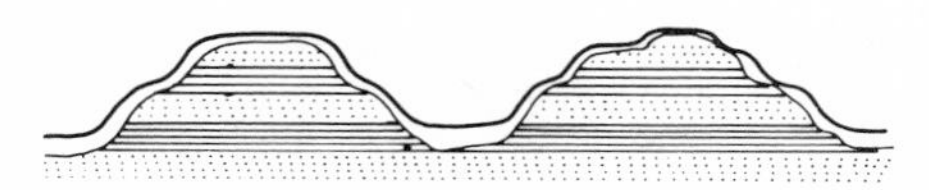 Hillsides in thick, interbedded sedimentary rock regions appear terraced; hilltops are at the same elevations. In sandstone-shale combinations the more resistant sandstone remains as a cap rock with steep sideslopes; the underlying shale has a more gradual sideslope. In limestone-shale combinations the limestone occupies the hilltops and uplands and may have solution features.	Dendritic: Medium to coarse 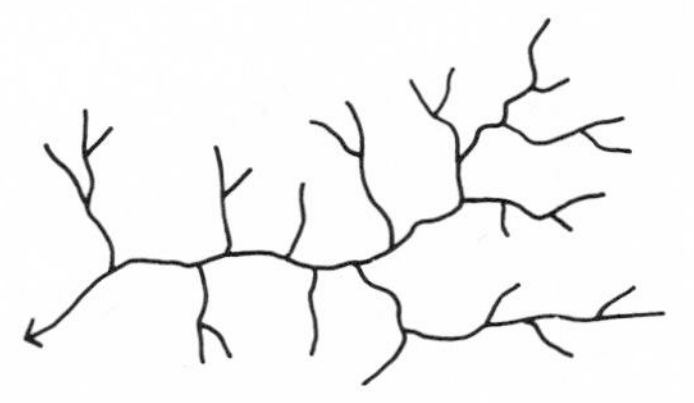 The drainage pattern is commonly a medium dendritic system and tends to be controlled by the most resistant rock in the series, commonly sandstone. Thick, limestone cap rocks may be characterized by solution sinkhole topography, with major streams following angular alignments.	Subdued banding Some banding may be apparent, especially between sandstones and shale, since sandstone is lighter as a result of the better drainage of its residual soil and rock structure. *Gullies* Vary Gully shapes vary but are usually of the sag-and-swale type, indicating cohesive soils. White gullies indicate limestone.	Cultivated and forested In sandstone and shale areas, the sandstone remains in forest cover and the shale is cultivated. Here, contour farming should not be confused with natural banding from different rock materials. In limestone and shale combinations, most of the land is cultivated and follows the natural patterns of sinkholes and the major drainage system.

Arid

Topography	*Drainage*	*Tone*	*Vegetation and land use*
Terraced hillsides  As in humid climates, hillsides appear terraced and hilltops are at the same elevations. The more resistant rocks occupying table rock positions, such as limestone and sandstone, maintain steep escarpments. Shale, usually underlying more resistant rock, is characterized by more gradual slopes.	Dendritic: Medium to fine  A dendritic drainage system of medium to fine texture is common, reflecting the nature of the upper strata of bedrock. Limestone does not develop solution cavities but may control surface drainage and valley alignments through its angular jointing system.	Banding Light and dark bands encircling hills and following contours indicate interbedded sedimentary rocks. Vegetation may follow sandstone layers and appears dark in tone. *Gullies* Few The lack of a significant residual soil does not allow for gully development.	Barren Because of the arid climate, rugged topography, and thin soil cover, interbedded sedimentary deposits in arid climates do not commonly have agricultural or other land use development. Most regions are nearly barren, their shrubby growths following bands on hillsides associated with seepage zones. Some areas are used for ranching and grazing purposes.

Table 5.10. Sedimentary rocks: Thin interbedded (humid and arid)

Humid

Topography	*Drainage*	*Tone*	*Vegetation and land use*
Uniform slopes	Dendritic: Medium	Medium gray	Cultivated and forested
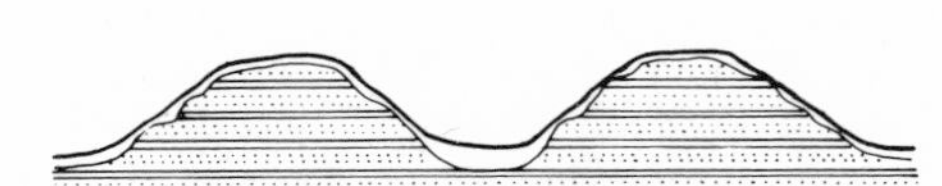 Thin, interbedded sedimentary materials do not have terraced slopes, although faint banding may be observed. Sandstones have steeper slopes than shale or limestone masses. Hilltops throughout such regions have approximately the same elevations.	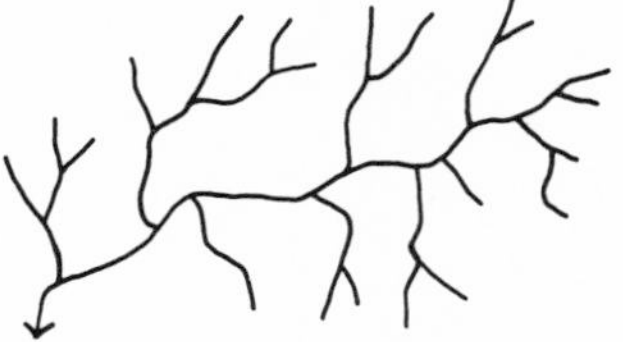 The drainage pattern in thin, interbedded sedimentary rocks is dendritic and of medium texture, similar to shale drainage in humid climates. The presence of shale in the interbedding, however, increases surface runoff, resulting in a more dissected topography.	Since the beds are thin and usually covered with residual soil which masks their tonal differences, an overall medium-gray tone results. *Gullies* Sag and swale. Gullies found in these regions are of the sag-and-swale type, indicating cohesive soils.	Gentle slopes, hilltops, and valleys are cultivated, while steeper slopes facing the drainage system are in forest cover. The beds are too thin for differences in vegetation preference to be easily observed. Highways tend to follow major valleys and ridgelines, to be curvilinear, and are controlled by the drainage system.

Arid

Topography	*Drainage*	*Tone*	*Vegetation and land use*
Minor terracing	Dendritic: Fine	Faint, narrow bands	Barren
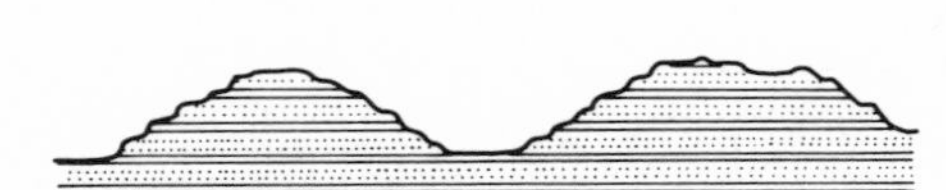 The well-dissected topography shows very narrow terracing, following hillside contours. Sandstones and limestones have vertical cliffs, while shale forms more gradual slopes.	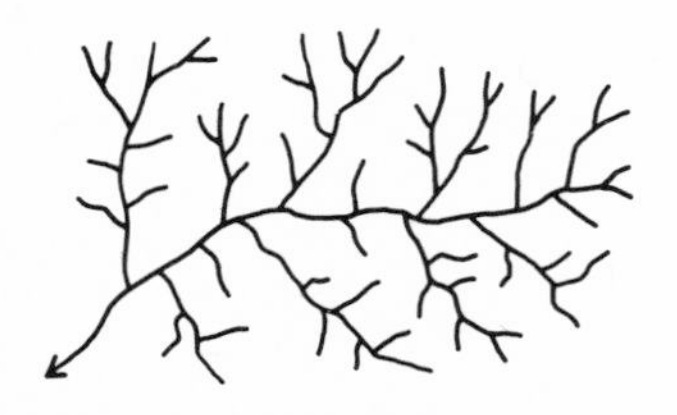 Interbedded sedimentaries in arid climates form dendritic patterns with fine textures. If shales are present, they provide impervious layers which increase surface runoff and dissection.	Since little residual soil is present, the different banded tones of the interbedded materials can be distinguished. *Gullies* Few. The lack of significant residual soil does not allow gully development.	The land is generally highly dissected, rugged, and has little soil cover. Land uses are not apparent. Scrub and grass growth are predominant, and the area may serve as grazing land or range land. Highways and other manmade structures are rarely present.

Soil Depth to Bedrock

Soil depth to bedrock varies according to the types of rock present but generally averages 2 to 10 feet in humid climates. Arid climates have little residual soil cover, for it is easily eroded by wind and water processes.

Issues of Site Development

Sewage Disposal

On-site sewage disposal systems, such as septic tank leaching fields, can be used in humid climates, but there are some constraints depending upon the specific rock types and residual soils. For example, percolation rates in shale are slow, and soil depths over limestone and sandstone may not be sufficient to protect the water resources from contamination. However, thin interbedded landforms develop predominantly

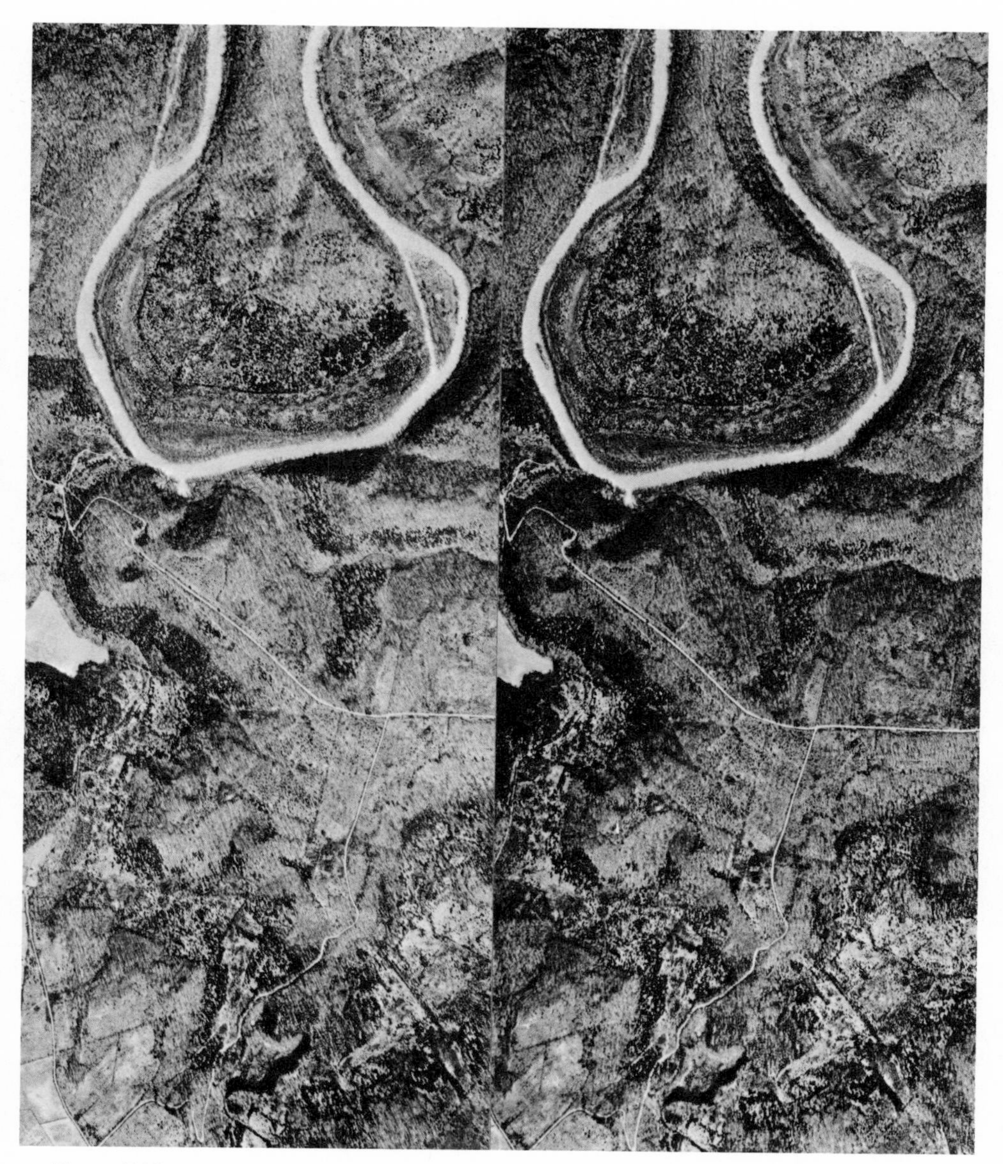

Figure 5.33. (A) Thick interbedded limestone-sandstone in Edmonson County, Kentucky, shows elements characteristic of each; the few sinkholes and angular drainage indicate limestone, while the coarse dendritic drainage and massive hills indicate sandstone. Photographs by the U.S. Geological Survey, Kentucky 3-ABC, 1:23,000 (1967′/″), March 19, 1953.

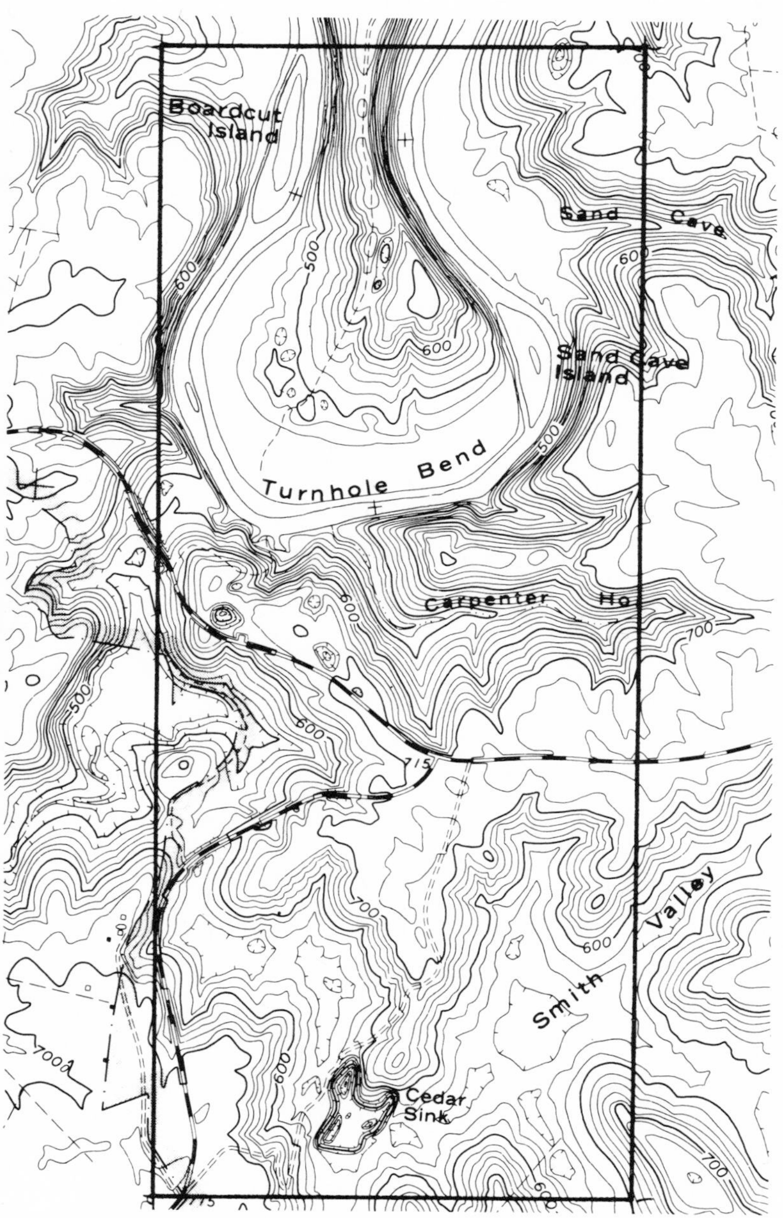

Figure 5.33. (B) U.S. Geological Survey Quadrangle: Rhoda (7½′).

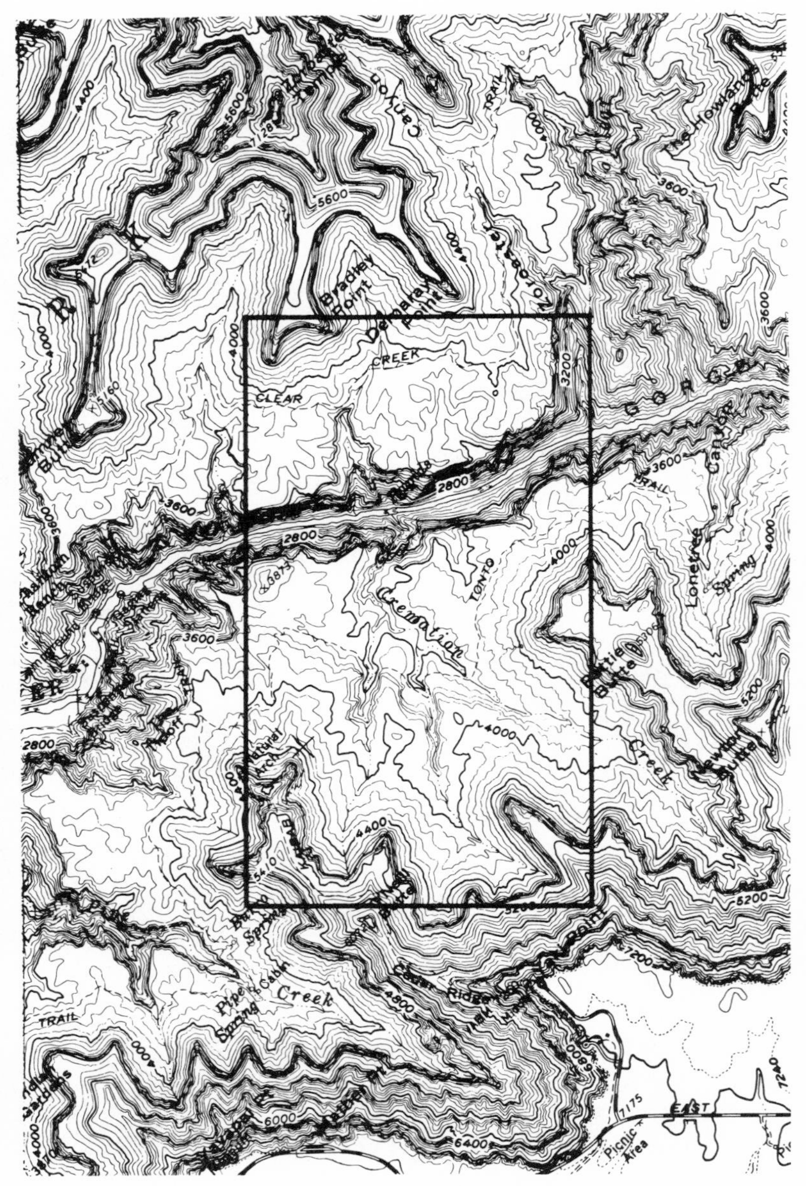

Figure 5.34. (A) Thick interbedded sedimentary rocks in an arid climate in Coconino County, Arizona, along the Grand Canyon. The bedded sedimentary rocks are underlaid by schist and granite. Photographs by the U.S. Geological Survey, Arizona 5-ABC, 1:36,000 (3000′/″), August 19, 1960.

Figure 5.34. (B) U.S. Geological Survey Quadrangle: Bright Angel (15′).

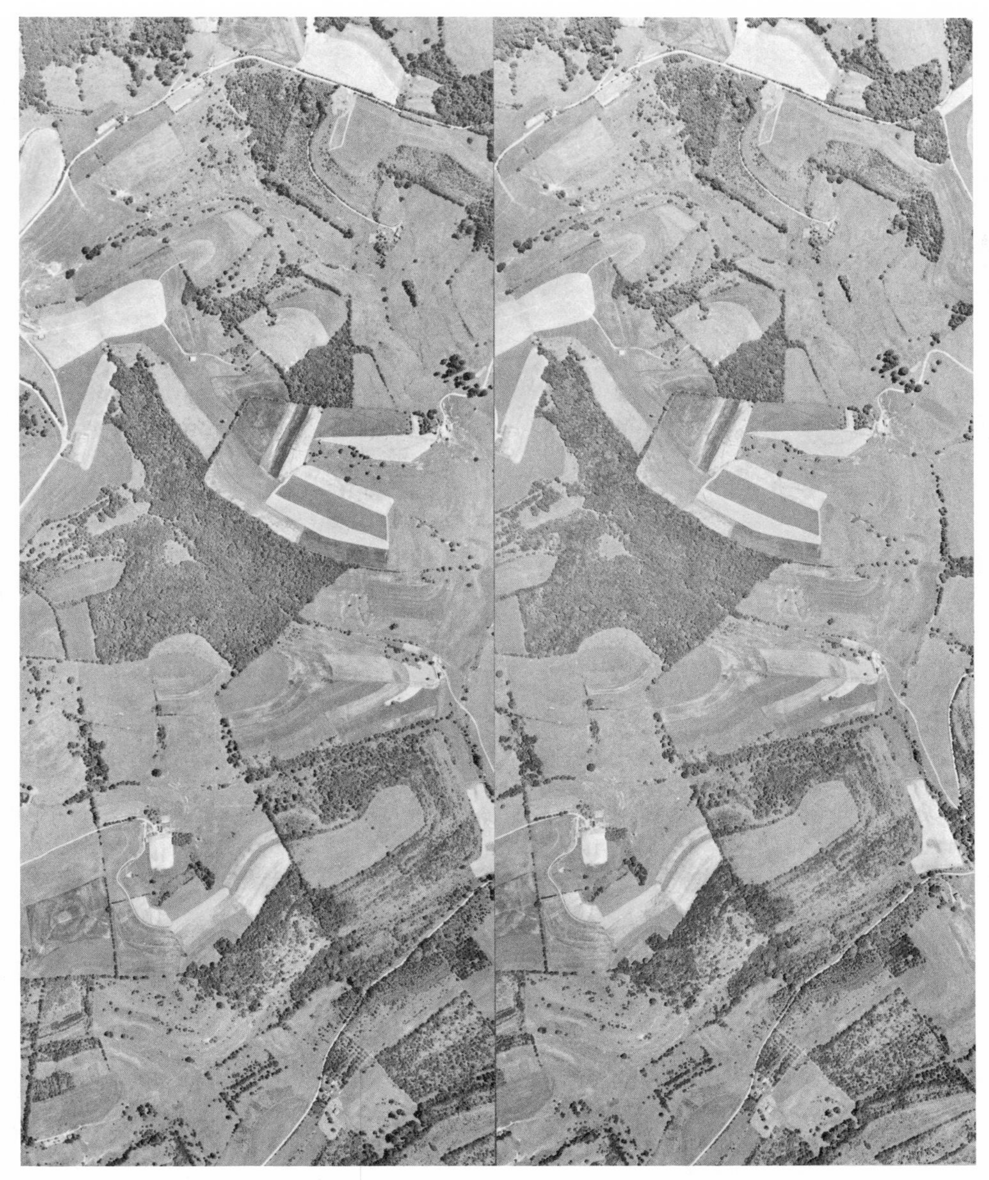

Figure 5.35. Thin interbedded sedimentary rock in a humid climate in Washington County, Pennsylvania, is characterized by even sideslopes, the tendency of hilltops to occur at equal elevations, and indications of thin, faint bands along hillsides. Photographs by the Agricultural Stabilization Conservation Service, APT-IV-58,59, 1:20,000 (1667′/″), June 14, 1958.

silty soils which may provide sufficient percolation and depth to bedrock for septic tank leaching fields. In planning intensive development the capabilities for on-site sewage disposal should be investigated, but it is most likely that a treatment system will have to be constructed.

Solid Waste

Sanitary landfill operations may be feasible in interbedded sedimentary deposits, depending upon the specific rock composition and bedding sequences. Thinly bedded rocks in humid climates develop sufficient soil depths and suitable soil textures for a sanitary landfill operation, but the depth to the seasonal high water table must be greater than 10 feet. Care should be taken not to contaminate the groundwater resource in limestone and sandstone regions.

Trenching

Trenching operations involve costs and difficulties related to the specific rock types encountered and their bedding sequence. Thickly bedded deposits of sandstone and limestone require the use of heavy equipment and blasting for rock removal, and per-cubic-yard costs of rock removal are four to five times those for trenching the same amount of deep, dry, moderately cohesive soil. Shale layers are relatively soft and can be excavated by power equipment without blasting. Typical costs for shale removal are 1.2 to 1.8 times the costs associated with removing the same amount of deep, dry, moderately cohesive soil. Thinly bedded rock structures can generally be removed without blasting if strong sandstones are not present. Costs of trenching in thin beds are commonly two to three times those for removing deep, dry, moderately cohesive soil. Seepage zones are encountered in deep trenching operations in all types of interbedding, resulting in additional costs.

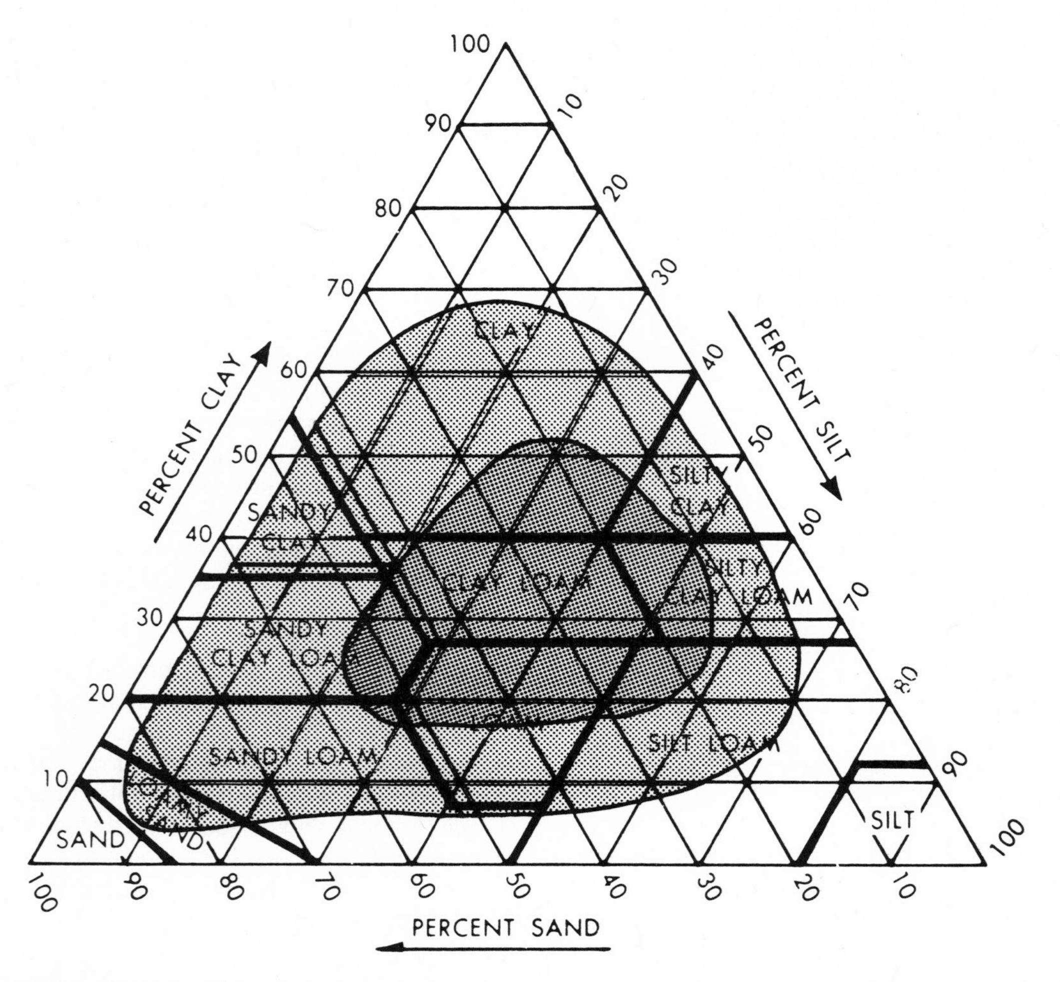

Figure 5.36. The U.S. Department of Agriculture soil texture triangle indicates a wide range of soil texture classification depending upon the specific rock type.

Excavation and Grading

Excavation and grading of large sites in humid and arid climates involve moving vast quantities of material, since the topography is usually rugged. In humid climates, up to 10 ft of soil may cover the bedrock, allowing some flexibility in shallow grading operations. Massive sandstones and limestone require blasting, resulting in costs four to five times those of associated with excavating the same amount of deep, dry, moderately cohesive soil. Thinly interbedded rock structures may not need blasting if they are shaly in nature, but costs are two to three times excavation costs for deep, dry, moderately cohesive soil. In sandstone or limestone beds, seepage zones may be encountered during excavation operations; these zones increase the potential for landslides, and stability studies should be undertaken by competent engineers or geologists.

Construction Materials (General Suitabilities)

Topsoil: Fair. The residual soils of interbedded sedimentary rocks are predominantly silty in texture but do not contain sufficient organic matter.

Sand: Poor. Some sands may be present in sandstone residual soils but they are mixed with fines.

Gravel: Not Suitable

Aggregate: Good to Poor. Suitability as a potential source of aggregate material is dependent upon the specific types of rock present. Limestones and sandstones offer the best potential sources.

Surfacing: Good to Poor. Road surfacing materials derived from limestone or sandstone are good, but shale materials are weak and wear rapidly.

Borrow: Fair to Poor. Residual soils from interbedded sedimentary rocks are naturally fine-textured and are hard to work when wet. Compaction of moderately cohesive soils for fills is accomplished through the use of pneumatic-tired rollers; more cohesive soils require strict control of moisture at the optimum moisture content and the use of sheepsfoot rollers. Specific materials under question should be tested and evaluated.

Building Stone: Excellent to Not Suitable. Limestones and sandstones are commonly utilized as building stone because of their resistive nature. Shale is very weak and weathers quickly and is therefore not suitable as a building material.

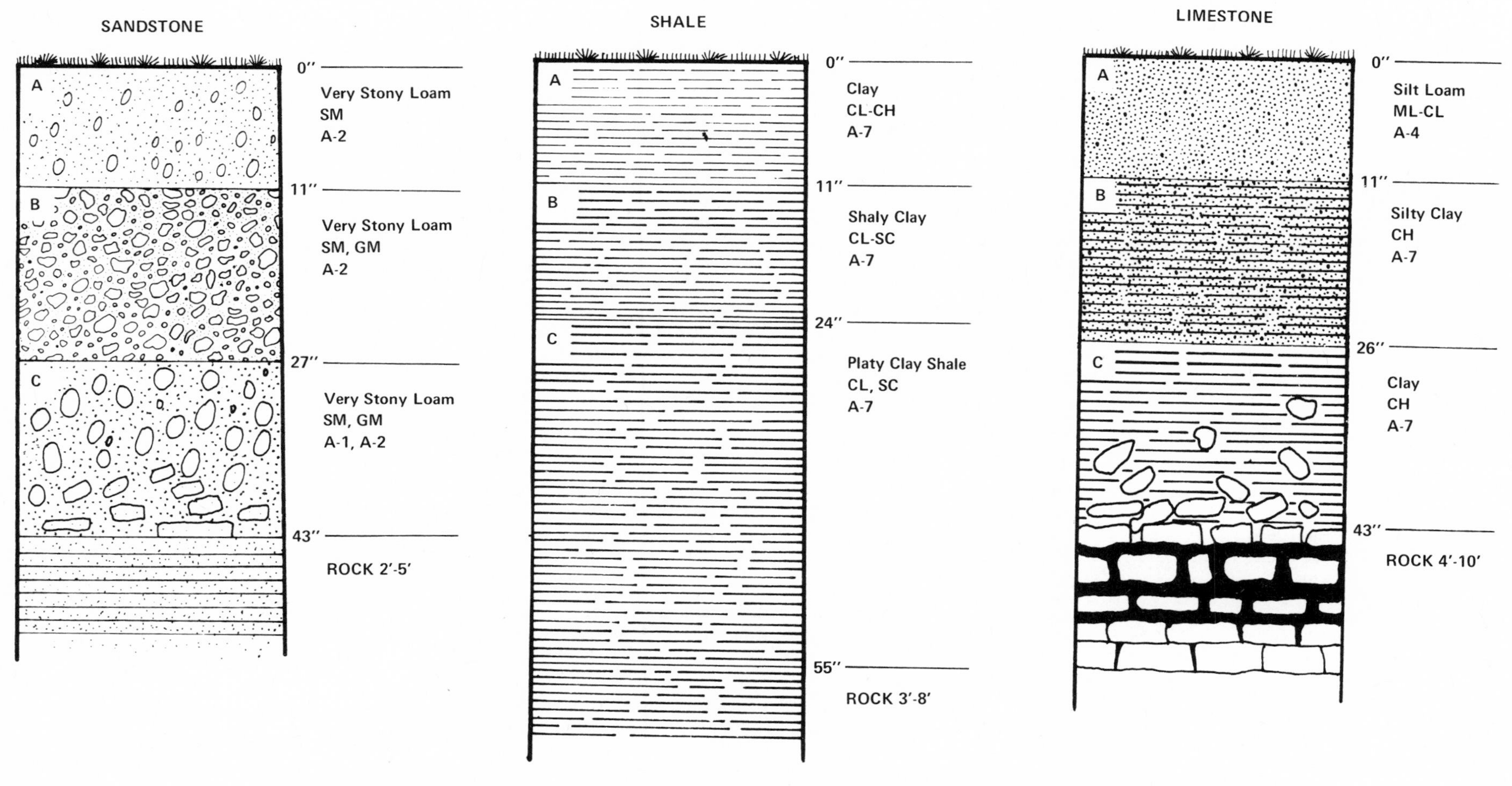

Figure 5.37. Soil profiles of residual soils from sandstone-limestone-shale as developed in a humid climate.

Mass Wasting and Landslide Susceptibility

Interbedded sedimentary rocks are susceptible to landslides if the interbedded sequences contain both pervious and impervious materials. Thus clay shales bedded with limestones or sandstones are unstable. When landslides do occur, they are small-scale and may be difficult to interpret from aerial photographs. Flat beds of rock are of course more stable than those that are tilted. Streams or excavations undercutting weak rock strata create very unstable conditions which are likely to slide. Soil creep is common.

Groundwater Supply

Interbedded sedimentary rocks have excellent potential for high-yield groundwater resources. The sandstone and limestone layers are porous and permeable and store and carry large amounts of underground water. There is generally no difficulty in locating sufficient water supply for intense developments and/or urban areas. Geologists and other consultants should make hydrological studies to determine the potential yield and quality of the groundwater. The presence of limestone,

Table 5.11. Flat-lying interbedded sedimentary rocks: aerial photographic and cartographic references

State	County	Photo Agency	Photograph no.	Scale	Date	Corresponding quadrangle and geologic maps from USGS	Corresponding soil reports from SCS
Arizona	Navajo	USGS*	Arizona 1-AB	1667′/″	9/15/51	Agathla Peak (15′)	–
Arizona	Coconino	USGS*	Arizona 5-AB	3000′/″	8/19/60	Bright Angel (15′)	–
Colorado	Montezuma	USGS*	Colorado 7-ABCDE	5000′/″	11/9/54	Mesa Verde National Park (1:31,250)	–
Kansas	Chase	ASCS	AXZ-2LL-127-128	1667′/″	7/5/70	None	–
Kansas	Chautauqua	ASCS	CWV-2GG-228-229	1667′/″	10/20/66	Cedar Vale East (7½′)	–
Kansas	Pottawatomie	USGS*	Kansas 2-AB	1667′/″	9/18/37	Manhattan (1:250,000)	–
Kentucky	Edmonson	USGS*	Kentucky 3-ABC	2000′/″	2/19/53	Rhoda (7½′)	–
Pennsylvania	Washington	ASCS	APT-9HH-109-111	1667′/″	10/4/67	Claysville (7½′)	1910†
Pennsylvania	Washington	ASCS	APT-1V-57-58	1667′/″	6/14/58	Canton (1:250,000)	1910†
Tennessee	Marion	ASCS	ALB-2HH-122-126	1667′/″	12/3/66	Whitwell (7½′)	1958
Texas	Blanco	ASCS	DME-1GG-174-175	1667′/″	10/20/65	Llano (1:250,000)	–
Texas	Bexar	ASCS	BQQ-1HH-230-233	1667′/″	10/20/66	Bulverde (7½′)	1966
Texas	Bexar	ASCS	BQQ-1HH-247-248	1667′/″	10/20/66	Bulverde (7½′)	1966
Texas	Coleman	ASCS	AWY-1HH-122-125	1667′/″	12/1/66	Brownwood (1:250,000	1922
Texas	Gillespie, Blanco	USGS*	Texas 6-ABCD	1583′/″	3/8/58	Cave Creek School (7½′), GQ-219, Rocky Creek, Stonewall, Liye, (7½′)	–
Utah	Wayne	USGS*	Utah 9-AB	2640′/″	7/12/39	Factory Butte (15′)	–
Utah	San Juan	USGS*	Utah 13-ABC	1667′/″	9/21/51	Upheaval Dome (15′)	1962
Utah	San Juan, Garfield	USGS*	Utah 15-ABC	2640′/″	7/23/38	Hite (15′), Prof. Paper No. 228	1962
Wisconsin	Iowa	ASCS	WT-1JJ-7-8	1667′/″	5/12/68	Spring Green (15′)	1962
Wisconsin	Iowa	ASCS	WT-4CC-137-138	1667′/″	8/28/62	Spring Green (15′)	1962

*From *Selected Aerial Photographs of Geologic Features in the United States*, U.S. Geological Survey Prof. Paper No. 590.
†Out of print.

which contains many dissolved minerals, makes the water hard. Seepage zones are common along hillsides where strata of water-bearing rock are exposed.

Pond or Lake Construction

Investigations are needed to determine capabilities and suitabilities for constructing ponds and lakes at specific sites. Potential seepage and leakage exist in sandstone and limestone areas. The topography in humid climates is favorable to the formation of impoundments, and the residual soils, upon compaction, are sufficiently impervious for bottom materials. In arid climates sufficient residual soils may not exist to line the bottom of the proposed water body, making it necessary to utilize bentonite or other commercially available liners. Shale is excellent as a bottom material in both climates, owing to its impervious nature. The fine residual soils associated with interbedded forms require good management techniques in the watershed; protection of the drainage system and first-order gullies slows siltation.

Foundations

Recommendations for foundations in interbedded sedimentary materials depend upon the specific compositions and sequences of bedding of each site. The residual soils are generally not suited to raft foundations because of their low load-carrying ability, but gravel pads may be used. Small residential structures with basement slabs commonly utilize continuous footings, since the materials deeper in the soil profile have higher load-bearing capacities. Where the soil depth is thin, as in sandstone deposits, foundations can transfer loads to the rock structure, using piers or footings. Large, massive loads can usually be suitably transferred, since soil depths are relatively thin and the bearing capacity values of these rocks are usually high. Thickly bedded limestone may contain solution cavities requiring filling and/or capping.

In arid climates foundations are no problem, since the rock structure is very close to or exposed at the ground surface. Basements may be expensive to construct in arid climates, unless they are located in shale deposits, for they involve excavation of massive bedrock materials.

Highway Construction

Highway construction through regions of interbedded sedimentary rock is expensive, owing to the amount of grading and excavation usually necessary. Blasting and heavy equipment are required in massive limestone and sandstone deposits, but shale and some thinly bedded deposits can be more easily excavated. Where the topography is extremely rugged, as in arid climates, tunneling may be desirable and many bridges, fills, and culverts may be needed. Major cuts in thinly bedded deposits require terracing to minimize the danger of slides and to hold down maintenance costs, and seepage is to be expected. Construction materials are not difficult to locate because of the variety of rock types present.

Interbedded Sedimentary Rocks: Tilted

Introduction

Most sedimentary rocks are originally bedded parallel to the horizon, but tectonic activity warps, twists, and folds the bedded structures into complex formations which are no longer parallel to but dip away from the horizontal plane. All tilted sedimentary rocks can be described by the direction of their dip and its angle. In such a description the strike indicates the direction of the bedding plane that intersects the horizon and is also perpendicular to the dip. For example, a structure of interbedded sedimentary rocks can dip to the southwest at 60 degrees with a strike running north-south (Figure 5.38).

Many complex structural forms of sedimentary rocks result from folding, including monoclines, anticlines, and synclines. (Note that these are not forms of surface topography but are geological structures.) *Monoclines* are bedded formations that contain a steplike bend; that is, the beds follow a gradual slope, become steeper, and then become more gradual. *Anticlines* are folds of rock strata in which the beds dip away from one another and the oldest rocks are in the center of the fold (Figure 5.39). *Synclines* are the opposite of anticlines; the beds dip toward each other, and the youngest rocks are in the center of the fold. Plunging anticlines and synclines (Figure 5.40) are those whose axes intersect the horizontal plane at an angle.

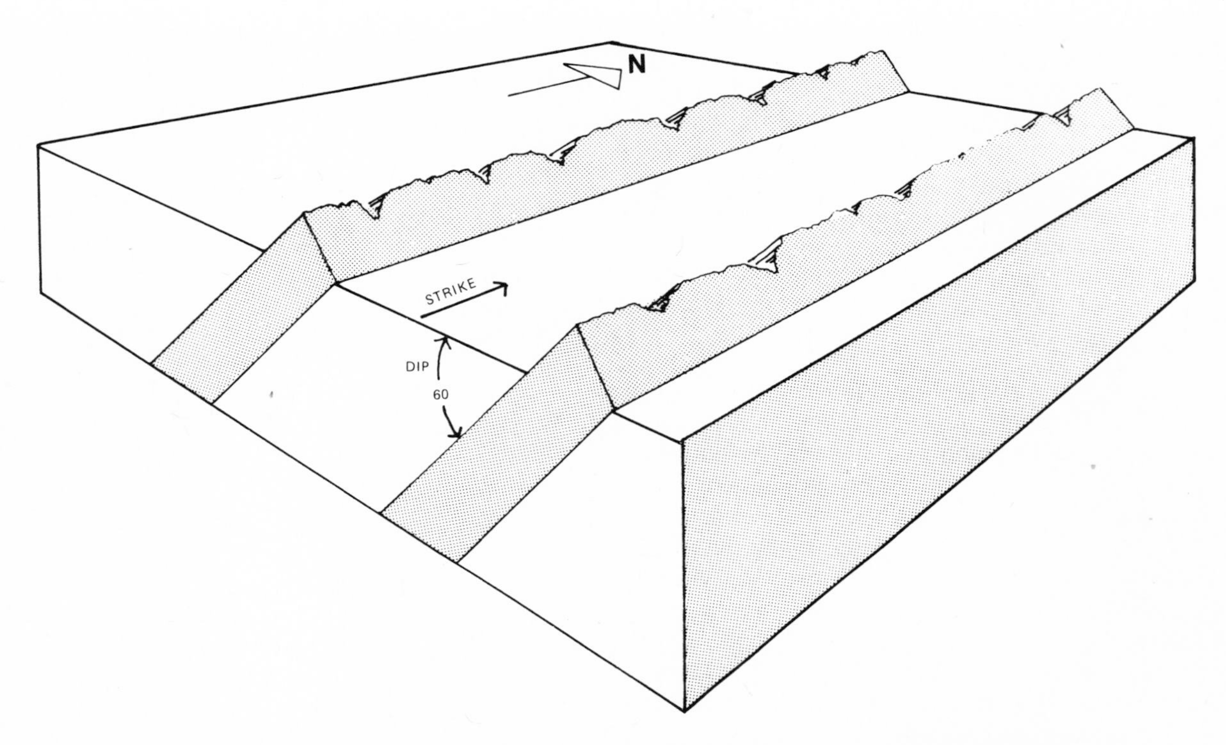

Figure 5.38. A strike and dip of sedimentary strata shown here dipping toward the west at 60 degrees with the strike running north-south.

Interpretation of Pattern Elements

Tilted, interbedded sedimentary rocks have significant pattern and engineering characteristics which are not similar to flat interbedded sedimentaries. Since different rock types have different resistances to weathering, the more resistant rocks dominate in the topography as uplands while the softer, less resistant rocks form lowlands. Among the distinctive types are sandstone, which develops sharp parallel ridges with a steep slope along one side and a more gradual slope on the other; shales, which are represented by smooth, rounded hills in the lowlands; and limestone, which develops karst topography in humid climates and also occupies lowlands. Tilted sedimentary rocks usually occur in any region of sedimentary deposits and present some major engineering and planning difficulties. The photographic interpretation key discusses the characteristics of tilted, interbedded sedimentary deposits in humid and arid climates.

Soil Characteristics

In humid climates residual soils from interbedded sedimentary rocks are typically shallow and silty and contain many rock fragments within the subsoil. Ridges of sandstone develop thin profiles, and limestone-shale valleys contain deeper, fine-textured, more plastic soils. Thick deposits develop profiles having the characteristics associated with each individual rock type.

In arid climates soil profiles either are very thin or may not exist. There is generally an accumula-

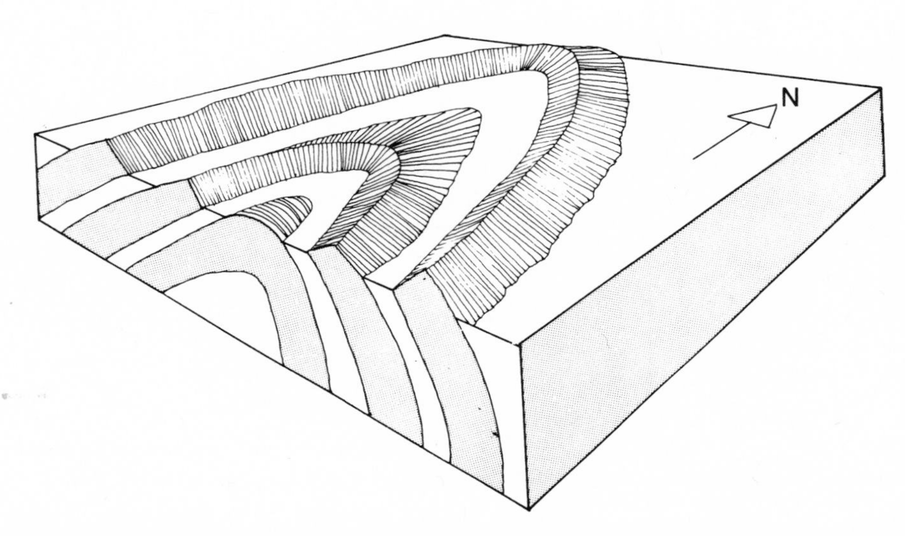

Figure 5.39. Plunging anticline. The folded, sedimentary structure has dips pointing away from one another and plunges toward the north. The oldest beds are near the center.

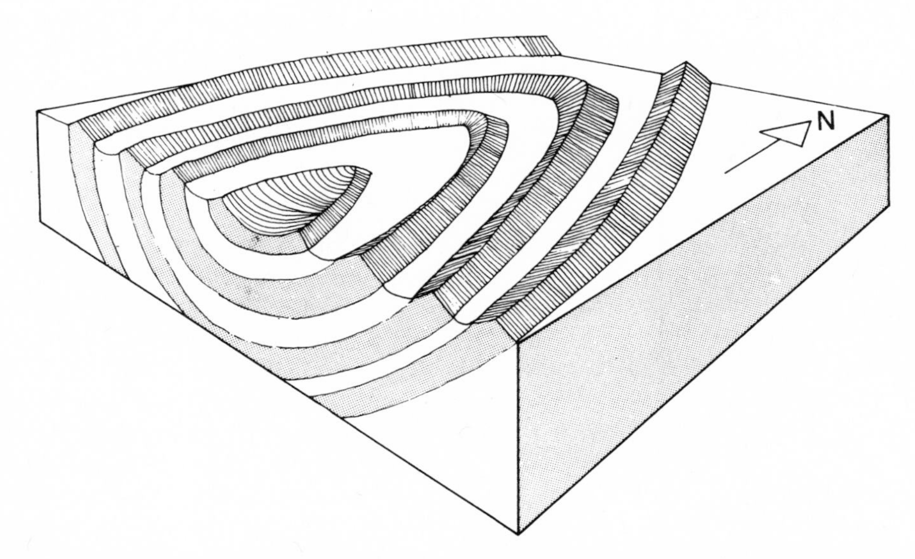

Figure 5.40. Plunging syncline. The folded, sedimentary structure has beds dipping toward one another. The youngest beds are in the center of the fold. The direction of the strike varies, and the syncline is plunging toward the south.

tion of talus debris along lower slopes and alluvial deposits in the valleys.

U.S. Department of Agriculture Classification

The soil characteristics described in the individual sedimentary rock sections should be consulted for specific rock types. Thin interbedded sedimentary rocks typically develop silty residual soils. Sandstones develop sandy loam, gravelly sandy loam, and loamy sand, which are nonplastic. Shales form silty clay, silt loam, silty clay loam, shaley clay, and platy clay, all of which have moderate to high plasticity. Limestone residual soils are commonly clay, silty clay, and silty clay loam with moderate to high levels of plasticity.

Unified and AASHO Classifications

Thinly interbedded rocks develop silty residual soils commonly classified as ML, MH, or ML-CL. Sandstones are categorized as SM, SM-SC, or GM in the Unified system and as A-4, A-2-4, and A-2 in the AASHO system. Limestone contains more fines, being described as CL, CH, MH, ML CH-MH, or occasionally ML-CL in the Unified system, and as A-6 and A-7 in the AASHO system. Shale is typically ML, CL, MH, CH, and CL-CH, classified by AASHO as A-6, A-4, and A-7.

Water Table

The depth to water table in residual soils is variable, being dependent upon the rock types and sequence of bedding. Sandstone and limestone residual soils are well drained; shale is not. Limestone residual soils become more impervious if disturbed by construction or grading. In humid climates interbedded residual soils derived from thin beds have water tables typically 4 to 10 feet beneath the surface. The water table is typically

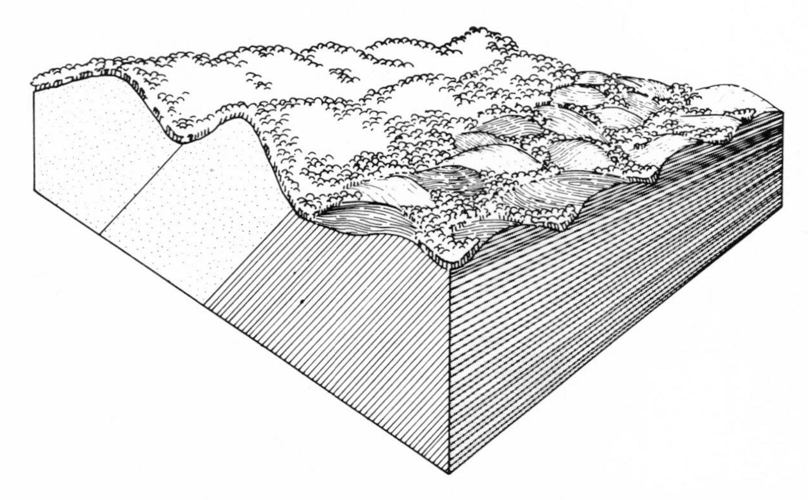

Figure 5.41. Tilted sandstone-shale in a humid climate. Tilted, interbedded sedimentary sandstone and shale in humid climates form a linear ridge-and-valley topography. The river and drainage system is typically located within the shale valleys along with most highways and settlement patterns. The landscape is visually absorptive, with regional views potentially existing along the elevated ridges of sandstone. The most sensitive visual area is the rim where the valley slopes meet the crest of the ridges; cutting of vegetation or development on the rims is readily observed from the adjacent valleys.

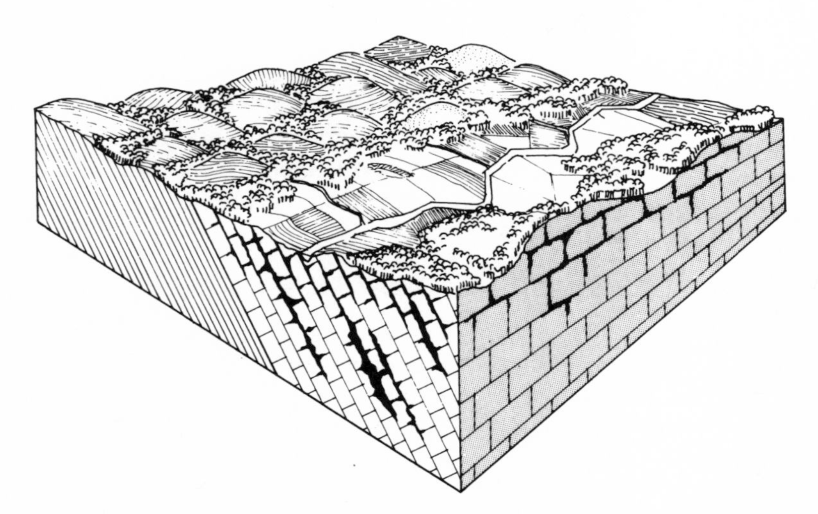

Figure 5.42. Tilted limestone-shale in a humid climate. The landscape of interbedded limestone and shale in a humid climate is not as visually absorptive as that of sandstone-shale. Few regional views can be obtained from the low shale hills. The undulating topography breaks up large spaces but is not rugged enough to give a real feeling of spatial enclosure.

closer to the surface in the valleys than in the uplands.

Drainage

The types of rock present and their interbedded sequence determine the internal drainage character of the rock structure and its impact upon the moisture within the residual soil. Sandstone and limestone soils are well drained, but shaly soils are fairly impervious and slow to drain. If limestone soils are disturbed, they will become impervious and poorly drained.

Soil Depth to Bedrock

Soil depths to bedrock are dependent upon the rock types but typically average 2 to 10 feet in humid climates. In comparison, limestones vary greatly in depth to bedrock but average 10 to 15 feet, and shales are typically 6 to 8 feet over weathered rock in humid climates. In arid climates all types of rock develop very thin residual soils which are quickly eroded by wind and water processes, exposing the parent rock material.

Issues of Site Development

Sewage Disposal

The use of on-site sewage disposal systems, such as septic tank leaching fields, is dependent upon the specific rock type and the residual soils. In humid climates the depth to bedrock, percolation, depth to water table, and slopes all provide constraints. Valley areas with greater soil depths may be found suitable, but accurate tests and analyses are needed to determine specific capabilities. Arid climates have very thin residual soils in the uplands, but deeper alluvial deposits are found in valleys. In both climates care should be taken that sewage disposal systems do not

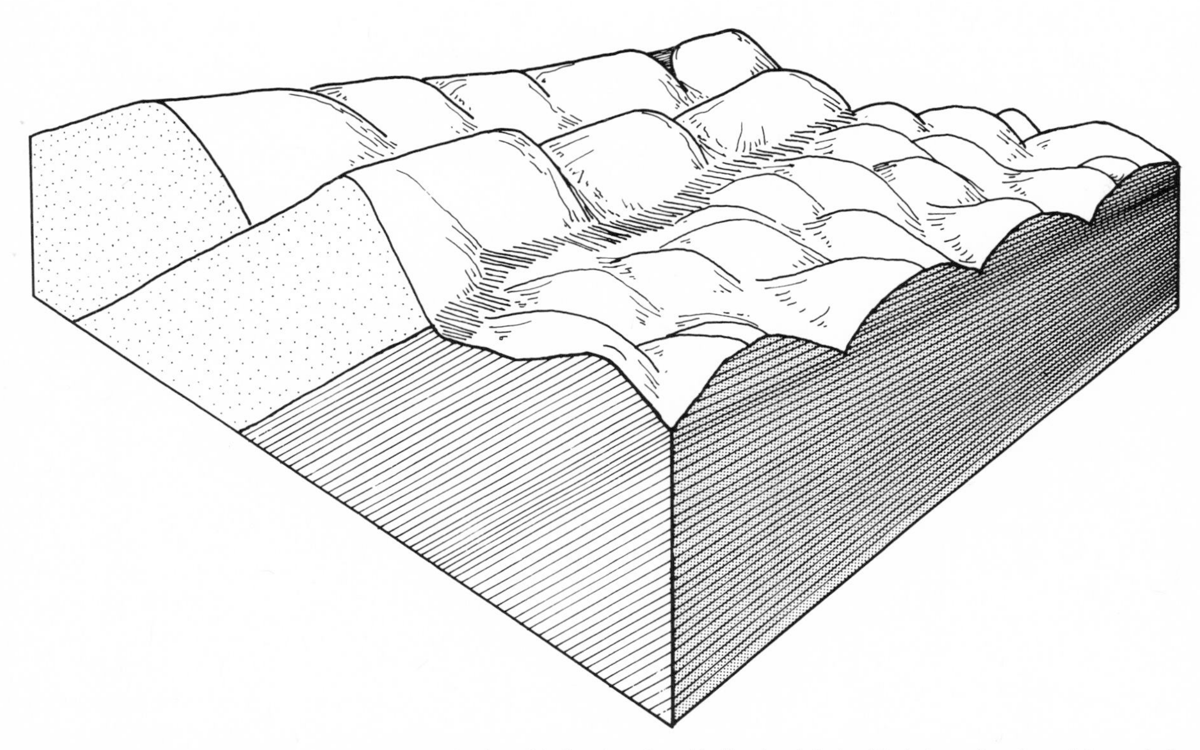

Figure 5.43. Tilted, interbedded sedimentaries in arid climates. In arid climates interbedded deposits of sandstone-shale or limestone-shale have the same visual expression. Both sandstone and limestone form resistant, sharp-edged ridges; the shale forms lowlands of undulating topography or soft, rounded hills. Little vegetation exists, and the land surface is covered with scattered grasses and shrubs.

contaminate the groundwater resources associated with sandstones and limestones. In areas of intense development and in urban areas sewage treatment systems normally must be constructed in order to insure the quality of the surface-subsurface water resources.

Solid Waste

Sanitary landfill operations may be feasible, depending upon the rock types and bedding sequences. Sites near the bases of slopes and in valleys are suitable if sufficient materials are available for covering and the depth to the water table is greater than 10 feet. Investigations by competent engineers or geologists should be made to determine the possibility of polluting the groundwater resources with leachate.

Trenching

Linear trenching alignments are difficult to route in sandstone-shale regions in humid climates, and in any rock combinations in arid climates. In shale valleys alignments that follow the natural grain of the topography involve fewer of the costs associated with rock removal and grade changes. Rock removal in sandstone is typically four to five times as costly as removing the same amount of deep, dry, moderately cohesive soil. Limestone-shale regions in humid climates have deeper residual soils, entailing costs only 1.2 to 1.8 times higher than for deep, dry, moderately cohesive soil trenching. Seepage zones between beds may be encountered, and deep trenches may need to be shored against collapse.

Excavation and Grading

Large, flat sites in regions of interbedded tilted sedimentaries require extensive excavation and grading. Rugged ridges of sandstone in humid and arid climates require blasting and heavy equipment for excavation, as does limestone in arid climates. Removal of massive rock is four to five times as expensive per cubic yard as removing the same amount of deep, dry, moderately cohesive soil. Shales can be excavated with power shovels without blasting, typically costing 1.2. to 1.5 times deep, dry, moderately cohesive soil excavation. Vertical or steep cuts may encounter seepage zones, and the rock mass has a high risk of sliding along joints and bedding planes where there is a much lower level of shearing resistance. This condition is extremely hazardous if the bedding planes parallel the slope and pervious-impervious materials are interbedded. In this case the accumulation of water increases the weight and shearing stresses in the pervious strata. Many failures have occurred in cuts under these conditions. Vertical cuts should be terraced and drained to minimize maintenance and to lower the potential for rock slides.

Construction Materials (General Suitabilities)

Topsoil: Fair. The residual soils of interbedded sedimentary rocks are predominantly silty in tex-

Table 5.12. Sedimentary rocks: Tilted, interbedded (humid and arid)

Humid

Topography

Parallel ridges

In sandstone-shale combinations sandstone forms resistant, sharp, parallel ridges; shale forms soft, rounded hills in the lowlands. Limestone-shale forms low, rounded hills in shale formations, and karst topography with rounded or oval sinkholes in limestone areas. Thin beds give the ridge topography a saw-toothed appearance.

Drainage

Trellis and dendritic

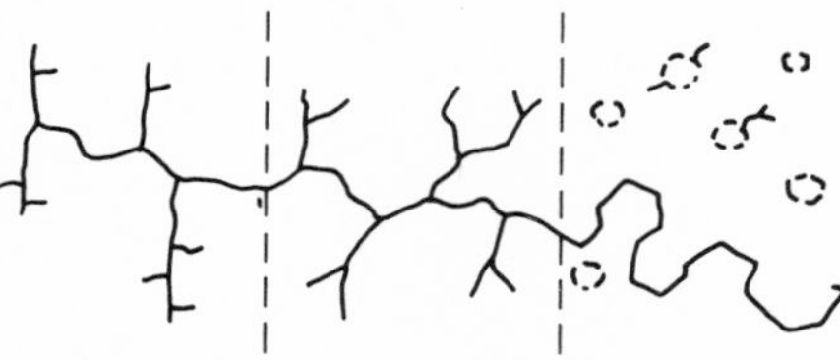

In most situations a trellis drainage pattern of medium to fine texture exists, controlled by the tilted rock structure. Interbedded limestone and shale may have a dendritic pattern and solution topography with internal drainage. Stream courses are generally located in shale lowlands; if they are located in limestone lowlands, some angularity should occur as a result of jointing influence.

Tone

Faint banding

Faint bands of tonal difference relate to the tones of the different rock materials. These tones may be masked by the residual soil.

Gullies

Vary

Gully cross sections relate to the surrounding residual soils. Generally, sag-and-swale gullies are found in shale and limestone. Gullies with steeper sides may indicate sandstone.

Vegetation and land use

Forested and cultivated

Vegetation and land use relates to the types of interbedded rock materials. Sandstones are very rugged, with steep slopes, and have forest cover; shale, with softer hills, has forest cover and/or cultivation; limestone is typically cultivated. Roads and major development patterns are concentrated in the valleys.

Arid

Topography

Saw-toothed ridges

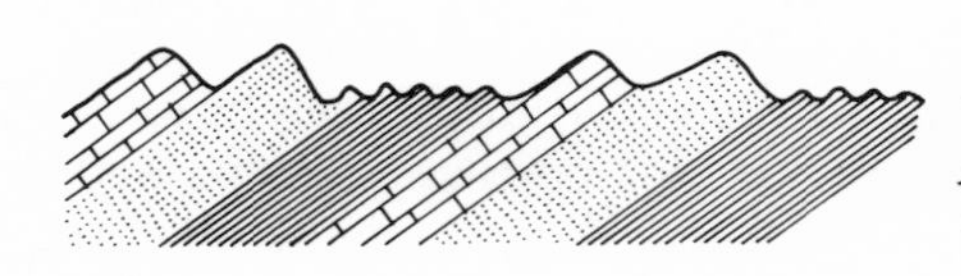

Ridge crests in arid climates are much sharper, and the ridges develop a saw-toothed appearance. The V's formed between the teeth point down the dip of the rock strata. Sandstone and limestone both form resistant ridges; shale forms conical lowland hills. Thin beds give the ridges a finer saw-toothed appearance.

Drainage

Trellis: Fine

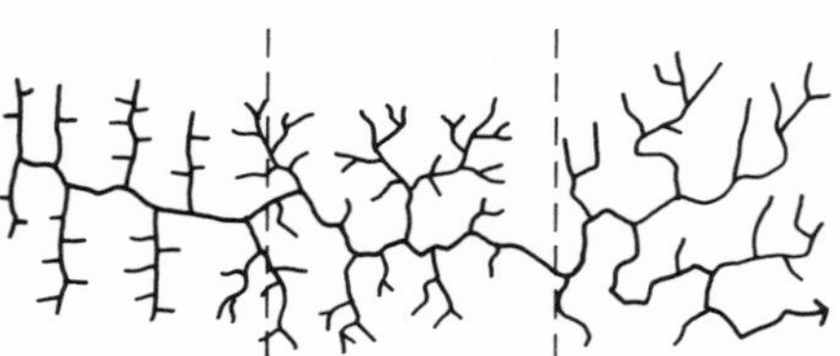

In arid climates tilted sedimentary rocks are identified by a trellis drainage pattern which is normally fine-textured. If there is a regional uniform slope, the trellis pattern may be modified, showing some parallelism.

Tone

Banded

Definite banding can be observed. Exposed limestone is the brightest, followed by sandstone and shale. Vegetation on slopes may confuse the appearance of natural tones.

Gullies

Few

Few gullies are present, since little residual soil exists. White gullies indicate limestone.

Vegetation and land use

Barren

Because of the arid climate, the ruggedness of the topography, and the thin soils, traditional land uses are not seen in these regions. Vegetation consists of scattered grasses and scrub growth, occasionally concentrated along outcrops of water-bearing rock. Few highways cross these regions which are used primarily as rangeland.

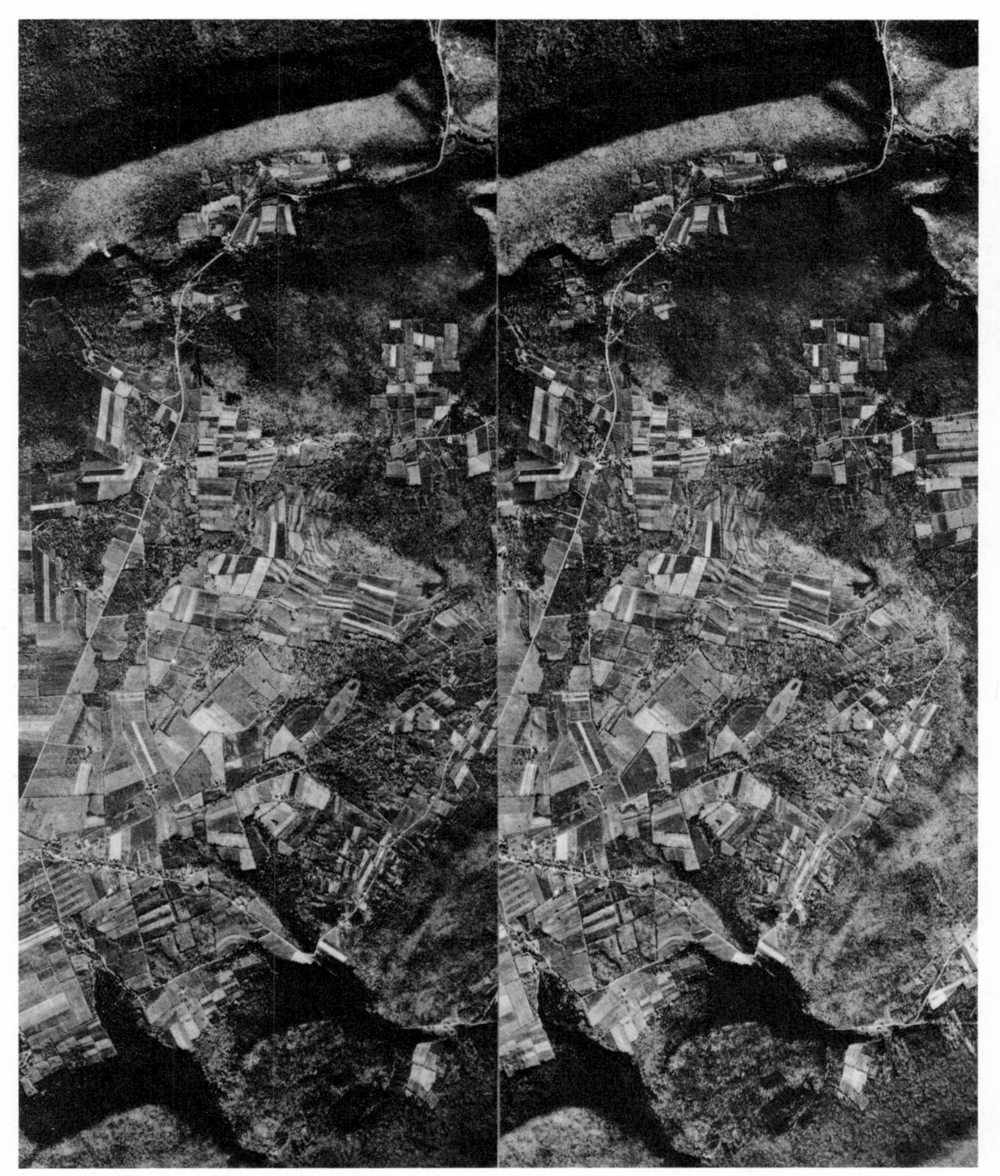

Figure 5.44. (A) Tilted, interbedded sandstone-limestone anticline formation in Lycoming and Clinton Counties, Pennsylvania, where the sharp ridges are the sandstones and the central valley is the limestone. Photographs by the U.S. Geological Survey, Pennsylvania 1-ABCD, 1:60,000 (5000′/″), 9/25/65.

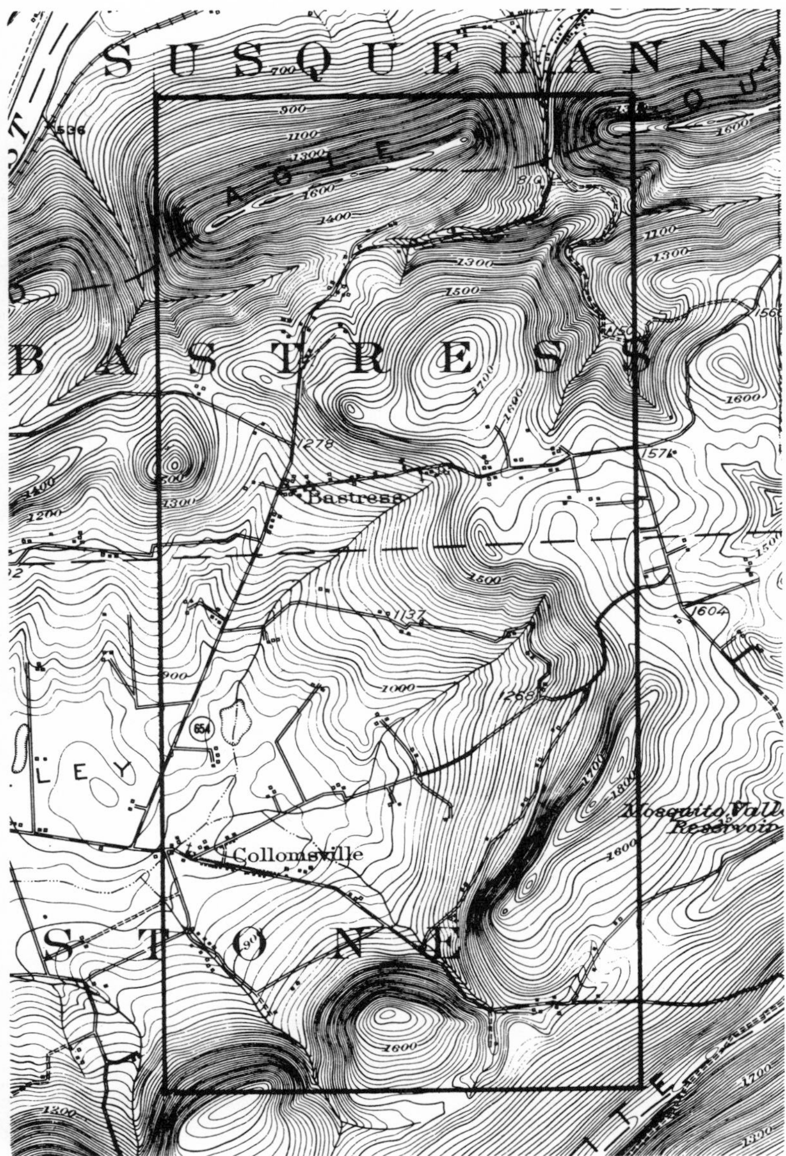

Figure 5.44. (B) U.S. Geological Survey Quadrangle: Williamsport (15′).

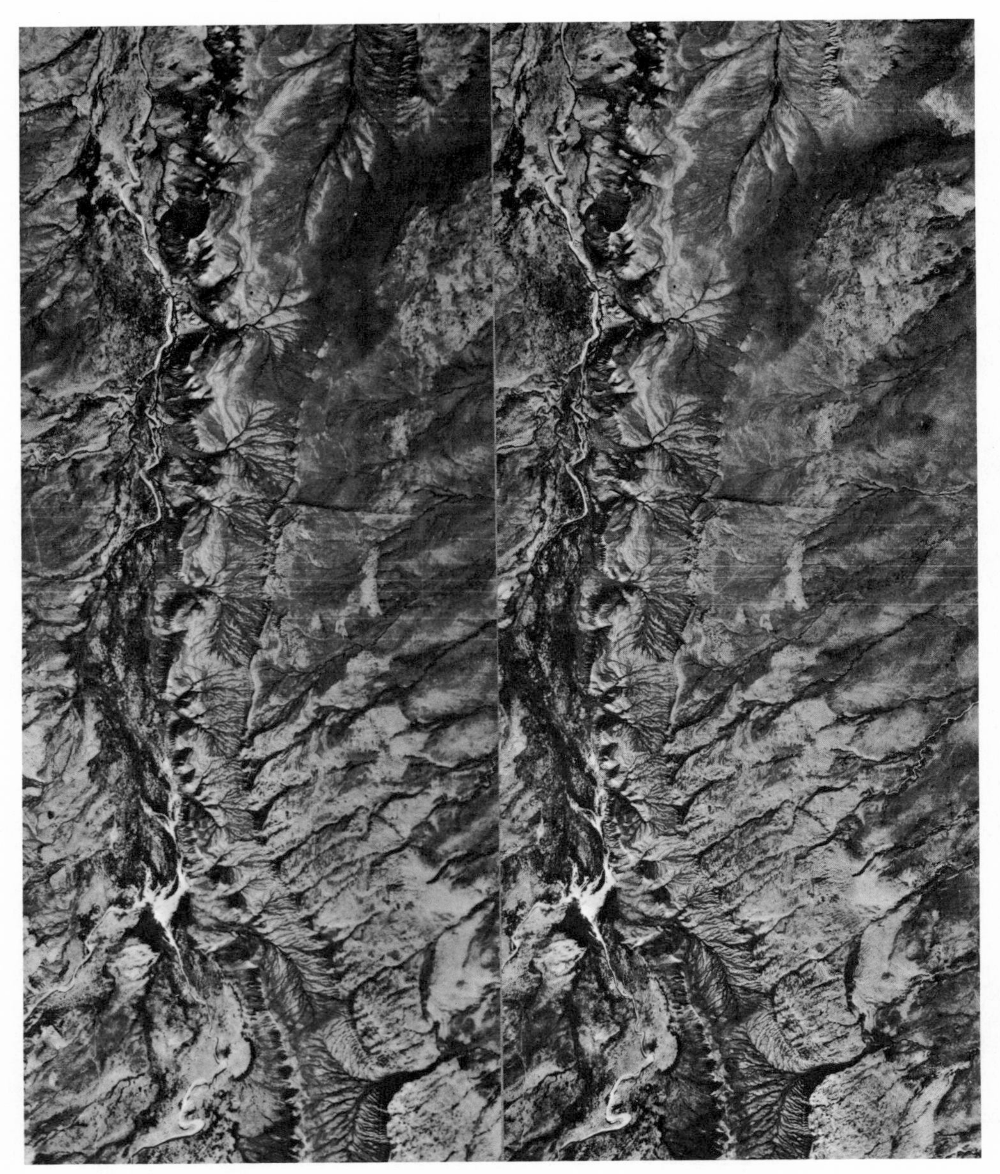

Figure 5.45. (A) Tilted, interbedded sandstone-shale in San Juan County, New Mexico, where the eastward-dipping sandstones form resistant cuestas which have a typical saw-toothed or zigzag outline. Photographs by the U.S. Geological Survey, New Mexico 1-ABCD, 1:54,000 (4500′/″), February 5, 1954.

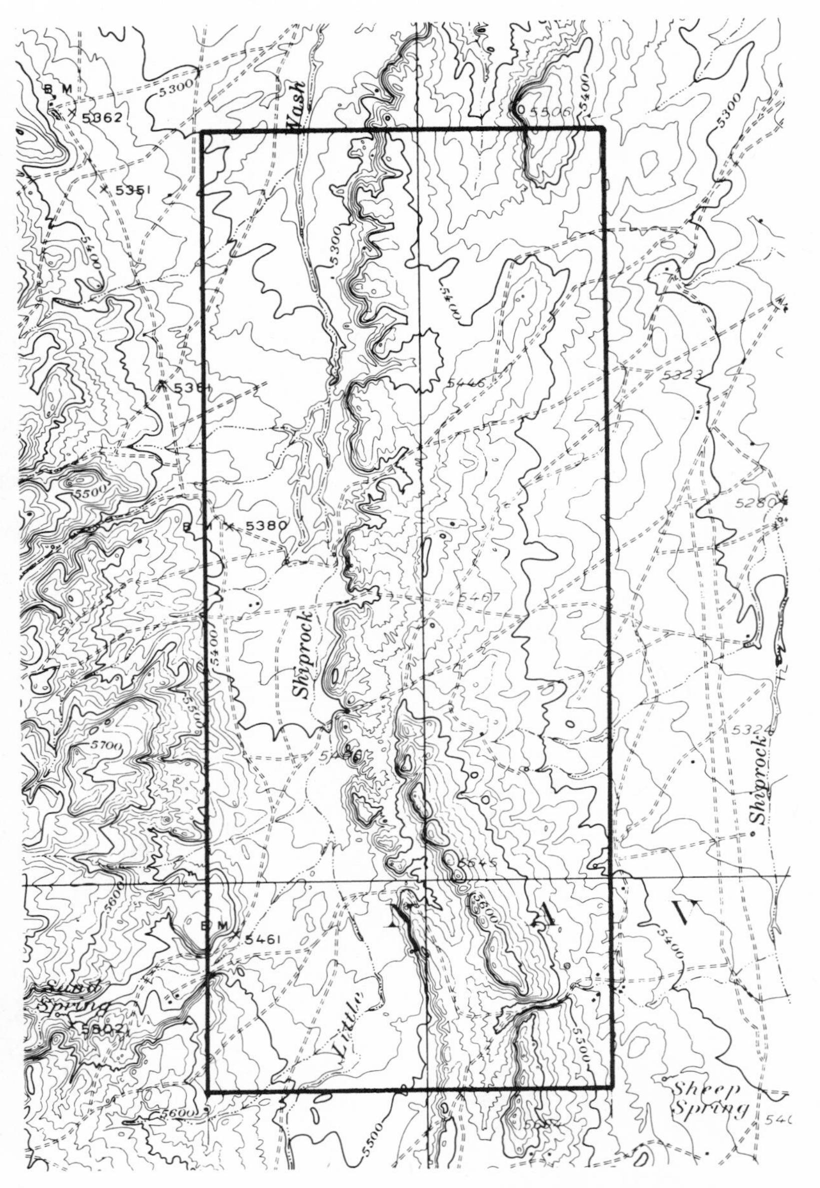

Figure 5.45. (B) U.S. Geological Survey Quadrangle: Ship Rock (15′).

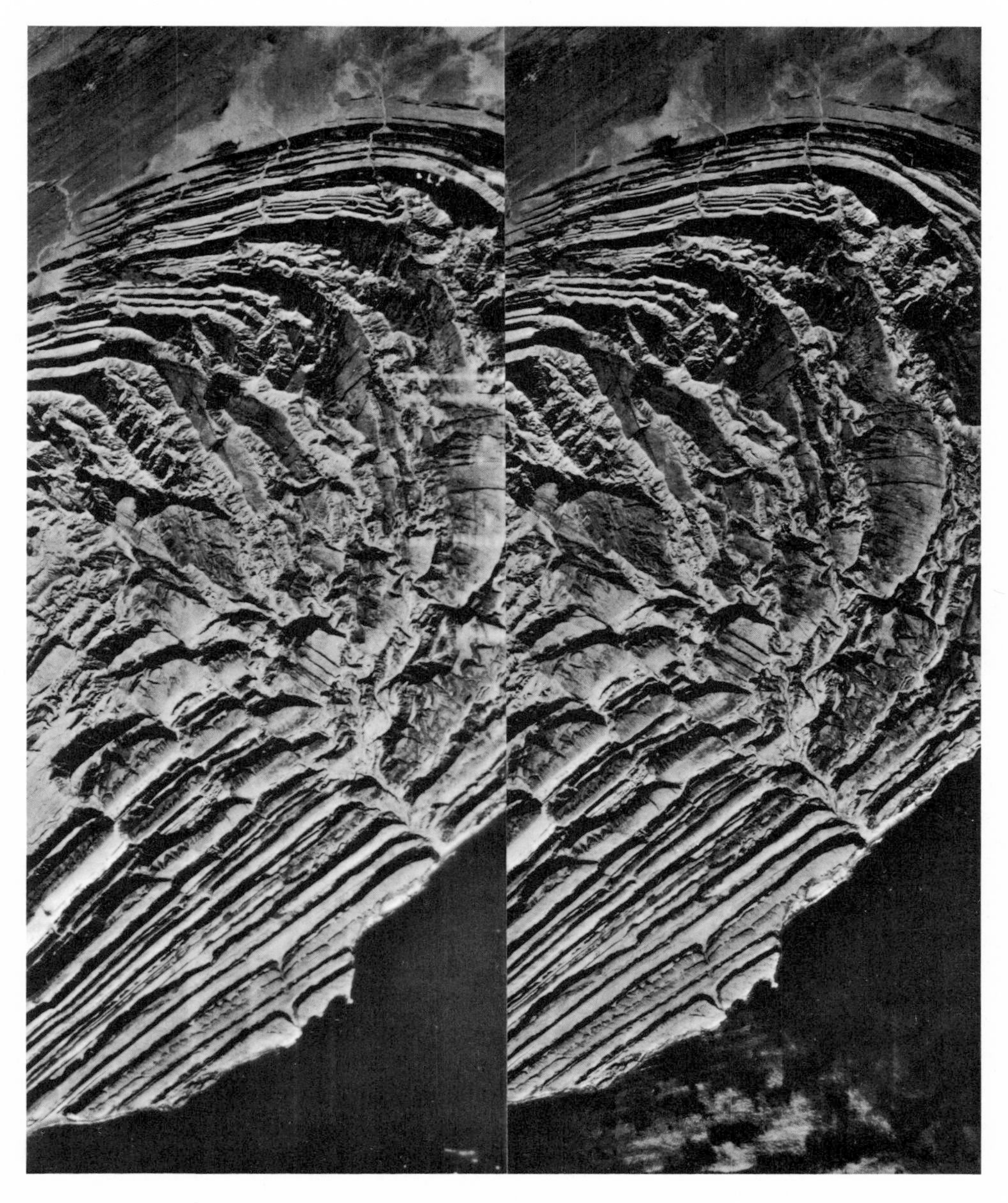

Figure 5.46. Plunging anticline of interbedded sandstone-mudstone in an arid climate in Pakistan. Photographs by the U.S. Geological Survey, Pakistan 2-ABCDE, 1:46,000 (3833′/″), 1953.

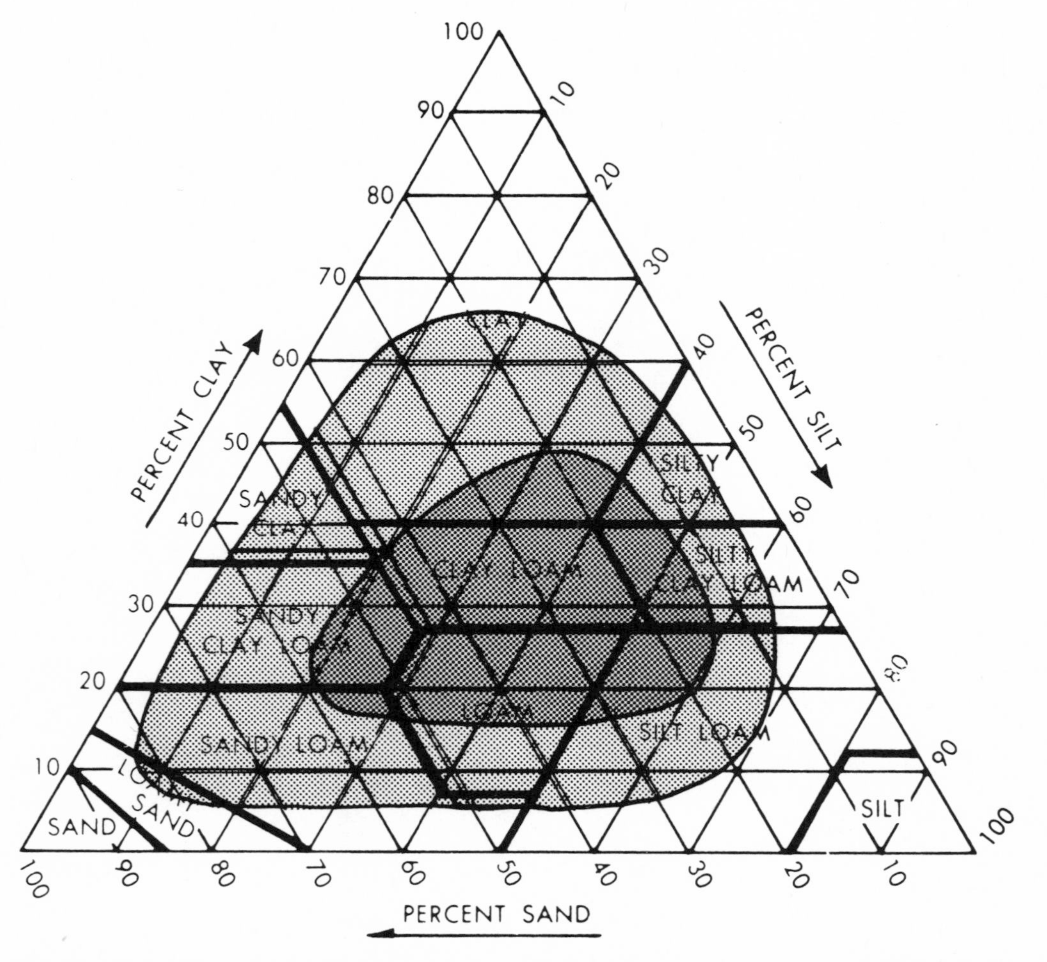

Figure 5.47. The U.S. Department of Agriculture soil texture triangle indicates a wide range of soil texture classifications depending upon the specific rock types.

ture but contain insufficient organic matter to have high value as topsoil.

Sand: Poor. Sands are found within the residual soils of sandstones but are mixed with fines.

Gravel: Not Suitable. Water-associated landforms in valleys may have gravelly materials.

Aggregate: Good to Poor. Suitability of materials for aggregate is dependent upon the specific rock types. Limestones and sandstones have the best potential.

Surfacing: Good to Poor. Road surfacing materials derived from limestone or sandstone are good; shale materials are weak and wear rapidly.

Borrow: Fair to Poor. Residual soils from interbedded sedimentary rocks are naturally fine-textured and hard to work when wet. Compaction of moderately cohesive soils for fills is accomplished through the use of pneumatic-tired rollers; more cohesive soils require strict control of moisture at the optimum moisture content and the use of sheepsfoot rollers. Specific materials under consideration should be tested and evaluated.

Building Stone: Excellent to Not Suitable. Limestones and resistant sandstones are commonly utilized as building stone because of their resistive nature. Shale is not suitable as a building material.

Mass Wasting and Landslide Susceptibility

Tilted, interbedded sedimentary rocks present a very hazardous high sliding potential in almost all situations. Moisture-bearing rocks such as limestones and sandstones increase in weight and mass as they absorb water and thus increase shearing stresses. Slides result when the shearing resistance is exceeded; the danger is increased by sudden shocks of an earthquake or blasting. Many types of mass wasting phenomena take place in tilted, interbedded regions, their particular form depending upon the angle of dip, the jointing system, and the climate. These include rockslides, rockfalls, debris slides, debris falls, and earthflows. Undercutting by streams or construction operations also creates very hazardous conditions.

A disastrous rockslide occurring in tilted, interbedded sedimentary rocks took place in 1963 at the Vaiont Reservoir near Longarone, Italy. The rockslide was attributed to a mass of pervious limestone, not having lateral support, which had become saturated with water. This

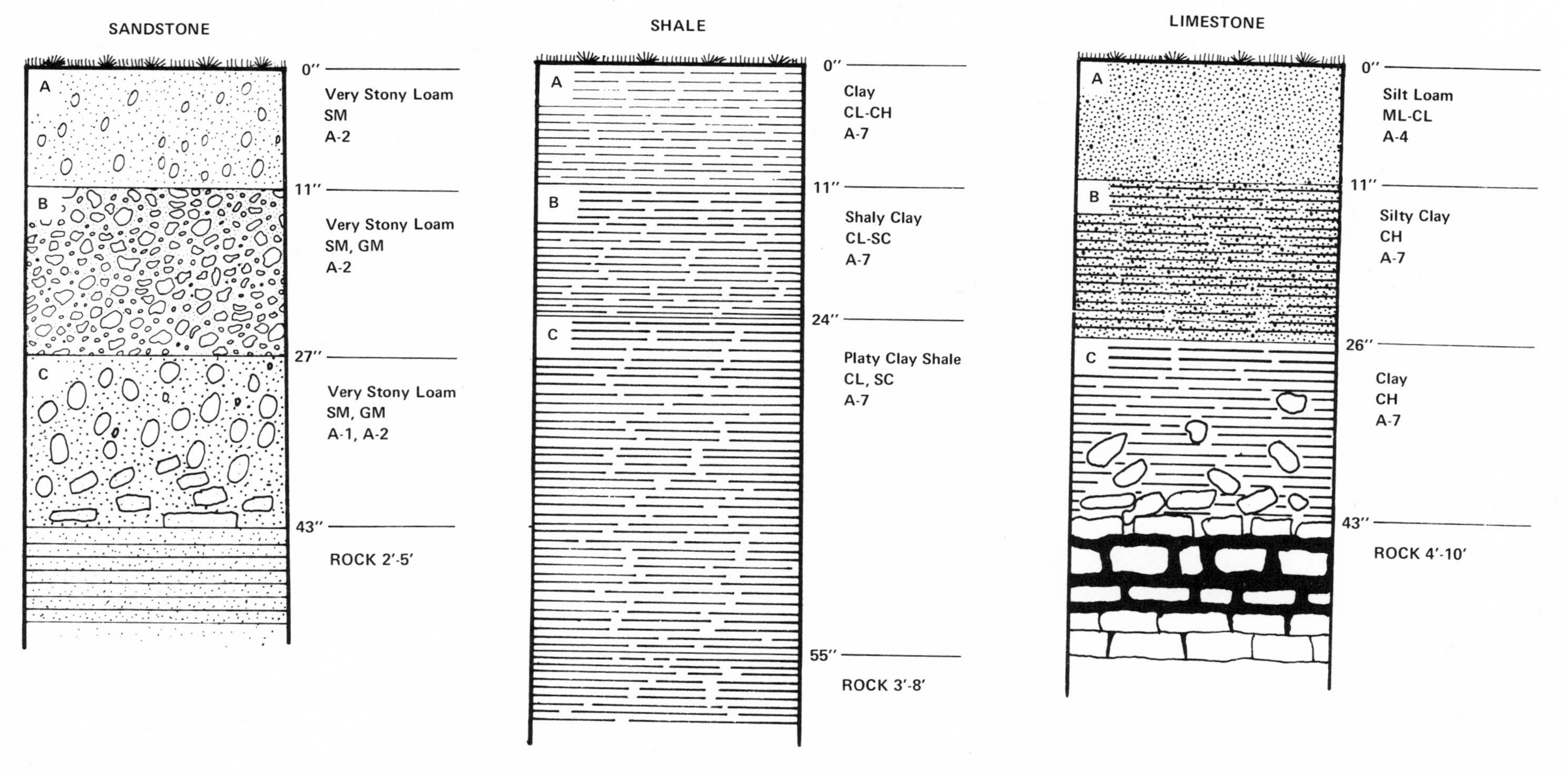

Figure 5.48. A residual soil profile of limestone-shale, interbedded sedimentary rock as found in a humid climate.

increased the mass of the rock and of the interbedded clay layers, and thereby the available shearing resistance was exceeded, causing the failure. A massive rockslide fell into the reservoir, creating a tidal wave which devastated villages located below the dam and claimed 3000 lives. (In Chapter 11, this and other similar events are described in more detail.)

Competent engineers and geologists should be consulted in the planning stages of all structural projects requiring excavation and foundations in tilted, interbedded sedimentary rocks.

Groundwater Supply

Interbedded sedimentary rocks provide excellent potential high-yield sources of groundwater. The sandstone and limestone layers are porous and permeable, storing and carrying large amounts of underground water. There is generally no difficulty in locating sufficient water supplies for large developments or urban areas in these regions. Geologists and other consultants can make hydrological studies to determine the potential yield and quality of the groundwater.

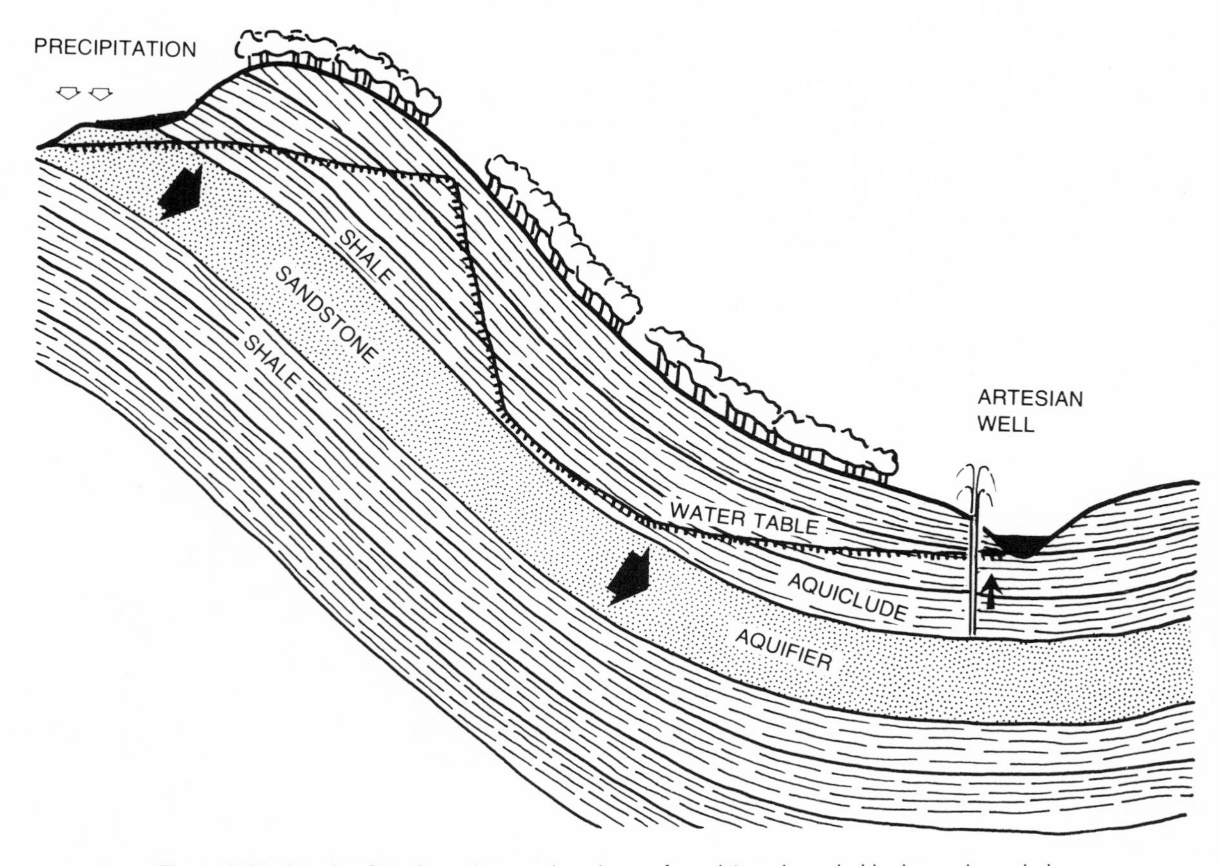

Figure 5.49. Artesian flow through a pervious layer of sandstone bounded by impervious shale.

In some regions tilted layers of sedimentary materials allow the flow of artesian water, thus forming springs and wells. In these cases the sandstone or limestone vein gathers water at a high elevation and pipes it through a vein bounded by impervious materials to an outcrop at a lower elevation, creating an artesian well or spring (Figure 5.49).

Pond or Lake Construction

Investigations are needed to determine capabilities for lake and pond construction in specific areas. Seepage and leakage of water are likely through sandstones and limestones. It is not difficult to find suitable sites for reservoirs in the ridge and valley topography. When compacted, the residual soils found in these areas provide an impervious layer for the bottom of the water feature. The fineness of the residual soils, however, requires preservation of the vegetation along the drainage system feeding the watershed, including first-order streams and gullies, in order to minimize silting.

Reservoirs in valleys where tilted sedimentaries parallel one or both of the valley walls, have increased landslide hazards. If some of the rock layers are unsupported on their lateral downslope side, the impounded water will help to saturate and lubricate the rock structure between the joints and beds, creating a very unstable condition. The Vaiont Reservoir disaster resulted from similar conditions.

Foundations

Suitabilities for different types of foundations depend upon the specific rock types and residual soils present. Residential units commonly use basement slabs and continuous footings; heavy structures rely upon piers or footings. Excavation for basements in arid climates is difficult, since the soil depth is thin and blasting is required except in shale formations.

Highway Construction

Highway construction is not difficult through tilted limestone-shale interbedded regions, but major excavation and grading operations occur when crossing tilted sandstone-shale regions. Highway alignments that follow the natural grain of the ridge-valley topography can minimize earthwork in shale deposits. Alignments perpendicular to the natural grain may require tunneling, bridging, rock excavation, and massive fills. Tunneling may be more economical where the topography is extremely rugged. Major cuts require terracing and drainage, especially in shale and in thin, interbedded deposits in order to minimize sliding and maintenance. Highway construction and alignments are more difficult in arid climates since the topography is more rugged.

Table 6.13. Tilted, interbedded sedimentary rocks: aerial and cartographic references

State or Country	County	Photo Agency	Photograph no.	Scale	Date	Corresponding quadrangle and geologic maps from USGS	Corresponding soil reports from SCS
Alaska	3rd Judicial Division	USGS*	Alaska 21-AB	3333′/″	9/14/55	Middleton Island (15′)	–
Arkansas	Garland	USGS*	Arkansas 3-ABC	1500′/″	4/9/65	Hot Springs (15′)	–
Arkansas	Logan	ASCS	II-1HH-65-66,	1667′/″	10/16/66	Little Rock (1:250,000)	–
Arkansas	Logan	ASCS	II-6BB-17-19	1667′/″	1/9/61	Little Rock (1:250,000)	–
Arkansas	Polk, Howard	USGS*	Arkansas 2-ABC	1967′/″	2/17/55	Umpire (15′)	–
Arkansas	Scott	USGS*	Arkansas 1-ABC	2010′/″	11/9/57	Waldron (15′), OM-192	–
Colorado	Jefferson	USGS*	Colorado 3-ABC	5280′/″	9/21/53	Ralston Buttes, Golden (7½′), GQ-103 Eldorado Springs, Louisville, MF-179	–
Kentucky	Meade, Hardin	USGS*	Kentucky 1-ABC	2000′/″	1/25/60	Flaherty (7½′), GQ-229	–
New Mexico	San Juan	USGS*	New Mexico 1-ABCD	4500′/″	2/5/54	Ship Rock (15′)	–
Oklahoma	Atoka	USGS*	Oklahoma 2-ABC	1416′/″	2/29/56	Stringtown (7½′)	–
Oklahoma	LeFlore	USGS*	Oklahoma 3-ABCD	1967′/″	2/9/55	Heavener (15′)	1931
Oklahoma	LeFlore	USGS*	Oklahoma 4-ABC	1967′/″	2/12/55	Page (15′)	1931
Oklahoma	Pushmataha	ASCS	CKH-3N-13-16	1667′/″	10/21/55	Tuskahoma (1:250,000)	–
Pakistan	West Pakistan	USGS†	Pakistan 2-ABCDE	3833′/″	1953	Army Map Service U501, G41-R	–
Pennsylvania	Carbon	ASCS	AQT-1KK-48-51	1667′/″	10/4/69	Palmerton, Pohopoco (7½′)	1962
Pennsylvania	Centre	ASCS	AQG-26-15-16	1667′/″	10/23/38	Harrisburg (1:250,000)	–
Pennsylvania	Lycoming, Clinton	USGS*	Pennsylvania 1-ABCD	5000′/″	9/25/56	Williamport	1923‡ 1966
Pennsylvania	Dauphin	ASCS	AHE-69-7-8	1667′/″	11/12/38	Harrisburg (1:250,000)	1972
Pennsylvania	Fulton	ASCS	AQB-5HH-17-19	1667′/″	9/4/67	Needmore (15′)	1969
Pennsylvania	Fulton	ASCS	AQB-4HH-32-34	1667′/″	9/4/67	Needmore (15′)	1969
Pensylvania	Fulton	ASCS	AQB-5HH-64-67	1667′/″	9/4/67	Clearville (15′)	1969
Pennsylvania	Huntingdon	ASCS	AQI-4HH-128-132	1667′/″	9/4/67	Tyrone (15′)	1944
Pennsylvania	Huntingdon	ASCS	AQI-8HH-59-61	1667′/″	9/4/67	Orbisonia (15′)	1944
Puerto Rico	Ciales, Morovis	USGS*	Puerto Rico 6-ABC	1667′/″	1/16/64	Ciales (7½′), Prof. Paper No. 501-B	–
Texas	Brewster	USGS*	Texas 1-ABC	5280′/″	5/15/54	Marathon (15′), Prof. Paper No. 187	–
Virginia	Lee	USGS*	Virginia 1-ABC	1667′/″	3/9/53	Rose Hill (7½′)	1953
Virginia	Shenandoah	ASCS	DJO-1FF-87-90	1667′/″	9/15/67	Strasburg (7½′)	–
Virginia	Shenandoah	USGS*	Virginia 5-ABC	2200′/″	3/9/45	Edinburg (15′)	–
Wyoming	Carbon	USGS	Wyoming 9-ABC	2308′/″	6/14/47	Saddleback Hills (15′), Como Ridge (15′)	–
Wyoming	Fremont	USGS*	Wyoming 6-ABC	1967′/″	10/22/48	Thermopolis (1:250,000)	–
Wyoming	Fremont	USGS*	Wyoming 7-AB	1967′/″	10/20/48	Thermopolis (1:250,000)	–

*From *Selected Aerial Photographs of Geologic Features in the United States,* U.S. Geological Survey Prof. Paper No. 590.
†From *Selected Aerial Photographs of Geologic Features outside the United States,* U.S. Geological Survey Prof. Paper No. 591.
‡Out of print.

CHAPTER 6

IGNEOUS ROCKS

INTRODUCTION

Igneous rocks are formed by the solidification of magma, or molten rock material, on or within the surface of the earth. These rocks are generally crystalline and have the qualities of hardness, durability, strength, density, low porosity, and low permeability. Igneous rocks are classified as either intrusive (formed beneath the surface of the earth) or extrusive (formed on the earth's surface) and are described by their mineral composition and crystalline texture.

Table 6.1 lists the major igneous rocks described in the following discussion, their chemical and mineral compositions, and the structures that typify them. This section concentrates upon widely distributed igneous rocks—the coarse-textured intrusive granitics and the glassy, surface-flow extrusive basaltics. The other group of igneous intrusive rocks, the porphyritics, fine-grained rocks containing at least 25% large crystals (phenocrysts), occur as localized deposits in small areas and are not discussed here.

Intrusive Igneous Rocks (Granitic Materials)

Intrusive igneous rocks were solidified from molten rock material beneath the surface of the earth and are commonly referred to as plutonic (after the god Pluto, ruler of the depths of the earth). All igneous intrusive rocks are plutons, regardless of size, shape, or composition.

The crystalline structure of igneous plutonic rocks is well developed owing to their slow process of solidification. During this process the rock

Table 6.1. Classification of igneous rocks*

			Basaltics (extrusive)	Granitics (intrusive)			
Rock-forming elements		Rock-forming minerals	Glassy surface flows	Fine-grained porphyritic dikes, sills, laccoliths		Coarse-textured batholiths	
Oxygen	O						↑ Acid
Silicon	Si	Quartz SiO_2	Rhyolite porphyry	Rhyolite-porphyry	Granite-	Granite	
Aluminum	Al	Orthoclase $KAlSi_3O_8$	Trachyte	Trachyte-porphyry	Syenite-porphyry	Syenite	
Potassium	K						
Sodium	Na	Plagioclase $CaAl_2Si_2O_8$ $NaAlSi_3O_8$	Andesite	Dacite-porphyry	Diorite-porphyry	Diorite	
Calcium	Ca						
			Basalt	Basalt-porphyry	Gabbro-porphyry	Gabbro	
Iron	Fe	Ferro-magnesian silicates					
		Biotite-mica Hornblende	Augitite	Augitite-porphyry	Pyroxenite-porphyry	Pyroxenite and periodotite	
Magnesium	Mg	Pyroxene Olivine	Limburgite				Basic ↓

*After A.K. Lobeck, *Geomorphology,* McGraw-Hill, New York, 1939.

mass surrounding the intrusion acts as an insulator against rapid heat loss and permits the formation and growth of large crystals. Because different minerals have unequal rates of crystalline growth and different solidification temperatures, a wide range of crystal sizes in a rock mass indicates the presence of many different minerals. Rocks with uniform crystal sizes generally contain a more uniform mineral composition.

Plutonic rocks underlie all rock types, forming a platform or basement supporting the surface rocks. Exposure to the earth's surface takes place if the overriding materials are weathered or eroded away. Plutonic rock occurs on only 15% of the earth's surface.

It is characteristic of igneous intrusive rocks to develop fractures or breaks which occur randomly throughout the rock, not aligned with normal mineral cleavages. These are formed during solidification of the rocks or when the overriding materials are removed through weathering and erosion, thus causing pressure release. Intrusive rocks are often highly fractured, and their engineering characteristics are correspondingly affected.

The family of plutonic igneous rocks comprises a variety of mineral compositions and relative acidities (see Table 6.1). They vary from granite which, since it has a high percentage of quartz, is the lightest and most acidic of the plutonic rocks, to gabbro, which is the darkest and most basic, having a high percentage of plagioclase feldspar and the ferromagnesian minerals.

For our purposes, the important qualities of plutonic igneous rocks are their engineering characteristics. The rock is of intrusive igneous origin, having a massive crystalline structure and very low porosity and permeability. A knowledge of specific differences in mineralogical structure is useful, but these differences can be identified only by close field examination.

Granite

Granite is the plutonic rock most commonly encountered. It generally contains a large amount of orthoclase feldspar in combination with quartz, with scattered accessory minerals such as hornblende, biotite, and mica. Its mineral composition gives it a light color tone and an acidic nature. Its strength, durability, and attractive appearance make granite valuable as building stone.

Syenite

Syenite is similar to granite in appearance, but close examination reveals important differences in mineral composition. Like granite, it contains orthoclase feldspar but, instead of quartz, has a high concentration of ferromagnesian minerals, such as mica, biotite, and hornblende. It is not as acidic as granite nor as light-colored.

Diorite

Diorite is a medium- to coarse-grained igneous rock consisting primarily of plagioclase and orthoclase feldspars. It contains little quartz, but dark ferromagnesian minerals such as hornblende, biotite, and pyroxene are abundant. Diorite is not as acidic as syenite nor as basic as gabbro. It is usually a dark-toned rock, owing to the higher concentration of dark minerals.

Gabbro

This variety of intrusive igneous rock is one of the darkest and most basic. It is composed mainly of plagioclase feldspars and dark ferromagnesian minerals such as pyroxene, olivine, and hornblende. Gabbro is commonly found extending over broad regions.

Extrusive Igneous Rocks (Basaltic Materials)

Extrusive igneous rocks are of two types. One type is formed by volcanic eruptions which pour molten lava onto the earth's surface where it solidifies. The other type includes fragmental rocks of all sizes which have solidified at the surface of the earth.

Volcanic magma does not develop a large crystalline structure, for the cooling of the material is rapid and the resulting crystalline texture is so fine that it is not apparent without magnification. Most extrusive rocks are dense and glassy in appearance, but they can be vesicular (filled with gas bubbles or even frothy).

Extrusive rocks occur throughout the world but account for only about 3% of the total, exposed, continental land surface, as opposed to 15% for intrusive rocks. The following discussion includes only those extrusive rocks that are often found and that cover extensive land areas.

Rhyolite

Rhyolite deposits occasionally cover extensive areas, but they are generally found as small, localized formations surrounded by other extrusive rocks. Some opinions hold that rhyolite is formed from extrusive materials of a viscous nature, thus accounting for the flow-banding that is frequently conspicuous. It is the lightest-toned and most acidic of the extrusive rocks, consisting primarily of quartz and orthoclase feldspar, with small concentrations of dark ferromagnesian minerals such as hornblende, biotite, and mica. The composition of rhyolite is similar to that of intrusive granite, but the crystalline structure is not as well developed because of the relatively faster cooling.

Trachyte

Trachyte is formed in localized deposits where magma flow had become quite viscous; this condition is indicated by conspicuous flow-banding. It is not as acidic as rhyolite nor as light in tone, and its mineral composition does not include as much quartz. It consists primarily of orthoclase feldspar and several of the dark ferromagnesian minerals such as hornblende, biotite, mica, pyroxene, or olivine. Its intrusive equivalent is syenite.

Andesite

These rocks are usually formed by the solidification of a magma more viscous than the material associated with the dark, basic rocks. Andesite is typically dense, hard, and durable, but extensive fracturing may exist, depending upon local conditions. This material consists of plagioclase feldspars and several of the dark ferromagnesian minerals such as olivine, pyroxene, or hornblende. Its intrusive equivalent is diorite.

Basalt

There are many definitions of basalt, but the term usually encompasses all dark, basic, surface-flow materials formed from lava flows. It may be either very dense or else vesicular, depending upon the conditions of its formation. It is the darkest and most basic of the commonly occurring extrusive rocks and consists of plagioclase feldspar and ferromagnesian minerals such as pyroxene, olivine, and hornblende. Its intrusive equivalent is gabbro.

Distribution of Igneous Granites and Basalts

Igneous intrusive rock forms are exposed over approximately 15% of the continental land surface of the world. Granite is predominant, providing the foundation for most continental masses and the central core of many mountainous structures. Basaltic and volcanic forms occur on approximately 3% of the earth's land surface and are scattered across most of the continents in small deposits. The significant worldwide granitic and

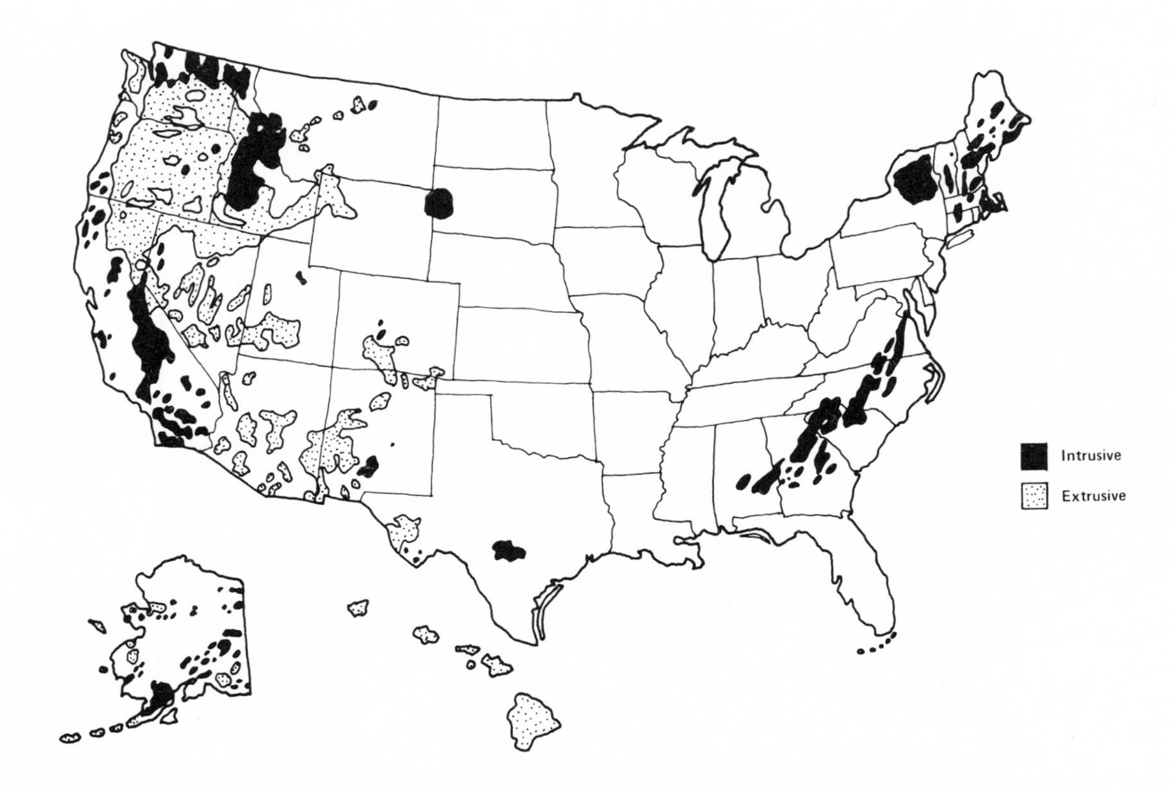

Figure 6.1. Distribution of igneous landforms in the United States. (After Belcher, D. J., and Associates, Inc., Ithaca, New York, "Origin and Distribution of United States Soils," 1946.)

basaltic occurrences are listed below; not included are the many small local outcrops also found throughout the world.

North America

United States. The New England states of Massachusetts, Vermont, New Hampshire, and Maine contain massive granitic forms which have been intruded into folded, twisted metamorphic rocks. Other granitic areas include the Adirondack Mountains of northern New York, a large batholith in central Idaho, the Black Hills of South Dakota, the Sierra Nevada region of California, and the northern Cascades in Washington. Many smaller, scattered deposits are found throughout the western states of Arizona, Colorado, Montana, Nevada, New Mexico, Oregon, and Wyoming. Basaltic formations are significantly distributed across the western United States, covering all of Hawaii, most of Washington, Oregon, Idaho, northwest Wyoming, northern California, and the Aleutian Island chain of Alaska. Scattered deposits are found in Nevada, Arizona, New Mexico, Utah, south central Colorado, and southwestern Texas.

Canada. Most of the eastern half of Canada is an exposed granitic region. Basaltic formations are found in south central British Columbia.

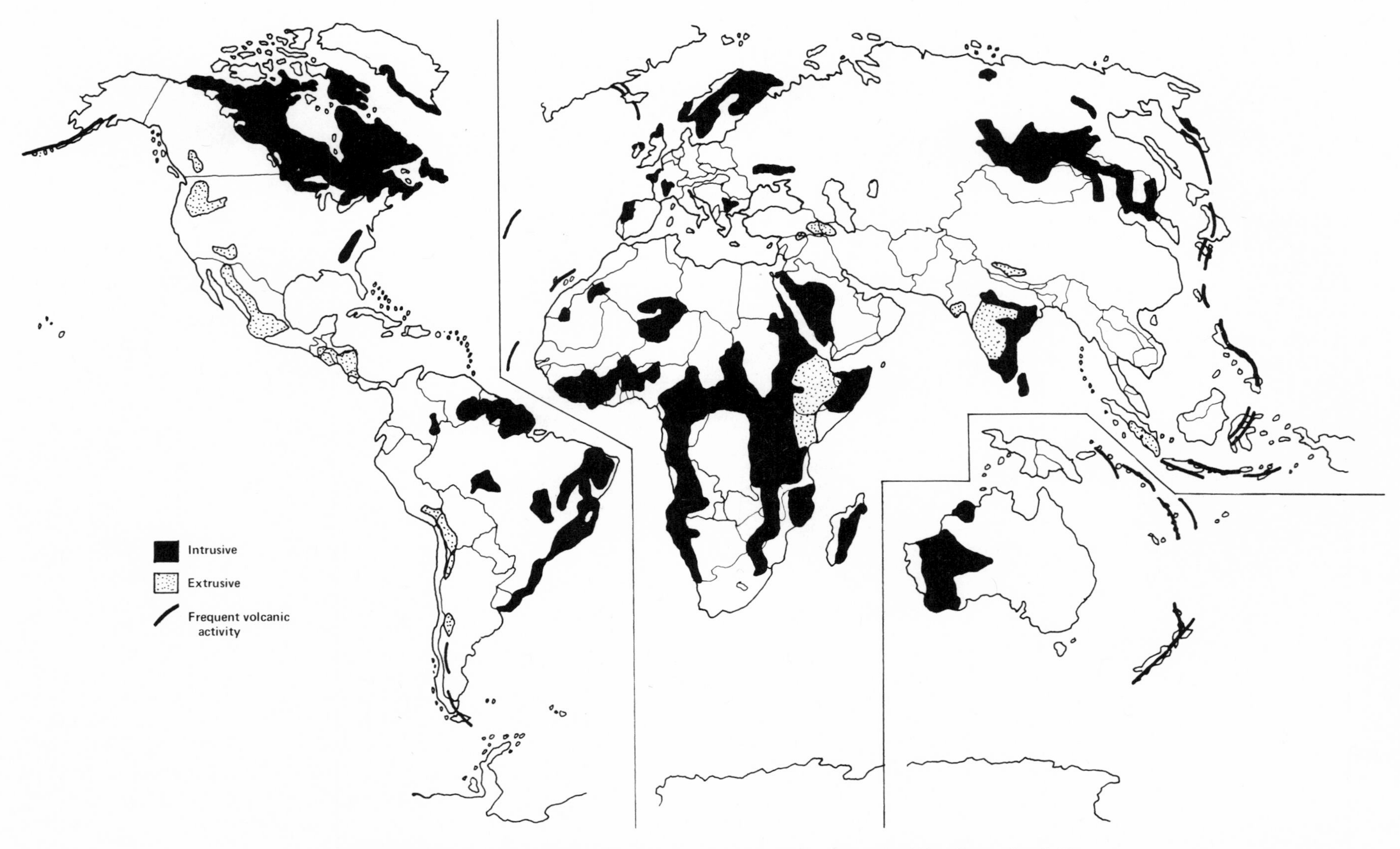

Figure 6.2. Distribution of major igneous formations throughout the world. (After Trewartha, G. T., Robinson, A. H., and Hammond, E. H., *Elements of Geography,* Mc-Graw-Hill, New York, 1967.)

South and Central America

Igneous granitic intrusions and associated metamorphic rocks are found in southern Venezuela, southern Guiana, Surinam, French Guiana, northern Brazil, the southeast coast of Brazil, and southern Uruguay. Basaltic formations are prominent through much of Central America, being found in central Mexico, El Salvador, western Honduras, western Nicaragua, and northern Costa Rica. South America contains a basaltic formation extending through eastern Chile into south central Peru.

Table 6.2. Igneous rocks: summary chart

Landform	Topography	Drainage*	Tone	Gullies	Vegetation and land use
Intrusive granite (large masses)					
Humid	Bold, domelike hills	Dendritic curved ends, M	Uniform, gray	U-Shaped	Forested and cultivated
Arid	A-shaped hills	Dendritic curved ends, F	Light, fractures	Few to none	Barren, grass cover
Intrusive granite (linear dikes)					
Humid and arid	Narrow, linear ridges†	None	Light or dark	None	Natural cover
Extrusive volcanic	Cinder cones	Radial, C-F	Dark gray	Vary	Barren, natural cover
Extrusive basalt flows	Level plain	Regional parallel, C to F	Dark, flow marks	Soft, U-shaped	Cultivated or barren
Extrusive interbedded	Terraced hillsides	Parallel dendritic	Banded	Vary	Cultivated or natural
Extrusive tuff	Sharp-ridged hills	Dendritic, F	Dull, light gray	Vary	Natural cover

*C, coarse; M, medium; F, fine.
†Ring dikes, a variation of this landform, are found with a circular narrow, sharp ridge.

Africa

Igneous granitic formations occupy large areas in Africa and are also scattered throughout almost all countries, including Madagascar. The broadest coverage extends across central Africa, then southward along eastern coastal sections. A large basaltic region extends across central Ethiopia into western Kenya.

Europe

Norway, Sweden, Finland, and bordering Russian territories consist mainly of granitic materials. Igneous intrusions with associated metamorphic rocks occupy northern Scotland, central France, and northwestern Spain. No significant basaltic or volcanic formations are found in Europe, but scattered deposits are found in northern Ireland, eastern Greenland, Iceland, and associated islands.

Asia

Exposed granitic intrusions occur over most of India, Ceylon, the northern portion of the Mongolian Republic, northern Manchuria, and the bordering regions of Siberia (Yablonovy Khrebet). A large basaltic deposit covering over 200,000 square miles occurs in west central India. Smaller formations are found in eastern Turkey, southern Tibet, and southern Sumatra.

Australasia

Most of Western Australia is granitic; there are smaller outcrops in the Northern Territory and Southern Australia. No significant basaltic formations are found in Australia.

Pacific Region

No significant regional granitic deposits are found in the Pacific Island region. Most of the islands of the Pacific central basin are basaltic, including the Hawaiian Islands, the Aleutians, and the foundations of many lagoon-forming islands.

Caribbean Region

Small, scattered granitic outcrops exist throughout Puerto Rico and the Virgin Islands. The Windward Islands consist of basaltic materials, and scattered deposits are found in the Dominican Republic.

REFERENCES

American Society of Photogrammetry, *Manual of Photo Interpretation,* American Society of Photogrammetry, Falls Church, Va., Chapter 4, 1960.

Buddington, A. F., "Granite Emplacement with Special Reference to North America," *Geological Society of America Bulletin,* Vol. 70, pp. 671–748, 1958.

Bullard, F. M., *Volcanoes of the Earth,* University of Texas Press, Austin, Texas, 1962.

Chapman, R. W., and M. A. Greenfield, "Spheroidal Weathering of Igneous Rocks," *American Journal of Science,* Vol. 247, 1949.

Cotton, C. A., *Volcanoes as Landscape Forms,* Whitcombe and Tombs, Christchurch, New Zealand, 1944.

Daly, R. A., *Igneous Rocks and the Depths of the Earth,* McGraw-Hill, New York, 1933.

Ernest, W. G., *Earth Materials,* Prentice-Hall, Englewood Cliffs, N.J., pp. 92–109, 1969.

Fenneman, N. M., *Physiography of the Western United States,* McGraw-Hill, New York, 1931.

Fenneman, N. M., *Physiography of the Eastern United States,* McGraw-Hill, New York, 1938.

Green, J., and N. M. Short, eds., *Volcanic Landforms and Surface Materials,* Springer Verlag, New York, 1971.

Hamblin, W. K., and J. D. Howard, *Physical Geology: Laboratory Manual,* 2nd ed., Burgess, Minneapolis, 1967.

Hamilton, W., and W. B. Myers, "The Nature of Batholiths," U.S. Geological Survey, Professional Paper No. 554-C, pp. C1–C30, 1967.

Hirashima, K. B., "Highway Experience with Thixotropic Volcanic Clay," *Proceedings,* Highway Research Board, Washington, D.C., 1948.

Knopf, A., "Igneous Geology of the Spanish Peaks Region, Colorado," *Geologic Society of America Bulletin,* Vol. 47, pp. 1727–1784, 1936.

Lo, K. Y., "Shear Strength Properties of a Sample of Volcanic Material of the Valley of Mexico," *Geotechnique,* No. 12, 1962.

Lobeck, A. K., *Geomorphology, An Introduction to the Study of Landscapes,* McGraw-Hill, New York, pp. 42–47, 1939.

Lobeck, A. K., *Things Maps Don't Tell Us,* Macmillan, New York, 1956.

Longwell, C. R., R. F. Flint, and J. E. Sanders, *Physical Geology,* Wiley, New York, Chapters 19 and 20, 1969.

Oliver, C., *Volcanoes,* M.I.T. Press, Cambridge, Mass., 1969.

Shelton, J. S., *Geology Illustrated,* W. H. Freeman, San Francisco, pp. 14–27, 1966.

Smalley, I. J., "Contraction Crack Networks in Basalt Flows," *Geological Magazine,* Vol. 103, 1966.

Stearns, H. T., *Geology of the Hawaiian Islands: Honolulu,* Hawaii Division of Hydrography, 1946.

Strahler, A. N., *Introduction to Physical Geography,* Wiley, New York, Chapters 25–27, 1965.

Terzaghi, K., "Dam Foundations on Sheeted Granite," *Geotechnique,* Vol. 12, No. 3, 1962.

Thorarinsson, S., *Surtsey: The New Island in the North Atlantic,* Viking, New York, 1967.

Thornbury, W. D., *Regional Geomorphology of the United States,* Wiley, New York, 1965.

Williams, H., "Volcanoes," *Scientific American,* Vol. 185, No. 5, pp. 45–53, 1951.

Intrusive Igneous Rocks: Granitic Forms

Introduction

Of all the different types of exposed igneous intrusive rock, the granites are the most common. This section concentrates upon granitic forms, which include—in addition to granite—syenite, diorite, and gabbro.

The form of granitic intrusive structure depends upon the extent of the deposit and the surrounding rock conditions. Structures commonly encountered are batholiths, laccoliths, sills, stocks (or bosses), and dikes. Although the appearance of these structures varies, their mineral and crystalline compositions are similar, as a result of their having had similar conditions during solidification.

Batholiths

Batholiths are massive intrusions which cover areas of more than 40 square miles. They are generally dome-shaped and have been forced up into the surrounding rock (Figure 6.3). Contact metamorphism may occur along the boundary between the intrusive mass and the surrounding rock material.

Daly (1933) has suggested the following objective criteria for the description of a batholith:

(1) It should be located in an area of mountain building.
(2) Its areal extent should be over 40 square miles (a body less than 40 square miles in extent, having the other characteristics of a batholith, is a stock).
(3) It should be discordant, cross-cutting the structure of the surrounding rock.
(4) It should possess an irregular but massive domed roof.

Figure 6.3. Batholith.

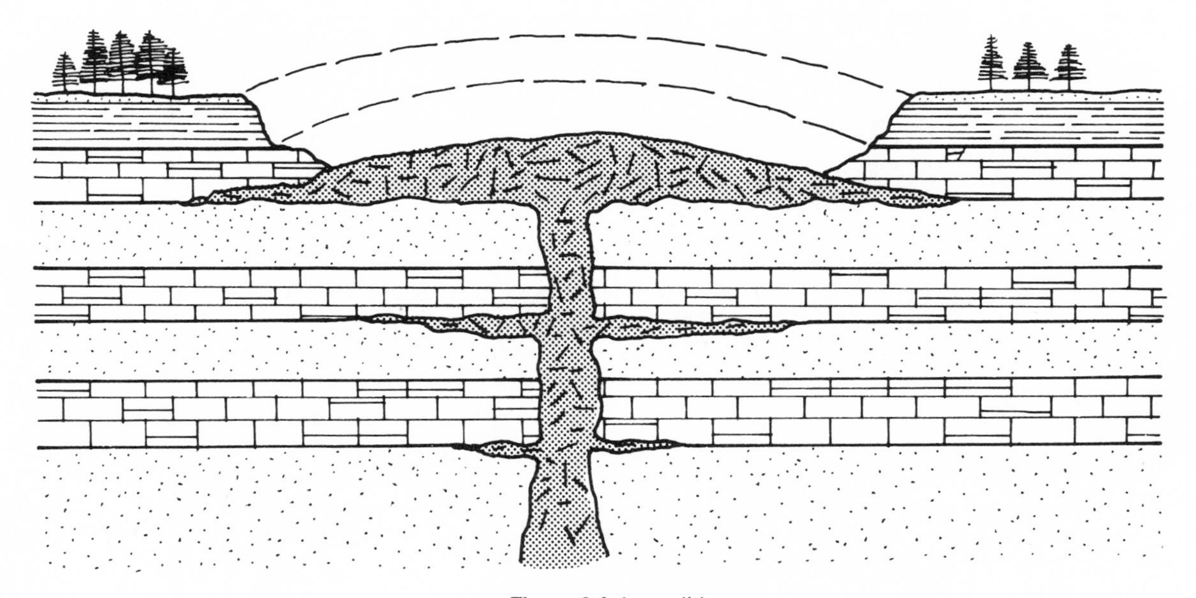

Figure 6.4. Laccolith.

(5) The sidewalls of the intrusive dome should be steep and smooth, dipping outward so that the body enlarges downward with no visible or inferable floor or underlying layer.
(6) It should occur as a composite of several intrusions.

Many of the batholiths currently described by cartographers and geologists do not fulfill all of the criteria specified by Daly, but his system does serve as a good guide. As in any classification system, other interpretations and modifications exist.

Laccolith

Laccoliths are intrusive forms which are lens-shaped and approximately concordant with the surrounding rock structure (Figure 6.4). The intrusion causes an uplifting of overlying rock, creating a distinct anticline of the strata. No rigorous guidelines have been established for the classification of these forms, which are very similar to sills. Some geologists classify forms as laccoliths if the ratio of the diameter to the thickness of the intrusion is less than 10 to 1.

Sills

Sills are sheets of intrusive igneous material which have been forced between layers of stratified rock (Figure 6.5). The resulting layer may be from several inches to hundreds of feet in thickness. These forms are very similar to laccoliths and are classified as such when the overlying rock material is slightly uplifted or domed.

Stocks (or Bosses)

Stocks are intrusive forms similar to batholiths but less than 40 square miles in area. They form

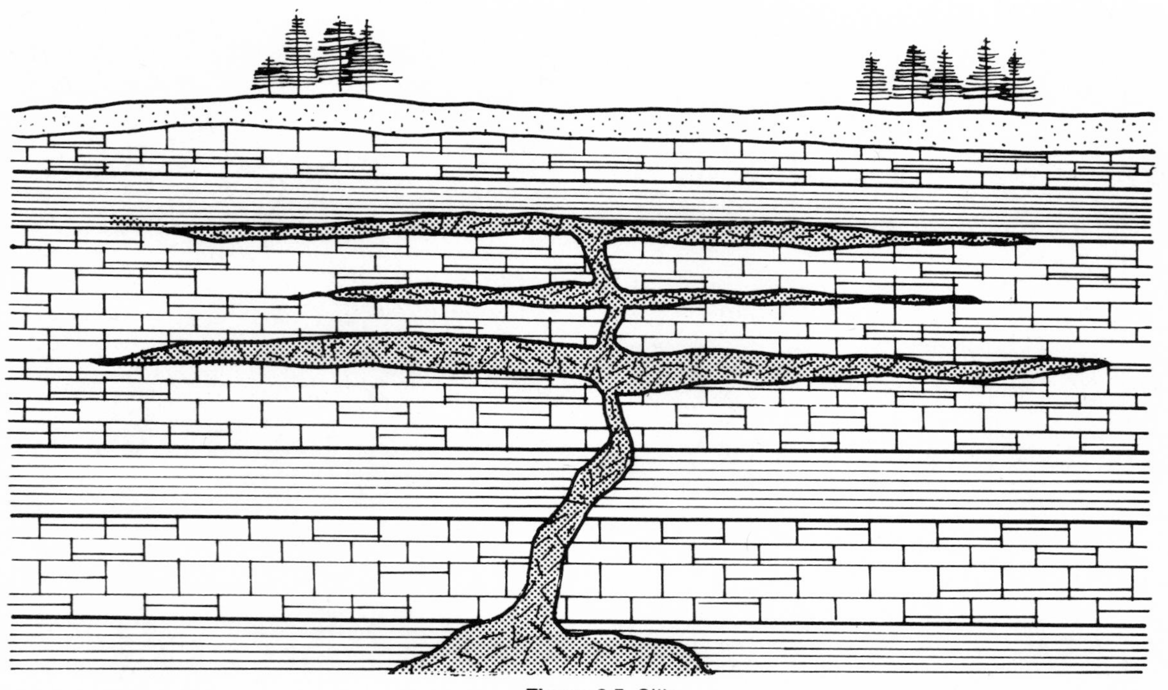

Figure 6.5. Sills.

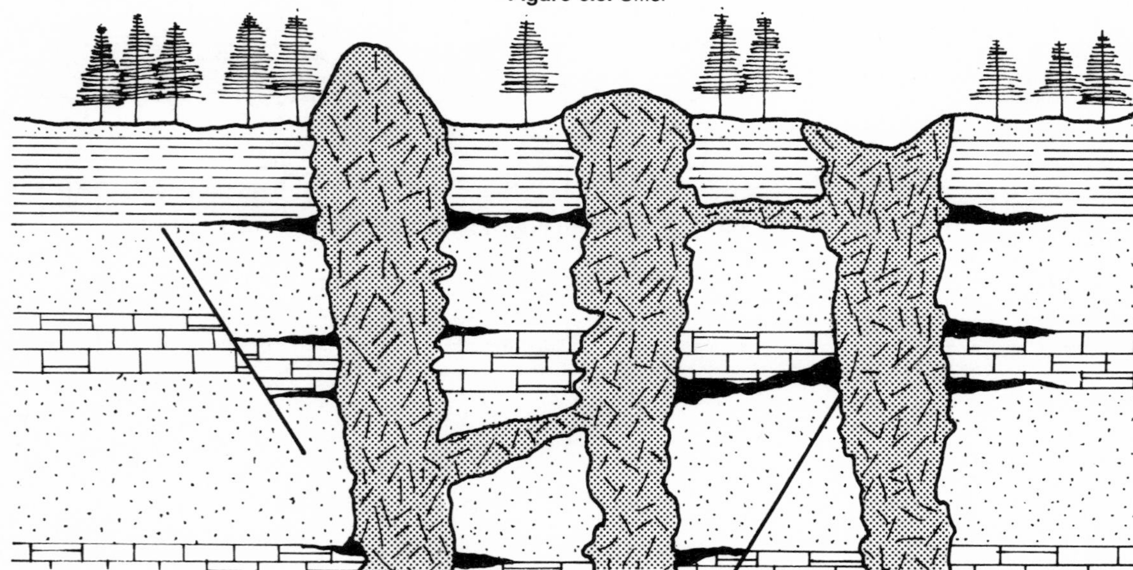

Figure 6.6. Dikes.

Figure 6.7. Igneous granitic dikes. Many radial, igneous granitic dikes occur in the Spanish Peaks region of Huerfano County, Colorado, and are characterized by their straight, resistant, thin linear form. Photographs by the Agricultural Stabilization Conservation Service, DYT-6P-85,86, 1:20,000 (1667′/″), October 7, 1955.

domes and knobs in the topography and usually have the other characteristics of batholiths.

Dikes

Dikes are formed when igneous intrusive material enters a crack between rocks. The intruded zone may be from several inches to several miles in width (Figure 6.6). Thus dikes may be either small and insignificant or large and influential in the regional physiography. (An example of the latter is the Spanish Peaks region of Colorado.)

Interpretation of Pattern Elements

Since batholiths, laccoliths, sills, and stocks are characterized by large, massive forms, while dikes are characterized by straight, linear forms, the interpretive characteristics of granitic forms are presented in these two catagories.

In humid climates granitic masses exhibit bold, massive, domelike hills and a medium-textured dendritic drainage pattern with curvilinear or sickle-shaped alignments. In arid climates these same forms exhibit a higher degree of dissection, in which the hills form small domes and ridges and are more A-shaped in section, containing many rounded boulders and exfoliated, exposed, bedrock outcroppings.

The domelike appearance of granitic hills, found in all climates, reflects a characteristic of granitic materials, that of being modified by the weathering processes known as *exfoliation*. This takes place when thin concentric shells, flakes, or scales of rock break off from the parent rock mass as a result of a combination of temperature, frost, and perhaps minor chemical effects.

In all climates linear forms, such as dikes, exhibit straight, narrow ridges having tones and

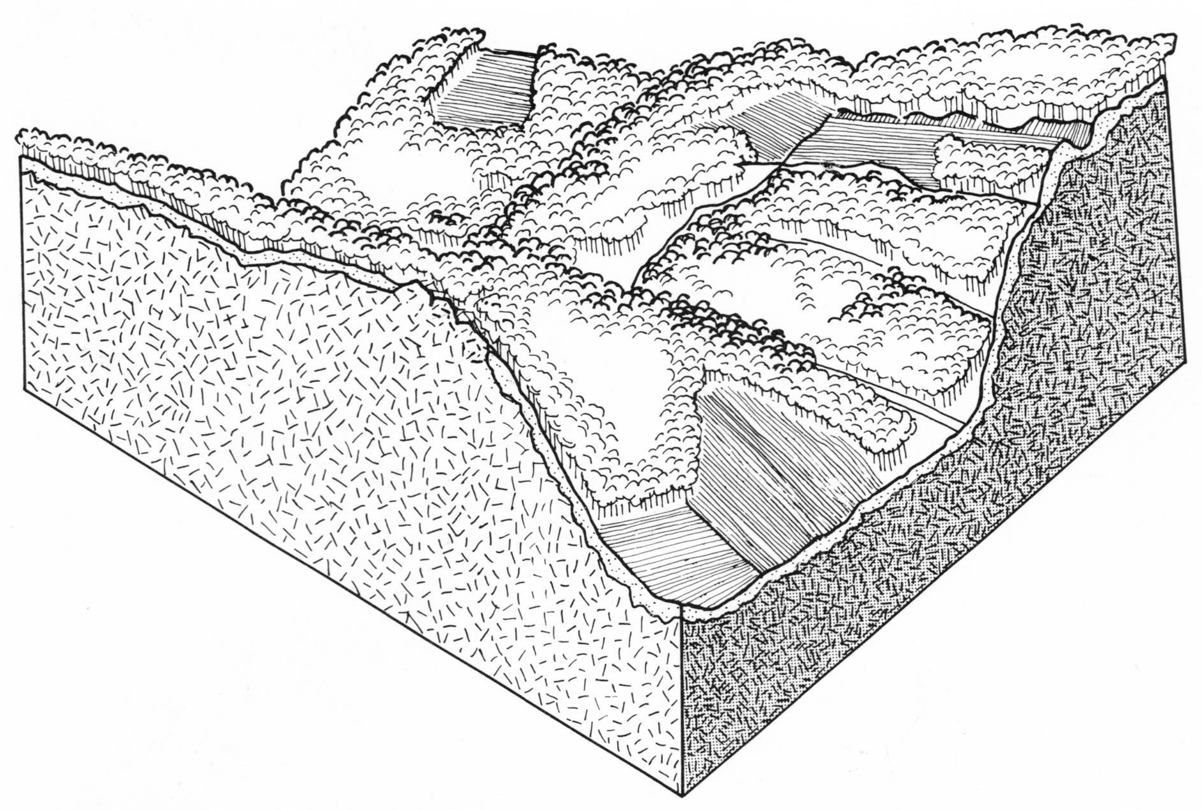

Figure 6.8. Massive granitic formations in a humid climate. The massive, rounded topography of granitic landforms in humid climates creates a visually absorptive landscape. Man's development patterns are confined to either the large valleys or the broad, smooth uplands. Large sites for airports are difficult to locate because of the relief of the topography. Road patterns are curvilinear in order to minimize rock removal and bridging. Agricultural field patterns are irregular on the hilltops but may be gridded in the valleys. The wooded slopes and hilltops are the most exposed areas when viewed from the valleys; the valleys and opposite slopes are the most exposed when viewed from hillsides and hilltops. The wooded, steep slopes facing most valley towns maintain the rural visual characteristics of these towns, and removal of the vegetation by development creates a severe visual impact.

vegetation cover different from the adjacent land surfaces.

Soil Characteristics

The residual soils developed from granitic intrusive rocks form relatively thin, sterile profiles over fractured bedrock. The surface layers are commonly silty sands; the subsurface is generally a sandy clay. Profile development is greater in unglaciated humid climates and is deep enough to allow some cultivation. In arid climates the soils are more sandy and thin and contain many rock fragments near the surface. Dike formations, being relatively small, do not develop a significant residual soil; they are very rocky and generally have a thin cover of sandy clay and rock fragments.

U. S. Department of Agriculture Classification

The U.S. Department of Agriculture classifies the granitic soil surface horizons as silty sand, silt loam, or sandy loam. Subsurface soils, which are slightly plastic, especially in humid climates, are classified as sandy clay, clay loam, or sandy clay loam. Rock fragments in the subsoil may be indicated in the soil series as rocky, stony, very rocky, or very stony. Figure 6.12 shows the USDA classification of typical granitic residual soils.

Unified and AASHO Classifications

Surface layers of granitic soils are commonly designated SM, with ML, SC, or ML-CL occasionally being used. The subsurface contains a greater concentration of fines which have been leached from the A horizon, resulting in an SC or a more plastic CH description. In some cases the subsurface is described by CL or CL-CH. Figure 6.13 shows a Unified classification of a typical residual soil profile weathered from granite in a subhumid climate. Categories within the AASHO classification system typically include A-4, with A-2 and A-6 occasionally occurring.

Water Table

In humid climates the seasonal high water table in residual granitic soils is near the surface. Because of the impervious nature of the rock material and the plastic subsoil, moisture retention is high and internal drainage is low. Because of these characteristics many areas of standing water are encountered, especially in small depressions during the wettest seasons of the year. Much of the water retained by the surface soil slowly percolates into fractures of the bedrock. No seepage of moisture takes place through the actual mass of the rock material, owing to its low permeability.

Drainage

Sandy-silty surficial soils developed in humid climates provide good initial surface drainage, but clayey subsoils have moderate to poor percolation. This latter condition is commonly described in soil reports as a "hard fragipan condition," that is, one in which (1) surface drainage is impeded by a layer of relatively impervious or poorly drained soil, and (2) standing water can therefore be expected during heavy rainfalls until the water runs off or slowly penetrates the soil. The impervious rock base further aggravates this situation.

Soil Depth to Bedrock

Because of the massive and uniformly resistive nature of granitic forms, relatively thin soil profiles are developed over the bedrock. Humid climates commonly have soil depths of 4 to 6 feet, at which point rock fragments are encountered, grading into the more massive bedrock. In arid climates thinner profiles are developed over partially weathered rotten rock, and soil depths are typically several inches to 2 feet in thickness. In all climates large, exfoliated boulders may be found within the soil profile.

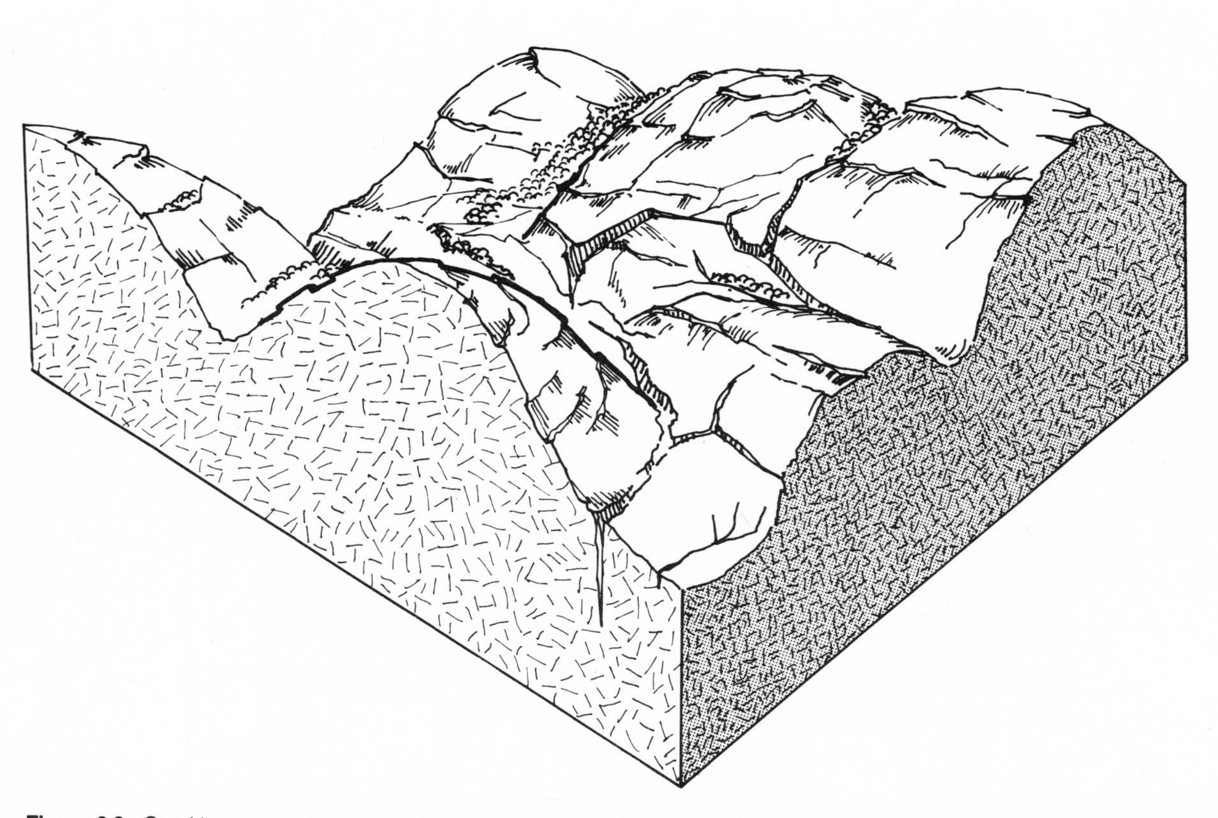

Figure 6.9. Granitic masses in an arid climate. Granitic landforms in arid climates appear bleak, dry, and desert-like. The dissected topography creates a wide variety of spatial change, but the scattered vegetation is uniform and monotonous. Hilltops occur at different elevations and offer a variety of regional and semi-regional views. Highways cross the broader valleys and offer rather closed visual experiences.

Issues of Site Development

Sewage Disposal

The thin soil depths, rock fragments, and plastic nature of the subsoil of granitic formations do not offer suitable conditions for septic tank leaching fields as a method of sewage disposal. Cesspools are used in some areas, but their maintenance is high since they require pumping several times a year. Large tract developments should consider installation of a sewage treatment system or linkage to existing community treatment systems to minimize potential pollution of the existing hydrological ecosystem.

Solid Waste

The thin soil depths and the plastic nature of the subsoil make it unsuitable for large sanitary landfill operations, and the seasonal high water table and impervious rock conditions present the further danger of ground and surface water contamination. Disposal of solid wastes in granitic regions can usually be acccomplished in other associated landforms; for example, major valleys

Table 6.3. Intrusive igneous rocks: Granitic forms: Large, massive (humid and arid)

Humid

Topography

Bold, domelike hills

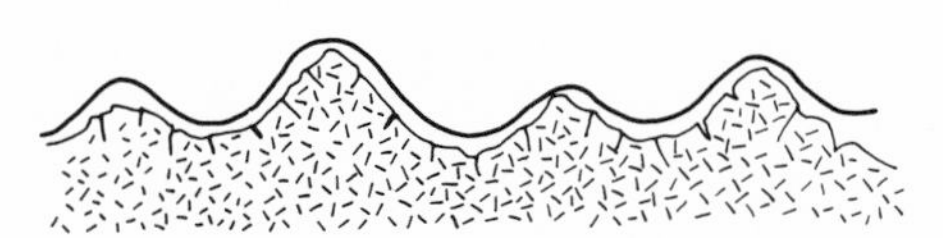

The topographic relief of granitic intrusions typically shows as massive, rounded, domelike hills. The tops of the hills are softly rounded; the sideslopes are steeper. The weathering processes of pressure-release and exfoliation tend to maintain the domelike appearance. Debris and large boulders weathered into rounded shapes accumulate in drainage courses and depressions.

Drainage

Dendritic: Medium

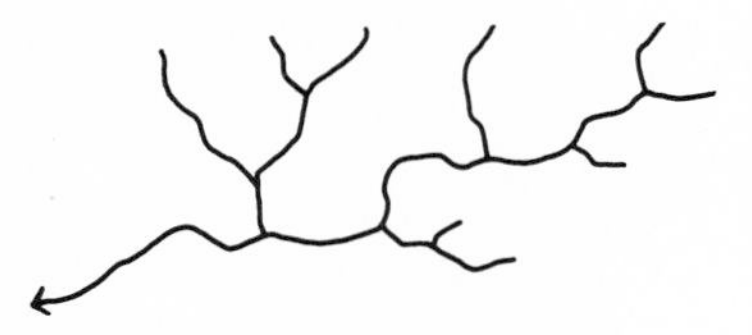

A dendritic drainage pattern of medium tex ture is common in humid climates and indicates the uniform materials of which the forms are made. The domelike hills cause curvilinear alignments to develop, and these are important evidence in the identification of granite. Intersections of tributaries occur at right angles or may be slightly acute upstream.

Tone

Uniform dull gray

The uniform composition of granite is reflected in its overall uniform soil tone which varies in color from dull to light gray. Fracture zones may be identified by linear, dark-toned areas.

Gullies

U-shaped

The residual soils developed from granite in humid climates show a U-shaped gully cross section, indicating a moderately cohesive, sandy-clay soil.

Vegetation and land use

Cultivated and forested

Nonglaciated regions are characterized by cultivation on the gently rolling hilltops and forests on the steeper slopes. Settlement patterns occur either on hilltops or in the major valleys. Glaciated regions typically have thin soil profiles over bedrock; their rugged topography is too severe for traditional, intense land uses. These areas are generally forest-covered.

Arid

Topography

A-shaped hills

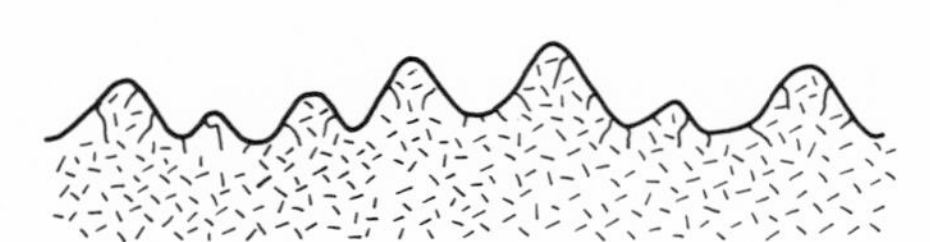

Bold, massive, domelike hills also occur in arid granitic regions. Many rounded boulders and rock outcrops can be observed as microfeatures, but these are not concentrated in depressions and drainage courses as in humid climates but rather tend to form large mounds as the surrounding residual soil is removed by wind erosion. Highly dissected areas appear, having A-shaped hills with steep sideslopes.

Drainage

Dendritic: Fine

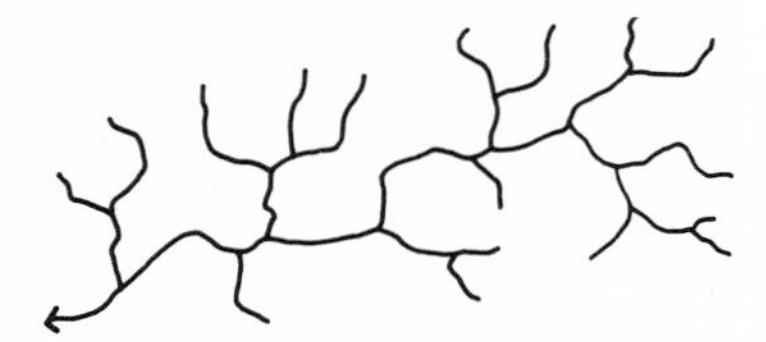

Drainage patterns in arid granitic regions are commonly dendritic and fine-textured. Granitic rock materials are massive and impervious to water, so a large amount of runoff can be expected. Curvilinear alignments are formed by the domelike hills and resemble sickle shapes in the drainage system. Intersections of tributaries occur at right angles or may be slightly acute upstream.

Tone

Light, banded

The general tone in these regions is very light and bright, owing to the thin soil cover and the light tones of the bare rock. Fractures appear as dark lines.

Gullies

Few or none

Gullies are rarely present since erosive forces do not result in the formation of the deep soil profiles needed for gully development.

Vegetation and land use

Barren, rangeland

Granitic landforms in arid climates develop very thin soil profiles punctured by many rock exposures and exfoliated residual boulders. Little vegetation is present because of the thin soil and arid conditions. Fractures, by accumulating greater soil and moisture, may have concentrations of vegetation. The typical land use is for grazing or as rangeland.

Table 6.4. Intrusive igneous rocks: granitic forms: Linear: Dikes (humid and arid)*

Topography	*Drainage*	*Tone*	*Vegetation and land use*
Straight, narrow ridges	None	Variable	Forested or barren
Igneous material injected into fault zones is generally more resistant than the surrounding rock and therefore forms sharp or rounded ridges. In rare instances the material composing the dike is less resistant and in this case linear depressions result. Occasionally, circular ring dikes are found; these are formed by the intrusion of igneous material along the faces of a circular fault. Commonly, the interior area has dropped down during the faulting process.	The forms are generally too small to develop any discernible drainage patterns. The steep slopes facilitate runoff to surrounding landforms.	The tone varies but is different than the tone of surrounding materials. In some regions dikes appear lighter in tone; in others they appear darker. *Gullies* None Because these forms are small in area, discernible gullies are not developed.	In humid climates dikes generally appear as a forested ridge bisecting the region. In arid regions little vegetation is found, owing to the elevated positions of this form and the thin soil cover.

*Because their characteristics are so alike, the pattern elements presented in this table apply to both humid and arid climates.

with large waterlaid deposits offer more suitable sites.

Incineration is the commonly used method of solid waste disposal in urban areas in granitic regions. To maximize economies it is recommended that future incineration plants be designed in combination with power production plants utilizing fossil fuels. In such a process combustible materials are separated from the solid wastes and burned with coal for heat generation and power production, and the expenses of incinerating the solid wastes are thereby returned through the power-generation facility. It should be noted that fossil fuels are not directly available in granitic landforms and must be transported from other regions.

Many urban areas in granitic regions do not dispose of their solid wastes by approved sanitary landfill operations. Since the supply of refuse far exceeds the soil material needed for cover, large, exposed, rat-infested mounds of garbage and waste are created. Other urban areas along coastal sections simply dispose of their refuse by dumping it into the ocean, thus relocating the problem so that it is no longer noticed. More creative thought is needed in this area, so that economically feasible and politically acceptable waste disposal systems can be designed for both large and small urban areas.

Trenching

In granitic regions excavation of material, when constructing trenches for pipelines, is generally expensive. Even in humid climates the thin soil profile does not permit a standard, 6-foot-deep trench to be excavated without encountering either rock fragments or bedrock, which must be blasted. Highly fractured granitic regions, however, facilitate blasting and removal operations. In other areas boulders may be scattered throughout the soil profile and cause higher-than-normal removal expenses. In humid climates the removal of massive bedrock is four to five times as expensive as removal of the same amount of deep, dry, moderately cohesive soil. Costs for removal of a combination of rock and soil in a standard 6-foot-deep trench, including blasting, are increased by a factor of three to four. As the depth of the trench increases, costs become higher; increasing the trench depth at 6-foot intervals increases the per-cubic-yard cost each time by 50% over the price for the first six feet.

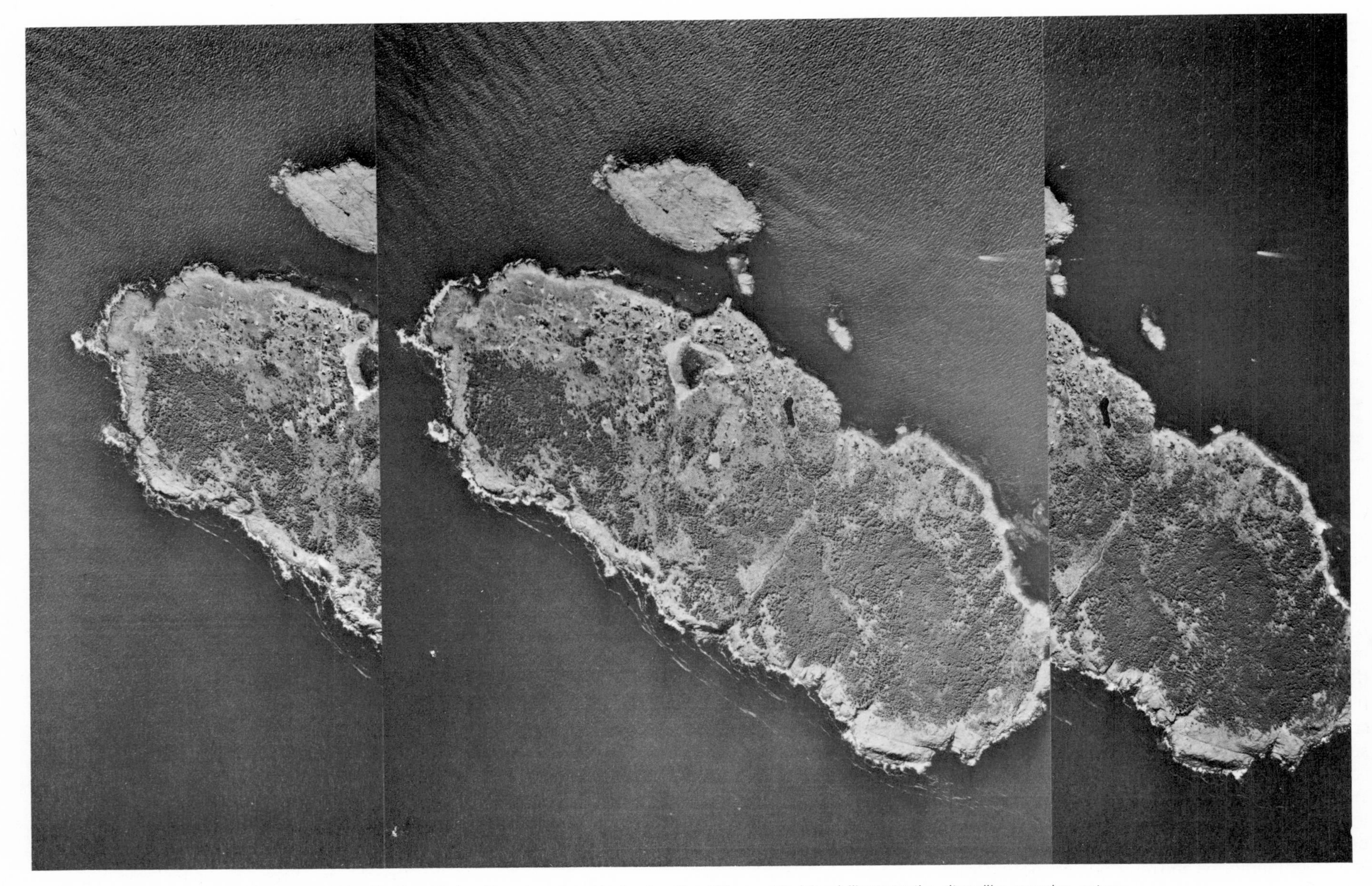

Figure 6.10. (A) Granitic landform in a humid climate. The hills on this island illustrate the domelike massive nature of granitic materials. Some major fracture patterns can be seen in the rock structure. Stereo-reversal as shown in the right stereopair can facilitate fracture, gully, or drainage pattern analysis. Photographs by the Soil Conservation Service, EPY-5GG-59,61, 1:20,000 (1667′/″), May 14, 1966.

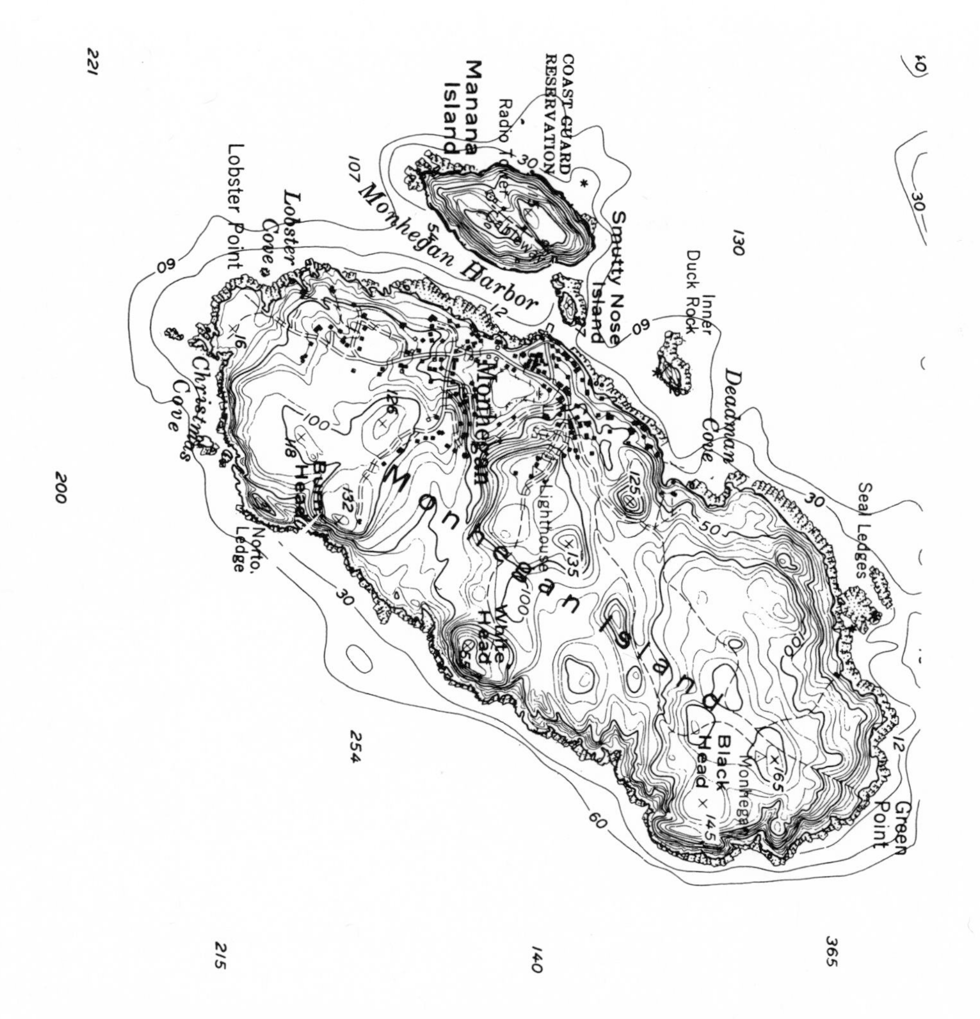

Figure 6.10. (B) U.S. Geological Survey Quadrangle: Monhegan Island (7½′).

Excavation and Grading

Grading and excavation to create large, level sites are difficult because of the thin soil depths, the resistant nature of the bedrock, and the massive, rolling character of the topography. Residual, rounded boulders can be found at all levels in the soil profile and often require blasting. Massive bedrock also requires blasting and heavy equipment for removal. Typical costs per cubic yard are four to five times higher than those encountered in deep, dry, moderately cohesive soils. The plastic subsoils in humid climates present grading and excavation difficulties during wet seasons and require compaction. Most grading operations in these regions emphasize fills over cuts because of the high costs of rock removal.

Construction Materials

Topsoil: Poor. The silty-sand surface soil horizon is too thin to provide a high potential source of topsoil.

Sand: Not Suitable. The sand that exists within the soil profile is mixed with silt and clay which are difficult to separate and remove.

Gravel: Not Suitable

Aggregate: Poor to Good. Aggregate made from granite breaks down quickly under chemical weathering. Therefore it is not recommended for use as aggregate in concrete. Granitic aggregate is suitable for use when it is not exposed to weathering. Thus it is a good source of aggregate material for road subbase and base-course material.

Surfacing: Good. The sandy-clay nature of the residual granitic subsoils makes them good materials for secondary road surfacing. The mix of sand and clay particles provides the necessary binder for stabilizing the surface and providing good wear.

Borrow: Poor to Good. Depending upon the specific purpose of a fill, granitic residual soils and rock material offer poor to good suitabilities. The plastic subsurface soil found in humid climates has limited uses because of its low load-bearing capacity although it becomes more suitable when compacted. Blasted rock fragments from large cuts and excavations can be used satisfactorily for large earthfill operations. Compaction of the moderately cohesive residual soils for fills is accomplished by pneumatic-tired rollers.

Building Stone: Excellent. Granite is commonly quarried for use as curbing and building stone; the relative content of quartz and feldspar determines its sensitivity to pitting and weathering.

Figure 6.11. (A) Granite in an arid climate. Granite in an arid climate in Inyo County, California, is characterized by a sickle-shaped dendritic drainage system and exposed rounded rock masses along the sides and tops of the hills. Photographs by the U.S. Geological Survey, California 19-ABC, 1:37,400 (3117′/″), July 17, 1947.

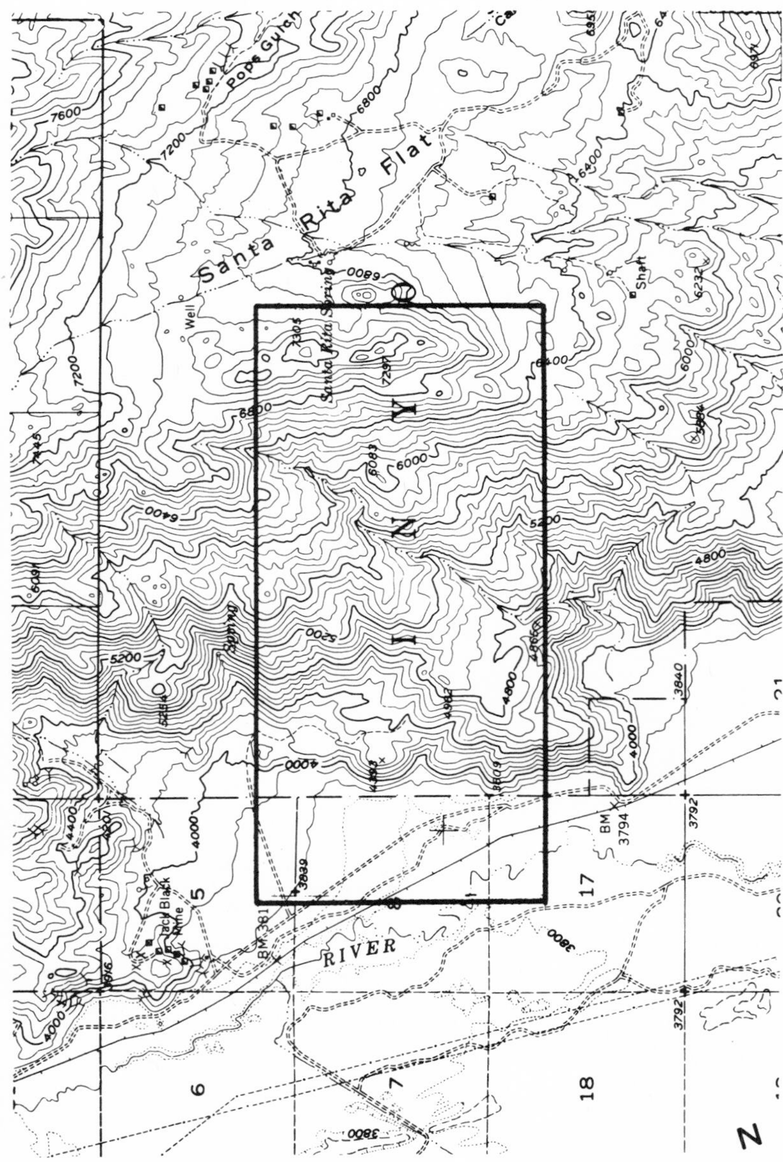

Figure 6.11. (B) U.S. Geological Survey Quadrangle: Independence (15′).

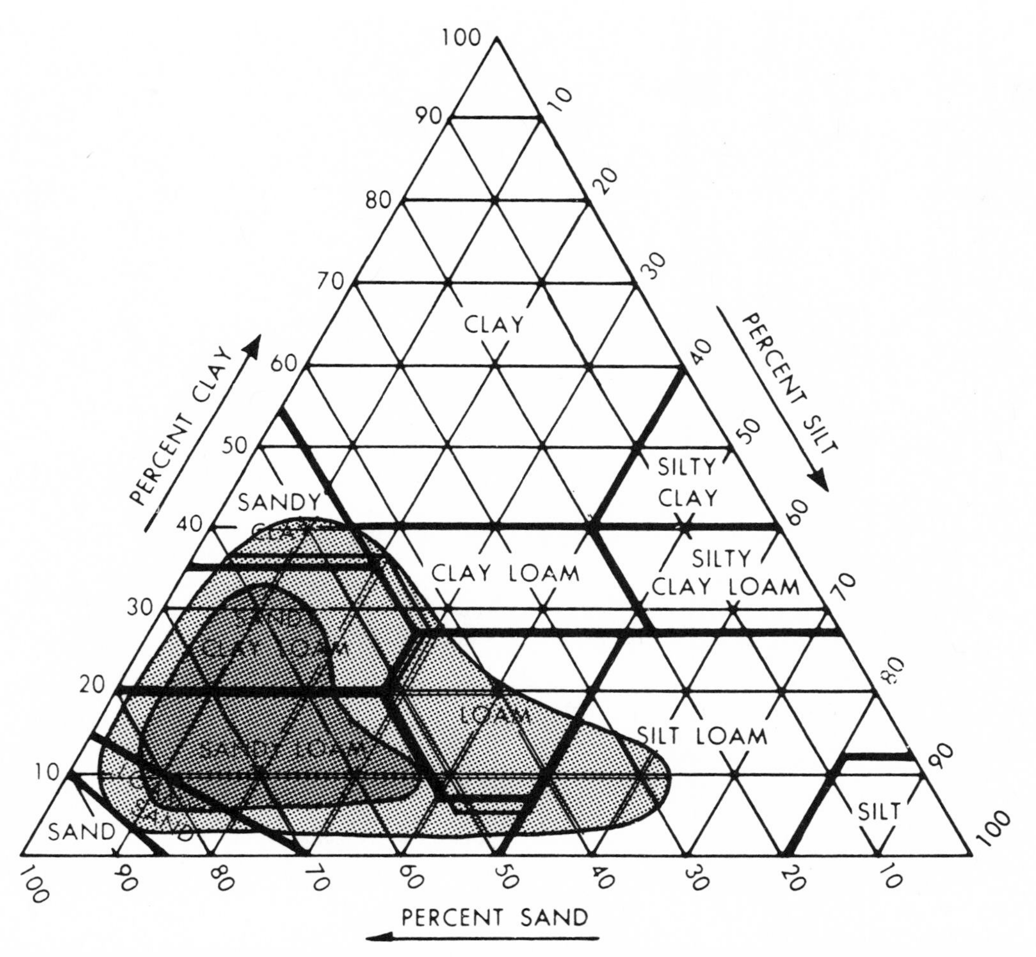

Figure 6.12. The U.S. Department of Agriculture soil texture triangle indicates commonly occurring granitic residual soils as sandy clay loam, sandy clay, and clay loam.

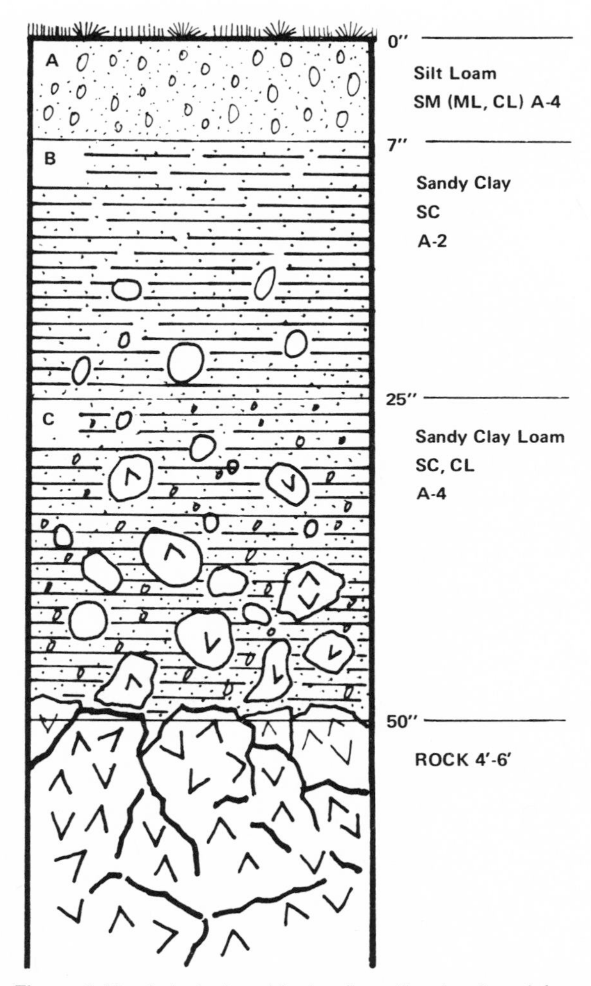

Figure 6.13. A typical residual soil profile developed from granite in a humid climate.

Mass Wasting and Landslide Susceptibility

Granitic regions are not generally susceptible to landslides. However, slides may occur along cuts on hillsides if the rock is highly fractured and contains sufficient water, especially when aided by frost forces. Soil slumps occasionally occur on hillsides during wet seasons as a result of high water content and the impervious nature of the bedrock; these conditions cause lateral movement of water through the subsoil and may provide sufficient lubrication for slumping. (See Chapter 11, "Mass Wasting.") In regions of highly fractured rock, high water tables, or high rainfall, field investigations must be carried out to determine the risks of mass wasting.

Groundwater Supply

In humid climates small water supplies, suitable for individual residences or small communities, can usually be obtained from shallow wells. Such wells, penetrating the shallow residual soil and fractured bedrock, capture water that flows and

Table 6.5. Granitic formations: aerial and cartographic references

State	County	Photo Agency	Photograph no.	Scale	Date	Corresponding quadrangle and geologic maps from USGS	Corresponding soil reports from SCS
California	Inyo	USGS*	California 19-ABC	3117′/″	7/17/47	Independence (15′), MF-254	–
California	Mariposa	USGS*	California 6-ABC	3917′/″	8/26/55	Hetch Hetchy Reservoir, Yosemite, Tuolumne Meadows, Merced Peak (15′), Prof. Paper No. 160	–
California	Mariposa	USGS*	California 7-AB	3917′/″	8/26/55	Hetch Hetchy Reservoir, Yosemite (15′), Prof. Paper No. 160	–
Georgia	DeKalb	USGS*	Georgia 1-AB	2467′/″	3/9/55	Stone Mountain (7½′)	1914†
Maine	Lincoln	SCS	EPY-5GG-59-61	1667′/″	5/14/66	Monhegan Island (7½′)	–
Maine	Piscataquis	USGS*	Maine 1-ABC	5250′/″	6/30/52	Katahdin, Harrington Lake, Telos Lake, Traveler Mt. (15′)	–
Minnesota	Cook	ASCS	CIT-3-56-58	1667′/″	7/6/40	Grand Portage (7½′)	–
Missouri	Iron	USGS*	Missouri 1-ABC	1667′/″	10/18/54	Ironton (15′)	–
New Hampshire	Coos	ASCS	DXU-6N-27-29	1667′/″	8/4/55	Lewiston (1:250,000)	1943
New Hampshire	Coos	ASCS	DXU-4N-41-43	1667′/″	8/4/55	Lewiston (1:250,000)	1943
New Hampshire	Grafton	USGS*	New Hampshire 1-AB	1667′/″	7/11/55	Franconia (15′)	1939
New York	Essex	ASCS	EKC-3CC-182-185	1667′/″	6/16/62	Au Sable Forks, Willsboro (15′)	–
New York	Essex	ASCS	EKC-1CC-200-203	1667′/″	6/16/62	Mt. Marcy (15′)	–
South Dakota	Custer	ASCS	BNW-6CC-80-81	1667′/″	8/28/62	Hot Springs (1:250,000)	–
South Dakota	Custer	ASCS	BNW-7CC-73-74	1667′/″	8/28/62	Hot Springs (1:250,000)	–
Wyoming	Park	USGS*	Wyoming 1-ABCD	3117′/″	7/7/48	Deep Lake (15′)	–

*From *Selected Aerial Photographs of Geologic Features in the United States,* U.S. Geological Survey Prof. Paper No. 590.
†Out of print.

Table 6.6. Granitic dikes: aerial and cartographic references

State	County	Photo Agency	Photograph no.	Scale	Date	Corresponding quadrangle and geologic maps from USGS	Corresponding soil reports from SCS
Colorado	Huerfano	ASCS	DYT-6P-83-86	1667′/″	10/7/55	Spanish Peaks (30′)	–
New Hampshire	Rockingham	USGS*	New Hampshire 2-AB	5667′/″	4/30/51	Mt. Pawtuckaway (15′)	1959
New Mexico	San Juan	USGS*	New Mexico 1-ABCD	5280′/″	2/5/54	Ship Rock (7½′), I-345	–

*From *Selected Aerial Photographs of Geologic Features in the United States,* U.S. Geological Survey Prof. Paper No. 590.

seeps along the transitional rock zone and small fractures. In arid climates location of suitable water supplies is more difficult. Intersecting fracture zones offer the greatest potential, owing to the seepage of moisture along these planes for great distances. However, it should be noted that the stability of the associated vegetation and wildlife in arid regions would be critically threatened by removal of water resources.

Large urban areas have difficulty in locating sufficient water supplies within granitic regions because of the impervious nature of the bedrock. Deep wells do not provide a high yield unless regional fracture zones that gather large amounts of subsurface water are encountered. Many areas must develop regional reservoirs or rely upon other sources of surface water for a sufficient water supply.

There are many examples of large urban areas in granitic regions that have severe difficulties in obtaining water supplies. Hong Kong, for example, has very little water available for public consumption. Even small towns in granitic regions may have problems in obtaining suitable water supplies. For instance, the summer tourist populations of fishing villages on many islands along the coast of Maine are constrained by the lack of available drinking water. During periods of drought or low rainfall, fresh water must be shipped to these areas from the mainland.

Pond or Lake Construction

The massive, rolling nature of granitic landforms presents difficulties in the location of dam structures, since valleys are wide. Because of these problems, power generation by hydroelectric means is not feasible in many granitic regions. Where damming conditions are favorable, the residual soil and the impervious rock can support the ponded water without significant loss through seepage. Only if the rock is highly fractured will losses through seepage be significant. In these regions impervious soil or synthetic materials can be used to coat the bottom of the area to be flooded.

Foundations

Granitic rocks have extremely high load-bearing capacities and, unless highly weathered and fractured, provide suitable platforms for any proposed loads. Large, heavy structures in shallow soil areas usually transfer their bearing loads directly to the rock via piers; however, in deep soil regions footings or rafts are typical. Smaller structures commonly use footings or rafts. Residential units are typically constructed with continuous footings, with or without basements, depending upon the depth of the associated soils. In regions where the rock is highly fractured, foundation problems are more severe, and in these cases fractured, rotten rock material may necessitate removal or the use of piers or deep footings to penetrate the weak zone. In thin soil areas, such as exist in Nova Scotia, indigenous construction of residential structures does not include excavated basements. Basements are created by placing the first-floor elevation 4 to 6 feet above the original grade and filling earth around the foundation after its construction. Fills are preferred to cuts because of the high costs of rock blasting and removal.

Highway Construction

Highways in granitic regions generally have raised grade lines and broad curvilinear alignments, following and linking large valleys or the lower slopes of major hills. Cuts are minimized because of the associated high costs of rock excavation. Highly dissected regions need more drainage structures and earthwork and perhaps bridging or tunneling, thus greatly increasing construction costs. Cold regions require thicker, granular subbases to prevent frost heaves of the road surface. Road cuts may encounter unpredictable seepage zones where rock fractures are found, and some sliding and slumping can be expected along these cut faces if the rock is highly fractured. Most materials needed for construction purposes are available within short distances.

Extrusive Igneous Rocks: Basaltic and Volcanic Forms

Introduction

Extrusive igneous forms are grouped into two categories: those associated with lava flows, and those resulting from explosive action. Lava flows originate from fissures, dikes, or cones and form broad plains, plateaus, domes, fields, and lava tongues. Mineralogically, lava contains basalt, andesite, trachyte, and rhyolite, but basalt is the most common. In this section the term *basalt* is used to describe extrusive rocks that are dark, basic, and fine-grained, and which typically develop columnlike structures along escarpments. The igneous formations resulting from explosive actions are fragmental and consist of volcanic breccia and tuff. The materials may be consolidated or unconsolidated and are commonly found in regions of present or recent volcanic activity.

Volcanic forms and lava flows have different characteristics at different stages of erosion (Figure 6.14). Young volcanoes are conical formations of cinder cones and associated flows. Mature volcanic cones are dissected by streams and glaciers which expose the internal structure of dikes. Old volcanoes no longer have conical forms but appear as volcanic necks, perhaps with radiating dikes. Young basalt flows appear as broad plains having a flat topography (viscous flows, however, are very rough in surface texture and have a twisted, ropy appearance). Mature basalt flows are dissected and show many lava or basaltic plateaus of equal elevation. Old basalt flows develop a peneplain topography of scattered mesas or buttes which are remnants of the original lava plain.

Figure 6.14. Erosion cycle of volcanic forms and basalt flows.

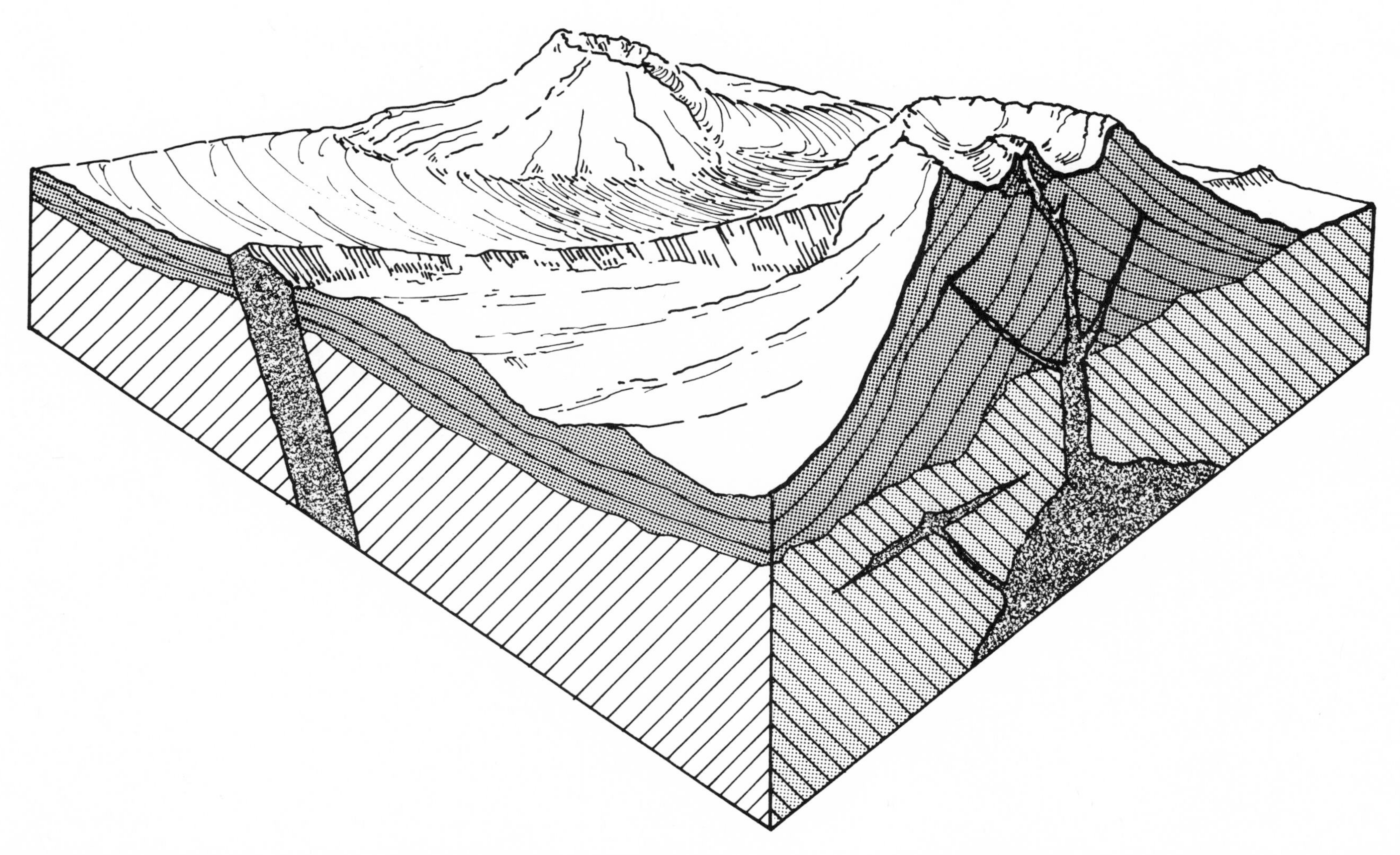

Figure 6.15. Volcanic cones. The upper slopes of volcanic cones are visually sensitive, owing to their elevated position above the lowlands. Construction of roads on these slopes requires cuts which potentially could have a high visual impact. Many cinder cones and volcanic structures are regionally significant in size and scale and provide a regional identity, for example, Mt. Shasta in California or Mt. Fujiyama in Japan.

Volcanic formations and flows are typically found in complex combinations, but the sequence of formation of these deposits can be quite simply described as follows. The central feature of the volcanic formation is the actual cinder cone or fissure from which the lava flow originated. Away from the cone, near the front of the flow, viscous materials solidified, forming rhyolite and andesite. Near the edge of the flow, interbedded deposits are common, consisting of fragmental rocks and flows; the coarse, fragmental rocks found here grade into fine, fragmental rocks such as tuffs.

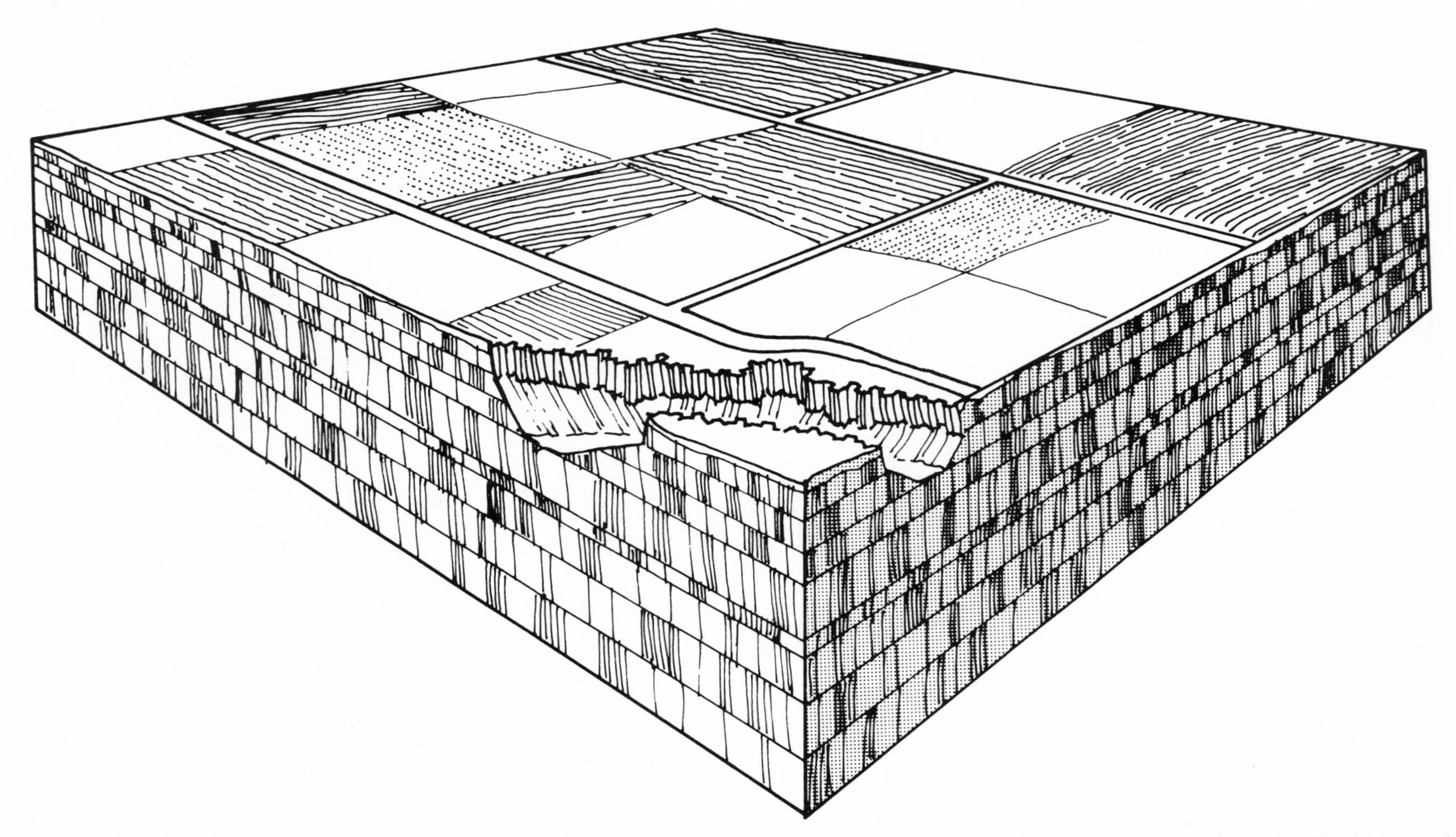

Figure 6.16. Young, nondissected basaltic plains. Young basaltic plains are not spectacular visually except where major drainage courses have disected the plateau and formed deep, vertical canyons. It is characteristic of vertical rock faces of basalt to have a columnar structure in which the columns appear as six-sided polygons. Devil's Tower near the Black Hills of Wyoming is an excellent example of this columnar structure. Young basaltic plains are typically treeless, which makes these regions visually sensitive to high towerlike structures or city skyline profiles.

Interpretation of Pattern Elements

Many complex local interbeddings, variations, and gradations of volcanic and flow forms can be identified, thus making it difficult to categorize the landform types strictly. For purposes of photographic interpretation and general terrain analysis,the key identifies the most common formations; however, for specific investigations more detailed classification is recommended. The key has been divided into two tables of two parts each, defining the characteristics and patterns associated with young volcanic forms and young basalt lava flows, and dissected interbedded flows and dissected fragmental tuffs. Except where noted, the landforms are similar in both humid and arid climates.

Young volcanic regions are characterized by a conical topography with (if exposed by erosion) associated radiating dikes. Basalt flows have flat surfaces with a parallel regional drainage system. These flows can be easily identified if an escarpment, exhibiting the typical saw-toothed outline associated with basalt columnar jointing, is present. Fragmental tuff regions have a fine-textured dendritic drainage similar to clay shale in appearance, but the hilltops occur at many different elevations and the ridges are sharp and have steep sideslopes. Basalt flows interbedded with ash have a terraced topography along the hillsides where the more resistant bands of basalt occur. The columnar structure and details of the drainage pattern are of value in distinguishing basalt materials from flat, interbedded sedimentary rocks.

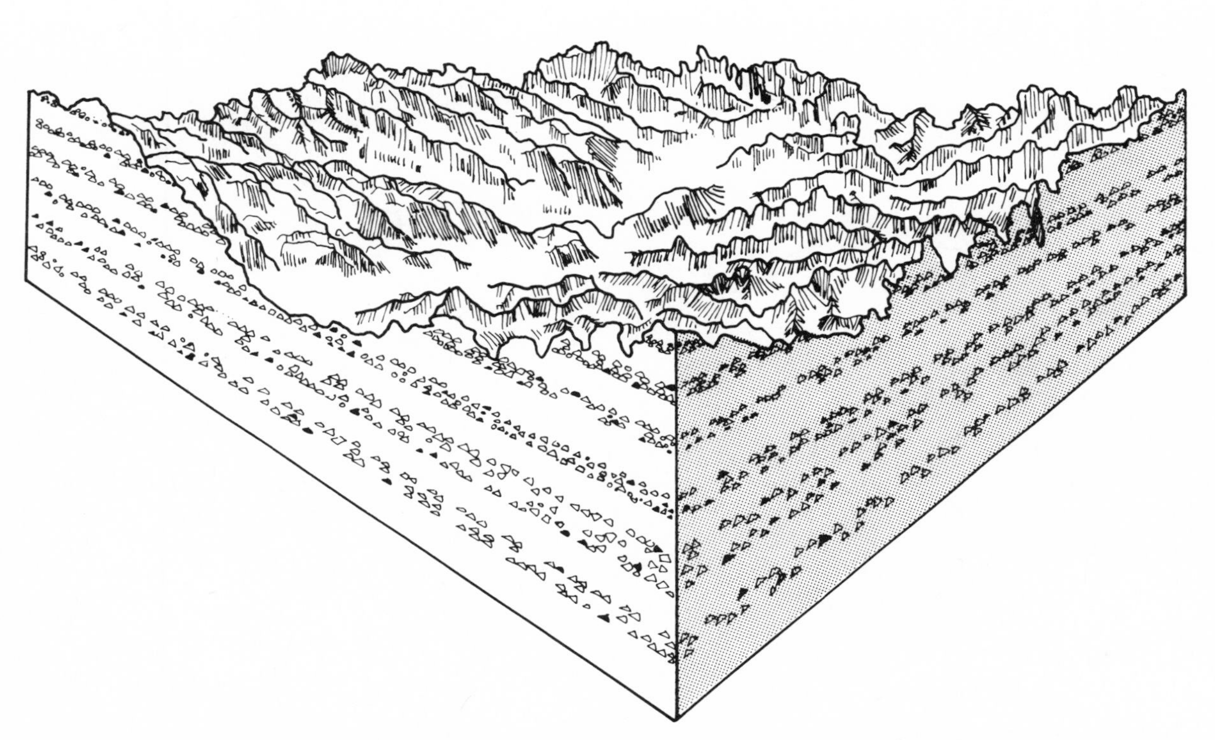

Figure 6.17. Volcanic tuff. Volcanic tuff regions offer one of the most visually closed landscapes in the world. The many peaks and pits can effectively screen most uses located within the valleys. Land uses and structures located on hilltops are observable from surrounding hilltops. Extreme variations in spaces and a large viewing potential exist within these regions.

Soil Characteristics

Depending upon the age of the deposit and the climate, either bare rock or a moderately deep residual soil is found. The residual soil consists of several inches of silty clay underlaid by clays or clay-associated soils. Humid and tropical climates typically have profiles in young formations that average 4 to 6 feet over bedrock. Arid and semiarid climates often develop less than 2 feet of residual soil cover and have occasional outcrops of bare rock.

U.S. Department of Agriculture Classification

On-site investigations should be undertaken in extrusive igneous regions because of the variety and complexity of materials that may be found. The U.S. Department of Agriculture commonly classifies basaltic residual soils as silty clay or silty clay loam in the surface horizons, and clay loam, silty clay, or clay in the subsurface (Figure 6.24). In arid climates rock fragments are common throughout the profile and the soil is described as stony. Interbedded flows can have a wide variation of compositions and rock fragments but are generally slit clays. Fragmental tuffs develop a clay subsoil under a surface soil of silty clay.

Unified and AASHO Classifications

Basaltic and tuff residual soils in humid or tropical climates develop profiles commonly categorized as ML or CL in the surface horizon and MH, or typically CH, in the subsurface horizon. Arid residual soils of basalt are commonly classified as ML-CL, ML, or GM in surface soils, and CH, CL, MH, ML-CL, or GC in the subsurface. The AASHO system categorizes basaltic and tuff residual soils as A-6, A-4, and A-7.

Water Table

Residual soils formed from basaltic materials are semi-impervious, but they are underlaid by highly fractured rock or tuff deposits which are well drained. The poorly drained characteristic of

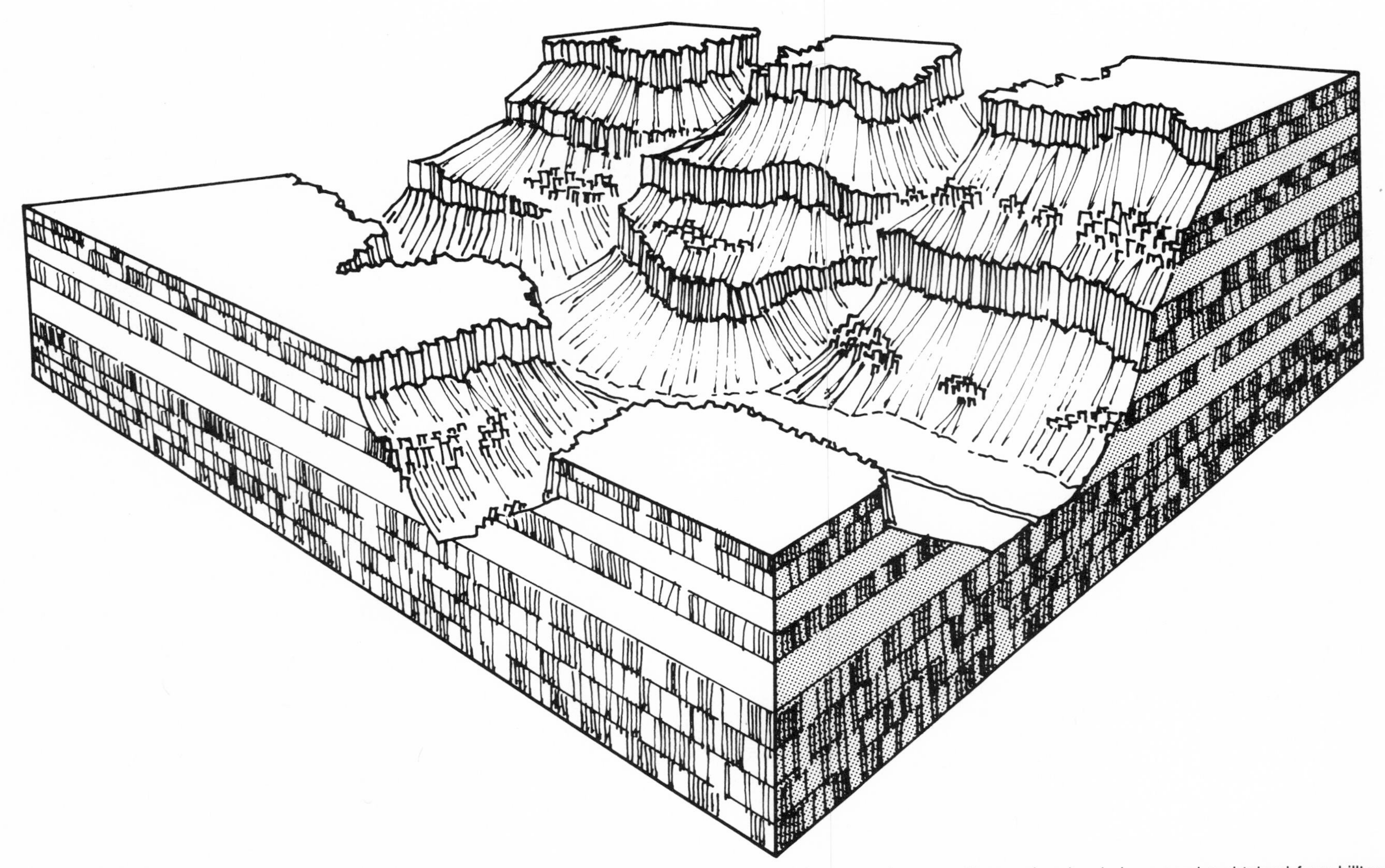

Figure 6.18. Interbedded basalt flows dissected basalt regions, or interbedded flows, produce a landscape that is visually absorptive. The many depressions and valleys create the feeling of spatial closure, and many valleys and regional views can be obtained from hilltops and ridgelines. Wide variation in spaces and a large viewing potential exist within these regions.

the residual soil creates a seasonally high water table in tropical and humid climates. The residual soils in arid climates are thinner and somewhat coarser and do not develop significant water tables.

Drainage

The fine-textured residual soils developed from basalt and tuff are not well drained through their surface horizons, and both basalt and tuff develop a heavy, plastic-clay subsoil which is semi-impervious. However, the basalt soils are underlaid by highly fractured rock with many joints, and this compensates for the clayey surface soil and helps it to drain. Tuff regions in turn are

Table 6.8. Extrusive igneous rocks: Young volcanic forms and basaltic flows (humid and arid)

Volcanic forms

Topography

Cinder cones

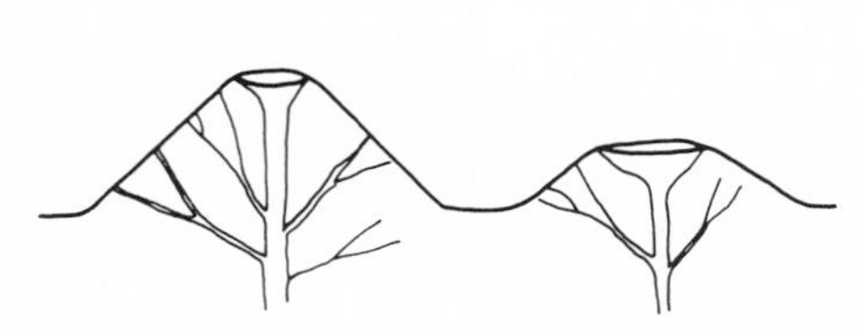

Rounded cinder cones are clear indicators of recent volcanic activity. Both mature and old dissected cones have some or all of the internal structure exposed, including the volcanic neck and radiating dikes.

Drainage

Radial

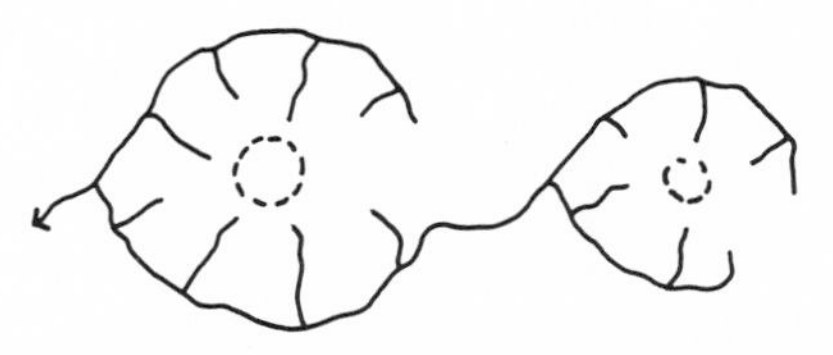

As the formation is dissected over time, a radial drainage pattern is developed around the circular volcanic cone. The texture of the pattern is dependent upon the climatic zone; the finest textures are found in arid climates. Many tributaries along the slopes appear parallel.

Tone

Dark gray

Dark gray tones are common, owing to the dark tone of the predominant basalt materials.

Gullies

Sag and swale

Many parallel gullies can be observed, especially in humid or tropical regions. Sag-and-swale cross sections indicate fine, cohesive materials. These gullies may be dark in tone.

Vegetation and land use

Natural cover

Young volcanic forms are potentially active and have steep sideslopes, neither of which qualities makes them attractive to traditional forms of land use. Most appear in their natural state except for cultivated lower slopes. Dissected cones are more rugged and are also generally undeveloped.

Basaltic flows

Topography

Level plain

Young basaltic flows form level or gently sloping, broad plains as a result of the fluid nature of the material. Mature and old dissected areas have many mesas or plateaus, with some pear-shaped appendages. Microfeatures of flow lines or gas pocket depressions may be observed on very young deposits. Vertical escarpments with columnar jointing develop along major drainage courses.

Drainage

Regional parallel: Coarse

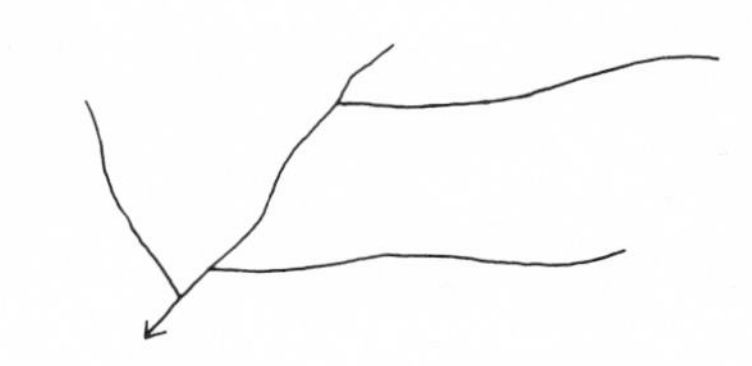

Basaltic rock itself is impervious to internal drainage, but the many joints and fractures create high permeability. Because of its gentle, uniform slope, the lava flow develops a coarse, regional, parallel drainage pattern.

Tone

Dark: Scattered spots

Tones in basaltic areas are typically dark, reflecting the coloring of the material. Scattered light spots are common and indicate differences in soil cover. Flow marks or blisters may also be visible.

Gullies

None

Few gullies are apparent in these formations because of the pervious nature of the rock material resulting from jointing.

Vegetation and land use

Cultivated or barren

Young flows in arid or semiarid climates are generally barren, owing to their thin soil cover. Older flows and those found in subhumid, humid, or tropical climates are usually intensively cultivated in rectangular and square field patterns. Highways are free to follow straight lines and grid patterns, since little constraint is presented by the topography.

Table 6.9. Extrusive igneous rocks: Dissected, interbedded flows and fragmental tuff

Fragmental tuff

Topography

Sharp-ridged hills

Tuff formations in all climates erode quickly and form sharp, knife-edged ridges within a highly dissected topography having characteristics similar to clay shale. Unlike the shales, however, hilltops do not occur at equal elevations. Sideslopes are very steep and of uniform slope, indicating homogeneity of materials.

Drainage

Dendritic: Fine

An extremely fine dendritic drainage pattern develops in tuff formations, indicating a uniform composition. Welded tuffs are more resistant and do not show this extremely fine pattern.

Tone

Dull, light gray

If the tonal pattern can be seen through the vegetative cover, it will usually appear as a uniform dull, light gray tone.

Gullies

Vary

Sizes of particles of tuff vary widely, from 4 millimeters in diameter to dust. V-shaped gullies are common, indicating unconsolidated materials of a granular texture.

Vegetation and land use

Natural cover

Tuff regions are so highly dissected and rugged that they are either in natural forest or grass cover, depending upon the local climate.

Interbedded flows

Topography

Terraced hillsides

A dissected topography exposes stratified layers and interbeddings along the sides of hills. Parallel ridges with convex sideslopes are common. In those areas not highly dissected, flow lines, similar to those associated with basalt, may be apparent.

Drainage

Parallel dendritic

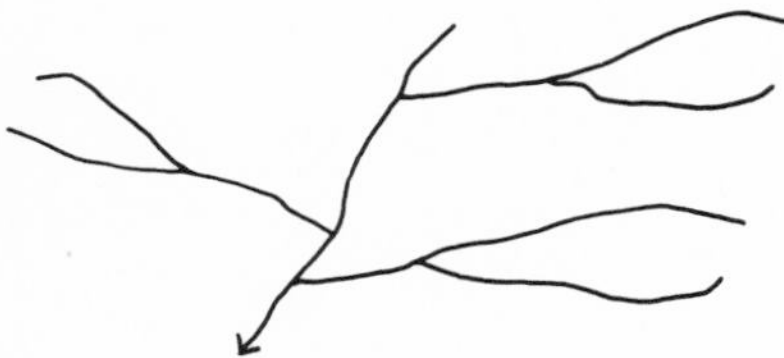

Interbedded deposits commonly develop dendritic patterns with parallel tributaries. The drainage systems in rhyolitic and andesitic materials tend to be parallel, with sicklelike, first-order tributaries.

Tone

Light and dark bands

Light and dark bands along hillsides indicate interbedding. The coloring relates to the type of materials; that is, basalts are dark, rhyolite and andesite are lighter, and tuffs have a light-gray tone.

Gullies

Vary

The thickness and sequence of interbedding influences the types of gullies found in these areas.

Vegetation and land use

Cultivated or natural

Humid climates are generally cultivated, depending upon the ruggedness of the topography and the state of development of the residual soil. Arid regions are typically barren. In both climates areas that are too rugged for cultivation are forest-covered.

Figure 6.19. (A) Young volcanic forms in Craters of the Moon National Monument, Butte County, Idaho, show the typical conical shape and associated lava flows. Photographs by the Agricultural Stabilization Conwervation Service, CVP-5W-193,194, 1:20,000 (1667′/″), July 23, 1959.

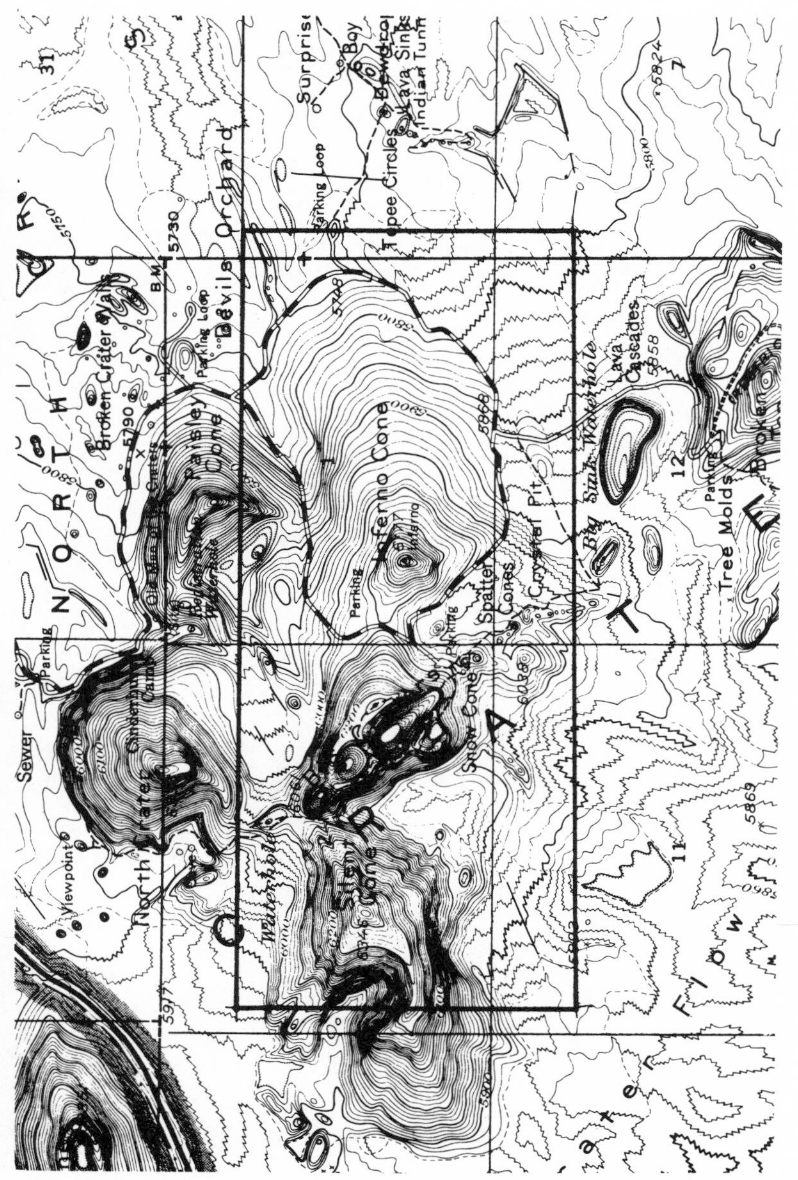

Figure 6.19. (B) U.S. Geological Survey Map: Craters of the Moon National Monument, 1:31,680.

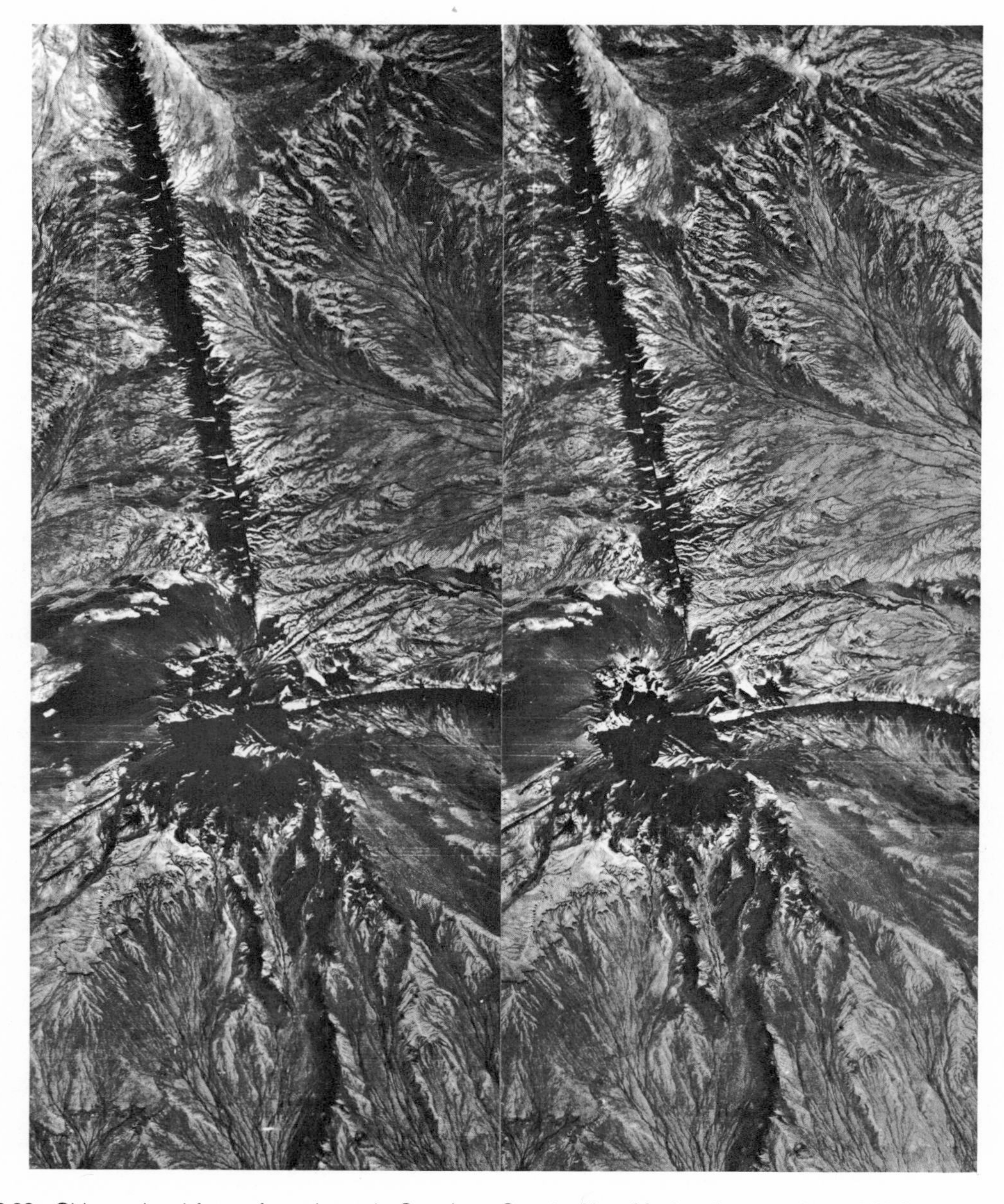

Figure 6.20. Old, erosional form of a volcano in San Juan County, New Mexico, is more than 1500 feet high and has many radiating dikes. Photographs by the U.S. Geological Survey, New Mexico 1-ABCD, 1:54,000 (4500′/″), February 5, 1954.

underlaid by materials of a gravelly texture which are extremely pervious and help to drain the surface layers.

Soil Depth to Bedrock

Soil depth to bedrock depends upon the age of the deposit and the climatic zone. Humid and tropical climates generally develop moderately deep soil profiles from young basalt and tuff, averaging 4 to 6 feet. Older deposits or soils on lower slopes are deeper as a result of the accumulation of soil transported by erosion, creep, and/or frost action. Arid or semiarid regions develop thin soil profiles, averaging 1 or 2 feet in thickness over weathered, fractured rock.

Issues of Site Development

Sewage Disposal

Even in humid or tropical climates, where sufficient soil depths occur, septic tank leaching fields are not generally a satisfactory method of sewage disposal in basaltic soils. The plastic clays found in the subsoil, or B horizon, form a semi-impervious layer with insufficient percolation. In arid regions soil depths are not great enough to warrant consideration of leaching fields. In isolated instances in both humid and arid regions, small pockets of soils can be found where conditions may be satisfactory for on-site disposal utilizing leaching fields, leaching beds, or seepage pits. Intense development or urban areas should construct a treatment system in order to protect the surficial water system and groundwater resources against contamination.

Solid Waste

Sanitary landfill operations can be undertaken in regions having sufficient soil cover and depth

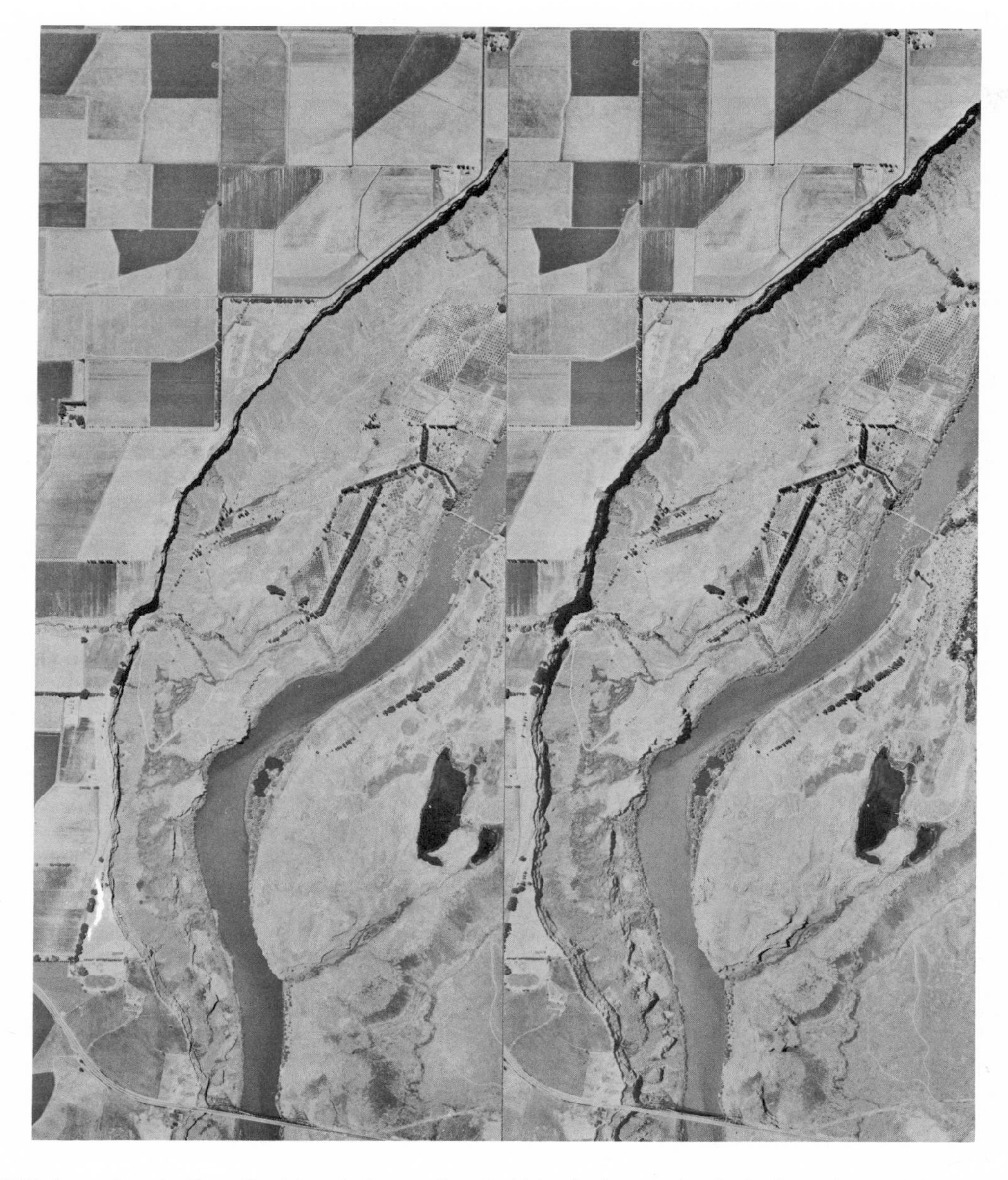

Figure 6.21. A nondissected basaltic plateau in Jerome County, Idaho, is characterized by its flat surface and the saw-toothed edge along the canyon walls. Photographs by the Agricultural Stabilization Conservation Service, YR-4T-21,22, 1:20,000 (1667'/"), July 15, 1957.

to water table. The silty clay soils typically developed are satisfactory for cover material and can be compacted to provide a fairly impervious layer over the waste materials, thus guarding against percolation of rainwater and the formation of leachate. Landforms of volcanic tuff may be less suitable for solid waste disposal, since the parent material is highly pervious and the risk of leachate percolation and contamination of groundwater resources is high. Soil studies and investigations should be initiated by competent engineers or geologists to determine the risks and feasibilities.

Trenching

Trenching costs in these regions relate to the depth of bedrock, hardness of bedrock, and degree of dissection existing within the topography. Young basalt flows are flat and create no difficulties in alignment, but in arid climates the thin soil profiles necessitate rock removal and blasting. However, the columnar structure and high fracturing characteristic of these flows facilitate the placement of charges and removal. Typical costs per cubic yard in these formations are 2.5 to 4 times those encountered in removing the same amount of deep, dry, moderately cohesive residual soil. Volcanic tuff, normally unconsolidated, does not require blasting, so trenching costs are relatively lower. Typical per-cubic-yard expenses are similar to those associated with deep, dry, moderately cohesive soil excavation; linear trench alignment location, however, is difficult because of the rugged terrain. Deep trenches in nonwelded or noncohesive tuff require sheeting. Interbedded flows and dissected basalt flows present alignment difficulties and require rock removal. Seepage zones may be encountered in faults and jointing patterns.

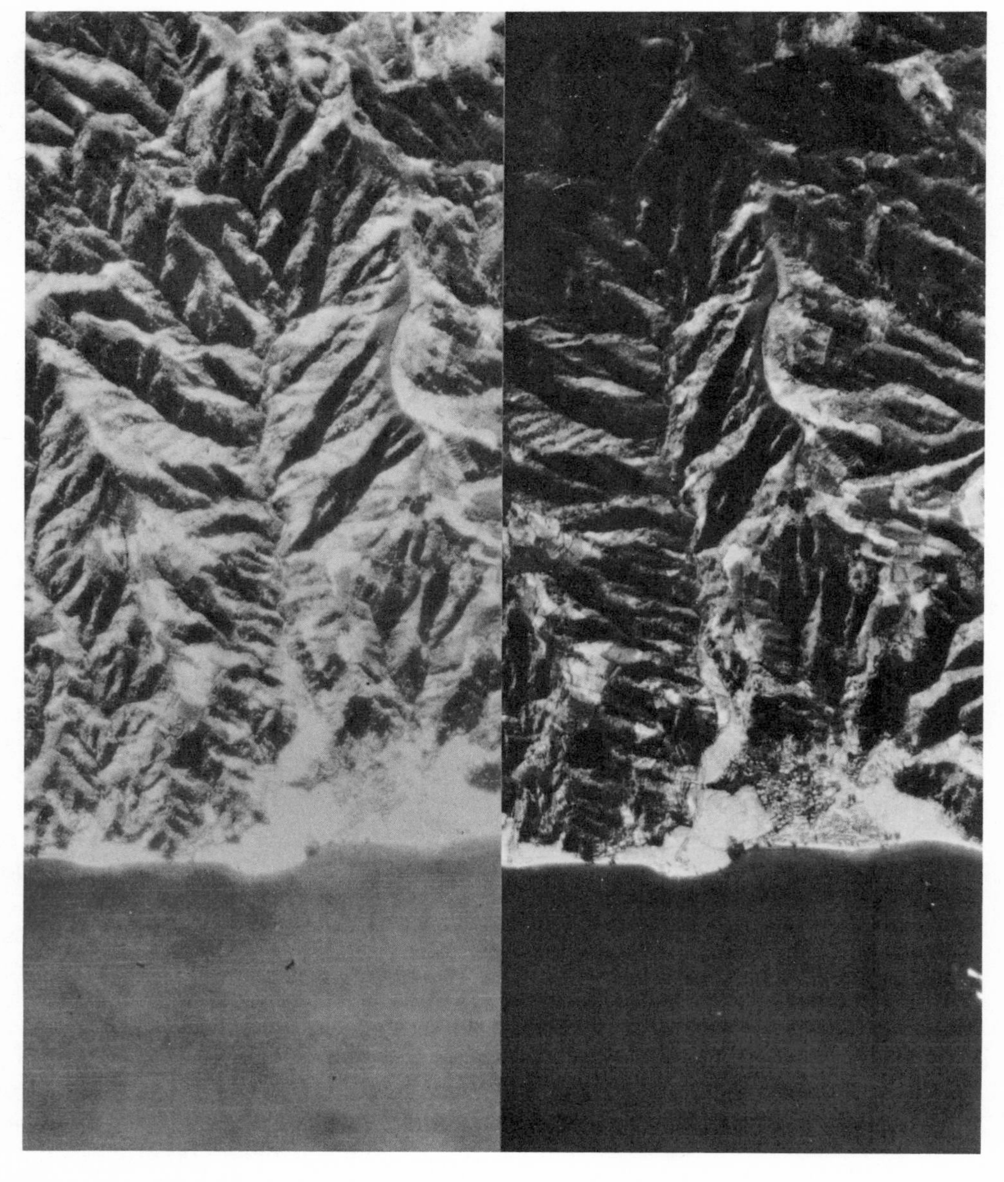

Figure 6.22. Volcanic tuff on Kyushu Island, Japan, shows the characteristic high degree of dissection typical of these weak materials. Photographs by the U.S. Geological Survey, Japan 7-ABC, 1:30,000 (1500′/″), December 4, 1947.

Excavation and Grading

The amount of excavation and grading needed to site large-area land uses in mature basaltic regions is minimal, owing to the flat topography. Very recent flows are very hummocky and require surface leveling. Interbedded flows and dissected basalt regions necessitate major rock cuts which require some blasting. Costs are typically three to four times those associated with deep, dry, moderately cohesive soil removal. The jointing and fracture patterns can be used for the placement of charges and ripping. Some seepage should be expected to occur in fracture and joint zones.

Volcanic tuff, being finely dissected, requires major excavation and cutting in order to obtain a large, flat site. The parent material is typically unconsolidated, does not require blasting, and therefore can be removed at costs similar to those associated with deep, dry, moderately cohesive soil excavation; the rugged topography, however, necessitates moving a much greater amount of material. Welded tuffs require blasting and increase costs by a factor of 4 to 5. Large quantities of water may be encountered, along with noxious gases or steam, during major excavation or tunneling operations (Belcher et al., 1951).

Construction Materials (General Suitabilities)

Topsoil: Excellent to Poor. Residual soils found in tropical climates serve as excellent topsoils for agricultural production. Arid climates typically develop thin profiles which are easily eroded.

Sand: Good to Not Suitable. Few sand-sized particles are created during the weathering process of these materials. Volcanic tuffs may contain sand-sized particles, but these may be soft.

Gravel: Excellent to Poor. The quality of

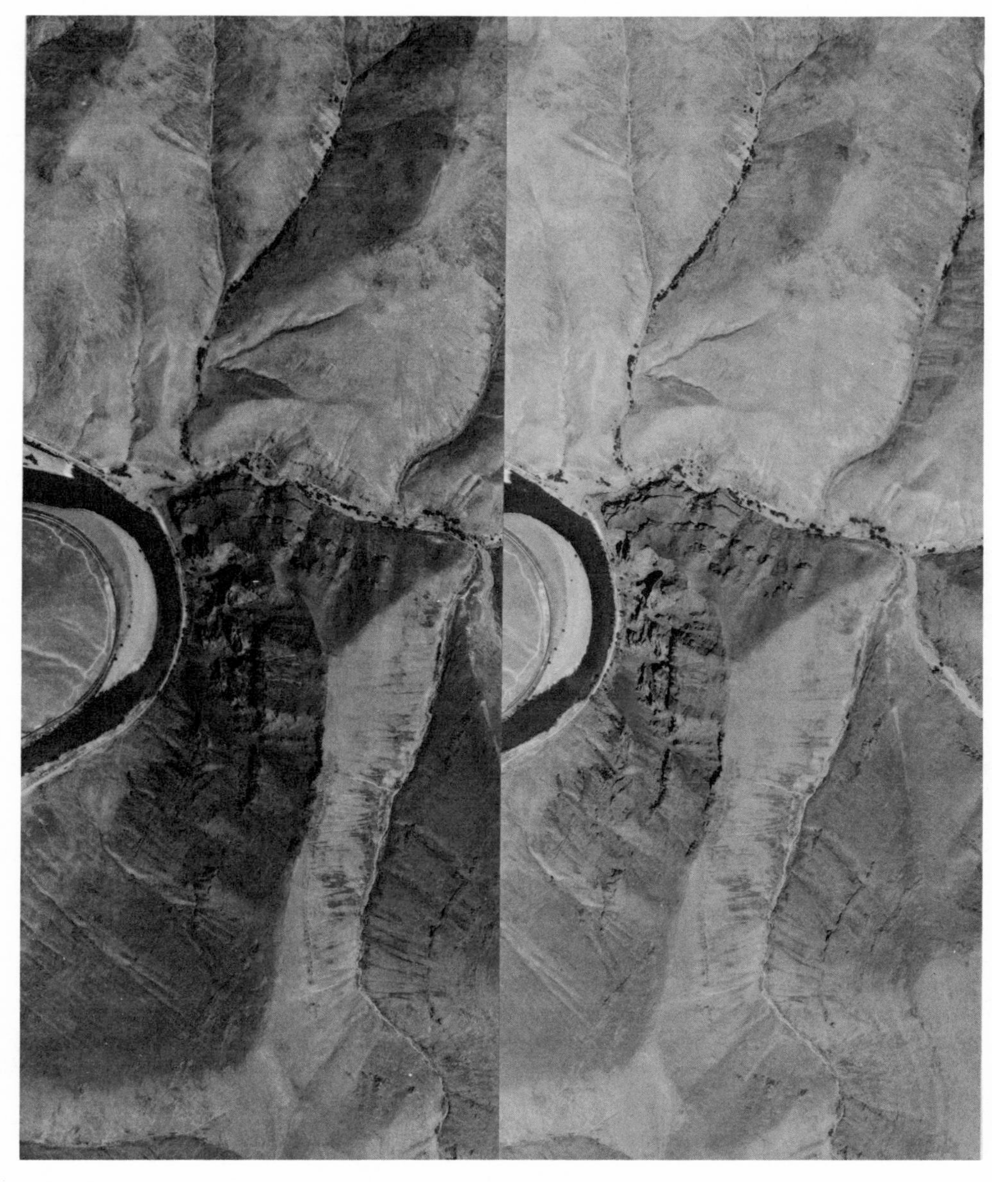

Figure 6.23. Interbedded basalt flows in Sherman County, Oregon, are identified by the terraced hillsides and the saw-toothed edges of the more resistant basalt flows. Photographs by the Agricultural Stabilizaition Conservation Service, AAE-3FF-4,5, 1:20,000 (1667′/″), June 3, 1965.

gravel from these regions depends upon the climatic zone, the amount and time of weathering, and the natural strength of the parent material. In arid climates several feet of stony gravel are part of the basaltic residual soil profile. Little if any gravel is found in tropical climates. Volcanic tuff may naturally exist as coarse sand or gravel, offering a good potential source, but the materials may be soft and wear quickly.

Aggregate: Excellent to Poor. The denser varieties of rock supply excellent aggregate materials, but the soft tuffs and weaker rocks wear quickly.

Surfacing: Good to Poor. Suitabilities for surfacing materials must be determined by analysis and evaluation, owing to the wide variety of materials. Generally, soils offering a range of textural sizes and having sufficient binder serve as good sources of surfacing material. Soils that are predominantly of one texture are not suitable unless mixed.

Borrow: Good to Poor. Tropical soils, when disturbed, acquire the characteristics of highly plastic materials, making them unsuitable for borrow unless compacted within narrrow moisture limits. Residual soils found in humid and tropical climates are hard to work, and it is difficult to maintain the moisture content within suitable limits. The gravelly materials found in tuff deposits are suitable sources of borrow. Compaction of the moderately cohesive residual soils for fills is accomplished through the use of pneumatic-tired rollers; more cohesive soils require strict control of moisture at the optimum moisture content and the use of sheepsfoot rollers.

Building Stone: Good to Not Suitable. The strong, dense rock varieties are valuable as sources of building stone for structures. There are many ornamental uses for lavas, such as

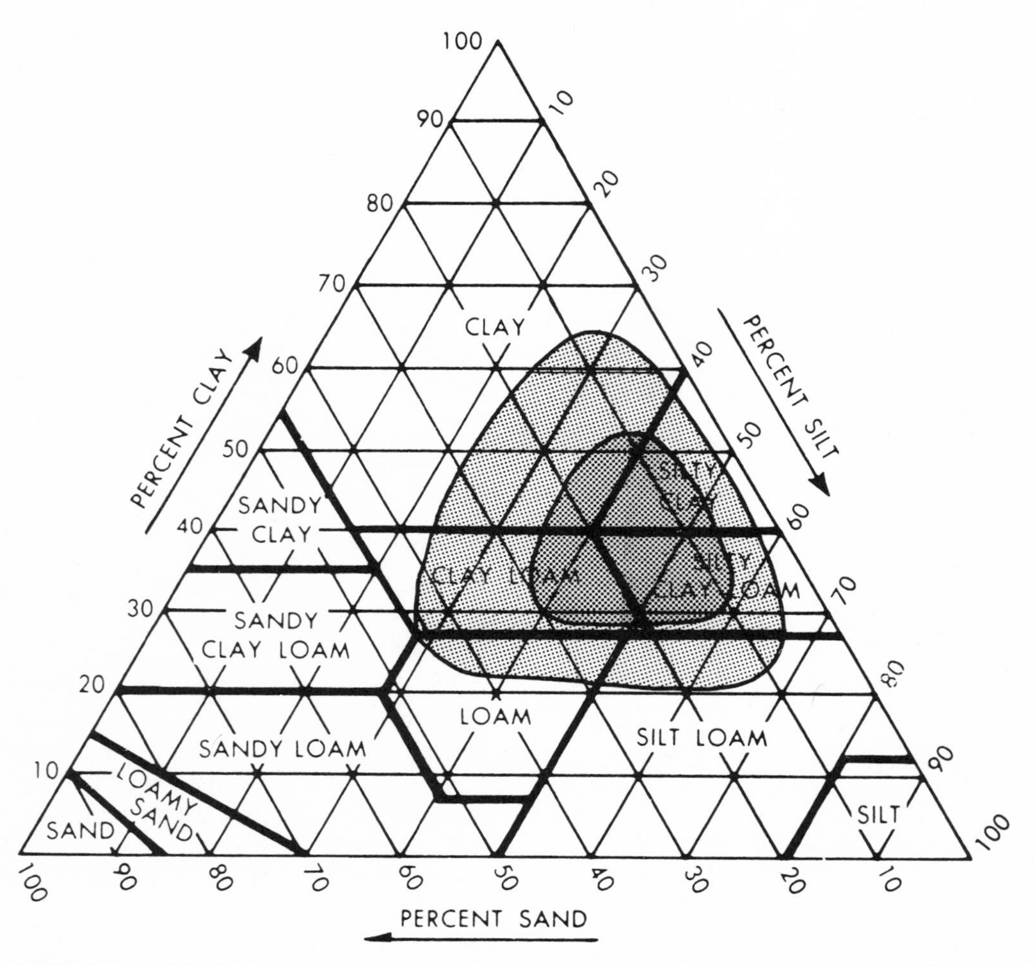

Figure 6.24. The U.S. Department of Agriculture soil texture triangle indicates silty clay, silty clay loam, and clay loam as the most commonly occurring residual soils developed from basaltic materials.

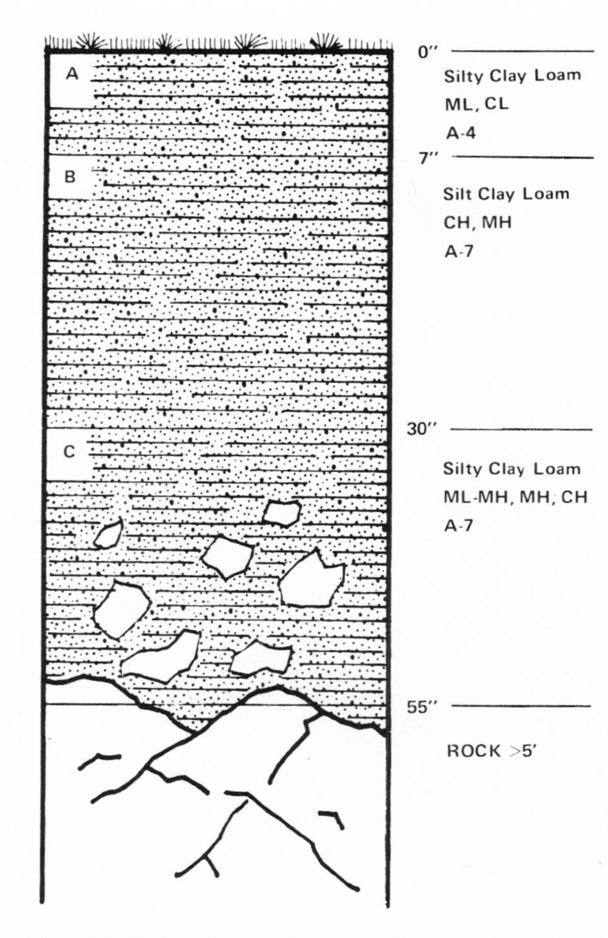

Figure 6.25. A typical profile of basaltic residual soil found in a subtropical climate.

exhibiting blister scars, gas pockets, or flow lines, in walls, building facades, and so on.

Mass Wasting and Landslide Susceptibility

Basaltic landforms, interbedded flows, and tuff are all extremely susceptible to sliding under numerous conditions. For example, where basaltic flows occupy upland or plateau positions in the topography, materials at the base of plateaus are easily eroded, undercutting the hillside. The vertical jointing, fractures, and seepage further contribute to the potential of massive slides along the faces of cliffs in these regions (Figure 6.26). Such slides can be easily identified on aerial photographs. Interbedded flows are highly unstable, since they are commonly underlaid by unconsolidated, easily eroded materials. Volcanic tuff is unstable because of the steepness of slopes and the movement of groundwater. Planners of projects requiring major grading, excavation, or construction should consult competent engineers or geologists early in design planning to determine

Figure 6.26. Landslide in basaltic formation in Twin Falls County, Idaho, in 1939. Photograph by W. C. Mendenhall, U.S. Geological Survey.

design standards and recommendations that reflect the stability of the proposed site.

Groundwater Supply

Even though basaltic materials are naturally impervious, their associated jointing and fracturing make the rock structure highly pervious and permeable, so groundwater is available in these formations. The depth to water is primarily dependent upon the drainage system and the climatic zone. Volcanic, unconsolidated tuffs can act as huge subsurface reservoirs because of their porous nature. Interbedded flows provide more complex groundwater systems since rocks of varying permeability are combined in one area; here, seepage zones may occur along hillsides where pervious materials overlie more impervious strata.

Pond or Lake Construction

Construction of ponds and lakes in extrusive volcanic regions encounters many difficulties. The

Table 6.10. Basalts, interbedded flows, volcanic features, and tuff: aerial and cartographic references

State or country	County or region	Photo Agency	Photograph no.	Scale	Date	Corresponding quadrangle and geologic maps from USGS	Corresponding soil reports from SCS
Volcanic formations and associated deposit							
Arizona	Coconino	USGS*	Arizona 10-AB	1333'/"	5/28/60	Flagstaff (1:250,000), water supply paper No. 1475-J	—
Arizona	Navajo	USGS*	Arizona 2-AB	1667'/"	9/15/51	Agathia Peak (15')	—
Chile	Andes	USGS†	Chile 4-ABC	3750'/"	4/27/61	USAF Aerographic Chart, Point Angamos (1:1,000,000)	—
Chile	Antogagasta	USGS†	Chile 5-ABC	5000'/"	4/7/61	USGS Bull. 1219	—
Guatemala	Sacatepequaz	USGS†	Guatemala 1-AB	4175'/"	1/29/54	USC and USGS ONC K-25 (1:1,000,000)	—
Hawaii	Hawaii	ASCS	EKL-4CC-120-124	2000'/"	2/1/65	Hawaii North (1:250,000), Hawaii South (1:250,000)	1955
Hawaii	Hawaii	USGS*	Hawaii 1-ABCDEF	3750'/"	10/14/44	Honokane, Kamuela (7½')	1955
Idaho	Butte	USGS*	Idaho 5-ABC	1667'/"	7/23/59	Craters of the Moon (1:31,680)	—
Idaho	Madison	USGS*	Idaho 7-ABC	2083'/"	10/8/50	Menan Buttes (7½')	—
Japan	O-Shima	USGS†	Japan 2-ABC	2500'/"	9/27/47	Army Map Service L774, 5951 II, O-shima (1:50,000)	—
Japan	O-Shima	USGS†	Japan 3-ABCDE	2500'/"	9/27/47	Army Map Service L774, 5951 II, O-Shima (1:50,000)	—
Japan	Kyushu	USGS†	Japan 7-ABC	2500'/"	12/4/47	Army Map Service L772, 4241 IV, Kagoshima (1:50,000)	—
Pagan	Mariana	USGS†	Pagan 1-ABC	2500'/"	2/12/46	Army Map Service W843, 3376 IINW, Shomushon (1:25,000)	—
New Mexico	Union	USGS*	New Mexico 5-AB	1667'/"	2/21/56	Dalhart (1:250,000)	—
Oregon	Klamath	USGS*	Oregon 4-ABCDE	4500'/"	8/9/53	Medford (1:250,000)	1908‡
Basalt flows							
Idaho	Canyon	ASCS	DHV-3T-77-78	1667'/"	7/5/57	Boise (1:250,000)	—
Idaho	Twin Falls	ASCS	YR-5T-178-180	1667'/"	7/16/57	Twin Falls (1:250,000)	1921
Idaho	Twin Falls	ASCS	YR-4T-21-22	1667'/"	7/15/57	Twin Falls (1:250,000)	1921
New Mexico	Sandoval	USGS*	New Mexico 8-ABCD	2640'/"	1935	Santa Ana Pueblo (7½'), OM-157	—
Oregon	Malheur	USGS*	Oregon 5-AB	2308'/"	9/14/46	Boise (1:250,000)	—
Interbedded basalt flows							
Idaho	Adams	ASCS	DYU-3P-91-94	1667'/"	8/22/55	Cuprum (15')	—
Idaho	Blaine	ASCS	CVO-1EE-243-245	1667'/"	7/3/64	Idaho Falls (15')	—
Oregon	Morrow	ASCS	AAG-7FF-208-210	1667'/"	9/20/65	Pendleton (1:250,000)	—
Oregon	Sherman	ASCS	AAE-3FF-4-5	1667'/"	6/3/65	The Dalles (1:250,000)	—
Washington	Douglas	ASCS	AAQ-6FF-48-50	1667'/"	7/18/65	Malaga (15')	—
Fragmental rocks: tuff							
LL-LL	Honshu Island	USGS†	Japan 1-ABC	3333'/"	11/6/47	Army Map Service L774, 5856 III, Karuizawa (1:50,000)	—
Japan	Honshu Island	USGS†	Japan 6-ABC	2500'/"	12/4/47	Army Map Service L772, 4241 IV, Kagoshima (1:50,000)	—
Japan	Kyushu Island	USGS†	Japan 7-ABC	2500'/"	12/4/47	Army Map Service L772, 4241 IV, Kagoshima (1:50,000)	—

*From *Selected Aerial Photographs of Geologic Features in the United States,* U.S. Geological Survey Prof. Paper No. 590.
†From *Selected Aerial Photographs of Geologic Features outside the United States,* U.S. Geological Survey Prof. Paper No. 591.
‡Out of print.

most severe relate to leakage through pervious, fractured bedrock. Openings in rocks and fractures do not enlarge themselves as in limestone. Denser, less-fractured materials offer more suitable sites. Volcanic fragmental rocks are soft, permeable, and weak, and may be unconsolidated; therefore they do not offer conditions suitable for the construction of reservoirs. Residual soil materials are clayey and provide excellent impervious bottoms after compaction.

Foundations

Suitabilities for the different types of foundations depend upon the specific rock types and residual soils present. Structures in thin-soil basaltic regions can utilize footings or rafts on gravel pads, whereas in regions of deep soils piers or deep footings may be needed for heavy loads. The residual soils found in tropical regions are lateritic, having high bearing capacities and good internal drainage if they are not disturbed during construction. Most residential structures utilize continuous footings with basement slabs; some are constructed with rafts on gravel pads. Volcanic tuff is adaptable to footing or raft foundations, but its surface clay soils, which are susceptible to volume change, must be considered.

Highway Construction

Highway construction is not difficult through basaltic plain areas. Since the topography is flat, little excavation and grading are necessary, and suitable construction and fill materials are available.

In tropical regions the residual clay soils of basalt have high bearing capacities and are suitable for subgrades if they are not disturbed. Construction operations and equipment must be careful not to destroy the existing soil structure. Interbedded flows and dissected basalt regions necessitate major cuts, bridging, massive fills, and many culverts. Although these escalate construction costs, they prevent potential sliding hazards.

Highway construction in volcanic tuffs is extremely difficult because of the rugged terrain and the highly dissected drainage network. Major cuts and tunneling may be necessary, but many zones of seepage and unstable soils may be encountered. Alignments are curvilinear, in an attempt to minimize earth work by following the lowlands around the small conical hills.

CHAPTER 7

METAMORPHIC ROCKS

INTRODUCTION

Metamorphic rocks are formed from sedimentary or igneous rocks through permanent physical or chemical alterations caused by heat, pressure, or the infiltration of other materials. These processes generally take place at depths below those associated with weathering and cementation. Metamorphic rocks can be classified as either foliated or nonfoliated; foliation refers to the existence of a parallel or nearly parallel structure along which the rock flakes or splits into thin slabs.

Table 7.1 lists the major types of foliated and nonfoliated metamorphic rocks and their mineral compositions. The felsic-mafic distinctions refer to a predominance of feldspar-silica minerals (felsic) or of magnesium-ferric minerals (mafic). Many of the rocks shown appear to have the same mineralogical composition, but they differ in their crystalline structure and particle size. Identifica-

Table 7.1. Classification of metamorphic rocks and their mineral compositions *

Foliates		Nonfoliates		
Felsic	Mafic			
Slate	–	Hornfels†	Quartzite	Marble
Phylite	Chlorite phylite			
Schist	Chlorite schist			
Gneiss	Amphibolite			

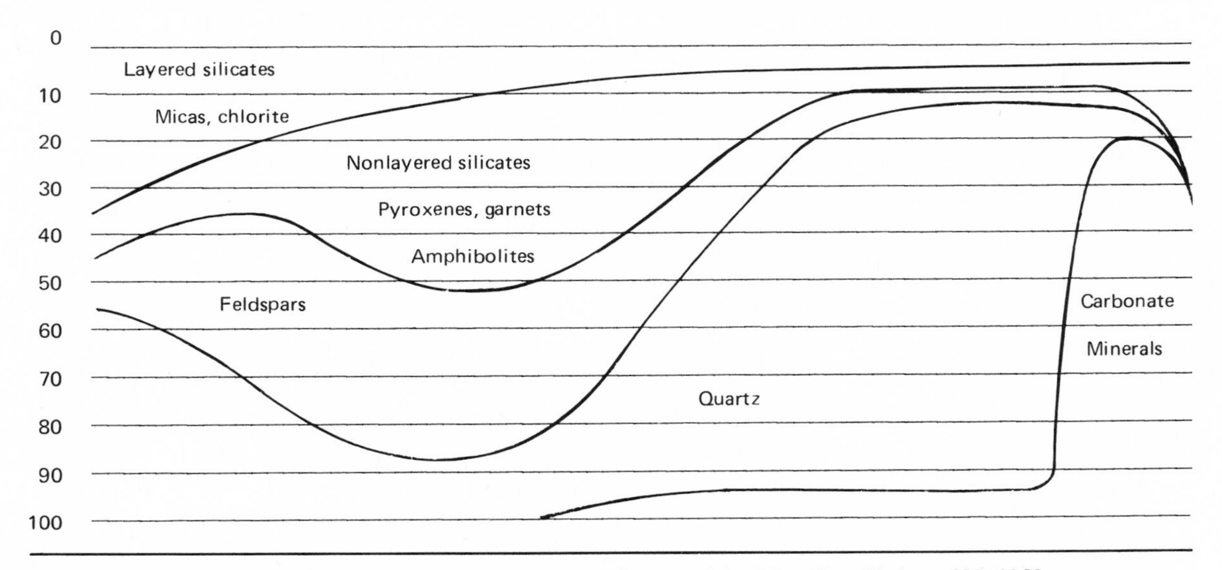

*After C. R. Longwell, R. F. Flint, and J. E. Sanders, *Physical Geology,* John Wiley, New York, p. 629, 1969.
†Contact metamorphism.

tion and minerals analysis can be made in field investigations.

It is difficult to construct a key for metamorphic formations, for they are usually interbedded and grade quickly from one material to another. Only the three major metamorphic rock types—slate, schist, and gneiss are described in the following landform sections because these are the only ones that occur to any significant extent. The definable visual characteristics of other metamorphic rocks are included in the following discussion. All the keys in the metamorphic rock section should be consulted when working in metamorphic regions.

Foliated Metamorphic Rocks

Foliated metamorphic rocks include slate, phylite, schist, gneiss, and their mafic equivalents. The rock types grade from slate to gneiss according to low and high degrees of metamorphism.

Slate

Slate is formed when shale is metamorphosed, forming a homogeneous, very fine-grained rock with no visible mineral grains. The color is typically blue-black, or "slate-colored." Slate splits with thin, perfect foliation, or slaty cleavage, is very smooth, and is commonly used for roofing materials. Slates grade from shale to slate to phylites, depending upon the degree of metamorphism. Slate can be distinguished from shale by its luster and its ring when dealt a sharp blow.

Phylite

Additional metamorphism of slate and/or fine-grained sedimentary rocks creates phylites. These are fine-grained and foliated, grading from slates through phylite into schist. Phylite is generally not as resistant as slate, and the cleavage has characteristics of both slate and schist. The parting of foliations can form smooth or irregular surfaces with an extremely lustrous, almost silvery appearance resulting from the parallel micaceous minerals. Chlorite phylite, the mafic variety, contains a greater composition of the mineral chlorite. The pattern elements of phylite as viewed from aerial photographs are very similar to those of slate.

Schist

Schists are metamorphic rocks which are well-foliated, that is, they have a visible crystalline structure in thin, closely spaced layers. Schists have undergone greater metamorphism than phylites and grade into gneiss. When weathered, the rock breaks easily along the foliation planes, forming a rough surface of parallel ridges and depressions which is reflected in the regional topography. The different, thin, mineral layers generally occur throughout the rock structure in a mixed composition. There are many different types of schists, and they are classified by their major mineral constituents, for example, biotite schist, mica schist, hornblende schist, graphite schist, and chlorite schist. Micaceous minerals tend to predominate in the composition of schist materials. More complex varieties are described as mica-hornblende schists, or other such combinations and may contain high-temperature minerals such as garnets. Schist is easily identified in the field by its closely spaced mineral foliations and its crystalline structure.

Gneiss

Gneiss is formed from sedimentary or igneous rocks that have undergone a high degree of metamorphism. The rock is foliated by thick or thin crystalline bands of different predominant minerals. The foliations can be layered and very regular, or twisted and contorted. The boundary between gneiss and schist is very hard to define; however, gneiss consists of the minerals commonly associated with granite, such as quartz and feldspars, whereas schist contains very small amounts of feldspars. Regions of metamorphic rock contain mixtures of all types of metamorphic materials, but gneiss tends to predominate. Gneiss is classified by its major mineral constituents, for example, granitic gneiss, quartzite gneiss, mica gneiss, conglomerate gneiss, hornblende gneiss, and gabbro gneiss. Complex varieties containing two or more significant minerals are listed under both, the one most predominant being first—for example, quartz-plagioclase gneiss.

Amphibolite

Amphibolite is similar to gneiss but contains abundant amphibole, a magnesium-ferric mineral. Amphibolite is darker in tone than gneiss, appearing dark green or black. It typically contains minor minerals such as micas, feldspars, garnets, and quartz. Its interpretive characteristics are similar to those of gneiss or granite. Field inspection is necessary for positive identification.

Nonfoliated Metamorphic Rocks

Nonfoliated metamorphic rocks include hornfels, quartzite, and marble. These usually occur in small, scattered deposits. Quartzite is the most common.

Hornfels

Hornfels are created by contact metamorphism, that is, by mineral alteration caused by heat and

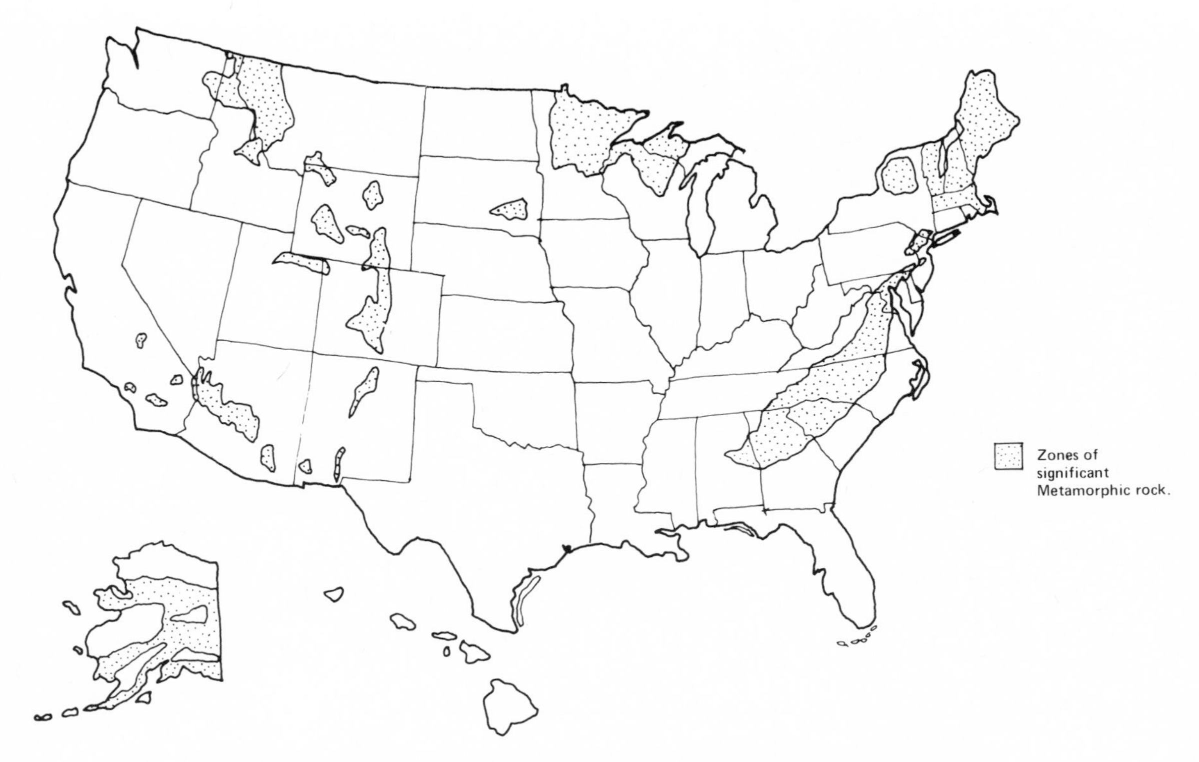

Figure 7.1. Distribution of major metamorphic rock formations in the United States. (After Belcher, D. J., and Associates, Inc., Ithaca, New York, "Origin and Distribution of United States Soils," 1946.)

pressure along fault plains or in contacts of intrusions and tectonic movements. High-temperature minerals are present, such as garnet and andalusite, scattered through a fine-grained, dark-colored rock material. These mineral formations do not constitute landforms and therefore do not have related pattern characteristics.

Quartzite

Quartzites are rocks derived from sandstones by crystallization or cementation with silica, forming a homogeneous, nonfoliated structure. These materials are very dense and break across the original sand grains, exposing a smooth, vitreous surface. Quartzite is generally light in color and completely impervious; it is the most resistant of all rocks that form sharp ridges and uplands with associated very shallow soil depths. Its pattern characteristics are similar to those of very hard sandstone.

Marble

Marble is formed by the metamorphosis of limestone and limestone-associated materials. The calcite or dolomitic grains are intergrown and/or cemented, then interlocked by the addition of calcite. Marble is readily recognized in the field by its characteristic effervescence when touched with dilute hydrocloric acid and by its white bandings and mottles which are caused by impurities such as iron oxides and organic matter. Marble weathers easily and forms lowlands and valleys in the topography.

Some metamorphic rocks not described are anthracite and graphite, which result from the metamorphosis of coal. These can be expected to be found adjacent to other metamorphic materials.

Photographic interpretation of metamorphic rocks is difficult because of the variety of rock types, their wide range of degree of metamorphism, and the small areal extent of single deposits. Rock types found typically occurring over large areas are included in the key and on the following distribution maps.

Distribution

Metamorphic rocks are widely scattered over the surface of the earth, especially in mountainous regions. It is difficult to define and map their distribution since metamorphism grades from igneous and sedimentary materials into highly metamorphosed complex forms. The following maps illustrate only the major areas of the world where metamorphic rock types are found.

North America

United States. Most of the major mountainous areas of the United States contain significant deposits of metamorphic rock. The Appalachian Mountains from New England to Georgia contain metamorphic rocks, including prominent formations of slate, gneiss, and schist. The Rocky Mountains contain many rocks, as does the Alaskan Range. Smaller scattered deposits can be found in Texas, Minnesota, and the Dakotas.

Canada. The Canadian Rockies, southeastern Canada, and New Brunswick are the regions where significant metamorphic deposits are found.

South and Central America

Principal metamorphic deposits are found along the Andes Mountain chain from Colombia through Chile. Smaller deposits are located in eastern Brazil, northeastern Venezuela, and northern Guyana. Central America has small, scattered deposits in southern Mexico and through the mountainous regions of Guatemala.

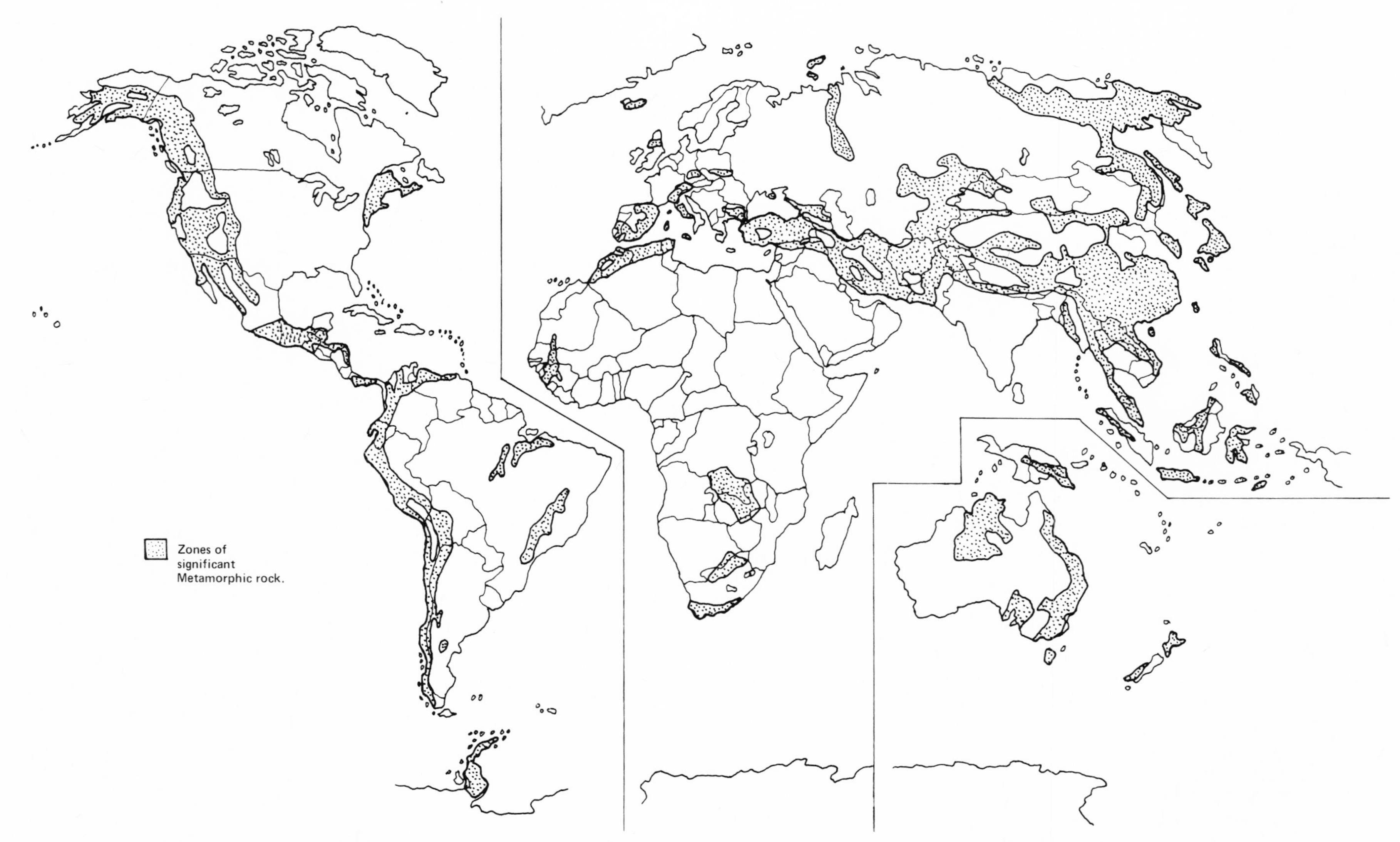

Figure 7.2. World distribution of major metamorphic rock formations. (After Trewartha, G. T., Robinson, A. H., and Hammond, E. H., *Elements of Geography,* Mc-Graw-Hill, New York, 1967.)

Table 7.2. Metamorphic rocks: summary chart

Landform	Topography	Drainage*	Tone	Gullies	Vegetation and land use
Slate					
Humid and arid	Many sharp ridges	Rectangular, F	Gray	Short, parallel	Natural cover
Schist					
Humid	Rounded, steep hills	Rectangular, M to F	Uniform light gray	Parallel, U-shaped	Cultivated and forested
Arid	Parallel laminations	Rectangular, M to F	Banded	Few, parallel, U-shaped	Barren and grass
Gneiss					
Humid and arid	Steep, parallel ridges	Angular, dendritic, M to F	Light uniform	U-shaped, few	Natural cover

*F, fine; M, medium.

Africa

Complex mixtures of metamorphic rocks and associated igneous rock structures are shown together throughout Africa, since distinctions are difficult to make between crystalline and metamorphic structures. Major deposits are found throughout the Near East, in the United Arab Republic, and in Saudi Arabia.

Europe

Metamorphic rocks are found in the Urals and the Alps, and there are other, extensive deposits in Sweden, Finland, and Norway.

Asia

Mixed, complex rock structures and ancient metamorphic and associated igneous rock structures are shown together throughout Asia, since distinctions are difficult to make between crystalline and metamorphic structures. Metamorphic rocks are not found in all of the areas shown but may occur in small deposits in some of them.

Australasia

Major metamorphic deposits are found in northern regions. Scattered deposits exist throughout the entire country.

Pacific Region

No significant deposits are found in the Pacific Islands region.

Caribbean Region

Small, scattered deposits can be found throughout the Caribbean region, including Cuba, Puerto Rico, and the West Indies.

REFERENCES

American Society of Photogrammetry, *Manual of Photo Interpretation,* American Society of Photogrammetry, Falls Church, Va., Chapter 4, p. 415, 1960.

Fenneman, N. M., *Physiography of the Eastern United States,* McGraw-Hill, New York, 1938.

Fenneman, N. M., *Physiography of the Western United States,* McGraw-Hill, New York, 1931.

Hamblin, W. K., and J. D. Howard, *Physical Geology: Laboratory Manual,* 2nd ed., Burgess, Minneapolis, 1967.

Harker, A., *Metamorphism,* Methuen, London, 1950.

Lobeck, A. K., Geomorphology, *An Introduction to the Study of Landscapes,* McGraw-Hill, New York, 1939.

Longwell, C. R., R. F. Flint, and J. E. Sanders, *Physical Geology,* Wiley, New York, Chapter 4 and Appendix C, 1969.

Shelton, J. S., *Geology Illustrated,* W. H. Freeman, San Francisco, Chapter 4, 1966.

Strahler, A. N., *Introduction to Physical Geography,* Wiley, New York, Chapter 16, 1965.

Thornbury, W. D., *Regional Geomorphology of the United States,* Wiley, New York, 1965.

Turner, F. J., and J. Verhoogen, *Igneous and Metamorphic Petrology,* 2nd ed., McGraw-Hill, New York, 1960.

Slate

Introduction

Slate is a foliated metamorphic rock formed by heat and pressure acting on sedimentary rock shale and thus changing its physical composition. The resulting rock has thin, parallel foliation planes (slaty cleavage) and mineral grains so small that they cannot be seen. The exposed rock usually appears dark gray or gray-black, although there are varieties that are red or greenish. Slates can be distinguished from shales by their dull luster and their metallic ring when given a sharp blow. Slates are formed by a low degree of metamorphism; they grade into phylites which are formed by a higher level of metamorphism. Any large zone of metamorphism may contain slates, but the slates themselves do not usually cover large areas. Slates are commonly interbedded with anthracite, a metamorphosed variety of coal.

Interpretation of Pattern Elements

Slates are characterized by a fine-textured, rectangular drainage pattern flowing around many small, sharp-ridged hills, and by a cover of natural vegetation since the topography is too rugged for cultivation.

Because their pattern characteristics and engineering capabilities are similar for both humid and arid regions, slate formations are discussed within one photographic key. Another metamorphic rock, phylite, is similar to slate in its interpretive and engineering characteristics and therefore is not discussed separately.

Soil Characteristics

Slate formations develop thin residual soils over partially weathered bedrock and contain many rock fragments. Silty clays are typically weathered from the fine-grained, slaty materials.

U. S. Department of Agriculture Classification

Slate materials develop thin soil profiles which grade into weathered and partially weathered slate bedrock. Soil textures are typically silt clays mixed with rock fragments and are described as stony (Figure 7.5). Many regions where slate occurs are rugged and still have their original vegetation; published soil surveys are usually not available, since there is little demand for the land for agriculture or development.

Unified and AASHO Classifications

Unified categories used to describe slate residual soils are commonly GM, GC, or GM-GC, indicating the presence of weathered slate fragments within the soil profile. AASHO categories are typically A-2, but may range to A-4 if only small numbers of fragments are found.

Water Table

Surficial water tables exist within the residual soil profile, owing to the impervious nature of the

Table 7.3. Metamorphic rocks: Slate (humid and arid)

Topography	*Drainage*	*Tone*	*Vegetation and land use*
Many sharp ridges	Rectangular: Fine	Gray	Natural cover
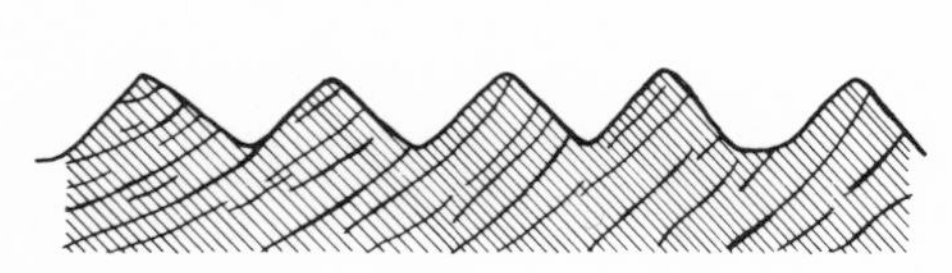 Slate weathers quickly by mechanical means, developing a very rugged topography with sharp ridges and steep hillsides. The small ridges and valleys tend to be parallel to one another. Elevations of the hilltops and valley bottoms tend to repeat over the region.	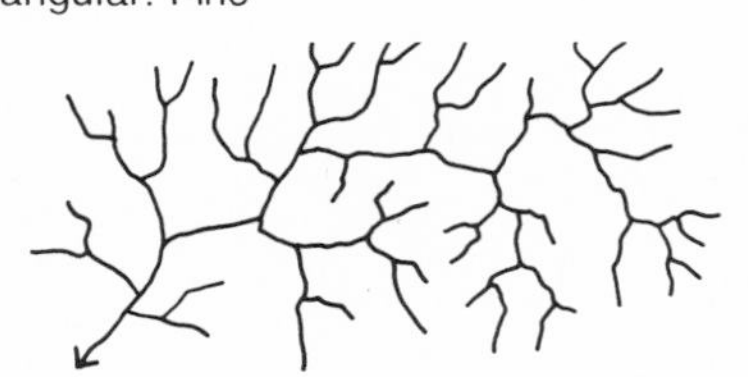 The drainage pattern developed in slate regions is a rectangular dendritic system, very fine in texture. The thin foliations or cleavages of the slate provide initial planes of weakness which control the drainage system. The drainage pattern found in slates is finer than the similar pattern found in schist.	In humid climates the natural tone is a uniform light gray, indicating homogeneous materials. In arid climates the hilltops are darker and the valleys lighter.	In humid climates little land use development is found, owing to the thinness of residual soils and the ruggedness of the topography. Most humid regions are covered with forest growth. Interbedding with anthracite may be indicated by strip mine operations following contours of the hillsides if beds are flat. Arid climates are barren, with scattered scrub and grasses only.
		Gullies	
		Short, parallel	
		Many short, parallel gullies are common, indicating an impervious bedrock and a high runoff percentage.	

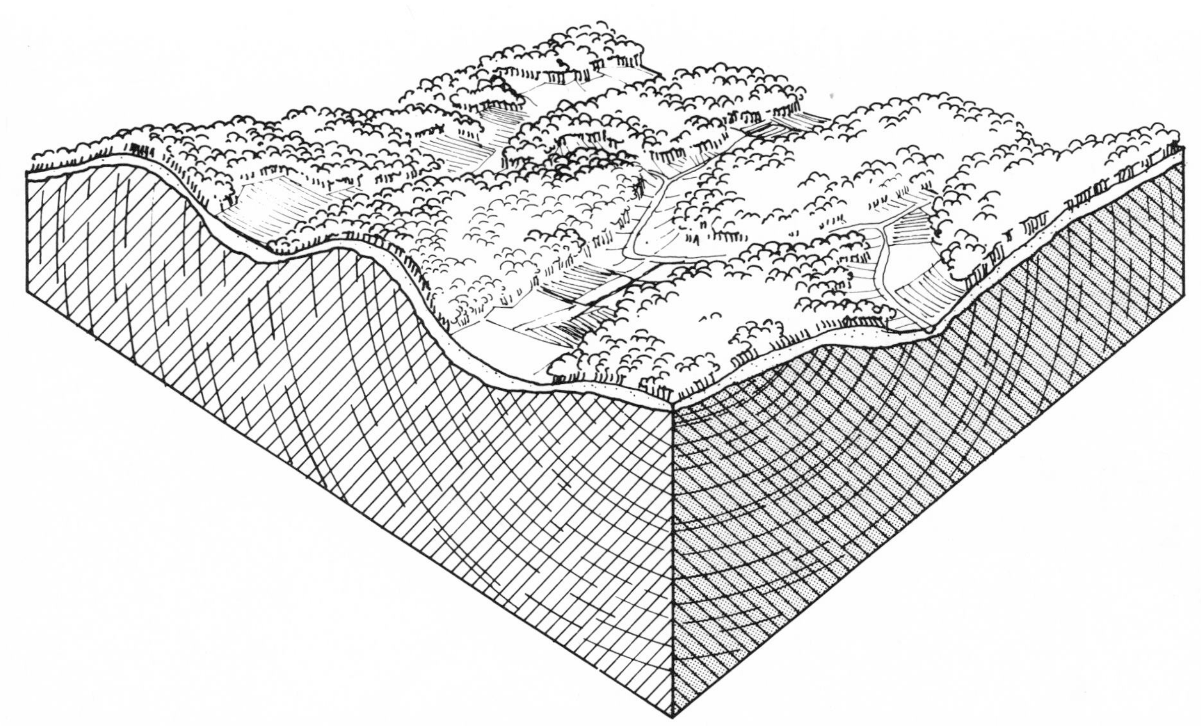

Figure 7.3. Visually, slate regions are very absorptive because of their rugged topography and fine dissection. Land uses hidden in depressions cannot be easily viewed from the surrounding region, and uses located on hilltops can be observed only from the immediate surrounding hilltops. Arid climates with slate formations are not as absorptive, owing to the lack of a heavy forest cover. Many spatial and viewing sequences exist in these regions. Depressions offer spatial closure. Hilltops offer valley rather than regional views, since all hilltops tend to occur at the same elevation.

bedrock and the fine textures developed within the surface horizons. In arid climates a water table exists only after heavy rainstorms. In humid climates the water table fluctuates according to the seasons of heaviest rainfall.

Drainage

Residual soils developed from slaty materials are moderately well drained, depending upon their depth and upon the amount of slate fragments contained within the subsoil. Rock fragments help to break up the fine-textured silt and clay residuum. Also, a high degree of rock fracturing beneath the soil surface facilitates seepage and aids surface drainage.

Soil Depth to Bedrock

Depths to bedrock are not significant in either humid or arid climates. In humid climates the profile is relatively thin, providing an average of 1 to 4 feet of cover over weathered, fractured bedrock. In arid climates little residual soil is found on hilltops, for it is moved by creep, frost action, or erosion to valleys and lowlands where depths are greater.

Issues of Site Development

Sewage Disposal

Utilization of septic tank leaching fields for sewage disposal is not feasible because of the rugged topography, steep slopes, and thin soils. Valleys may have deeper soil profiles, but the seasonal high water table is a constraint. Intense developments or urban areas must rely upon treatment systems if water quality is to be maintained.

Solid Waste Disposal

Sites for sanitary landfill operations are difficult to locate in slate formations because of the steep topography, its ruggedness, and the thin soil cover. Deep deposits of residual soil provide suitable cover if slaty fragments do not predominate.

Trenching

Linear trenching alignments are difficult to route in slate regions because of the rugged topography. Alignments that follow the natural grain of the parallel valleys encounter less rock removal and grade changes. Normal trenching operations involve some rock removal, but trenches less than 6 feet in depth may not require blasting. Trenching across hilltops and ridges encounters more rock which requires blasting. Rock removal is three to four times as expensive as removing the same amount of deep, dry, moderately cohesive soil. The fractures, weathering, and cleavages of this rock facilitate blasting and removal operations. Seepage zones are commonly encountered.

Excavation and Grading

Construction of large, flat sites in slate formations requires extensive excavation and grading. The highly dissected topography requires blasting

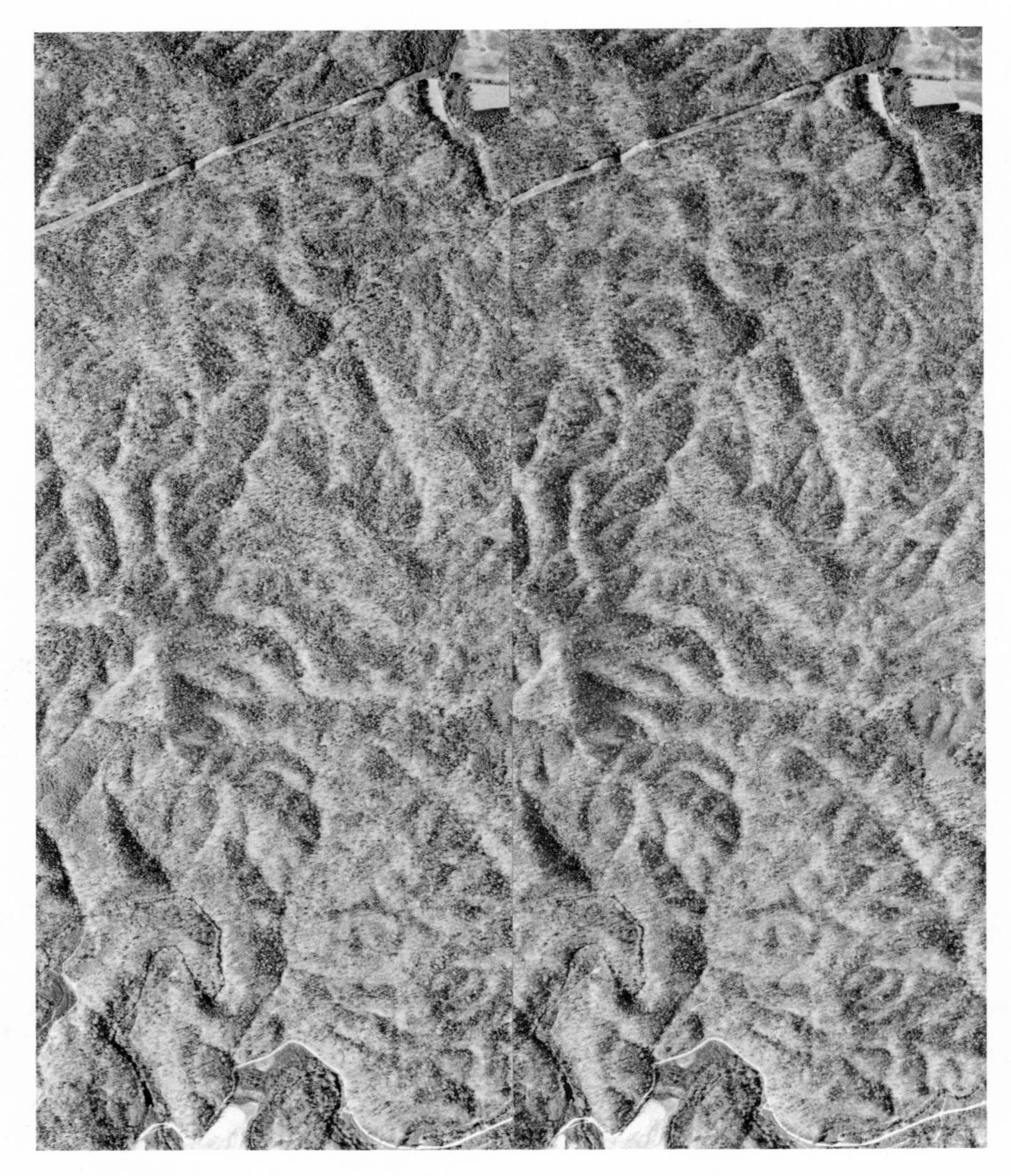

Figure 7.4. (A) Slate in a humid climate in Calhoun County, Alabama, is characterized by a finely dissected topography, fine rectangular drainage pattern, and lack of developed land. Photographs by the Agricultural Stabilization Conservation Service, GR-2LL-144,145, 1:20,000 (1667′/″), November 20, 1969.

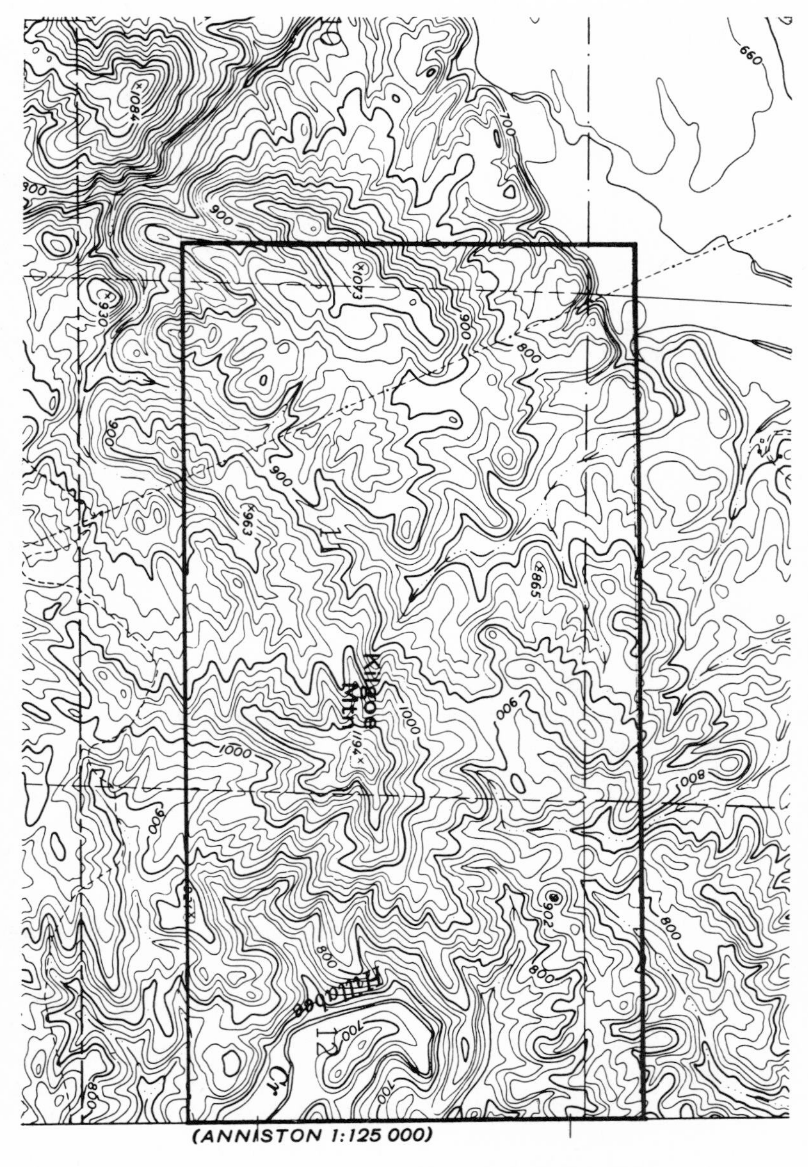

Figure 7.4. (B) U.S. Geological Survey Quadrangle: Oxford (7½′).

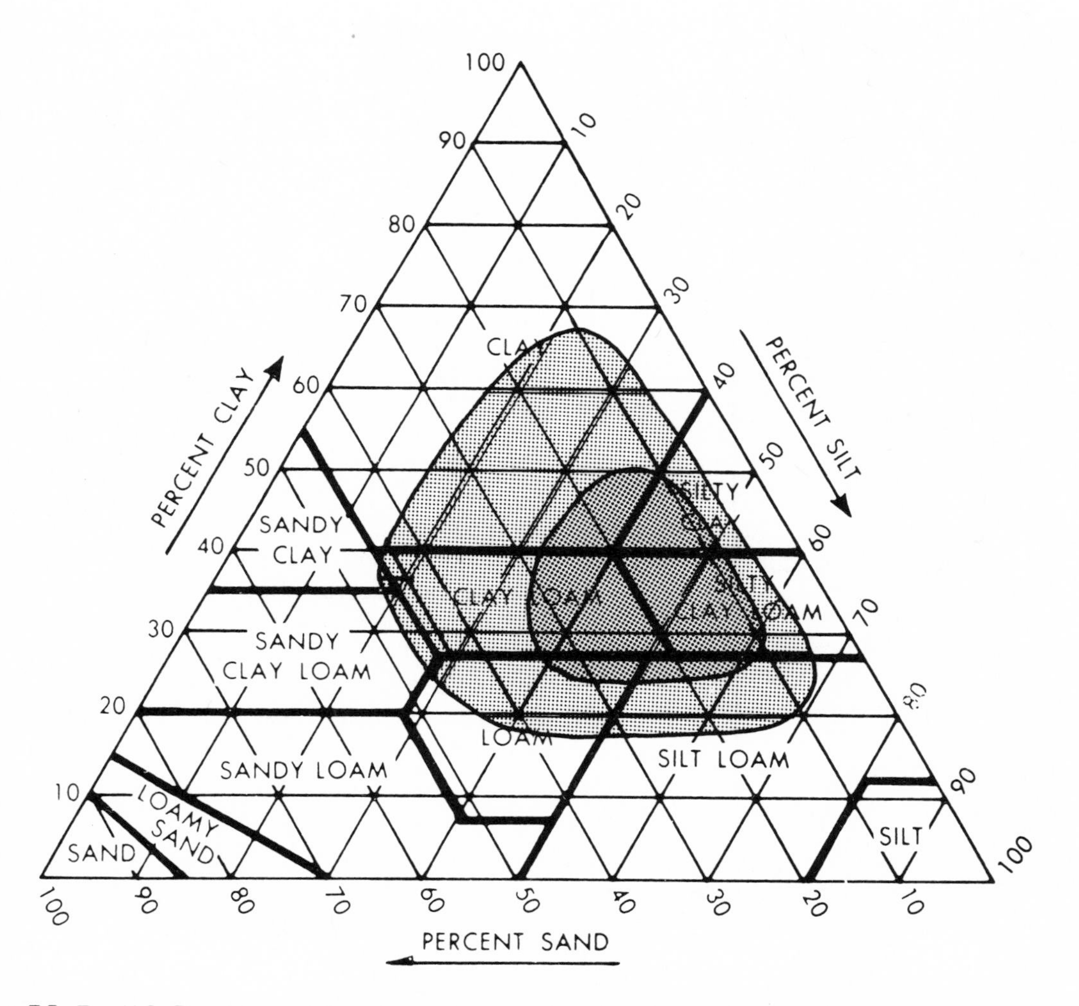

Figure 7.5. The U.S. Department of Agriculture soil texture triangle indicates silty clay, clay loam, and silty clay loam as the most commonly occurring residual soils developed from slate.

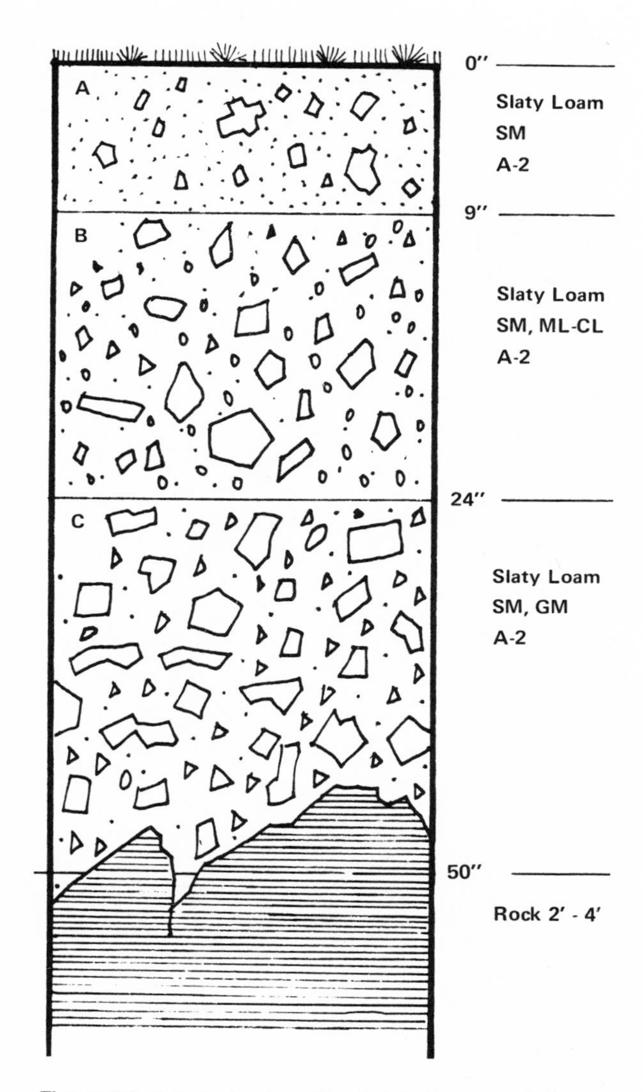

Figure 7.6. A typical soil profile of slate in a humid climate.

for rock removal, and seepage zones may be encountered. Rock removal costs are typically three to four times those for removing the same amount of deep, dry, moderately cohesive soil. The fractures, weathering, and cleavages facilitate blasting and ripping. Rockslides should be anticipated from slips along cleavage planes where shearing resistance is lowest. Excavated materials have a swell factor of approximately 20%. To minimize maintenance, vertical cut faces may be terraced and drained so as to minimize accumulation of debris in the cut and to lower shearing stresses.

Construction Materials (General Suitabilities)

Topsoil: Poor. The thin residual soil, which consists of silty clays mixed with rock fragments, has little organic matter and is not a valuable source of topsoil material.

Sand: Not Suitable

Gravel: Not Suitable. Even though gravelly textures are described in the Unified classification system, gravel-sized materials are thin and platy and easily break into smaller particles.

Aggregate: Not Suitable. The thin natural cleavage does not make this material suitable for aggregate materials. It wears quickly and breaks down into smaller particles.

Surfacing: Not Suitable

Borrow: Good to Fair. Slate bedrock materials removed from cuts can be utilized as fills with little difficulty. These materials expand approximately 20% when excavated and utilized as fill. The residual soils have textures ranging from gravel-sand through silts and clay, and compaction of these materials for fill is best accomplished through the use of pneumatic-tired rollers. Highly weathered residual soils which are more cohesive and contain less coarse fraction can be compacted at the optimum moisture content by using sheepsfoot rollers.

Building Stone: Good. Slate is not used as building stone but as roofing material. The thin cleavage allows the slate to be broken into shinglelike pieces which wear well for this purpose.

Mass Wasting and Landslide Susceptibility

The steep slopes and impervious nature of the bedrock provide conditions highly susceptible to soil creep and downslope soil movement. Landslides can occur when slippage takes place along cleavage planes where the shearing resistance is lowest, and if the cleavage planes are parallel to the hillside slope, a highly unstable condition results. Seepage of water through fractures and joints also contributes to sliding tendencies by increasing the shearing stress. Thorough investigations should be initiated wherever major cuts are to be made.

Groundwater Supply

Some groundwater resources may be available from within slate formations, owing to the well-developed jointing and fracturing, but the rock material itself is impervious. Intensive developments and urban areas must rely upon other underlying sources or surface supplies. Engineering or geological consultants can determine water supply plans for a particular demand and region.

Pond or Lake Construction

Ponds and lakes can be constructed in slate regions with little difficulty if the extent of rock fracturing and jointing is thoroughly investigated. If the area contains significant seep zones, leakage can potentially take place and cause dam or embankment failures. Slaty residual soils, even with compaction, may not be suitable as liner for a reservoir bottom, for compaction is not complete if there is a high content of slaty fragments. First-order gullies and streams within the watershed feeding the reservoir should be kept in vegetative cover to protect against siltation, since the gully system in these regions normally follows steep gradients and runoff from heavy rainfalls can acquire velocities sufficient to move a great deal of material.

Table 7.4. Slate: aerial and cartographic references

State	County	Photo Agency	Photograph no.	Scale	Date	Corresponding quadrangle and geologic maps from USGS	Corresponding soil reports from SCS
Alabama	Calhoun	ASCS	GR-2LL-144-145	1667'/"	11/20/69	Atlanta (1:250,000)	1961
Alabama	Calhoun	ASCS	GR-2LL-204-205	1667'/"	11/20/69	Atlanta (1:250,000)	1961
Alabama	Randolph	ASCS	HT-1EE-50-53	1667'/"	12/8/64	Wedowee (15')	1967
Alabama	Randolph	ASCS	HT-1EE-79-82	1667'/"	12/8/64	Atlanta (1:250,000)	1967
Alabama	Talladega	ASCS	HW-2LL-274-275	1667'/"	11/20/69	Birmingham (1:250,000)	–
Georgia	Unknown	SCS	DQH-3-211-212	1667'/"	1/18/54	–	–
Georgia	Gilmer	ASCS	BUR-3EE-161-163	1667'/"	10/16/63	Tickanetley (7½')	–
Georgia	Rabun	ASCS	CZF-1EE-10-12	1667'/"	10/6/63	Rainey Mountain (7½')	1920†
Tennessee	Sevier, Swain	USGS*	Tennessee 2-ABC	2333'/"	3/30/63	Great Smoky Mt. (East) (15'), Prof. Paper No. 349-B	1956

*From *Selected Aerial Photographs of Geologic Features in the United States,* U.S. Geological Survey Prof. Paper No. 590.

†Out of print.

Foundations

Shallow foundations can be used for both large and small structural loads in slate formations, since the bedrock provides a high bearing stratum and soil depths are shallow. Residential units commonly utilize continuous footings with basement slabs. Very shallow soil conditions facilitate the use of individual footings and framing systems which allow floor grades to be broken; this in turn minimizes grading operations and disruption of the natural landscape, allowing structures to step up or down hillsides. Major structural loads can also utilize footings or piers.

Highway Construction

Highways are difficult to construct across rugged, dissected slate formations because of the alignment problems and the high volume of earthwork needed. If connecting valleys can be followed, costs can be minimized, but routes that bisect ridges and valleys involve very high grading costs and may necessitate the use of massive fills, bridging, or tunneling. Major cuts require terracing to minimize maintenance of rock debris created by future weathering and soil creep. Slate formations are generally not extensive in their coverage, and in developing or planning in areas with slate deposits, it may be more feasible to consider bypassing the entire formation.

Schist

Introduction

Schist is a metamorphic rock of either igneous or sedimentary origin having the arrangement of the mineral structure highly foliated, and platy in appearance. The term schistosity refers to the splitting property of schist rocks, in which thin layers readily flake from the foliated rock; this is the result of its having alternating layers of different minerals, which can be easily observed without magnification. Schists are usually composed of layered silicate minerals, such as mica, but without a significant feldspar content. Mafic schists are formed where magnesium-ferric minerals such as chlorite are abundant. The classifications of schists are determined by the predominant mineral composition or by the mineral most responsible for causing foliation; they include biotite schist, mica or muscovite schist, hornblende schist, talc schist, graphite schist, and chlorite schist. Schists grade into phylites, which represent less metamorphism, and gneiss, which represents the next highest order. Schists commonly occur in metamorphic rock regions mixed with gneiss and phylite.

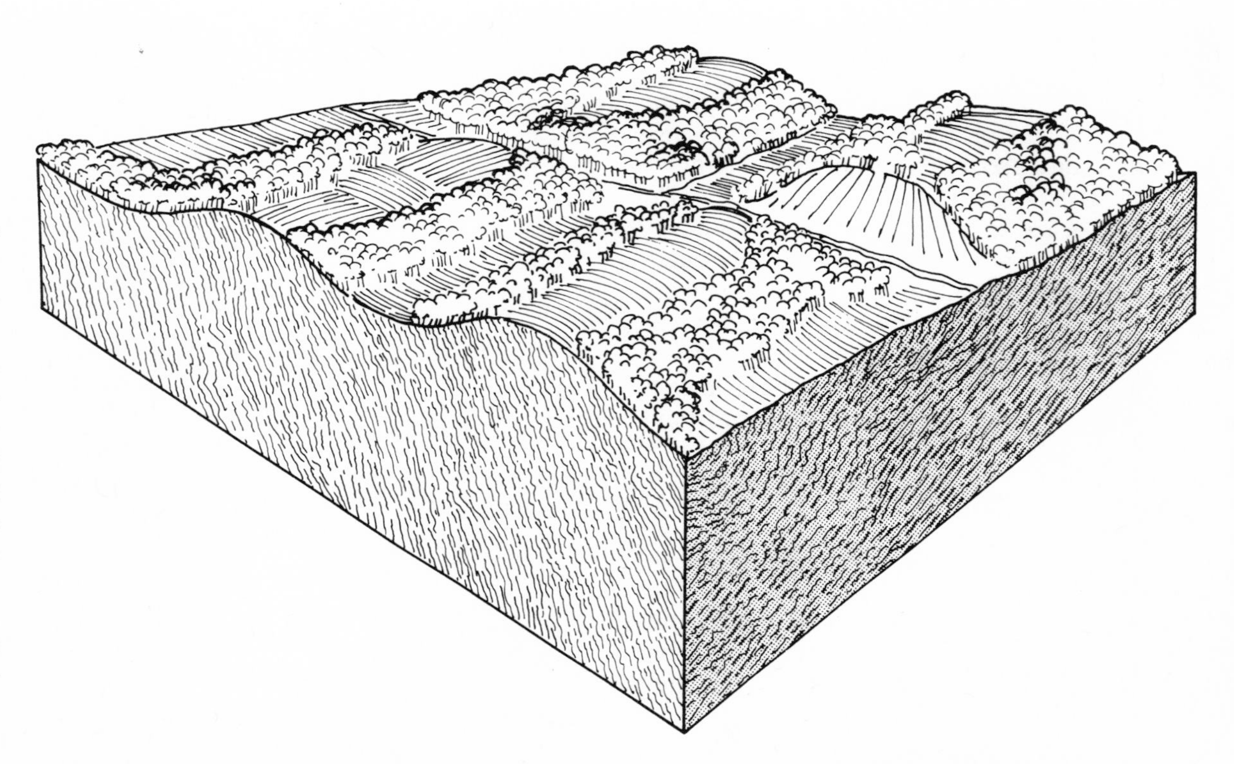

Figure 7.7. Schist in a humid climate. The visual character of schist formation in humid climates is typified by an undulating landscape which, because of its mixture of agriculture and forest growth, offers much variation in local views and scenes. The fact that land uses are located along the hilltops provides for a visually exposed landscape with local valley views.

Interpretation of Pattern Elements

Reflecting their foliated structure, schist formations develop a topography whose appearance is similar to a small sample of the rock. Parallel ridges and valleys form, which control and influence the drainage system. In humid climates, where a deeper residual soil is formed, hilltops are rounded and sideslopes are steep. In arid climates the topography may show faint tonal bandings caused by the laminated structure. For purposes of the photographic key and terrain analysis, schist is described here for both arid and humid climates. Field investigations can easily identify and further analyze the mineral composition of schist for detailed studies.

Soil Characteristics

Residual soils from schist materials develop moderately deep profiles of fairly well-drained sandy silts and clays. The soil may contain many fragments of mica, which significantly affect its engineering characteristics; the subsurface soil grades into weathered and partially weathered bedrock. The B horizon of the residual soil contains higher clay concentrations and has slower percolation rates, thus contrasting with the C horizon, or parent material, which is rather well drained since it consists of silts and clays. Deposits tend to accumulate in lowlands, having been moved from the uplands by creep, erosion, and frost action.

U. S. Department of Agriculture Classification

The U. S. Department of Agriculture commonly classifies residual schist soils as varieties of sandy silt or sandy clay. The soil triangle in Figure 8.10 shows the textural distributions most commonly found in humid climates. The B horizon of the

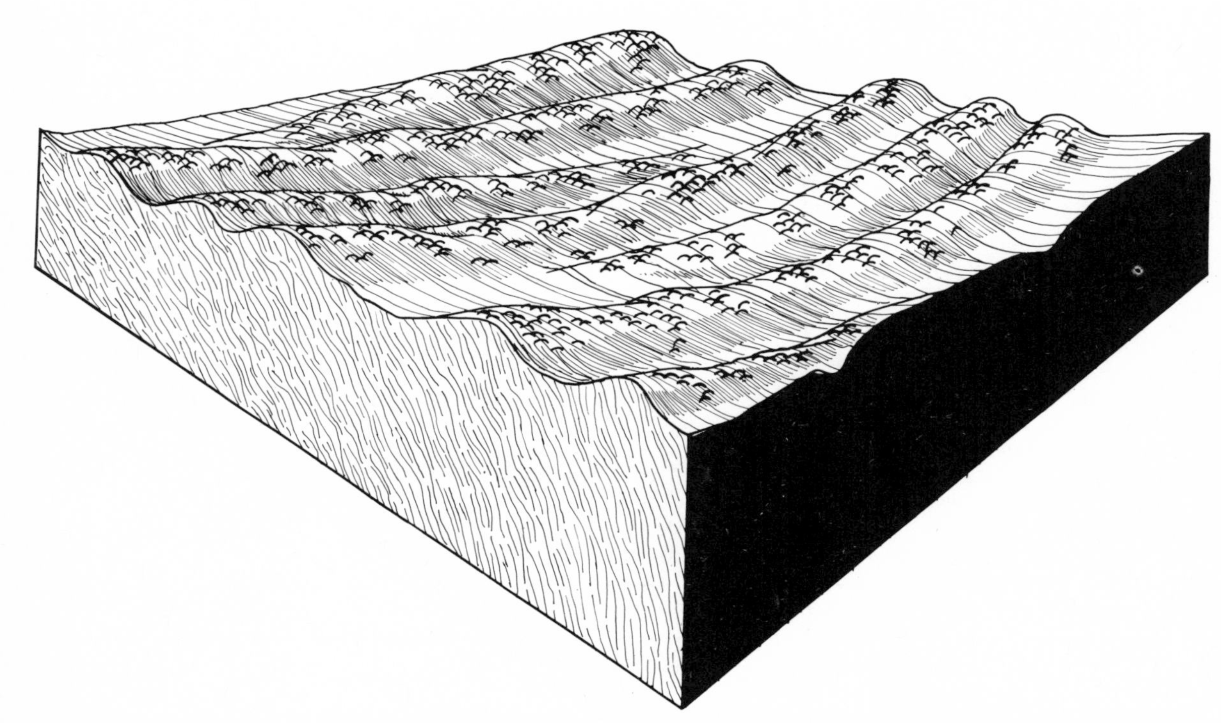

Figure 7.8. Schist in an arid climate. The landscape of schist formations in arid climates is moderately undulating, offering some spatial and viewing variation. Valleys are not generally tight enough to give a feeling of spatial closure; the hilltops offer local valley views but are not high enough to provide regional viewing.

soil develops a fairly impervious clay layer which affects the drainage characteristics of the soil. In arid climates thinner profiles are found, and these have a higher concentration of rock fragments.

Unified and AASHO Classifications

Surface residual soils of schist in humid climates are commonly classified as SC, SM, ML-CL, or SC-SM. Subsurface soils are SM, ML-CL, ML, SC, or CL. The B horizon contains a higher concentration of clay and is commonly described as MH-CH, CL, ML-CL, or MH. The AASHO system categorizes surface soils as A-4 or A-6, the B horizon as A-7 or A-6, and the C horizon or parent material as A-4 or A-7. In arid climates the coarser classes prevail where soils are present.

Water Table

In humid climates a seasonal high water table may exist after heavy rainfalls or spring thaws, reflecting slow percolation through the B horizon of the soil. In arid climates a water table may be found near the surface in deeper valley deposits.

Drainage

The internal drainage of schist residual soils does not constrain land uses or development. The B horizon is poorly drained but is underlaid by a well-drained C horizon of sufficient depth.

Soil Depth to Bedrock

Schist formations in humid climates usually develop deep soil profiles averaging 10 to 15 feet but are occasionally as deep as 50 feet in older formations. The parent material grades into weathered rock material which eventually passes into massive bedrock. Arid climates have many rock outcrops and exposures, but thicker soils can be found in valleys and depressions protected from wind erosion.

Issues of Site Development

Sewage Disposal

Septic tank leaching fields are not recommended in schist residual soils in humid climates because of the low percolation rates, caused by the clay content of the soils, and the steep slopes along hillsides. Percolation rates range from less than 2 to 4 inches per hour, the lower rates being most frequently encountered. Mica schists and residual soils containing many rock fragments have higher percolation rates, in some cases very rapid ones. In these zones care should be taken not to contaminate the water table. Intensive developments or urban areas can utilize leaching fields, but it may be better to develop a treatment system that maintains the quality of the water resources.

Solid Waste Disposal

Sanitary landfill operations in residual schist soils in humid climates can be satisfactorily estab-

Table 7.5. Metamorphic rocks: Schist (humid and arid)

Humid

Topography	*Drainage*	*Tone*	*Vegetation and land use*
Rounded crests, steep sideslopes 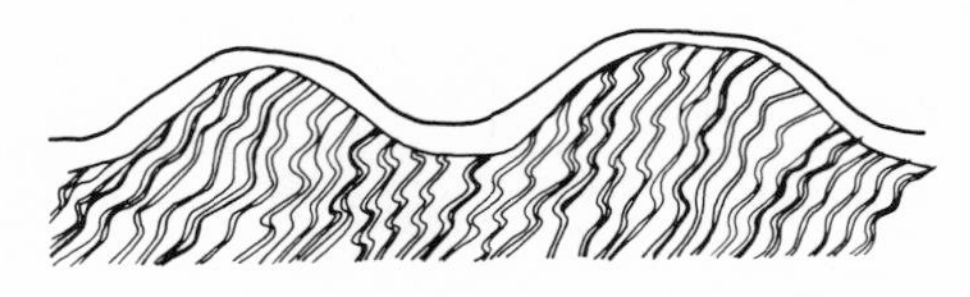 In humid climates schist formations develop deep residual soils reflected by rounded hills with steep sideslopes. The overall appearance of the topography is that of smooth, undulating hills.	Rectangular dendritic: Medium to fine  A rectangular, dendritic drainage system of medium to fine texture, having many long, deep, parallel gullies, develops in humid climates. The drainage pattern is developed and controlled by the parallel foliation structure of the parent material, whose weaker materials are more easily eroded to form valleys and whose more resistant materials form ridges.	Uniform light grays In humid climates uniform light grays are observed unless obscured by vegetative cover. *Gullies* Parallel, U-shaped Many parallel, U-shaped gullies occur, indicating sandy or mica fragments within the residual soil. Gullies are dark, contrasting with tones in surrounding fields.	Cultivated and forested In humid climates schist formations may occur entirely in forest cover or in a combination of cultivation and forest cover, depending upon the degree of dissection of the topography. Typically, the rolling hilltops are cultivated and the steeper slopes facing the drainage system are forested. Contour farming may be practiced, indicating highly erosive residual soils.

Arid

Topography	*Drainage*	*Tone*	*Vegetation and land use*
Parallel laminations 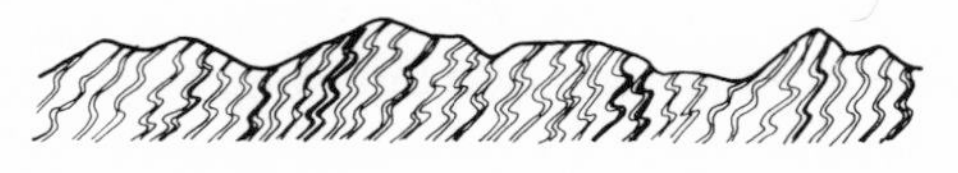 In arid climates schist formations develop topography that appears fairly rugged, the form of the ridges and valleys being controlled by the obvious foliation of the material. When residual soils are thin, the characteristic banding may be seen.	Rectangular dendritic: Fine  A rectangular, fine-textured drainage system develops in arid climates. Many gullies exist, and tributaries are angular, reflecting the structural control of the rock.	Light, faint banding Arid climates have predominantly light tones, but faint banding, emphasized by vegetation, may be discernible. The bandings are parallel and do not follow contours along hillsides. *Gullies* Parallel, U-shaped Many steep-sided gullies are found, but they do not necessarily reflect the characteristics of the residual soil since it is very thin.	Natural cover The rugged topography and thin soil cover do not make these areas attractive to intensive land use or development. They are typically covered with scattered grasses and scrub growth with some concentrations of vegetation occurring where deeper soils have developed or more moisture is available. The foliation within the rock structure is usually emphasized by vegetation.

lished, but other materials for covering may be needed if residual soils are micaceous. Micaceous soils are very difficult to compact and have high levels of internal drainage which allow rainwater to penetrate the landfill and allow leachate to percolate through the profile, potentially contaminating groundwater resources. The depth to water table is the most severe constraint in locating a suitable potential site.

Trenching

Normal-sized, 6-foot trenches should not encounter bedrock in humid climates. Typical removal costs for the residual soil are approximately the same as for those involved in deep, dry, moderately cohesive soils. In arid climates rock is encountered, requiring blasting and heavy equipment for removal. Typical costs are three to four times the costs associated with removal of the same amount of deep, dry, moderately

Figure 7.9. (A) Schist formations in a humid climate in Montgomery County, Maryland, along the Potomac River, are characterized by a rectangular drainage pattern. Photographs by the U.S. Geological Survey, Maryland 1-ABCD, 1:19,000 (1583′/″), March 18, 1963.

Figure 7.9 (B) U.S. Geological Survey Quadrangle: Seneca (7½′).

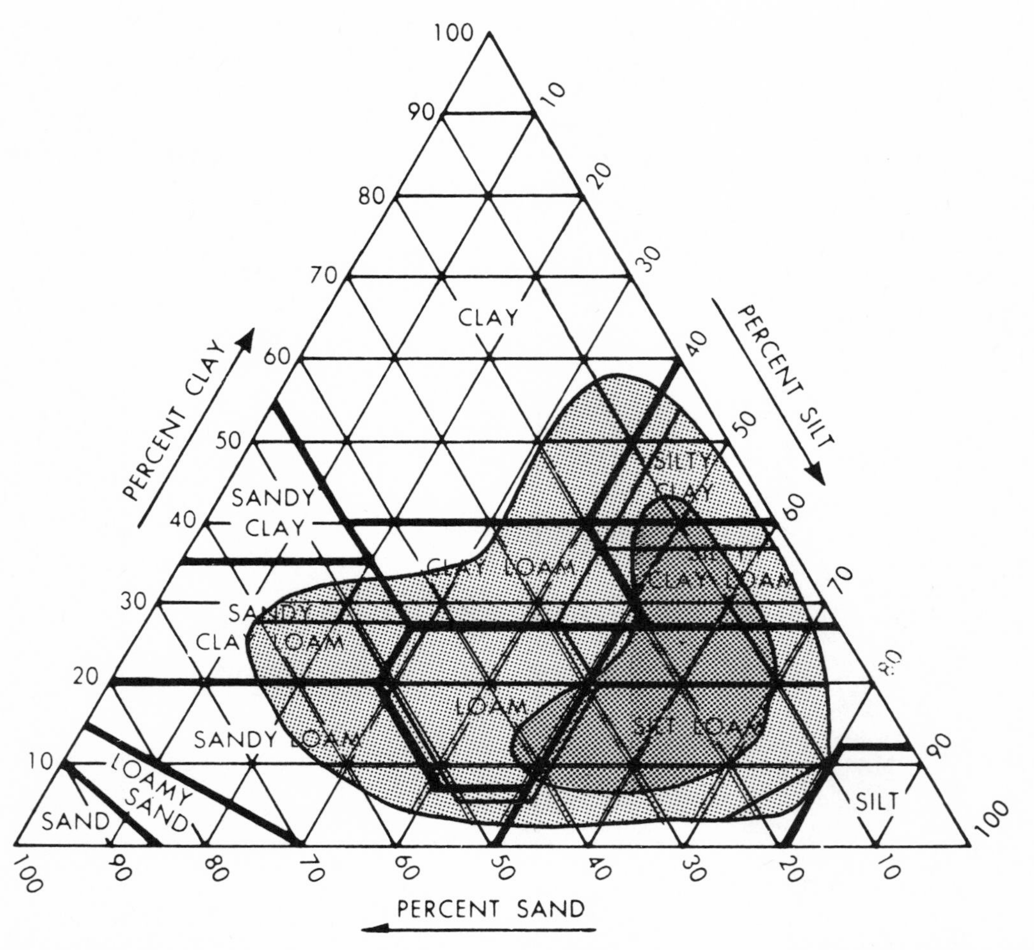

Figure 7.10. The U.S. Department of Agriculture soil texture triangle indicates sandy silts or sandy clays as commonly occurring in residual soils developed from schists.

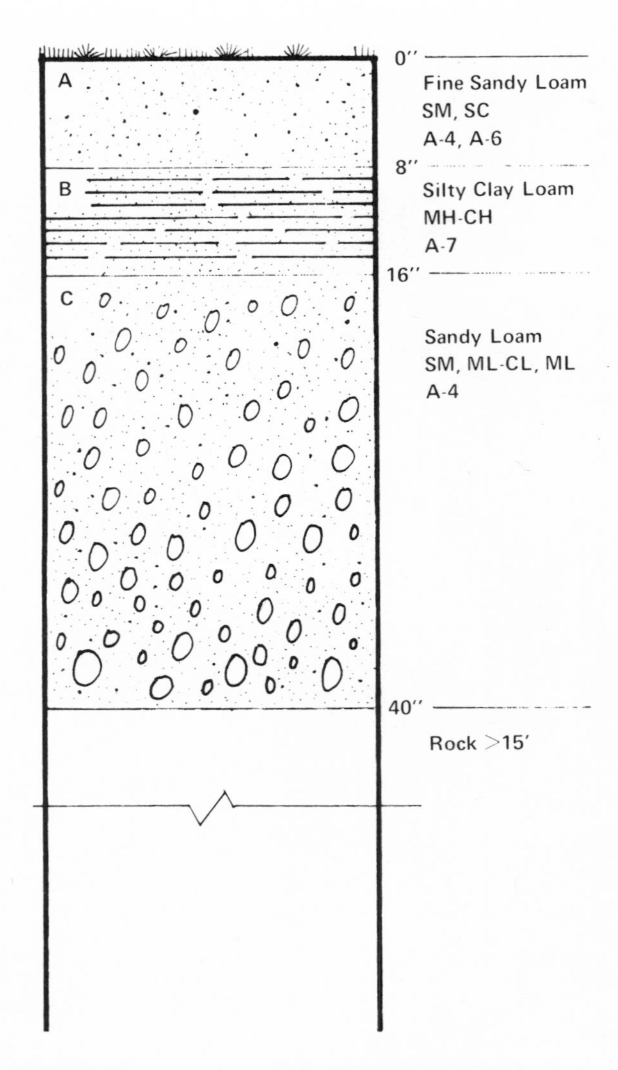

Figure 7.11. Soil profile. A typical schist residual soil developed in a humid climate.

cohesive soil. In deep trenching operations seepage zones may be encountered, and these aggravate the danger of slides along foliation planes where shearing resistance is low.

Excavation and Grading

Excavation operations to obtain large, level sites in arid or glaciated regions require significant movement of materials. Rock is encountered in these areas since little residual soil is present, and this results in higher costs. Rock excavation requires blasting and heavy equipment, at three to four times the costs involved in removing the same amount of deep, dry, moderately cohesive soil. Seepage along foliation planes may aggravate sliding potential. In nonglaciated regions in humid climates, some grading is necessary in order to create a large, flat site, but bedrock is generally not encountered owing to the greater depth of residual soil. Excavation of residual soils costs approximately the same as removing the same amount of deep, dry, moderately cohesive soil unless the water table is encountered, in which case costs will be increased 35 to 40%.

Construction Materials (General Suitabilities)

Topsoil: Good to Fair
Sand: Not Suitable
Gravel: Not Suitable. Unified categories indicating gravelly fragments refer to partially weathered rock materials in the profile, but these break and wear easily along foliation planes.

Aggregate: Not Suitable. The natural foliation of the rock structure provides planes of weakness which absorb moisture, then expand and decompose into smaller particles, making them unsuited for use as aggregate material.

Surfacing: Fair to Poor. Residual soils may not contain sufficient fines to act as the necessary binder in surfacing materials. Soils containing a high concentration of mica fragments are not suitable, since they cannot be easily compacted to provide a wearable firm base.

Borrow: Good to Fair. Schist residual soils provide suitable materials for borrow, but high mica content causes compaction difficulties. Compaction of nonmica, moderately cohesive residual soils for fills is accomplished through the use of pneumatic-tired rollers; nonmica, more cohesive soils require strict control of moisture at the optimum moisture content and the use of sheepsfoot rollers.

Building Stone: Not Suitable. The foliation in schist causes planes of weakness which absorb water, then expand and decompose, making this material unsuitable for building stones.

Mass Wasting and Landslide Susceptibility

Schist residual soils are moved downslope by soil creep, erosion, and frost action and accumulate in valleys and lowlands. Rock slides are not common in the natural topography because of the gently rounded hills, but excavation or tunneling creates cut surfaces which may be unstable. Seepage zones and fracturing may allow water to enter foliation planes, increasing the shearing stress as water accumulates. Swelling in the foliation planes caused by water increases the potential for slippage along the planes where shearing resistance is lowest. Hillsides or cuts parallel to foliation planes are extremely susceptible to failure.

Groundwater Supply

Schist materials are not naturally permeable to water except along fractures and split foliation planes. Water seeps from the surface-subsurface soil layers into the weathered rock, where it flows and seeps through the fractures. Large urban areas in these regions usually obtain their water from surface sources or from surface soil aquifers related to other types of landforms. Suitable water supplies can generally be found in schist formations owing to the deep, weathered rock and the fractured nature of the materials.

Pond or Lake Construction

Ponds or lakes can be constructed within schist formations, but several difficulties may be encountered. The soil materials are naturally well drained, necessitating treatment of the reservoir bottom to provide a more impervious layer. If a significant amount of micas is present, the existing materials will be extremely difficult to compact and utilize. Thus any use of the residual materials requires both compaction and strict control of the moisture content. Fracture zones within the rock may present leakage potential and should be thoroughly investigated. Since the topography is dissected in an angular dendritic pattern, it is well suited to the placement of dams, and the reservoirs thus created will have a high linear footage of shoreline.

Foundations

Suitabilities for different foundation types depend upon the specific composition of the residual soils. Design should be based upon field surveys and test samples from the construction site. Past construction within humid regions has generally utilized continuous footings for small structures with basement slabs. Soil depths are suitable in humid climates for basement excavation, but grading can be minimized if individual footings are used to allow floor grades to terrace the hillsides. Rafts on gravel pads may be used, but volume-change soils may be present and require detailed analysis. Large structural loads may utilize shallow foundations, such as piers or footings, where the soil depths are thin; deeper deposits may require the use of deep footings or rafts.

Highway Construction

Highway construction in these regions is not difficult in either humid or arid climates, since the relief is not significant. Humid climates having deep soil profiles allow alignments to be established that encounter little rock excavation. The material removed from cuts can be utilized as road fill if mica content is not excessive. The residual soils are slightly susceptible to volume change by frost action or wetting and drying. In arid climates little excavation is necessary since the topography is not rugged, but any excavation encounters rock. The direction of foliation planes, fractures, and seepage zones should be analyzed to determine the best facing for rock surfaces, so as to protect against sliding potential and to minimize future maintenance of weathered debris.

Table 7.6. Schist: aerial and cartographic references

State	County	Photo Agency	Photograph no.	Scale	Date	Corresponding quadrangle and geologic maps from USGS	Corresponding soil reports from SCS
Georgia	Clark	ASCS	ATG-5HH-141-143	1667'/"	2/13/67	Athens East (7½')	1968
Georgia	Douglas	ASCS	JU-3GG-4-5	1667'/"	2/20/66	Campbelltown (7½')	1961
Georgia	Douglas	ASCS	JU-3GG-6-8	1667'/"	2/20/66	Campbelltown (7½')	1961
Maryland	Montgomery, Fairfax	USGS*	Maryland 1-ABCD	1583'/"	3/18/63	Seneca, Falls Church, Rockville, Vienna (7½')	1961
Maryland	Montgomery	ASCS	NV-4LL-166-170	1667'/"	9/1/70	Sandy Spring (7½')	1961
Maryland	Montgomery	ASCS	NV-4LL-129-132	1667'/"	9/1/70	Sandy Spring (7½')	1961

*From *Selected Aerial Photographs of Geologic Features in the United States,* U.S. Geological Survey Prof. Paper No. 590.

Gneiss

Introduction

Gneiss is a metamorphic crystalline rock, originating from sedimentary or igneous materials, whose original mineralogical composition was modified by heat and pressure. It is a massive crystalline rock similar in appearance and topography to granite but containing foliated layers or bands of many different minerals. These bands may be a fraction of an inch to several inches in width; the cyrstalline structure is visible without magnification. Its predominate minerals determine the classification of the different types of gneiss. These include simple combinations such as quartzite gneiss, mica gneiss, and hornblende gneiss, as well as more complex types such as biotite-quartz-plagioclase gneiss (in which plagioclase is the predominate mineral). Gneiss represents a high order of metamorphism; it is graded into by schists, which represent a lower level of metamorphism. Occurrences of gneiss are common in metamorphic rock regions, often intermixed with schist or other metamorphic rock types.

Interpretation of Pattern Elements

Gneiss formations in mature landscapes develop a topography reflecting the same foliated characteristics observable in a hand sample. Thus ridges and valleys develop where differences in resistance to weathering occur in the rock, and the topography is rugged, having parallel, sharp-edged ridges and steep sideslopes. In the following key, distinctions between the patterns of humid and arid climates are not made, since they are similar unless specifically indicated.

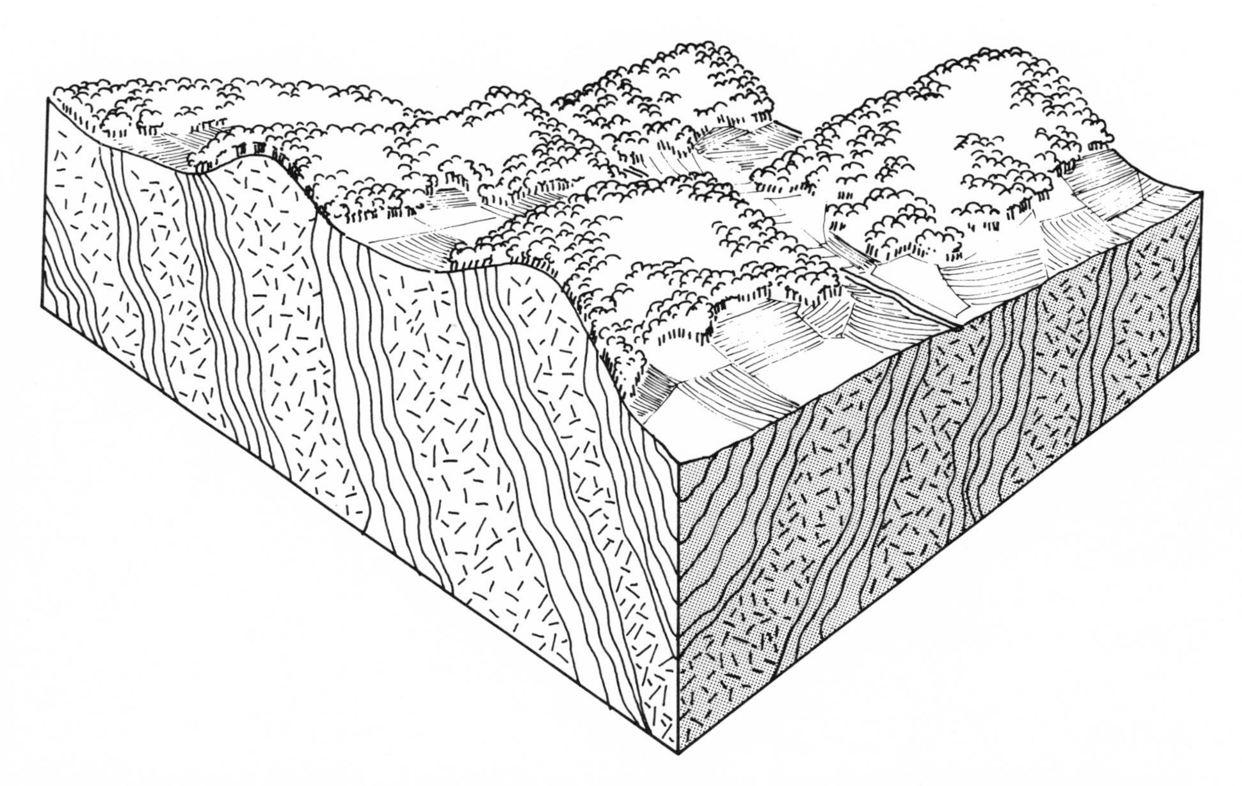

Figure 7.12. Gneiss regions have a high level of visual diversity. Valleys are tightly enclosed by steep-sided slopes, providing a feeling of visual closure. The ridgetops are narrow and permit views on both sides. Hilltop elevations vary and therefore provide both valley and regional views. Man's settlement patterns and transportation systems are generally located in the valleys, where construction costs are lower. The most sensitive visual zone is where the sideslopes of valleys meet the ridges; any disruption of the vegetation or development along this rim zone will have a maximum visual impact on the valley. Development in the valleys, however, is well absorbed, being visible only from surrounding ridges. In arid climates the lack of vegetative cover causes the landscapes to be slightly more exposed than those of humid climates, but they are nonetheless essentially visually absorptive, having many valleys and depressions where development can be hidden and screened. Many varieties of spatial closure and viewing capacity occur within the many valleys and along the ridgelines.

Soil Characteristics

Gneiss residual soils contain sandy silts or sandy clays and have a plastic subsoil (B horizon) in humid climates. In mature or young landscapes the soils on ridges are thin, and materials are moved downslope by creep, erosion, and frost action. In humid climates the profiles average 5 to 10 feet in depth; in arid climates the profile is thinner and coarser. Rock fragments are found in the subsoil, which grades from partially weathered rock into massive bedrock. Older formations develop more undulating landscapes with soils of greater depth and higher clay content.

Table 7.7. Metamorphic rocks: Gneiss (humid and arid)

Topography	*Drainage*	*Tone*	*Vegetation and land use*
Sharp, steep-sided, parallel ridges	Angular dendritic: Fine to medium	Uniform light	Natural cover
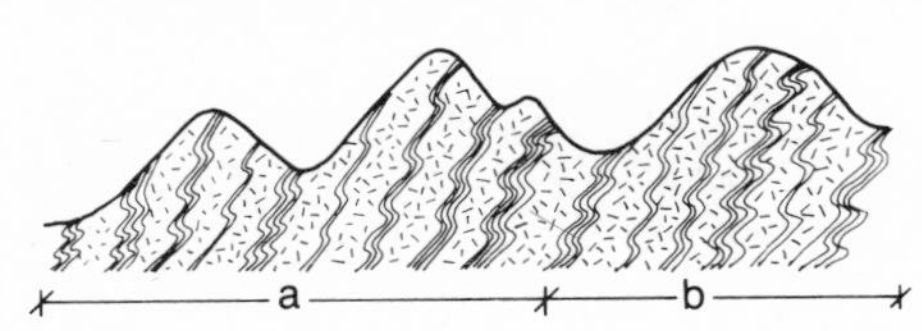 Gneiss formations develop parallel, sharp-ridged hills with steep sideslopes; this topography is the result of differential weathering of the foliations within the rock structure (a). Glaciated regions (b) develop the same topography, except that ridgetops may be rounded as a result of glacial scouring.	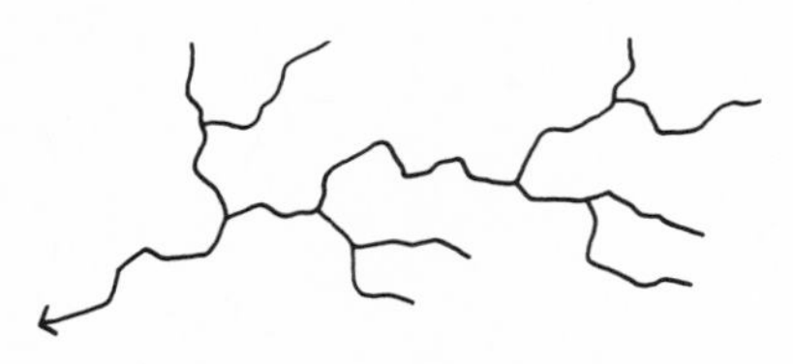 The weathering of the rock foliation initiates and controls the placement of the drainage system which is angular and dendritic. In regions where parallel ridges and valleys are developed, the system appears rectangular. The texture is generally fine or medium.	If the ground surface is not concealed by vegetation, a uniform light tone develops. No banding is apparent. *Gullies* U-shaped Where deep residual soils have developed, gullies may be present, indicating a moderately cohesive, sand-clay mixture.	In both climatic regions the thinness of the residual soils and the rugged topography do not present conditions attractive to the establishment of land uses. Humid climates are forested. In arid climates a scrub and grass cover develops, which occurs at greater densities in the valleys where soil depths are greater and moisture is available.

U.S. Department of Agriculture Classification

The U.S. Department of Agriculture classifies gneiss residual soils in humid climates as sandy loam, sandy clay loam, and clay loam. The B horizon usually has a more impervious, finer-textured composition, typically classified as clay or clay loam. Significant rock fragments in the surface or subsurface soils are described as stony or rocky. In arid climates thinner profiles develop, and soil textures are coarser. Sandy loams predominate.

Unified and AASHO Classifications

Residual gneiss soils in humid climates typically develop SM, SM-SC, ML-CL, or ML in the surface soils; MH and CH in the B horizon; and SM, ML, CL, MH-CH, and ML-CL in the parent material or C horizon. The AASHO system categorizes surface soils as A-2 or A-4, the B horizon as A-5 or A-7, and the C horizon as A-4, A-6, or A-7.

Water Table

Residual soils of gneiss have moderate to very slow percolation rates and high levels of runoff. During wet seasons in humid climates, there may be a seasonal perched water table. The rather impervious soil structure and steep slopes contribute to rapid runoff.

Drainage

Residual gneiss soils in humid climates may be either practically impervious or moderately drained, depending upon the specific composition of the parent material. The soil structure may aid the internal drainage, and it should be noted that disruption of the structure by construction creates a more impervious material. Soils in arid climates are well drained because of the steep slopes with their associated high velocities of runoff.

Soil Depth to Bedrock

In humid climates moderately deep to deep residual profiles develop, which average 5 to 10 feet in thickness but may be as deep as 60 feet in some areas. The surface soils grade into rock fragments, partially weathered rock, and finally massive bedrock. Soils on the tops of ridges tend to be thinner, since weathered materials readily move downslope through soil creep, erosion, and/or frost action. Glaciated regions may have little soil cover along the ridges, and these will have been carved and scoured into elongated formations. In arid climates gneiss formations develop thin profiles. There are many rock exposures, boulders, and ridgelines of bare rock.

Issues of Site Development

Sewage Disposal

The more granular residual soils may have suitable percolation rates for use as septic tank leaching fields, but typically the clay content of the soil and the impervious B horizon cause percolation rates to be below the required minimum. Topo-

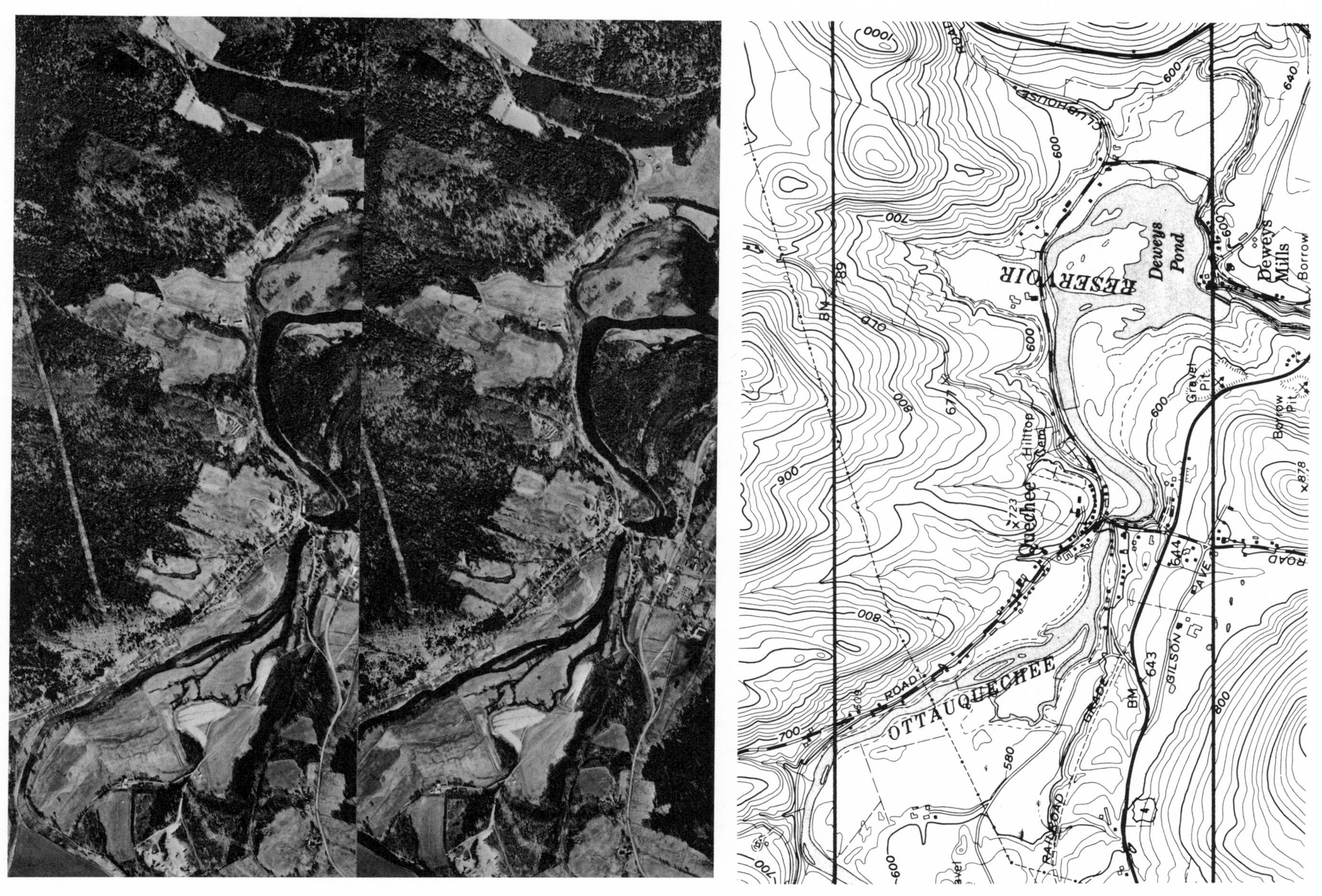

Figure 7.13. (A) Glaciated gneiss in a humid climate in Windsor County, Vermont, is characterized by parallel ridges and valleys. Photographs by Raytheon-Autometrics, VBM-6824-6-168,169, 1:24,000 (2000′/″).

Figure 7.13. (B) U.S. Geological Survey Quadrangle: Quechee (7½′).

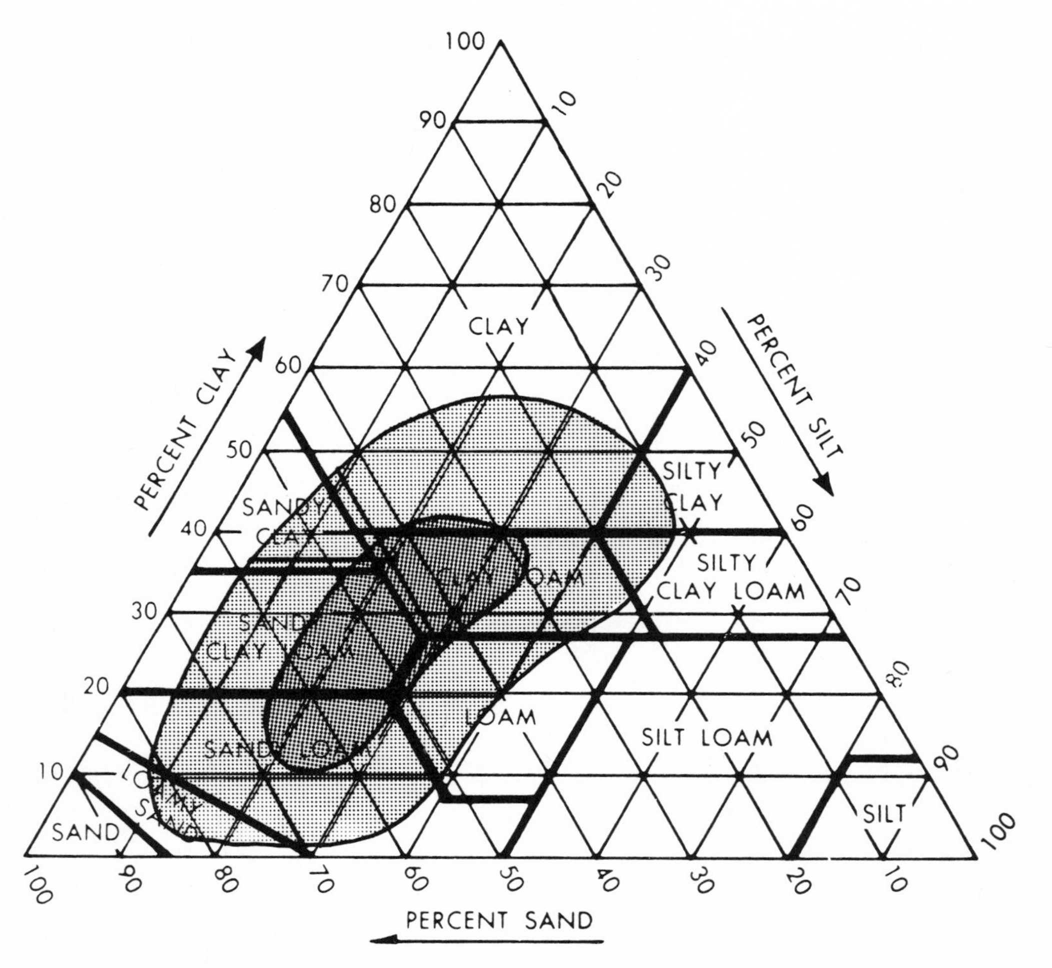

Figure 7.14. The U.S. Department of Agriculture soil texture triangle indicates sandy loam, sandy clay loam, and clay loam as the residual soils most commonly occurring from gneiss formations.

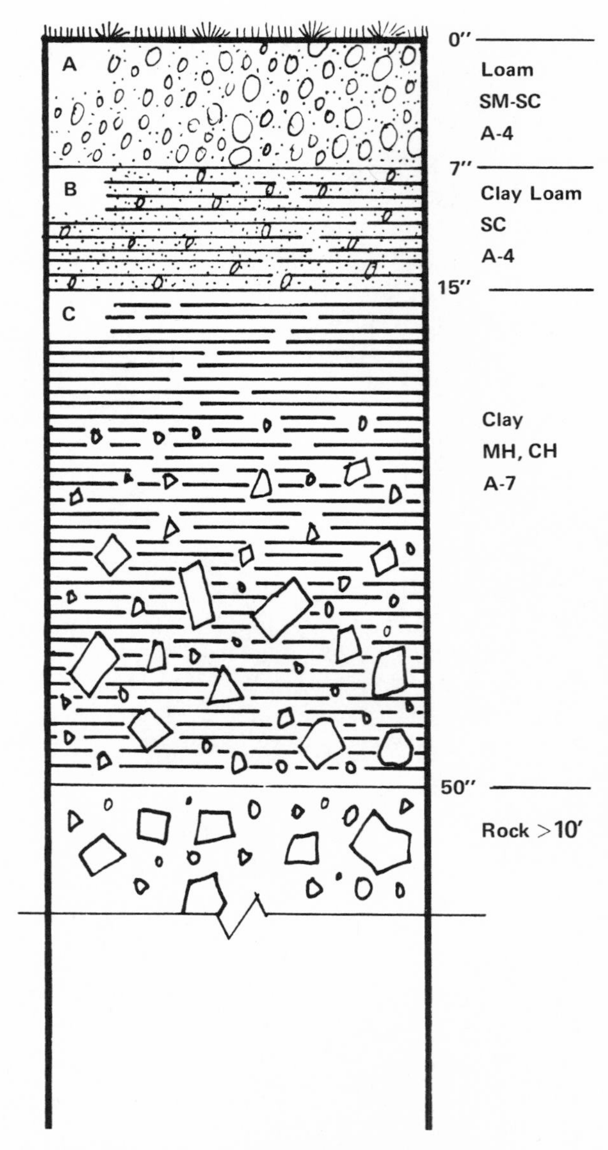

Figure 7.15. A typical soil profile developed from gneiss in a humid climate.

graphic relief is also a severely limiting factor in locating suitable leaching field sites in these regions. The depths to bedrock or the water table should not be serious constraints. Hilltops, however, especially in glaciated landscapes, have thinner soil profiles which may not be sufficiently deep for this type of disposal system. Both intensive developments and uban areas may find it possible to use some on-site septic disposal systems, but treatment is recommended to maintain the quality of the surficial water resources.

Solid Waste Disposal

Sites for sanitary landfill operations can be located with little difficulty in regions of gneiss residual soils. The rugged topography is an obstacle, but the edges of major valleys can be seriously considered for this use, as long as sufficient depth to water table exists (greater than 10 to 15 feet). The fine-textured soils are suitable for compacted cover in the landfill operation and prevent rainwater from penetrating the fill and percolating leachate into the groundwater table. Mica gneiss residual soils contain mica fragments, however, making compaction difficult. If sufficient cover materials are available, old rock quarries where gneiss has been excavated for building stone or aggregate should be considered for sites.

Trenching

Linear trenching alignments are difficult to route in gneiss regions, owing to the high degree of dissection and the massive bedrock. In order to minimize earthwork, when parallel ridges and valleys occur the natural grain of the topography should be followed. Trenching across ridges encounters massive, crystalline bedrock, very similar to granite, which requires the use of heavy equipment and blasting. Typical per-cubic-yard costs for removing bedrock are four to five times those for removing the same amount of deep, dry, moderately cohesive soil. Trenching in arid climates involves the same difficulties and requires still more rock removal, since residual soils are thinner. In all climates seepage zones exist in highly fractured or foliated rock areas, but they are not commonly found since the rock tends to be massive.

Excavation and Grading

The creation of large, flat sites in both humid and arid climates necessitates significant earthwork operations. Bedrock removal is required, at costs four to five times those for removing the same amount of deep, dry, moderately cohesive soil. Blasting and heavy equipment are necessary because of the massive, crystalline, granitelike nature of the rock. Where highly fractured or foliated rocks occur, seepage zones may be encountered. Grading of residual soils is difficult, owing to the problems of controlling the moisture content. Furthermore, earthwork operations are usually limited to the drier seasons of the year because the plastic nature of the subsoil causes it to become sticky and hard to work when wet. Micaceous residual soils are difficult to compact properly.

Construction Materials (General Suitabilities)

Topsoil: Good to Fair. Residual soils are suitable as sources of topsoil, but their organic content is low.

Sand: Not Suitable. Even though there are sands within the residual soils, they are mixed with fines and are difficult to separate.

Gravel: Not Suitable

Aggregate: Good to Fair. Aggregate materials originating from gneiss are suitable, although not as durable as those originating from granite.

Surfacing: Good to Poor. The residual soils contain suitable binders for surfacing use, but they may be lacking in course-grained materials. Micaceous residual soils are rated poor for surfacing since they are not easily compacted.

Borrow: Good to Fair. Moisture limits must be carefully observed, but these soils are generally suitable for road fill. Micaceous residual soils are not as attractive, since they cannot be compacted. Pneumatic-tired rollers are used for the compaction of moderately cohesive fills, and sheepsfoot rollers are used for more cohesive soils.

Building Stone: Excellent to Good. The more massive, crystalline varieties of gneiss are very attractive for building stone. Also, the foliated layers of gneisses of different mineralogical compositions provide interest and variation when used in polished facades.

Mass Wasting and Landslide Susceptibility

Gneiss formations are structurally stable except where highly weathered and strongly foliated zones occur within the rock structure. Soil creep and slumping tendencies occur along the sides of steep hills. If vegetation has been removed along hillsides, gullying is common.

Groundwater Supply

Gneiss rock is impervious to water, having the low porosity and low permeability characteristics of crystalline rock. Water is carried within the rock structure but through fractures and partially weathered material. Major fracture zones may intersect in such a way as to provide a regional, subsurface drainage network, and wells drilled in fracture intersections may obtain good water supplies. Most intensive developments and urban areas, however, must rely upon surface water sources or those available from soil aquifers in associated adjacent landforms.

Pond or Lake Construction

The strength of the rock material, its massiveness, and its impervious nature are conditions favorable to the construction of reservoirs. However, formations containing highly foliated or fractured rock may allow seepage and leakage and should be thoroughly investigated. The residual soils are suitable for use as bottom materials, since they become impervious when compacted. Gneiss regions that have a residual soil high in micaceous particles present compaction difficulties. The dendritic topography of gneiss regions provides many sites for dams; these have

Table 7.8 Gneiss: aerial and cartographic references

State or country	County	Photo Agency	Photograph no.	Scale	Date	Corresponding quadrangle and geologic maps from USGS	Corresponding soil reports from SCS
Brazil		USGS*	Brazil 1-ABC	1667'/"	11/29/64	Mapa geológico de Distrito Federal (1:87,000)	–
Maine	Oxford	SCS	ENM-2EE-213,214	1667'/"	5/2/64	Bethal (15')	–
Maine	Oxford	USGS	GS VARQ 2-109, 110	1667'/"	5/6/65	Bethal (15')	–
Vermont	Windsor	USGS	VBM-6824 6 166, 167	2000'/"	4/20/68	Quechee (7½')	1916†

*From *Selected Aerial Photographs of Geologic Features outside the United States*, U.S. Geological Survey Prof. Paper No. 591.
†Out of print.

very angular shorelines with many linear feet of water contact.

Foundations

Gneiss rock materials, unless they are highly fractured and/or foliated, make an excellent foundation base having a high bearing capacity. Residential units are typically constructed with basement slabs and continuous footing foundations. Soil depths in nonglaciated, humid climates are suitable for basement construction, but in arid climates or glaciated regions rock removal may be necessary. The use of raft foundations requires gravel pads on the residual soils, owing to their plastic nature, but specific tests can determine suitabilities. Foundations for large structural loads can utilize deep footings or rafts in deep soil areas. Shallow soil zones facilitate the use of piers or footings to transfer the applied load to the rock stratum.

Highway Construction

Highway construction across gneiss formations encounters alignment problems in trying to minimize earthwork, bridging, and tunneling. Alignments following the natural grain of the topography through parallel valleys can minimize costs in comparison with routes that bisect ridges and valleys. However, for most routes large cuts are necessary and require blasting and removal of rock. Massive fills, large bridges, and tunnels are occasionally economically feasible as alternatives. In massive structures vertical faces resulting from rock "presplitting" hold well without flaking and weathering. Highly foliated or fractured structures need terracing along the cut face to minimize future maintenance or weathered debris. Problems of highway construction are similar in arid climates, although intensified by the fact that more rock excavation is necessary because of the thinner residual soil.

CHAPTER 8

GLACIAL LANDFORMS

INTRODUCTION

Glaciers are formed when snow and ice, having accumulated to a certain weight, lose their familiar crystalline structure and become plastic. As the accumulation continues, the structure begins to flow outward under its own weight, and the destructive process of glaciation is begun. During its progress the glacier flows with a tremendous mass; it scours, scrapes, and transports soil and rock materials across hundreds of miles, depositing them along the entire extent of its flow. Glacial deposits both influence and disrupt postglacial drainage systems.

Glacial landforms are the erosion and deposition formations left on the earth's surface as a result of glacial activity. These formations occur to a significant extent on the earth's surface, since approximately 30% of its land area has been glaciated at some period. At present, 10% of the earth's land surface is occupied by glaciers, the main ice sheets being in Greenland and Antarctica.

Two types of glaciation occur, at two different scales, and each creates a distinct kind of landscape. Alpine glaciation is the result of the actions of valley glaciers and is found in mountainous regions. Continental glaciation was caused by glacial activity that affected or covered all or a significant portion of a continental landmass.

Alpine Glaciation

Alpine glaciation and its associated features can be observed in most of the world's mountainous areas. Valley glaciers, similar to stream tributaries, flow toward lower elevations, carving and scouring rock materials, creating steep-walled valleys and jagged ridgelines. The flow of ice continues until the rate of melting, known as ablation, equals the rate of advancement. When the ice melts, the glaciated topography is exposed and exhibits such features as cirques, aretes, U-shaped valleys, hanging valleys, and moraines (**Figure 8.1**). Cirques are amphitheater-shaped depressions found in higher mountainous areas near ridgelines. Often, the front of a cirque is blocked by moraines or rock debris, and a tarn or steep-sided lake having no apparent inlet is thus created. When two or more cirques intersect a jagged ridgeline, an arete is formed. U-shaped valleys are glacial troughs carved by the main ice flows; they have rounded bottoms and very steep walls. Sometimes the smaller intersecting valleys have floors well above that of the major valley, and hanging valleys occur. Indeed, it is common to find entire interconnecting systems of hanging valleys which have been created by interconnected valley glaciers; these appear treelike in plan (Figure 8.2). Where ice flows terminate, depositing their contained materials and those pushed in front or between ice lobes, moraines are formed (Figure 8.3). Some high mountainous regions of the world still have active alpine glaciers; other regions such as the White Mountains of New England show many features of previous alpine glacier activity.

Continental Glaciation

Continental glaciation occurs when great masses of ice accumulate and cover all or a significant portion of a continental surface. Several theories have attempted to explain the causes of continental glaciation. It is attributed to (1) a significant decrease in the solar energy reaching the earth, which lowered temperatures; (2) an increase in the elevation of major land masses during the Pliocene and early Pleistocene epochs, which acted to intercept moisture and thereby developed major snowfields; (3) a reduction of carbon dioxide in the atmosphere, which lowered temperatures; (4) the presence of large amounts of volcanic dust in the atmosphere, which blocked solar radiation and thereby lowered temperatures; (5) shifts in the position of the earth's poles, which enhanced the conditions necessary for vast quantities of ice growth on the continental landmasses; and (6) changes in oceanic currents, which affected moisture flows and temperatures and created conditions favorable to massive ice growth upon continents. No matter what the exact cause, the concept of glaciation is now universally accepted.

In North America four major glacial stages, which covered significant portions of the land, have been identified. The oldest deposits now found come from the Nebraskan glaciation, which extended through the upper Mississippi Valley. Few of these materials are now exposed to the ground surface, since they were covered over in subsequent glacial periods. The Kansan glaciation period which followed extended southward into northern Kansas and Missouri, and here some exposed drift still exists. The next period of American continental glaciation was the Illinoian. Drift from this period averages 30 feet in thickness across the Midwest region of Ohio, Illinois, and Indiana, and small bands of it can be found in southern Illinois, Indiana, and Ohio where it was not covered by later deposits. The most recent period of glaciation was the Wisconsin, which occurred 10,000 to 80,000 years ago. It covered most of the previous glacial deposits in the northern Midwest, all of New England and Canada, and most of Alaska.

Glacial Erosion

Field investigations commonly encounter evidences of glacial erosion recorded on exposed

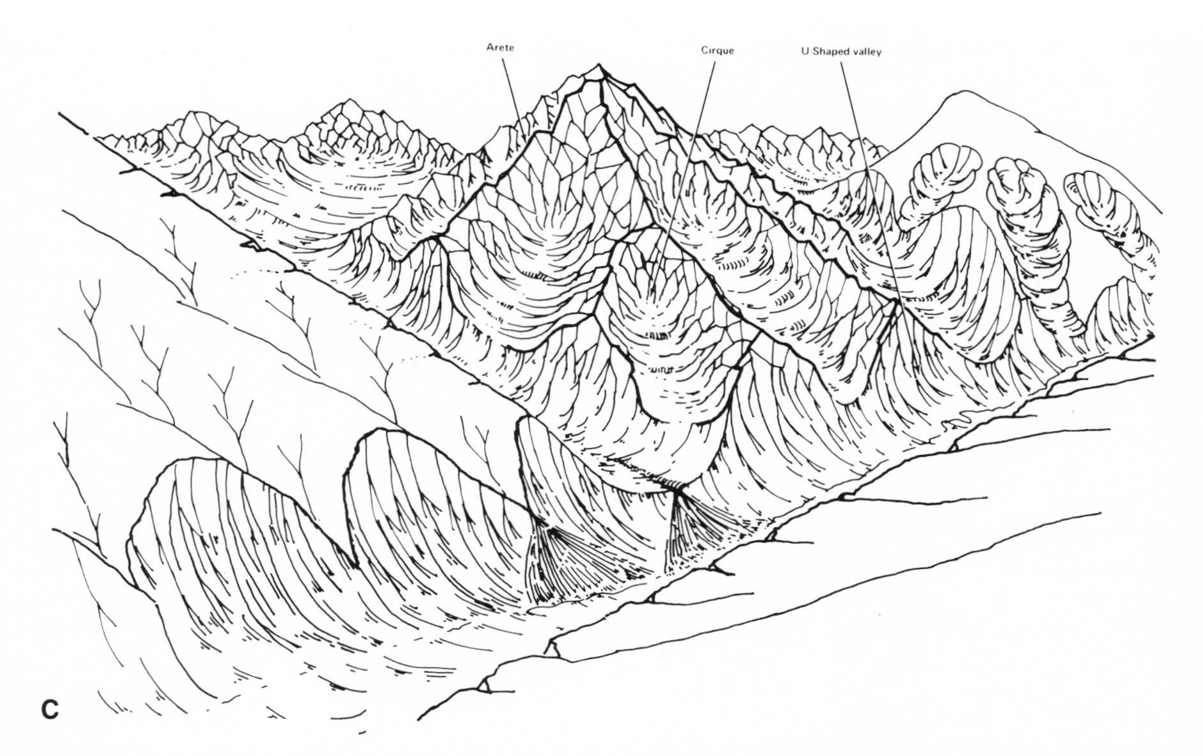

Figure 8.1. Effects of alpine glaciation. (A) Before glaciation, the region has rounded slopes and ridges with V-shaped valley bottoms. (B) The glacial erosive process carves new forms out of the mountain masses and creates sharp divides. (C) Upon disappearance of the glacial ice, the topography appears with sharp ridges or divides and gentle, U-shaped valleys. (After W. M. Davis and A. K. Lobeck.)

hard rock or on the surfaces of stones found within the residual soils. As the massive layers of glacial ice moved across the land, both solid bedrock and residual soils were captured in the frozen mass and helped to polish and grind the underlying rock surfaces. Large blocks of resistant rock, dragged by the ice, created the scratches or striations that now indicate the direction of the ice movement. When sharp-pointed rocks were dragged over bedrock, they formed crescent-like gouges, called chatter marks, whose curvature is opposite to the direction of flow. Angular edges or scratches on rocks and pebbles indicate rasping by glacial ice; rounded edges on these materials indicate transportation and erosion by fluvial processes. When the moving ice sheets encountered a knob of bedrock, they overrode the front-facing side and plucked rocks from the lee side. The resulting cigar shaped rock masses are called roches moutonnees and are similar to drumlins in form, one end having a gradual slope and the other a steeper, more jagged slope. The glacial ice lobes that followed preglacial valleys occasionally eroded and formed elongated depressions; many of these valleys are now filled with water, for example, the Finger Lakes in New York and many lakes in Minnesota and Canada.

Glacial Deposits

Glacial ice is heavily loaded with accumulated rock and soil debris known as glacial drift. Glacial drift refers to all types of debris transported by glacial ice, without any sorting or stratification. Rocks and soil textures of all sizes are transported, ranging from boulders as large as houses, called eradics, to very fine silts and clays. The crushing and grinding of the rock materials creates very fine, uniform, silt-sized particles (rock flour) which have angular, unweathered characteristics. As the ice advanced, then receded and melted, all of these materials were eventually dumped or scattered or sometimes stratified to create a variety of landforms (Table 8.1). The following list includes the glacial landforms of till, glaciofluvial, and glaciolacustrine origin, which have major differences in interpretive and engineering characteristics.

Glacial Till Deposits

As a glacial ice flow becomes heavily loaded with soil and rock mixtures, materials may be deposited and compressed under the ice flow itself. Deposits of such lodgment till contain heterogeneous mixtures of clay, silt, sand, gravel, and boulders and form unconsolidated, unsorted, and unstratified plains. Other deposits of till originate from materials previously suspended in glacial ice, which drop onto the ground surface

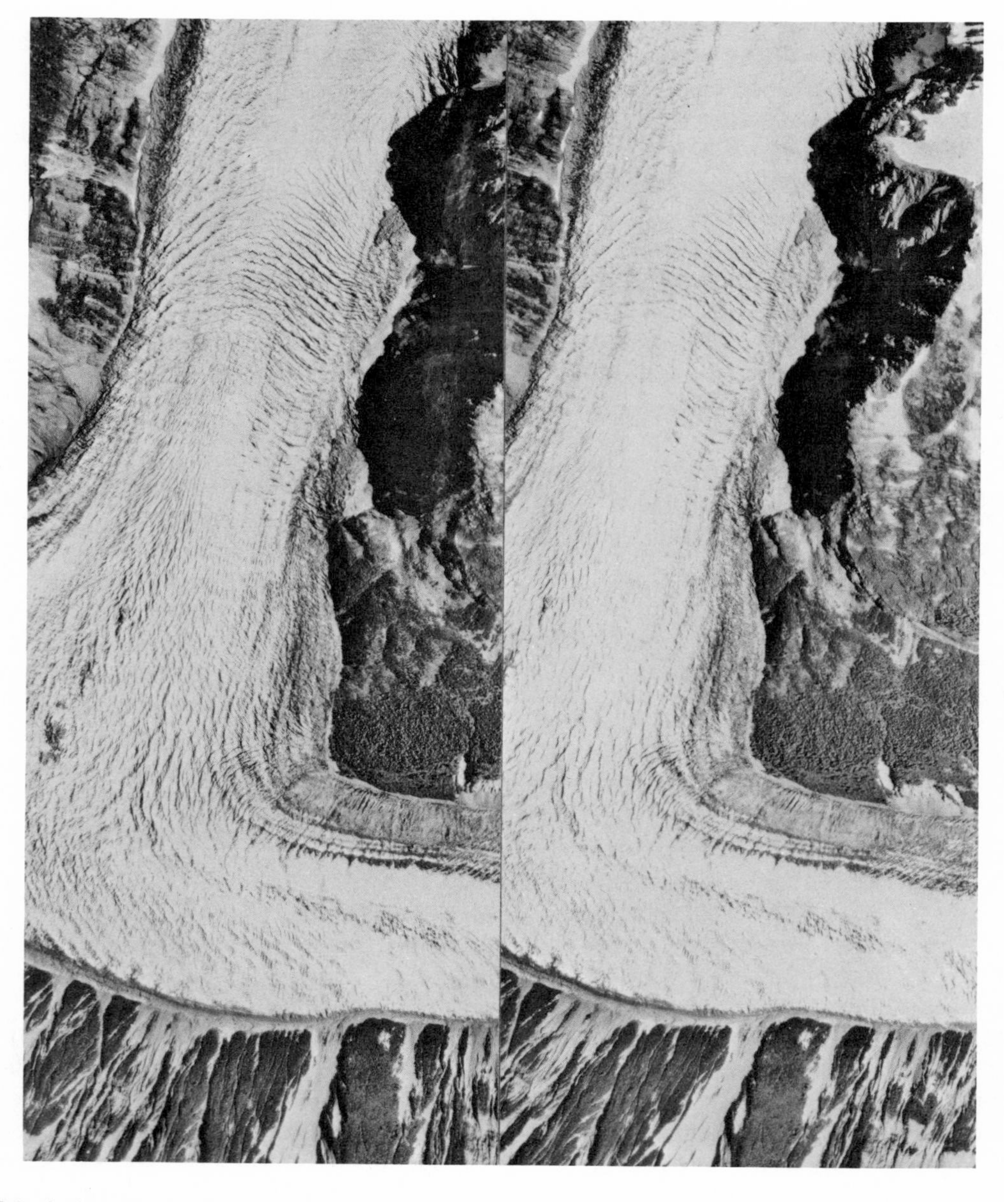

Figure 8.2. Active valley glaciers in Alaska. The South Crillon Glacier is shown. The linear bands of material carried by the ice are readily observed. Photographs by the U.S. Geological Survey, Alaska 28-ABC, 1:25,000 (2083′/″), August 11, 1959.

as it melts. These materials are commonly referred to as ablation tills or ground moraine.

Till Plain. Till plains are made of glacial till and have a flat, broad topography. Many of the deep till deposits in the midwestern United States are described as till plains.

Ground Moraine. These formations originate from till materials, either those deposited under glacial ice or those deposited by the process of ablation (in which debris suspended in the glacier is dropped onto the land surface as the ice melts). Many geologists do not make a sharp distinction between till and ground moraine except to say that the latter contains a higher percentage of ablation till. On many maps the terms till and ground moraine are used interchangeably.

Moraine. Advancing ice lobes push before them large quantities of heterogeneous debris, and this forms moraines after the ice melts. *End moraines* refer to those forms created by ice lobes which periodically retreated and advanced; they appear as a series of concentric formations behind a terminal moraine. *Terminal moraines* are formations which indicate the furthest advance of any one ice lobe during its glacial period. *Interlobate moraines* are formed between intersecting ice lobes and are commonly observed in areas of alpine glaciation.

Drumlin. The exact process by which drumlins were formed is not clear, but they are definitely of glacial origin, being composed of till which was compressed into streamlined, cigar-shaped formations under advancing ice flows. The drumloidal shape is steeper on the end from which the ice advanced; the lee side tapers gradually to the ground surface. The axes of the drumlins indicate the direction of ice movement.

Glaciofluvial Deposits

As glaciers melt, large quantities of water are

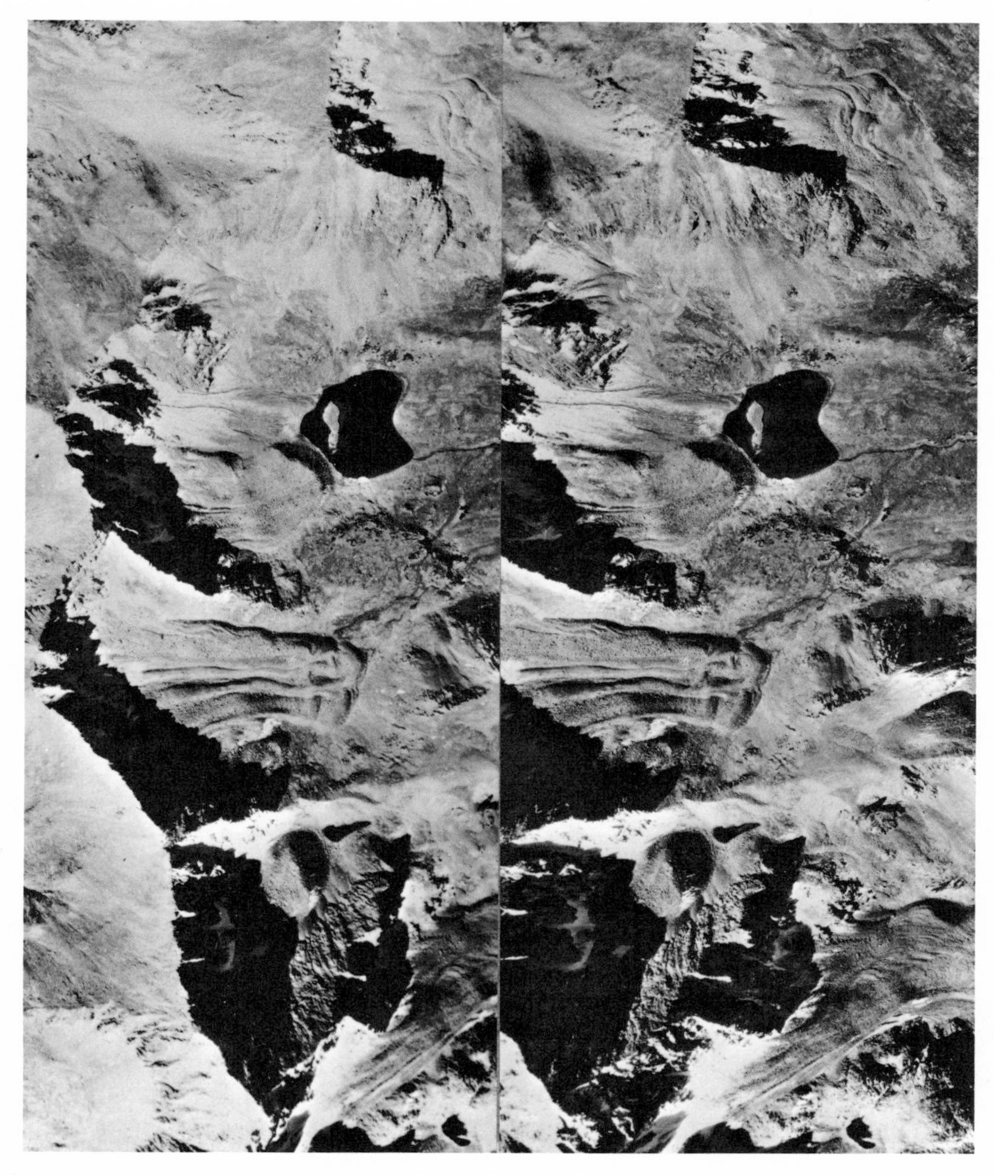

Figure 8.3. Features of alpine glaciation, including cirques, tarns, and rock glaciers, are apparent in Chaffee and Gunnison Counties, Colorado, along the north side of the Continental Divide. Photographs by the U.S. Geological Survey, Colorado 5-AB, 1:18,400 (1533'/"), September 24, 1956.

released, and these streams carry and sort the glacial debris into a variety of stratified formations known as eskers, kames, kame terraces, outwash plains, and valley trains. Some of these landforms, such as eskers, kames, and kame terraces, were created with some or all of the formation in contact with the glacial ice, whereas outwash plains and associated varieties were formed completely independent of contact with the ice.

Eskers. Eskers are glaciofluvial, stratified materials which were deposited from the bedload of streams or rivers flowing on, in, or under glacial ice. They consist primarily of sands and gravels and show as a long, narrow, winding ridge.

Kame. Crevasses or depressions on the surface of glaciers accumulate surface runoff and transported soil particles and, upon the melting of the ice, form conical or sharp-ridged hills. Sands or gravels are predominate in poorly stratified layers.

Kame Terrace. Kame terraces are formed by stratified sands and gravels which were deposited between glacial ice sheets and associated higher ground. They appear as elevated terraces along the edge of valley floors.

Outwash. Large flood plains are created when vast quantities of meltwater are overloaded with soil and rock debris. These flood plain or outwash formations consist primarily of sands and gravels in well-stratified layers.

Valley Train. In regions of undulating or rugged topography, outwash deposits were contained within valleys. These deposits are called valley trains. Many of the outwash deposits found in the northeastern United States are valley trains.

Glaciolacustrine Deposits

Glaciolacustrine deposits are stratified materials whose deposition or formation sequences occurred in relation to lakes. Materials that were

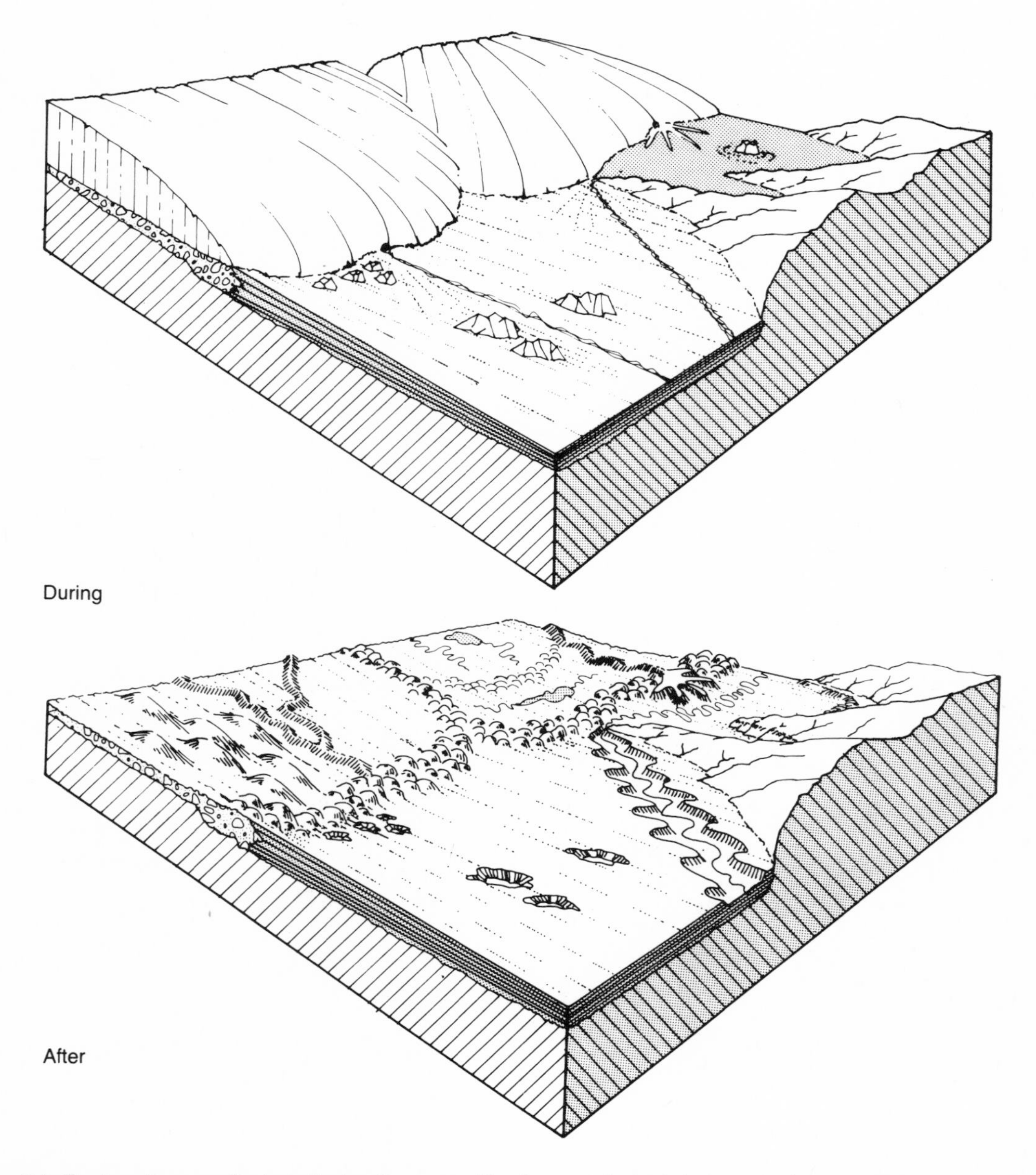

Figure 8.4. The two diagrams illustrate the landforms resulting from continental glaciation. (After A. N. Strahler, *Introduction to Physical Geography,* John Wiley, New York, 1965, p. 328.)

Table 8.1. Glacial landforms—parent material—origin

Parent material	Landform	Origin
Glacial drift (till), nonstratified	Till plain, Ground moraine, End moraine, Drumlin	Glacial
Stratified drift, ice contact	Esker, Kame, Kame terrace	Glaciofluvial
Stratified drift	Outwash, valley Valley train	
Stratified drift	Lake bed, Beach ridge	Glaciolacustrine

deposited on lake bottoms are classified as lake beds; those formed by wave action on lake shorelines are beachridges.

Lake Bed. Glacial lake bed formations consist of sediments that were deposited on a lake bottom during and/or following glaciation. The lake itself is often a temporary formation, formed by ice dams in valleys or between the ice front and moraines or high ground.

Beach Ridge. Beach ridges are formed on lake shorelines by the action of waves which transport, sort, and deposit sand and gravel. If the water level decreases, the deposits are often found as concentric ridges, each representing a previous shoreline.

Distribution

Glacial deposits occur over 30% of the exposed land surface of the earth. It is difficult to give a general description of the distribution of individual glacial landforms because of their small size and scattered occurrence. The major glaciated regions and their most important associated deposits are described in the following discussion.

North America

United States. Most of the northern third of the United States was glaciated. The younger, Wisconsin age materials are predominant. Till deposits of the older Kansan age are located in eastern Nebraska, northeast Kansas, northern Missouri, and southern Iowa; Illinoian age tills are exposed along a narrow strip passing through southern Illinois, Indiana, and Ohio. Moraines are scattered throughout the Midwest in central Iowa, Minnesota, eastern South Dakota, central North Dakota, Wisconsin, northern Illinois, Indiana, Ohio, northwestern Pennsylvania, and most of Michigan. The New England states have a few end moraines in sourthern Rhode Island and along the northeast coast of Maine. Drumlin fields, which are concentrated in groups and rarely found isolated, occur in Minnesota, eastern Wisconsin, Iowa, Illinois, Michigan, western New York, Massachusetts, and Connecticut. Outwash plains and valley trains are widely distributed throughout all the glaciated regions; major deposits are scattered across the Midwest, northern Indiana, Michigan, Long Island, southeastern Massachusetts, and scattered valleys in New England and New York. Eskers and kames are commonly found throughout the New England states, Minnesota, Wisconsin, Michigan, Indiana, Ohio, and New York. Glacial lake beds and beach ridges are found in central and eastern North Dakota, western Minnesota, northern South Dakota, south central Montana, and around the Great Lakes Basin; many scattered deposits occur throughout New England. Major formations and deposits are shown in Figure 8.5.

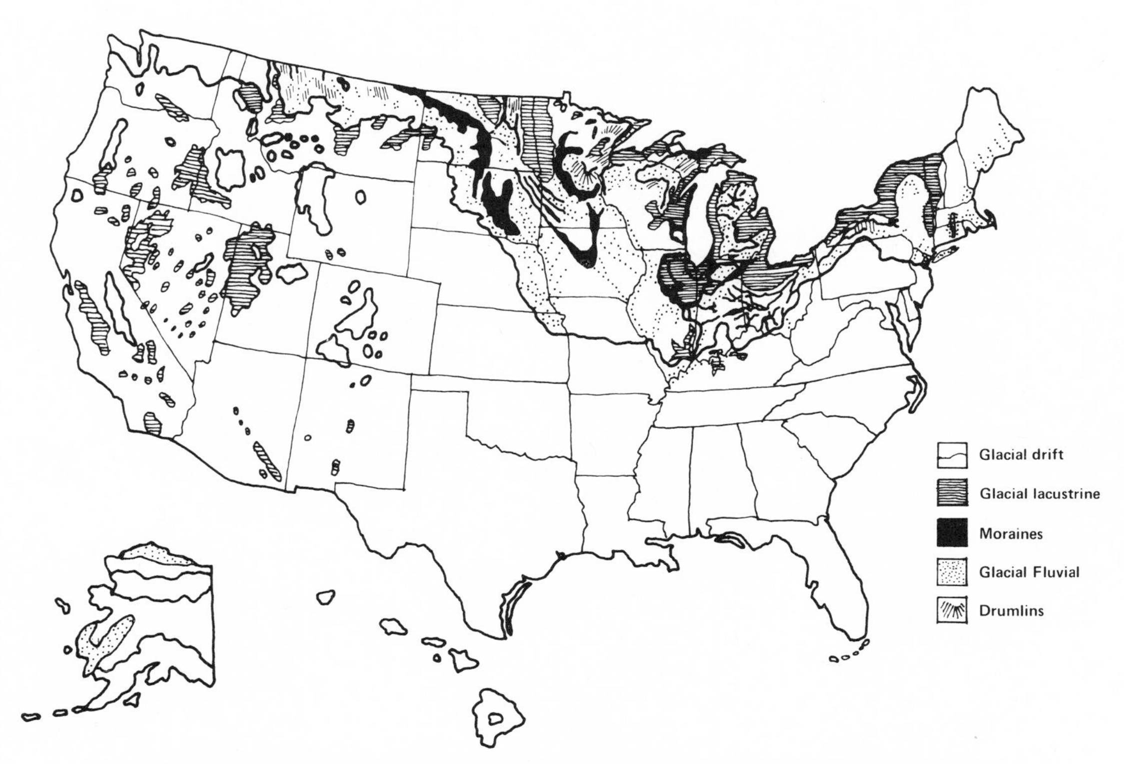

Figure 8.5. Distribution of major groups of glacial landforms across the United States. (After "National Atlas," U.S. Geological Survey, Washington, D.C., 1971.)

Canada. Virtually all of Canada has been glaciated, and all regions of the country have a Wisconsin age till cover of varying thicknesses. Moraines are well developed across Ontario, Quebec, and western Canada; extensive drumlin fields occur in northern Saskatchewan, Manitoba, and sourthwestern Nova Scotia; vast outwash deposits are found across most of central Canada; eskers are scattered through northern Canada, western Canada, and southwestern Quebec and Ontario; and prominent lake beds and beach ridges are found in Manitoba and Saskatchewan.

Central and South America

No significant portions of Central America were glaciated, but mountain chains in Peru and Chile in South America contain landscapes formed by alpine glaciation.

Africa

The African continent shows no features of continental glaciation.

Europe

Most of northern Europe was glaciated, creating formations that cover more than 6 million square kilometers. Tills cover significant portions of Nor-

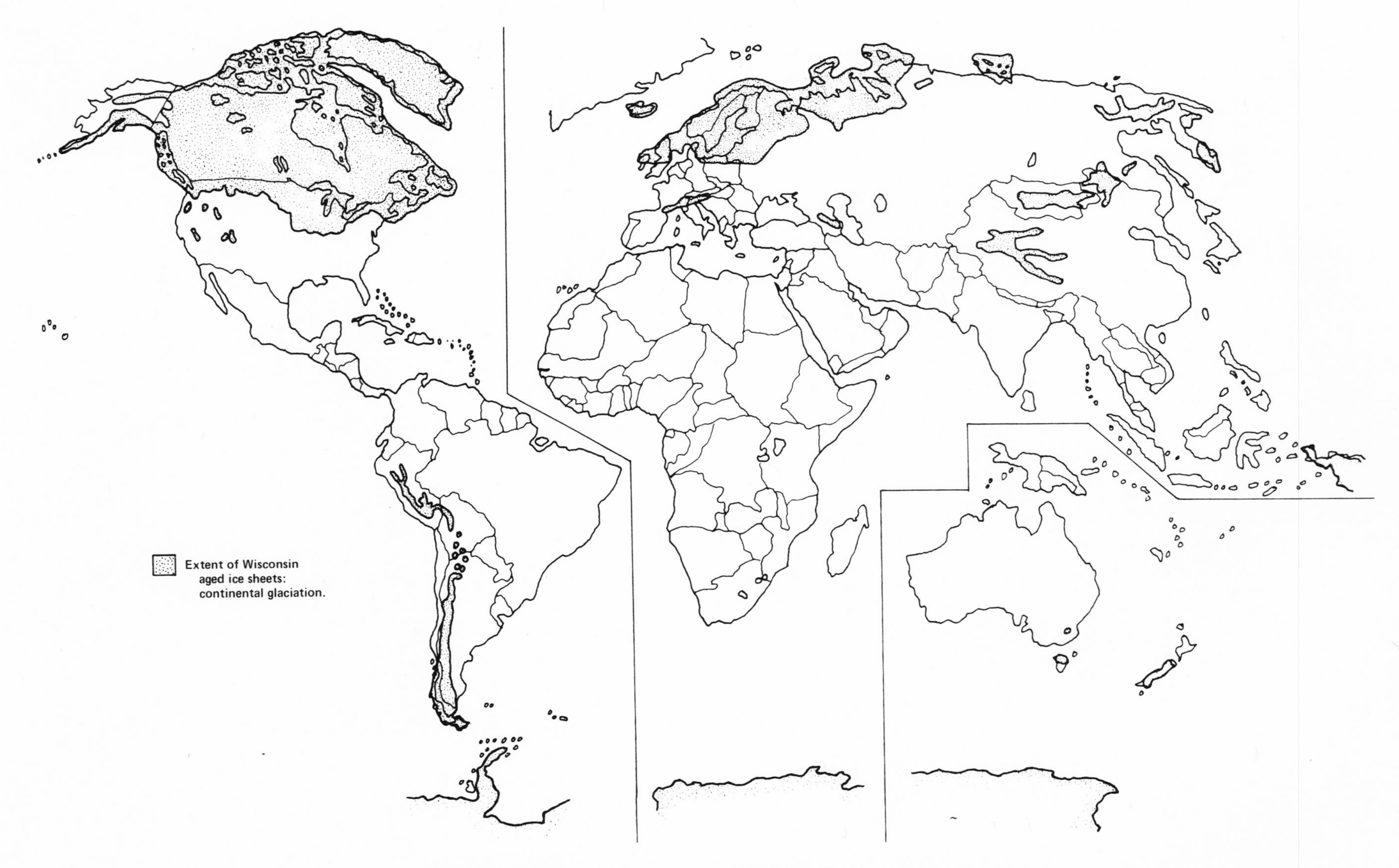

Figure 8.6. Distribution of major groups of glacial landforms around the world.

way, Finland, Sweden, the northern British Isles, Denmark, northeastern Germany, northern Poland, and northwestern Russia. Smaller till deposits occur in the Alps and Pyrenees mountain chains. End moraines are found in southern Finland, Denmark, southern Sweden, and northern Germany; drumlins occur in Switzerland and Germany; and outwash is found throughout northwestern Russia, Germany, and around the Alps and Pyrenees. Other glaciofluvial deposits, such as eskers and kames, are scattered across most of glaciated Europe. Lake bed (lacustrine) forma-

Table 8.2. Glacial landforms: summary chart

Landform	Topography	Drainage	Tone	Gullies	Vegetation and land use
Till					
Thick					
Young	Flat plains	Deranged	Mottled	Few to soft	Cultivated, gridded*
Old	Dissected plateaus	Dendritic, medium	Subdued mottles	Box-shaped†	Cultivated and forested*
Thin					
Young	Rock-controlled	Rock-controlled	Light gray	Vary	Cultivated and/or forested*
Old	Rock-controlled	Rock-controlled	Dull gray	Vary†	Cultivated and forested*
End moraine	Undulating to rugged	Deranged	Light to dark	Vary	Cultivated and/or forested*
Drumlin	Drumlin-shaped	None	Light gray	None to few	Cultivated and/or forested*
Glaciofluvial					
Esker	Snakelike ridges	None	Light	None	Natural cover
Kame	Cone or ridged hills	None	Light	Few V-shaped	Natural cover
Outwash					
Plain	Flat plains	Internal, channel scars	Light	None	Cultivated or natural*
Pitted	Pitted plains	Internal	Light	Few V-shaped	Cultivated or natural*
Valley train	Flat valley bottom	Internal	Light	None	Cultivated or natural*
Glaciolacustrine					
Lake beds					
Sandy	Flat plains	Internal and ditches	Light dull gray	None	Cultivated
Clay	Flat plains	Broad meanders	Dull gray	Ditches	Cultivated

*Semiarid or arid climates may have some grain crops but land is typically in natural cover of scattered grasses.
†White-fringed gullies are direct indicators of Illinoian age till.

Table 8.3. Alpine glaciation: aerial and cartographic references

State or country	County or area	Photo Agency	Photograph no.	Scale	Date	Corresponding quadrangle and geologic maps from USGS	Corresponding soil reports from SCS
Alaska	—	USGS*	Alaska 30-AB	3333′/″	7/5/48	Petersburg (1:250,000)	—
Colorado	Chaffee, Gunnison	USGS*	Colorado 5-AB	1533′/″	9/24/56	Mt. Harvard (15′)	—
Ecuador	Azuay area	USGS†	Ecuador 3-AB	3333′/″	6/22/63	Army Map Service, 1301, SA17, Quito (1:1,000,000)	—
Maine	Piscataquis	USGS*	Maine 1-ABC	5250′/″	6/30/52	Katahdin, Harrington Lake, Telos Lake, Traveler Mountain (15′)	—
Montana	Glacier	USGS*	Montana 2-ABCD	5000′/″	8/25/58	Chief Mountain (30′)	—
Utah	Wayne	USGS*	Utah 6-ABC	3117′/″	10/3/52	Mt. Olympus, Mt. Tom (15′)	—
Washington	Snohomish	USGS*	Washington 2-ABC	2308′/″	10/7/44	Glacier Peak (15′)	1947

*From *Selected Aerial Photographs of Geologic Features in the United States*, U.S. Geological Survey Prof. Paper No. 590.
†From *Selected Aerial Photographs of Geologic Features outside the United States,* U.S. Geological Survey Prof. Paper No. 591.

tions are found throughout the Scandinavian countries and the northern British Isles.

Asia

In Asia approximately 8 million square kilometers of land were glaciated, including northern Siberia, the central Siberian plateau, New Siberian Island, Wrangell Island, the Koryak Mountains, the Kamchatka Peninsula, the Transbaykal and Altai highlands, the Himalaya Mountains of central Asia, and the Caucasus Mountains in Asia Minor.

Australasia

Only the western half of Tasmania and a portion of South Island in New Zealand were glaciated to a significant extent. The glaciation was primarily in the form of alpine glaciers.

Pacific Region

No significant glacial deposits are found in the Pacific Island region.

Caribbean Region

No glacial formations occur in the Carribean region.

REFERENCES

American Society of Photogrammetry, *Manual of Photo Interpretation,* American Society of Photogrammetry, Falls Church, Va., Chapter 5, 1960.

Bloom, A. L., *The Surface of the Earth,* Prentice-Hall, Englewood Cliffs, N.J., Chapter 7, 1969.

Bretz, J. H., "Keewatin End Moraines in Alberta, Canada," *Geological Society of America Bulletin,* Vol. 54, pp. 31–52, 1943.

Capps, S. R., "Glaciation in Alaska," U.S. Geological Survey, Professional Paper Number 170, pp. 1–8, 1931.

Chadwick, G. H., "Adirondack Eskers," *Geological Society of America Bulletin,* Vol. 39, pp. 923–929, 1928.

Chorley, R. J., ed., *Water, Earth and Man,* Methuen, London, 1969.

Coates, D. R., ed., *Glacial Geomorphology,* State University of New York, Binghamton, N.Y., 1974.

Dyson, J. L., "Ice-ridged Moraines and Their Relation to Glaciers," *American Journal of Sciences,* Vol. 250, pp. 204–211, 1952.

Embleton, C., and C. King, *Glacial Geomorphology,* Halsted Press, N.Y., 1975.

Evans, I. S., B. A. Kennedy, M. G. Marcus, and J. Rooney, "Snow and Ice," in R. J. Chorley, ed., *Water, Earth and Man,* Methuen, London, Chapter 8, 1969.

Fenneman, N. M., *Physiography of the Western United States,* McGraw-Hill, New York, 1931.

Fenneman, N. M., *Physiography of the Eastern United States,* McGraw-Hill, New York, 1939.

Flint, R. F., "Eskers and Crevasse Fillings," *American Journal of Science,* Vol. 15, pp. 410–416, 1928.

Flint, R. F., "Growth of the North American Ice Sheet during the Wisconsin Age," *Geological Society of America Bulletin,* Vol. 54, pp. 325–362, 1943.

Flint, R. F., *Glacial and Pleistocene Geology,* Wiley, New York, 1957.

Flint, R. F., et al., "Glacial Map of the United States East of the Rocky Mountains," *Geological Society of America,* New York, 1959.

Gilbert, G. K., "Crescentic Gouges on Glaciated Surfaces," *Geological Society of America Bulletin,* Vol. 17, pp. 303–314, 1906.

Hamblin, W. K., and J. D. Howard, *Physical Geology: Laboratory Manual,* 2nd ed., Burgess, Minneapolis, 1967.

Hobbs, W. H., *Characteristics of Existing Glaciers,* Macmillan, New York, 1911.

Holmes, C. D., "Kames," *American Journal of Science,* Vol. 245, pp. 240–249, 1947.

Jahns, R. H., "Sheet Structure in Granites: Its Origin and Use as a Measure of Glacial Erosion in New England," *Journal of Geology,* Vol. 51, 1943.

Johnston, W. A., "Glacial Lake Agassiz, with Special Reference to the Mode of Deformation of the Beaches," *Canada Geological Survey Bulletin,* No. 7, 20 pp., 1946.

Leverett, F., "Moraines and Shore Lines of the Lake Superior Basin," U.S. Geological Survey, Professional Paper No. 154, pp. 1–72, 1929.

Lobeck, A. K., *Geomorphology, An Introduction to the Study of Landscapes,* McGraw-Hill, New York, pp. 42–47, 1939.

Lobeck, A. K., *Things Maps Don't Tell Us,* Macmillan, New York, 1956.

Longwell, C. R., R. F. Flint, and J. E. Sanders, *Physical Geology,* Wiley, New York, Chapter 12, 1969.

Maclaren, C., "The Glacial Theory of Professor Agassiz," *American Journal of Science,* Ser. 1, Vol. 42, pp. 346–365, 1842.

Price, R. J., *Glacial and Fluvioglacial Landforms,* Hafner, New York, 1973.

Moore, R. C., *Historical Geology,* McGraw-Hill, New York, 1933.

Salisbury, N. E., "Relief: Slope Relationships in Glaciated Terrain," *Annals of Association of American Geographers,* Vol. 81, 1962.

Shaler, N. S., "On the Origin of Kames," *Boston Society of Natural History Proceedings,* Vol. 23, pp. 36–44, 1884.

Sharp, R. P., "Glacier Flow," *Geological Society of America Bulletin,* Vol. 65, pp. 821–838, 1954.

Shelton, J. S., *Geology Illustrated,* W. H. Freeman, San Francisco, Chapter 18, 1966.

Smalley, I. J., and D. J. Unwin, "The Formation and Shape of Drumlins and Their Distribution and Orientation in Drumlin Fields," *Journal of Glaciology,* Vol. 7, 1968.

Smith, H.T.U., "Giant Glacial Grooves in Northwest Canada," *American Journal of Science,* Vol. 246, pp. 503–514, 1948.

Strahler, A. N., *Introduction to Physical Geography,* Wiley, New York, Chapter 22, 1965.

Till (Ground Moraine)

Introduction

Glacial till or ground moraine consists of glacial drift or till materials in heterogeneous mixtures of clay, silt, sand, gravel and boulders, unstratified, unconsolidated, and unsorted. Till accumulates both from materials lodged beneath glacial ice and from materials dropped onto the land surface upon melting of glacial ice (Flint, 1957). The textural distribution of till varies in relation to the rock type from which the glacier acquired its load. Tills in the Midwest consist primarily of silts and clays originating from soft sedimentary rocks such as shale, limestone, or weakly cemented sandstone; tills in New England are typically coarser, being derived from crystalline rocks such as granite, gneiss, and schist. Exceptions to this are the clayey tills in southern New York and northern New Jersey.

Glacial till or ground moraine is the most common of all glacial landforms, being found wherever glaciation has occurred. Many geographers and geologists classify ground moraine and till plains as types of till deposits. For purposes of the photographic key and terrain analysis, however, we have arbitrarily classified till plains as a flat, topographic form of ground moraine.

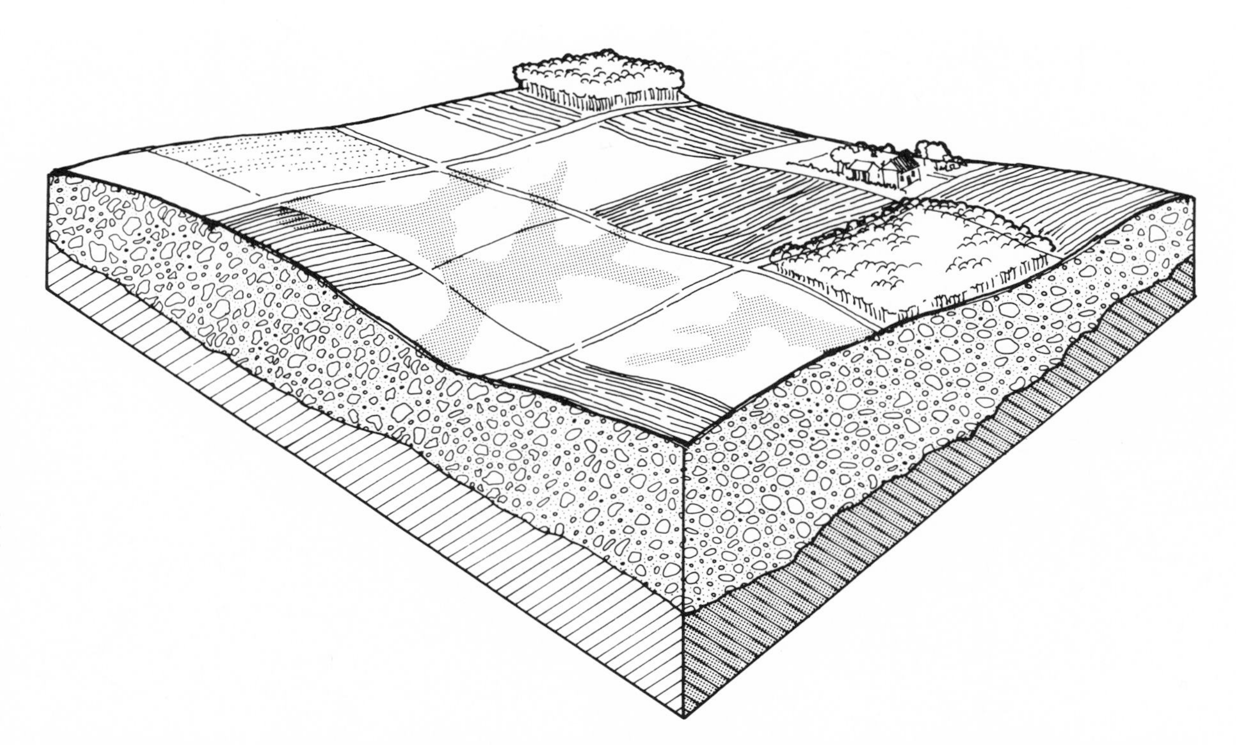

Figure 8.7. Young, thick till is characterized by its flat, topographic plain, lack of integrated drainage, and mottling of photographic tone. Visually, young, thick tills in the Midwest appear as monotonous, flat plains offering little viewing potential or spatial variation. Little variation in land use is found. The landscape is rather open and, because of the lack of vegetative cover, visually sensitive to structural development. Older, dissected areas offer more diversity. Visual closure occurs along drainage courses. The rock-controlled topography, covered with a thin veneer of till, is more undulating or rugged and offers a greater diversity of visual pattern.

Interpretation of Pattern Elements

There are many variations of till or ground moraine landforms, and it is difficult to include within a photographic key all the major categories. For example, in the Midwest young, thick till deposits of the Wisconsin glacial stage appear as broad, flat plains and are described as till plains, while in southern Iowa, eastern Nebraska, northeastern Kansas, and northern Missouri, where till formations were formed during older glacial periods (Kansan), they originally occupied flat plains but now have integrated surficial drainage systems. Again, Illinoian glacial deposits found in southern Illinois, southern Indiana, and southern Ohio can be 30 feet thick over sedimentary bedrock, while in New England relatively thin till deposits are found over very complex metamorphic and igneous rock structures. In order to give full exposure to the great variety of glacial till formations in the United States, the photographic key is divided into the following categories:

(1) Thick, young till deposits: tills of Wisconsin age found in the midwestern United States, of sufficient thickness to obliterate any pattern signatures of the underlying rock type or structure.

(2) Thick, old till deposits: formations occurring in the central midwestern United States, deposited during the Kansan and Illinoian glacial stages, of sufficient thickness to hide all patterns of the underlying rock type and structure.

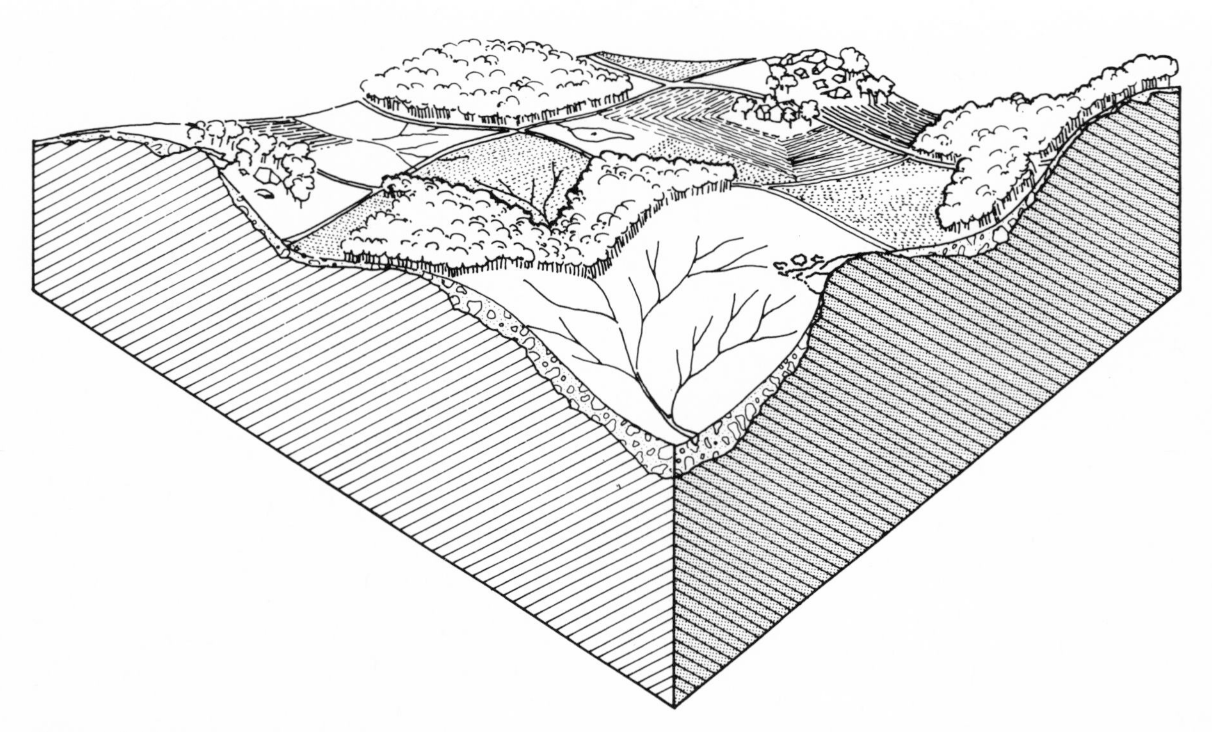

Figure 8.8. Thin, young tills show topographic and drainage characteristics associated with the underlying rock materials. The hilltops and the rims, where the valley slopes meet the ridgelines, are visually sensitive when viewed from the valleys. The landscape is visually absorptive and closed, owing to the heavy vegetative cover and topographic relief. Town centers and most development are located within lowlands and valleys.

(3) Thin, young till deposits: tills of Wisconsin age overlying bedrock, in which the patterns and topographic control of rock types and structure are apparent and/or predominant.

(4) Thin, old till deposits: tills of Kansan and Illinoian age, thinly veneered over bedrock, in which patterns and topographic control of rock type and structure are apparent and/or predominant.

Pattern characteristics of till deposits are generally the same in both humid and arid climates; therefore, climatic distinctions are not made unless required.

Soil Characteristics

The composition of till or ground moraine landforms varies, being dependent upon the source of the glacial drift. Midwestern tills are typically deep and contain a predominance of fine soil textures. Young, Wisconsin age tills originated from sedimentary rocks in northern Wisconsin, Minnesota, and Michigan, and consist of silty or clay loams; Kansan or Illinoian tills are more weathered and finer in texture, developing deeper profiles over a silty-clay hardpan 8 to 10 feet below the surface. Tills found in New England are of Wisconsin age but are coarser in texture, since the parent material originated from crystalline rocks in Canada, northern Vermont, New Hampshire, and Maine. Tills in southern New York and northern New Jersey contain more clay. The following ratings of these soils are presented only as guidelines; field investigations and samples are necessary to determine design standards.

U.S. Department of Agriculture Classification

Wisconsin Till: Sedimentary Rock Origin (Midwest). Midwestern glacial tills of Wisconsin age originate from sedimentary rocks and typically contain surface horizons of silt loam, silty clay loam, or clay loam. The B soil horizon contains more fines or clays which have been leached from the surface layers. The parent material or C horizon contains silty clay loam, clay loam, silt loam, or loam (Figure 8.13). Depressions contain more fines than slightly elevated positions.

Wisconsin Till: Crystalline Rock Origin (New England). Wisconsin age tills formed from crystalline rock are common in New England and are fairly coarse in texture, having surface horizons of sandy loam. B horizons contain more fines, and the parent material is typically a sandy loam, fine sandy loam, sandy clay loam, or silt loam (Figure 8.14). Tills in northern New Jersey and southern New York contain more clay.

Old Till (Kansan and Illinoian Age): Sedimentary Origin (Midwest). Older tills develop thick, weathered profiles with a semi-impervious silty clay hardpan 8 to 12 feet below the surface. Higher concentrations of silt and clay are found in the surface soil, and these are commonly classified as silty clay loams and silty clay.

Unified and AASHO Soil Classifications

Wisconsin Till: Sedimentary Rock Origin (Midwest). Surface soils are commonly described

Figure 8.9. (A) Thick, young till: Wisconsin age. Thick deposits of young Wisconsin age till are common throughout the midwestern United States and are usually referred to as till plains. Photographs by the Agricultural Stabilization Conservation Service, BWI-1BB-27,28, 1:20,000 (1667′/″), May 27, 1961.

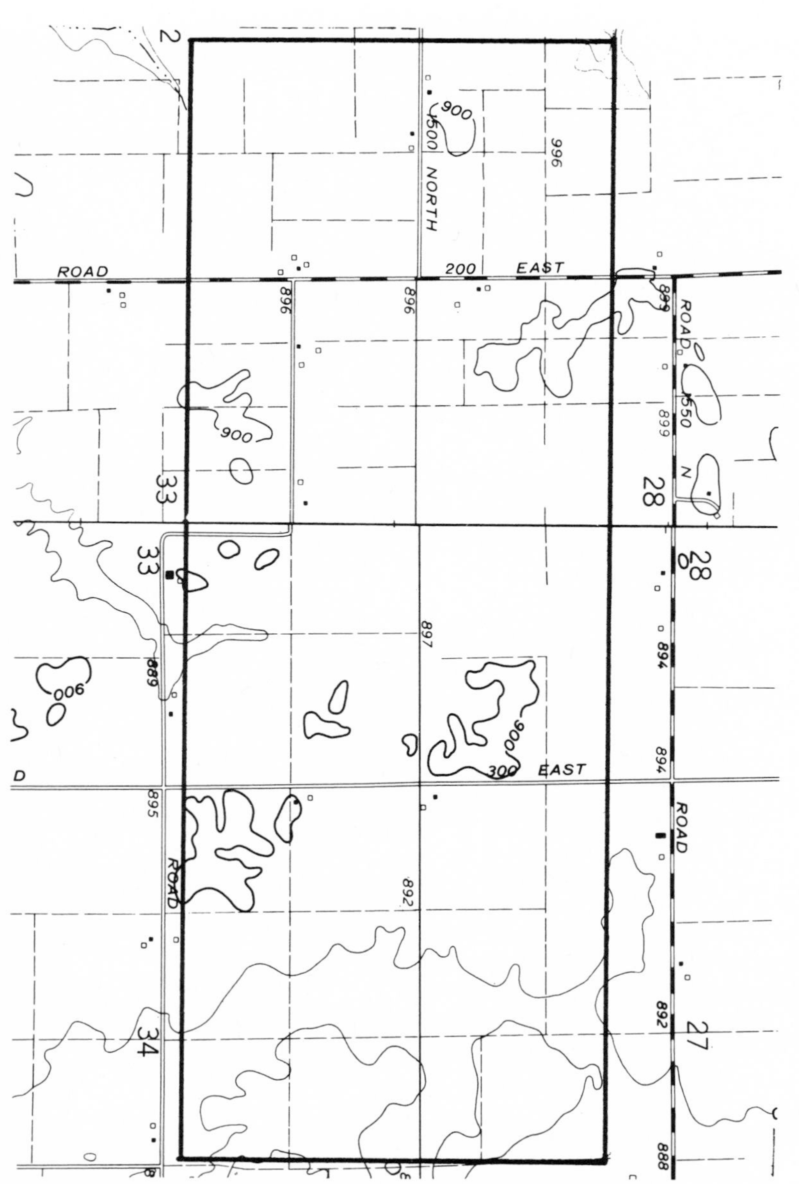

Figure 8.9. (B) U.S. Geological Survey Quadrangle: Alexandria, Gaston (7½′).

Table 8.4. Glacial till (till plains): Thick, young till (Wisconsin) and thick, old till (Illinoian, Kansan, Nebraskan) (humid and arid)

Thick, young till

Topography	*Drainage*	*Tone*	*Vegetation and land use*
Flat plains 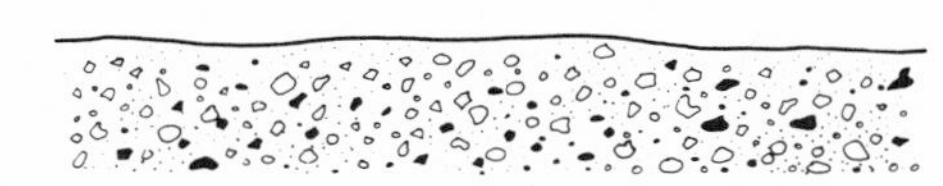 Thick, young till deposits of Wisconsin age are usually described as till plains but may also be mapped as ground moraine. The topography is very flat and undulating, typically not changing more than a few feet in elevation over an area of several square miles.	Deranged 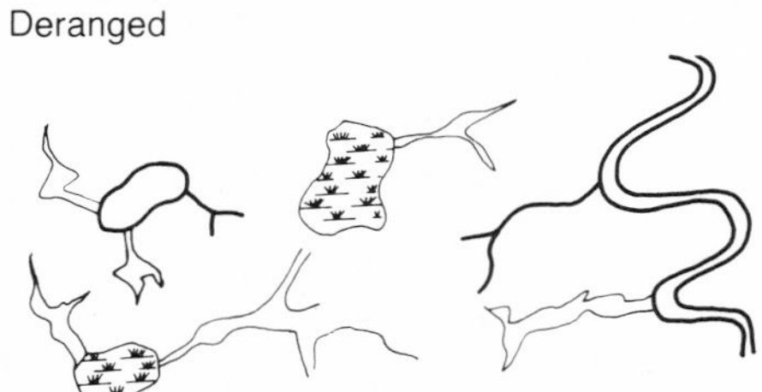 Since this formation is geologically young, runoff systems have not yet been established in well-defined, integrated dainage networks. Many depressions and swamps are found along with a few, meandering regional rivers.	Mottled Mottled soil tones indicate moisture differences. A slow gradation of light to dark tones between mottles indicates fine materials; sharp changes indicate coarser materials. *Gullies* None or few Generally, drainage has not developed sufficiently for gullying. Any gullies found are of the sag-and-swale type, indicating fine, cohesive materials.	Cultivated The thick Midwestern till deposits of the Wisconsin age are typically under agricultural use. Field and road patterns are square or rectangular and gridded, since there are few topographic or drainage constraints. Arid climates have grass cover.

Thick, old till

Topography	*Drainage*	*Tone*	*Vegetation and land use*
Dissected plains 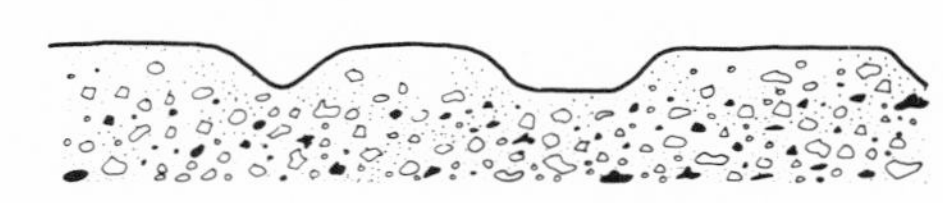 Old, thick deposits of Kansan and Illinoian age till were originally flat plains in form, but age has allowed erosion to dissect the topography. Undissected portions of the topography appear as flat plains.	Dendritic: Medium 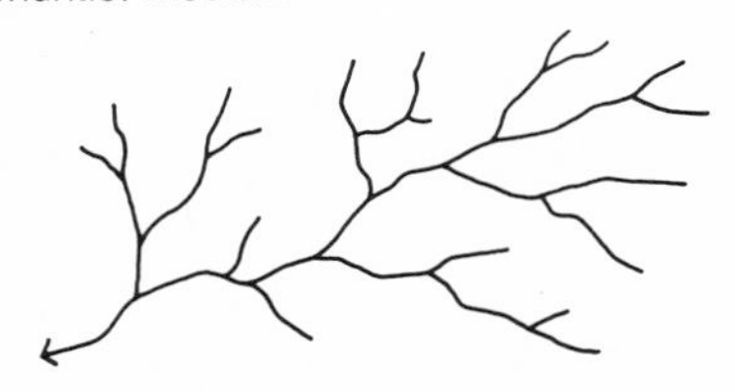 Illinoian and Kansan tills develop integrated dendritic drainage systems of medium texture. Illinoian tills can be identified by the occurrence of white gullies feeding the drainage pattern; these are caused by exposure of the light-colored topsoil.	Varies Kansan tills are a dull light-gray tone with few apparent mottles. Illinoian tills are light with white gullies. *Gullies* Box- (U)-shaped The high silt content is indicated by box-shaped gullies. White edges of gullies are indicators of silty Illinoian age till.	Cultivated Most of these deposits are cultivated, forming square fields and gridded road systems unless adjacent to drainage courses. Eroded areas adjacent to drainage courses may have tree cover, thus emphasizing the dendritic form. Arid climates have grass and scattered scrub cover.

as ML, CL, and ML-CL, with some OL soils occurring in bottomlands. The more impervious B horizon is typically CL, CL-CH, or CH; the parent material or C horizon is CL or ML. The AASHO system classifies the surface soils as A-6, the B horizon as A-6 or A-7, and the C horizon, or parent material, as A-4, A-6, or occasionally A-7 (Figure 8.15).

Wisconsin Till: Crystalline Rock Origin (New England). The Unified surface soils are typically SM, SC, or ML, the B horizon SM or ML, and the parent material or C horizon SM, ML, and

Table 8.5. Glacial till: Thin, young till (Wisconsin) and thin, old till (Illinoian, Kansan, Nebraskan) (humid and arid)

Thin, young till

Topography	*Drainage*	*Tone*	*Vegetation and land use*
Varies	Rock-related	Light gray	Cultivated and forested
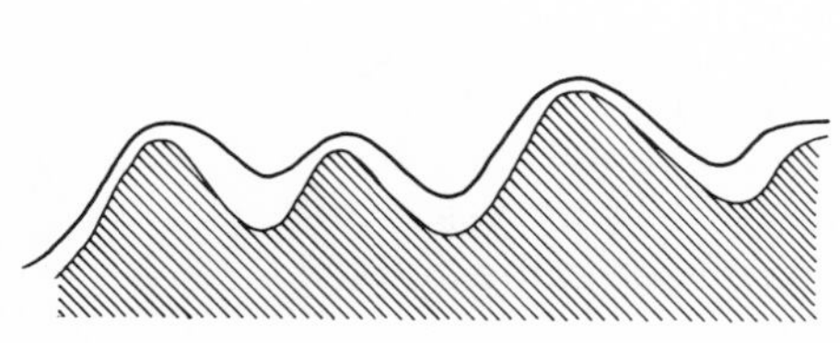 The land surface should show characteristics of form or drainage related to the underlying rock strata. In New England the topographic form generally reflects the massive character of the crystalline rock.	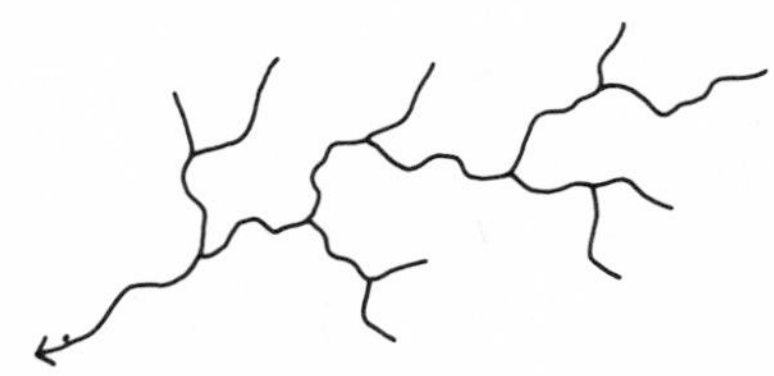 The drainage system is a significant indicator of shallow till deposits over rock structures, and angular alignments or other characteristics of rock-controlled drainage are apparent. Unlike deep, young till formations, these deposits have integrated drainage systems resulting from their slightly greater relief.	Tones of underlying, unrelated materials are subdued by the thin cover of till. Slight mottling may be apparent. *Gullies* Vary. Gullies indicate the textural composition of soils. Thus they are commonly V-shaped in New England and of the sag-and-swale type in the Midwest.	The ratio of cultivation to forest cover depends upon the topographic relief and the thinness of the soil cover. In southern New England and western New York, an undulating topography occurs, providing an approximately even mix of agriculture and forest cover. In northern New England the topography is more rugged, with thinner soil depths, which is reflected by a greater percentage of forest cover.

Thin, old till

Topography	*Drainage*	*Tone*	*Vegetation and land use*
Varies	Rock-related	Dull gray	Cultivated and forested
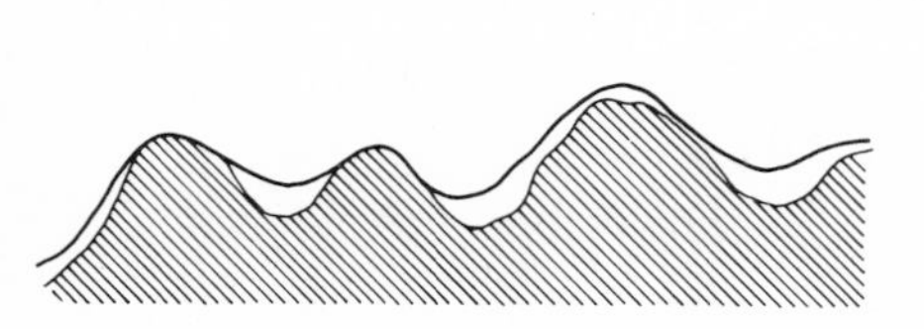 The sedimentary rocks found in southern Illinois, Indiana, and Ohio under Illinoian till do not control the topographic relief significantly. The topography is moderately undulating and dissected.	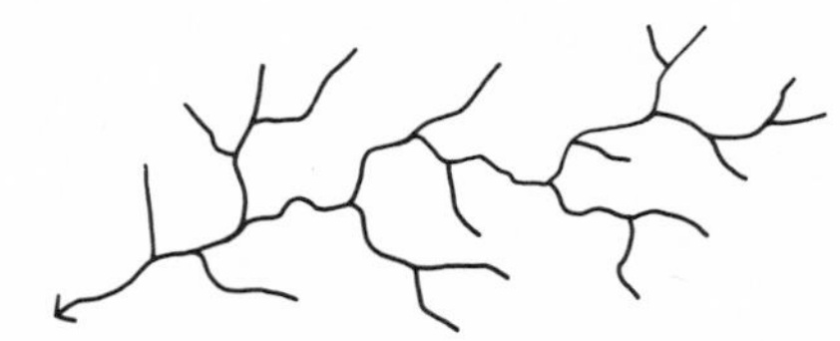 Drainage systems with angular alignments or other rock control patterns indicate till cover of insufficient depth to mask characteristics of the underlying rock structure. White gullies and gray tones indicate Illinoian or Kansan tills, respectively.	Kansan tills have dull gray tones; Illinoian till tones are lighter, and there are white-fringed gullies. *Gullies* Box- (U)-shaped. Box-shaped gullies are common in Illinoian tills, indicating a high silt content. Other gully sections that occur indicate the different textures of their enclosing soils.	These formations support mixtures of agriculture and forest cover. Field patterns, being constrained by the drainage system and the topographic influence of the rock structure, are a combination of rectangular and irregular shapes. Tree growth occurs along banks of the drainage system where slopes are too steep for cultivation.

GM if gravelly. The AASHO system lists A-4 or A-2 for surface soils, A-4 or A-2 in the B horizon, and A-2 or A-4 for the parent material or C horizon (Figure 8.16).

Old Till (Kansan and Illinoian Age): Sedimentary Rock Origin (Midwest). The parent material is typically CL or CH. The AASHO classification is A-6 or A-7.

Water Table

In deposits of glacial till, the seasonal high water table is typically found near the surface during

Figure 8.10. Thick, old till: Illinoian age. The older till of Illinoian age is more dissected and shows the characteristic white-fringed gullies. Photographs by the Agricultural Stabilization Conservation Service, RN-1P-167,168, 1:20,000 (1667'/"), September 12, 1955.

all or part of the year. Midwestern tills, finer in texture, commonly have compact subsoils, and their seasonal high water tables are 1 to 3 feet below the surface. New England tills also have compact subsoils, creating seasonal high water tables 3 to 5 feet beneath the surface. The older tills, which have more weathered profiles, have dense, silty clay layers 10 to 12 feet beneath the surface and the seasonal high water table at approximately 5 feet. In all cases depressions have water tables closer to the surface.

Drainage

Young glacial tills in the Midwest have slow permeability and are typically tiled when in agricultural use, for the fine soil texture and firm substratum do not facilitate good internal drainage. Young New England tills are coarser in texture but still have compact subsoils with poor internal drainage and slow permeability and percolation rates. Older tills are very poorly drained, since they have weathered longer and have developed a compact subsurface hardpan.

Soil Depth to Bedrock

Soil depth to bedrock in till deposits varies a great deal, and sophisticated photographic interpretation and/or field analysis is required in order to map it. Tills greater than 7 to 10 feet in depth begin to mask completely all pattern traces of underlying rock structure. A thin till condition is interpreted by identifying dark tones created by subsurface rock fractures or bedding planes and the influence of the underlying rock on the drainage pattern.

Issues of Site Development

Sewage Disposal

Septic tank leaching fields are difficult to site in fine-textured till formations, since the seasonal

Figure 8.11. (A) Thin, young till: Wisconsin age. Wisconsin age till is thinly veneered over metamorphic rock (gneiss) in Windsor County, Vermont. The main valley contains deeper deposits of glaciofluvial materials, indicated by the soil depths observed within the road cuts. Photographs by Raytheon-Autometrics, VBM-6824-6-170,171, 1:24,000 (2000′/″).

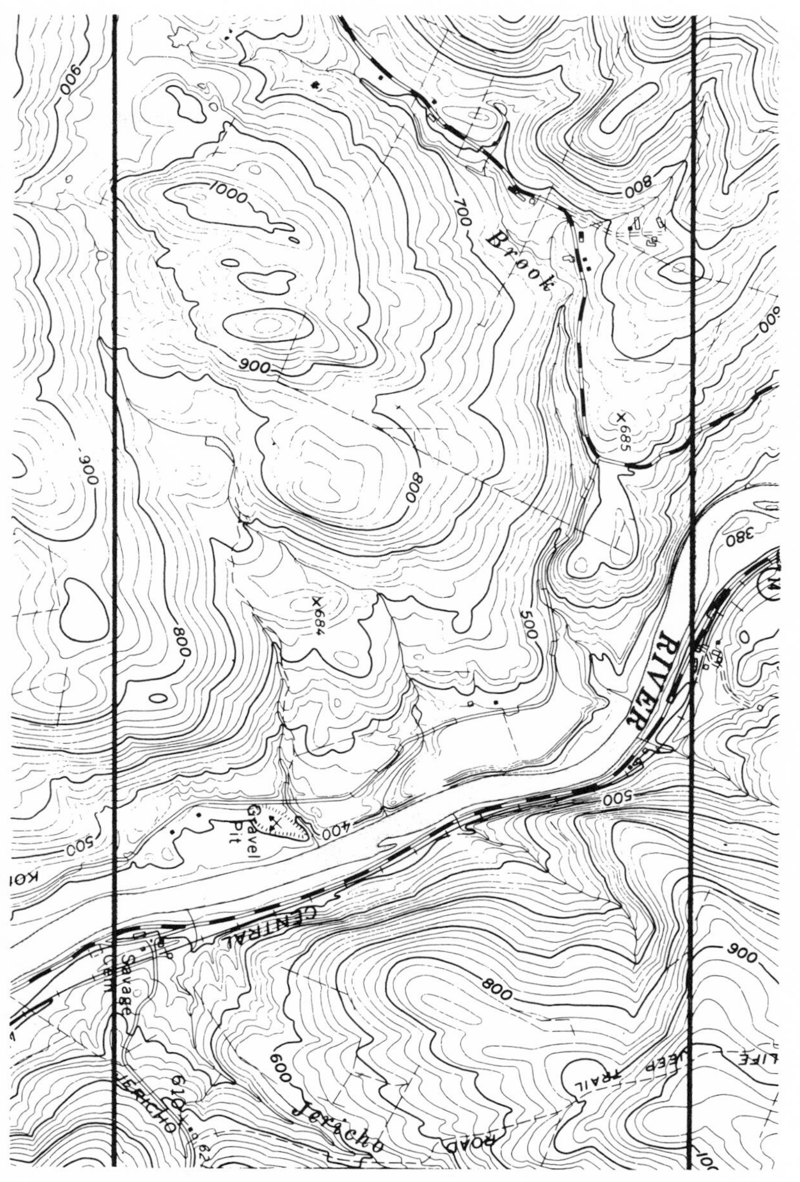

Figure 8.11. (B) U.S. Geological Survey Quadrangle: Quechee (7½′).

Figure 8.12. Thin, old till: Illinoian age. Illinoian age till is thinly veneered over limestone in Ripley County, Indiana. The angular drainage pattern is characteristic of limestone, while the fine, white-fringed gullies are characteristic of the Illinoian till. Photographs by the Agricultural Stabilization Conservation Service, RK-2P-1,2, 1:20,000 (1667′/″), September 12, 1955.

high water table is near the surface and percolation rates are slow. Coarse tills may provide suitable sites if the depth to water table and percolation rates are above the minimums, but thin till formations may have insufficient soil depths. Intensive developments and urban areas should develop sewage treatment systems to maintain the quality of water resources.

Solid Waste Disposal

Because of the variability of the materials, onsite investigation and detailed analysis are necessary to determine the capabilities of these soils for solid waste disposal. In the Midwest, sanitary landfill operations are difficult to site in glacial till formations because of the high seasonal water table. However, if sited, soil textures are suitable for cover and will prevent potential penetration of rainfall and percolation of leachate which could contaminate groundwater resources.

In New England sanitary landfill operations can be located in till if a site has sufficient depth to water table and sufficient fines in the soil to provide a cover that does not allow percolation of rainwater and leachate. Suitable locations are usually not difficult to identify.

Trenching

Trenching in finer-textured midwestern tills encounters high water tables and plastic soils during the spring or after heavy rainfalls. Trenching costs are slightly higher than but similar to those encountered in removing the same amount of deep, dry, moderately cohesive soil. If the water table is encountered, costs will increase by 35 to 40%. Trenching operations in New England tills may encounter many boulders, cobbles, and even bedrock. Linear alignments may be difficult, owing to the rugged topography, wet areas, and rock outcrops, Typical average costs per cubic

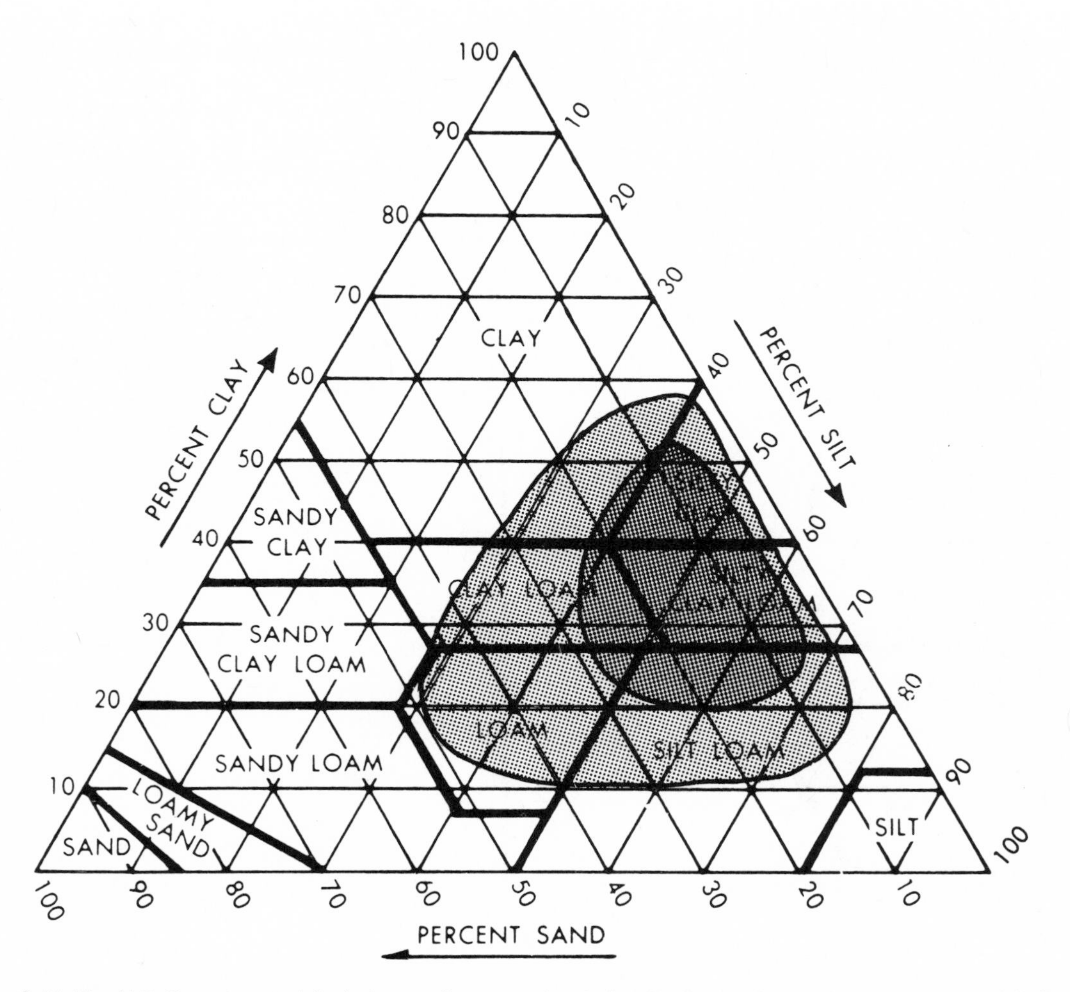

Figure 8.13. The U.S. Department of Agriculture soil texture triangle for tills derived from soft sedimentary rock indicates silty clay, silty clay loam, and silt loam as dominant materials.

yard are 1.2 to 2.2 times those associated with removing the same amount of deep, dry, moderately cohesive soil. Removal of large boulders over 1 cubic yard in size or of bedrock require blasting, and costs increase by a factor of 4 to 5.

Excavation and Grading

In young and old till deposits in the Midwest, grading to construct large, flat sites does not involve any significant difficulties or expenses, since the topography is naturally flat. Pit excavations require dewatering since the water table is typically near the surface. Low areas are difficult to work since they are wetter and the moisture content exceeds the plasticity limit. Typical per-unit grading and excavation costs are approximately the same as those associated with removal of a deep, dry, moderately cohesive soil, unless the water table is encountered, which will increase costs by 35 to 40%. In New England the topography is more rugged and the soils are thinner, containing many stones and boulders; per-cubic-yard excavation and grading costs are 1.2 to 2.5 times those associated with removing the same amount of deep, dry, moderately cohesive soil. Also, bedrock and large boulders in New England are generally crystalline, requiring blasting and heavy equipment, and rock removal increases costs by a factor of 4 to 5.

Construction Materials (General Suitabilities)

Topsoil: Good to Poor. The high water table and low organic content are constraints in the Midwest; the stony tills found in New England are not desirable.

Sand: Not Suitable. Little sand is found in midwestern tills. That found in New England tills is mixed with fines and difficult to separate. However, suitable sources of sand can easily be found in other, associated glacial landforms.

Gravel: Not Suitable. Other landforms can generally be found in glaciated regions which offer suitable sources of gravel materials.

Aggregate: Not Suitable. Other, associated landforms may supply suitable materials.

Surfacing: Excellent to Fair. Glacial deposits provide many ranges of textural sizes which can easily be mixed to provide materials suitable for surfacing.

Borrow: Good to Poor. Midwestern tills are predominantly fine-textured and present problems

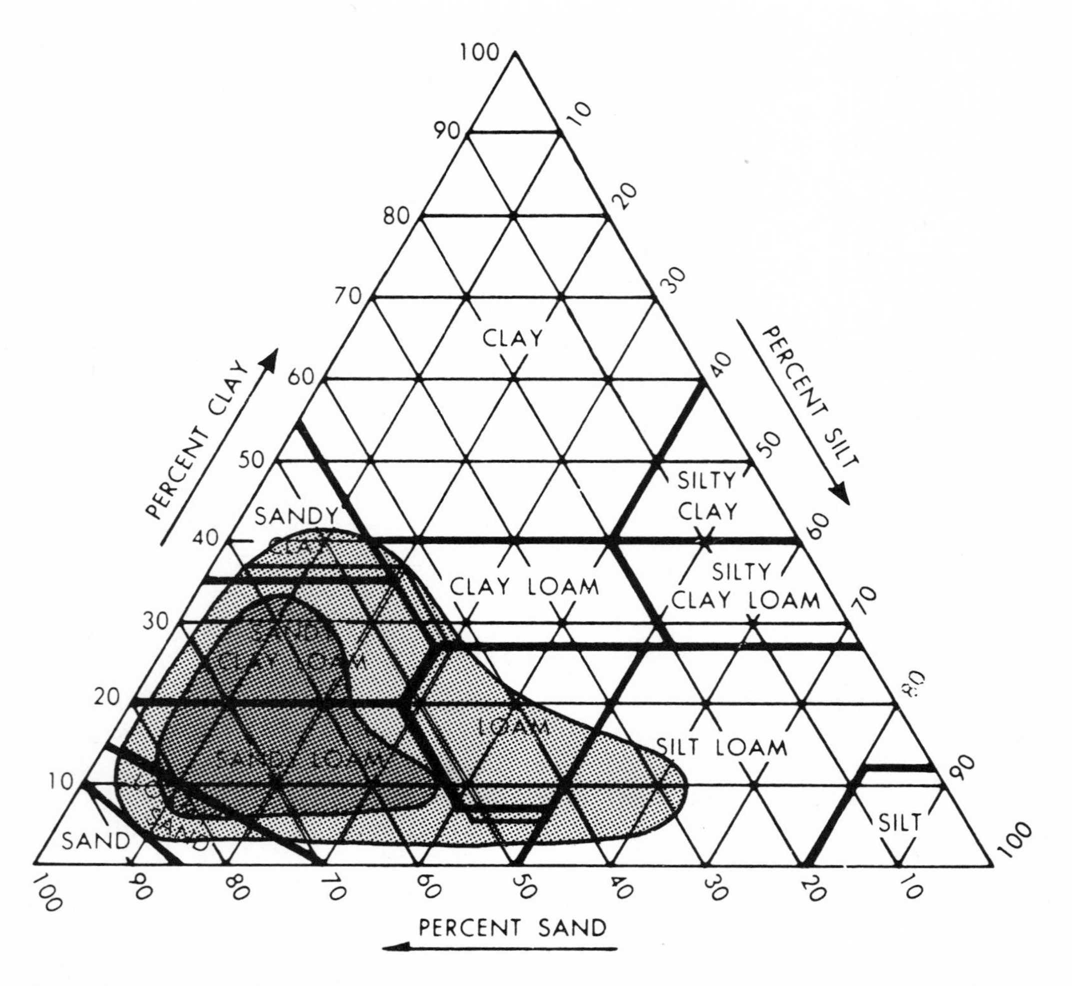

Figure 8.14. The U.S. Department of Agriculture soil texture triangle for tills derived from igneous rock indicates sandy loam, sandy clay loam, and loam as dominant materials.

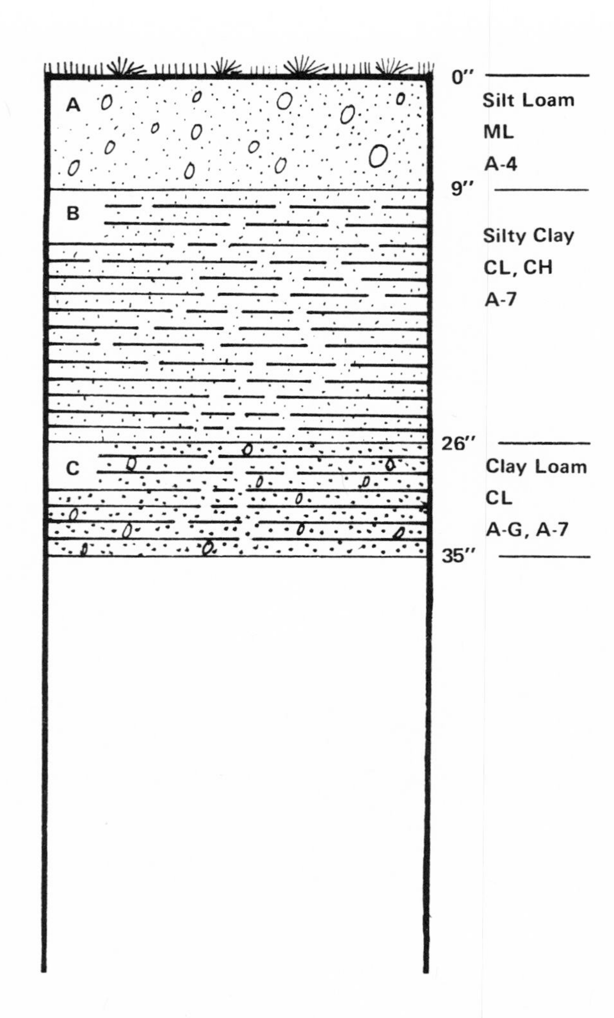

Figure 8.15. A typical soil profile developed in till derived from soft sedimentary materials in a humid climate.

involving high volume change, high plasticity, high water table, and frost action, thus making them poor sources of road fill and general borrow materials. These materials are more suitable if dried and compacted, but the moisture content is difficult to control. However, they can be suitably compacted for fills at the optimum moisture content by using sheepsfoot rollers. New England tills are better suited for borrow materials since they are coarser and easier to work; they are compacted with penumatic-tired rollers.

Building Stone: Not Suitable. Many of the boulders and cobbles are utilized for stone walls and walkways, and for some housing construction.

Mass Wasting and Landslide Susceptibility

In flat midwestern tills there are few occurrences of mass wasting and landslide activity unless

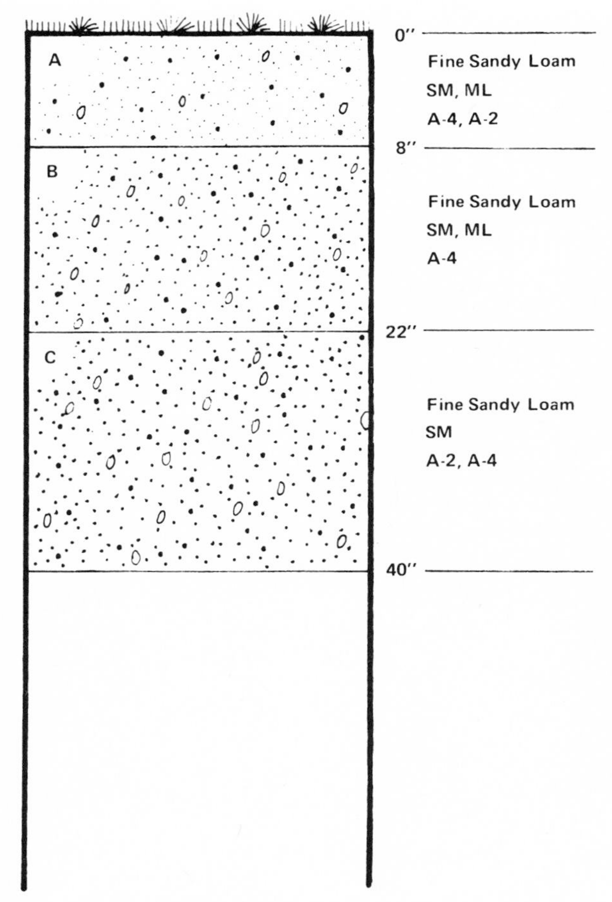

Figure 8.16. A typical soil profile developed in till derived from igneous-metamorphic rock in a humid climate.

steep faces are created by excavation or stream undercutting. (The last-mentioned areas may potentially slump if high shear stresses are present.) New England tills are more susceptible since they are deposited on underlying materials which have a variety of slope conditions. Tills along steep slopes have a tendency to slip along the bedrock plane if moisture and slope gradients are sufficient to create little shear resistance. Soil creep occurs along all hillsides and moves particles slowly downslope. If they are not well enough anchored and drained, artificial fills will occasionally fail along the bedrock plane. In till regions the type, attitude, character, fracturing, and so on, of the underlying rock must be considered along with the nature of the overlying materials to determine the best construction and design techniques. Specific site investigations are always necessary early in the planning process to determine soil stabilities.

Groundwater Supply

The semi-impervious nature of till does not provide a high yield of groundwater. Interbedded or buried sand and gravel layers may exist that provide excellent yields, but they must be located by test borings and field investigations. Also, water-bearing rock aquifers may underlie the till, or other water-bearing surficial deposits may be found in associated landforms. In New England, for example, glacial outwash deposits commonly occupy large valleys and provide excellent sources and supplies of groundwater.

Pond or Lake Construction

The creation of small ponds in Midwestern tills is not difficult, owing to the impervious nature of the soil materials upon compaction and to the many, naturally occurring, small depressions. Large reservoir sites are difficult to locate since there is little topographic relief. Older, more dissected tills provide alternatives for damming a drainage system or tributary. Kansan and Illinoian tills provide excellent bottom linings upon compaction because of their impervious nature. New England tills, commonly thin over bedrock, need detailed investigations to determine their specific suitabilities. In these regions leakage can be expected if the coarser tills are used for bottom materials, but the more rugged topography should provide many alternative dam and reservoir sites.

Foundations

Suitabilities for different foundation types depend upon the specific composition of the soils, the topographic slope and soil depths to bedrock. Young till formations contain many variable soils with different plasticity levels and limits of supporting power. Continuous footing foundations with basement slabs are typical of residential structures in deep till, but rafts are also used occasionally. Large structural loads can be transferred by deep footings or rafts. In thin till regions small structures use footings; larger structural loads can use piers or footings. Specific field investigations are necessary to determine actual design standards.

Highway Construction

In thick, young Wisconsin tills, problems of highway construction are related to the suitability of the local materials for subbase and base courses, the high water table, and the frost heave potential. Depressions and low areas have finer soils and higher levels of plasticity, and fill must be found in other, associated landforms. Old Kansan, Illinoian, or Nebraskan till formations are more weathered and have poor load-bearing capabilities. The drainage system presents some alignment difficulties and may necessitate minor bridging and the use of culverts.

Thin till deposits over rock, as typified in New England, encounter alignment difficulties and involve increased excavation and grading costs. The massive bedrock requires blasting and heavy equipment; the more rugged, rock-controlled topography and integrated drainage system may

Table 8.6. Glacial till: aerial and cartographic references

State	County	Photo Agency	Photograph no.	Scale	Date	Corresponding quadrangle and geologic maps from USGS	Corresponding soil reports from SCS
Indiana	Dearborn	ASCS	RN-1P-166-167	1667'/''	9/12/55	Cincinnati (1:250,000)	—
Indiana	Madison	ASCS	BWI-1BB-27-30	1667'/''	5/27/61	Alexandria, Gaston (7½')	1967
Indiana	Montgomery	ASCS	BWJ-1DD-43-44	1667'/''	5/31/63	Darlington (7½')	1912†
Indiana	Montgomery	ASCS	BWJ-1DD-10-11	1667'/''	5/31/63	Darlington (7½')	1912†
Indiana	Montgomery	ASCS	BWJ-2DD-22-23	1667'/''	5/31/63	Darlington (7½')	1912†
Indiana	Ripley	ASCS	RK-2P-1-2	1667'/''	9/12/55	Cincinnati (1:250,000)	—
Indiana	Tipton	ASCS	BWM-4R-134-137	1667'/''	9/21/56	Kempton (7½')	1912†
Iowa	Buenavista	USGS*	Iowa 1-ABC	5833'/''	9/23/50	Fort Dodge (1:250,000)	—
Massachusetts	Hampshire	ASCS	DPB-4LL-218-220	1667'/''	10/19/70	Mt. Tom (7½')	1928
Massachusetts	Worcester	ASCS	DPV-2LL-34-36	1667'/''	10/18/70	Ashby (7½')	1922
New York	St Lawrence	USGS*	New York 5-AB	1667'/''	5/4/60	Pope Mills (7½')	1925
New York	St. Lawrence, Jefferson	USGS*	New York 4-ABC	1583'/''	5/4/60	Muskellunge Lake, Hammond, Redwood, Chippewa Bay (7½')	1925
North Dakota	Emmons	USGS*	North Dakota 1-AB	5000'/''	9/28/52	Jamestown, Bismark (1:250,000) I-331	—
Ohio	Logan	ASCS	DU-3G-85-88	1667'/''	6/5/50	Marion (1:250,000)	1939
Wisconsin	Kenosha	ASCS	XD-100-136-137	1667'/''	6/24/63	Silver Lake, Racine (15')	1971

*From *Selected Aerial Photographs of Geologic Features in the United States*, U.S. Geological Survey Prof. Paper No. 590.
†Out of print.

require the use of massive fills, bridges, and tunnels. Seepage zones may be encountered in cuts along the zone where the soils are adjacent to impervious bedrock. These seepage zones may facilitate slumping during the spring or after heavy rainfalls which act to increase the shearing stress. Erosion is critical on steep slopes, and artificial fills must be properly anchored and drained when placed on sloping rock surfaces.

Moraines

Introduction

Glacial moraines are formed by the accumulation of drift materials as a result of the direct action of deposition and thrust deformation of glacial ice (Flint, 1957). Moraines usually form at the edge or margin of a glacial ice sheet and have an associated belt of hummocky topography. They consist of heterogeneous, unconsolidated to moderately consolidated, unstratified to partly stratified mixtures of clay, silt, sand, gravel, cobbles, and boulders. The specific composition of moraines is dependent upon the parent glacial drift. Fine-grained drifts are generally derived from soft, sedimentary rocks, as typified in the Midwest; coarse-grained drift originates from crystalline rocks and is common in New England. Most moraines contain unstratified till, but some small areas of stratification occur as a result of localized fluvial actions when the ice underwent ablation. Large quantities of meltwater create a high percentage of well and poorly stratified deposits within the moraine.

Moraines are classified by their depositional relationship to the glacial ice sheets as end, terminal, lateral, interlobate, and ground moraines. *End moraine* is a general classification for hummocky drift deposits that have been pushed and carried in front of or lateral to an advancing glacial ice lobe. *Terminal moraines* record the furthest downslope advance of a specific ice lobe. *Lateral moraines* are formed along the edges of ice lobes and ideally grade into a terminal moraine. Interlobate moraines are built along the junction of two adjacent glacial lobes. *Ground moraine* consists of both lodgment and ablation till having low relief and no transverse linear elements (Flint, 1957). Its characteristics are similar to those of the till deposits discussed in the previous section.

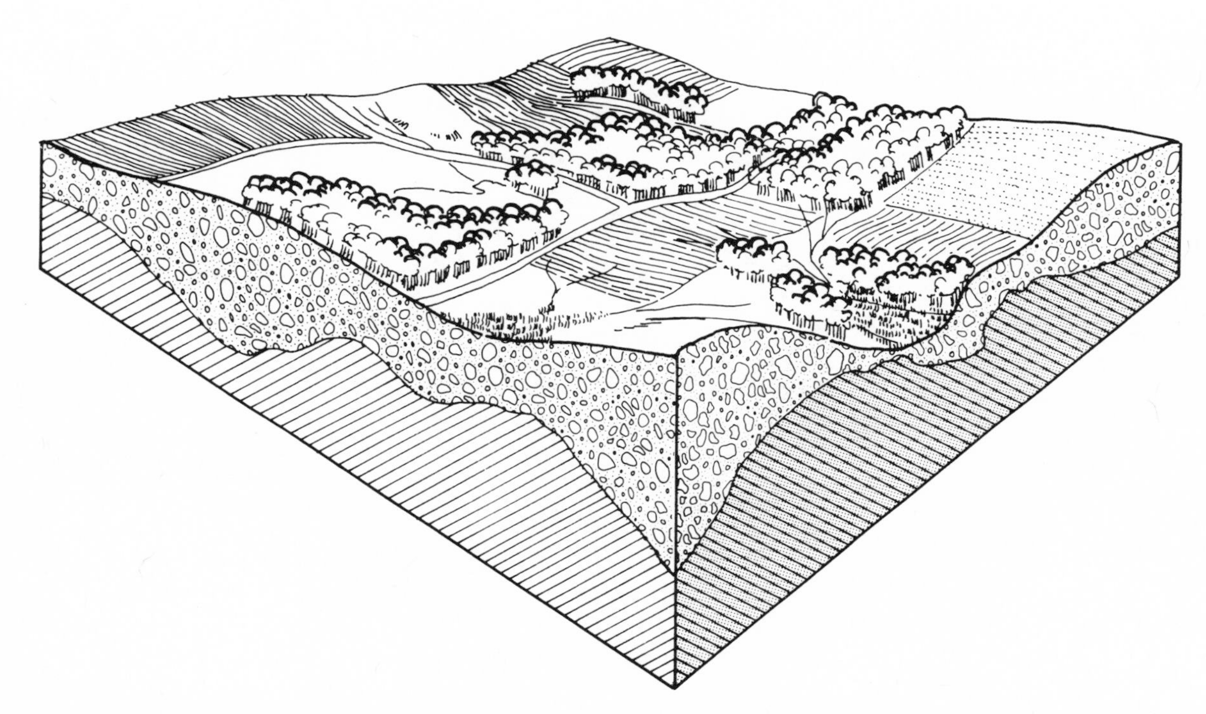

Figure 8.17. Slightly undulating moraine typical in the Midwestern United States. Glacial end moraines have visual and pattern characteristics distinctly different from their surrounding landforms. Moraines of slightly undulating topography are similar to till plains, except that there is more forest cover and spatial variation. Depressions are usually occupied by ponds and swamps. In arid climates there may be a mixture of agriculture and grass cover. Wetter depressions may be cultivated, uplands may be grass covered.

The topographic characteristics of moraines are modified by erosion after initial deposition. The effectiveness of the erosion processes varies, depending upon the climate, the textural composition of the soils, and the topographic slope. The initial morainic form varies from undulating to very rugged, depending on the amount and distribution of drift within the glacier and at its margin, the rate of movement, the rate of ablation, and the amount and concentration of meltwater. For example, massive end moraines are formed when the load is heavy and the movement and ablation rates are high.

End moraines are distributed across the glaciated regions of the United States, the largest formations being found in the Midwest. Some valley-contained moraines are found in central western New York and on the northern edge of Long Island, and New England has deposits along the northeastern coast of Maine, in southern Rhode Island, at the base of Cape Cod, and on Martha's Vineyard Island in Massachusetts.

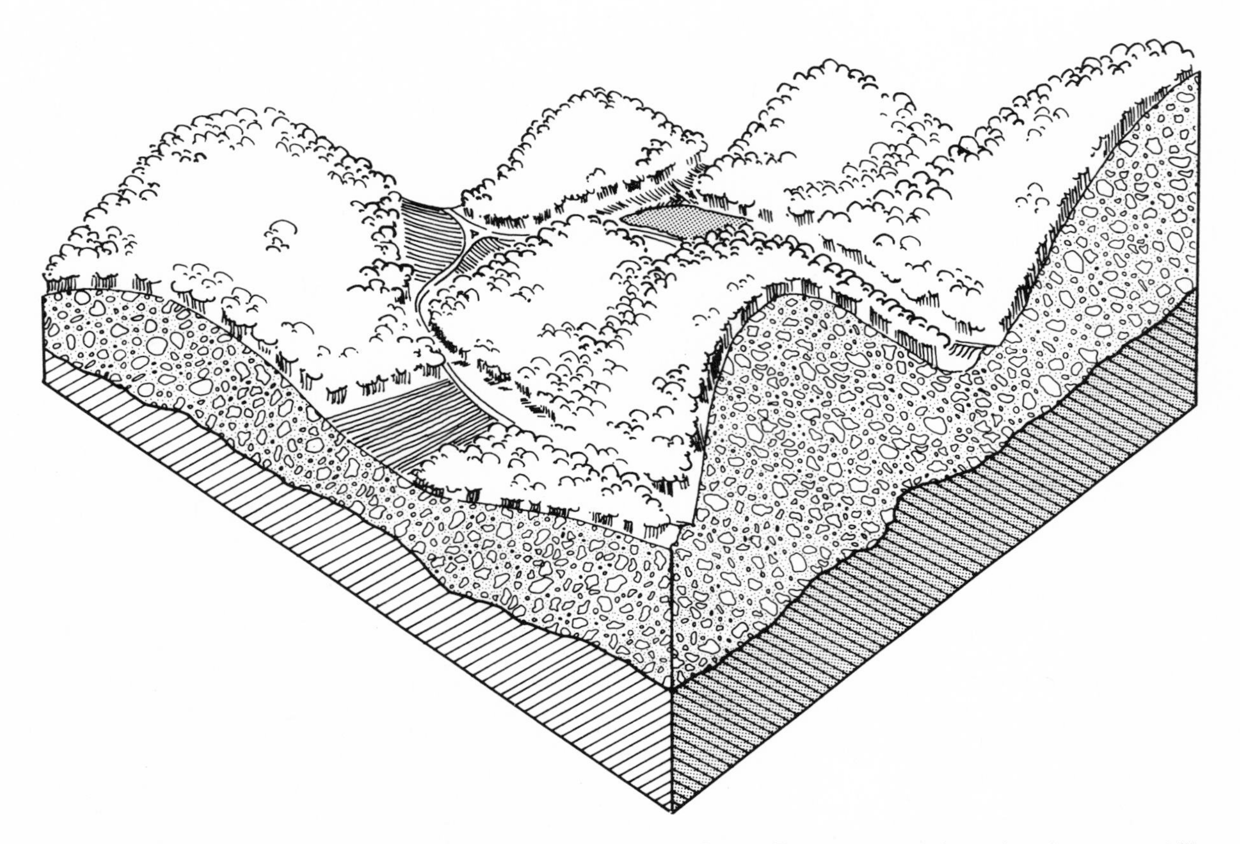

Figure 8.18. Rugged moraines typical in the northeastern United States. The more rugged moraines have many hills and depressions of different elevation, offering a wide diversity of spatial and viewing potential. Visually, rugged, massive moraines create an absorptive landscape because of the many depressions; any land use or structural development located in these depressions can be viewed only from the surrounding hilltops. The hills and valleys are not continuous, so the most sensitive visual zones are the tops of the highest hills or the slopes and tops of the hills on the edge of the morainic deposit adjacent to other lowland landforms. Regional viewing potential exists on hilltops along the edge of the moraine formation; these provide observation points over the adjacent, typically flat till plain or outwash. Rugged, massive moraines are predominately forested. Few roads cross the formation, and highways have curvilinear alignments. Some of the less rugged land surface between hills may be cultivated. In summary, the spatial variation and the viewing capacities offered by this formation make it valuable for recreation sites, since it typically adds diversity to the flat, monotonous surrounding landscape.

Interpretation of Pattern Elements

The topographic key in Table 9.7 is primarily concerned with moraines in general and does not specify type. Detailed observation of the surrounding region is needed to determine whether a moraine is specifically end, terminal, lateral, or interlobate. Moraines as landforms are characterized by a linear, sometimes crescent band of discontinuous hills and depressions occurring at different elevations. An undulating topography indicates fine soil textures, while rugged, steep-sided hills indicate coarser, more permeable soils. There tends to be a rather distinct boundary between an end moraine and its adjacent landforms. The moraine is also probably more rugged or undulating than its surrounding formations unless they are valley moraines. More detailed observation is necessary to determine soil textures, even though general distinctions are presented in the key.

Soil Characteristics

The transported soils comprising end moraines are very difficult to interpret and classify, for significant changes in texture, plasticity, and stratification can occur within very short linear distances and depths. Generally, the soils of moraines are related to the origin and composition of the glacial drift. In the Midwest many moraines contain fine soils derived from sedimentary rocks, but those that originate from crystalline rock are coarse in texture. In the latter case erosion during and after deposition may have removed some of the finer soil particles and redeposited them in depressions. The following discussion illustrates typical general conditions, but specific field investigations are needed in order to obtain design criteria.

U.S. Department of Agriculture Classification

The glacial drift comprising moraines, which originates from sedimentary rocks, typically contains predominantly fines such as silty clay loam, silt loam, and/or clay loam. Depressions are likely to contain a greater concentration of clays and to have a higher organic content. Glacial drift

Table 8.7. Glacial landforms: End moraine (humid and arid)

Topography	*Drainage*	*Tone*	*Vegetation and land use*
Undulating and rugged	Deranged	Light and dark	Cultivated and forested
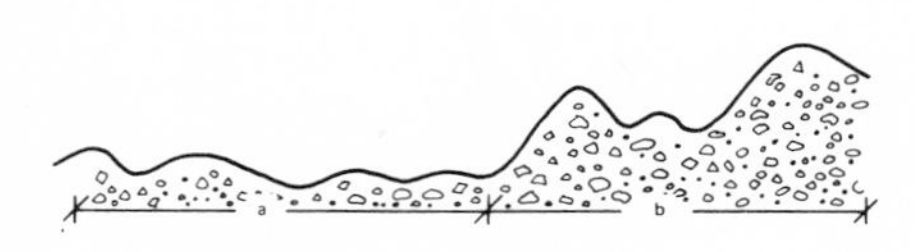 End moraines create a topography with a linear, discontinuous band of hills and depressions, whose elevation differences seldom exceed 500 feet. Softly rounded, undulating hills (a) indicate fine soils (common in the Midwest); rugged, steep-sided, sharp-ridged hills (b) indicate coarse soils. Hill masses are relatively small when compared to bedrock formations.	 Unless soils are very coarse or the climate is arid, many ponds and swamps are found in depressions at varying elevations. Small dendritic patterns drain into these depressions. Fine soils may have greater drainage development with a finer pattern texture, indicating higher runoff coefficients.	Coarse materials have light tones, but wet depressions appear dark. Fine materials have dull gray tones grading into darker tones in depressions. *Gullies* Vary Sag-and-swale gullies indicate fine materials; U- or V-shaped gullies indicate sand-clay or sand mixtures. Fine soils produce many gullies and/or sheet erosion.	The amount and mix of forest cover and cultivation depend upon soil composition, climate, and relief. The more rugged and coarse-textured moraines are forest-covered in humid climates and grass-covered in arid climates. Undulating moraines have a mix of cultivation and forest cover in humid climates and cultivation and grass cover in arid climates. Many ponds and swamps are found.

formed from crystalline rocks is typically coarser in texture and contains gravelly sandy loams or other predominantly gravelly or sandy mixtures. Depressions also contain more fines and are poorly drained.

Unified and AASHO Classifications

Glacial drift comprising moraines derived from soft sedimentary rocks is typically classified as ML, CL-CH, CL, and/or CH, with the parent material commonly being CL or CH. The AASHO classification of the same materials is typically A-6 or A-7, and occasionally A-4. Coarser drift is predominantly sand and gravel mixtures and is classified as SM-SP, GW, GC-GP, SC-SP, SM-SC, SW-SP, GM, GC, SM, or SC. AASHO describes these soils as A-1, A-2, and A-4. Depressions in both fine and coarse drift contain higher levels of clay, silt, and organic matter and may be described as CH, CL, ML-CL, OL, and/or OH. AASHO describes depression soils as A-7, A-5, A-4, and/or A-6. Clearly, there is sufficient variation to necessitate field investigations and sampling for detailed mapping.

Water Table

The depth to the seasonal high water table is difficult to interpret, owing to the variability of the materials. Typically, the water table averages 5 feet from the surface, but on-site investigations are needed for detailed mapping. Where the water table occurs at the ground surface, swamps and ponds can be observed.

Drainage

The internal drainage of moraines is variable, since a range of materials may be encountered. Stratification of silt and/or clay lenses inhibits drainage and creates a perched water table. Fine materials are naturally poorly drained; coarse soils are moderately to well drained.

Soil Depth to Bedrock

Morainic deposits are generally of sufficient thickness so that bedrock outcrops do not occur. Most construction procedures should not encounter bedrock, but large boulders may occur, randomly distributed throughout the formation.

Issues of Site Development

Sewage Disposal

Because of the variability of end moraine materials, on-site investigation and detailed analysis are necessary to determine its capabilities. In coarse-textured soils percolation rates are generally suitable for septic tank leaching fields, but slopes and seasonal high water tables act as constraints for their location.

Figure 8.19. (A) Undulating moraine typical of the midwestern United States. This moraine in Rock County, Wisconsin, is not sufficiently rugged to constrain agricultural use of much of the land. Some of the depressions are ponded, and the heavy forest growth in the woodlots indicates that the area originally had heavy forest cover rather than grass. Photographs by the Agricultural Stabilization Conservation Service, XA-2R-63,64, 1:20,000 (1667′/″), May 18, 1956.

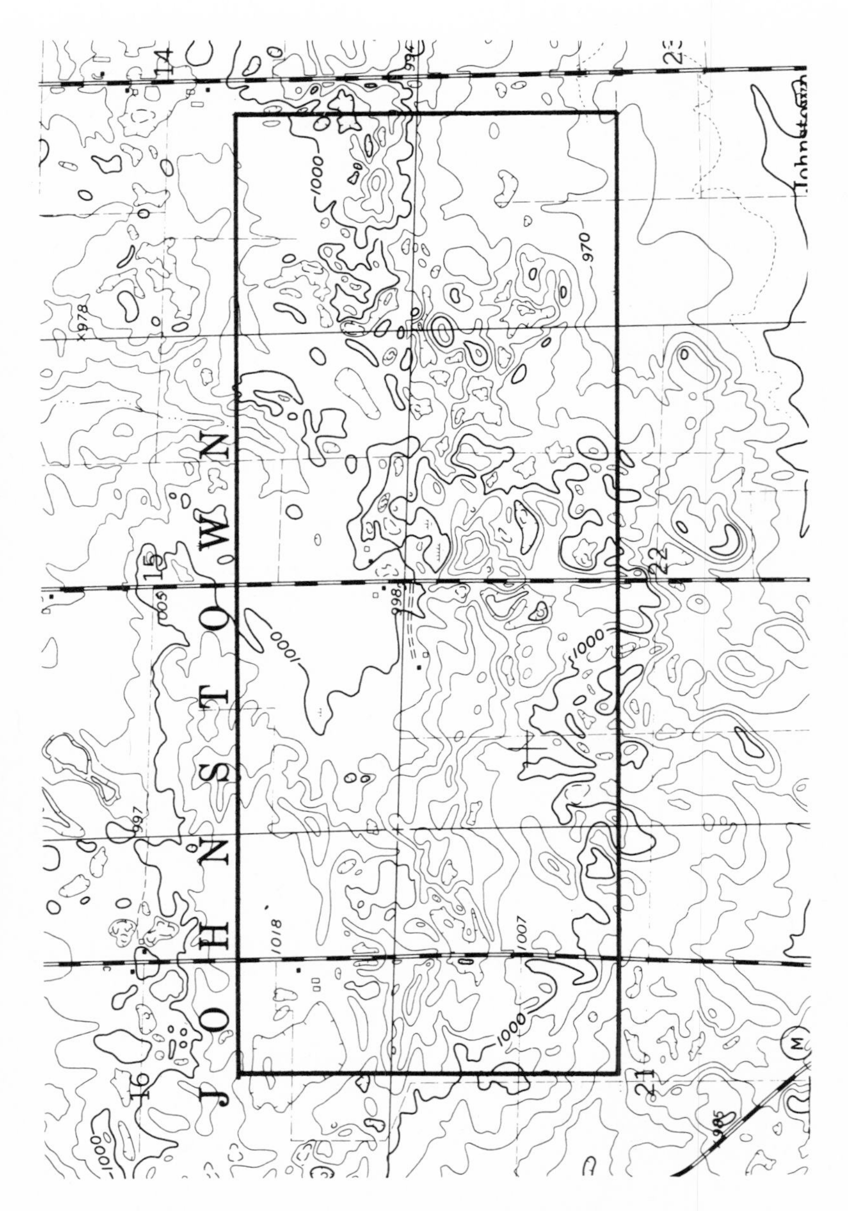

Figure 8.19. (B) U.S. Geological Survey Quadrangle: Avalon (7½′).

Figure 8.20. (A) Rugged end moraine as typical of the northeastern United States. This rugged moraine in Washington County, Rhode Island, is largely forest covered with little established land use development. Photographs by the Agricultural Stabilization Conservation Service, DPK-4H-125,126, 1:20,000 (1667′/″), November 12, 1951.

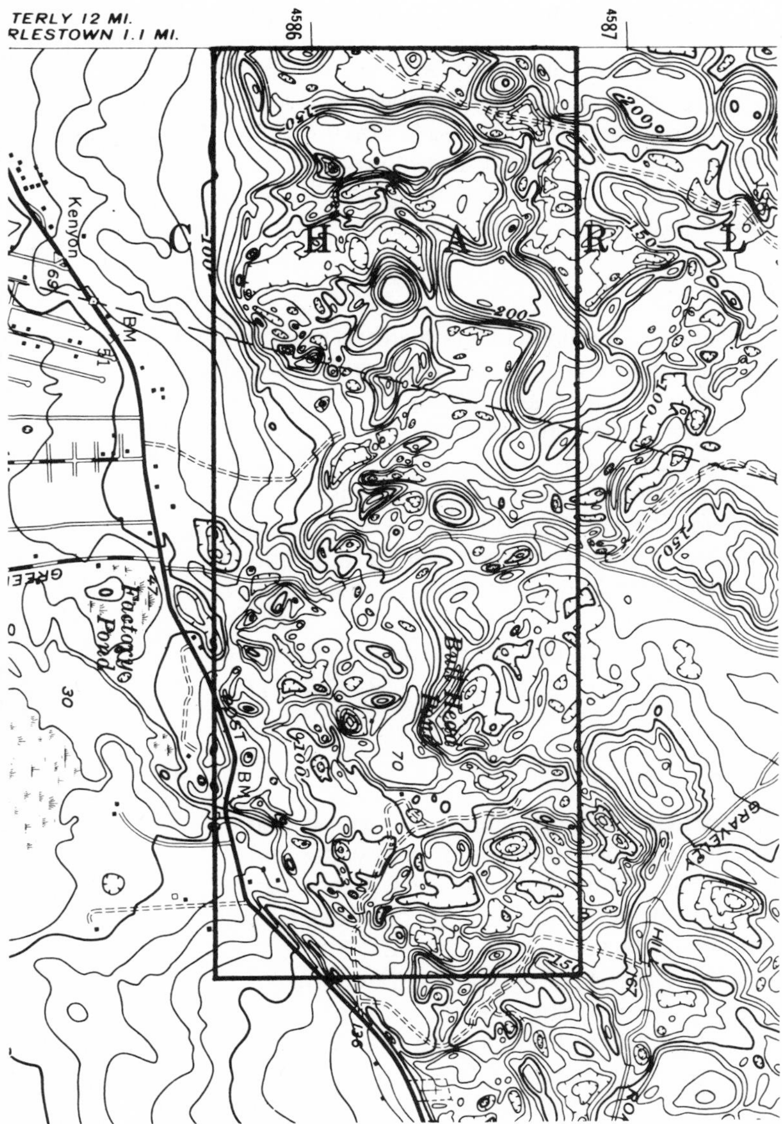

Figure 8.20. (B) U.S. Geological Survey Quadrangle: Kingston, Carolina (7½′).

Solid Waste Disposal

Suitable sites can usually be found for sanitary landfill operations, although detailed analysis is needed to determine actual capabilities. Suitable cover materials should be used to prevent percolation of rainwater leachate, and sites should have 10 to 15 feet to the seasonal high water table to prevent groundwater contamination.

Trenching

Trenching operations in fine-textured moraines encounter high water tables and standing water in depressions, increasing costs by 35 to 40%. Typical per-cubic-yard costs are 1.2 to 2.2 times those encountered when removing the same amount of deep, dry, moderately cohesive soil. More granular moraines do not present significant problems unless large boulders are found, in which case blasting and heavy equipment is required, thus increasing costs. Rugged, massive moraines present linear alignment difficulties and may require pumping stations for liquid-carrying pipelines.

Excavation and Grading

Excavation and grading in fine-textured morainic formations encounter high water tables and standing water in depressions. These conditions make it very difficult to carry out these operations when the soils are frozen, and the rugged topography requires vast earthwork operations if a large, flat site is to be graded. The more rugged the topography, the greater the total volume of materials to be moved. Typical per-cubic-yard costs are 1.2 to 2.5 times those associated with removing the same amount of deep, dry, moderately cohesive soil; drainage of the water table, if necessary, increases costs 35 to 40%. Large boulders may be encountered, requiring some blasting and heavy equipment, and organic soils are commonly found in depressions. Granular morainic formations are typically well drained; the only potential cost increase factor would be the occurrence of many large boulders.

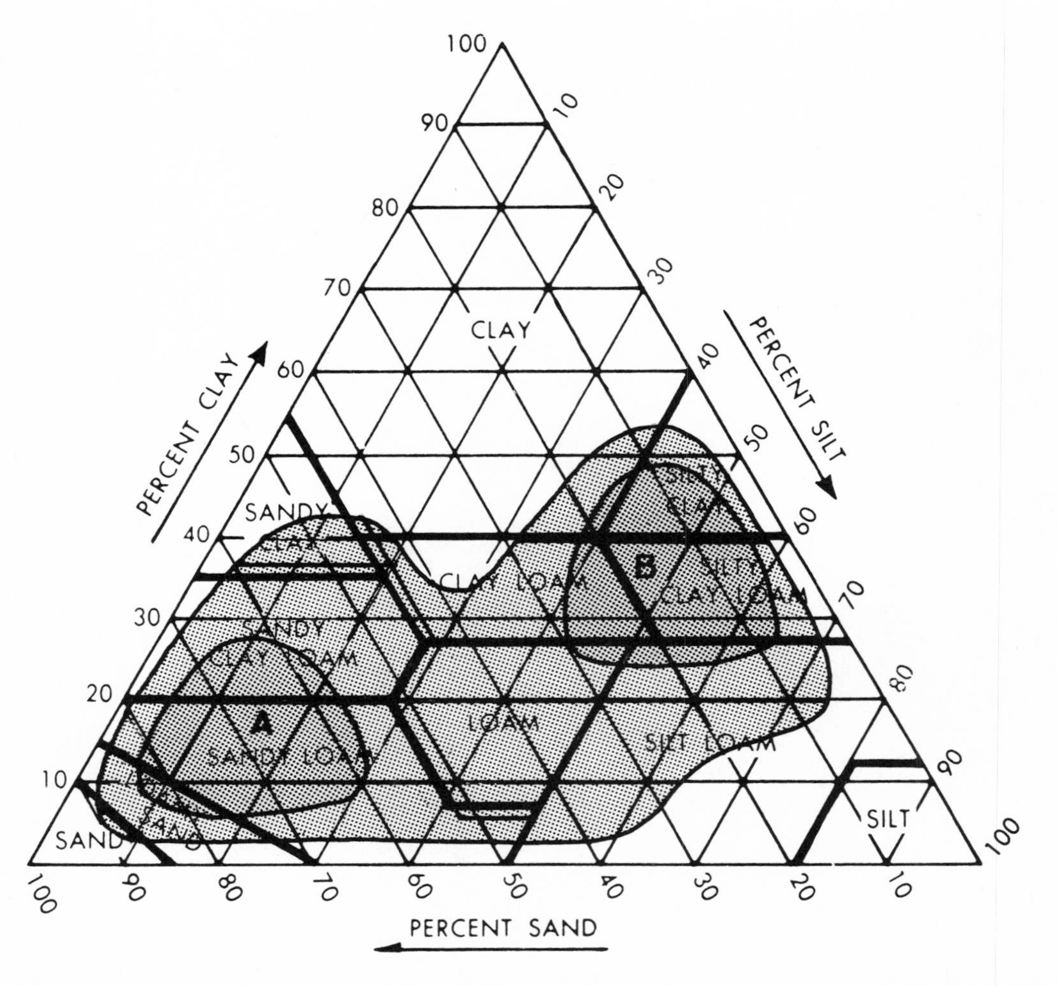

Figure 8.21. The U.S. Department of Agriculture soil texture triangle indicates sandy clay loam and sandy loam for tills derived from igneous-metamorphic rock (A) and silty clay, silty clay loam, and silt loam for tills derived from softer sedimentary rocks (B).

Construction Materials (General Suitabilities)

Topsoil: Good to Unsuitable. Fine-textured moraines may provide good sources of topsoil material, but coarse deposits may be unsuitable.

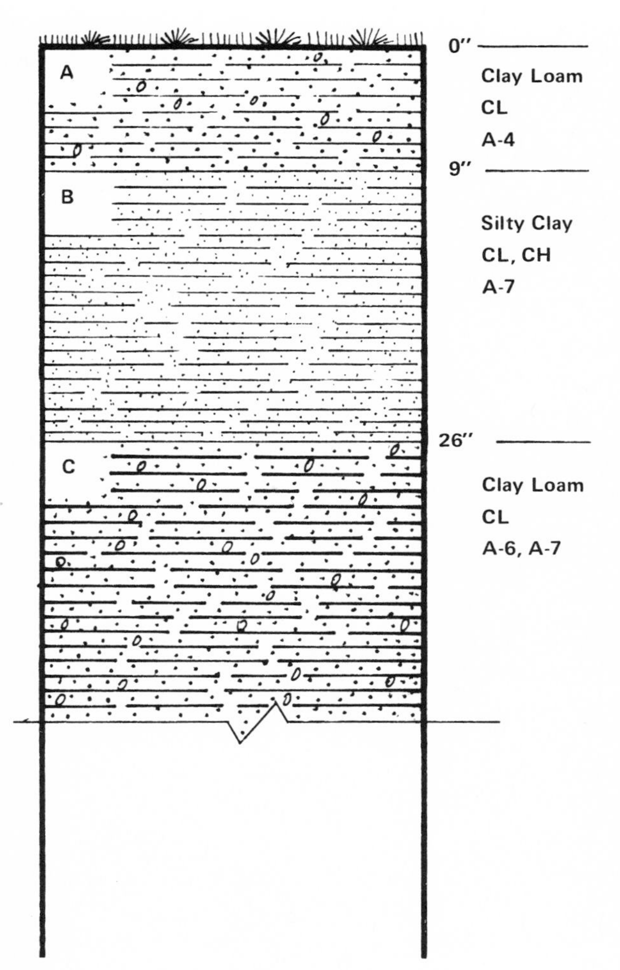

Figure 8.22. A typical soil profile of glacial till developed from soft sedimentary rock in a humid climate.

Depressions may contain soils containing a greater number of fines and higher organic content.

Sand: Good to Unsuitable. Fine-textured moraines are not suitable sources of sand materials, but coarse-textured formations may provide good supplies, especially where stratified. Potential sources must be investigated in the field.

Gravel: Good to Unsuitable. Sources are unsuitable in fine-textured moraines except where localized stratified deposits occur. Good sources are commonly found in coarse-textured moraines, especially where stratified.

Aggregate: Good to Unsuitable. Coarse mixtures may offer good sources of aggregate, but local field experience is helpful in determining suitabilities.

Surfacing: Excellent to Fair. Glacial deposits provide many ranges of textural sizes which can easily be mixed to provide materials suitable for surfacing.

Borrow: Excellent to Poor. Fine-grained deposits may present problems relating to high volume change, organic content, high water table, and frost action, thus making them poor sources of road fill and general borrow. These fine-grained materials are suitable for fills but require compaction by sheepsfoot rollers at optimum moisture content. Coarser tills are excellent sources of borrow for fills, and compaction is accomplished with pneumatic-tired rollers.

Building Stone: Not suitable. Many of the boulders and cobbles are used for the construction of stone walls, walkways, and exterior housing walls.

Mass Wasting and Landslide Susceptibility

Moraines are moderately susceptible to landslides under numerous conditions. Fine-textured soils and steep, rugged, massive formations have the highest hazards. Past debris flows, slides, and slumps may be observed from aerial photographs. Steep embankments created by stream undercutting or excavation may be highly unstable. Shearing stresses can build up if silt or clay seams within the subsoils cause differentials in drainage. Specific on-site studies are needed to determine actual hazards and potentials.

Groundwater Supply

The glacial drifts comprising moraines vary widely in texture and degree of stratification and present subsoil conditions ranging from very pervious to practically impervious. On the average, the deep glacial drift comprising moraines can supply a moderate yield of groundwater resources.

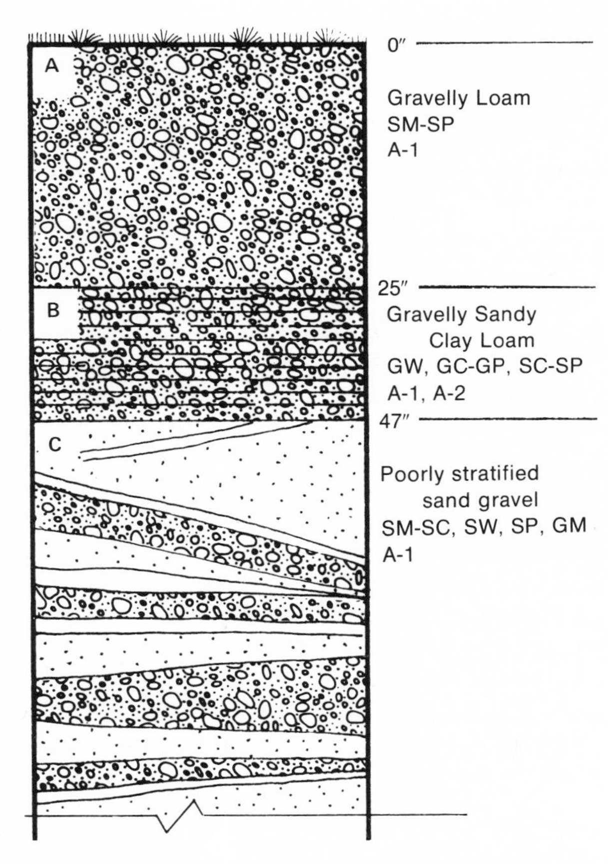

Figure 8.23. A typical soil profile of glacial till developed from igneous-metamorphic rock in a humid climate.

Table 8.8. Glacial end moraines: aerial and cartographic references

State	County	Photo Agency	Photograph no.	Scale	Date	Corresponding quadrangle and geologic maps from USGS	Corresponding soil reports from SCS
Alaska	–	USGS*	Alaska 11-AB	3333'/"	8/27/52	Mt. McKinley (15') Prof. Paper No. 373	–
Alaska	–	USGS*	Alaska 15-ABCD	3333'/"	7/54; 9/49	Healy (15')	–
Alaska	–	USGS*	Alaska 20-ABCD	3333'/"	8/12/50	Valdez (15')	–
Alaska	1st Judicial Division	USGS*	Alaska 25-ABC	2083'/"	8/11/59	Mt. Fairweather (1:250,000)	–
Alaska	1st Judicial Division	USGS*	Alaska 29-ABC	2083'/"	8/11/59	Mt. Fairweather (1:250,000)	–
Massachusetts	Barnstable	ASCS	DLP-1LL-118-120	3333'/"	9/20/70	Dennis (7½')	1920†
Massachusetts	Barnstable	ASCS	DLP-4LL-4-7	3333'/"	10/29/70	Sandwich (7½')	1920†
Massachuseets	Dukes	ASCS	DPO-1LL-27-29	3333'/"	10/20/70	Vineyard Haven (7½')	1925
Minnesota	Clay	ASCS	BXR-4V-63-64	1667'/"	11/12/58	–	–
New York	Tompkins	ASCS	ARU-3N-70-71	1667'/"	10/25/54	West Danby (7½')	1965
North Dakota	Kidder	USGS*	North Dakota 2-ABC	5000'/"	7/31/52	New Rockford (1:250,000), 1-331	–
North Dakota	Kidder, Logan, Stutsman	USGS*	North Dakota 3-AB	5000'/"	7/28/52	Jamestown (1:250,000), 1-331	–
Pennsylvania	Northampton, Monroe	USGS*	Pennsylvania 5-ABC	1967'/"	4/24/51	Stroudsburg, Saylorsburg, Wind Gap, Bangor (7½')	–
Rhode Island	Washington	USGS*	Rhode Island 1-ABCD	1667'/"	11/12/51	Kingston, Carolina (7½')	1939
Wisconsin	Dane	ASCS	WU-3CC-197-198	1667'/"	9/7/62	Rutland (7½')	1970
Wisconsin	Fond du Lac	USGS*	Wisconsin 2-ABCD	1967'/"	4/30/52	Kewaskum (15'), Prof. Paper No. 106	1911†
Wisconsin	Jefferson, Walworth	USGS*	Wisconsin 3-ABCD	1415'/"	4/8/55	Whitewater (15')	1912† 1920†
Wisconsin	Kenosha	ASCS	XD-1DD-136-137	1667'/"	6/24/63	Silver Lake, Racine (7½')	1971
Wisconsin	Rock	ASCS	XA-2R-63-64	1667'/"	5/18/56	Avalon (7½')	1917†

*From *Selected Aerial Photographs of Geologic Features in the United States,* U.S. Geological Survey Prof. Paper No. 590.
†Out of print.

Pond or Lake Construction

Specific, on-site investigations are needed because of the variability of the materials. Fine-textured soils offer excellent bottom materials upon compaction, but coarser soils may have high levels of permeability. Many natural ponds and swamps are found in depressions in humid and subhumid climates. Natural ponds are generally shallow and are difficult to maintain against algae growth and siltation.

Foundations

Enough variability of soil conditions generally exists to provide suitable sites for large and small structural loads with few problems. Uplands or the tops of small hills are better drained and typically contain soils that are coarser than deposits found in depressions. On high load-bearing capacity soils, footings and rafts can be utilized for both small and large structural loads. Continuous footing foundations with basement slabs are typical of residential structures in morainic formations.

Highway Construction

In fine-textured moraines the high water table, associated frost potential, and major excavation required in rugged topographies provide constraints reflecting higher costs for highway construction. Rugged, massive moraines present problems of alignment, and major earthwork is necessary. Cuts through depressions encounter wet soils which have higher levels of organic content. It may also be necessary to drain ponds and swamps or to make use of low bridges and/or partial fills and culverts. In fine-textured moraines suitable construction materials and fills may be available from adjacent landforms; in coarse-textured formations suitable materials generally exist within the moraine but may be difficult to locate.

Drumlins

Introduction

Drumlins are landforms composed of heterogeneous glacial drift which may include some lenslike layers of sand and gravel. Typically, these forms have a smooth, oval shape and an elongated axis parallel to the flow of glacial ice. Typically, drumlins measure ½ to 1 mile in length, 500 to 1500 feet in width, and 60 to 200 feet high. The end facing the glacier or stoss end is generally steeper and blunter than the tail or lee side.

There are two theories about the origin of glacial drumlins, but there is general agreement that they were formed under advancing glacial ice. In a few cases drumlin fields may have been formed by an ice sheet overriding previously deposited moraines and creating many single or interlocking drumlins whose axes radiate away from the center of the ice lobe. The drumlin field in southeastern Wisconsin between Fond du Lac and Madison appears to have been formed by this action. Evidence also indicates that some drumlins were formed when ice overrode materials previously deposited in ice fissures. Probably, both theories are correct.

Drumlins are classified by their soil relationship with rock. A typical drumlin is composed of compacted glacial drift laid down in deep deposits over bedrock. Other drumlins, classified as crag-and-tail drumlins, were formed around rock obstructions and have a rock nose with drift trailing behind. Glacial ice, since it places tremendous loads and pressures on the surface bedrock, frequently carves and molds drumloidal forms out of the rock structure itself. Solid bedrock formations with drumlin shapes are known as roches moutonnées; bedrock formations with a thin veneer of till over the rock are rock drumlins.

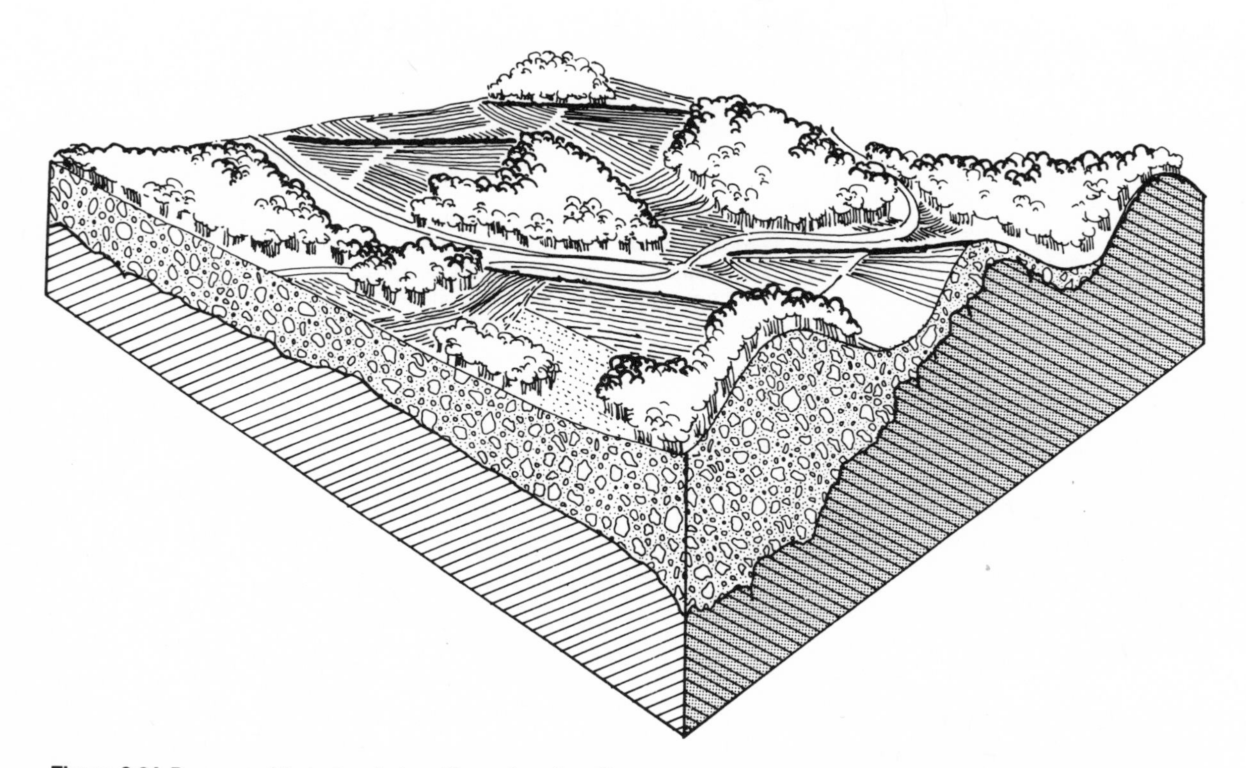

Figure 8.24. Because of their elevated positions, drumlins offer a greater viewing capacity than their typically flat surrounding landscapes. Drumlin hills are relatively low, however, and therefore do not provide significant regional views. Drumlin hilltops and ridges are visually sensitive when viewed from surrounding lowlands, and if vegetation on them is cleared, any land use development along the ridges will be exposed.

Interpretation of Pattern Elements

Drumlins can be readily identified by their topographic form, and it is sometimes possible to distinguish their predominant soil textural compositions by observing their shapes and side slopes. Smooth, broad shapes and slopes may indicate a predominant content of fines, while steeper slopes and long, narrow shapes may indicate coarser textures. However, other factors, such as gullies, tone, vegetation and land use patterns, and microfeatures, should be carefully studied along with the topography in determining composition. Table 8.9 shows the patterns associated with drumlin formations in both humid and arid climates. No distinctions are made between deep drumlin deposits, rock drumlins, or roche moutonnees, for knowledge of the region and/or field investigations may be necessary to make these distinctions accurately.

Table 8.9. Glacial landforms: Drumlins (humid and arid)

Topography	*Drainage*	*Tone*	*Vegetation and land use*
Drumlin-shaped	None	Light gray	Natural or cultivated
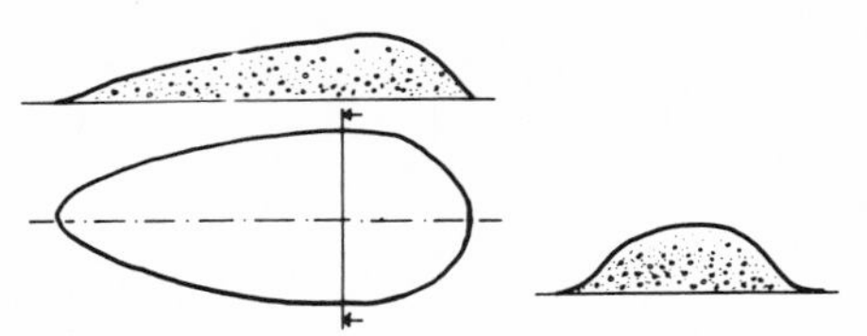 Drumlins are characterized by smooth, oval, cigar-shaped hills whose axes show the direction of glacial flow. Ideally, the side facing the glacier is steeper than the more tapering lee side. Lengths are typically ½ to 1 mile, widths, 500 to 1500 feet, and heights 60 to 200 feet.	These formations are relatively small and do not develop their own characteristic drainage patterns. Some gullying may be apparent along the base edge and helps in identifying the composition of the soils.	Light tones indicate the well-drained, elevated position of this landform.	Drumlins in humid climates may be cultivated. The steeper slopes have forest cover or pasture. The presence of orchards indicates the well-drained character. Rock drumlins and *roche moutonnees* in Canada are commonly forest-covered. Drumlins in arid climates are usually covered with grass.
		Gullies	
		Few or none	
		A few gullies may be found along the base edge of a drumlin. These are useful in identifying the composition of the soils.	

Soil Characteristics

The soil textures and characteristics of drumlins are similar to those of the surrounding glacial drift, although drumlin soils have a shallower surface profile, owing to sheet erosion, and are more compact. The subsurface may be so compacted as to inhibit root penetration. Many stones, boulders, and cobbles may be present if these occur naturally in the drift of the region. Soft, rounded, drumlin shapes typically indicate a predominantly fine-textured till, whereas steep-sided, sharp-crested shapes indicate coarser materials. Groups of drumlins found in a drumlin field usually have the same textures. Specific field investigations and/or knowledge of the region may be desirable in order to map soils accurately.

U.S. Department of Agriculture Classification

Glacial drift originating from soft sedimentary rocks typically contains predominantly fines forming clay loam, silty clay loam, and silt loam. Glacial drift formed from crystalline rocks is generally coarser in texture and contains gravelly sandy loam, gravelly loam, sandy loam, and loam, and other predominantly gravelly and sandy mixtures. Those soils containing large numbers of stones are described as stony, very stony, rocky, and so on.

Unified and AASHO Classifications

Glacial drift derived from soft sedimentary rocks is typically classified as ML, ML-CL, and CL (which is commonly the parent material). The AASHO description of these materials is A-6 and/or A-4. The plastic soils found in glacial drift, such as CH (A-7), do not commonly occur in drumlin landforms. Coarser drifts originating from more resistant sedimentary or crystalline rocks form predominant sand and gravel mixtures, including SM, SC, SM-SC, GM, GC, and GM-GC. The AASHO system commonly describes these soils as A-2, A-4, and occasionally A-1.

Water Table

Water tables usually are not near the surface, since a large portion of rainfall runs off the land surface because of its elevated position and the compacted nature of the materials. Depths are typically greater than 10 feet.

Drainage

Typically, drumlin formations are well drained as a result of their elevated positions. The glacial till deposits between drumlins, however, are usually poorly drained.

Soil Depth to Bedrock

Field investigations and/or knowledge of the region are important in obtaining accurate information about soil depth to bedrock. Many drumlins

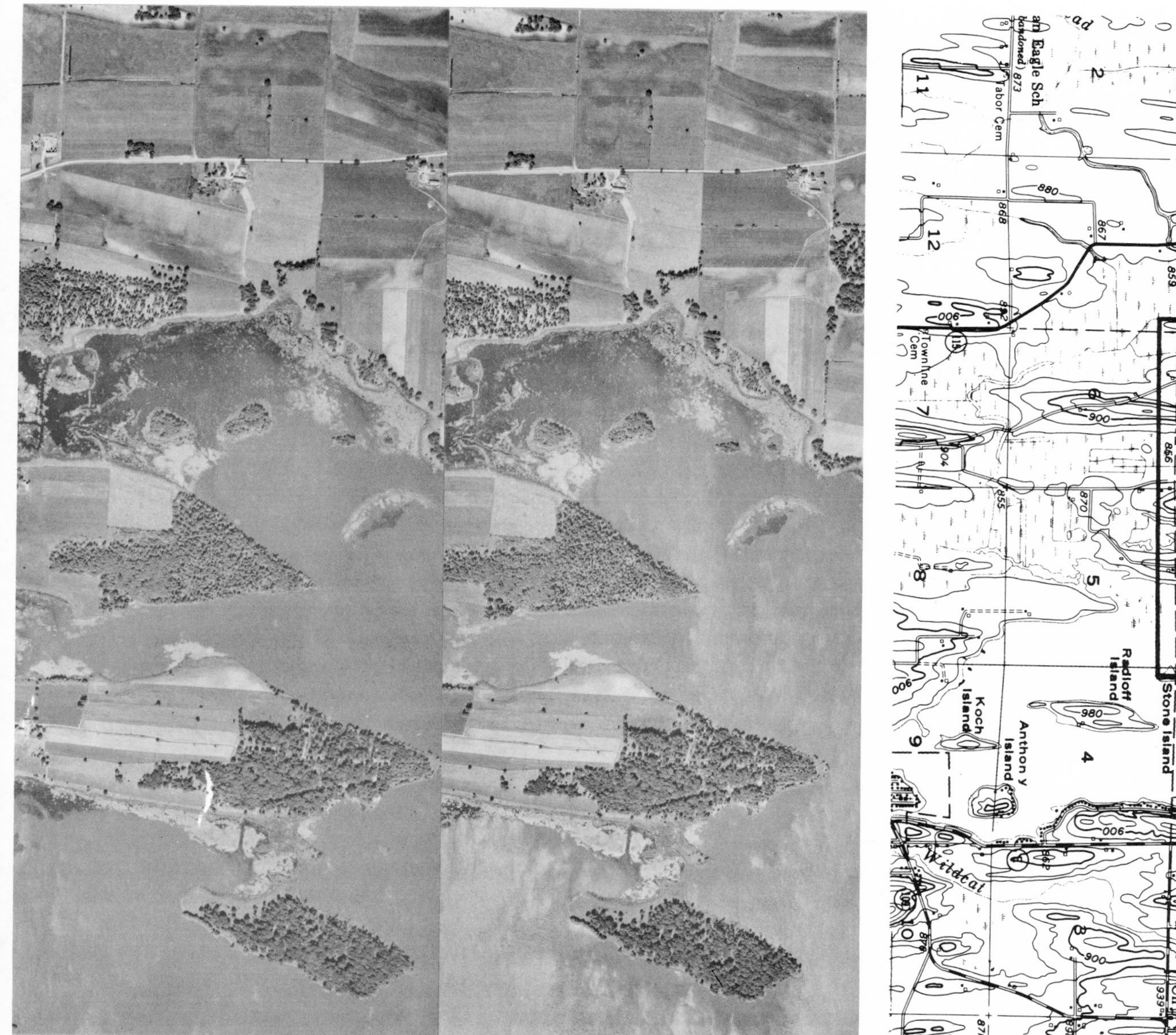

Figure 8.25. (A) Drumlins in Dodge County, Wisconsin, show the typical topographic shape associated with these formations. Photographs by the Agricultural Stabilization Conservation Service, AX-5R-90,91, 1:20,000 (1667′/″), June 27, 1956.

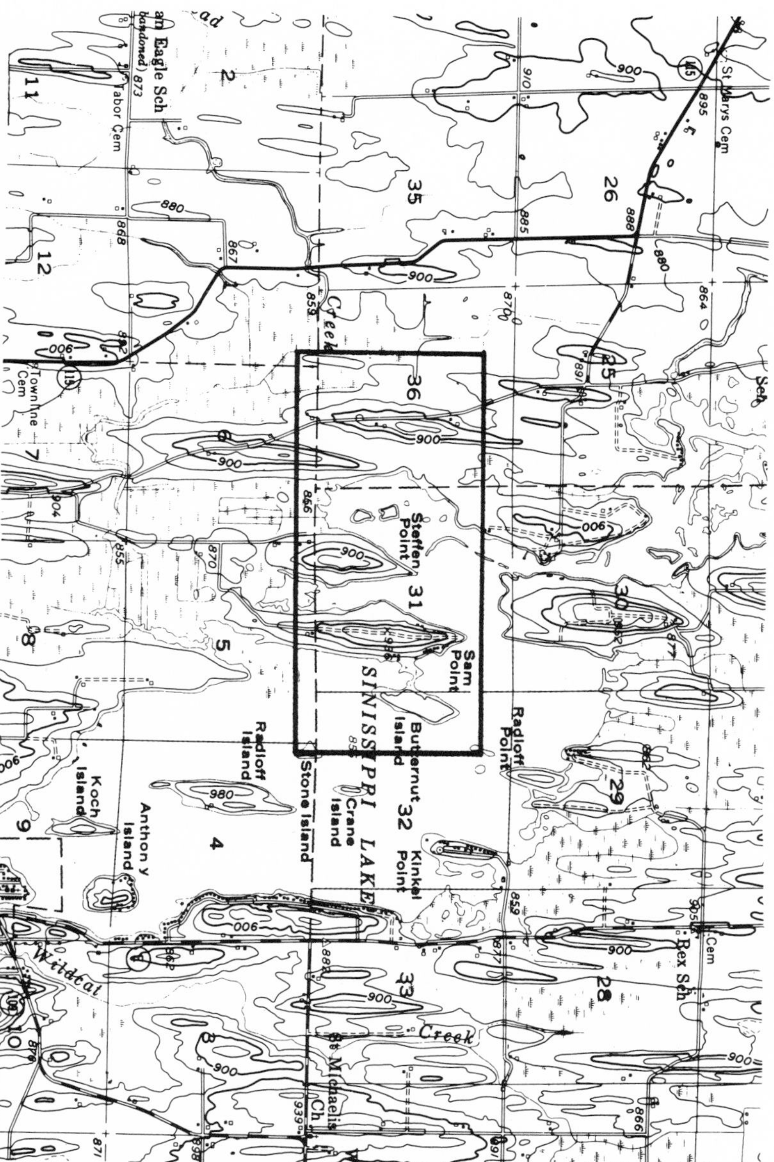

Figure 8.25. (B) U.S. Geological Survey Quadrangle: Horicon (15′).

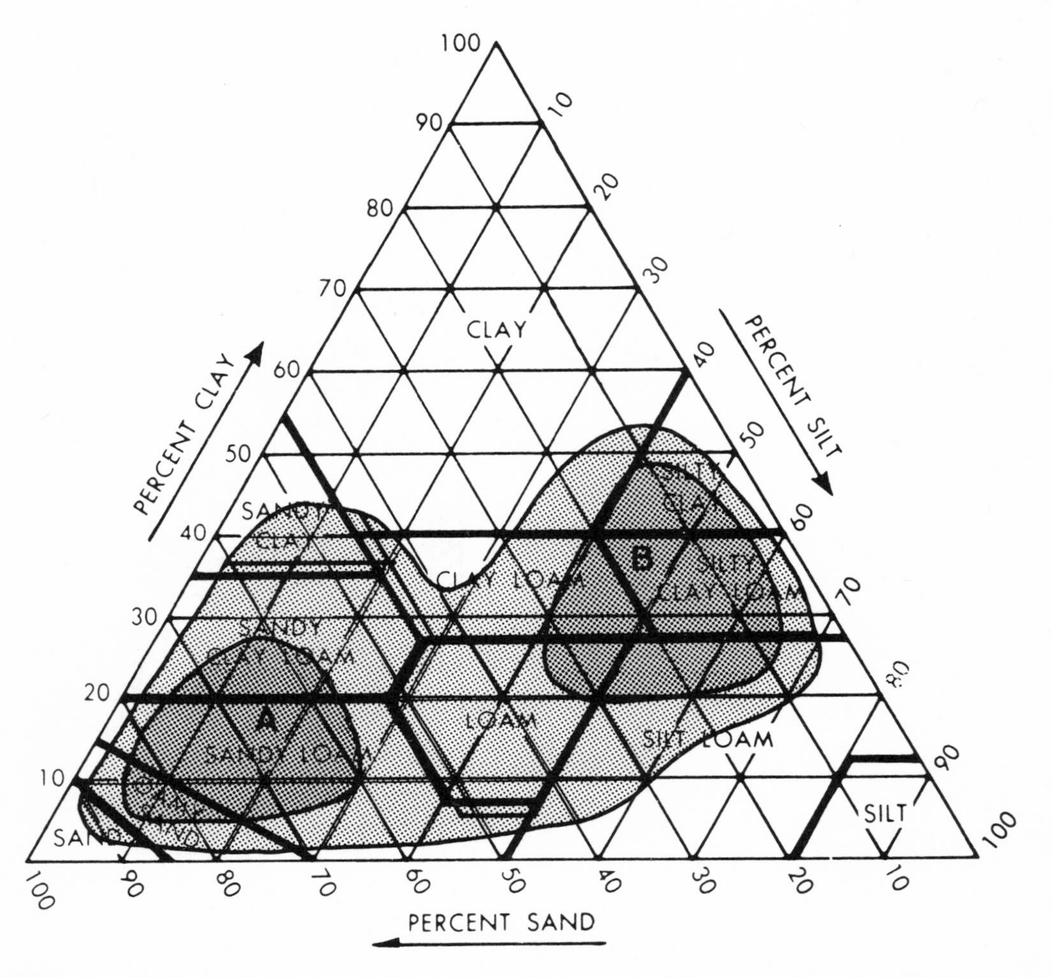

Figure 8.26. The U.S. Department of Agriculture soil texture triangle indicates sandy clay loam and sandy loam for tills derived from igneous-metamorphic rock (A) and silty clay, silty clay loam, and silt loam for tills derived from softer sedimentary rocks (B).

in the Midwest consist almost entirely of deep tills, whereas many in New England have rock cores (rock drumlins) or are of the crag-and-tail variety and have rock noses. Many drumloidal formations in Canada are solid rock or roches moutonnees.

Issues of Site Development

Sewage Disposal

Because of the variability of the glacial drift soils comprising drumlins, detailed analyses are necessary in order to determine their capabilities. In coarse-textured soils percolation rates are moderate and sites are usually suitable for septic tank leaching fields, although the sideslopes, which commonly exceed 30%, cause constraints. Housing lots and septic disposal fields located along ridgelines should encounter favorable conditions. The fine-textured, compacted drumlin soils typical of the Midwest and northwestern New York do not have suitable percolation rates.

Solid Waste Disposal

The elevated topography of drumlin formations provides sites that are well drained and have no significant water table problems. Large formations, consisting of sandy soils mixed with fines, offer the highest potential for disposal use. Siting, however, is difficult, owing to the steep sideslopes, and excavation for cover material may present problems because of the compacted and semi-cemented nature of the soils. Along the edges of drumlins, old borrow pits can sometimes be found, and these may be used for suitable sanitary landfill sites.

Trenching

Trenching operations through drumlins involve relatively higher costs, because of the steep sideslopes and the compact nature of the subsoil material. Typical costs are 1.8 to 3.0 times those encountered in excavating the same amount of deep, dry, moderately cohesive soil. Large boulders or bedrock may be encountered in drumlins, rock drumlins, or roches moutonnees, thus escalating removal costs. Large boulders and massive bedrock require blasting and heavy equipment and result in a cost increase by a factor of 4 to 5. Dense drumlin fields present linear alignment difficulties for liquid-carrying pipelines and may require pumping stations for grade changes.

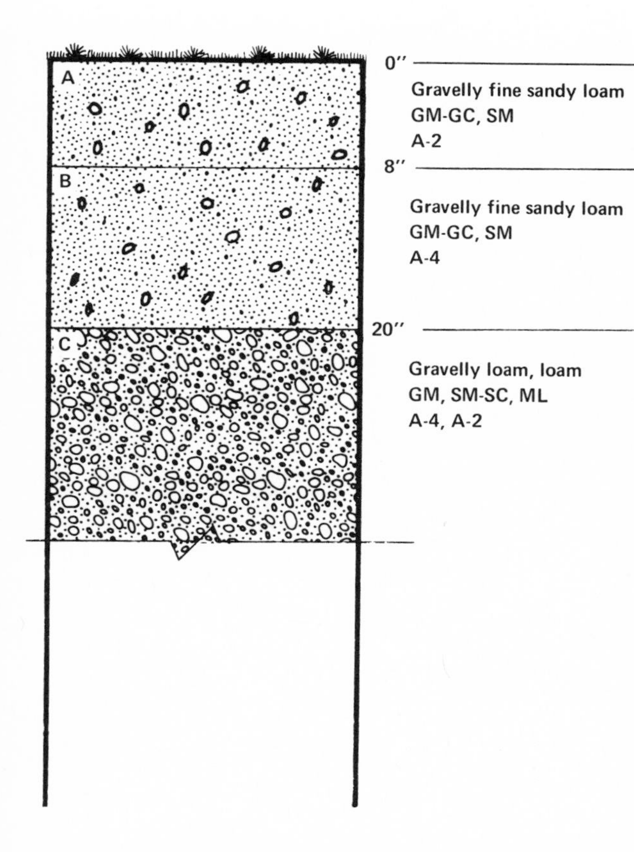

Figure 8.27. A typical soil profile developed in a drumlin form in New England. The till comprising the drumlin is of igneous-metamorphic origin.

Excavation and Grading

The creation of large, flat sites in dense drumlin fields requires extensive grading operations because of the many changes in elevation. Typical costs for excavation and grading are 1.5 to 2.8 times those involved in removing or moving the same amount of deep, dry, moderately cohesive soil, the increased costs being related to the compact and dense nature of the subsoil. Large boulders and/or the occurrence of massive bedrock require blasting and heavy equipment, raising costs by a factor of 4 to 5. Typically, naturally occurring, large, flat sites may be found in landforms adjacent to drumlin fields.

Construction Materials (General Suitabilities)

Topsoil: Poor. Drumlin soils are generally thin, as a result of sheet erosion. Surface soils better for agricultural purposes are usually found in the formations between drumlins.

Sand: Not Suitable

Gravel: Not Suitable. Coarse materials may exist within drumlins, but they are compact, mixed with fines, and difficult to separate.

Aggregate: Not Suitable

Surfacing: Good to Fair. Drumlin materials contain well-graded materials with sufficient fines to provide suitable binder for surfacing materials.

Borrow: Good to Fair. Drumlin soils are generally good sources of borrow material since the soils are well graded. Fine, compacted soils may not be as desirable as coarse materials.

Building Stone: Not Suitable. Many of the boulders and cobbles are used for the construction of stone walls, walkways, and exterior housing walls.

Mass Wasting and Landslide Susceptibility

Because of their compact subsoils, drumlin formations are not typically highly susceptible to sliding. Materials are moved downslope by a combination of creep, frost heave, and erosion. The thin soil veneers found on rock drumlins may, under wet conditions, slip along sideslopes.

Groundwater Supply

Drumlins have very small groundwater yields, since the compact nature of the subsoil resists the force of pumping. Associated lowland landforms may provide more suitable groundwater resources.

Pond or Lake Construction

Drumlin forms are elevated and rounded and therefore do not offer many suitable sites for reservoir construction. Although small ponds can be created for agricultural use, they will have little catchment area due to the convex nature of side-slopes where water runs away from all points and does not form a collector system. The compact soils are suitable for holding water.

Foundations

The compacted nature of drumlin soils provides load-bearing capacities suitable for most foundations. Small structural loads may utilize raft foundations or continuous footings with basement slabs; large structural loads may also utilize footings or rafts, depending upon the specific site conditions and structural system. In cold climates the occurrence of water-bearing silt lenses presents a potential for frost heave.

Highway Construction

It is difficult to minimize cut and fill in locating highway alignments through dense drumlin fields. Fine-textured, dense drumlin soils are more difficult to excavate than coarser materials, but they are useful for fill. Rock drumlins, crag-and-tail formations, and roches moutonnees may necessitate rock removal and blasting, escalating construction costs. Landforms between drumlins may be poorly drained and require drainage and/or roadbed elevation.

Table 8.10. Glacial drumlins: aerial and cartographic references

State	County	Photo Agency	Photograph no.	Scale	Date	Corresponding quadrangle and geologic maps from USGS	Corresponding soil reports from SCS
Massachusetts	Plymouth	ASCS	DPT-2LL-165-167	3333′/″	10/6/70	Hull (7½′)	1969
Massachusetts	Suffolk	ASCS	DPU-5LL-118-120	3333′/″	10/29/70	Boston South, Hull (7½′)	–
Minnesota	Todd	USGS*	Minnesota 1-AB	1667′/″	7/21/63	Brainerd (1:250,000)	–
Montana	Phillips	USGS*	Montana 5-AB	1667′/″	7/12/58	Havre (1:250,000), I-327	–
New York	Wayne	USGS*	New York 3-AB	5208′/″	5/6/60	Ontario, Williamson, Macedon, Palmyra (7½′)	1919†
New York	Wayne	ASCS	ARO-1N-78-79	1667′/″	7/26/54	Savannah (7½′)	1919†
Wisconsin	Dodge	ASCS	AX-5R-89-92	1667′/″	6/27/56	Horicon (15′)	–
Wisconsin	Dodge	ASCS	AX-6R-204-205	1667′/″	7/10/56	Beaver Dam (15′)	–
Wisconsin	Fond du Lac	USGS*	Wisconsin 1-ABCD	2000′/″	4/24/52	Fond du Lac (15′), Prof. Paper No. 106	1911†
Wisconsin	Jefferson	ASCS	WV-4R-124-126	1667′/″	7/6/57	Watertown (15′)	1912†

*From *Selected Aerial Photographs of Geologic Features in the United States,* U.S. Geological Survey Prof. Paper No. 590.
†Out of print.

Eskers

Introduction

As glacial ice melts, large river systems of meltwater carrying bedloads of soil and rock debris flow on, in, or under the ice sheet. When the ice retreats, the bedload and bottom deposits of these streams form snakelike ridges, or eskers, of poorly to moderately well-stratified sands and gravels. Eskers are classified as glaciofluvial, ice-contact landforms since they were deposited by water that was in contact with the ice. These formations are found both on top of and partially buried by other glacial landforms. They are found crossing till plains and drumlins and may be partially buried by lake beds or outwash. They are found in uplands as well as lowlands with little regard to the underlying topographic characteristics. Many eskers in the Midwest and New England cross lakes and, where they occasionally protrude through the water surface, form snakelike peninsulas or island chains (Figure 8.28).

Eskers are found in a wide variety of forms and sizes, but their dimensions are commonly ½ to 1 mile in length, less than 200 feet in width, and less than 100 feet above adjacent elevations, and they have sideslopes of approximately 30 degrees and a snakelike ridge. Variations include beaded eskers, crevasse eskers, and esker trains. Beaded eskers were formed by the same process as other eskers but have segments missing, probably eroded, and thus a discontinuous linear form. Crevasse eskers have a similar appearance to beaded eskers but were formed in ice crevasses, where stratified sands and gravels accumulated; after melting of the glacial ice, these forms appeared as straight, sharp-ridged landforms which are not continuous over long distances. Esker trains are formed by the presence of heavily

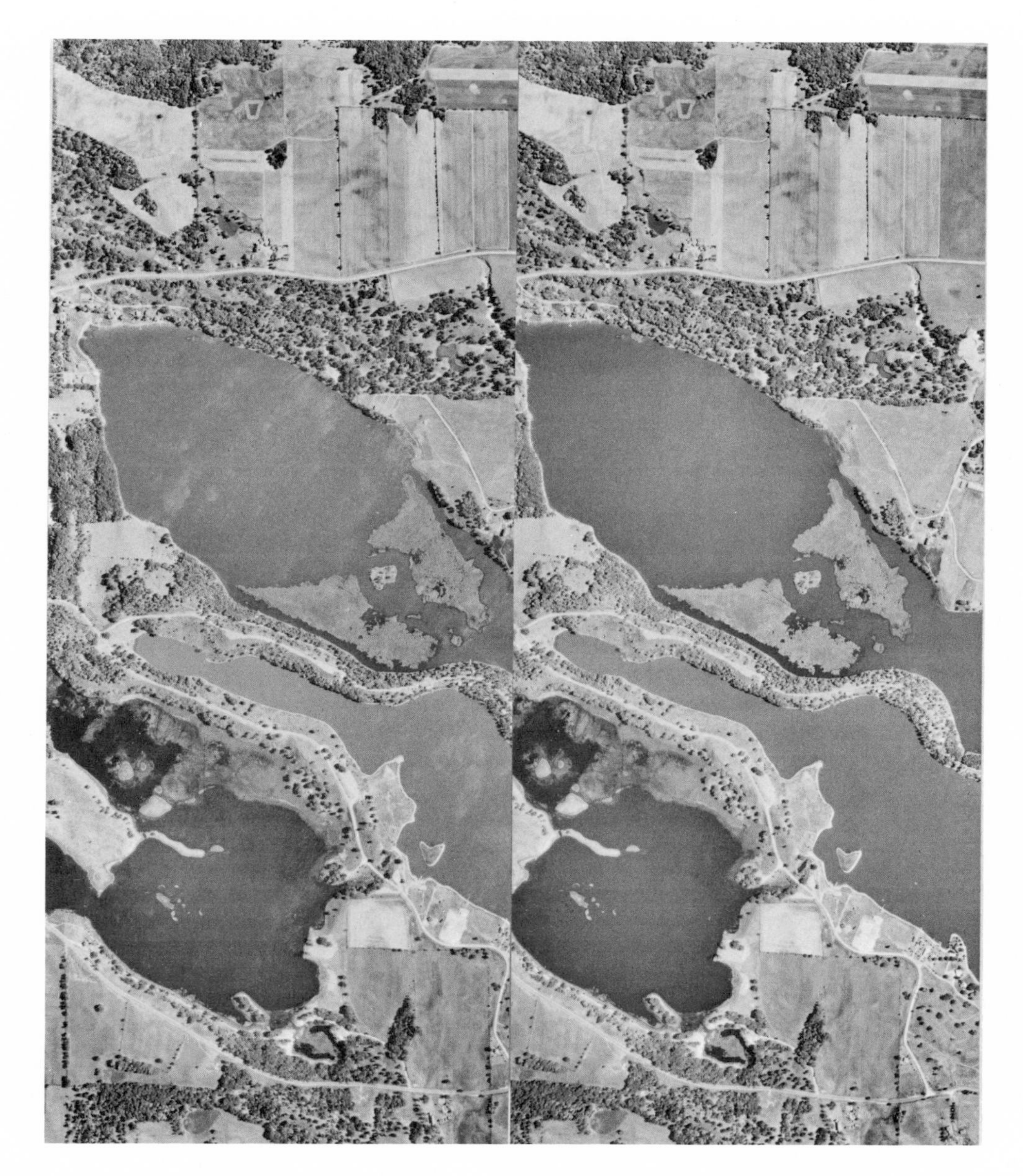

Figure 8.28. An esker in Walworth County, Wisconsin, near Whitewater, extends as a peninsula into a large pond. Photographs by the Agricultural Stabilization Conservation Service, XB-3R-97,98, 1:20,000 (1667′/″), October 2, 1956.

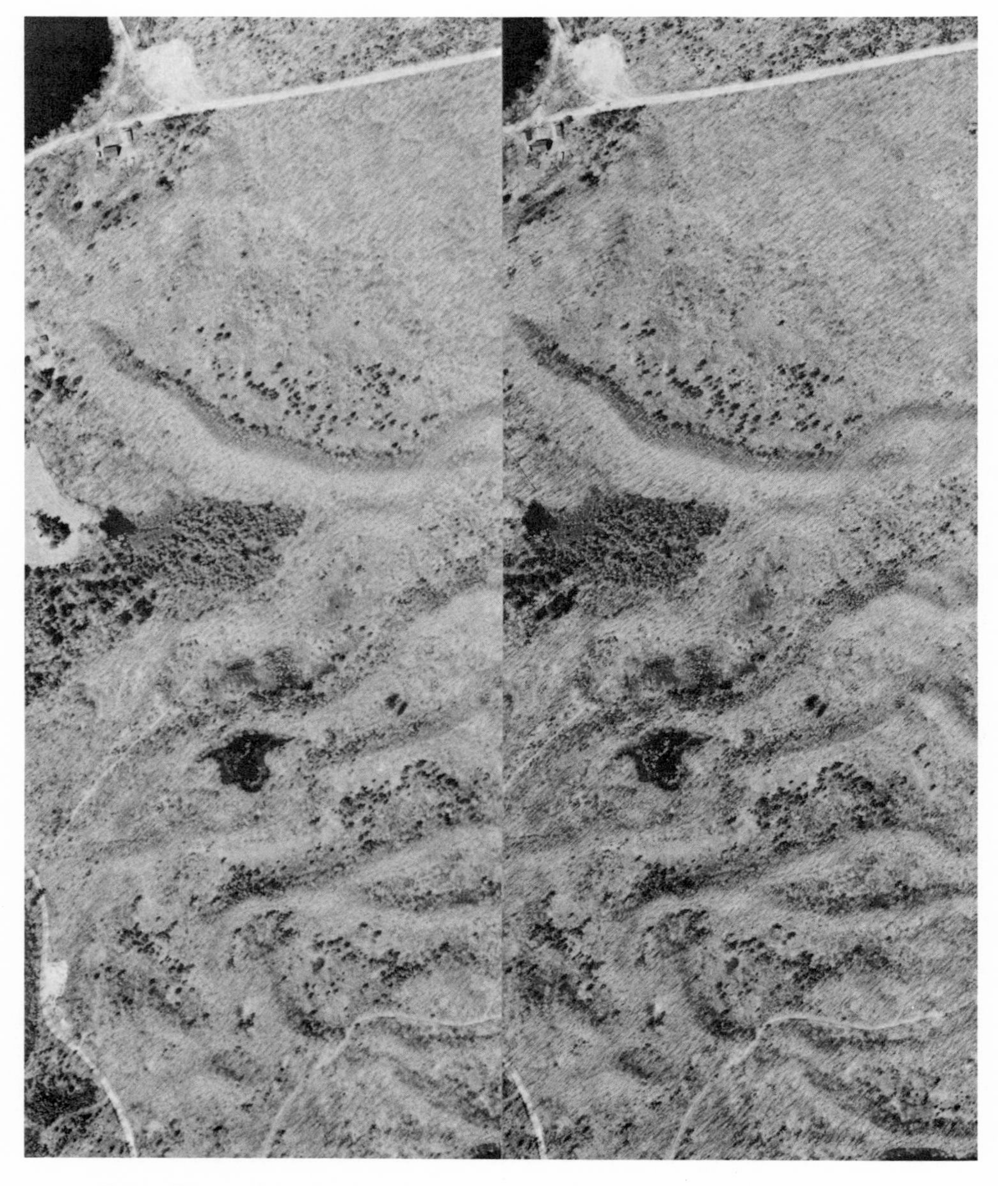

Figure 8.29. An esker train in Norfolk County, Massachusetts. Source unknown.

loaded, multi- or braided stream channels flowing on, in, or under the glacial ice; the resulting deposits take the form of many meandering, interconnected eskers (Figure 8.29).

Interpretation of Pattern Elements

Eskers occur commonly in all glaciated areas and can be easily recognized by their snakelike, linear topographic form. The photographic key shows the characteristics of eskers in general and is not specific as to type or climatic location unless stated.

Soil Characteristics

Eskers have very granular, coarse-textured soils which are useful and valuable for construction purposes. Most formations are moderately to poorly stratified and contain poorly graded gravels and sands mixed with some fines. The central area of the esker may be better stratified than the edges which usually were disturbed during deposition. Finer surface soils sometimes develop; these are typically rather shallow but are thickest along the lower edges of the esker.

U.S. Department of Agriculture Classification

Eskers contain predominantly mixtures of coarse soils typically described as gravelly sands, very gravelly sands, or stratified sands and gravels.

Unified and AASHO Classifications

Predominantly coarse soils are categorized as GM-GC, GP, SP, and occasionally GW or SW. AASHO describes the same materials as A-1, A-2, and A-4.

Water Table

Because of their elevated position, steep slopes, and very coarse soils, water tables do not occur within eskers. However, adjacent land-

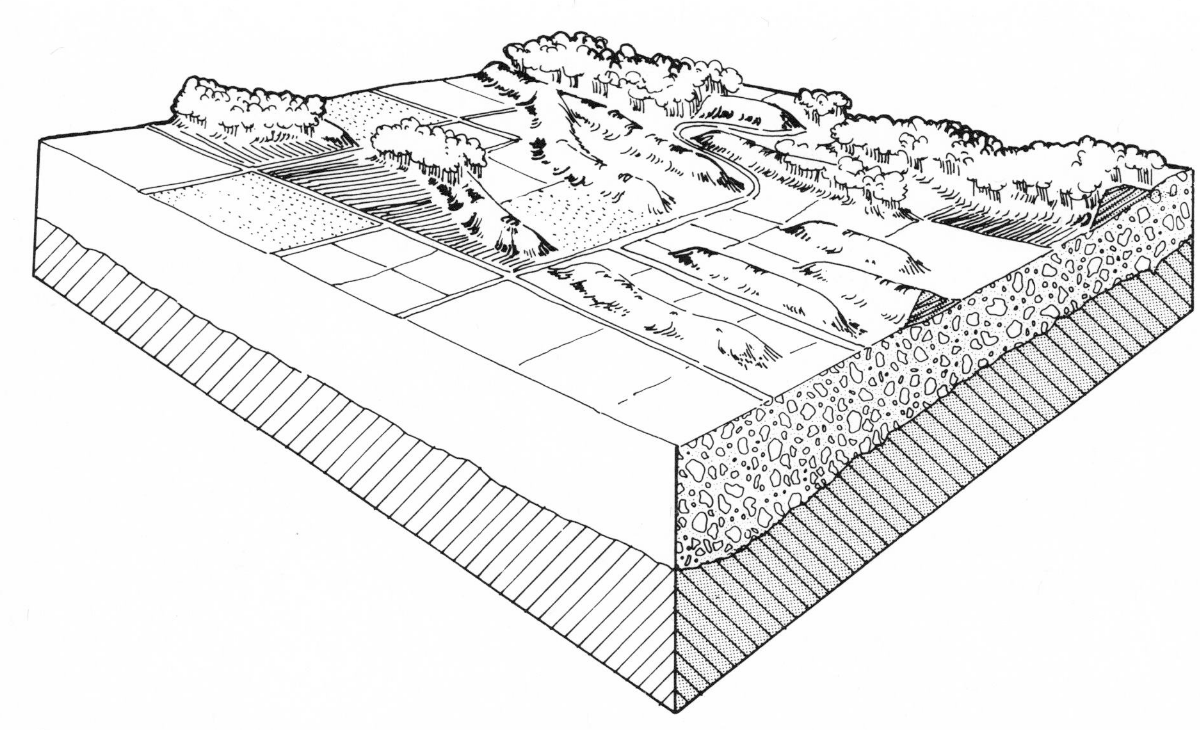

Figure 8.30. Eskers are almost unique among landforms in offering exciting viewing platforms which cross the landscape above the surrounding terrain and are continuous for significant distances. The ridgelines are sharp and narrow, allowing views from both sides; these views are not regional but are views of the immediate surroundings. Eskers provide a natural base for recreation circulation systems, such as hiking, riding, bicycling, or pleasure driving. When viewed from surrounding lowlands, the elevated esker ridges are visually very sensitive to any development or timber cutting.

forms occupying lower topographic positions may have water tables near their surfaces.

Drainage

Percolation and internal drainage is very rapid because of the coarse, rather poorly graded soils. Esker soils are generally too droughty for agricultural use.

Soil Depth to Bedrock

The entire esker formation consists of soil materials, and the soil profile is therefore deep. Bedrock may be located directly beneath the esker, however.

Issues of Site Development

Sewage Disposal

Most state regulations for septic tank leaching fields are satisfied by esker formations, since they have suitable percolation rates. However, it should be noted that many states do not specify a maximum percolation rate, and in these soils there is a danger that the sewage effluent will percolate too quickly, before it is sufficiently acted upon by the aerobic bacteria, and contaminate the water tables in the lowlands. Some states require setback distances from steep slopes for the location of seepage fields to prevent leakage outbreaks along the hillsides, but the sharp ridgelines of eskers make it difficult to comply with these standards without grading. Local health codes and associated water resources should be investigated before leaching field disposal systems are utilized on these formations.

Solid Waste Disposal

Eskers contain mixtures of coarse soils which make them unsuitable for the location of sanitary landfill operations, for the granular soils allow percolation of rainwater and seepage of leachate into the groundwater table.

Trenching

Trenching through eskers encounters difficulties relating to the steep sideslopes. Liquid-carrying pipelines that cross eskers require either excavation of a deep trench, or pumping stations when the pipe is laid over the esker. Trenching is not difficult, for the soils are unconsolidated, and costs are slightly less than those encountered in trenching deep, dry, moderately cohesive soils.

Excavation and Grading

The unconsolidated granular materials can be easily excavated by light equipment. Land uses requiring large, flat sites are not typically located on esker formations because of their narrow form; such a use would require that the esker be completely graded out and used as fill. Costs of such an operation are slightly less than those involved in removing the same amount of deep, dry, moderately cohesive material. Dewatering is generally not necessary for excavations in eskers.

Table 8.11. Glacial landforms: Eskers (humid and arid)

Topography	*Drainage*	*Tone*	*Vegetation and land use*
Snakelike Ridge	None	Light	Natural cover
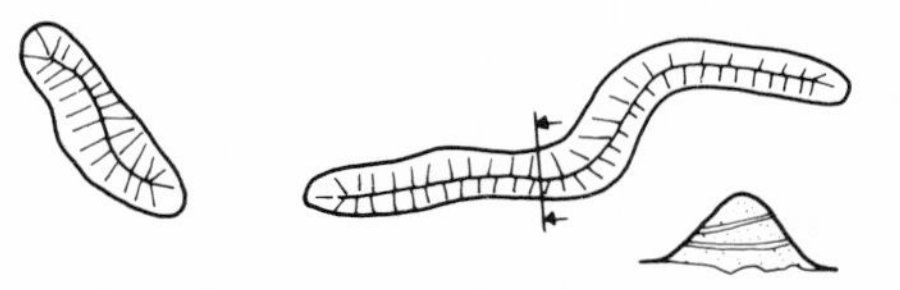 Eskers are primarily identified by the presence of snakelike ridges which do not follow a constant elevation. Typically, these formations are ½ to 1 mile in length, less than 200 feet in width, less than 100 feet above adjacent elevations, and have sideslopes of approximately 30 degrees.	Drainage patterns do not develop on eskers because of their small areal extent. Rainfall runs off the steep slopes and/or percolates into the granular subsoils. Eskers perpendicular to drainage courses develop seepage zones on their opposite sides which continue the drainage system. Eventually, breaks may develop where the esker is eroded.	The exposed soils, if dense tree cover is not present, have light tones, indicating droughty soils. Dense vegetation cover has a dark tone. *Gullies* Few or none Some short, V-shaped gullies may be found near the base of esker slopes if sufficient erosive runoff occurs during rainstorms, for instance, in arid climates.	Eskers are forest-covered in humid climates and grass-covered in arid climates. The droughty, well-drained nature of the soil causes them to be excessively drained for agricultural use, but, for the same reason, they are attractive for highway route location in humid climates.

Construction Materials (General Suitabilities)

Topsoil: Not Suitable. The very thin surface soils are generally not suitable as sources of topsoil material.

Sand: Good to Poor. Eskers contain a large percentage of gravels with some mixed sands which are difficult to separate.

Gravel: Excellent

Aggregate: Excellent

Surfacing: Not Suitable. Surfacing materials need sufficient fines to act as a binder so that corrugated surfaces will not develop with wear. Materials from eskers are suitable only if mixed with fines.

Borrow: Excellent. The stratified layers may be excavated against their open face to mix the interbedded sands, gravels, cobbles, and fines that may occur. These fill materials are generally excellent for all granular earthfill uses, which explains why many commercial borrow pits can be found on eskers. Compaction of the cohesionless sands and gravels for fills is best accomplished with vibrators or a combination of vibration and rolling. Sands and gravels mixed with fines (silt and clay) are compacted for fills by using pneumatic-tired rollers.

Building Stone: Not Suitable. Many of the stones and cobbles are used for the construction of walkways, stone walls, and exterior housing walls.

Mass Wasting and Landslide Susceptibility

Because of their granular composition and excellent drainage, esker formations are stable. The steep sideslopes are at the natural angle of repose of the materials comprising the esker. Surface soils that develop are moved downslope by creep, frost action, and erosion.

Groundwater Supply

Except in rare instances, eskers are too well drained and granular to contain any groundwater resources. However, when they are buried in relatively impervious, glacial lake bed deposits beneath the water table, they provide a very high-yield capacity.

Pond or Lake Construction

Their small areal extent and granular soils make these forms unsuitable for siting reservoirs.

Foundations

Eskers contain granular soils and have very high load-bearing capacities, suitable for almost any foundation system. Small structures can utilize raft foundations or continuous footings with basement slabs. Larger structures can also use rafts or footings as well as piers. Seams of silt or fine sand are occasionally found in these soils, and these could present a frost heave potential in cold climates. Eskers in their natural state contain unconsolidated soils. If an earthquake or

Figure 8.31. An esker in Oconto County, Wisconsin, appears to have been eroded where it crosses the small stream. Photographs by the Agricultural Stabilization Conservation Service, BIC-IV-37,38, 1:20,000 (1667′/″), May 16, 1958.

similar type of shock occurs, the soils may consolidate and cause structural failure. Therefore, before any large structural loads are applied, these soils should be compacted in place by vibrations produced by pile drivers or vibroflotation techniques.

Highway Construction

Highway designers are fortunate if esker formations occur near or along potential corridors or center-line locations. The granular soils are a valuable source of construction materials. The esker itself can be used as a natural roadbed for light-duty or scenic roads, and it offers viewing potential from both sides of the roadway because of its narrow width. In regions of permafrost a high priority is placed upon eskers for highway route location. Some eskers in the Northwest Territories of Canada continue for tens of miles and provide excellent highway foundations. In all regions eskers adjacent to highway construction are commonly utilized as borrow pits for construction materials.

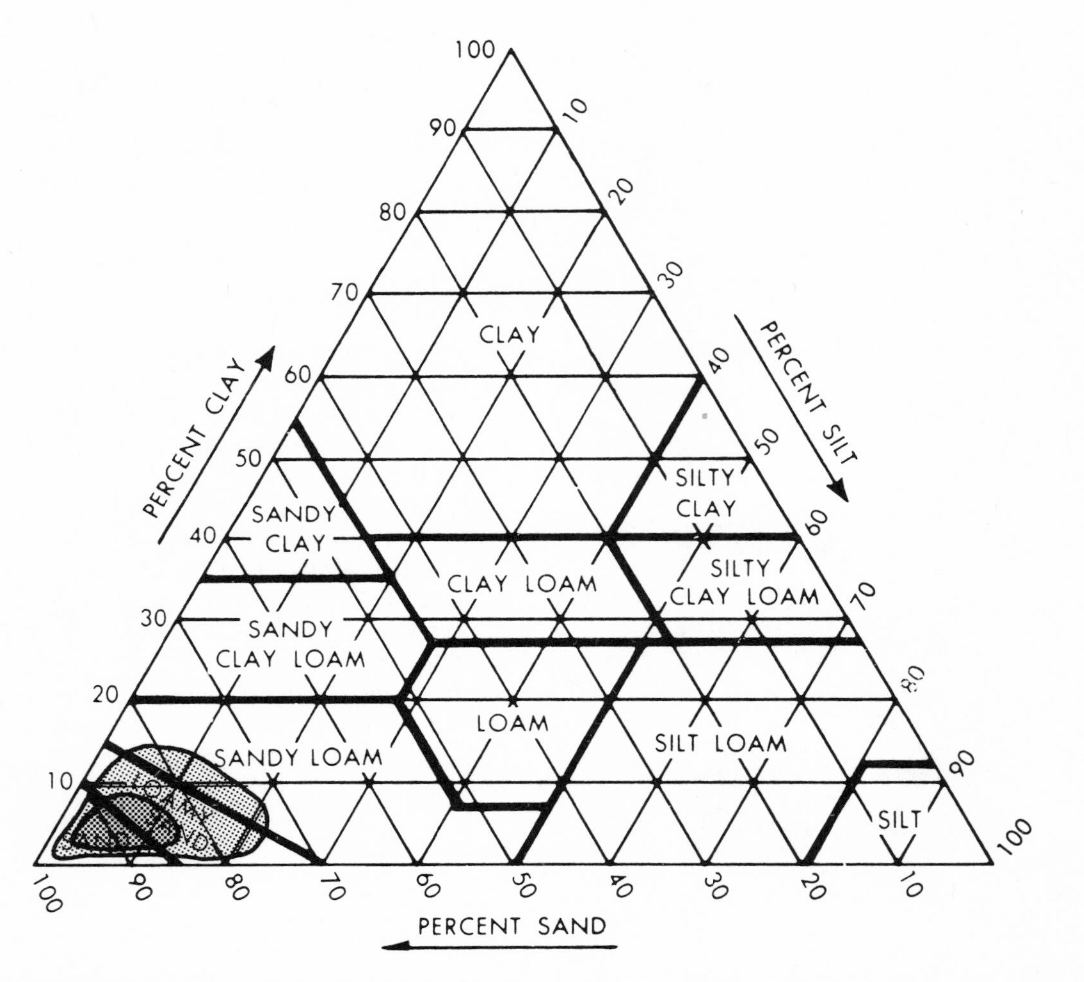

Figure 8.32. The U.S. Department of Agriculture soil texture triangle typically shows sandy materials for eskers. The subsurface materials are stratified sands and gravels.

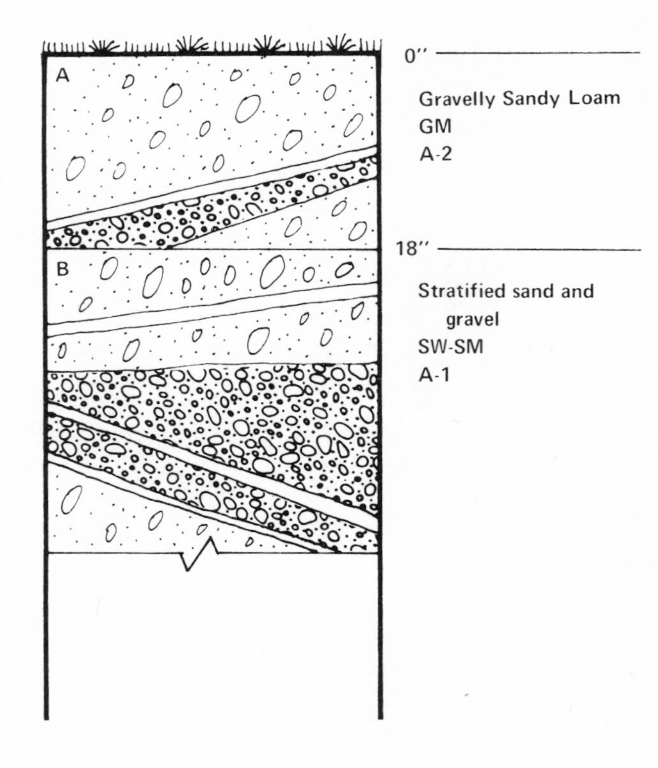

Figure 8.33. A typical profile for soil materials found in an esker. Aroostook County, Maine, Colton gravelly, sandy loam.

Table 8.12. Glacial eskers: aerial and cartographic references

State	County	Photo Agency	Photograph no.	Scale	Date	Corresponding quadrangle and geologic maps from USGS	Corresponding soil reports from SCS
Maine	Aroostook	ASCS	AHZ-4CC-105-106	1667'/"	8/24/62	Sherman (15')	1964
Maine	Hancock	USGS*	Maine 2-ABC	2000'/"	5/1/56	Great Pond (15')	–
Massachusetts	Norfolk	ASCS	DPS-6K-124-125	1667'/"	7/7/52	Wrentham (7½')	1920†
Michigan	Cheboygan	USGS*	Michigan 5-AB	1416'/"	5/3/53	Tower (15')	1939
Minnesota	Itasca	USGS*	Minnesota 2-AB	1416'/"	5/2/47	Calumet (7½'), 1-403	–
New York	Tioga	ASCS	ASM-1P-157-158	1667'/"	6/28/55	Owego, Endicott, Hartford (15')	1953
Wisconsin	Oconto	ASCS	BIC-IV-37-38	1667'/"	5/16/58	Oconto (7½')	–
Wisconsin	Walworth	ASCS	XB-3R-97-98	1667'/"	10/2/56	Whitewater (7½')	1920†

*From *Selected Aerial Photographs of Geologic Features in the United States,* U.S. Geological Survey Prof. Paper No. 590.
†Out of print.

Kames

Introduction

Kames are glaciofluvial, ice-contact deposits very similar to and closely associated with eskers. They are formed by the deposition of poorly stratified sands and gravels in potholes and depressions on or in the glacial ice, which leaves isolated conical hills, mounds, and knobs when the ice melts. Kames are common in any glaciated area and can be found on any topographic surface, deposited on tills and drumlins, or partially or completely buried by lake beds or outwash. They are generally found associated with eskers and in areas where stagnant edges of glacial ice occurred.

Other formations of glacial origin, classified as types of kame deposits, are not included in this section since they are in fact different landforms and have different origins. Kame terraces, for example, are not related to kames but are terrace formations of stratified drift which were deposited by glacial meltwater between glacial ice and adjacent areas of high ground. Technically, a kame terrace is a constructional outwash terrace left after the disappearance of the ice, and it is very similar in characteristics and engineering applications to outwash (see the next section, "Outwash"). Kame moraines are really moraines with many hummocky hills and depressions and so appear kame-like. Groups of kames are classified as kame fields; they should not be referred to as kame moraines.

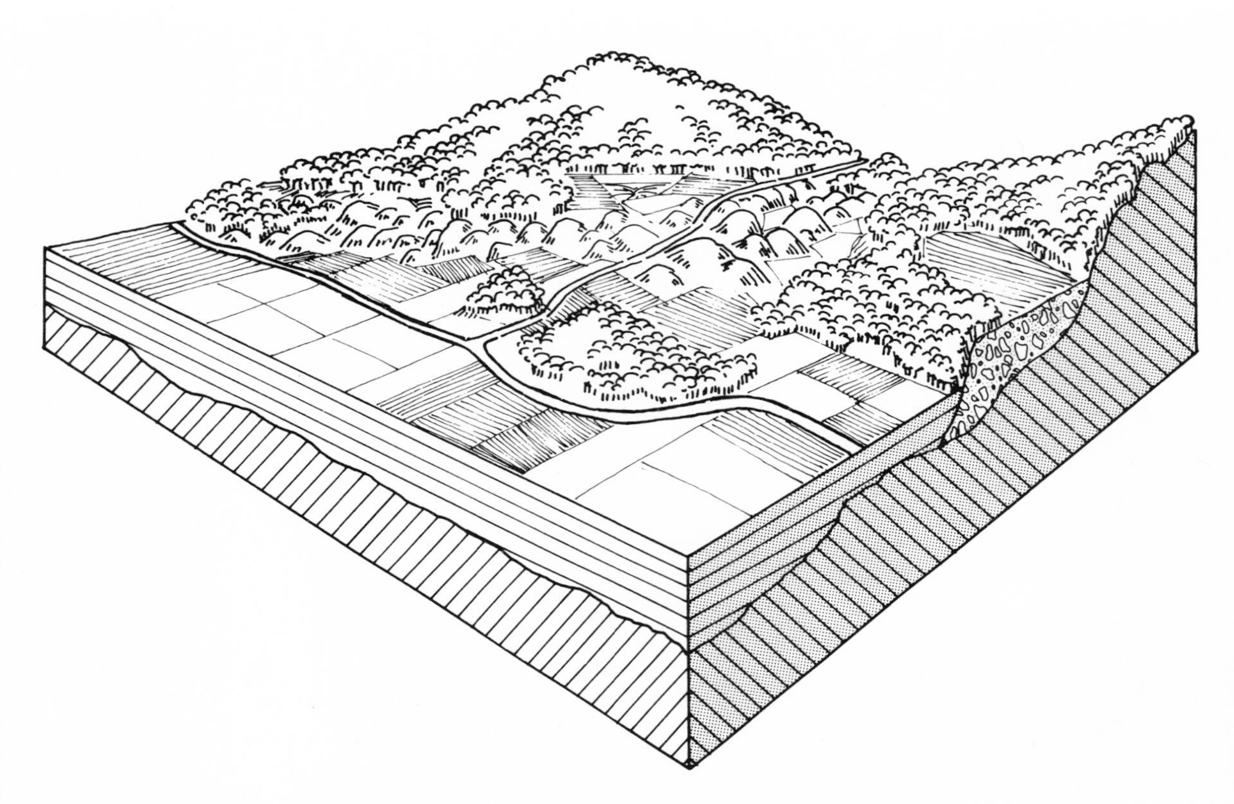

Figure 8.34. Kames can act as exciting viewing platforms or pedestals, since they are elevated above typically flat or undulating topographies and are small in size, offering complete 360-degree views from their tops. The elevations do not allow regional views. When viewed from the lowlands, the tops of kames and the sideslopes are visually sensitive to any development or timber cutting.

Interpretation of Pattern Elements

Kames can occur individually or in groups or fields. In all cases they exhibit definite rounded outlines or boundaries. The photographic interpretation key shows the characteristics of kames for both humid and arid climates, making distinctions between the two only when it is pertinent.

Kames are characterized by their unique topography of conical or elongated, sharp-ridged hills. Drainage patterns do not develop on kames, owing to their small areal extent and to the granular subsoils which give the surface a photographic tone which is light in comparison with surrounding materials. The steep slopes and droughty soils do not facilitate cultivation, and these forms are therefore typically forest-covered and have little undergrowth.

Soil Characteristics

Kames contain mixtures of poorly stratified sands and gravels but generally contain more fines than eskers. Lenses of fine sand and silt are common, along with many directional variations in bedding and textural sequences.

Table 8.13. Glacial landforms: Kames (humid and arid)

Topography	*Drainage*	*Tone*	*Vegetation and land use*
Conical hills	None	Light	Natural cover
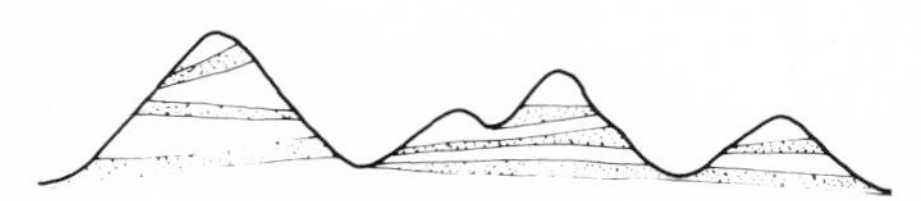 Single kames appear as conical or elongated hills; their dimensions are usually less than 400 feet in any one direction, and less than 50 feet above adjacent elevations. They have sideslopes of approximately 30 degrees, or close to the natural angle of repose for granular materials. Many kames are irregular in outline and form hooked shapes.	Drainage patterns do not develop on kames because of their small areal extent and because their steep slopes do not develop a watershed (that is, rainfall runs off the slopes and/or percolates into the granular subsoils). Kames are small formations and do not, as is common with eskers, block other drainage systems.	Light tones appear through the vegetation cover, reflecting the droughty soils and lack of undergrowth. Surrounding landforms are typically wetter, showing darker tones. *Gullies* Few or none. Short, V-shaped gullies may occur near the bases of kame slopes if there is sufficient erosive runoff during rainstorms. This occurs especially in arid climates.	In humid climates the typical cover is forest, owing to the droughty soils and the steep sideslopes. In arid climates kames are grass-covered. The rapid internal drainage makes these forms too dry for agricultural use, but they are attractive as sources of construction materials, as shown by the occasional presence of borrow pits.

U.S. Department of Agriculture Classification

Kames contain predominantly mixtures of coarse soils which are typically described as gravelly sands, very gravelly sands, or stratified sand and gravels.

Unified and AASHO Classifications

The predominantly coarse soils are categorized as GP, SP, GP-GM, GM-GC, SM-SC, or SP-SM, but very rarely GW or SW. AASHO describes the same materials as A-1, A-2, and occasionally A-4.

Water Table

Because of their elevated position, steep slopes, and very coarse soils, kames do not commonly hold water tables. If there are silt lenses, they may be water-bearing and may create some associated engineering problems. Adjacent landforms occupying lower topographic positions are generally wet and have water tables near the ground surface.

Drainage

Percolation and internal drainage are very rapid because of the coarse, somewhat poorly graded soils. These soils are generally too droughty for agricultural use.

Soil Depth to Bedrock

The entire formation consists of deep soil materials. Bedrock may be located directly beneath the kame, however.

Issues of Site Development

Sewage Disposal

Conditions at the tops of kames satisfy most state regulations for the location of septic tank leaching fields, since there are suitable percolation rates. However, it should be noted that many states do not specify a maximum percolation rate, and therefore there is a danger that sewage effluent will percolate too quickly and will contaminate the water table in lowlands. Some states require seepage field setback distances from steep slopes to prevent leakage outbreaks along the hillsides. It is difficult to satisfy this criterion on sharp-ridged kames, but flat-topped kames sometimes qualify.

Solid Waste Disposal

Kames contain mixtures of granular materials, thus making them undesirable for the location of sanitary landfill operations. The coarse soils allow percolation of rainwater and the seepage of leachate, which may contaminate the groundwater table.

Trenching

Trenching through kames encounters linear alignment difficulties because of the steep side slopes. Liquid-carrying pipelines must either excavate a deep trench or provide a pumping station if the line is laid over the kame itself. Generally, kames are small enough to allow alignments to

Figure 8.35. Kames in Cortland County, New York, appear with somewhat irregular conical shapes with steep sideslopes. Photographs by the Agricultural Stabilization Conservation Service, ASK-1P-48,49, 1:20,000 (1667′/″), May 20, 1955.

be routed around them. The actual trenching operations are not difficult in the unconsolidated soils; costs are slightly less than those associated with trenching the same amount of deep, dry, moderately cohesive soil.

Excavation and Grading

The unconsolidated granular materials can be easily excavated and graded with light equipment, and the resulting costs are slightly less than those encountered in removing the same amount of deep, dry, moderately cohesive soil. Land uses requiring large, flat sites are not typically located on kames, because of their small, rugged forms, unless the kames are completely graded.

Construction Materials (General Suitabilities)

Topsoil: Not Suitable. Kames may develop thin surface soils, but they are not suitable sources of topsoil materials.

Sand: Excellent

Gravel: Excellent

Aggregate: Good. Because kames are rather small formations and have poorly stratified layers, they may present some problems in locating large supplies of aggregate materials.

Surfacing: Not Suitable. Surfacing materials need sufficient fines to serve as binder, so that corrugated surfaces will not develop under traffic wear. Materials from kames are suitable only if mixed with sufficient fines.

Borrow: Excellent to Fair. Kames are excellent sources of granular borrow materials but, if a well-graded borrow is desired, they are not suitable. Noncohesive granular materials used for fills can be compacted by vibrators or by a combination of vibration and rolling.

Building Stone: Not Suitable. Many of the stones and cobbles are used for the construction

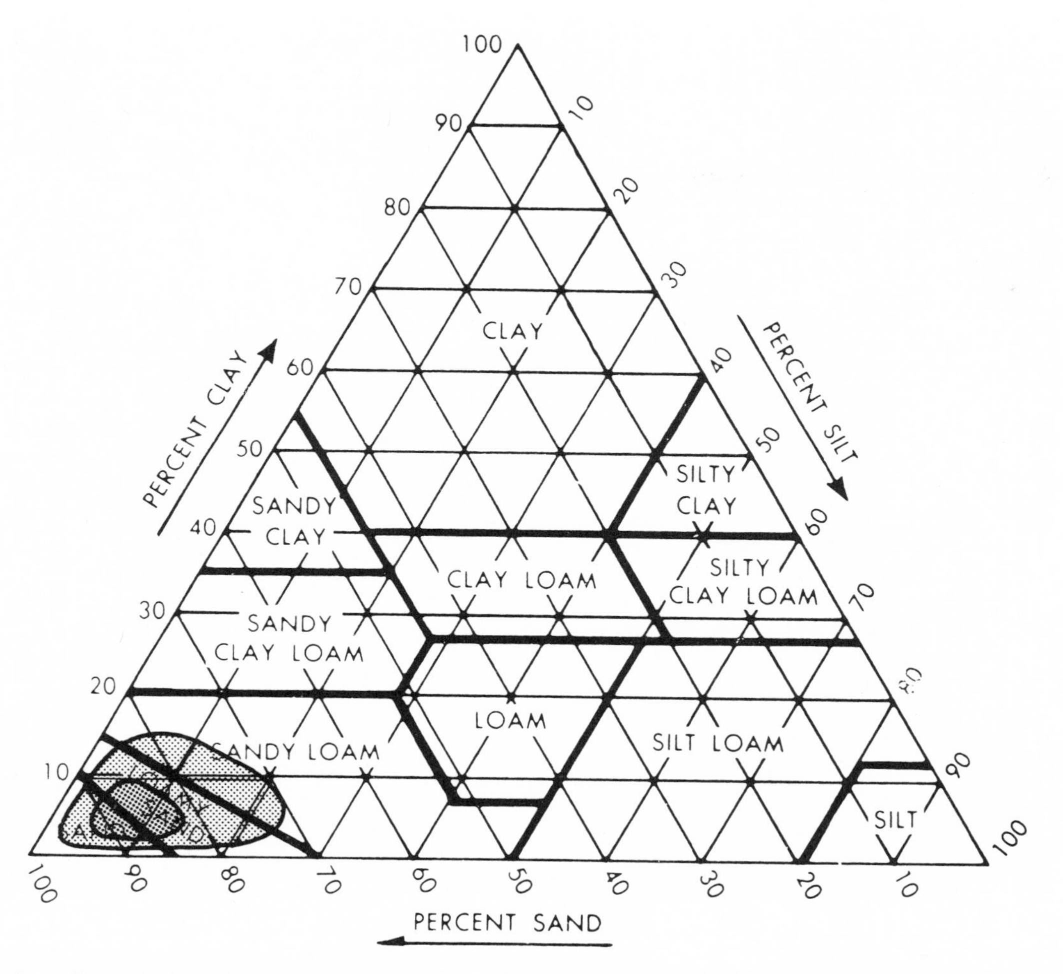

Figure 8.36. The U.S. Department of Agriculture texture triangle indicates sandy materials, but kames contain stratified or poorly stratified sands and gravel.

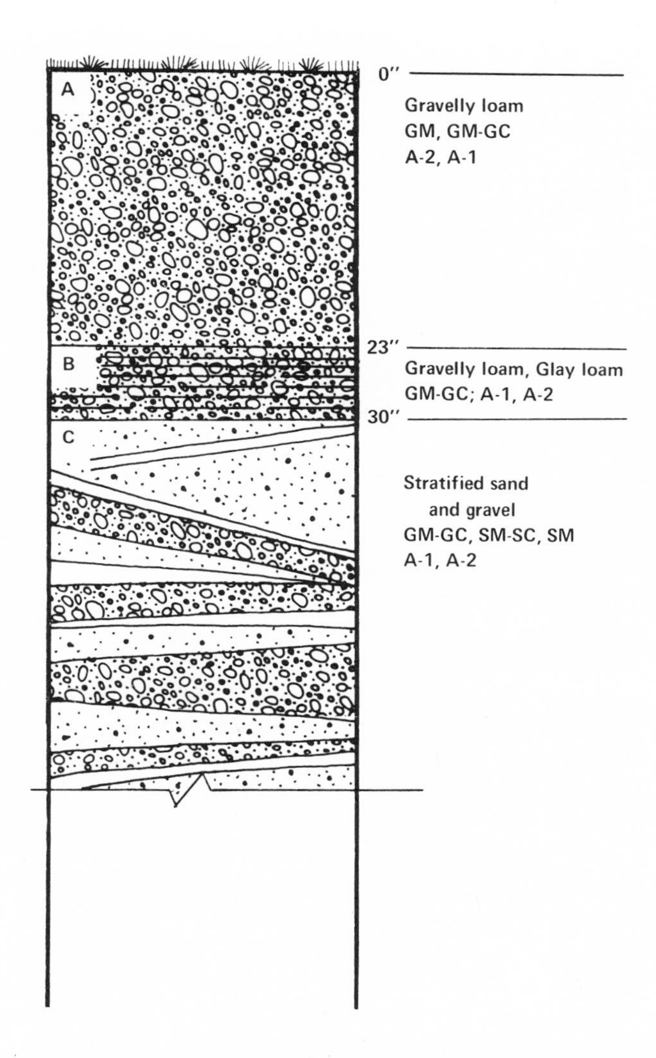

Figure 8.37. A typical soil profile of kame materials in Cortland County, New York. Several feet of sandy, silty materials are found over the stratified sand and gravel.

of walkways, stone walls, and exterior house walls.

Mass Wasting and Landslide Susceptibility

Kame formations are relatively stable, owing to their granular composition and excellent internal drainage. The steep sideslopes are close to the natural angle of repose of the materials comprising them. Surface soils that develop are moved downslope by soil creep, frost action, and/or erosion.

Groundwater Supply

Kames are too well drained and granular to contain groundwater resources except in rare instances. However, when kames are buried in impervious lake beds, they provide a high water yield capacity since they occur beneath the water table.

Table 8.14. Glacial kames: aerial and cartographic references

State	County	Photo Agency	Photograph no.	Scale	Date	Corresponding quadrangle and geologic maps from USGS	Corresponding soil reports from SCS
New York	Cortland	ASCS	ASK-1P-48-50	1667'/"	5/20/55	Cortland (7½')	1961
New York	Erie	ASCS	ARF-3GG-192-193	1667'/"	7/3/66	East Aurora (7½')	1929
New York	Erie	ASCS	ARF-3GG-132-137	1667'/"	7/3/66	Springville (15')	1929
New York	Warren	ASCS	EHI-2JJ-85-88	1667'/"	9/15/68	Glens Falls (15')	—
New York	Washington, Warren	ASCS	EHI-2JJ-74-77	1667'/"	9/15/68	Glens Falls (15')	—

Pond or Lake Construction

The small areal extent and granular soils of kames do not provide satisfactory sites for the placement of reservoirs.

Foundations

Kames contain very granular soils which have very high load-bearing capacities and are suitable for most foundations. Small structures can utilize rafts or continuous footings with basement slabs; large structural loads can also use rafts or footings, as well as piers. Seams of silt or fine sand should be located, since these have potential for frost heave movement in cold climates. Kames in their natural state contain unconsolidated granular soils. If an earthquake or other similar shock occurs, the soils may consolidate and cause structural failure. Therefore, before major structural loads are applied, these soils should be compacted in place by vibrations produced by pile drivers or vibroflotation techniques.

Highway Construction

Highways are not typically located on kame landforms because of their small areal extent and their usefulness as sources of construction materials. Many kame materials are used as borrow during nearby highway construction operations.

Outwash

Introduction

Outwash landforms consist of alluvium containing stratified silts, sands, and gravels which were deposited by glacial meltwaters; outwash is usually formed beyond the furthest advance of the glacier.

In analyzing the soils of an outwash formation, it can be seen that the coarsest particles are deposited nearest the origin or apex. Theoretically, the materials grade outward from the center, from gravels into gravels and coarse sands, and finally into sand and silt plains. Typically, clays are not deposited in outwash formations but are carried away; eventually, they may be deposited in lake beds, ocean bottoms, or large deltas. The coarse soil fragments found near the apex of the outwash include rocks with angular surfaces; stones found further from the apex have been abraded and rounded by the transporting process. Gravel-sized fragments are predominant and are located within the upper and midrange of the outwash fan.

The gravelly materials of outwash landforms cause, as they are deposited, many braided, meandering channels, typically 1 to 5 feet deep and approximately 25 feet in width. The braided pattern starts when the glacial meltwater becomes overloaded with debris. Then as each stream channel becomes filled with the sediments, it develops levees; these cause the channel to overflow, and eventually a new channel is started adjacent to the old. However, the new channel is at a slightly lower elevation and with a higher gradient. The depressed topographic position of the abandoned channel scars allows more moisture to collect, increasing weathering, and therefore developing a deeper soil profile containing a higher content of clays. The channel depressions then act to hold water and slowly release or infiltrate it into the gravelly subsoils. The infiltration basins that indicate gravelly materials are seen on aerial photographs as darker tones, created by the higher moisture and organic content of the soils contained in them.

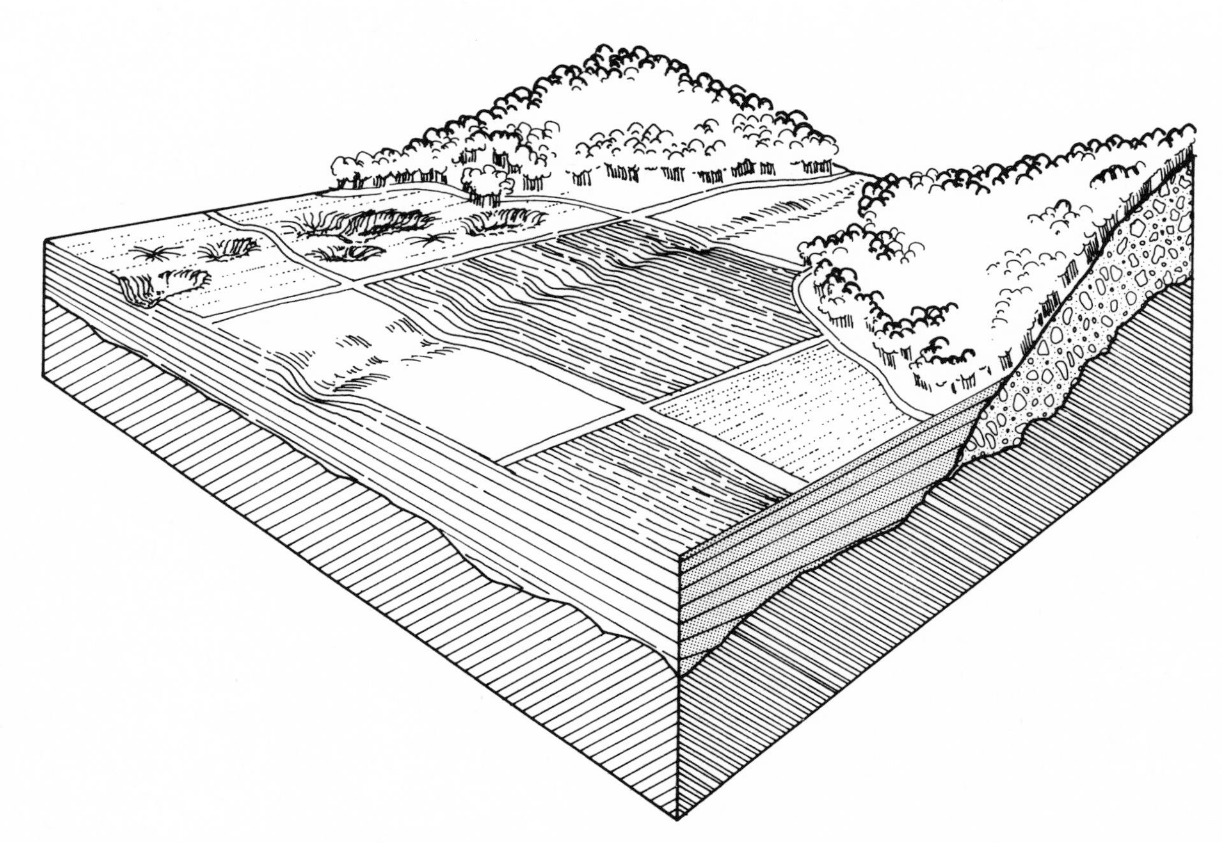

Figure 8.38. Gravel and sand outwash plains are very flat and do not present any spatial or viewing variation. When traveling through these areas, which typically contain little forest cover, the intense agricultural development becomes monotonous. Tower elements have a high visual impact in such regions because of the lack of any topographic or vegetation screening. Pitted outwash deposits provide more spatial variety, since they are visually absorptive within the pits or depressions. Land uses located within these pits can be observed only from the immediate depression edge, and this is commonly forested.

The lower reaches of an outwash fan develop into broad sand-silt plains. These do not have braided scars, because the water velocities were lower and significant channels were not created and because any channels that were created were readily erased by wind erosion.

Outwash deposits occur either within valleys, as is common in the northeastern United States,

Table 8.15. Glacial landforms: Outwash (plains and pitted) (humid and arid)

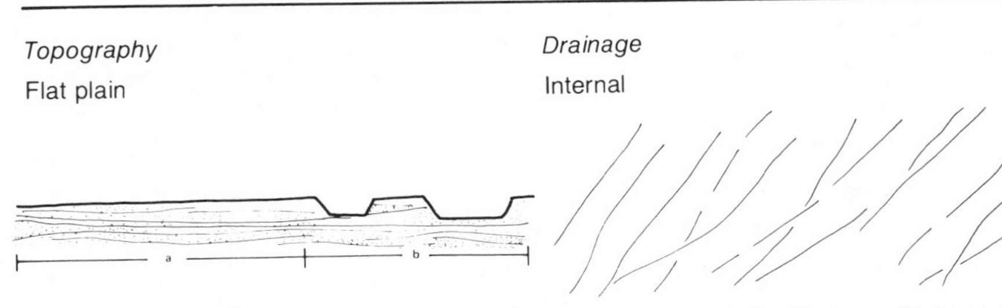

Topography	*Drainage*	*Tone*	*Vegetation and land use*
Flat plain	Internal	Light-banded	Natural and cultivated
Surfaces are very flat (a), sloping as little as ¼ to ½%. Gravel deposits have surface irregularities in the form of ancient channel scars; sand plains are uniformly flat. Pitted outwash (b) contains depressions or pits which may be filled with water and may have steep sideslopes.	The well-graded granular soils have very high percolation rates and very rapid internal drainage and therefore do not allow sufficient runoff for the creation of an integrated drainage system. Some ditching may be observed, indicating efforts to lower the high water table. Pits are typically filled with water.	The well-drained soils have light tones. Channel scars or infiltration basins in gravelly deposits have dark tones, indicating moist, organic materials.	In humid climates outwash plains are intensively cultivated. Pitted outwash is usually cultivated around the pits, although in New England the pits themselves are used as cranberry bogs. Highways are typically gridded. Fields are rectangles or squares. Arid climates have grass cover, indicating the droughty, unproductive nature of the granular soils.
		Gullies	
		None	
		Because of the high permeabiity of the soils, sufficient runoff to create gullying does not occur, even in arid climates.	

or on flat plains formed by large, coalescing alluvial fans. They can be pitted or nonpitted. Most outwash plains have flat, smooth surfaces, but near the outwash apex, where ice blocks were buried and subsequently melted, pits occur; these are steep-sided depressions commonly filled with water or organic materials. Outwash deposits having many pits and depressions are classified as *pitted outwash.* Outwash deposits contained within valley walls are referred to as *valley trains*; such deposits may show terracing, indicating postglacial erosion. Valley trains are common in most major valleys in rugged landscapes such as those in the northeastern United States. *Outwash plains* developed in the Midwest where the surrounding and underlying topography is flat; here, several large outwash fans come together or coalesce into broad, flat plains which may extend over hundreds of square miles.

Interpretation of Pattern Elements

Outwash deposits are easily identified by their flat or pitted topographic forms, lack of drainage systems, and channel scars. The most difficult formations to distinguish are sand plains, which occasionally grade into sandy lake bed deposits. The photographic key distinguishes between outwash plains and pitted outwash plains. Valley trains are easily identified by their occurrence in valleys. The effect of climate, humid or arid, is apparent only in the vegetation and land use patterns.

Soil Characteristics

The soils associated with outwash landforms are waterlaid, stratified deposits horizontally bedded in layers of quite uniform texture. The surface soils typically develop a profile of fine sands and silts; these grade into the stratified silts, sands, and gravels of the subsoils. Gravelly outwash deposits contain infiltration basins at slightly lower ground elevations.

U.S. Department of Agriculture Classification

The U.S. Department of Agriculture commonly classifies surface soil horizons as silt loam, silty clay loam, loam, and sandy loam. These surface horizons grade into subsurface soils described as stratified sands or as stratified silts, sands, and gravels.

Unified and AASHO Classifications

Symbols indicating coarse-grained soils are typical, including GW, SW, GP, SP, and their combinations, with GM and SM occasionally occurring. Typical AASHO categories are A-1 and, occasionally, A-2, A-3, and A-4.

Water Table

The soils are highly permeable and well drained internally, but water tables are usually within 10 feet of the surface in valley deposits. Outwash plains in the Midwest may have a greater depth to water table. Drainage of water is dependent upon subsurface groundwater seepage and flow. Depressions, excavations, and borrow pits may be filled with water, indicating the elevation of the groundwater table.

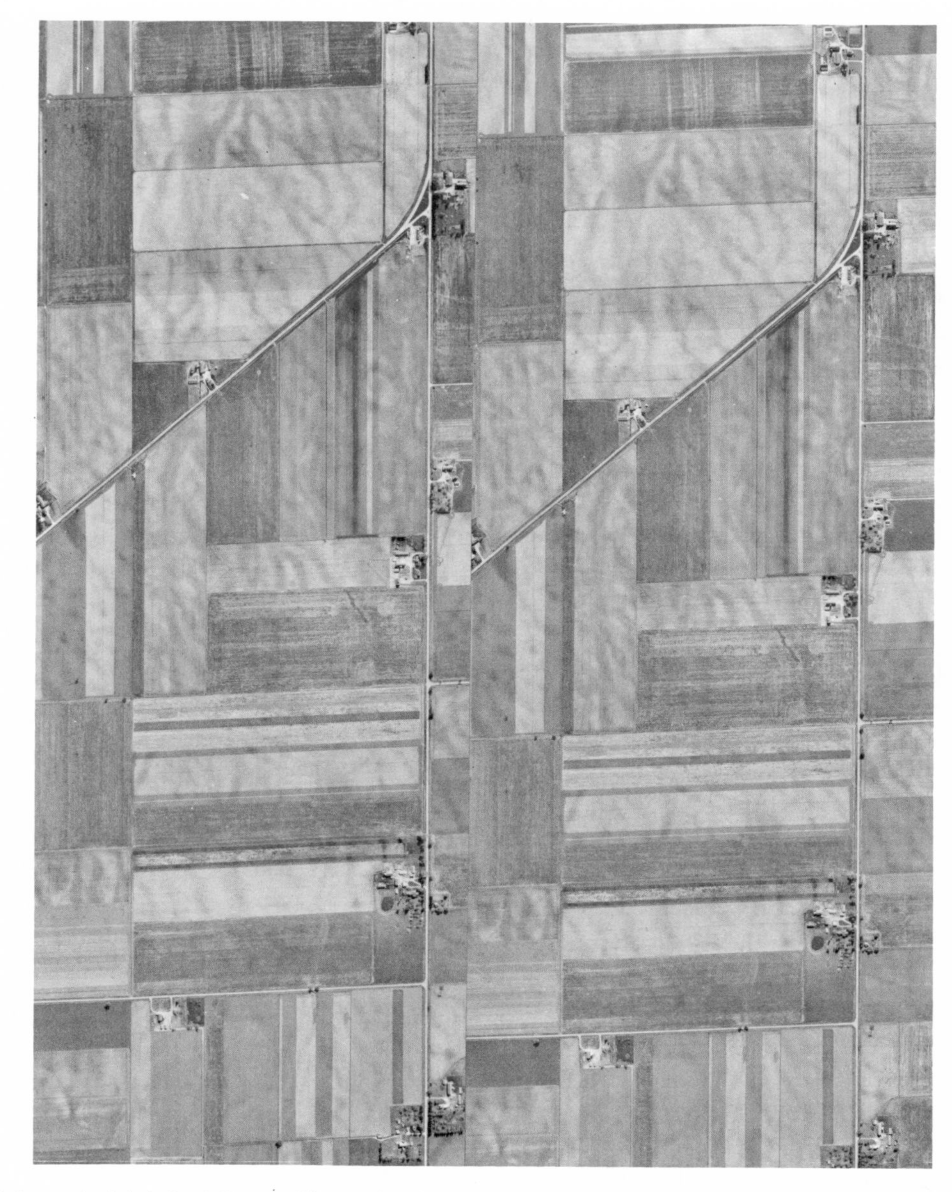

Figure 8.39. Outwash plain in Rock County, Wisconsin, with characteristic flat topography and ancient channels. Photographs by the Agricultural Stabilization Conservation Service, XA-2R-66,67, 1:20,000 (1667′/″), May 5, 1956.

Gravelly outwash deposits may develop seasonal, perched, high water tables in more poorly drained channel scars or infiltration basins.

Drainage

Percolation and internal drainage are very rapid because of the coarse soils. However, old channel scars occurring in gravelly deposits have more fines and higher organic contents and are therefore drained slowly. If the groundwater table is near the surface, the outwash will be poorly drained, but ditching and other methods of artificial drainage can, depending upon the surrounding topography and underlying materials, quickly lower the high water table.

Soil Depth to Bedrock

The entire formation consists of deep soil materials, and bedrock is not commonly encountered during normal construction operations. In New England outwash may be less than 20 feet in thickness and is underlaid by till or occasionally lake bed materials.

Issues of Site Development

Sewage Disposal

Most state regulations concerning the location of septic tank leaching fields are satisfied on outwash plains if there is sufficient depth to the groundwater table. However, it should be noted that many states do not indicate a maximum percolation rate, and therefore there is a potential danger of the sewage effluent percolating too quickly and contaminating the groundwater table. Local health codes and sanitary engineers should be consulted to determine suitabilities.

Solid Waste Disposal

Outwash plains contain mixtures of granular materials, making them undesirable for the loca-

Figure 8.40. (A) Valley train of outwash material in Cortland County, New York, containing some water-filled pits. Photographs by the Agricultural Stabilization Conservation Service, ASK-1P-86,87, 1:20,000 (1667′/″), May 20, 1953.

Figure 8.40. (B) U.S. Geological Survey Quadrangle: Homer (7½′).

tion of sanitary landfill operations. The coarse soils allow the percolation of rainwater and the seepage of leachate, which may contaminate the groundwater table.

Trenching

Little difficulty is encountered in outwash formations during trenching operations unless the water table is very close to the surface. When the water table is not encountered, costs are slightly less than those associated with removing the same amount of deep, dry, moderately cohesive soil. Deep trenching operations within the water table require drainage by pumping from sumps or well points until the construction is completed.

Excavation and Grading

Few problems are encountered in excavating outwash materials above the water table; costs are equivalent to those associated with removing the same amount of deep, dry, moderately cohesive soil. Excavation of materials beneath the water table encounters a condition in which hydrostatic uplift causes sand and water to boil upward from the bottom of the excavation. Such soils require dewatering by the use of methods such as sumps and well points.

Construction Materials (General Suitabilities)

Topsoil: Fair. Generally, these soils are rather droughty, owing to their predominant composition of granular materials and their low organic content.

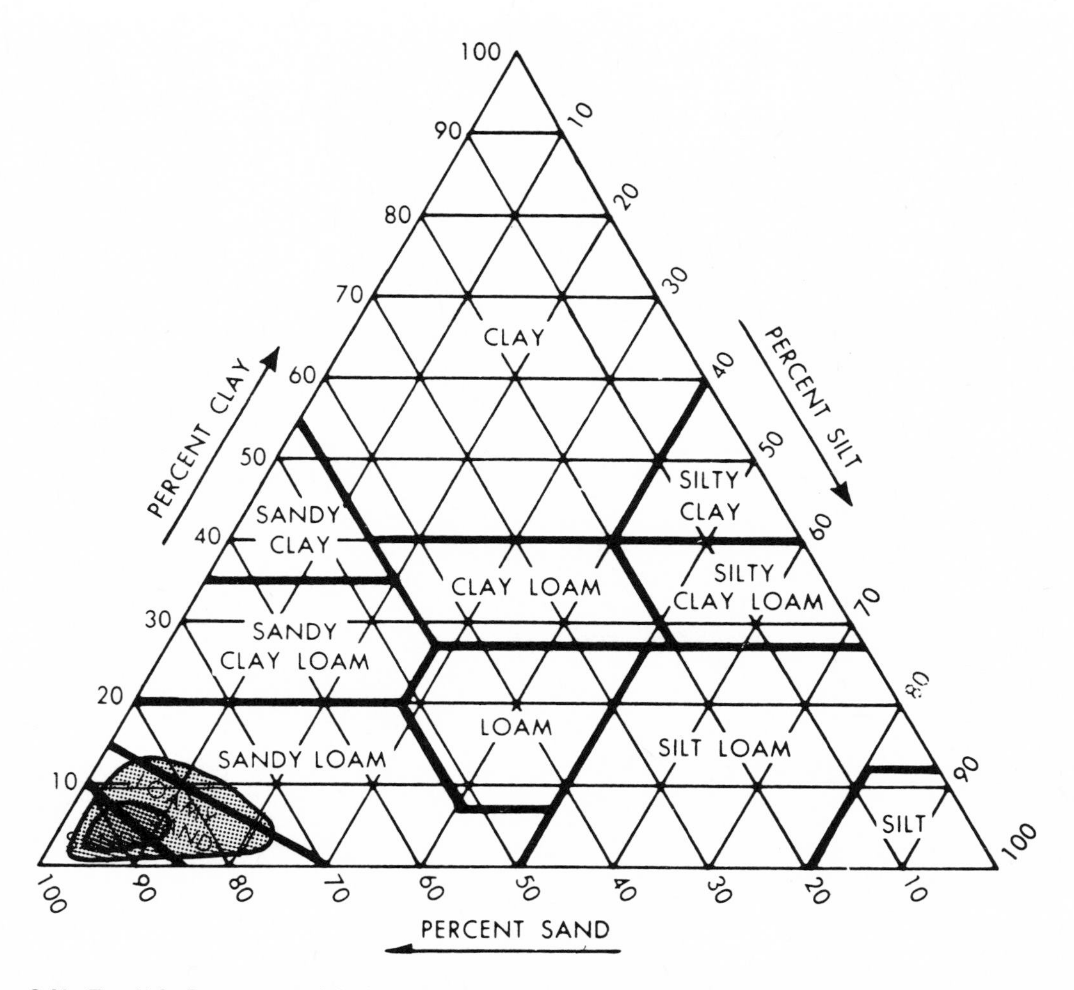

Figure 8.41. The U.S. Department of Agriculture soil texture triangle typically indicates sandy materials, but they are stratified and mixed with gravels.

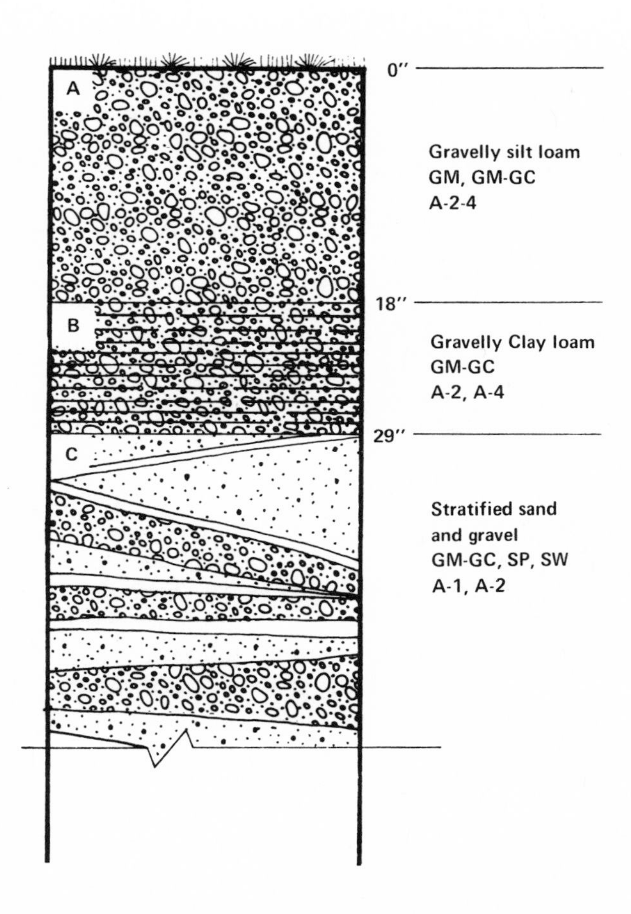

Figure 8.42. Typical outwash materials are shown in the soil profile of Palmyra cobbly loam in Cortland County, New York.

Sand: Excellent. Suitability depends upon the specific composition of the outwash. Sufficient sandy materials can usually be found along the lower third of the deposit.

Gravel: Excellent. The suitability of these materials depends upon the specific composition of the outwash and the distance of a specific site from the apex. Channel scars indicate the infiltration basins associated with gravelly materials; these are usually found in the middle and upper thirds of the deposit.

Aggregate: Good. Suitable aggregate materials can generally be found in these stratified, well-graded formations.

Surfacing: Not Suitable. Surfacing materials need sufficient fines (clays) to serve as binder, or corrugated surfaces will develop with wear. Materials from outwash formations are suitable if mixed with sufficient fines.

Borrow: Excellent. Outwash deposits generally offer excellent granular borrow materials, but excavation should take place along vertical faces to facilitate mixing of the clean, stratified layers.

Compaction of materials for fills is accomplished by the use of vibrators or by a combination of vibrators and rollers.

Building Stone: Not Suitable.

Mass Wasting and Landslide Susceptibility

Outwash formations are relatively stable except under extreme, not commonly occurring conditions. One such condition results when embankments are oversteepened by excavation or stream erosion. If the outwash materials are underlaid by more impervious deposits, such as till, unstable conditions will occur along cuts where the till is exposed and where lateral support of the outwash is removed. Also, in such a situation, seepage of groundwater through outwash increases its weight and mass and acts to lubricate the impervious till surface, thus enhancing its potential for sliding. When such interbedding of pervious-impervious layers occurs, specific field investigations should be undertaken and consultants engaged to determine the actual conditions and hazards.

Groundwater Supply

Outwash formations offer excellent groundwater resources, since they have a very high yield capability and recharge potential. In New England these deposits are very important as water supply resources, since the bedrock is crystalline and nonwater-bearing. Care should be taken not to contaminate such groundwater resources, since the pollutant will easily be carried through the subsurface aquifer and eventually affect a large part of it.

Pond or Lake Construction

The flat surfaces and high rates of percolation of outwash deposits do not present suitable conditions for the construction of reservoirs. The natural soil materials require coating with bentonite or other commercially available materials in order to create a suitable impervious bottom layer. Excavation of soils below the water table creates water bodies, but their water level fluctuates in relation to the elevation of the of the groundwater table. Pitted outwash depressions are typically ponded, also reflecting the elevation of the groundwater table. If the water levels of several ponds are compared, they will be found to be nearly equal, indicating that the surface elevation of the water table reflects the granular nature of the materials.

Foundations

Outwash formations contain soils with moderate load-bearing capacities which provide fair to good foundation surfaces. Small structures can utilize rafts or footings; residential units typically use continuous footings with basement slabs. Moderate-load structures may also utilize rafts or footings. Heavy structures should utilize friction piles or, if necessary, point-bearing piles. Channel scars in gravelly outwash deposits may require special treatment because of the higher concentration of fines or if there is a perched water table.

Highway Construction

Highways can be constructed through outwash

Table 8.16. Outwash plains, pitted outwash, and valley trains: aerial and cartographic references

State	County	Photo Agency	Photograph no.	Scale	Date	Corresponding quadrangle and geologic maps from USGS	Corresponding soil reports from SCS
Alaska	—	USGS*	Alaska 19-ABC	3250′/″	7/17/50	Kenai (A-2) (1:63,360)	—
Alaska	—	USGS*	Alaska 24-AB	3333′/″	6/21/48	Yakutat (1:250,000)	—
Massachusetts	Dukes	ASCS	DPO-1LL-19-21	3333′/″	10/20/70	Vineyard Haven (7½′)	1925
Massachusetts	Dukes	ASCS	DPO-1LL-6-8	3333′/″	10/20/70	Tisbury Great Pond (7½′)	1925
Montana	Sheridan	USGS*	Montana 12-ABC	1667′/″	7/31/40	Wolf Point (1:250,000)	—
New York	Cortland	ASCS	ASK-1P-48-49	1667′/″	5/20/59	Cortland (7½′)	1961
New York	Cortland	ASCS	ASK-1P-86-87	1667′/″	5/20/53	Homer (7½′)	1961
New York	Tompkins	ASCS	ARU-3N-184-185	1667′/″	5/20/53	Speedsville (7½′)	1965
Rhode Island	Washington	USGS*	Rhode Island 1-ABCD	1667′/″	11/12/51	Carolina, Kingston (7½′)	1939
Wisconsin	Adams	ASCS	AJA-1T-135-136	1667′/″	6/29/57	Briggsville (15′)	1920†
Wisconsin	Rock	ASCS	XA-1R-100-101	1667′/″	5/16/56	Avalon (7½′)	1917†
Wisconsin	Rock	ASCS	XA-2R-65-66	1667′/″	5/18/56	Janesville, Shopiere (15′)	1917†

*From *Selected Aerial Photographs of Geologic Features in the United States,* U.S. Geological Survey Prof. Paper No. 590.
†Out of print.

formations with little difficulty. Local materials are suitable for the highway foundation and are sometimes susceptible to volume changes where water-bearing silt lenses are found. Few drainage structures are needed because of the lack of surface runoff. Some alignment difficulties may be encountered in heavily pitted outwash deposits, and such situations require more grading. In New England towns where highways have been constructed on outwash deposits, serious problems involving infiltration of salt runoff into the groundwater from highway winter salting have arisen, and many of these towns are attempting either to stop highway salting or to provide suitable trapment and filtering of the salt runoff before the usefulness of their groundwater resources is destroyed. The town of Burlington, Massachusetts, has recently barred the use of salt on roads within the town area because of an increase over the past several years in the chloride content in the municipal water supply. This increase has been traced to salt runoff contamination of the glacial outwash soil aquifer. A more serious source of chloride contamination is the storage areas where the salt is stockpiled. Heavy runoff following storms can carry vast amounts of dissolved chlorides into a drainage network which may feed a downstream soil aquifer.

Lake Beds

Introduction

Glacial lake beds are deep-water lacustrine deposits formed from glacier-caused lakes. Other lake bed deposits not related to glaciation are discussed in a separate section under waterlaid landforms.

Glacial lakes are formed either close to the ice sheet or many miles outside the area of glaciation. In the one case, temporary lakes are formed between the melting glacial ice mass and glacial moraines or some other adjacent higher ground. After the ice melts these lakes find or erode drainage channels and are emptied, leaving bottom and near-shoreline soil deposits as evidence of the lake's existence. The other type of lake is formed many miles from the actual glacier. Here, valleys draining meltwater are dammed by a geological structure or by a landslide. Later, erosion exposes the lacustrine soil deposits.

There are three types of glacial lacustrine deposits. All three types are the result of sediments flowing into the lake, encountering lower velocities, and being deposited. The first group includes shoreline-associated deposits, such as deltas and beach ridges. These landforms, which consist of sands and gravels, are related primarily to water or fluvial processes and are therefore discussed in Chapter 10. A second type of formation occurs adjacent to the shore edge. Here, fine sands and coarse silts are deposited and are classified as coarse lake bed sediments or sandy lake bed deposits. The third type of landform is created at and around the lake center, where water velocities are very low and allow deposition of the finest silts and clays. These formations, classified as fine lake bed sediments or clay lake bed deposits, are the most common and cover large areas, generally hundreds of square miles, and contain materials often 100 or even 200 feet in depth.

Figure 8.43. Pleistocene, varved silt and clay in Ferry County, Washington. Each horizontal layer of silt and clay represents 1 year's accumulation of materials. Photographs by F. O. Jones, U.S. Geological Survey.

Clay lake beds contain horizontally bedded layers of silts and clays, known as *varves*, which represent yearly accumulations of bottom sediments. During the summer months, when the water is blown by wind and storms and circulated by vertical currents, its slightly higher water velocities allow only silt-sized particles to be deposited; finer clays are held in suspension until winter when the surface is ice-covered and the internal circulation and velocity of the lake decrease (Figure 8.43). Typically, a yearly varve is ½ to ¼ inch thick; the summer silt is at the bottom, the winter clay at the top. Ages of glacial lakes are determined by counting these yearly

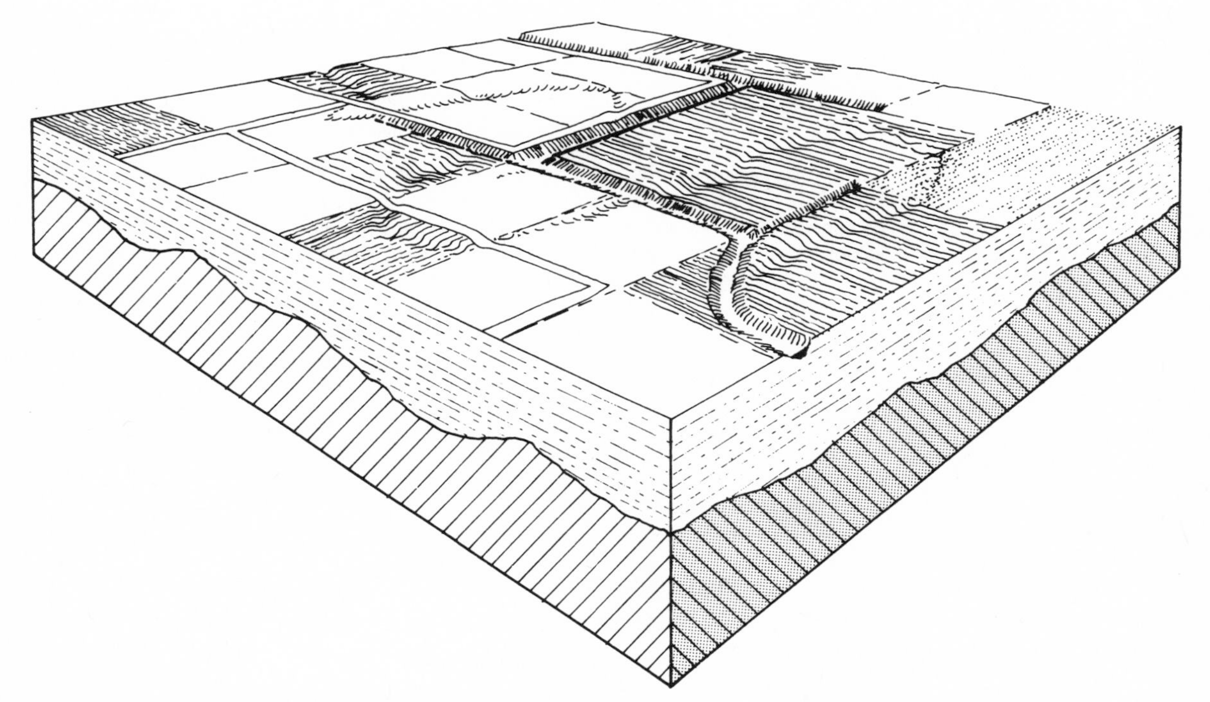

Figure 8.44. Glacial lake bed regions provide very monotonous, repetitious landscapes; they are extremely flat over broad areas and have no variation in land use pattern. There is little if any spatial variation or screening, owing to the lack of topographic features and vegetative cover. High tower elements have a high visual impact, as do city skylines; conversely, low, broad structures are visually absorbed. Shoreline deposits, such as beach ridges, provide some topographic variation and can act as elevated platforms for minor, localized views. Shoreline deposits are typically developed as town or village centers since they provide relatively well-drained sites and some view of the surrounding landscape.

varves. When a sufficient number of lakes occurs along the north-south line of a glaciated region, varve samples can be correlated to indicate accurately the rate of ice withdrawal. Such a study was undertaken in New England by Ernest Antevs, who determined that ice covered New England for approximately 5000 years before it withdrew to the Canadian border at an average rate of 0.05 miles (265 feet) per year. Similar studies in Scandinavia by Gerard De Geer indicated that the ice retreat there took place at a rate of approximately 7 miles per 100 years, or 0.07 miles (370 feet) per year. In both instances the retreat of the glacial ice has been shown to have been very rapid on the geological time scale.

Some glacial lake beds do not occur in valleys or lowland areas but can be found elevated several hundred feet above the lowlands. These temporary lakes were dammed by moraines or ice sheets which eventually melted or were eroded, allowing the water to drain. Lake beds of this nature are common throughout New York and the New England states of Vermont, New Hampshire, and Massachusetts.

Interpretation of Pattern Elements

Glacial lake beds are easily identified by their indistinct, transitional boundaries, broad, level plains, drab, dark-gray tones, intense cultivation, and artificial drainage structures (Figure 9.44). Coarse lake bed sediments, consisting of fine sands and coarse silts represent shallow or near-shore deposits and appear better drained and show lighter tones and fewer drainage features; deep, fine-textured deposits containing fine silts and clays have dark drab tones and many artificial drainage structures. The photographic interpretation key is organized in relation to the features associated with deep, fine-textured lake bed deposits; coarse lake bed sediments are discussed when their identifying patterns are significantly different.

Soil Characteristics

Most glacial lake bed deposits consist of varved, fine silts and clays. Surface soils typically contain silty clays, loams, and moderate to high amounts of organic matter; subsoils are typically plastic clays and clay loam. The water table is generally within 2 to 4 feet of the surface, except where the deposit is elevated, facilitating drainage.

U. S. Department of Agriculture Classification

The U. S. Department of Agriculture commonly classifies surface soil horizons as silt loam, silty clay loam, silty clay, or even clay. The subsoil contains plastic silty clay and clay. Sandy lake beds obviously contain more sands, whose parent material is commonly silt loam or sandy loam.

Table 8.17. Glacial landforms: Lake beds (humid and arid)

Topography	*Drainage*	*Tone*	*Vegetation and land use*
Flat, broad plains	Artificial	Dull gray	Cultivated
Fine silts and clays form flat, broad plains unless dissected along the edge of a major river. Coarser silts and fine sands may cause a slightly undulating landscape, reflecting wind erosion and deposition of sands.	 The flat topography and high water table do not allow the establishment of an integrated drainage system. Rainfall is temporarily ponded and then evaporates or is slowly absorbed by the soils. Artificial drainage features and a few broadly meandering streams may be present.	The fine-textured silts and clays are indicated by drab, dark-gray tones. The coarser silts and fine sands are indicated by lighter tones and some mottling. *Gullies* None No gullies are apparent unless the lake bed is dissected, in which case ideal sag-and-swale gullies indicating cohesive soils are seen.	Most of the land is intensively cultivated, reflecting the moist soils and high productivity. Some vegetative cover may occur in swamp areas where the water table is at the surface, and bare spots may be observed in dry sandy areas. Many artificial drainage structures are apparent, along with dead furrows resulting from attempts to lower the water table.

Unified and AASHO Classifications

Symbols indicating fine-textured silt-clay lake beds are typically CH, MH-CH, CL-CH, and occasionally CL and ML-CL. Typical AASHO categories include A-7, A-5, and occasionally A-4 or A-6. Sandy lake beds are commonly classified by the Unified system as SM, ML, ML-MH, or ML-CL; the AASHO designations are A-4, A-5, or A-6, and occasionally A-7.

Water Table

Because of the flat topography and the impervious nature of the soils, the water table occurs within 2 to 4 feet of the surface except where elevated. In lowland deposits, however, the seasonal high water table is practically on the surface over the entire landform. Sandy lake beds are better drained and may not have seasonal high water tables as near the surface.

Drainage

Fine-textured clay lake beds are poorly drained, and the soil materials are impervious. Sandy lake beds are better drained and do not require as many dead furrows and drainage ditches to provide sufficient drainage for agriculture.

Soil Depth to Bedrock

Fine-textured clay materials indicate the past location of the lake where waters were deep and formed deep soil layers up to 100 or even 200 feet thick. Bedrock is therefore not commonly encountered in these areas during construction operations. The shallower deposits of sandy lake beds may be directly underlaid by bedrock, but investigations are needed to determine its presence. Most lake bed deposits are underlaid by till.

Issues of Site Development

Sewage Disposal

The high water table and the impervious soils do not present suitable conditions for the use of septic tank leaching fields, and potential flooding constrains the use of sewage lagoons. Many farmers in these areas use cesspools for treatment of household wastes, but these require periodic pumping. Intensive development and urban areas require treatment facilities if the quality of the surface water resources is to be maintained. The sandy, coarse-textured deposits may offer better suitabilities for on-site disposal systems if the depth to water table is sufficient.

Solid Waste Disposal

Fine-textured lake bed deposits can supply suitable sites for sanitary landfill operations, but only if the depth to water table is 10 to 15 feet below the surface, insuring protection against groundwater contamination. However, it is difficult to find such depths to water table occurring in these landforms unless they are elevated. If conditions are suitable, the fine-textured soils will provide suitable cover over the landfill and will prevent penetration of rainwater or seepage.

Figure 8.45. (A) Glacial clay lake bed in Clay County, Minnesota, is characterized by a very flat topography, artificial drainage, and a broad, meandering stream. Photographs by the Agricultural Stabilization Conservation Service, BXR-4V-3,4, 1:20,000 (1667′/″), November 12, 1958.

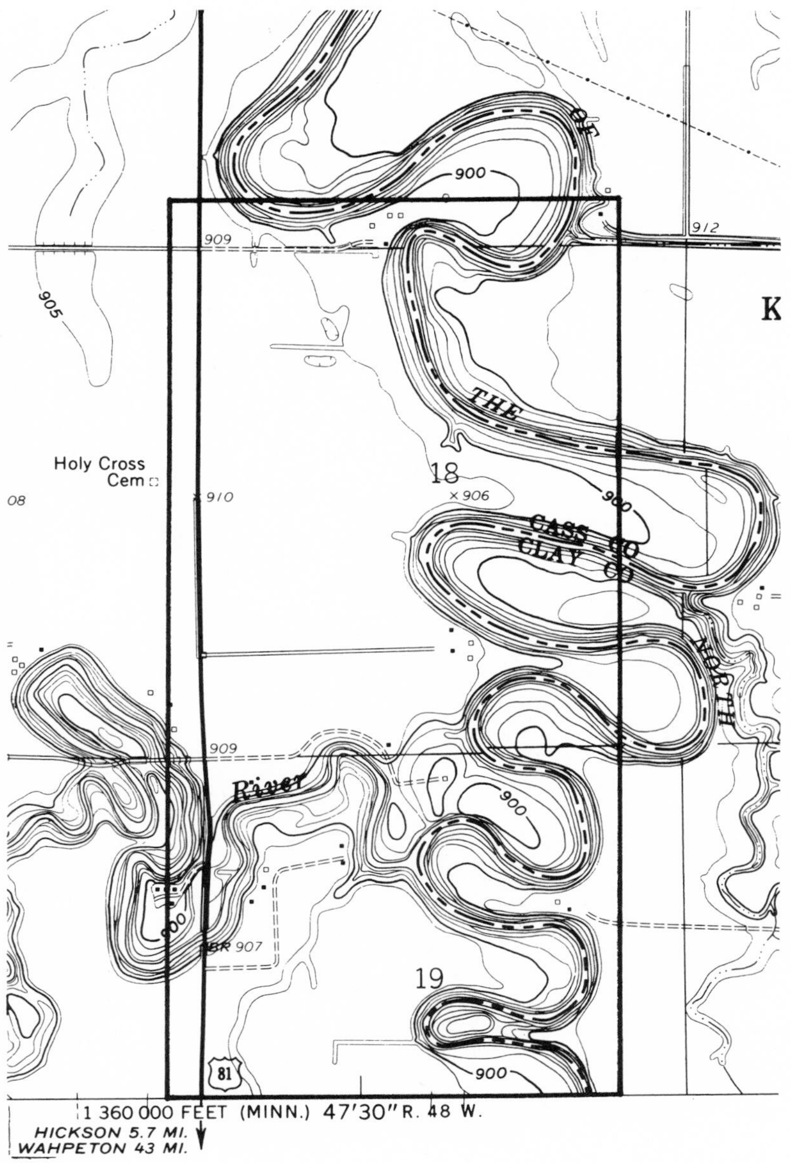

Figure 8.45. (B) U.S. Geological Survey Quadrangle: Fargo, South Dakota (7½′).

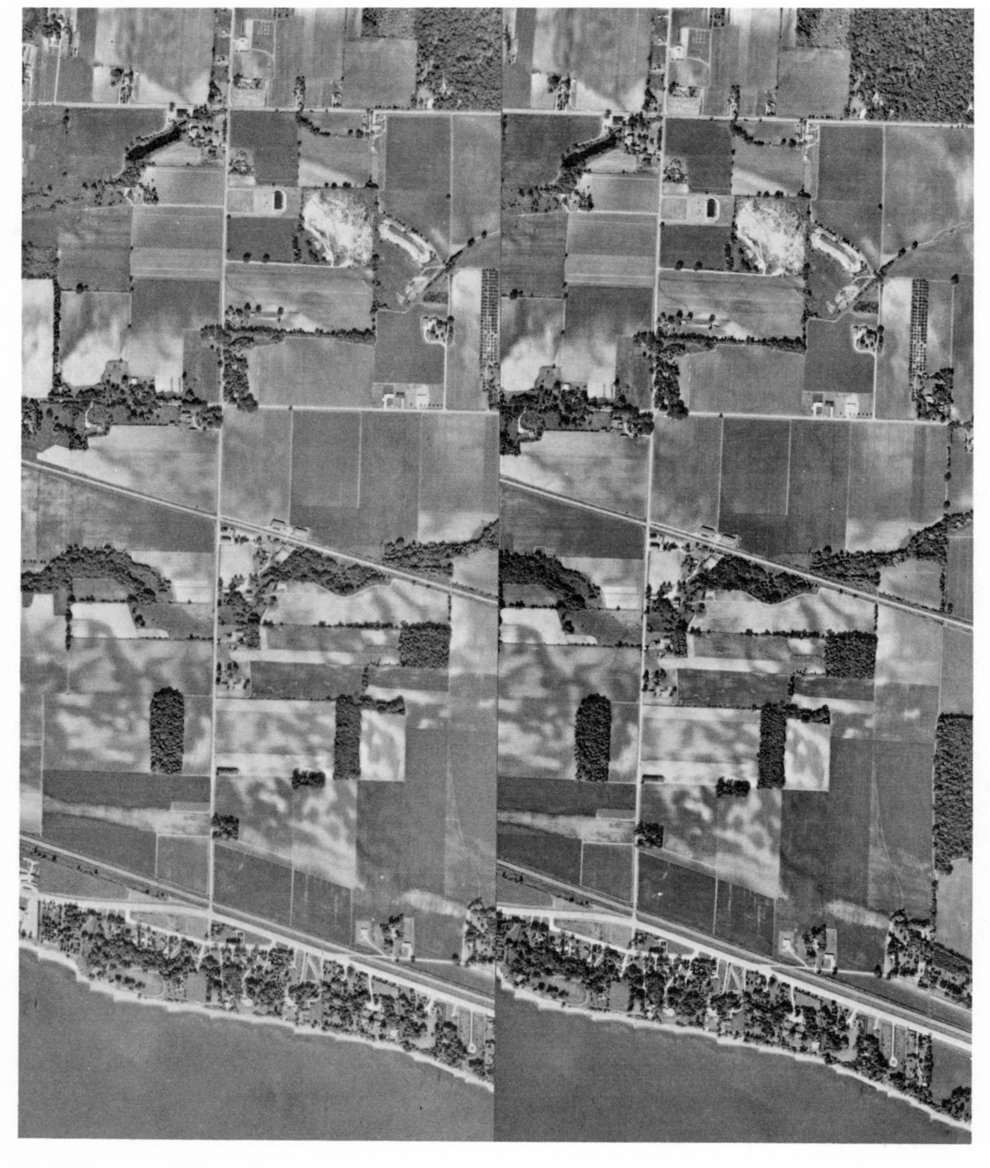

Figure 8.46. Sandy lake bed deposit in Racine County, Wisconsin, indicated by the mottled tones. Photographs by the Agricultural Stabilization Conservation Service, XC-1DD-41,42, 1:20,000 (1667'/"), June 6, 1963.

Trenching

Trenching problems occur in lake bed deposits if the water table is near the ground surface; drainage by electro-osmosis or vacuum pumps is required in these cases. The soils are difficult to handle because of their wet, sticky, plastic nature, and generally costs are 1.5 to 2.5 times those associated with removal of the same amount of deep, dry, moderately cohesive soils. Liquid-carrying pipelines covering long distances may involve problems in acquiring sufficient slope for gravitational flow, since the landscape is flat over large regions, and pumping stations may be necessary in order to raise line elevations periodically to provide sufficient flow. Lake bed formations in arid climates do not have shallow water tables and do not present many difficulties in trenching. Sandy lake beds may involve trenching problems similar to those encountered in sandy outwash deposits; these are discussed under "Trenching," in the section on "Outwash."

Excavation and Grading

The need for grading to create large, flat sites is at a minimum in lowland, lake bed formations, owing to the natural flatness of this landform. However, the high water table and sticky, plastic nature of the fine-textured clay materials are difficult to work with; the soils are more easily handled in arid climates. Deep excavations require drainage by electro-osmosis or vacuum pumps, and the depth of the cut in soft clays is controlled by failure of the sides of the excavation. Pit excavations in soft clays require the use of adequate sheeting and bracing to resist lateral pressures, or reduction of sideslopes from one on three to one on five. Clays in these formations create many construction problems related to their specific properties, and detailed site investigations and

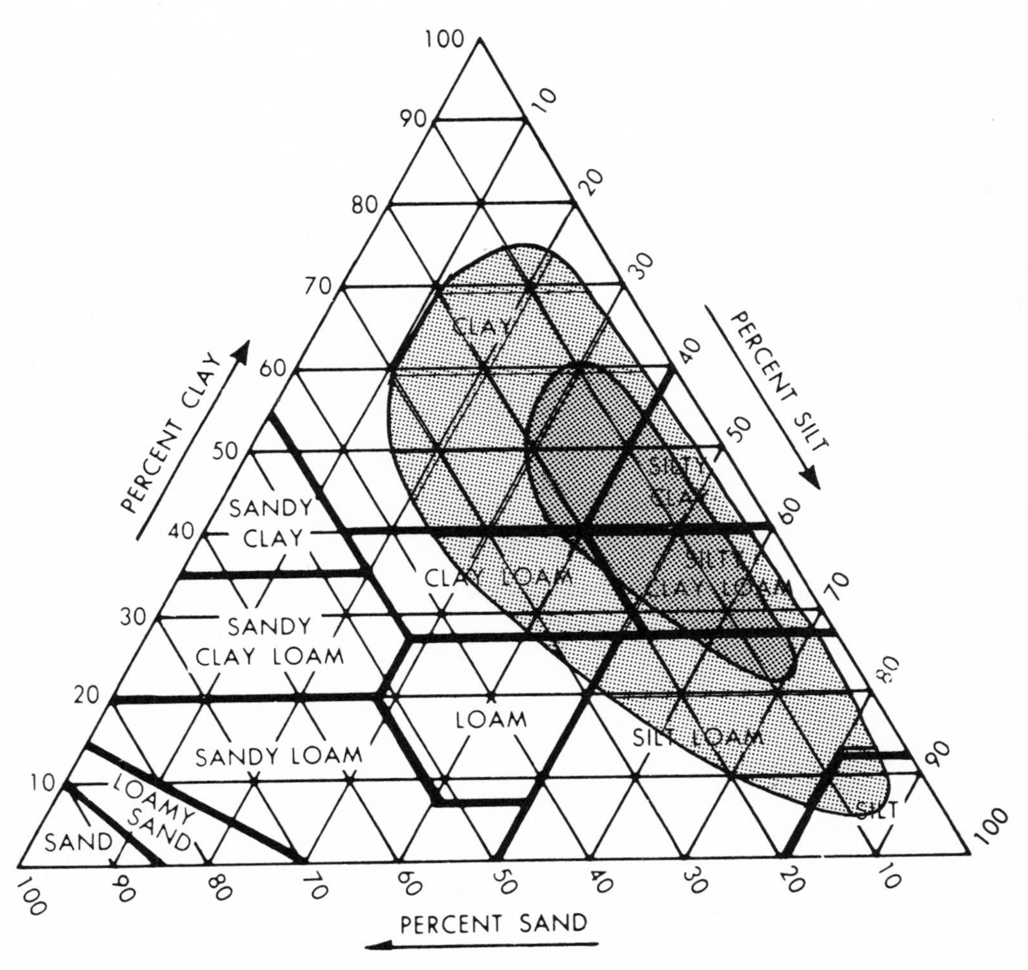

Figure 8.47. The U.S. Department of Agriculture soil texture triangle indicates silty clay and silty clay loam as dominant materials for silt-clay glacial lake beds. Sandy lake beds are not shown but contain sandy loam, loam, and silt loam.

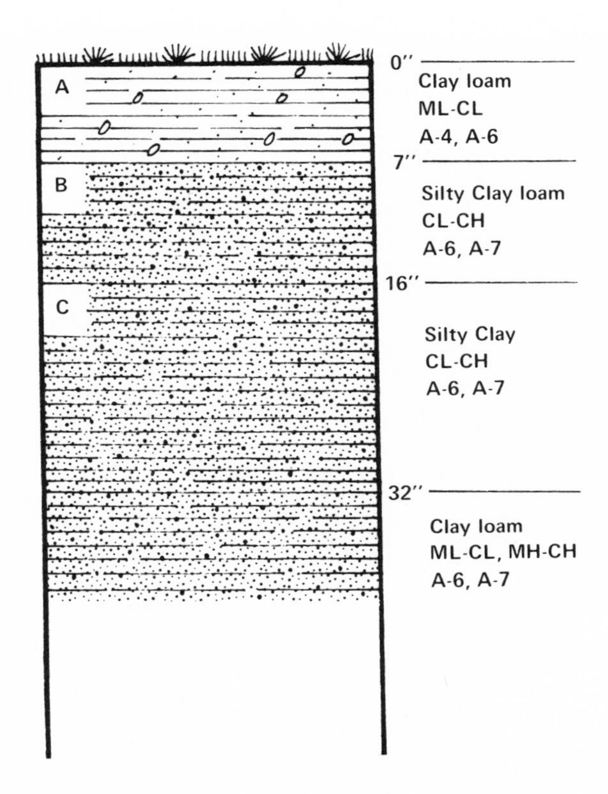

Figure 8.48. A typical soil profile of silt-clay lake bed materials.

borings are required in order to determine the most appropriate construction procedures.

Construction Materials (General Suitabilities)

Topsoil: Good to Fair. Depending upon their specific organic, moisture, and textural content, the surface soils generally provide rather productive agricultural soils.

Sand: Not Suitable. Sands are not common in deep lake bed deposits, although associated shoreline features such as beach ridges may provide a supply.

Gravel: Not Suitable. Gravels are not found in deep lake bed deposits, but associated shoreline features such as beach ridges may be a suitable source.

Aggregate: Not Suitable. Aggregate materials are not found within deep lake bed formations, but associated shoreline features such as beach ridges may provide a suitable source.

Surfacing: Fair. Fine-textured lake beds may provide suitable materials for surfacing if these are mixed with coarser-textured soils. Use of the deposits in their natural state results in dust when dry, and slippery surfaces when wet.

Table 8.18. Glacial lake beds: aerial and cartographic references

State	County	Photo Agency	Photograph no.	Scale	Date	Corresponding quadrangle and geologic maps from USGS	Corresponding soil reports from SCS
Michigan	Lenawee	ASCS	XV-2DD-211-212	1667'/"	1963	Saline (15')	1961
Minnesota	Clay	ASCS	BXR-IV-121-122	1667'/"	7/18/58	—	—
Minnesota	Clay	ASCS	BXR-4V-3-4	1667'/"	11/12/58	Fargo South (7½')	—
New York	Niagara	ASCS	ARE-16-51-52	1667'/"	9/6/38	—	1947
New York	Monroe	ASCS	ARK-3GG-87-89	1667'/"	7/1/66	Hamlin (7½')	1938
New York	Monroe	ASCS	ARK-3GG-53-55	1667'/"	7/1/66	Brockport (7½')	1938
North Dakota	Grand Forks	ASCS	ZZ-3CC-275-276	1667'/"	7/12/62	—	1904†
North Dakota	Grand Forks	USGS*	North Dakota 4-AB	5000'/"	9/10/52	Larimore (15'); I-331	1904†
North Dakota	Grand Forks	USGS*	North Dakota 5-AB	5000'/"	8/1/52	Larimore (15'); I-331	1904†
North Dakota	Grand Forks	USGS*	North Dakota 6-ABC	1667'/"	6/24/62	Emerado (15'); I-331	1904†
Ohio	Lorain	USGS*	Ohio 1-ABC	5000'/"	5/16/60	Avon, N. Olmsted (7½'), 1-316	—
Ohio	Lucas	ASCS	BUG-4-47-48	1667'/"	5/23/40	—	1943
Wisconsin	Racine	ASCS	XC-1DD-41-42	1667'/"	6/24/63	Racine South (7½')	1970

*From *Selected Aerial Photographs of Geologic Features in the United States,* U.S. Geological Survey Prof. Paper No. 590.
†Out of print.

Borrow: Poor. Materials contained in lake bed deposits are fine-textured, plastic, and too wet to provide suitable borrow. Moisture contents are extremely difficult to control even when the soils are used dry and compacted with sheepsfoot rollers. More suitable materials may be available in nearby beach ridge formations.

Building Stone: Not Suitable

Mass Wasting and Landslide Susceptibility

Landslides do not normally occur in lake bed formations under their natural, flat topographic conditions. However, if the lake bed is either dissected, exposed, and eroded by an ocean, lake, or stream; or vertically cut in excavations; unstable conditions may result. Slumps and slides in clay materials are often associated with a rapid downdraw of water from reservoirs and streams. In such cases the normally stable, sloping stream bank is left, still saturated with groundwater, at the original higher elevation and places heavy pressure (shear stress) on the slope surface; if the available shearing resistance is insufficient, sliding will occur.

Groundwater Supply

Fine-textured lake bed deposits are very impervious to groundwater flow and therefore supply very low yields. Even the forces of pumping are exceeded by the soil's ability to retain the water. Occasional coarse-textured deposits may be found that can supply suitable groundwater resources. For example, buried glacial eskers or kames or lacustrine beach ridges may act as high-yield soil aquifers; these have a high municipal value where they occur. Other water-bearing materials, such as soil aquifers or water-bearing rock, may also be found beneath the lake bed formation. Sandy lake beds, similar to sandy outwash plains, may provide a high-yield groundwater supply, depending upon the specific textures and interbedded sequences of the contained materials.

Pond or Lake Construction

Reservoir areas within fine-textured lake bed formations have no seepage difficulties, but some grading is usually necessary in constructing the impoundment basin since the topography is naturally too flat. Siltation presents a major high-maintenance problem, because of the easily eroded fine silts and clays that occur throughout such regions. Sediment catchment basins are rarely effective since they trap only those particles that settle quickly, leaving suspended particles to settle in the deep-water portion of the lake where lower velocities occur. The best control is to keep the entire watershed, especially the first-order drainage system (where channelized flow first begins), well covered with dense ground cover vegetation.

Foundations

Fine-textured lake bed soils present limitations for the different foundation types because of their

Figure 8.49. Landslides in varved clay materials along the Franklin D. Roosevelt Lake in Washington. Photograph by F. O. Jones, U.S. Geological Survey.

susceptibility to volume change and low load-bearing capabilities. Typical foundations for small structures are footings on compacted fill or the use of rafts on gravel pads. Large structural loads typically use point-bearing piles to transfer the applied load to more stable underlying materials. In some areas where the bearing capacities are high, footings or rafts may be feasible. In all cases detailed engineering analysis is needed early in the planning process to determine the properties of the clay materials at the site under the proposed structure loads.

Highway Construction

Highway construction through fine-textured lake bed formations necessitates the use of granular borrow materials to provide a suitable foundation under base courses or bituminous surfaces. A-1 or A-2 materials are needed above the moist lake bed parent materials to provide sufficient insulation to alleviate frost heave potential and the effect of volume changes resulting from moisture fluctuation. The sides of roads need ditching to encourage drainage and require maintenance to combat siltation. Alignments are not difficult and typically follow straight lines, not being constrained by topographic features and/or major drainage dissection.

CHAPTER 9

EOLIAN (WINDLAID) LANDFORMS

INTRODUCTION

The effects of the transportation of soil grains by eolian (wind) action are very similar to those of the transportation of materials by stream (fluvial) action. In eolian transportation rock particles are carried in the atmosphere, either close to the surface of the ground in a fashion similar to stream bedloads, or higher above the surface of the ground in a fashion similar to suspended stream loads. The lower layer, or bedload, consists of sands which scatter and bounce along the surface, perhaps never rising more than 4 feet above the surface. The upper layer, or suspended load, contains fine-grained, silt-sized particles which are carried for great distances at high altitudes.

Transportation of Sand Grains (Bedload)

Movement of sand grains is initiated when wind velocities are sufficient to start the rolling of grains across the ground surface. This continues until some of the grains hit other grains, bouncing them into the air. The grains are thus elevated into zones of higher air velocity, and this has the effect of moving them further and faster, bouncing them off still other sand grains and scattering other particles into the higher-velocity zone (Figure 9.1). Eventually, the ground is covered with sand particles jumping, bouncing, and scattering along the surface, and this process continues until wind velocities decrease. Sand grains transported by air movement are all remarkably close to the same size (0.3 to 0.15 millimeter) and require an effective wind speed of only 11 miles per hour to initiate movement. Smooth surfaces facilitate movement of particles; rough surfaces, which have many depressions and eddies, effectively lower wind velocities near the ground surface.

Transportation of Fine-Grained Particles (Suspended Load)

Transportation of fine-grained particles is a more complex process not directly initiated by the wind itself. Silt and clay particles are much smaller, more uniform, and more tightly packed than sand grains; they are part of a smooth soil layer which does not extend above a layer of dead air, very close to the ground, created by sand particles and other obstructions. Sand particles, however, are large and widely scattered and have much of their mass above the dead-air layer, which is only about 1/30 of their diameter. Transportation of the fine-grained particles is initiated by the movement of the sand grains, which dislodges the finer particles by scattering them out of the dead-air layer. Therefore, even though silt and clay particles are much smaller than sand particles, the wind velocities necessary for their erosion are the same as those associated with the movement of sand-sized particles. Once the fine particles enter the zone of air movement, they are easily suspended in vertical eddies. The fine particles are carried greater distances than sands since they do not settle until air velocities are

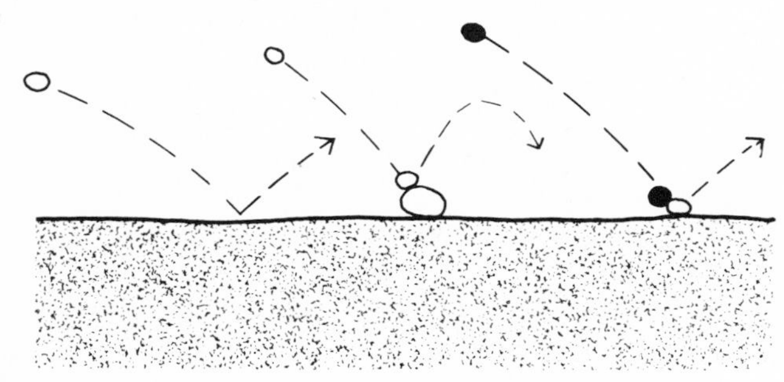

Figure 9.1. Sand particles move with a bouncing action which helps to scatter other particles from the ground surface.

less than 100 centimeters per second (compared to approximately 500 centimeters per second for sand).

Load-Carrying Ability of the Atmosphere

The amount of soil material that could be carried by the atmosphere if it were fully loaded would exceed the carrying capacity of the hydrological surface drainage system of the world by several thousand times. However, because of the wide spread protective vegetative cover on soils and the difficulties associated with initiating eolian erosion of fine-grained materials, only about 1% of this potential capacity is in effective operation.

Rising air having a velocity of 3 feet per second carries soil particles 0.1 millimeter in diameter; air having a velocity of 40 feet per second lifts particles larger than 1 millimeter. Measurements of suspended dust particles have shown that with sufficient wind velocities 1 cubic mile of air can support and carry over 4000 tons of soil particles. A typical duststorm 300 to 400 miles in diameter can carry 100,000,000 tons of dust—which could form a pile 100 feet high and 2 miles in diameter (Lobeck, 1939). The great dust storms of the 1930s are estimated to have exceeded these conservative figures and to have carried up to 166,000 tons of dust per cubic mile in clouds above 12,000 feet in altitude.

Volcanic eruptions also illustrate the tremendous sediment-carrying ability of the atmosphere. The eruption of Krakatoa in the Dutch East Indies in 1883 placed sufficient volcanic ash in the atmosphere to form deposits 2 to 4 inches thick as far as 1000 miles downwind from the eruption. In 1947 the volcano Hekla in Iceland erupted and within 1 hour deposited ash over 1500 square kilometers around the crater; 51 hours later finer ash materials were spread 3000 miles to Finland.

Wind Erosion

Erosion by wind occurs in two forms: the first is *deflation,* the removal of loose soil particles; the second is *abrasion,* defined as the cutting action of wind-borne particles. Of the two, the process of deflation is most conspicuous and provides most of the sediments carried by the wind.

Deflation is common in arid and semiarid climates where there is little vegetation cover to protect the soil surface. Sands and finer-textured silts and clays are removed by the winds, leaving larger pebbles and rock fragments. As this process continues, the ground surface is lowered irregularly until it develops a surface armor of larger rock fragments which protects it from further deflation.

Abrasion of rock surfaces by wind takes place when soil particles carried by the wind strike larger rock surfaces. Deflation armor shows many polished surfaces caused by abrasion by sand grains traveling across the ground. Other rocks develop ventifacts, that is, flat sides with angular edges where the rock was exposed to wind-borne particles (Figure 9.2).

Most of the soil materials significant in the formation of windlaid landforms originate from the process of deflation. The formations resulting from abrasion indicate the presence of wind erosion, but the soil particles removed do not contribute much to the total particle load of the wind.

Eolian (Windlaid) Landforms

Eolian deposits include (1) sand dunes, which occur near the source of the material, and (2) loess, or silt deposits, which are often carried great distances by the wind and which can cover large areas.

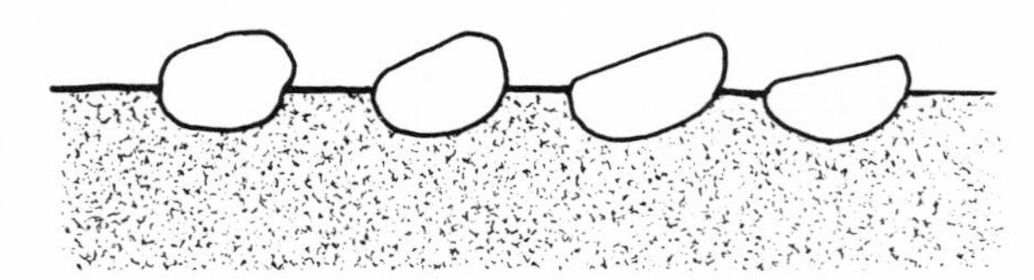

Figure 9.2. Ventifacts are formed from pebbles and stones of the deflation armor; they are exposed to blowing sand grains which slowly abrade the windward surface. Successive stages in the formation of a ventifacts are shown.

Sand Dunes

Windlaid sand formations occur near the sources of the sand, which may be old or present coastal shorelines, sandy lake beds, sandy glacial outwash plains, sandy river flood plains, or sandstone residual soils. Sand dunes are eolian (windlaid) deposits of sandy particles which have a topographic definition. The classification of sand dunes reflects their differing topographic shapes which result from variations in wind direction, amounts of available sand, and the wind velocity when they were formed. Sand dune formations include beach, barchan, transverse, U-shaped (wind-drift), and longitudinal. The sand particles comprising these forms are almost all the same size, varying from 0.3 to 0.15 millimeter.

Loess

Windlaid silt deposits known as loess are found at great distances from their sources which may be glacial lacustrine lake beds, outwash plains, desert areas, flood plains, or even residual soils of shale or limestone. Loess deposits cover extensive areas of hundreds of square miles and generally decrease in thickness as the distance from the source increases. Loess materials are composed of unconsolidated or weakly consolidated, silt-sized particles 0.01 to 0.05 millimeter in size. The mineral content varies according to the source of the parent material. Most loess particles are angular and are typically held together by calcareous cement or binder. Loess is a light-colored material, since it is very porous and well drained vertically.

Distribution

Eolian landforms are widely distributéd through all parts of the world but are especially common adjacent to glaciation zones and large flood plains, in arid climates, and along coastal areas. Major windlaid landforms are located as follows.

North America

United States. Belts of sand dunes are found along the coasts of the Atlantic and Pacific Oceans and the Great Lakes. Extensive inland dune formations are found in the Great Basin in Nevada, the deserts of southern California, the San Luis Valley of southern Colorado, and the White Sands National Monument near Alamogordo, New Mexico, which covers approximately 500 square miles. In north central Nebraska, in the Sand Hill region, there are approximately 18,000 square miles of stabilized sand dunes. Small dune areas occur near Saratoga, New York, north of Albany, New York, and in the plains northwest of Mt. McKinley, Alaska. Large river basins in semiarid climates typically have well-developed dune formations, and in the United States these are found adjacent to the Arkansas, Columbia, and Snake Rivers in the Northwest.

Loess deposits are found adjacent to the large rivers of the Mississippi Valley and cover significant portions of the states of Iowa, Illinois, northern Missouri, southwestern Wisconsin, western Tennessee, and western Mississippi. The loess deposits of southern Nebraska and northern Kansas originated in the semiarid regions of Colorado. In southeastern Washington and southeastern Idaho, the Palouse deposits form thick, rolling hillsides of transported loess soils.

Canada. Sand dune deposits in Canada are few and scattered. Loess is found in small, localized deposits through southern Saskatchewan.

South and Central America

The most prominent sand dune formations occur in the central basins of Peru and Chile. Extensive deposits of loess cover most of northern Argentina, southern Uruguay, and southern Paraguay.

Africa

Extensive areas in Africa are covered with sand dunes; in the Sahara these account for one-ninth of the total desert area, or 300,000 square miles.

Small areas of loess cover are found along the coastal sections of Morocco, Algeria, Libya, and the United Arab Republic. Scattered small deposits occur across north central Africa from Senegal to Ethiopia.

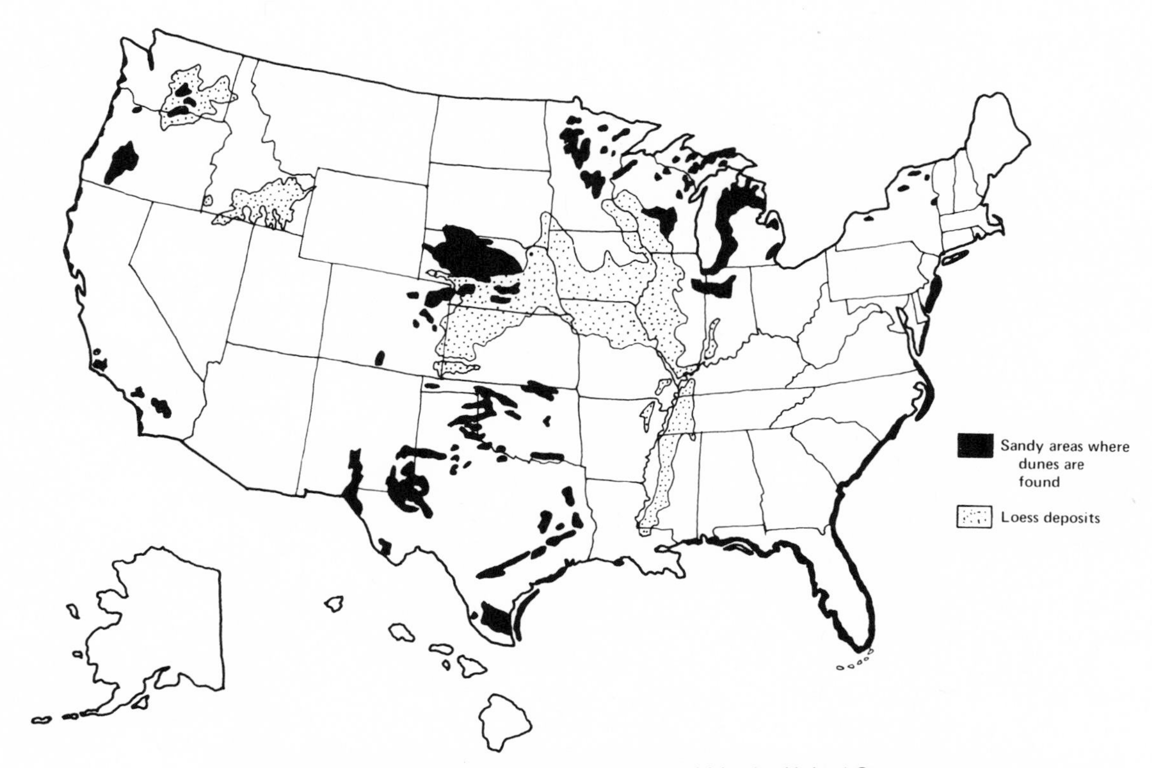

Figure 9.3. Distribution of eolian landforms within the United States. (After "National Atlas," U.S. Geological Survey, Washington, D.C., 1971.)

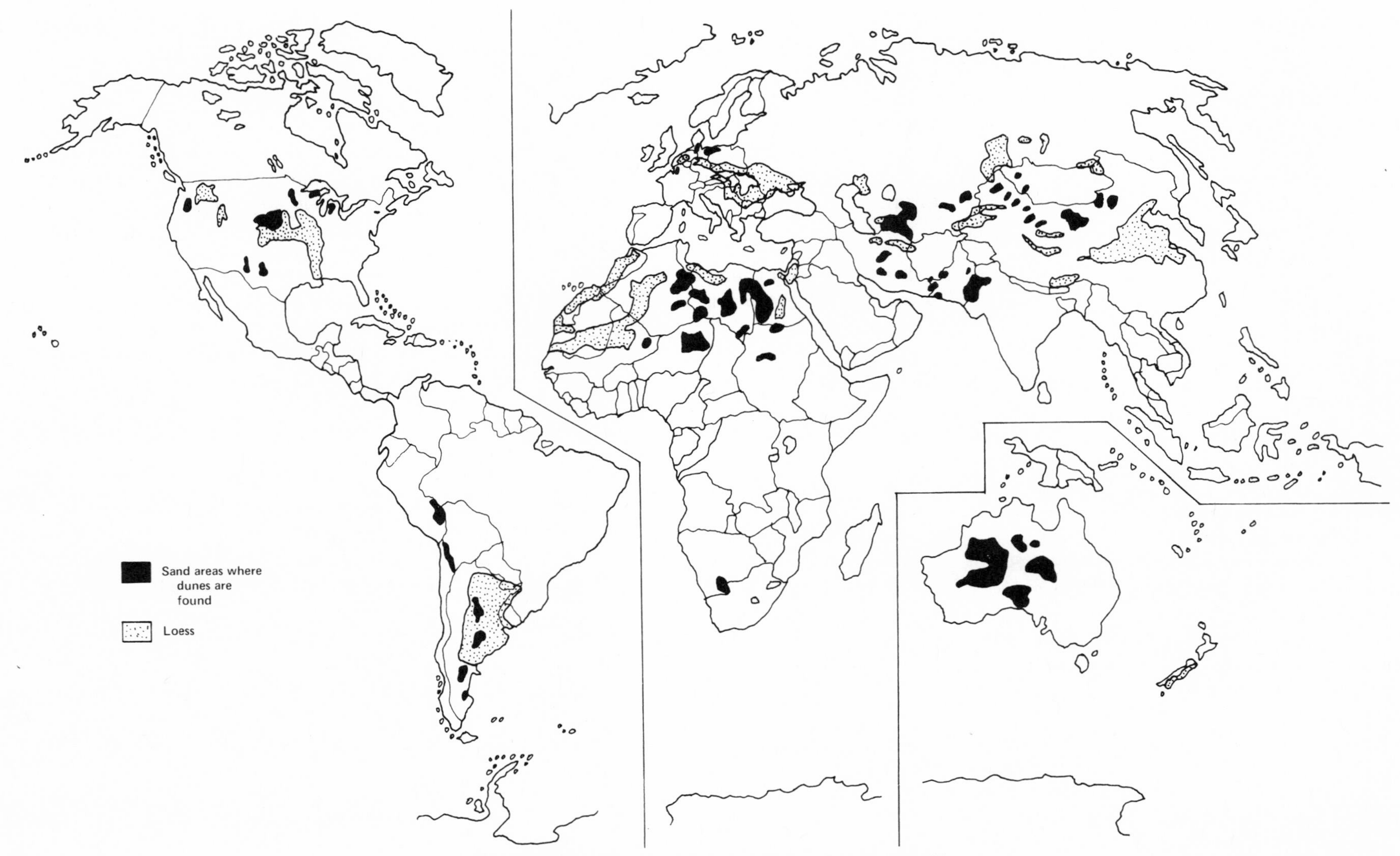

Figure 9.4. Distribution of eolian landforms around the world.

Europe

Sand dunes are dominant features along the French coast of the Bay of Biscay, and the coasts of Belgium, the Netherlands, Denmark, and eastern Russia along the Baltic Sea. Dune formations also occur along the major rivers in southern France, Spain, and southern Russia north of the Black Sea.

A large east-west belt of loess material is found just south of the limit of glaciation extending through northeastern France, Belgium, Poland, Czechoslovakia, Rumania, and southern Russia north of the Black Sea.

Table 9.1. Eolian landforms: summary chart

Landform	Topography	Drainage	Tone	Gullies	Vegetation and land use
Sand dunes					
Beach	Hummocky	None	Bright*	None	Barren or natural cover†
Transverse	Parallel ridges, perpendicular to wind	None	Bright*	None	Barren or natural cover†
Barchan	Crescent downwind	None	Bright*	None	Barren or natural cover†
Wind-drift	Crescent upwind	None	Bright*	None	Barren or natural cover†
Longitudinal	Parallel ridges, perpendicular to wind	None	Bright*	None	Barren or natural cover†
Star-shaped	Star-shaped hills	None	Bright*	None	Barren or natural cover†
Loess					
Deep, Young	Parallel, smooth hills	Pinnate to dendritic, fine	Uniform light	Box-shaped	Cultivated
Dissected (Old)	Rugged, steep hills	Pinnate, fine	Lights to darks	Box-shaped	Natural cover

*Bright tones are found where the soil is exposed; vegetation cover on stablizied dunes may mask tones.

†Stabilized dunes may exhibit grass or forest cover in humid climates, and grass cover in subhumid, semiarid climates. Forest cover in humid climates does not develop much undergrowth.

Asia

Sand dunes occur in many areas of Asia, including a central deposit running from Mongolia to northern India, the Gobi Desert, the region from Syria through Iran, and most of Saudi Arabia.

Major deposits of loess are found throughout the Hoang-Ho River Valley in China and along the northeastern edge of the Caspian Sea.

Australasia

Much of central Australia contains sand dune formations, including the Great Sandy, Gibson, Great Victoria, Tanami, and Simpson Deserts.

Thin loess deposits cover the central portion of Australia. Thicker deposits occur along the southeastern coast of South Island, New Zealand.

Pacific Region

Small, calcareous sand dunes of local origin are scattered through the Pacific Islands. No significant deposits of loess are found.

Caribbean Region

Small sand dunes, containing calcareous sands of local origin, are scattered through the Caribbean Islands. No significant deposits of loess are found.

REFERENCES

American Society of Photogrammetry, *Manual of Photo Interpretation,* American Society of Photogrammetry, Falls Church, Va., Chapter 5, 1960.

Bagnold, R. A., *The Physics of Blown Sand and Desert Dunes,* William Morrow, New York, 1941.

Bigarella, J. J., "Dune Sediments," *Bulletin of American Association of Petroleum Geologists,* Vol. 53, 1969.

Blackwelder, E., "Geomorphic Processes in the Desert," *California Division of Mines Bulletin,* No. 170, Chapter 5, pp. 11–20, 1954.

Byan, K., "Glacial versus Desert Origin of Loess," *American Journal of Science,* Vol. 143, 1945.

Chorley, R. J., ed., *Water, Earth and Man,* Methuen, London, 1969.

Clevenger, W. A., "Experiences with Loess as a Foundation Material," *Transportation,* American Society of Civil Engineers, 1958.

Cooper, W. S., "Coastal Sand Dunes of Oregon and Washington," *Geological Society of America Mem. 72,* 1958.

Cooper, W. S., "Coastal Dunes of California," *Geological Society of America Mem. 104,* 1967.

D'Appolonia, E., *Loose Sands: Their Compaction by Vibraflotation,* American Society of Testing Materials, Technical Publication No. 156, 1953.

Fenneman, N. M., *Physiography of the Western United States,* McGraw-Hill, New York, 1931.

Fenneman, N. M., *Physiography of the Eastern United States,* McGraw-Hill, New York, 1939.

Flint, R. F., *Glacial and Pleistocene Geology,* Wiley, New York, Chapter 10, 1957.

Flint, R. F., "Leaching of Carbonates in Glacial Drift and Loess as a Basis for Age Correlation," *Journal of Geology,* Vol. 57, pp. 297–303, 1949.

Holm, D. A., "Desert Geomorphology in the Arabian Peninsula," *Science,* Vol. 132, 1960.

Hooke, R. L., "Processes on Arid-Region Alluvial Forms," *Journal of Geology,* Vol. 75, 1967.

Leighton, M. M., and H. B. Williman, "Loess Formations of the Mississippi Valley," *Journal of Geology,* Vol. 58, pp. 599–623, 1950.

Lobeck, A. K., *Geomorphology, An Introduction to the Study of Landscapes,* McGraw-Hill, New York, Chapter 11, 1939.

Longwell, C. R., R. F. Flint, and J. E. Sanders, *Physical Geology,* Wiley, New York, Chapter 13, 1969.
Mabbutt, J. A., *Desert and Savana Landforms,* M.I.T. Press, Cambridge, Mass., 1970.
Mason, C. C., and R. L. Folk, "Differentiation of Beach, Dune and Aeolian Flat Environments by Size Analysis," *Journal of Sedimentary Petrology,* No. 28, 1958.
Matalucci, R. V., J. W. Shetton, and M. Abdelhady, "Grain Orientation in Vicksburg Loess," *Journal of Sedimentary Petrology,* Vol. 39, 1969.
Obruchev, V. A., "Loess Types and Their Origin," *American Journal of Science,* Vol. 243, pp. 256–262, 1945.
Page, L. R., and R. W. Chapman, "The Dust Fall of December 15-16, 1933," *American Journal of Science,* Ser. 5, Vol. 28, pp. 288–297, 1933.
Shelton, J. S., *Geology Illustrated,* W. H. Freeman, San Francisco, Chapter 17, 1966.
Shepard, F. P., and R. Young, "Distinguishing Between Beach and Dune Sands," *Journal of Sedimentary Petrology,* Vol. 31, 1961.
Smith, H.T.U., "Physical Effects of Pleistocene Climatic Changes in Nonglaciated Areas: Eolian Phenomena, Frost Action, and Stream Terracing," *Geological Society of America Bulletin,* Vol. 60, pp. 1485–1516, 1949.
Strahler, A. N., *Introduction to Physical Geography,* Wiley, New York, Chapter 24, 1965.
Thornbury, W. D., *Principles of Geomorphology,* Wiley, New York, Chapter 12, 1954.
Thorp, J., et al., *Pleistocene Eolian Deposits of the United States,* Geological Society of America, Boulder, Colo., map, 1952.
Turnbull, W. J., "Utility of Loess as a Construction Material," *Proceedings, 2nd International Conference on Soil Mechanics,* Rotterdam, 1948.

Sand Dunes

Introduction

A sand dune is a ridge or mound composed of predominantly sandy particles deposited by wind. The process of dune formation begins when sand grains are moved across the ground surface by air currents. If an obstruction, such as a small topographic irregularity or a clump of vegetation, is encountered, the air current is deflected and a downwind eddy is created. The eddy decreases the effective wind velocity, allowing the sand grains to be deposited. After a small mound of sand is formed, wind deflection is further increased and a larger downwind eddy results. The process continues, and eventually a large sand landform or sand dune is formed. Equilibrium of size is established when the sand erodes from the dune crest as rapidly as it is deposited; the height of the dune then becomes constant, while the dune mass moves slowly in a downwind direction (Figure 9.5).

Sand dunes conform to the dominant wind direction of their location; their windward slopes are gradual, and their lee slopes are steep, reflecting the angle of repose for sand grains (about 34 degrees). Dunes are typically 30 to 100 meters in elevation but some are as high as 200 meters (approximately 700 feet). Some dunes are formed from materials other than sand. Dunes in the White Sands National Monument consist of gypsum particles cemented into sand sizes; dunes throughout the Pacific and Caribbean islands contain calcareous materials.

If average wind velocities are low enough, vegetation may become established on sand dunes, protecting and stabilizing the sand particles. Eventually, the entire dune is covered with vegetation. Removal of the vegetative cover seriously jeopardizes the stability of the dune and allows the wind to erode the exposed sand.

The shapes assumed by dunes vary; they are controlled and affected by the volume of sand available, sand grain size, wind velocity and regularity, topography, and the amount of vegetative cover. Listed below are the common forms of dunes, classified by their topographic appearances.

Beach Dunes

Beach dunes occur along coastal areas. Their very hummocky form is the result of a sufficient supply of sand, variations in wind direction, and a wide variety of wind velocities. Further inland, beach dunes may be stabilized by vegetation.

Transverse Dunes

Transverse dunes are long and thin and form wavelike ridges perpendicular to the dominant wind direction. A constant wind develops a gentle windward slope and a steeper lee slope. A region of many transverse dunes is characterized by lines of parallel ridges, occasionally broken by wind blowouts. These dunes generally occur in areas that have abundant sand, little vegetation, and gentle winds from one major direction. These dunes are quickly stabilized in humid climates because of their low rate of movement.

Barchan Dunes

Barchan dunes develop crescent-shaped outlines, horns pointing downwind. The crest line is toward the inside of the crescent, indicating that the predominant wind direction is from the outside face. These dunes develop in areas of strong winds, limited sand supply, constant winds from one major direction, and flat topography.

Wind-Drift or U-Shaped Dunes

These dunes reflect the blowouts that result when other dune forms are subjected to very strong winds from a constant direction with little sand supply. Here the crest line is located toward the outside of the crescent, reflecting a wind direction from inside the U, or opposite the direction of the horns.

Longitudinal Dunes

Longitudinal dunes form long, narrow, parallel ridges of sand whose equal sideslopes indicate wind directions parallel to their long axes. These forms are a result of very strong winds, a

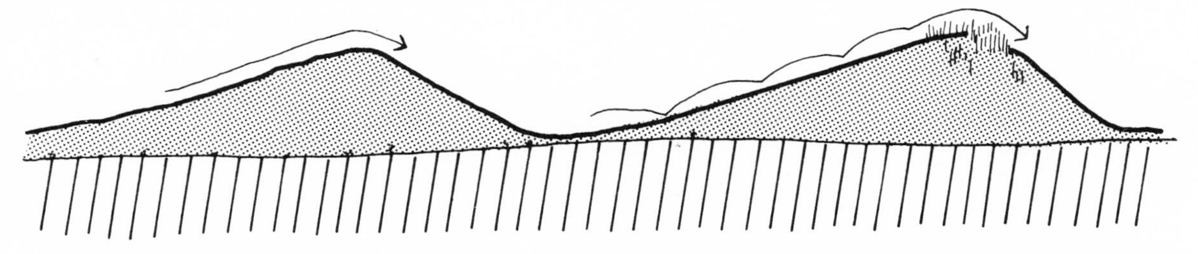

Figure 9.5. A sand dune in cross-section showing the unequal sideslopes reflecting the dominant wind direction.

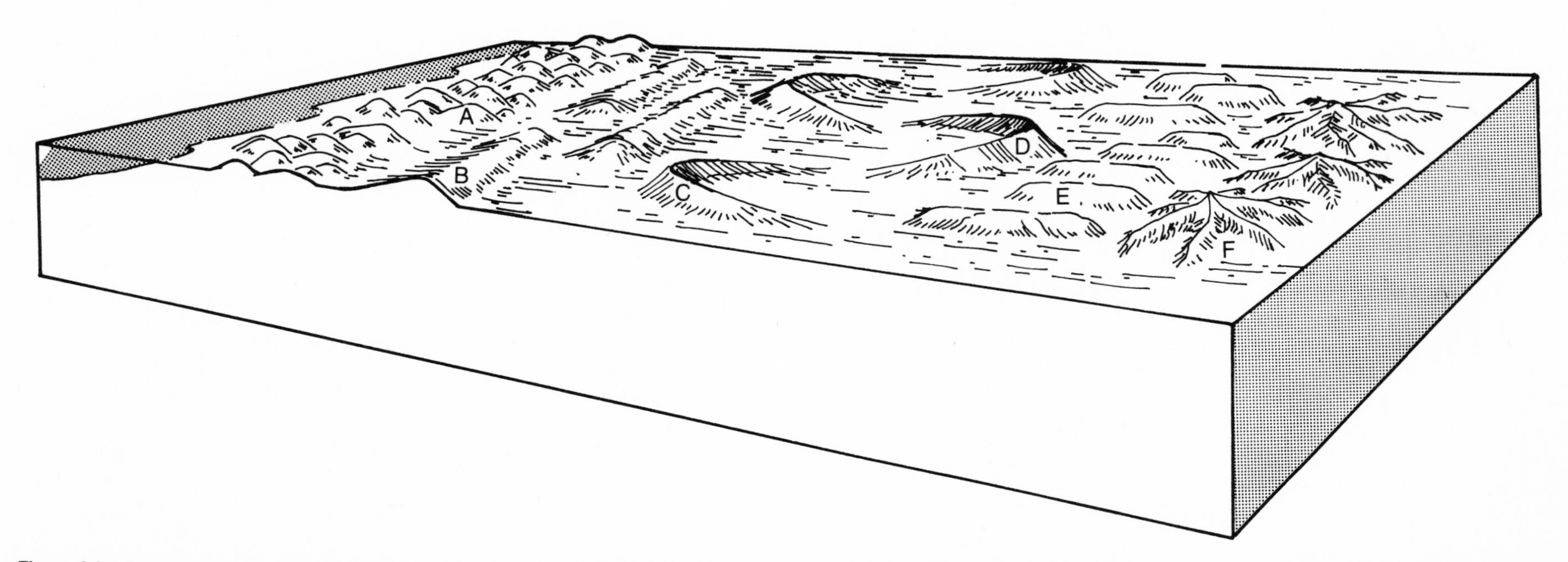

Figure 9.6. Types of sand dunes reflect the grain size, the volume of sand available, the wind velocity and regularity, the topography, and the amount of vegetative cover. (A) Beach dunes. (B) Transverse dunes. (C) Barchan dunes. (D) Wind-drift dunes. (E) Longitudinal dunes. (F) Star-shaped dunes.

Table 9.2. Eolian landforms: Sand dunes (humid and arid)

Topography	*Drainage*	*Tone*	*Vegetation and land use*
Dune forms	None	Light or bright	Barren and natural cover
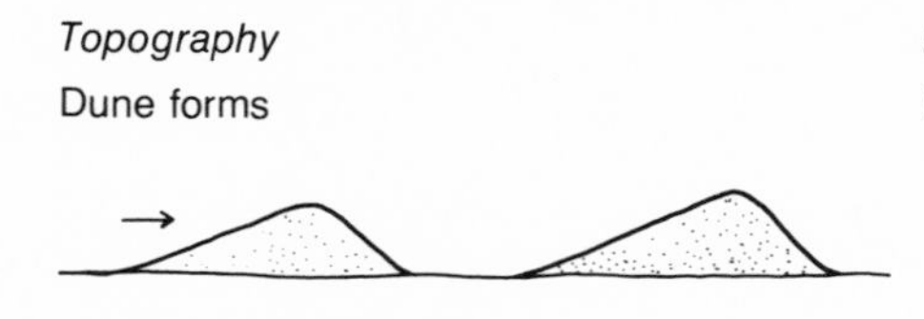 Sand dunes have streamlined shapes parallel and perpendicular to the dominant wind directions. The windward slope is gentle; the leeward slope is steep, reflecting the angle of repose for sand of approximately 34 degrees. Types of dunes include: beach, transverse, barchan, wind-drift, longitudinal, and star-shaped.	Because of the small areal extent of dune forms, their outward-facing slopes, and the porous soil materials, surface drainage patterns are not established. Percolation is very rapid, effectively draining rainfall through the subsurface.	Exposed sands have very bright white tones. Vegetation on stabilized dunes appears gray, with bare spots appearing white. *Gullies* None Normally, gullies are not developed because of the rapid internal drainage. Any that are present exhibit the V-shape characteristic of uncohesive, sandy soils.	In humid climates coastal dunes near the shoreline are typically barren; inland, where they are protected from the wind, they develop vegetative cover. Older, stabilized dunes may be covered with grass or scrubby trees. In arid climates dunes are barren as a result of their droughty condition.

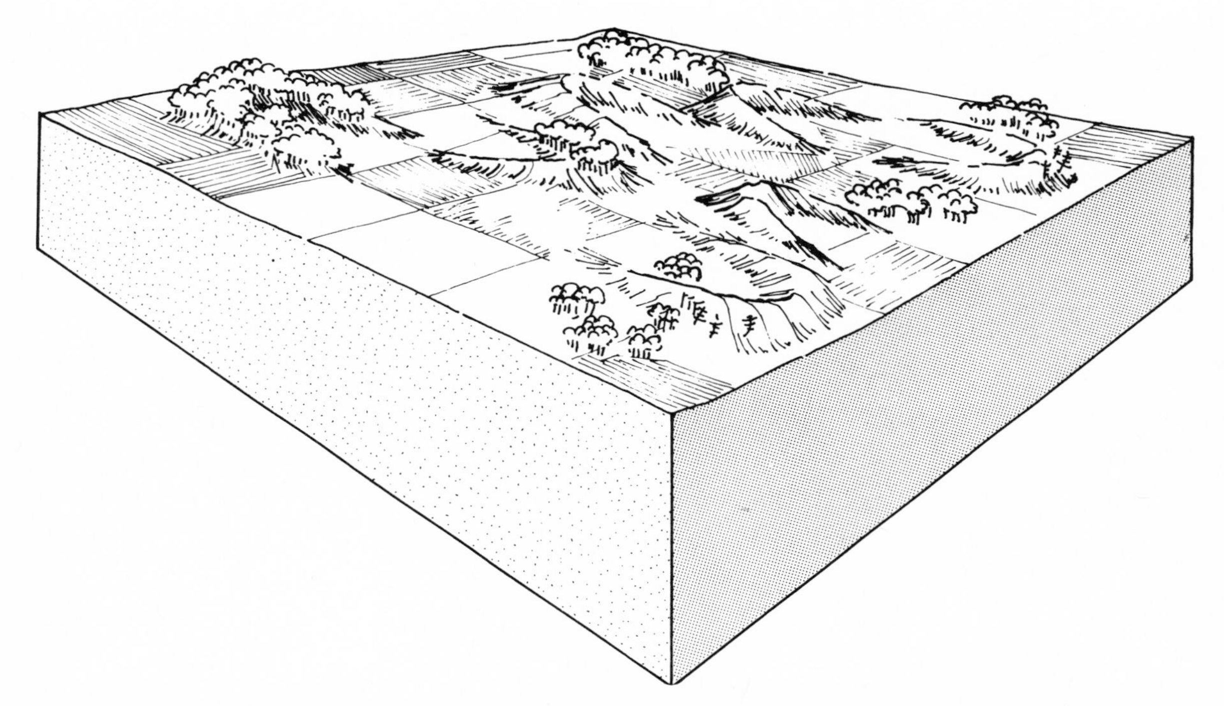

Figure 9.7. Active and stabilized sand dunes. Active sand dune formations have a great deal of spatial variation, depending on the size and spacing of the dunes. There is little land use development on dunes that are in the active formation stage, and these provide a natural, dynamic landscape. The lack of vegetation allows spatial variation and character to be controlled by the topography of the dunes. Any land uses that do occur have a high visual impact because of the lack of vegetative screening. Stabilized sand dune formations also have a high degree of spatial variation and viewing capacity. If the dunes are forest-covered, they are visually very absorptive and will effectively screen most low structures and land uses. Grass-covered dunes, such as those common throughout Nebraska, are visually absorptive for low structures, for these can be observed only from the immediately surrounding ridgelines and hilltops.

small supply of sand, and some variation in the dominant wind direction.

Star-Shaped Dunes

These dunes have a many-pointed-star shape, reflecting a limited sand supply and widely variable wind directions. For all practical purposes these dunes are stationary until a wind from a dominant direction develops.

Interpretation of Pattern Elements

Sand dune landforms are easily distinguished by their topography, which generally shows a directional tendency parallel or perpendicular to the dominant wind directions. Constant changes in wind direction create hummocky shapes. The boundaries of active sand dune landforms are very clear and distinct; stabilized dunes have indistinct transitions to surrounding landforms.

Soil Characteristics

The soils of sand dunes are poorly graded, consisting of many one-sized sand particles. Up to 98% of the material passes through a no. 40 sieve; up to 10% passes through a no. 200 sieve. The soils are noncohesive and typically contain rounded quartz grains. Dunes in some parts of the world, such as the Pacific Islands, consist of calcite or gypsum particles.

U.S. Department of Agriculture Classification

The U.S. Department of Agriculture typically classifies the parent materials as sand or fine sand. Older, stabilized dunes tend to develop thin surface soils of fine sand, loamy sand, or loamy fine sand.

Unified and AASHO Classifications

The Unified system typically classifies surface soils in stabilized dunes as SM and the parent material as SP. The AASHO system describes these soils as A-3.

Water Table

Because of the poorly graded nature of these materials and the associated rapid internal drainage, no seasonal high water table is found near the surface.

Drainage

The clean, even-sized, poorly graded sands provide very rapid internal percolation and a drainage aided by the elevated topographic position of the landform.

Soil Depth to Bedrock

These forms do not contain any bedrock, but they may rest directly upon rock surfaces. The

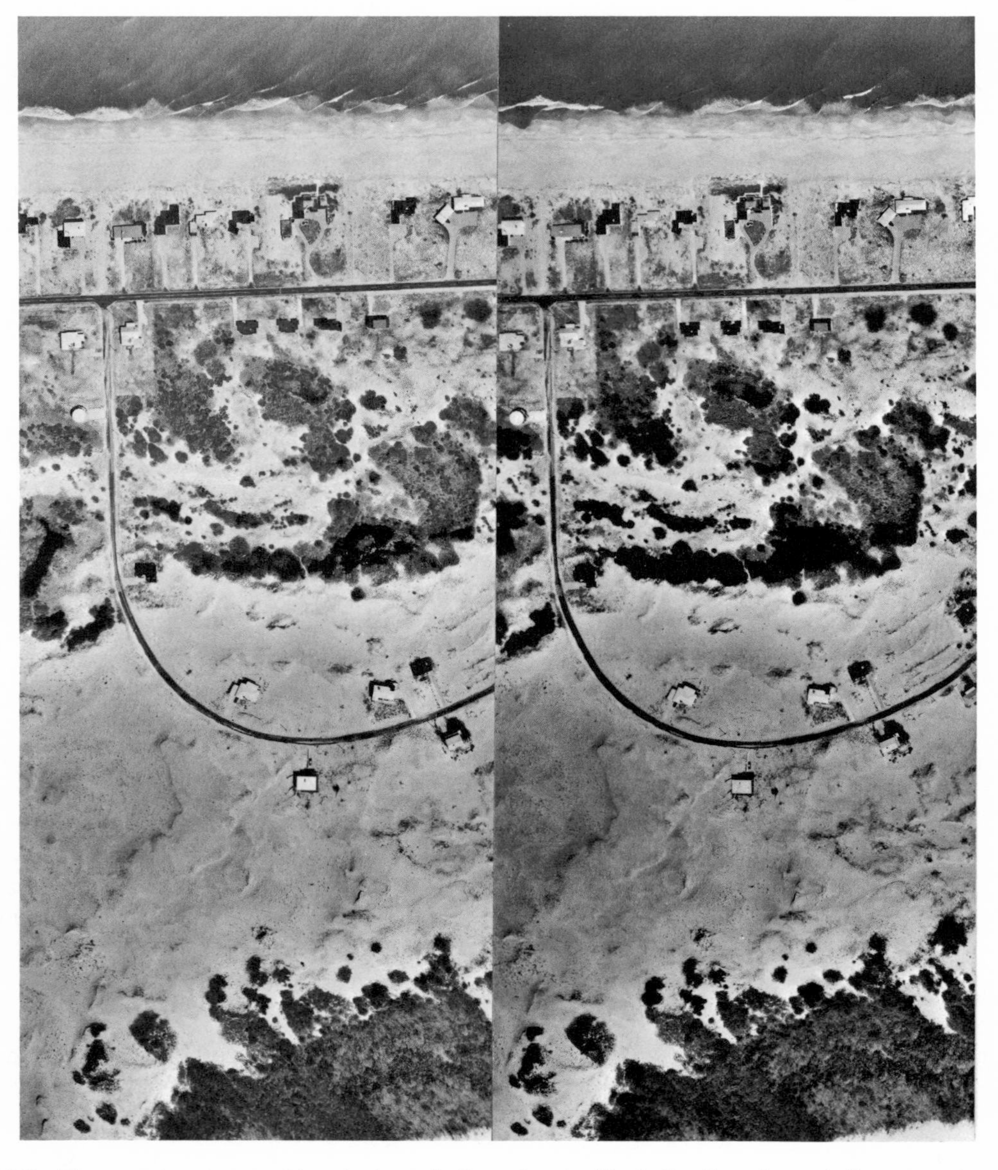

Figure 9.8. The complex shapes of beach dunes in Dare County, North Carolina, do not indicate any dominant wind direction. Photographs by Aerial Data Reduction Associates, Riverside, New Jersey, 1:4800 (400′/″), March 21, 1971.

surrounding and underlying formations need identification and analysis in order to determine bedrock depths.

Issues of Site Development

Sewage Disposal

Most state standards for septic tank leaching fields are satisfied by sand dune formations. However, most states do not have maximum percolation rates in their restrictions to protect against the groundwater contamination which can occur if sewage effluent percolates through the subsoil too rapidly and enters the groundwater table before the action of the aerobic bacteria has been completed. Pollution of this nature is very common in coastal areas where densely developed resort areas have been allowed to use septic systems, and it can occur even when the water tables are more than 10 feet below the surface. Intensive developments or urban areas should provide treatment systems to guarantee the quality of groundwater resources. Sewage lagoons can be utilized if their bottoms are coated with impervious materials.

Solid Waste Disposal

Sanitary landfill operations are not recommended in sand dune formations since the materials are of uniform size and allow rapid percolation of rainfall, the formation of leachate, and the percolation of the leachate to contaminate the groundwater table.

Trenching

The coarse-grained, clean, poorly graded sandy materials can be easily worked in all seasons under a variety of moisture contents. Trenching obviously requires sheeting, and costs are equivalent to those involved in excavating the same amount of deep, dry, moderately cohesive soil. Alignments may constrain themselves to lower areas adjacent to dunes, thus minimizing trenching. Pipelines through active dune areas are initially constructed on stable platforms away from or between sand formations, and the dunes are then allowed to migrate freely over the pipe. Pipelines laid on active dunes risk exposure as the dunes migrate and remove lateral support.

Excavation and Grading

The coarse-grained, clean, poorly graded, sandy materials can be easily worked in all seasons under a variety of moisture conditions. Large, flat sites require extensive grading, owing to the hummocky nature of the topography, but costs are equivalent to removing the same amount of deep, dry, moderately cohesive soil. Light equipment can be utilized for grading and excavation operations, but the abrasive sand grains cause excessive wear on the machinery. Deep pit excavation below the water table requires sheeting and dewatering.

Construction Materials (General Suitabilities)

Topsoil: Unsuitable

Sand: Excellent. Sand dunes contain natural, poorly graded, clean sand grains, all primarily the same size, and are an excellent source of sand materials.

Gravel: Not Suitable

Aggregate: Not Suitable

Surfacing: Not Suitable. Surfacing materials need fines to act as binder and prevent the formation of corrugated surfaces under wear.

Borrow: Good to Poor. The uniform sand grains should be mixed with fines to make suitable

Figure 9.9. Complex transverse dunes in Pakistan. The dominant wind direction is from the lower left. Photographs by the U.S. Geological Survey, Pakistan 1-AB, 1:40,000 (3333′/″), 1953.

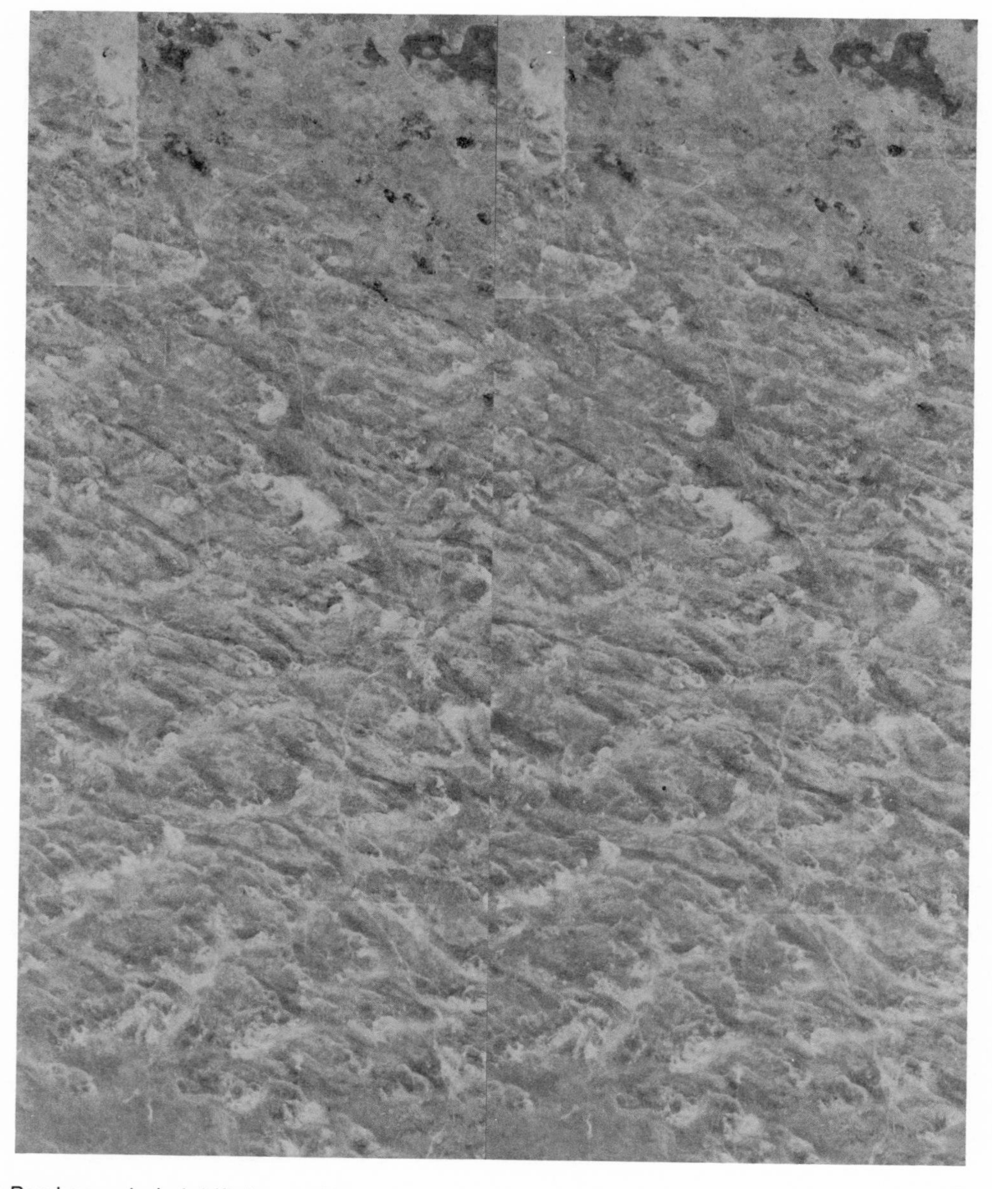

Figure 9.10. Barchan and wind-drift dunes which have been stabilized in Box Butte County, Nebraska. The dominant wind direction is from the upper left. Photographs by the Agricultural Stabilization Conservation Service, CAK-2AA-276,277, 1:20,000 (1667′/″), July 14, 1960.

borrow material. Fills should be laterally contained if possible and compacted with vibrators.

Building Stone: Not Suitable

Mass Wasting and Landslide Susceptibility

Sand dunes are very stable since they are uniform in composition and well drained. If the angle of repose of these materials is exceeded during excavation or stream undercutting, sliding will take place until the angle is restored.

Groundwater Supply

The coarse, clean, poorly graded soils of individual dunes do not have the capillary ability to retain moisture or a groundwater supply. Landforms underlying sand dune formations may offer suitable groundwater supplies. Coastal dune formations can act as giant sponges, holding a fresh groundwater supply slightly above sea level and the salt water table. Overpumping of the fresh water, however, allows the saltwater to contaminate the supply. This condition is common along the Atlantic coastal regions where withdrawal of water has exceeded its recharge ability.

Pond or Lake Construction

The small areal extent of individual dunes and their porous soils make them unsuitable for the siting of reservoirs. Small ponds can be created in dunes if impervious bottom materials are applied and the water content is refurbished by pumping to compensate for that lost by evaporation.

Foundations

Sand dunes contain granular, clean, poorly graded materials which are unconsolidated and potentially compressible. Sudden shocks from earthquakes or similar vibrations may cause compression and possible structural failure if corrective measures are not taken before construction.

Table 9.3. Eolian sand dunes: aerial and cartographic references

State or country	County or region	Photo Agency	Photograph no.	Scale	Date	Corresponding quadrangles and geologic maps from USGS	Corresponding soil reports from SCS
Alaska	–	USGS*	Alaska 10-AB	3333′/″	8/27/52	Mt. McKinley (1:250,000), Prof. Paper No. 373	–
Arizona	Coconino	USGS*	Arizona 11-ABC	4500′/″	2/19/54	Flagstaff (1:250,000)	–
California	Imperial	USGS*	California 39-AB	1667′/″	11/10/59	Kane Spring NE (7½′), Kane Spring NW (7½′)	1903‡
California	Santa Barbara	USGS*	California 35-AB	5280′/″	4/1/60	San Minguel (7½′)	1958
Chile	Tarapaca	USGS†	Chile 1-ABCD	5000′/″	4/7/55	USAF Aerographic chart, Point Angamos (1:1,000,000)	–
Colorado	Saguache, Alamosa	USGS*	Colorado 10-ABC	4416′/″	10/5/53	Trinidad (1:250,000)	–
Indiana	Lake	ASCS	BFJ-1V-34-36	1667′/″	6/26/58	Gary (7½′)	–
Indiana	Porter	ASCS	BFP-2V-55-56	1667′/″	9/12/58	Portage (7½′)	1916‡
Indiana	Porter	ASCS	BFP-2N-34-35	1667′/″	9/8/54	Dune Acres (7½′)	1916‡
Iran	Sarayan	USGS†	Iran 1-AB	4508′/″	8/7/56	Coast and Geodetic Survey ONC, G-5 (1:1,000,000)	–
Iran	–	USGS†	Iran 2-A	4508′/″	8/25/56	Coast and Geodetic Survey ONC, H-7 (1:1,000,000)	–
Iran	Shahr-e Lut	USGS†	Iran 3-AB	4508′/″	9/22/56	Coast and Geodetic Survey ONC, H-7 (1:1,000,000)	–
Iran	–	USGS†	Iran 4-AB	4583′/″	8/22/56	Coast and Geodetic Survey ONC, H-7 (1:1,000,000)	–
Massachusetts	Barnstable	ASCS	DPL-1LL-112-114	3333′/″	9/20/70	Provincetown (7½′)	1920‡
Massachusetts	Barnstable	ASCS	DPL-1LL-108-109	3333′/″	9/20/70	Provincetown (7½′)	1920‡
Michigan	Luce	ASCS	BVY-3K-87-90	1320′/″	7/10/53	Sault Saint Marie (1:250,000)	1929
Nebraska	Box Butte	ASCS	CAK-2AA-276-277	1667′/″	7/14/60	Alliance (1:250,000)	1916‡
Nebraska	Grant	USGS*	Nebraska 3-AB	1667′/″	8/16/55	Ashby (15′)	–
Nebraska	Logan	ASCS	CAX-2P-36-37	1667′/″	9/3/55	Cody Lake (15′)	–
Nebraska	Sheridan	USGS*	Nebraska 2-ABC	1667′/″	8/23/54	Antioch (15′)	1918‡
Nebraska	Sheridan	ASCS	CBD-7N-5-8	1667′/″	8/19/54	Lakeside (15′)	1918‡
Oregon	Morrow	ASCS	AAG-5FF-45-47	1667′/″	8/6/65	Pendleton (1:250,000)	–
Oregon	Umatilla	ASCS	NZ-7EE-39-42	1667′/″	7/25/64	Pendleton (1:250,000)	1948
Saudi Arabia	–	USGS†	Saudi Arabia 1-AB	4583′/″	11/28/54	Army Map Service K462; HF 39W NW Rub' al Khali (1:500,000)	–
Washington	Grant	USGS*	Washington 6-ABC	1667′/″	8/11/54	Moses Lake (15′)	–
Pakistan	West Pakistan	USGS†	Pakistan 1-AB	3333′/″	1953	Army Map Service U501; H-41X Kharan Calat (1:253,440)	–
Pakistan	West Pakistan	USGS†	Pakistan 3-ABC	2915′/″	1953	Army Map Service U502; NG42-11 Mirpur Khas (1:250,000)	–
Wisconsin	Door	ASCS	BHQ-2AA-12-13	1667′/″	9/18/61	Jacksonport (15′)	1916‡

*From *Selected Aerial Photographs of Geologic Features in the United States,* U.S. Geological Survey Prof. Paper No. 590.
†From *Selected Aerial Photographs of Geologic Features outside the United States*, U.S. Geological Survey Prof. Paper No. 591.
‡Out of print.

Figure 9.11. Stabilized longitudinal dunes in Logan County, Nebraska. The fine, dotted lines in the fields are windrows. Photographs by the Agricultural Stabilization Conservation Service, CAX-2P-36,37, 1:20,000 (1667′/″), September 3, 1955.

These materials can be suitably compacted in place by vibrations caused by pile drivers, by the process of vibroflotation, or even by dynamite. Small structural loads can utilize footings or rafts, and residential units are typically constructed with continuous footings and basement slabs although rafts and footings without basements are common. Heavy structural loads typically utilize deep foundations such as friction piles; the driving process compacts much of the surrounding soil material. Care should be taken during construction maintain the stability of the sand, either by the natural vegetation or by temporary mechanical means. Lack of a protective cover can initiate wind erosion and blowout of the formation, thus critically endangering the structure if footings or portions of a raft are undermined. Sand dunes in cold, permafrost climates are valuable building platforms since they are not susceptible to capillary moisture and therefore frost heave and permafrost do not occur.

Highway Construction

There are potential maintenance problems when highways are located through active dune areas, for dunes can move across the roadway. Stabilization of adjacent dunes by vegetative cover may be necessary; the roadbed itself can be stabilized by the addition of clay, oil, or asphalt.

Stabilized sand dune formations have a hummocky topography, but cuts and grading in them are not difficult, owing to the unconsolidated nature of the soil. The shapes and alignments of the specific site determine whether or not the dunes can be used for the location of the roadbed center line. Their topographic elevation, excellent drainage, and high bearing capacities make these formations ideal alignment corridors in regions of permafrost. Areas between dunes may need fill materials, depending upon the specific landforms present.

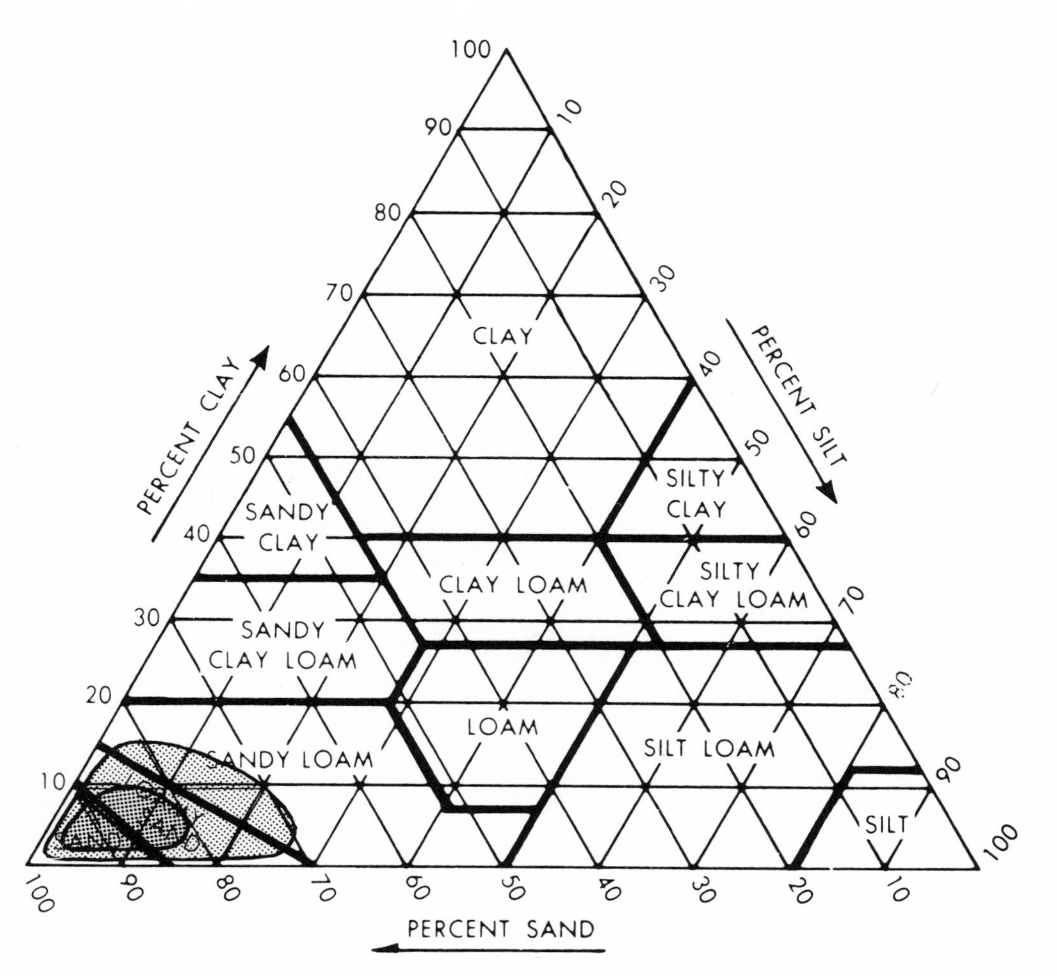

Figure 9.12. The U.S. Department of Agriculture soil texture triangle indicates sand or loamy sand for these landforms.

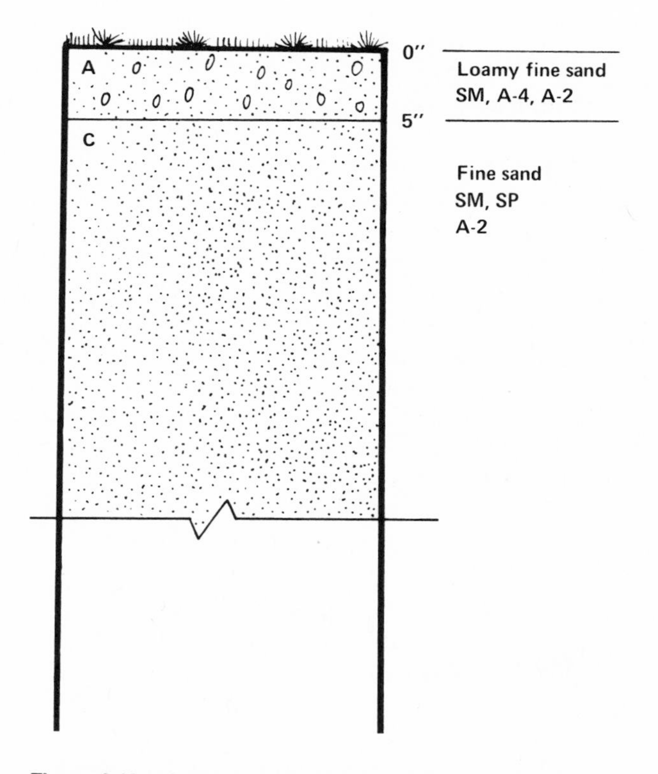

Figure 9.13. A typical soil profile developed on a sand dune in a semiarid climate.

LOESS

Introduction

Loess consists of fine-grained silts and clays that have been deposited by wind. These materials are unconsolidated or weakly consolidated and generally contain 50 to 70% silt, with most particles ranging from 0.0l to 0.05 millimeter in size. The mineral content varies, depending upon the source of the parent material, and most grains are angular in shape. Residual soils are very porous and light-toned and are homogeneous in nature, having no stratification. Loess is characterized by its ability to hold a vertical face, but sloping surfaces are highly susceptible to erosion.

The silts found in loess originate from large glacial outwash plains, fluvial flood plain formations, or desert areas where saltation by sand grains placed large amounts of silt and clay into the lower atmosphere. The materials are carried long distances by winds, and the deposits decrease in thickness with the distance from the source. It has been estimated that the loess deposits in the Midwest accumulated at a rate of 0.25 inch per year.

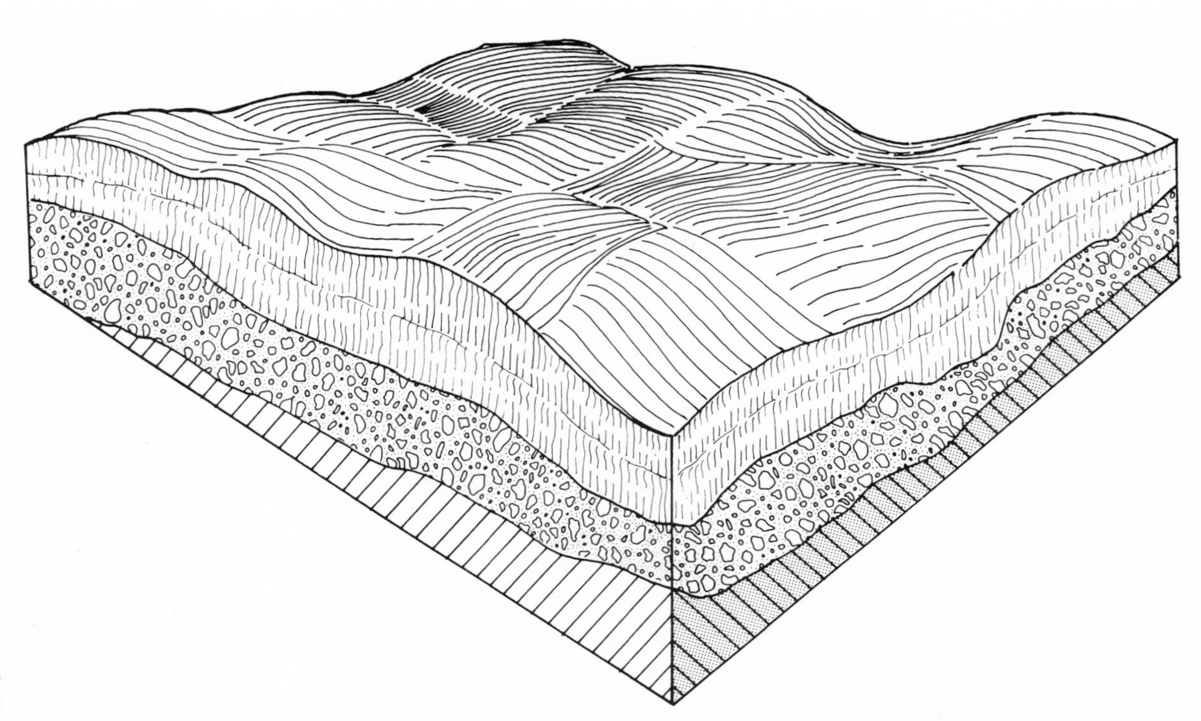

Figure 9.14. Nondissected loess landforms present a topography of gently rolling, somewhat parallel hills which are intensely cultivated. The hilltops and broad ridges are visually exposed, since no high vegetation is found to create significant screening. The hills are not high enough to offer regional views, but they do offer elevated positions for local valley views. The topographic changes do not give a feeling of spatial closure, but they provide some minor variation as one moves through the region. The land use pattern has little diversity and typically consists of broad, rolling agricultural fields.

Interpretation of Pattern Elements

Loess is easily identified by its characteristic pinnate drainage pattern and box-shaped gullies, which indicate the high silt content of the residual soil. The soils are found covering all types of landforms in varying thicknesses, either dominating the topography or, as a thin layer, masking the character of the underlying material. In some regions loess materials have been reworked by water and form broad river terraces, but their interpretive and engineering characteristics are essentially the same.

Soil Characteristics

Loess soils typically contain 50 to 70% unconsolidated to weakly consolidated silt which is vertically well drained unless disturbed by compaction or other operations. Loess is highly erosive along sloping surfaces but extremely stable in holding a vertical face.

U.S. Department of Agriculture Classification

The U.S. Department of Agriculture typically classifies loess soils as silt loam, silty clay loam, or silt. Poorly drained depressions may have heavier soils of the same texture which have higher levels of organic content and plasticity.

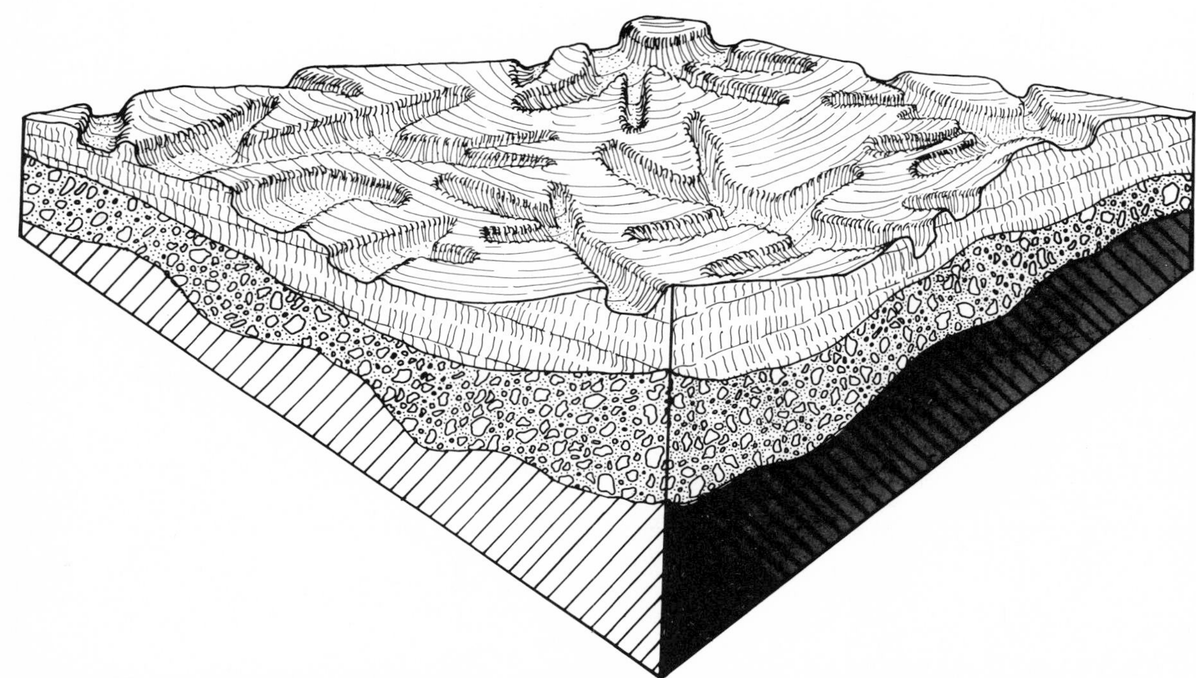

Figure 9.15. Dissected loess formations have very steep slopes and a rugged topography which is covered with grass in arid climates and forested in humid climates. The landscape is visually very absorptive because of the many tight valleys which can be observed from the immediate ridges. The ridges offer some regional viewing potential, while the valleys provide tight spatial closure owing to the steep, high, adjacent slopes. A great deal of spatial variation is encountered when traveling through these regions.

Unified and AASHO Classifications

The Unified system classifies these soils as ML, ML-CL, and occasionally CL. Depressions that are poorly drained develop CL or CH subsoils. AASHO symbols are commonly A-4, and occasionally A-7 or A-6.

Water Table

A seasonal high water table is generally found beyond 5 feet below the surface. Wetter conditions are encountered in depressions, where the seasonal high water table is typically 1 to 3 feet beneath the surface.

Drainage

Loess materials are well drained vertically but poorly drained horizontally. Disruption of the soil structure by compaction, grading, or filling changes the drainage characteristics to an impervious condition.

Soil Depth to Bedrock

During construction operations in deep loess formations, rock is not encountered, but operations in thin deposits may encounter rock if it is the underlying material. The identification of depths to bedrock requires investigation and analysis of the underlying material. In a situation in which there is thin loess over rock, the rock can be identified if its structure controls topographic and drainage characteristics.

Issues of Site Development

Sewage Disposal

Septic tank leaching fields generally operate satisfactorily in these formations because of the moderately well-drained condition of the subsoil. Depressions are less suited for leaching field location, since the water table may be near the surface. Soils should not be disturbed during construction or they will become impervious.

Solid Waste Disposal

Sanitary landfill operations do not generally encounter difficulties in site location within these landforms. The loess materials can be used for cover when compacted, for they form an impervious layer which prevents rainwater percolation and eventual contamination of the groundwater

Table 9.4. Eolian landforms: Loess (humid and arid)

Topography	*Drainage*	*Tone*	*Vegetation and land use*
Rugged and smooth hills	Pinnate and dendritic	Uniform light	Natural and cultivated
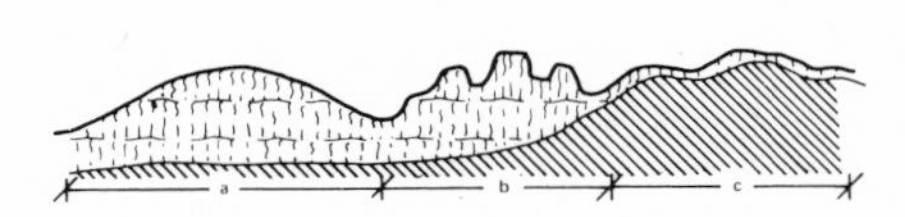	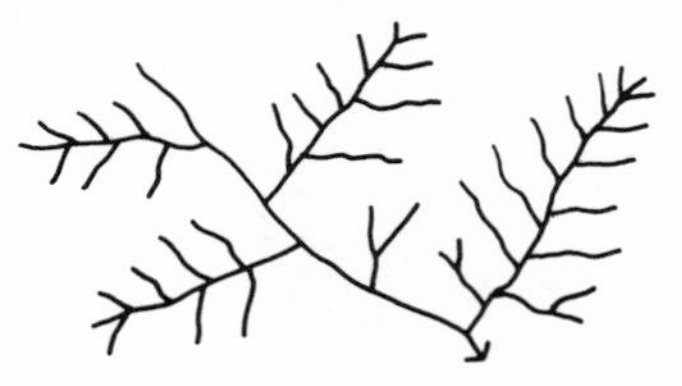	Deep, nondissected formations have uniform light tones; deep, dissected forms show gentle gradations from light to medium tones. Thin deposits mask the tones of the underlying materials.	Deep, gently rolling loess landforms are intensively cultivated; highly dissected landforms are in natural cover. Forest growth is found in humid climates; in arid climates grass cover is typical. Where highways cross these regions, steep, vertically stepped road cuts may be observed.
Deep deposits of loess are characterized by smooth, rounded, convex hills which may be parallel (a). Typically, highly dissected areas are adjacent to major streams (b). The topography of shallow loess deposits is controlled by the underlying material (c).	Highly dissected loess deposits have a fine-textured pinnate drainage pattern with flat-bottomed tributaries. In nondissected areas a regional, slightly parallel, dendritic pattern develops, with box-shaped gullies. Such erosional features as catsteps along hillsides or residual pinnacles in valleys indicate silty soils.	*Gullies* Box-shaped The box-shaped gully, indicating a high silt content in the surrounding soil, is best exemplified in silt landforms. Pinnacle remnants may be found in valleys.	

table. Depressions may not be suitable sites, since the water table is near the surface.

Trenching

Trenching in loess materials is not difficult, owing to the lack of rock or boulders and the ability of the materials to stand in vertical sections. The materials are hard to work when wet. Frost heave occurs only in places where the water table is near the surface. Trenching costs are equivalent to those encountered in removing the same amount of deep, dry, moderately cohesive soil. The highly dissected topographies present difficulties in selecting alignments that minimize grade changes. Liquid-carrying pipelines passing through these regions may require the use of pumping stations for grade changes; short segments are able to follow valley bottoms with little difficulty.

Excavation and Grading

Dissected loess areas require significant grading operations in order to obtain a large, flat site. Excavation is neither difficult nor costly; the unconsolidated material keeps costs at a level equivalent to those encountered in removing the same amount of deep, dry, moderately cohesive soil. Final grading and drainage must be carefully studied to minimize potential erosion; cut faces are stable if vertical, but drainage should not be allowed to cross sloping surfaces. If these soils are compacted, their structure is destroyed, and a material with a higher runoff coefficient and erosion potential results.

Construction Materials (General Suitabilities)

Topsoil: Good to Poor. The most suitable topsoil materials are found in the weathered surface horizons.

Sand: Not Suitable
Gravel: Not Suitable
Aggregate: Not Suitable

Surfacing: Not Suitable. Loess soils may be suitable for surfacing if other materials are added for stabilization.

Borrow: Fair. Several difficulties are encountered when loess materials are utilized for borrow. The moisture content is critical to proper compaction, and it is difficult to control. When the material is too dry, it is difficult to compact and easily eroded by wind action; when it is wet, it is difficult to work and compact. Because of the one-size characteristic of the soil, the addition of other materials may be required for stabilization. Pneumatic-tired rollers are commonly used for compaction of fills.

Building Stone: Not Suitable. Although they are not so used in the United States, loess forma-

Figure 9.16. (A) Dissected loess in a semiarid climate in Sherman County, Nebraska, where the characteristic pinnate drainage pattern and box-shaped gullies are easily recognized. Photographs by the Agricultural Stabilization Conservation Service BNO-2T-138, 139, 1:20,000 (1667′/″), June 2, 1957.

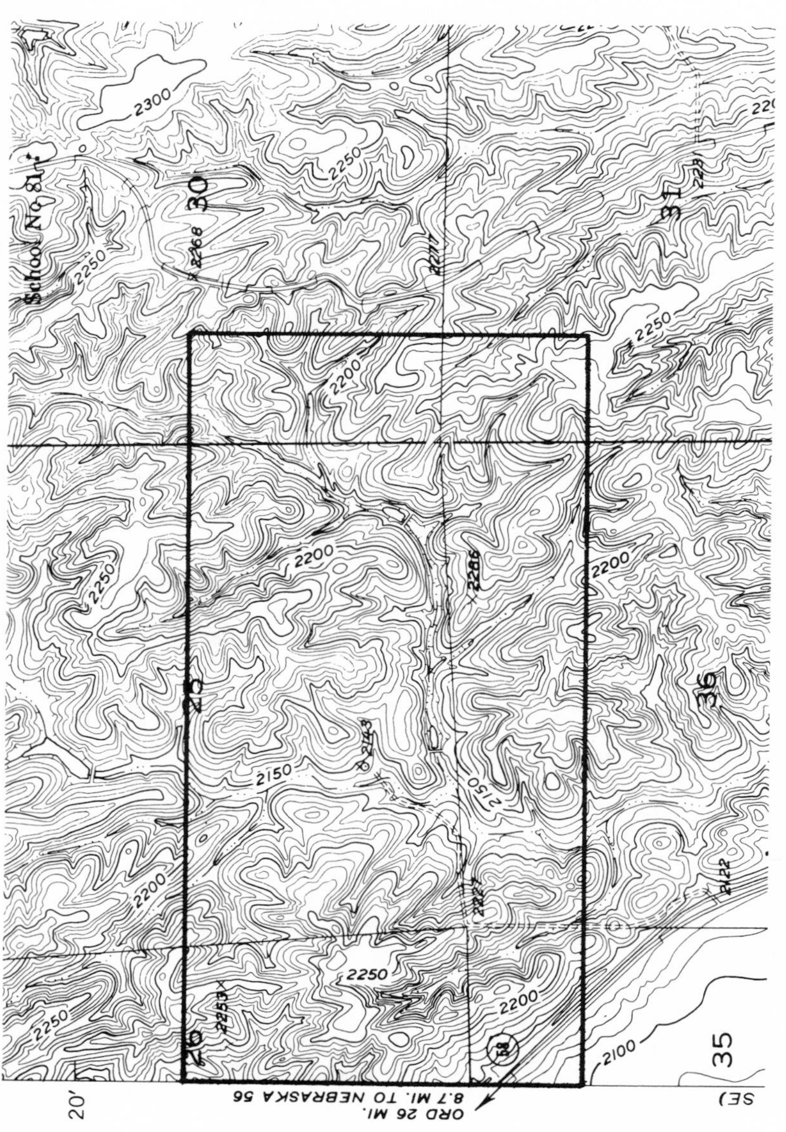

Figure 9.16. (B) U.S. Geological Survey Quadrangle: Loup City (7½′).

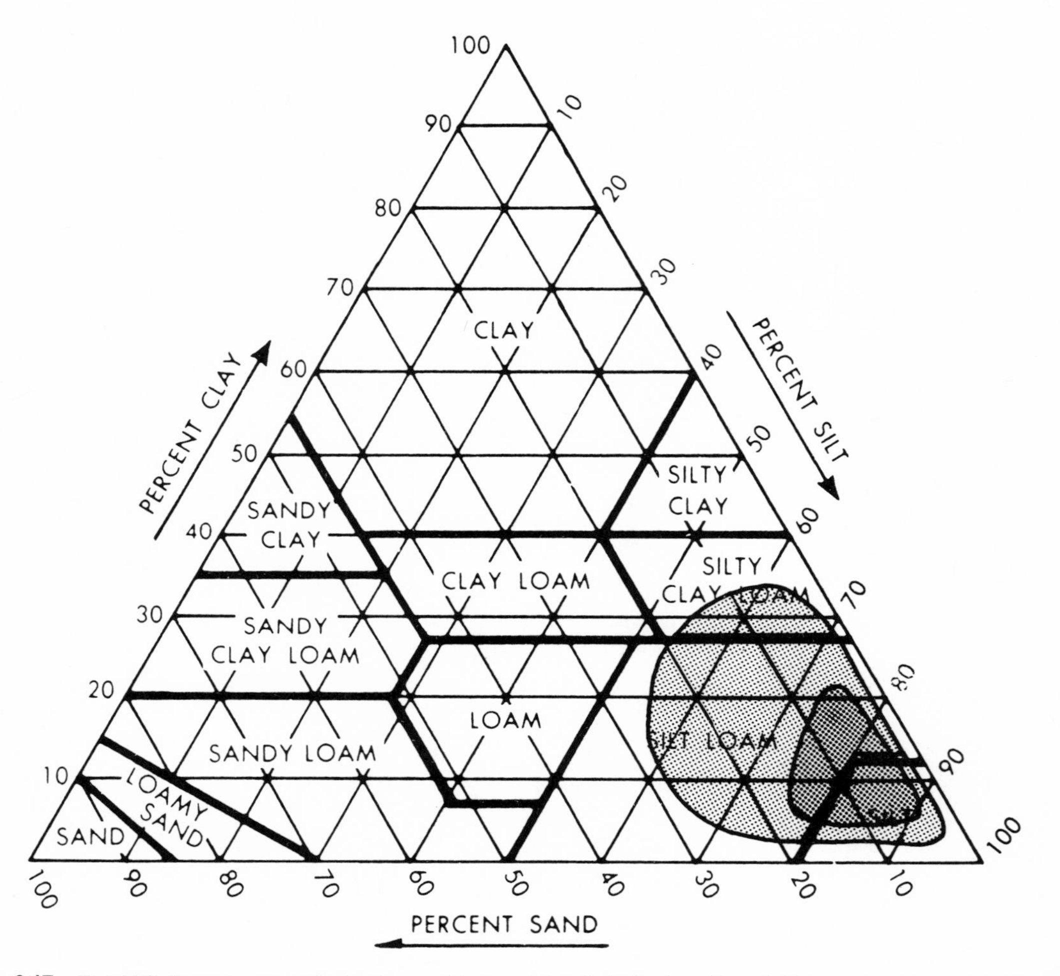

Figure 9.17. The U.S. Department of Agriculture soil texture triangle indicates silt and silt loam particles as the dominant materials.

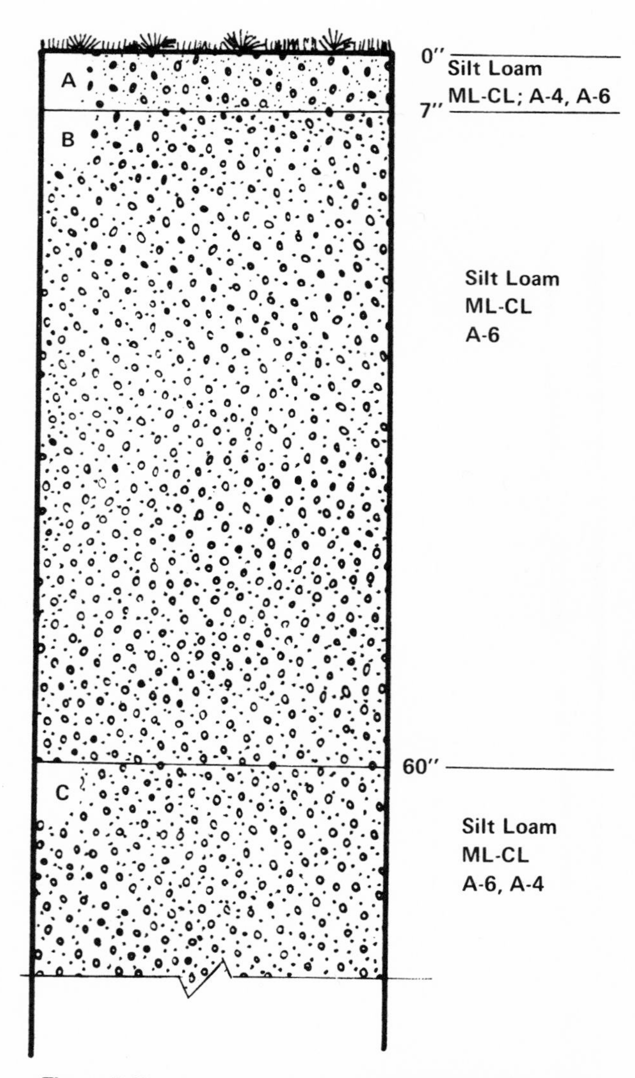

Figure 9.18. A typical soil profile developed in loess in a semiarid climate.

tions in many regions of the world are used as sources for construction materials and for shelter as well. In China and in some parts of Europe, loess bound with grass, sticks, straw, and so on, is used to build structures. In China vertical cliffs of loess deposits are often carved and hollowed to form underground rooms beneath tilled fields.

Mass Wasting and Landslide Susceptibility

Soil materials in loess formations are moved downslope mainly by erosion. Catsteps are common and indicate minor slumps along hillsides; these appear as small terraces several feet wide and several inches to several feet high. On aerial photographs catsteps show as faint, light lines which follow contours. As opposed to other soil materials, vertical cuts in loess are extremely stable, but sloping surfaces are very susceptible

Table 9.5. Eolian loess: aerial and cartographic references

State	County	Photo Agency	Photograph no.	Scale	Date	Corresponding quadrangle and geologic maps from USGS	Corresponding soil reports from SCS
Alaska	—	USGS*	Alaska 14-ABC	3333'/"	8/17/52	Mt. McKinley (1:250,000)	—
China	—	USGS†	China 1-ABC	2500'/"	4/27/47	Army Map Service; 1301, NJ49, Tang-Chu (1:1,000,000)	—
China	—	USGS†	China 2-AB	4175'/"	1943	Army Map Service; 1301, NJ50, Pei Ping (1:1,000,000)	—
Colorado	Dolores	USGS*	Colorado 6-AB	2270'/"	9/15/47	Cortez (1:250,000), Prof. Paper No. 475-C	—
Iowa	Benton	USGS*	Iowa 2-AB	1667'/"	8/14/63	Waterloo (1:250,000)	1921‡
Iowa	Benton	USGS*	Iowa 3-ABC	5833'/"	10/18/50	Waterloo (1:250,000)	1921‡
Nebraska	Buffalo	ASCS	BMO-1T-124-125	1667'/"	5/26/57	Alfalfa Center (7½')	1924‡
Nebraska	Sherman	ASCS	BNO-2T-137-138	1667'/"	6/2/57	Loup City (7½')	1931
Nebraska	Sioux	ASCS	CBE-5N-9-10	1667'/"	6/29/54	Alliance (1:250,000)	1919‡
Nebraska	Valley	ASCS	BNQ-2T-51-52	1667'/"	5/27/57	North Loup (7½')	1932
Washington	Adams	USGS*	Washington 8-AB	1667'/"	8/28/50	Ritzville (1:250,000)	1967
Washington	Adams	USGS*	Washington 9-AB	1667'/"	8/29/50	Washtucna (15')	1967
Wisconsin	Grant	ASCS	CI-4BB-192-193	1667'/"	10/17/61	Kieler (7½')	1961

*From *Selected Aerial Photographs of Geologic Features in the United States,* U.S. Geological Survey Prof. Paper No. 590.
†From *Selected Aerial Photographs of Geologic Features outside the United States,* U.S. Geological Survey Prof. Paper No. 591.
‡Out of print.

to erosion. Undercutting by streams is a common cause of slides in vertical faces.

Groundwater Supply

Some groundwater resources can be obtained from loess materials, but the yield is slow because of the slow horizontal percolation. More suitable water resources are commonly found in the underlying materials.

Pond or Lake Construction

Loess formations are generally suitable for the construction of reservoirs. When compacted, the residual soils are suitable for bottom materials, since they become impervious. Highly dissected formations provide many sites suitable for the placement of dam structures. Maintenance against siltation is difficult, since much of the runoff contains suspended silts and clay, and sediment catchment basins are rarely effective since they trap only the coarser materials. The fine-textured particles in the runoff slowly settle to the reservoir bottom during the winter or at other times when the water velocities and currents are low. The best management control is to keep the entire watershed, especially the first-order drainage system where channelized drainage first begins, well covered with dense vegetation.

Foundations

Loess soils generally provide suitable load-carrying support for raft and footing foundations for low and moderate loads. Heavier structures may have to rely upon piles, depending upon the specific nature of the materials, their depth, the underlying materials, and the framing system proposed.

Highway Construction

Highway construction involves a variety of problems in loess formations. The dissected landforms require a great amount of grading and earthwork, bridging, culverts, and perhaps tunneling, thus escalating construction costs. Highway cuts should be left holding a vertical face; if deep, they should be terraced and drained. The loess parent materials along embankments need stabilization through the addition of sands, cement, or other materials. Construction materials may have to be imported from great distances, since loess does not contain any variation in textural distribution and so does not supply aggregate, clay, sand, or gravel. Frost heave problems may occur in cold climates where the water table is near the surface. Final grading and drainage must be carefully planned and reviewed in order to minimize potentially severe erosion.

CHAPTER 10

FLUVIAL (WATERLAID) LANDFORMS

INTRODUCTION

Waterlaid materials are represented in many varied and complex landforms found throughout the world, and when examining landforms and undertaking terrain analysis, it is essential to have a solid understanding of fluvial processes. Since it is impossible to present all of the necessary information on the subject in this brief introduction, the references at the end of this section are strongly recommended for further study.

In the fluvial process water transports eroded sediments and rock debris, sorts the materials, and eventually deposits them. These erosive and transporting processes are dynamic, modifying and creating waterlaid landforms and utilizing the weathered and eroded materials of other landforms. Mass wasting processes interact with fluvial processes to denude the landscape. Thus mass wasting, in the form of creep, slumps, sliding, and so on, moves most of the soil debris downslope where it is then picked up and transported by the stream system (Figure 10.1). As the process continues, the landscape is lowered until the dissecting drainage system reaches a graded condition, or equilibrium, in which the average rate of soil debris entering the stream from all its tributaries and slopes is equal to the average rate at which the stream can transport the load. The erosion sequences discussed in Chapter 2 illustrate the stages associated with land mass denudation, from original uplift through peneplanation. These stages should not be confused with the stages of stream development by which a stream's sediment load-carrying ability and erosion potential are classified.

The formation of streams is initiated when either of two conditions occurs with rainfall: (1) the soil absorbs the water until it is saturated, and (2) rainfall takes place faster than it can be absorbed. Either of these conditions causes the formation of puddles in small depressions. When these are filled, the water begins to flow in swales between the puddles, and the basic ties of an integrated drainage system come into existence. Eventually, the flowing water begins to have velocities high enough to erode or to pick up and carry particles of soil, initiating the establishment of rills, gullies, and stream beds.

Fluvial Erosion

Fluvial systems have the energy to erode and transport large quantities of soil debris. The materials and debris carried within a system are of two kinds, the bedload and the suspended load. The bedload consists of particles that roll and slide along the channel bottom when the water velocity creates a drag over and around a particle sufficient to exceed its coefficient of friction. Collisions and bouncing place the particle into higher-velocity currents, and this process of saltation, as well as sliding and rolling, moves the bedload debris slowly downstream according to the increase and decrease of local velocities and turbulence.

The suspended load in a fluvial system consists of finer particles, such as silts, clays, and fine sands. These are held in suspension when turbulence is greater than the settling velocities of the materials. The movement of a bedload helps add to a suspended load, for the larger particles of the bedload can break up cohesive, finer particles and place them in reach of the higher velocities above the stream bottom. Once these fine particles are suspended, they can be carried great distances. Large river systems have a great ability

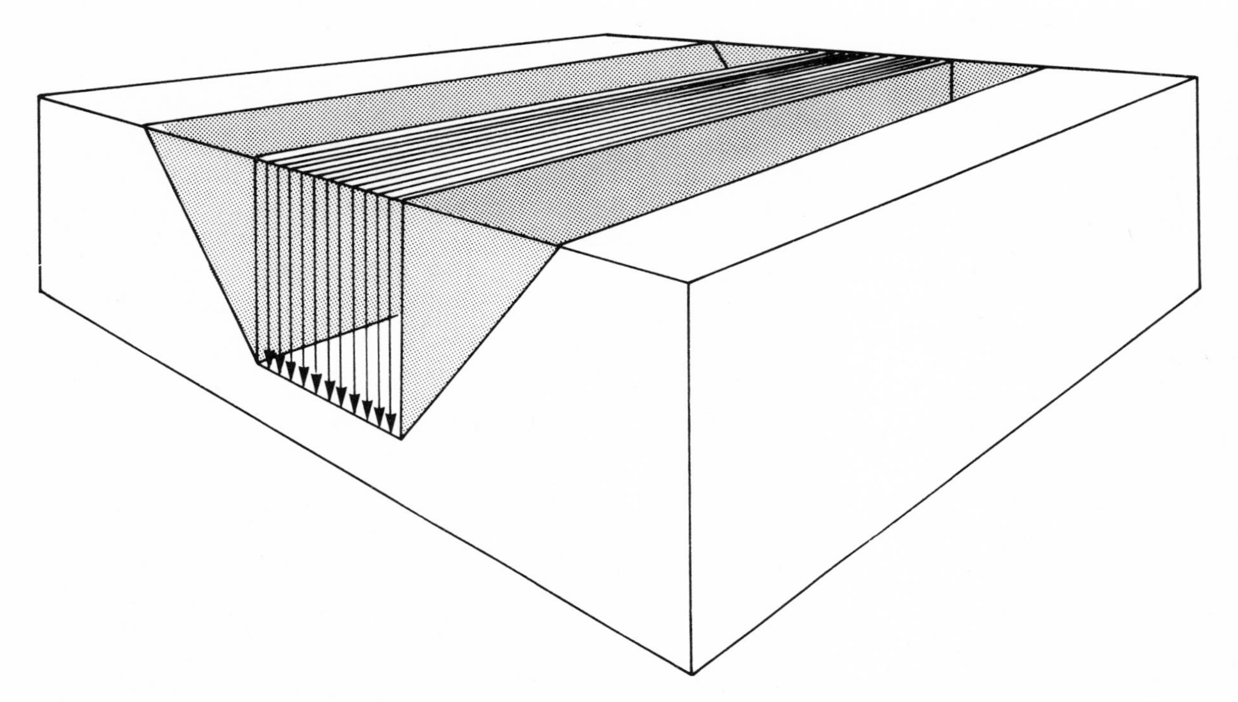

Figure 10.1. Mass wasting and fluvial erosion. The combined effects of mass wasting and fluvial erosion can be seen in this diagram. The valley was formed by removal of all materials by fluvial erosion. The shaded areas represent sections where mass wasting transported materials downslope into the stream bed for fluvial erosion.

to carry suspended sediments, especially during flood stage. As a yearly average, the Mississippi River carries approximately 500 to 600 parts per million of suspended sediment, but during flood stage its suspended load may increase to 2600 parts per million. During flood stage the Colorado River above Hoover Dam carries up to 40,000 parts per million of sediment; this is 1/25 the weight of the water. And the Yellow River in China has been measured when the weight of the suspended solids has been greater than the weight of the water. It is conjectured that the suspended load, rather than the bedload, of a stream system is the means whereby most materials are moved; the Mississippi River moves all but about 10% of its load by suspension.

The amount of turbulence within a stream system significantly influences its erosive and transporting powers. A smooth channel bottom of silt and clay has a layer of water directly above it of very low velocity and laminar flow, and this makes it very difficult to erode and transport the bottom particles even though the stream velocity may far exceed the settling velocity for these particles. Turbulent flows, however, are characterized by the existence of many eddies created by bottom roughness and/or a very high stream velocity. The more turbulent the flow, the greater the contained circulating velocities and the greater the load that can be eroded and transported by the system.

Stream Energy

The economy or energy flow of a stream system consists of its total input and consumption of energy and the changes caused by it. The rainfall distributed over the United States in 1 year has a great amount of potential energy, and this potential is described in hydrological terms as the weight of the water (1 cubic foot) times the height of the land above sea level; thus 1 cubic foot of water weighing 62.4 pounds has 312,000 foot-pounds of energy at an altitude of 5000 feet (62.4×5000). Much of this potential energy is converted to kinetic energy as the water flows down through drainage systems toward sea level; kinetic energy is responsible for the erosion and transportation of soil and rock debris suspended or in the bedload of the fluvial system.

Stream economy or energy flow is dependent upon the discharge (cubic feet per second), the size and shape of the channel (feet), the velocity (feet per second), the gradient (percent slope), and the load (sediment parts per million). In hydrological computations it has been found that the discharge of any stream system (Q) is directly proportional to the channel width (w), times the channel depth (d), times the water velocity (v). Thus

$$Q = wdv$$

As the formula indicates, a change in one of the quantities affects and changes the other variables. In humid climates the amount of discharge of any given stream system is greatest at the mouth, the furthest downstream portion of the system. Obviously, if the discharge is greatest at the mouth, the other variables of channel width, depth, and velocity are also greater at the mouth, and it is true that river systems near their mouths commonly have wider, deeper channels which both decrease friction and allow for a higher flowing velocity.

All stream systems flowing toward sea level are attempting to achieve a gradient where all erosive forces will cease. This level, if obtainable, is that of sea level and is referred to as the base level of any stream system. When a reservoir is created anywhere along a stream gradient, it acts as a local base level for all the drainage flowing into it.

The profile of stream gradients is typically concave upward toward the headwaters. The processes and relationships represented by the formula $Q = wdv$ are active as the waters descend through the drainage system, acquiring and depositing bedload. If a reservoir structure is placed across the stream profile, however, the energy process is interrupted; the stream velocity decreases and a great deal of the suspended load and bedload is deposited. When the water leaves the reservoir over the spillway, its erosive power has been relatively increased, since its bedload has been removed, and the channel below the dam is exposed to increased forces of erosion. For example, after construction of Hoover Dam across the Colorado River in 1935, the channel below the dam lowered its depth 2 to 6 feet for the first 13 miles. Dams constructed across stream systems must therefore compensate in their design for the deposition of sedimentary materials within the reservoir and for the increased forces of erosion downstream.

Stages of Fluvial Systems

Stream systems are commonly classified by geomorphologists as young, mature, or old. These categories describe the degree of dissection within a region, however, rather than the relative age of landforms or geological structures.

The young or youth stage of a stream is characterized by the ability of the fluvial system to erode its own channel. Regional dissection of the landforms has begun and is characterized by steep stream gradients and high velocities which are able to carry the load delivered by the tributaries. The drainage system, although integrated, is rather coarse in texture and does not exhibit flood plains or meanders. The valleys are steep-sided and V-shaped, and the stream occupies the valley bottom. Rapids and waterfalls and rocky channels containing many potholes are common. Lakes and swamps occupy large depressions.

The mature stage of a stream system has a reduced gradient whose velocities are sufficient only to transport the load contained within the system. Unless its load is reduced or its flow increased, the river does not erode the channel bottom. A well-integrated drainage system is established, which has accomplished maximum dissection of the topography. Flood plains and their associated features develop in major valley bottoms which are graded or in equilibrium. When all the tributaries become graded, the system reaches advanced maturity; when the first-order drainage and gullies become graded, the system enters old age.

Old-age stream systems are either completely graded or in equilibrium (i.e., erosion is equal to deposition). They have extensive flood plains, rivers with very low gradients, and rivers with a great deal of meandering. The topography is either flat or undulating and is usually characterized by extensive flood plains and occasional exposed monadnocks (isolated hills). This last stage of fluvial erosion is referred to as a peneplain and is characterized by a dominant, flat plane (Lobeck, 1939).

The climatic factors of different regions influence the physical appearance of the peneplain. In *humid climates* peneplains have a gently rolling topography and deep soil development. In *semiarid climates* peneplains are characterized by broad alluvial fans, which coalesce and fill valleys, and by scattered rock outcrops in the uplands. In *arid climates* peneplain landscapes are typically broad and flat, punctuated by scattered, flat deflation hollows developed by wind erosion and the formation of deflation armor. In *tropical climates* savanna, or inselberg, landscapes develop; domelike hills protrude from the flat wash or peneplain developed around them, formations created by the differential, deep rock weathering common in granitic formations in tropical climates. *Arctic and subarctic climates* develop peneplains similar to those found in semiarid climates; the valleys are filled and there are scattered rock outcrops in the uplands.

Most of the world's land surface has been dissected to the peneplain or old-age stage of fluvial dissection. Sometimes a region is rejuvenated—that is, uplifted—thus causing the stream system to be no longer graded. If rejuvenation occurs where there is a peneplain, the existing, broad meanders of its stream system become entrenched in the underlying rock material and acquire gradients not normally associated with meandering forms.

Types of Fluvial (Waterlaid) Formations

When carrying velocities decrease, the sediments transported by fluvial processes are deposited. These deposits are sorted, since different velocities cause settlement of different textural sizes. Thus deep-water deposits consist primarily of fine silts and clays, while regions with higher-velocity waters may accumulate gravels and cobbles. The fluvial landforms discussed in this chapter are grouped into categories by their process of deposition and by their planimetric topographic forms. They include river fluvial formations such as flood plains, terraces, and deltas; alluvial landforms including alluvial fans, valley fills, and continental alluvium; lacustrine formations including lake beds (playas) and organic deposits; and marine landforms such as coastal plains, beach ridges, and tidal flats. It should be noted, however, that these landforms can occur in places other than those mentioned here; beach ridges, for example, can be found along glacial or nonglacial lakes. The fluvial landforms are briefly defined below.

Group 1: River-Associated Forms: Flood Plains, Terraces, Deltas

Flood Plains. Flood plains are formed when sediments carried by rivers and streams are deposited during floods in slack-water areas where velocities are low. Many features are associated with flood plains, including old meander channels or meander scrolls, the water-filled, meander cutoffs known as oxbow lakes, natural levees, point bars, slack-water areas, and various levels of terracing. The whole range of soil types and depths, and moisture and organic content is encountered in these formations.

Terraces. Terraces are topographic forms carved in bottoms of deep alluvial deposits or extensive flood plains. Most often, they originate from postglacial erosion of vast alluvial deposits; as the load is reduced following glaciation, the streams cut deeply through the alluvial materials forming terraces. In the case of flood plains, the the deposits are built up thickly by slow-moving floodwaters but are later eroded during normal flow. Terraces are not discussed separately in the following landform sections but in relation to the landforms where they commonly occur, such as flood plains and alluvial materials.

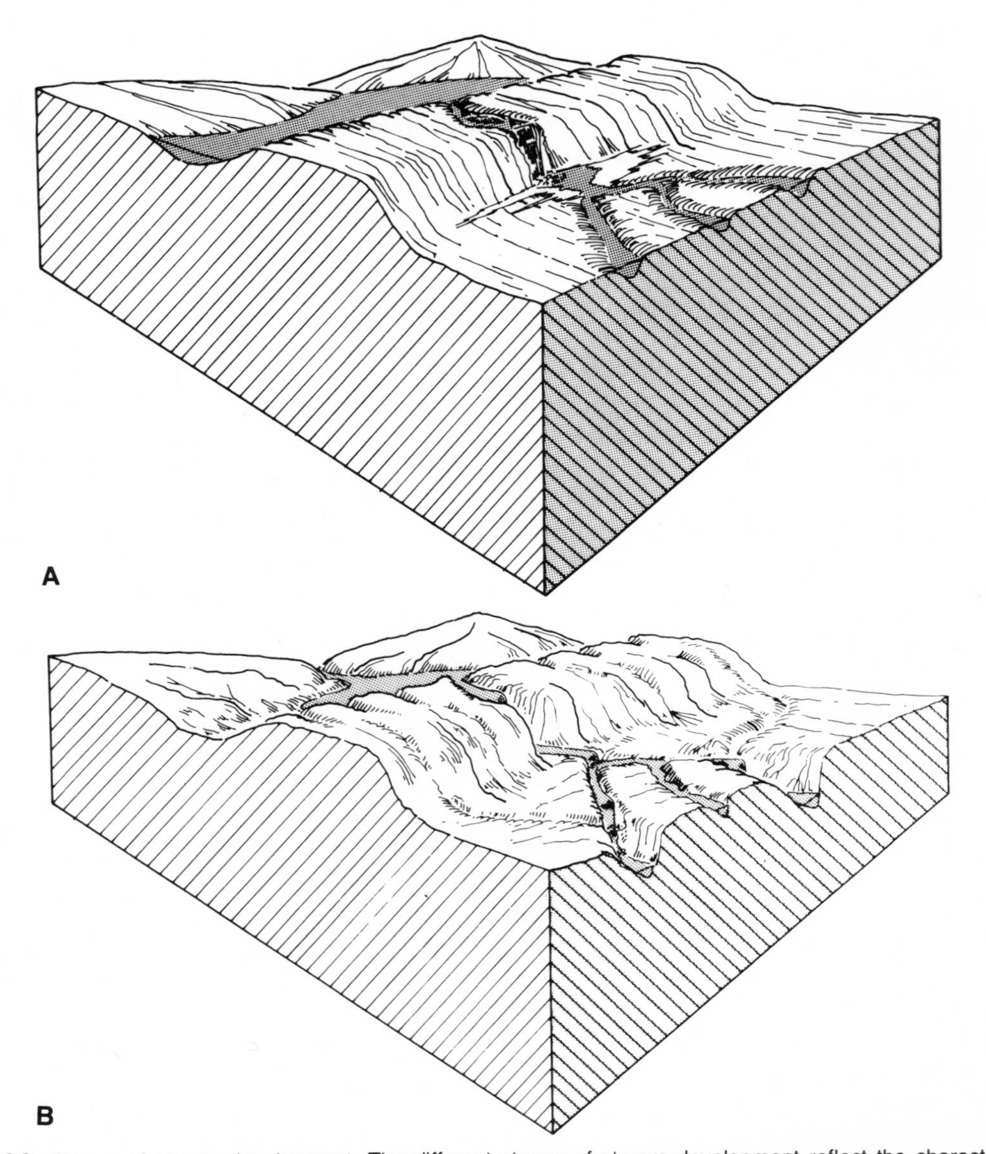

Figure 10.2. Stages of stream development. The different stages of stream development reflect the characteristics of the drainage system rather than the land mass. (A) In the initial stage many lakes, waterfalls, swamps, and rapids reflect the beginnings of an integrated drainage system. (B) In the middle youth stage, the drainage is integrated. Lakes disappear and rapids and small waterfalls are found within the narrow gorge. (C) As the fluvial system matures, it develops a narrow flood plain with wide, broad meanders. (D) Full maturity is characterized by the development of a broad flood plain with many associated features such as meander scars, meander scrolls, oxbow lakes, yazoo streams, and so on.

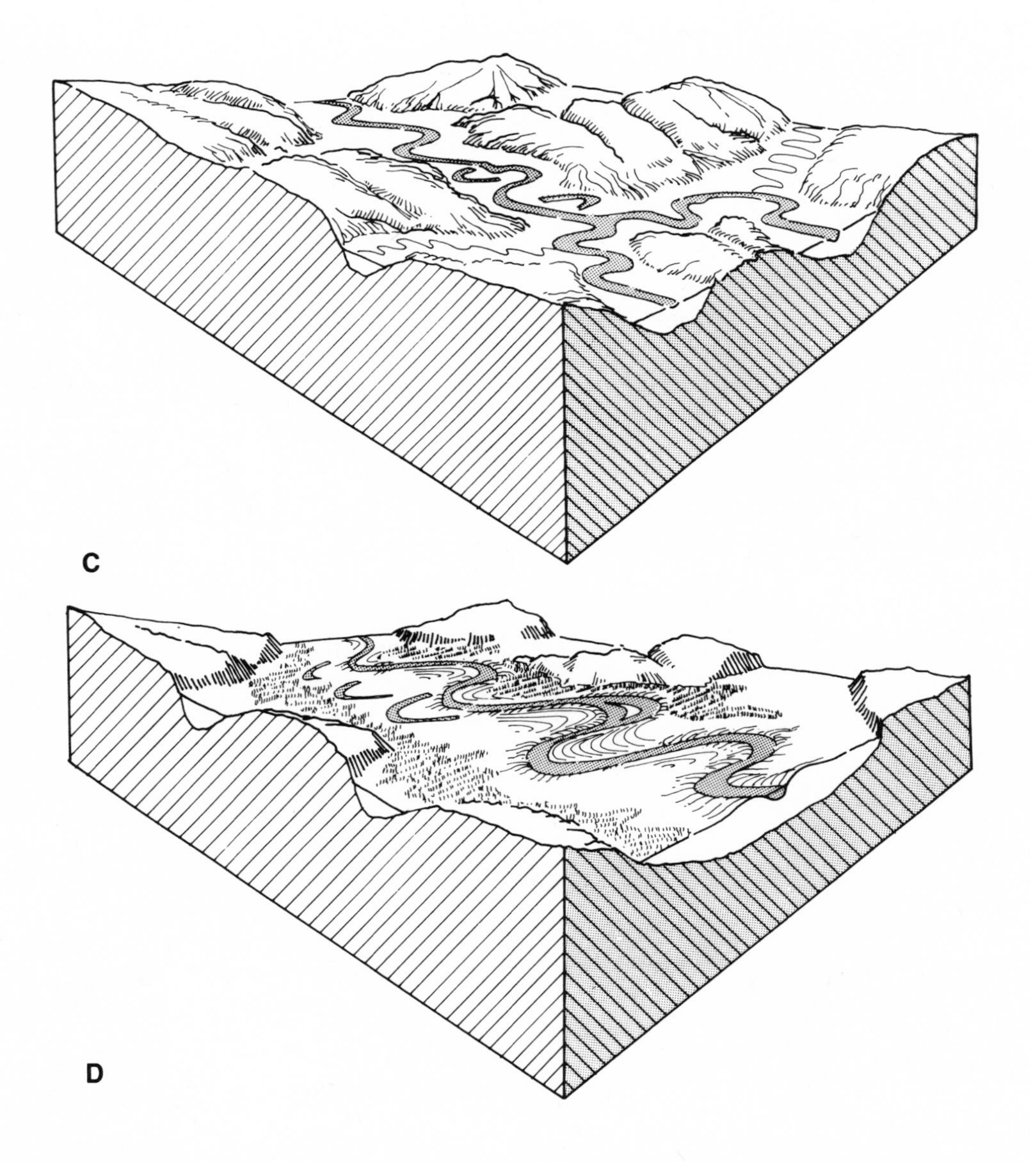

Deltas. Deltas are formed when a heavily loaded stream system encounters a large body of water, either an ocean or a lake. This immediately decreases both the gradient and the flow velocity, allowing sediments and debris to settle in sorted layers. The coarsest materials are found near the middle of the delta; the finer sands are found along the fringe. Many different types of deltas are classified, including birds-foot, estuarine, and arc, according to their planimetric shape or their position relative to the surrounding topography.

Group II: Alluvial Forms:
Alluvial Fans, Valley Fills, Continental Alluvium

Alluvial Fans. These formations are constructed by the sediments from loaded streams when they change gradient abruptly, as from a steep valley onto a relatively flat surface. The highest and coarsest portion of the fan is near the apex, where boulders and cobbles are found. The fan slopes toward its outer fringe, with slopes varying between 1 and 10 degrees. This landform is discussed as a type of alluvial deposit.

Valley Fill. Valley fills are alluvial materials transported from surrounding uplands which occupy valleys and basins. They occur in semiarid and arid climates. Valley fills are initiated by the formation of alluvial fans; these coalesce and fill the valley with alluvial materials. This landform is discussed as a type of alluvial deposit.

Continental Alluvium. Continental alluvial deposits occur where stream systems originating in mountains wash down large amounts of sediment onto adjacent plains. These materials are stratified and may be partially cemented. Continental alluvium landforms are discussed as types of alluvial landforms.

Group III: Freshwater Formations:
Lake Beds, Playas, Organic Deposits

Lake Beds. Lake beds are deposits of sediments laid down in lake bottoms which have since been drained and exposed. In North America most major lakes and lake bed formations are of glacial origin; lake beds, therefore, are discussed in Chapter 8.

Playas. Playas are lake bed formations found in semiarid or arid climates. Generally, they

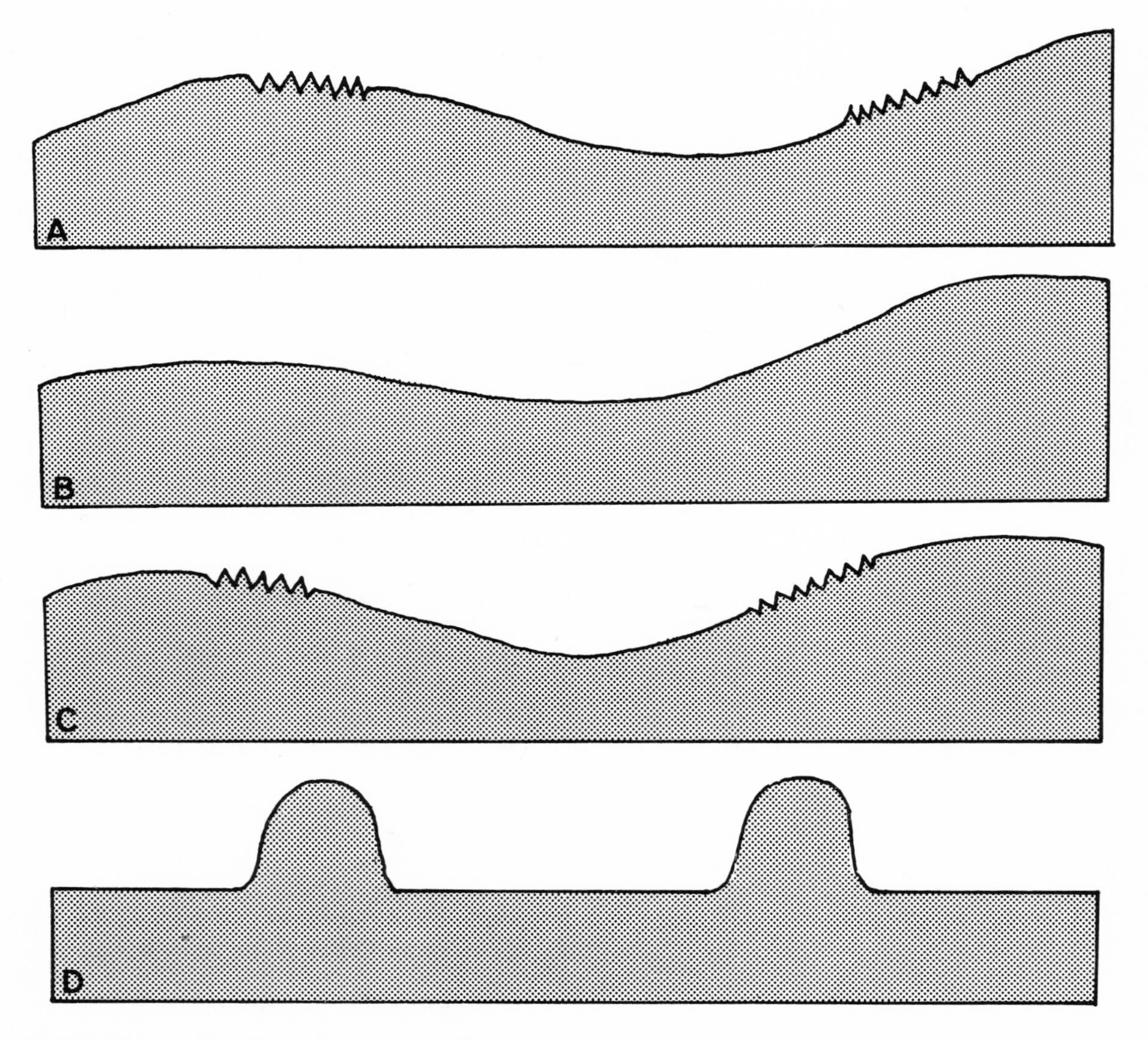

Figure 10.3. Types of peneplains. (A) Peneplain in an arid climate. (B) Peneplain in a humid climate. (C) Peneplain in a tropical climate. (D) Peneplain landscape in a subarctic climate.

occupy low basins which may have been former lakes and which may periodically be filled with water. These formations consist of stratified silts and clays which contain large quantities of soluble salts. Playas are also referred to as dry lakes, alkali flats, or salinas.

Organic Deposits. Organic materials accumulate in undrained depressions, eventually forming peat. Such deposits are generally classified by their predominate types of associated vegetation such as marshes, bogs, and swamps. When these formations are found in association with saltwater, they are classified as tidal flats.

Group IV: Marine Formations:
Coastal Plains, Beach Ridges, Tidal Flats

Coastal Plains. Coastal plains are formed when a sea floor is uplifted and exposed and becomes part of the earth's land surface. They are composed of unconsolidated to moderately consolidated sands with some silts and clays in stratified layers. The following two landforms, beach ridges and tidal flats, can in a broad sense be considered associated landforms of coastal plain formations.

Beach Ridges. Beach ridges are low ridges of stratified, unconsolidated sand and gravel which represent ancient shorelines. Each ridge was constructed by the wave action of lakes or oceans and represents a previous water elevation.

Tidal Flats. Tidal flats are stratified, unconsolidated sediments deposited along coastal areas under tidal influence and fluctuation. Different types are classified, according to the predominate composition, as sand flats, marsh flats, and mud flats.

Distribution

Since most landforms and regions are subject to erosion and the deposition of sediments from rainwater runoff, fluvial landforms can be found in all parts of the world. The following description highlights only some major occurrences of fluvial formations.

North America

United States. Many large rivers in the United States form large flood plains; examples are the Mississippi, the Missouri, the Ohio, and the Connecticut. The better known delta deposits include the Mississippi River, the Colorado River, and the St. Clair River deltas. Alluvial deposits are scattered through western valleys, including Great Valley, California, but are also rather extensive in the Midwest from the base of the Rocky Mountains to the Mississippi River. Many exposed glacial and nonglacial lake beds occur across the United States; those in arid or semiarid climates, such as the Great Salt Lake, are alkaline. Coastal plains, associated beach ridges, and tidal marshes are extensive along the Gulf and Atlantic Coasts, north to Massachusetts. Beach ridges are also found adjacent to ancient glacial lake beds, including the earlier shorelines of the Great Lakes. Organic deposits are very common landforms in all regions, distributed throughout humid, subtropical, and tropical climates.

Canada. Most of the fluvial landforms in Canada are flood plains, lake beds, and organic deposits. The MacKenzie River system is the largest, draining most of west central Canada and ending in an estuarine delta in the Northwest Territories. There are thousands of organic deposits, as a result of the recent glacial activity.

South and Central America

The Amazon River and the Parana River are the major drainage systems of South America. Major deltas are found at the mouth of the Amazon in Brazil and the Orinoco in Venezuela. Alluvial formations are found along the base of the Andes Mountains. Some small lake beds are found in Argentina, and lake beds and playas are abundant through Central America. Many large swamps and organic deposits are found in South America, especially in southwestern Brazil, Paraguay, northern Argentina, and northern and eastern Bolivia. Some coastal plain deposits are found along the coasts of the Guianas.

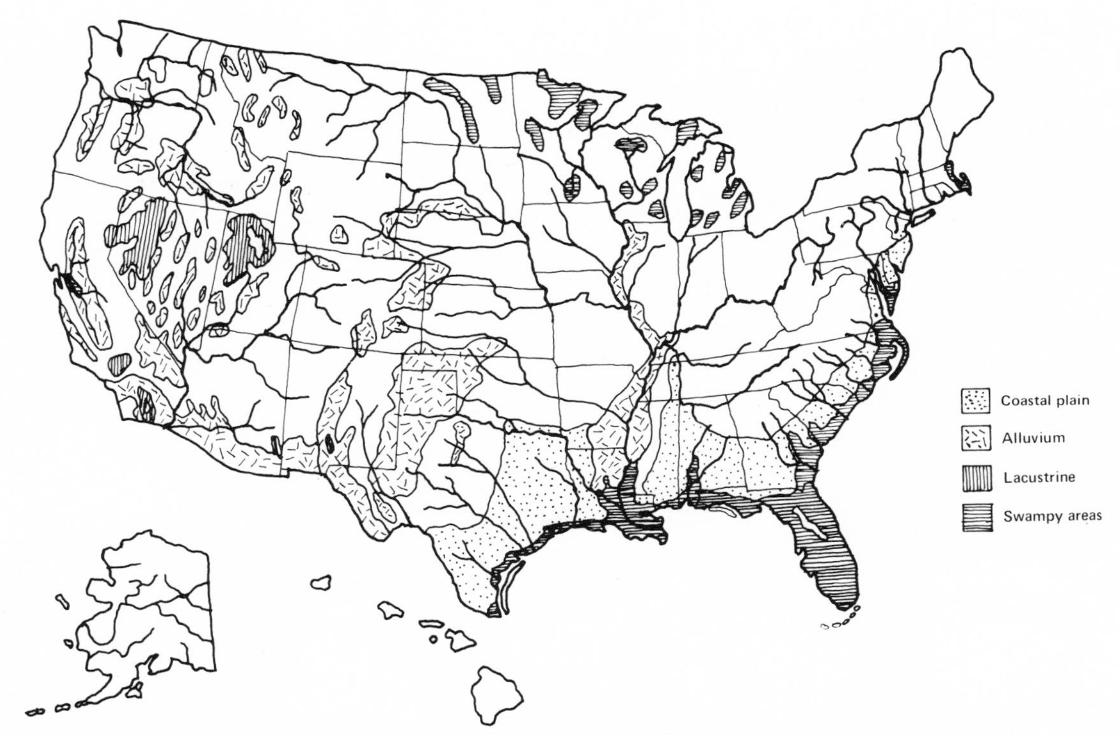

Figure 10.4. Distribution of fluvial forms throughout the United States. Note that swampy areas indicate only 10-50% coverage by actual swamps. (After "National Atlas," U.S. Geological Survey, Washington, D.C., 1971.)

Africa

Major river systems include the Nile, Zambezi, Congo, and Niger Rivers. Large deltas are found at the mouths of the Nile and Niger Rivers. Extensive portions of Africa are covered with alluvial deposits, with some scattered playa lake beds. Most significant marine formations are found along the Mediterranean, although there are scattered fluvial deposits along the South Atlantic and Indian Oceans.

Europe

There are many river systems in Europe, including the Rhine, the Danube, and the Volga in eastern Russia. The Danube and Volga Rivers have developed large deltas. Many lakes and lake beds are found throughout Europe, especially in Scandinavia. Local deposits of alluvium are found adjacent to the Alps. Organic deposits are common, scattered throughout Europe. Marine formations are found from the Netherlands east to Poland with some scattered features along the northern coast of eastern Russia.

Asia

Many large river systems are found in Asia, including the Ob, Yenisei, and Lena in Russia,

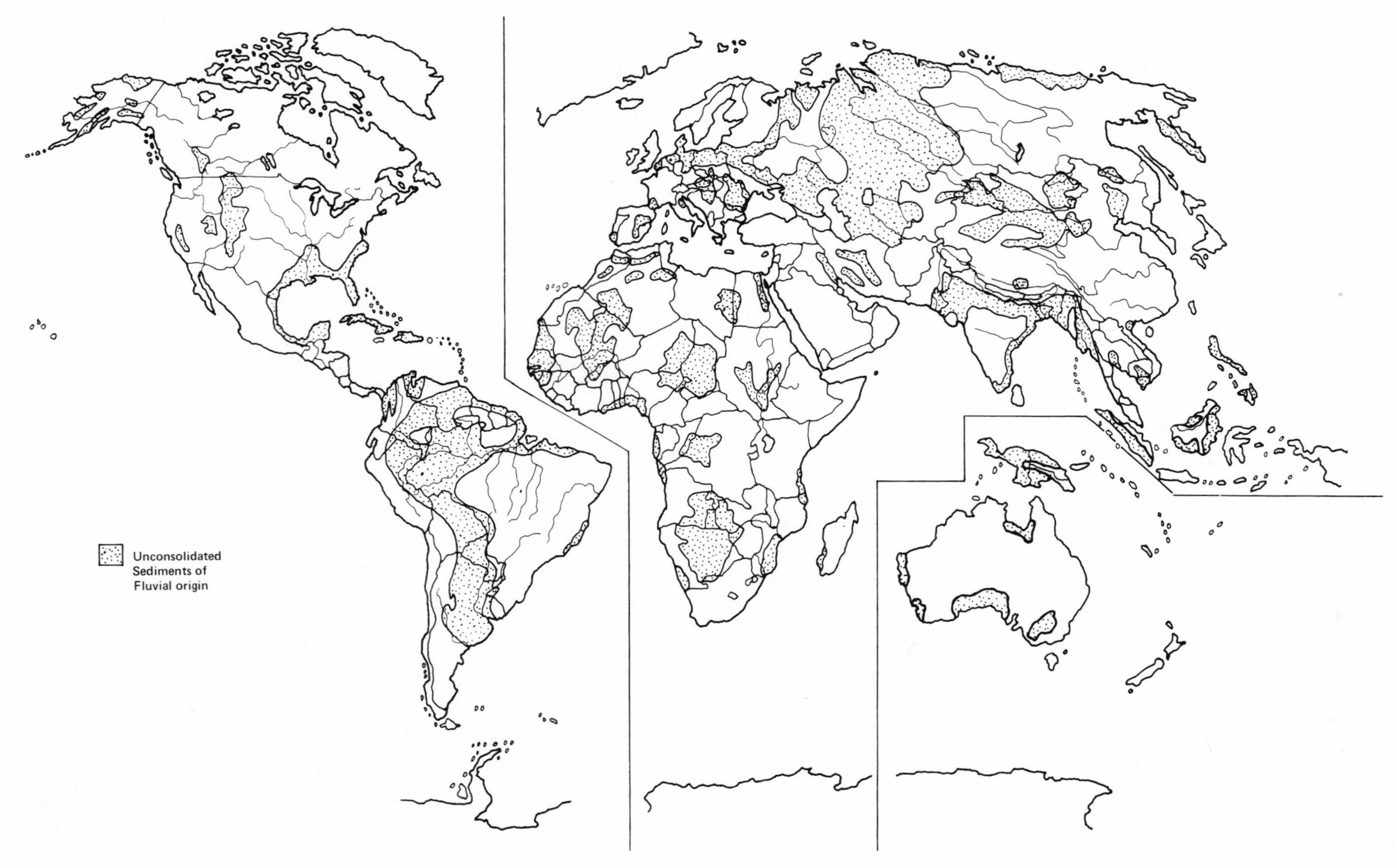

Figure 10.5. Distribution of fluvial landforms throughout the world. (After Trewartha, G. T., Robinson, A. H., and Hammond, E. H., *Elements of Geography,* McGraw-Hill, New York, 1967.)

the Hwang Ho and Yangtze Kiang in China, the Mekong in southeast Asia, and the Ganges in India. Major deltas include those found at the mouths of the Lena and Indigirka Rivers in Russia, the Indus River in Pakistan, the Cauvery, Godavari, and Mahanadi Rivers in India, the Irrawaddy in Burma, and the Mekong in South Vietnam. Alluvium is widespread throughout Russia, Tibet, and the Near East, along with many

Table 10.1. Summary chart: Fluvial landforms

Landform	Topography	Drainage	Tone	Gullies	Vegetation and land use
River-associated forms					
Flood plains					
Meander	Flat	Broad Meanders	Complex	Vary	Cultivated or natural
Covered	Flat	Flood plain features*	Complex	Vary	Cultivated or natural
Terraces	Terraced valley sides	Internal—dendritic	Light	V- or U-shaped	Cultivated or natural
Deltas	Flat, slight slope	Numerous channels	Light to mixed	None	Natural or developed
Alluvial forms					
Alluvium					
Fans	Fan-shaped	Radial—braided	Light gray	None	Scattered scrub and grass
Valley fill	Filled valley bottoms	Parallel—braided	Uniform light gray	None	Natural or cultivated
Continental	Flat level plains	Internal (depressions)	Uniform light gray	Few, U-shaped	Grain or natural cover
Freshwater deposits					
Lake beds (playas)	Flat plain, basin	None—evaporation	Light to scrabbled	None	Barren to cultivated†
Organic deposits	Flat, depressions	None—dendritic	Dark to light	None (ditched)	Vegetated to cultivated
Marine forms					
Costal plains					
Young	Flat, undulating	Parallel—dendritic, coarse	Mottled	Soft, U-shaped	Cultivated and forested
Old	Dissected, cuestas	Dendritic, medium to fine	Uniform light	U-shaped (V)	Forested, some agriculture
Beach ridges					
Sand	Soft, parallel ridges	Internal	Light	None, V-shaped	Natural cover to cultivated‡
Gravel	Sharp, parallel ridges	Internal	Light	None, V-shaped	Natural cover to cultivated‡
Tidal flats					
Marsh	Flat	Tidal, dendritic	Vary	None	Marsh grass
Mud	Flat	Tidal, dendritic	Vary	None	Little vegetation
Sand	Flat	Tidal, parallel channels	Light to dark	None	None

*Flood plain features include old meander channels, oxbow lakes, meander scrolls, natural levees, slack-water areas, bars, and so on.
†Cultivated if under irrigation.
‡Ancient glacial beach ridges are commonly cultivated in humid, subhumid climates.

organic deposits, lake beds, and playas. Marine formations are common along the coasts of northern Russia and Siberia, southeastern Asia, India, Iran, and Saudi Arabia.

Australasia

The Murray River is the largest system, draining most of southeast Australia. The predominant fluvial landforms in this region are the playas scattered through central, south central, and western Australia. Some scattered marine formations are found in areas protected from ocean erosion.

Pacific Region

Local marine formations are the predominate fluvial landforms.

Caribbean Region

Local marine formations are the predominate fluvial landforms.

REFERENCES

Adams, W., L. Lepley, C. Warren, and S. Chang, "Coastal and Urban Surveys with IR," *Photogrammetric Engineering*, Vol. 36, No. 2, pp. 173–180, 1970.

Allen, J.R.L., "A Review of the Origin and Characteristics of Recent Alluvial Sediments," *Sedimentology*, Vol. 5, 1965.

Allen, J.R.L., *Physical Processes of Sedimentation*, Allen and Unwin, London, 1970.

Anderson, R. R., and F. J. Wobber, "Wetlands Mapping in New Jersey," *Photogrammetric Engineering*, Vol. 39 (April) 1973.

Bagnold, R. A., "Some Aspects of River Meanders," U.S. Geological Survey, Professional Paper No. 282-E, 1960.

Bagnold, R. A., "Beach and Nearshore Process," in M. N. Hill, ed., *The Sea,* Vol. 3, Wiley, New York, 1963.

Bascomb, W., *Waves and Beaches,* Anchor Books, Doubleday, New York, 1964.

Bird, E.C.F., *Coasts,* M.I.T. Press, Cambridge, Mass., 1969.

Blissenbach, E., "Geology of Alluvial Fans in Semi-Arid Regions," *Bulletin of Geological Society of America,* No. 65, 1954.

Bogardi, J. L., "Some Aspects of the Application of the Theory of Sediment Transport to Engineering Problems," *Journal of Geophysical Research,* Vol. 66, 1961.

Bull, W. B., "Types of Deposition on Alluvial Fans in Western Fresno Co., California," *Geological Society of America Bulletin,* No. 71, 1960.

Bull, W. B., "Alluvial Fans and Near Surface Subsidence in Western Fresno County, California," U.S. Geological Survey, Professional Paper No. 437A, 1964.

Bull, W. B., "Geomorphology of Segmented Alluvial Fans in Western Fresno County, California," U.S. Geological Survey, Professional Paper No. 352E, 1964.

Chow, V. T., *Handbook of Applied Hydrology,* McGraw-Hill, New York, 1964.

Colby, B. R., *Effect of Depth of Flow on Discharge of Bed Material,* U.S. Geological Survey, Water-Supply Paper No. 1498-D, 1961.

Conkling, H., R. Eckis, and P.J.K. Gross, *Ground Water Storage Capacity of Valley Fill,* California Division of Water Resources, Bulletin No. 45, 1934.

Dolan, R., and L. Vincent, "Coastal Processes," *Photogrammetric Engineering,* Vol. 39 (March) 1973.

Eakin, H.M., and C.B. Brown, *Silting of Reservoirs,* U.S. Department of Agriculture, Technical Bulletin, No. 542, 1939.

El-Ashry, M. R., and H. R. Wanless, "Shoreline Features and Their Changes," *Photogrammetric Engineering,* Vol. 33, No. 2, pp. 184–189, 1967.

Emery, K. O., "Grain Size of Marine Beach Gravels," *Journal of Geology,* Vol. 63, 1955.

Fenneman, N. M., *Physiography of the Eastern United States,* McGraw-Hill, New York, 1939.

Friedkin, J. F., *A Laboratory Study of the Meandering of Alluvial Rivers,* U.S. Waterways Engineering Experiment Station, 1945.

Gilbert, G. K., *The Transportation of Debris by Running Water,* U.S. Geological Survey, Professional Paper No. 86, 1914.

Gregory, K. J., and D. E. Walling, *Drainage Basin Form and Process,* Wiley, New York, 1973.

Guilcher, A., *Coastal and Submarine Morphology,* Wiley, New York, 1958.

Hack, J. T., "Interpretation of Erosional Topography in Humid Temperate Regions," *American Journal of Science,* Vol. 258A, 1960.

Hayes, M. O., and J. C. Boothroyd, *Storms as Modifying Agents in the Coast Environment—Field Trip Guidebook,* Geology Department, University of Massachusetts, Amherst, 1969.

Helgenson, G. A., "Water Depth and Distance Penetration," *Photogrammetric Engineering,* Vol. 36, No. 2, pp. 164–172, 1970.

Holman, W. W., and H. C. Nikola, "Airphoto Interpretation of Coastal Plain Areas," *Highway Research Board Report No. 83,* National Academy of Science, Washington, D.C., 1953.

Hooke, R. LeB., Houng-Yi Chang, and P. W. Weiblen, "Desert Varnish: An Electron Probe Study," *Journal of Geology,* Vol. 77, 1969.

Horton, R. E., "Erosional Developments of Streams and Their Drainage Basins; Hydrophysical Approach to Quantitative Morphology," *Bulletin of Geological Society of America,* Vol. 56, 1945.

Hoyt, J. H., "Barrier Island Formation," *Bulletin of Geological Society of America,* No. 78, 1967.

Jahns, R. H., "Geologic Features of the Connecticut Valley, Massachusetts-Related to Recent Floods," U.S. Geological Survey, Water Supply Paper No. 996, 1947.

King, C.A.M., *Beaches and Coasts,* St. Martin's, New York, 1960.

Krumbein, W. C., and J. S. Griffith, "Beach Environment in Little Sister Bay, Wisconsin," *Bulletin of the Geological Society of America,* Vol. 49, 1938.

Lawson, A. C., "The Petrographic Designation of Alluvial Fan Formations," Department of Geological Sciences, University of California, 1925.

Leopold, L. B., and T. Maddock, Jr., *The Hydraulic Geometry of Stream Channels and Some Physiographic Implications,* U.S. Geological Survey, Professional Paper No. 252, 1953.

Leopold, L. B., and J. P. Miller, *Ephemeral Streams-Hydraulic Factors and Their Relation to the Drainage Net,* U.S. Geological Survey, Professional Paper No. 282-A, 1956.

Leopold, L. B., and M. G. Wolman, "River Channel Patterns," U.S. Geological Survey, Professional Paper No. 282B, 1957.

Leopold, L. B., M. G. Wolman, and J. P. Miller, *Fluvial Processes in Geomorphology,* W. H. Freeman, San Francisco, 1964.

Linsley, R. K., M. A. Kohler, and J.L.H. Paulhus, *Applied Hydrology,* McGraw-Hill, New York, 1949.

Lobeck, A. K., *Geomorphology, An Introduction to the Study of Landscapes,* McGraw-Hill, New York, Chapters 5–7, 10, and 13, 1939.

Pestrong, R., "Multiband Photos for a Tidal Marsh," *Photogrammetric Engineering,* Vol. 35, No. 5, pp. 453–470, 1969.

Porter, O. J., "Studies of Fill Construction over Mud Flats: Including a Description of Experimental Construction Using Vertical Sand Drains to Hasten Stabilization," *Proceedings of the 1st International Conference on Soil Mechanics,* Cambridge, Mass., 1936.

Reimold, R. J., J. L. Gallagher, and D. E. Thompson, "Remote Sensing of Tidal Marsh," *Photogrammetric Engineering,* Vol. 39 (May) 1973.

Reineck, H. E., and I. B. Singh, *Depositional Sedimentary Environments,* Springer-Verlag, New York, 1975.

Scherz, J. P., D. R. Graff, and W. C. Boyle, "Photographic Characteristics of Water Pollution," *Photogrammetric Engineering,* Vol. 35, No. 1, pp. 38–43, 1969.

Schumm, S. A., "Effect of Sediment Characteristics on Erosion and Deposition in Ephemeral Stream Channels," U.S. Geological Survey, Professional Paper No. 352C, 1961.

Seher, J. S., and P. T. Tueller, "Color Aerial Photos for Marshland," *Photogrammetric Engineering,* Vol. 39 (May) 1973.

Stafford, D. B., and J. Langfelder, "Air Photo Survey of Coastal Erosion," *Photogrammetric Engineering,* Vol. 37, No. 6, 1971.

Strandberg, C. H., *Aerial Discovery Manual,* Wiley, New York, Chapters 13–18, 1967.

Tuyahow, A. J., and R. K. Holz, "Remote Sensing of a Barrier Island," *Photogrammetric Engineering,* Vol. 39 (February) 1973.

Twenhofel, W.H., *Principles of Sedimentation,* McGraw-Hill, New York, 1950.

Van Lopik, J. R., A. E. Pressman, and R. L. Ludlum, "Mapping Pollution with Infrared," *Photogrammetric Engineering,* Vol. 34, No. 6, pp. 561–564, 1968.

Williams, P.F., and B.R. Rust, "The Sedimentology of a Braided River," *Journal of Sedimentary Petrology,* Vol. 39, 1969.

Wobber, F. J., and R. R. Anderson, "ERTS Data for Coastal Management," *Photogrammetric Engineering,* Vol. 39 (June) 1975.

Flood Plains

Introduction

Flood plains border streams and rivers and are formed by the deposition of sediments carried by streams and deposited during floods. For purposes of aerial photographic interpretation, flood plains have been placed in three categories: meander flood plains, covered flood plains, and composite flood plains.

Meander flood plains are commonly associated with the mature or old-age stages of the cycle of stream erosion. They are formed by streams which are subject to full loads but which do not overflow their banks during floods. The primary zone of erosion is along the outer edges of the meanders; the most active zone of deposition is along the inside of the meanders where sand or gravel bars are commonly formed. Many features which indicate migrating stream channels can be found, such as meander scrolls, cutoff meanders, oxbow lakes, and other features of abandoned channels. The meandering form of the flood plain indicates a low gradient. The soils are usually coarse-textured.

Covered flood plains are formed from sediments deposited when floodwaters overflow river banks and form natural levees and slack-water deposits. This type of flood plain is commonly associated with the old-age stage of the stream erosion cycle. Natural levees are formed adjacent to the stream channel and consist of relatively coarse sediments deposited as the river water overflowed the bank and the velocity of the water was significantly decreased. Silts and clays are carried by the remaining flood waters to the lower reaches of the flood plain where low velocities allow their deposition. Compared to meander flood plains, covered flood plains have a predominate composition of fines. Natural levees have coarser materials.

Composite flood plains contain features common to both meander and covered flood plains, since the stream systems in this case are subject to both full and overbank loads. The features of meander flood plains, such as meander scrolls, oxbow lakes, and abandoned channels, are commonly found mixed with features of covered flood plains, such as natural levees and slack-water deposits (Figure 10.6).

Associated with flood plain formations are *terraces,* which represent portions of a flood plain of some previous larger river system. Terraces are common in glaciated regions. Glacial meltwaters, heavily loaded with sediments, formed large, deep flood plains, and after glaciation, as climatic conditions changed and meltwater flow and sediments decreased, the stream system increased its ability to erode from its channel and removed much of the previously deposited alluvium, transporting it downstream. Often, over a large area, several levels of terraces may be observed; such a situation requires the use of small-scale photographs or photographic indexes to include all the significant marginal boundaries (Figure 10.7). Microfeatures such as gullies and miniature drainage patterns are critical for the interpretation of the predominant soil textures. Soils in glacial terraces are predominantly coarse and consist of sands and gravels. Characteristically, terraces have steep banks, a few V-shaped gullies, and occasional dark, spotty tones indicating infiltration drainage associated with gravel deposits. Coarse textures are also indicated by the disappearance of drainage channels from adjacent uplands when the terrace is encountered. In contrast, terraces of marine or lacustrine origin commonly contain finer-textured silts and clays; since they actually represent partially dissected variations of those landforms, they are discussed in the sections entitled "Lake Beds" and "Coastal Plains."

Interpretation of Pattern Elements

Meander flood plains are easily identified by the many meander features, scrolls, and abandoned channels which indicate lateral channel erosion and deposition. The topography is typically flat. The many complex variations in soil tone indicate differences in moisture content. In most regions, unless the water table is not on or very near the ground surface, these deposits are cultivated. Occasionally, ditching and other means of artificial drainage can be observed—results of attempts to lower the high water table. Continuous lines of spoil along the sides of ditches indicate the presence of coarse materials; that is, if the materials were fine, the spoil would act as a dam for surface water drainage into the ditch, but here drainage is obviously able to occur through the subsoil.

Covered flood plains are identified by their flat topography and features associated with overbank flooding and deposition, such as natural levees and slack-water deposits. Many complex tones are found, indicating differences in soil moisture and associated texture. Cultivation is generally the predominant form of land use unless the water table occurs very near or on the surface, in which case artificial drainage ditches may be used to lower the high water table. Alternating lines of spoil along each side of the ditches may indicate fine-textured soils; here, the alternating arrangement allows for surface drainage of water.

Composite flood plains are the most common type and have the characteristics of both meander and covered flood plains.

Terraces are identified by their flat, elevated topographic form occurring along valley walls;

their microfeatures identify the nature of the materials comprising them.

Soil Characteristics

A wide variety of soils of almost all textures and grades can be found in flood plains, and their distribution may vary on a small scale, both vertically and horizontally. Clean, coarse gravels are difficult to locate but may be found in terraces of glacial origin. Soil tone is useful for mapping and locating differences in soil texture; organic deposits are indicated by black tones, while coarser sands along point bars or natural levees appear light or bright. The associated landforms within the watershed determine the soil textural composition. Stream systems eroding fine-textured sediments develop fine-textured flood plains; coarse sediments form relatively coarse flood plains.

U. S. Department of Agriculture Classification

All textural types may occur within the soil triangle representing a flood plain deposit. Slack-water deposits are commonly clay, silt clay, and silty clay loam; point bars are sandy; natural levees are sandy, loamy sand, sandy loam, or silt loam; and depressions or swamps may contain organic material or peat. Even if surface soils are accurately mapped, subsurface conditions may vary widely from them.

Unified and AASHO Classifications

Almost any Unified category can be found, except for GP which generally does not occur. GW soils are difficult to locate but may be associated with braided channels or along point bars. Natural levees may contain SM or ML soils; slack-water areas, ML, CL, MH, CH, and MH-CH

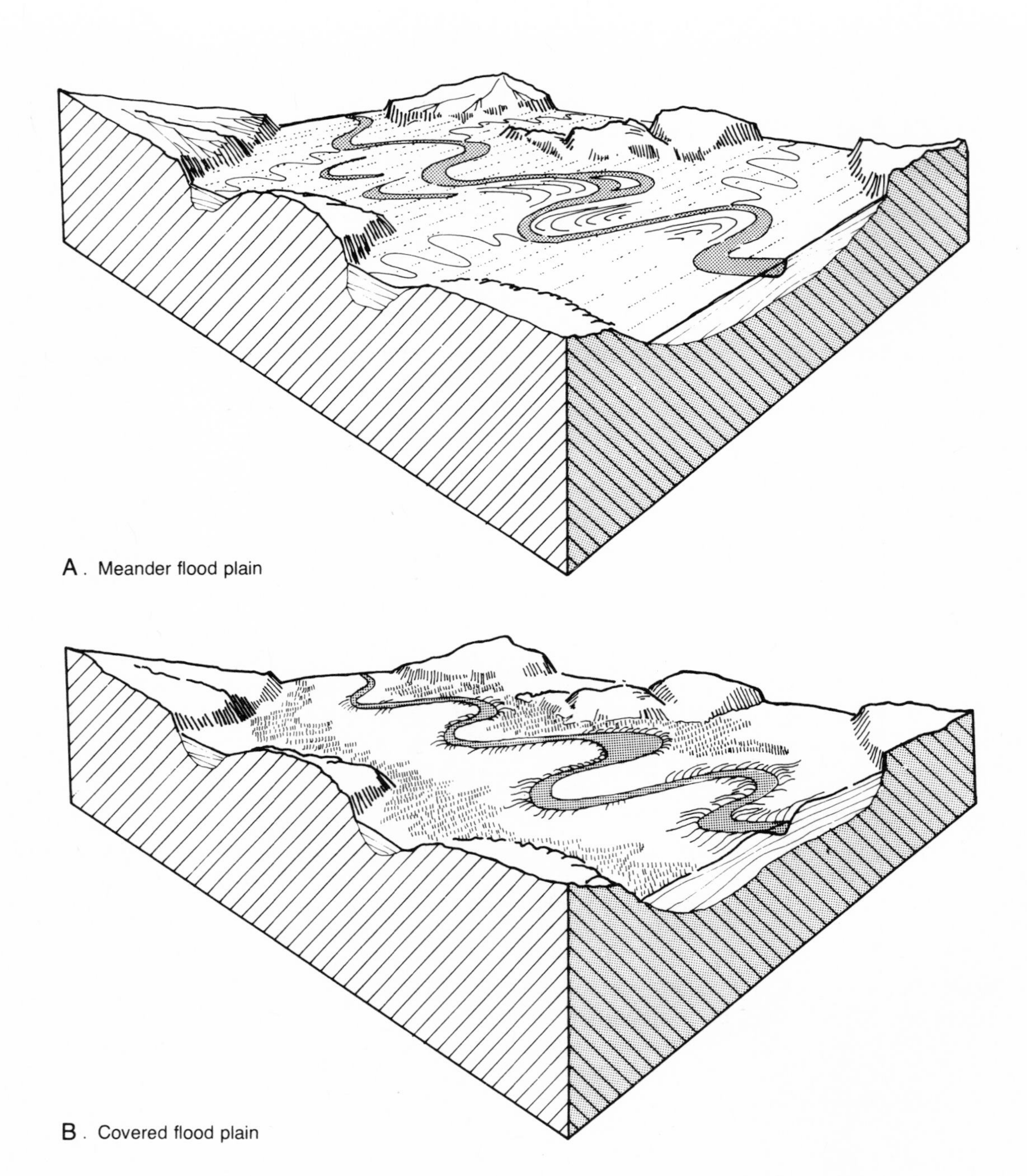

A. Meander flood plain

B. Covered flood plain

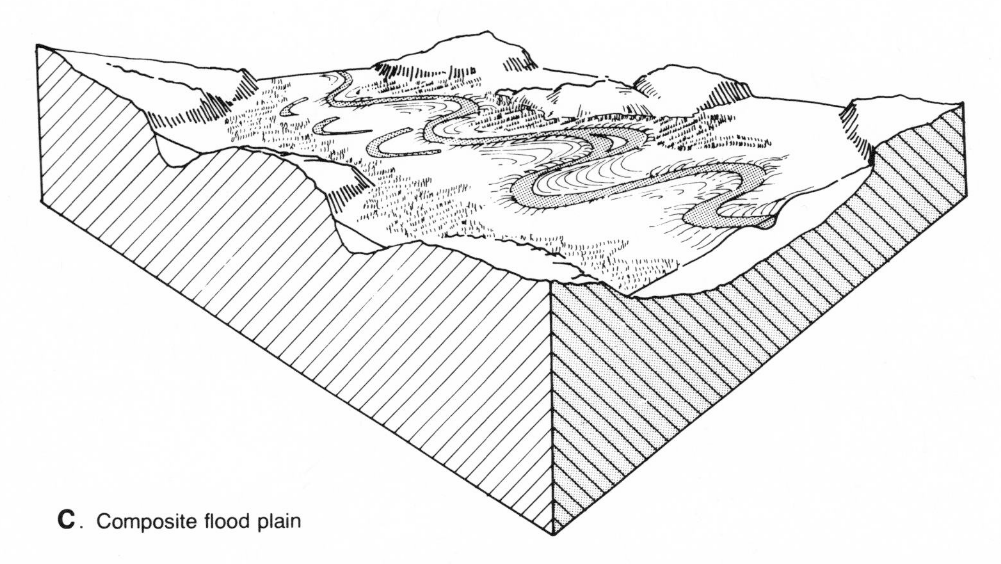

Figure 10.6. Meander, covered, and composite flood plains. Flood plains are typified by a rather flat, monotonous topography and land use pattern offering little spatial or viewing variation. Flood plains contained within wide, broad valleys may not have any feeling of enclosure but be visually similar to a flat plain. Minor surface irregularities may exist, but these do not offer any significant viewing capabilities. Flood plains in narrow valleys may provide a feeling of spatial enclosure as a function of their surrounding type of landform. Some diversity and variation may occur with the presence of levees and abandoned meander channels; these show changes in land use and vegetation and contrast with the surrounding conditions. Terraces along the valley walls provide elevated viewing platforms over the adjacent lowlands, especially when viewed from the terrace edge, but when a terrace occurs, its rim edge—the edge where the sideslope meets the edge of the plateau—is visually sensitive when viewed from the valley below. (A) Meander flood plains are characterized by lateral erosion which develops many meander scrolls, abandoned channels, and oxbow lakes. These systems are not subject to overbank flooding. (B) Covered flood plains are subject to overbank flooding and form characteristic slack-water and natural levee deposits. (C) Composite flood plains have features associated with both meander and covered flood plains.

soils; and depressions and swamps, OL, OH, and Pt soils. Clean sands, such as SW or SP, may be found in point bar deposits. A wide range of AASHO categories is also found in flood plains, A-1 being common in point bars, A-4, A-5, A-6, or A-7 in slack-water deposits, and perhaps A-2 in natural levees. Again, the surface soil pattern may be, and typically is, underlaid by different categories of materials.

Water Table

Water tables generally occur near the surface in humid and subhumid climates. The seasonal high is typically from 1 to 3 feet, or may even be at the surface in depressions. Artificial drainage may be attempted in order to lower the high water table so the land can be utilized for cultivation. Depressions contain swamps or standing water; if these are not included in the permanent pattern of cultivation, they represent areas that are wet throughout the year.

Drainage

Most flood plains are poorly drained, as a result of their low topographic position and the typical predominance of fine soils. Artificial channels are common, indicating efforts to lower the high water table.

Soil Depth to Bedrock

In normal construction operations in flood plain formations, bedrock is not encountered. Typically, wider flood plains indicate deeper deposits.

Issues of Site Development

Sewage Disposal

Septic tank leaching fields are usually difficult to locate in flood plains, owing to the associated high water table. Coarser soils which have sufficient depth to the water table may still present the hazard of groundwater contamination, since percolation rates may be too rapid for sufficient decomposition of the effluent by aerobic bacteria. Flooding and seasonal high water tables, where they occur, force effluent to the surface and create a health hazard. Specific tests are necessary to determine capabilities, since the underlying soils typically differ from the surface soils, making them difficult or impossible to interpret from aerial photographs.

Solid Waste Disposal

Sanitary landfill operations are not recommended in flood plains because of the high water table, flooding, and possible contamination of

Table 10.2. Fluvial landforms: Flood plains (meander, covered, composite)

Topography	*Drainage*	*Tone*	*Vegetation and land use*
Flat	Meanders or channels	Complex	Cultivated or natural
Flood plains are characterized by their predominantly flat topography, although there may be some surface irregularities caused by abandoned channels. Natural levees occupy slightly elevated positions; slack-water deposits are found in lower areas. Terraces may be found along valley walls.	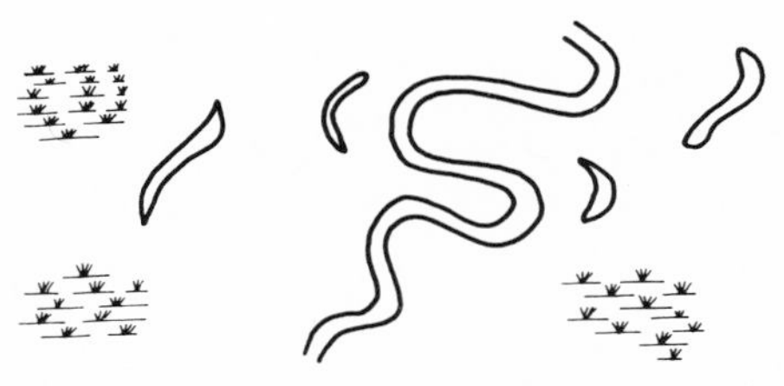 Most flood plains contain a major drainage channel which either meanders through the valley bottom or is braided. Many undrained swamps and ponds may exist, indicating a deranged pattern. In some arid and semiarid regions, drainage patterns may not even be apparent, but this does not necessarily indicate porous materials.	Photographic tones may vary widely from dark to light, dull to bright, indicating variations from well-drained to poorly drained soils and from fine to coarse textures. *Gullies* Few. Because of the flat topography, gullies are not normally found, but they are important microfeatures for identifying the composition of terrace soils.	Most flood plains contain rich agricultural soils which have high moisture availability relative to the surrounding region and are typically intensively cultivated. If the water table occurs near or on the surface, a natural cover of dense vegetation is typical, but in arid regions alkali deposits prevent dense growth.

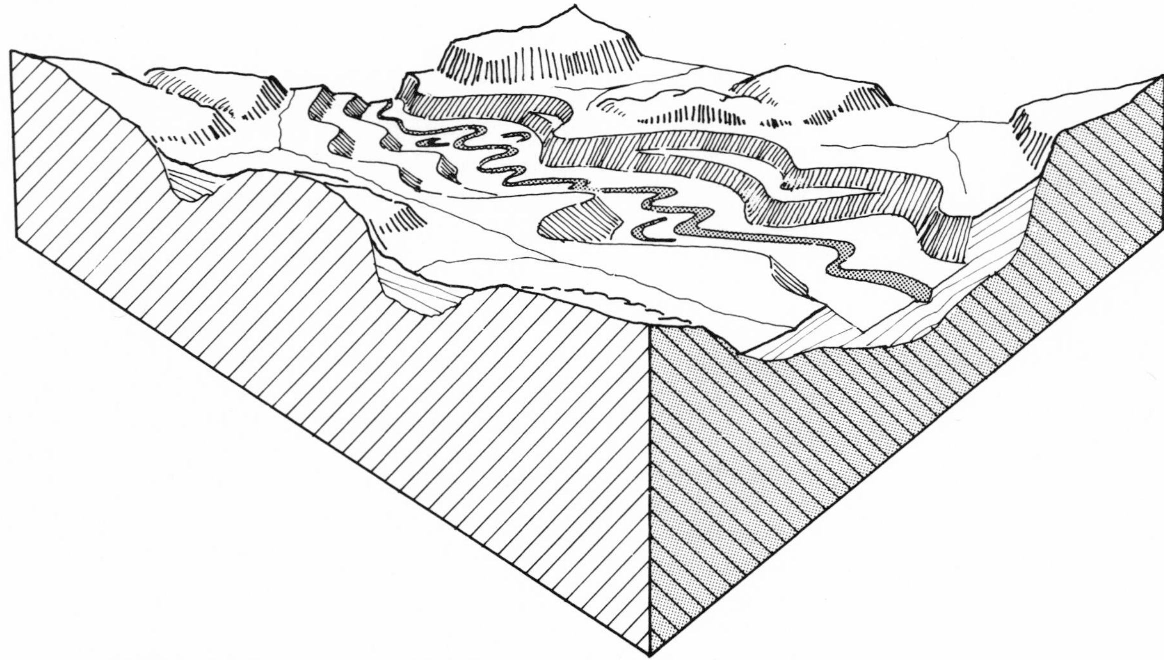

Figure 10.7. Terraces develop when a graded stream slowly erodes its previously deposited alluvium.

groundwater resource. Terraces may offer more suitable sites if they contain fines sufficient to prevent internal seepage of leachate.

Trenching

Trenching operations commonly encounter a high water table which may require draining or dewatering. Costs are typically 1.3 to 2.5 times those encountered in removing the same amount of deep, dry, moderately cohesive soil, the higher cost reflecting the high water table. Alignments are difficult around swamps, meander channels, abandoned channels, and so on.

Excavation and Grading

Excavation in active flood plains commonly encounters the water table, therefore requiring drainage or dewatering for pit-type excavation. Excavation or removal of materials can also be

Figure 10.8. Composite flood plain in Madison County, Idaho, has features of abandoned meander channels, oxbow lakes, meander scrolls, point bars, natural levees, and slack-water deposits. Photographs by the U.S. Geological Survey, Idaho, 7-ABC, 1:25,000 (2083′/″), October 8, 1950.

accomplished with a clam bucket or drag line apparatus. Costs are generally 1.3 to 2.5 times those encountered for the removal of deep, dry, moderately cohesive soils.

Construction Materials (General Suitabilities)

Topsoil: Excellent to Fair. In most regions flood plains provide excellent sources of topsoil materials containing a range of soil types which can be mixed for any desired composition. Highly organic materials can be found in swamps and depressions.

Sand: Excellent. Sands are generally found along stream beds, within point bars, or adjacent to braided streams. Natural levees may contain fine sands.

Gravel: Poor. Gravels do not commonly occur over large areas, but scattered deposits may be found within point bars or adjacent to braided channels. Outwash terraces originating from glacial meltwater may contain predominately sands and gravels, which are indicated by tonal and drainage characteristics.

Aggregate: Not Suitable. Coarse gravels generally do not occur, although small deposits may be found.

Surfacing: Excellent to Fair. In most flood plains a mix of fine and coarse-textured soils can be obtained which is suitable for the surfacing of secondary roads.

Borrow: Excellent to Poor. In most flood plains a variety of soil textures can be found, some of which provide excellent materials for borrow. The predominately fine-textured soils of covered flood plains are not a suitable source, since they require drying and compaction within moisture limits that may be difficult to control.

Building Stone: Not Suitable

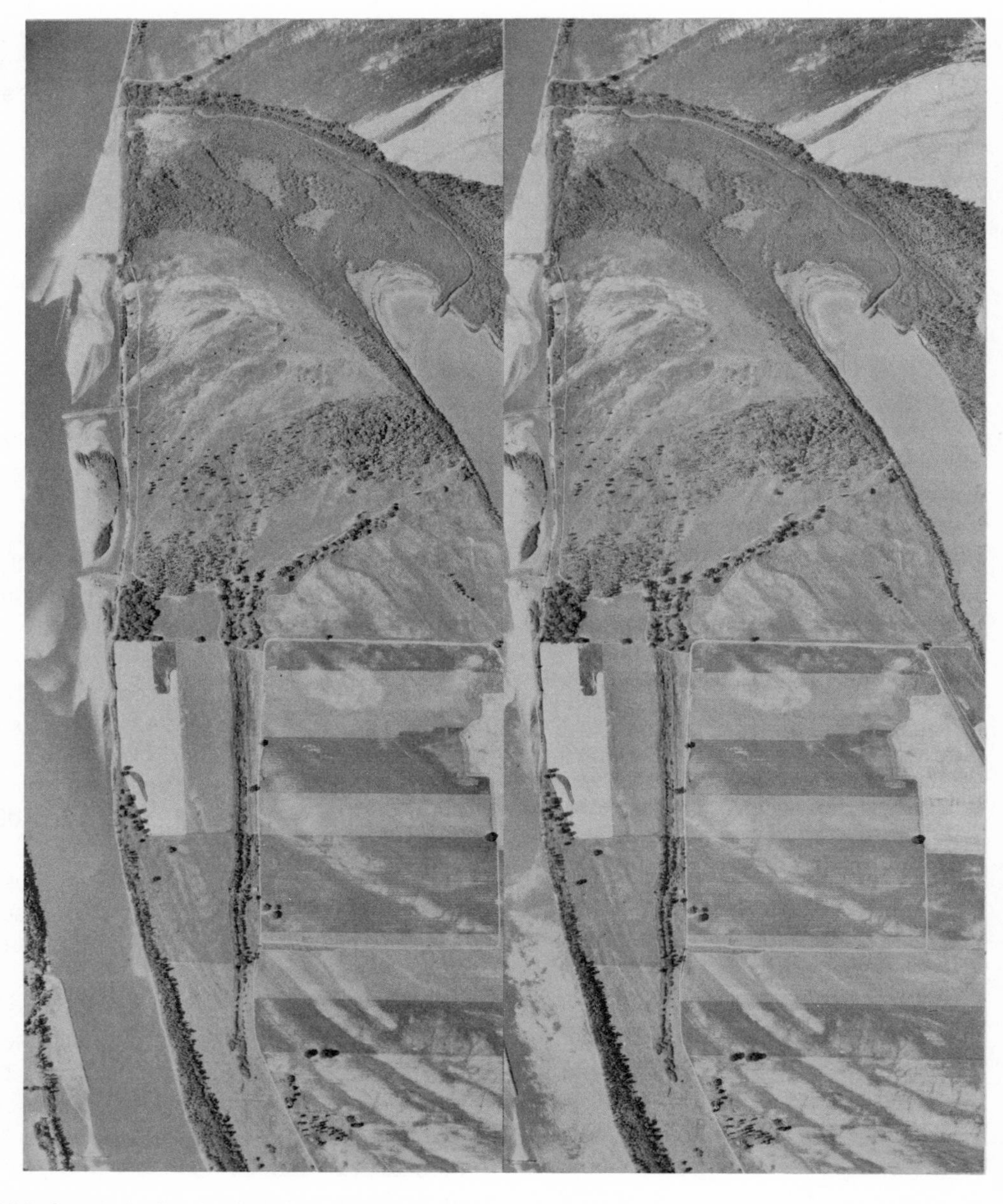

Figure 10.9. Composite flood plain in Arkansas clearly shows meander scrolls, natural levees, and abandoned meanders. Photographs by the Agricutural Stabilization Conservation Service, 1A-1HH-143,144, 1:20,000 (1667'/"), October 12, 1966.

Mass Wasting and Landslide Susceptibility

Flood plains generally occupy low topographic positions and are therefore not susceptible to landslides under normal conditions. Minor slumping is common along stream banks where undercutting has oversteepened slopes, especially along the outer edges of meanders in meander flood plains. Slumping hazards along stream banks are increased after floods, when the rapid drainage of flood waters temporarily creates an elevated water table in the surrounding flood plain and thus increases the mass along the edge of the river bank. Terraces of marine or lacustrine materials may contain varved materials and fine soil textures and therefore be extremely susceptible to sliding upon small increases in their moisture content.

Groundwater Supply

The groundwater table in flood plains is near the surface and typically fluctuates as flow rates increase or decrease. Enough mixtures of soil materials are encountered in a vertical section so that most wells have high yields. However, very fine-textured, shallow flood plain deposits have rather low yields, even if there is a high water table, for the ability of the soil to retain water may be greater than the ability of a pump to remove it.

Pond or Lake Construction

Reservoir areas sited in fine-textured flood plains do not encounter difficulties with seepage, but they probably require grading in order to create an impoundment basin. Subsurface exploration is necessary to determine whether or not granular substrata exist which could facilitate leakage.

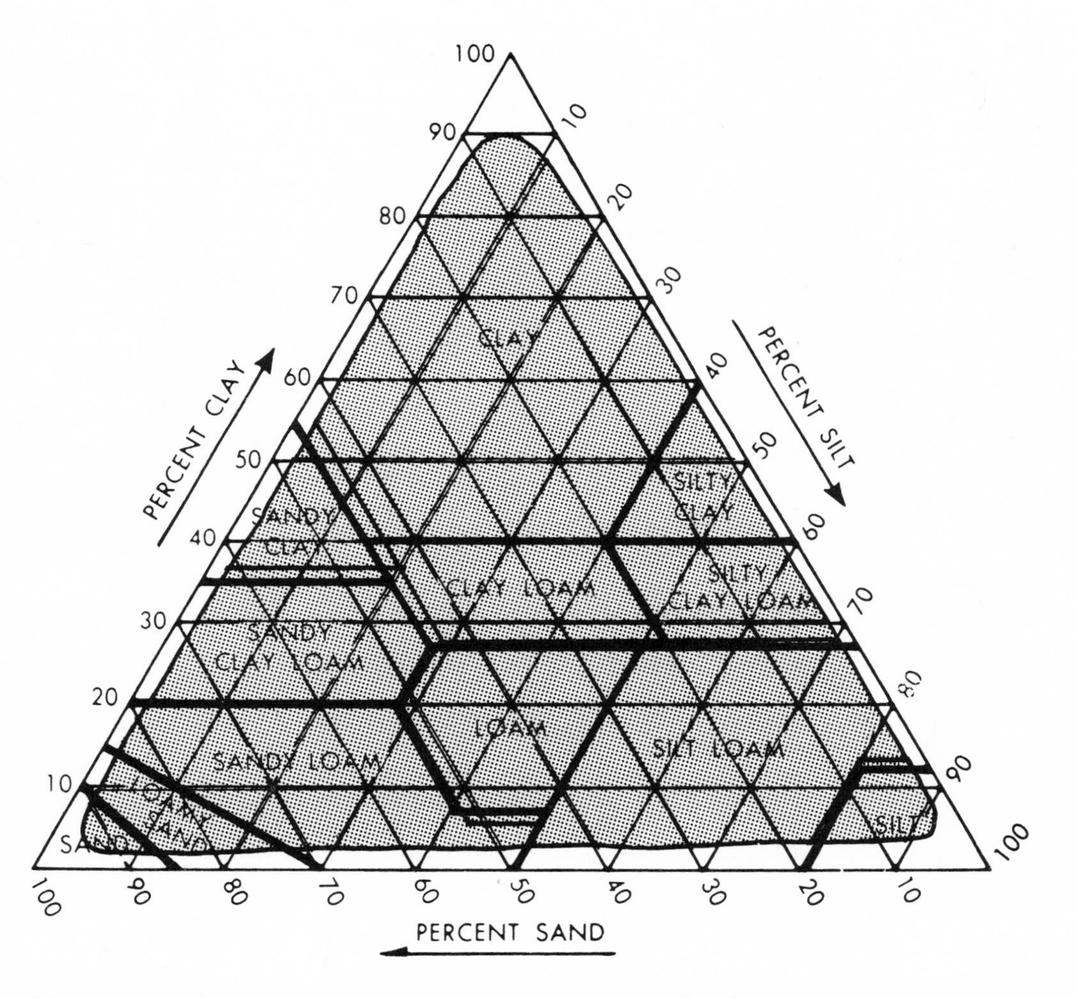

Figure 10.10. The U.S. Department of Agriculture soil texture triangle indicates the wide range of soil types plus gravel commonly found in flood plains.

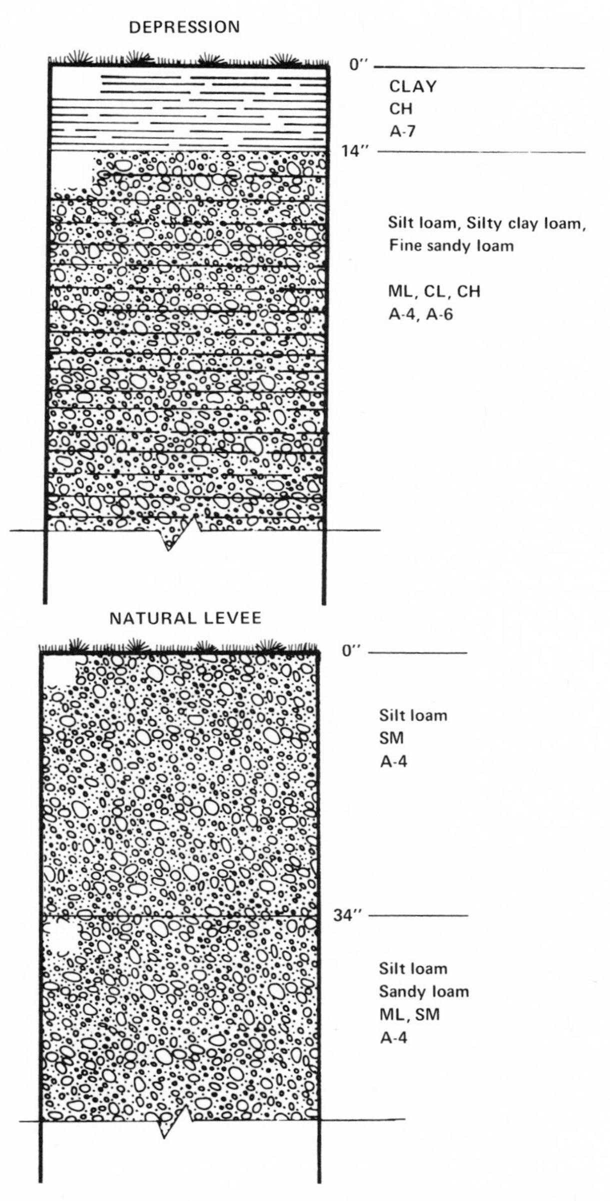

Figure 10.11. Typical soil profiles developed in a covered flood plain.

Maintenance against siltation is difficult because of the predominance of fine-textured materials throughout the immediate drainage basin. Small ponds can be easily constructed by excavating materials to and below the water table.

Foundations

In addition to the obvious flooding hazard, flood plains do not generally offer attractive building sites, owing to their great variations in soil conditions and the prevalence of high-volume-change

Table 10.3. Fluvial flood plains: aerial and cartographic references

State	County	Photo Agency	Photograph no.	Scale	Date	Corresponding quadrangles and geologic maps from USGS	Corresponding soil reports from SCS
Alaska	—	USGS*	Alaska 24-AB	3333'/"	6/20/48	Yakutat (1:250,000) I-271	—
Arkansas	Desha	ASCS	CFG-2HH-2-7	1667'/"	10/27/66	Big Island (15')	—
Arkansas	Conway	ASCS	IA-1HH-83-88	1667'/"	10/12/66	Morrilton West (7½')	1907†
Arkansas	Conway	ASCS	IA-1HH-73-81	1667'/"	10/12/66	Morrilton W, Morrilton E (7½')	1907†
Arkansas	Conway	ASCS	IA-1HH-140-145	1667'/"	10/12/66	Morrilton West (7½')	1907†
Arkansas	Conway	ASCS	IA-1HH-149-154	1667'/"	10/12/66	Morrilton West (7½')	1907†
Arkansas	Conway	ASCS	IA-1HH-201-204	1667'/"	10/12/66	Houstin (7½')	1907†
California	Inyo	USGS*	California 19-ABC	3116'/"	7/17/47	Independence (15'), MF-254	—
California	Inyo	USGS*	California 27-AB	4000'/"	12/2/48	Ash Meadows, Ryan (15'), Prof. Paper No. 424-D	—
Colorado	Montezuma	USGS*	Colorado 7-ABCDE	5000'/"	11/9/54	Mesa Verde National Park (1:31,250)	—
Colorado	Weld	USGS*	Colorado 1-AB	2308'/"	6/16/47	Greeley (1:250,000)	—
Idaho	Madison	USGS*	Idaho 7-ABC	2083'/"	10/8/50	Menan Buttes (7½'), Prof. Paper No. 450-E	—
Indiana	Vanderburgh, Henderson	USGS*	Indiana 2-ABCD	2367'/"	4/16/50	Wilson, Henderson, West Franklin (7½')	1944
Kansas	Morton	USGS*	Kansas 1-ABCDEF	1667'/"	8/23/36, 6/29/60	Dodge City (1:250,000), Prof. Paper No. 352-D	1963
Louisiana	Madison Parish	USGS*	Louisiana 2-ABC	1667'/"	5/17/56	Talla Bena (15')	—
Mississippi	Issaquena	USGS*	Mississippi 1-AB	1667'/"	11/30/49	Lorenzen (15')	1961
Mississippi	Tallahachie	USGS*	Mississippi 2-ABC	1667'/"	9/21/62	Philipp (15')	1970
Nebraska	Garden	USGS*	Nebraska 4-ABCDEF	1667'/"	7/12/39, 6/20/60	Scotts Bluff (1:250,000)	1924
Nebraska	Loup	USGS*	Nebraska 6-AB	1667'/"	7/10/54	Taylor SE, Sargent East (7½')	1937
Pennsylvania	Lycoming, Clinton	USGS*	Pennsylvania 1-ABCD	5000'/"	10/25/56	Williamsport (15')	1923†, 1966
Virginia	Shenandoah	USGS*	Virginia 5-ABC	2268'/"	3/9/45	Strasburg, Edinburg (15'), Prof. Paper No. 354-A	—
Wisconsin	Buffalo	ASCS	BHM-5FF-153-158	1667'/"	8/11/65	Alma, Cochrane, Winona (15')	1962
Wyoming	Carbon	USGS*	Wyoming 9-ABC	2308'/"	6/14/47	Saddleback Hills, Como Ridge (15')	—

*From *Selected Aerial Photographs of Geologic Features in the United States,* U.S. Geological Survey Prof. Paper No. 590.
†Out of print.

fine soils. Small structures may require friction piles or spread footings to elevate first floors above the seasonal high water table or potential flood levels. Rafts are commonly utilized, placed on a gravel pad or on naturally occurring, elevated, coarse soils. Heavier structures typically use friction piles, raft-type foundations, or bearing piles to carry loads to more stable underlying materials. In all cases detailed engineering analyses are needed early in the planning process to determine the properties of the underlying materials for the proposed structural loads.

Highway Construction

Depending upon the predominant soil textures of the landform, highways through flood plains may require granular borrow materials. Few lateral drainage structures are needed, nor are cuts or bridging, thus costs are relatively low. The required construction materials can commonly be found on-site, except in flood plains having fine-textured soils. Fine subsoils in cold climates require insulation from frost heave.

Deltas

Introduction

Deltas are formed when a stream system encounters a large body of water, either an ocean or lake; this immediately decreases both the gradient and the flow velocity and allows the sediments and debris to settle. The form and appearance of a delta is dependent upon the specific river and its load and on the characteristics of the water body in which the delta is formed. The several types of deltas can be classified by their planimetric shapes, which reflect differences in composition; these include arcuate or arc deltas, estuarine fillings or deltas, and bird's-foot deltas.

Arc deltas typically contain relatively coarse soil materials and have a characteristic fan-shaped outline, the convex edge facing the water body. Many shallow drainage channels are found on the fan, but the porous materials, especially in arid climates, facilitate subsurface flow. Small-scale photographs or photographic indexes may be necessary to observe the entire delta and its characteristic outline. Major arc deltas of the world include those of the Nile, the Mekong, the Rhine, the Hwang Ho, the Ganges, the Irrawaddy, the Volga, and the Lena.

Estuarine deltas are characterized by the uplands that surround the river mouth and delta. The entire delta is relatively narrow and usually has many islands and inlets. These deltas contain a complex variety of soil materials. Typical examples include the deltas of the Mackenzie, the Loire, the Seine, the Hudson, and the Elbe.

Bird's-foot deltas are very similar to covered flood plains, being subject to overbank floods and tending to develop natural levees. The natural levees help to stabilize and maintain stream flow in the channels which give the landform its characteristic, coarse-branching or clawlike bird's-foot pattern. Fine silts and clays are the predominate materials, even within the natural levees. Offshore irregularities, such as mud lumps, can be identified and are associated with larger bird's-foot formations. Typical, well-known examples can be seen at the mouths of the Mississippi and St. Clair Rivers.

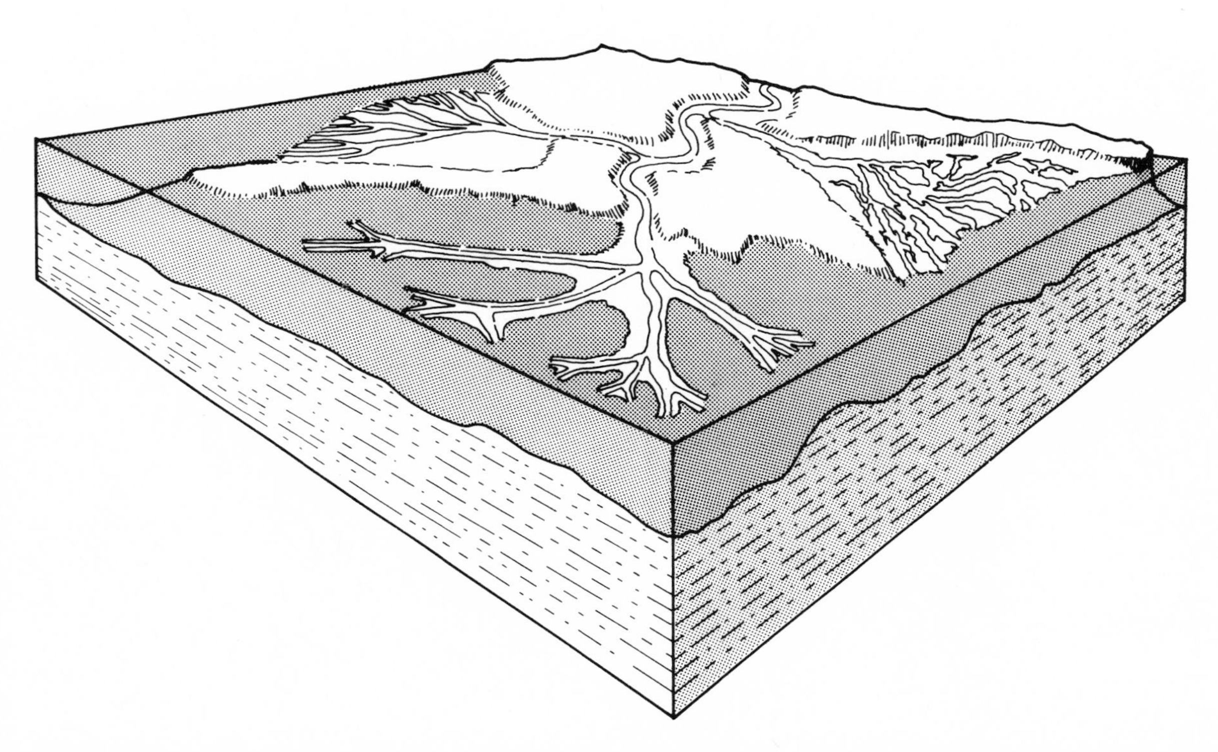

Figure 10.12. Types of deltas classified by their planimetric shape. (A) Arc delta. (B) Bird's-foot delta. (C) Estuarine delta.

The character of the water body in which a delta is formed determines its appearance and the bedding characteristics of its soils to a certain degree. Deltas formed in lakes or inland seas are more perfect in outline and more even in their bedding than those found in oceans, where waves and currents erode the delta forms, redepositing materials, forming bars, masking the form, and modifying the beds.

When examined in cross section, deltas show a relatively flat surface and several distinct zones of bedding. The bottomset and foreset beds underlie the delta formation and are exposed only if the delta has been uplifted or otherwise exposed and eroded. The topset beds occur on the surface and are therefore those commonly encountered.

Interpretation of Pattern Elements

Deltas are easily identified by their distinctive planimetric outlines. They are distinguished from

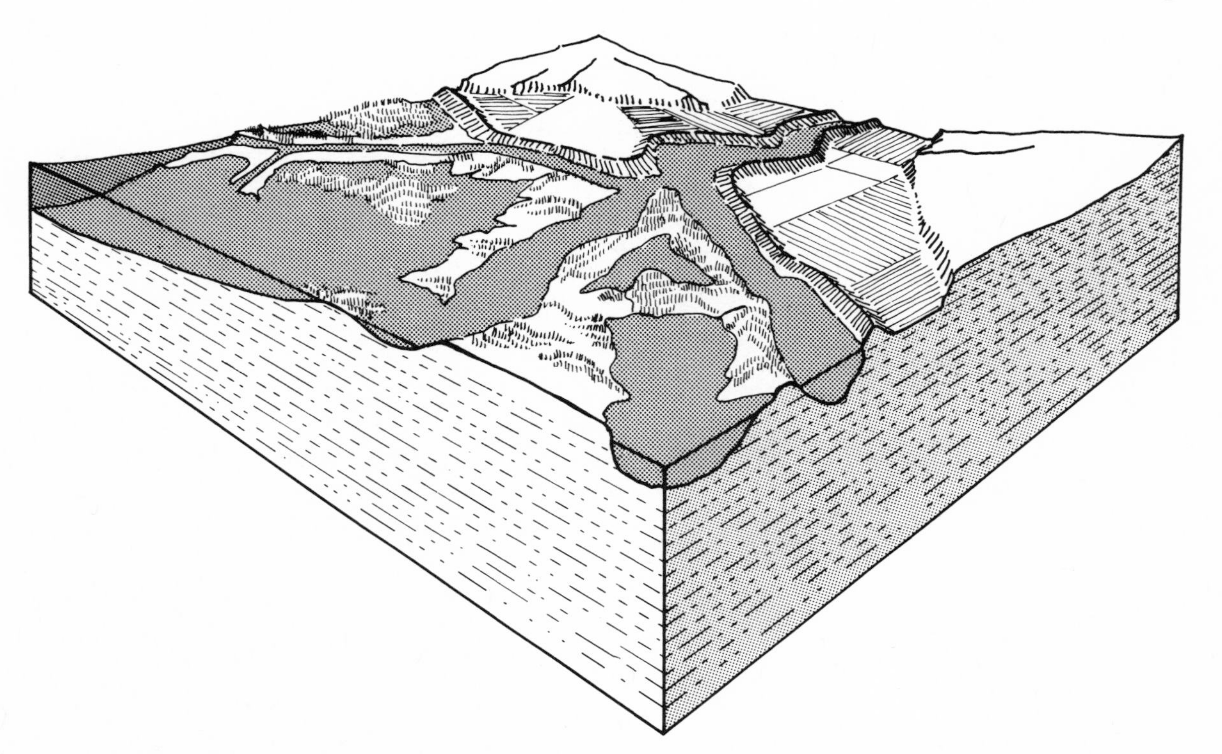

Figure 10.13. Visually, small delta formations may be exciting because of their relationship to water, for the flat topography of the delta occurs as a platform extending out into the adjacent water body. Deltas that are very large and broad in extent appear as broad, rather monotonous plains, lacking topographic relief. Granular, stable arc deltas generally fall under development pressure because of their attractiveness as water-related homesites, and the pleasure of being adjacent to water may compensate for the higher development costs incurred in overcoming the foundation and sewage disposal problems encountered on these formations.

alluvial fans by their relatively flat surfaces. Arc, bird's-foot, and estuarine deltas each have their own planimetric outlines.

Soil Characteristics

The following discussion of soils concentrates on arc deltas because they are the most common delta form. Typically, arc deltas contain relatively coarse, granular materials. Larger bird's-foot deltas contain more fines; these deltas are very similar to covered flood plains in soil and engineering characteristics. Estuarine delta soils are very complex and require a high level of detailed analysis and field investigations.

U. S. Department of Agriculture Classification

The U. S. Department of Agriculture commonly classifies surface soils of arc deltas as varieties of loam, silt loam, sand loam, and so on, and subsurface soils as crossbedded stratified sands and gravels. Bird's-foot deltas contain silts and clays in the subsurface, as opposed to the sands and gravels of arc deltas.

Unified and AASHO Classifications

The subsurface materials found in arc deltas are commonly categorized in the Unified system as poorly graded gravels and sands or as sands and gravels mixed with fines, such as GM, GC, SM, and SC. The most common categories are SP and GP. Surface soils are commonly fine textured and may be described as ML, CL, MH, and CH, with scattered deposits of peat and highly organic materials where the water table occurs at the surface. Common AASHO categories for subsurface materials in small arc deltas include A-1, A-2, and rarely A-3; these deltas offer better sources of well-graded sands and gravels.

Water Table

The water table occurs within a few feet of the ground surface in delta formations, owing to the flat topography and its association with water.

Drainage

Arc deltas, which contain fairly granular materials, are well-drained along the surface, since any rainfall percolates at once to the water table close beneath the surface. Bird's-foot deltas, which contain more fine soils, are poorly drained and develop many swamps and organic materials.

Soil Depth to Bedrock

Delta formations do not relate to any rock structure and are generally deep over underlying rock materials.

Table 10.4. Fluvial landforms: Deltas

Topography	*Drainage*	*Tone*	*Vegetation and land use*
Delta outline	Numerous channels	Mixed	Natural or developed
Deltas have planimetric shapes characteristic of their type: (a) arc, (b) estuarine, or (c) bird's-foot. Their surfaces, or topset beds, are generally flat, with a slight slope toward the water.	 Arc deltas (a) generally have many radial drainage channels which may show some braiding. Estuarine deltas (b) contain many interconnecting channels which carry the drainage around the many islands. Bird's-foot forms (c) typically contain two or three major channels in a branching alignment.	Tones are generally mixed in delta formations, indicating slight differences in materials, moisture or vegetation types. The fringes of deltas may be light gray, indicating suspended sediments.	Depending upon their soil materials and depths to water table, deltas are either covered with natural vegetative growth or are developed in a wide range of land uses, such as recreation,industrial park sites, or agriculture. For example, many small arc deltas in the Finger Lakes of New York state have been developed into marinas.
		Gullies	
		None	
		Gullies are associated only with exposed, eroded delta formations; their forms indicate the predominant textures of the soil materials.	

Issues of Site Development

Sewage Disposal

The high water table prohibits the use of septic tank leaching fields on arc deltas. Even if the delta were exposed and the water table at sufficient depth, the rapid percolation through the granular materials could still potentially contaminate the groundwater resources. Development located on the natural levees of bird's-foot deltas also encounters disposal difficulties, since the soils may have low percolation rates and water tables that fluctuate seasonally. Intensive development projects upon delta landforms should consider the use of treatment facilities if the quality of the water resources is to be maintained.

Solid Waste Disposal

Sanitary landfill operations are not recommended on deltas because of the high water table, danger of flooding, and possible contamination of the water resources. Ancient, exposed deltas are rarely suitable either, since they generally contain granular materials which facilitate percolation of leachate and do not provide a suitable source for cover materials.

Trenching

Trenching through delta formations encounters additional expenses, related to the high water table, which require temporary drainage or dewatering. Costs are therefore approximately 1.3 to 1.6 times those encountered in removing the same amount of deep, dry, moderately cohesive soil. Highly organic soils are more dificult to drain and escalate costs to as high as three to four times those for normal conditions. All these increases in costs are primarily related to the high water table.

Excavation and Grading

Excavation and grading operations of any significance encounter the water table, therefore necessitating the use of clam bucket, dredge, or drag line operations. Typical costs in granular and fine soils are 1.3 to 1.6 times those associated with removing deep, dry, moderately cohesive soils; in highly organic materials, costs are three to four times as high. Large pit-type excavations may require dewatering by pumping from sumps or well points.

Construction Materials (General Suitabilities)

Topsoil: Good to Poor. Some finer-textured deltas, such as the bird's-foot type, are good sources for topsoil and organic materials. Arc deltas are generally granular in composition but may have small, localized deposits of topsoil materials.

Sand: Good to Poor. Generally, only young arc deltas are good sources for well-graded sand materials. Some fine sands may be found in the natural levees of bird's-foot deltas.

Gravel: Good to Not Suitable. Only young arc deltas are good sources for either well-graded or poorly graded gravel deposits.

Figure 10.14. An arc delta in Tompkins County, New York, has been developed into a marina and recreation land. Photographs by the Agricultural Stabilization Conservation Service, ARU-1EE-74, 75, 1:20,000 (1667'/"), July 6, 1964.

Aggregate: Poor. Only young arc deltas provide potential suitable sources for aggregate.

Surfacing: Excellent to Fair. In most deltas a mix of fine and coarse materials can be obtained which is suitable for surfacing secondary roads. Bird's-foot delta sources containing fine soils, unless mixed with other, granular materials, create slippery surfaces when wet and are dusty when dry.

Borrow: Good to Poor. Deltas are generally poor sources for well-graded borrow materials. Young arc deltas have the highest potential, and when used for fill are compacted with pneumatic-tired rollers. Most materials are mixed with fines, and bird's-foot deltas typically contain more plastic soils which require drying and compaction by sheepsfoot rollers within strict moisture limits.

Building Stone: Not Suitable

Mass Wasting and Landslide Susceptibility

Deltas generally occupy low topographic positions with no steep, vertical escarpments and therefore are not susceptible to landslides under normally occurring conditions. Exposed, elevated deltas are commonly of glacial origin and are rather stable, as a result of their granular composition. Thin layers or beds of silt, however, can cause differences in internal drainage which can create a hazardous sliding potential.

Groundwater Supply

The groundwater table in delta formations is also the water table, which occurs near the ground surface. The elevation of the groundwater is fairly constant, reflecting the elevation of the surrounding body of water. Arc deltas of granular composition in freshwater lakes are excellent locations for wells, which have a high-yield capacity.

Figure 10.15. Bird's-foot delta in St. Clair County, Michigan, at the mouth of the St. Clair River. Photographs by the Agricultural Stabilization Conservation Service, XP-1EE-27,28, 1:20,000 (1667′/″), July 6, 1964.

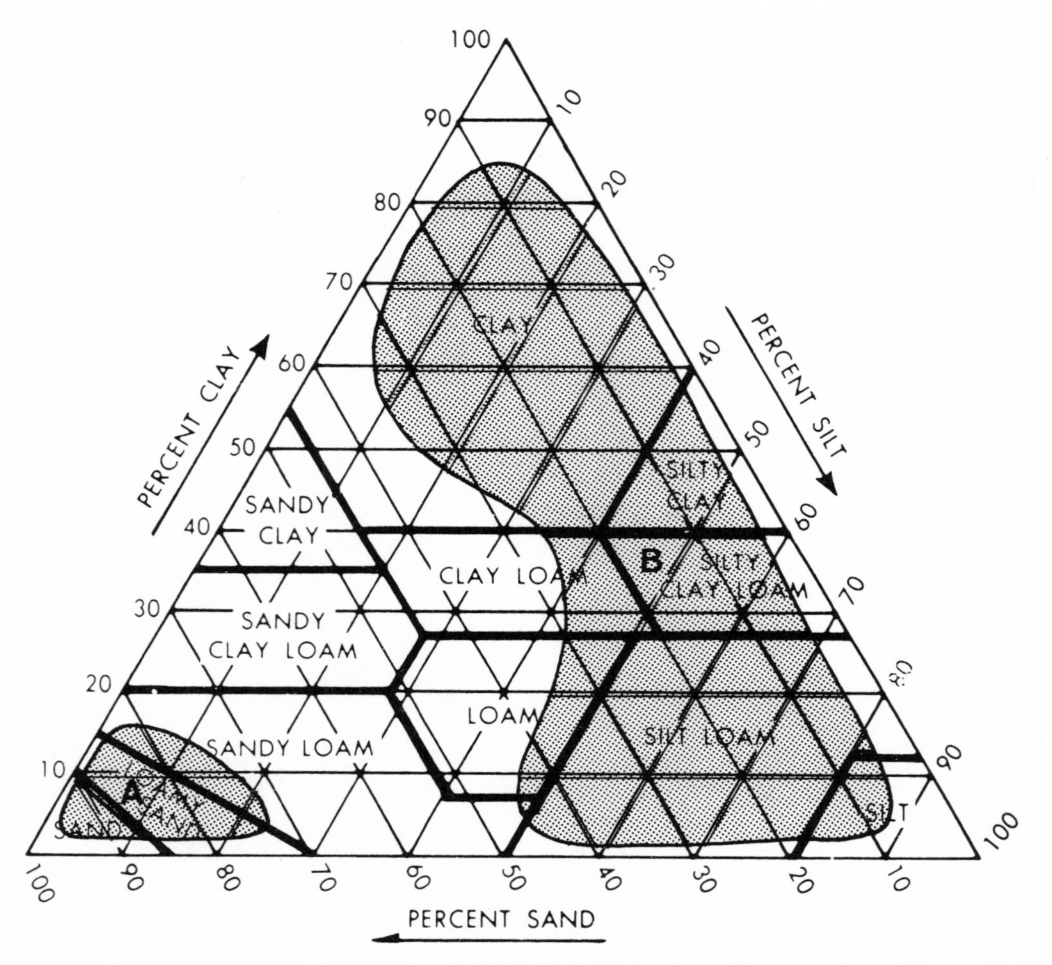

Figure 10.16. The U.S. Department of Agriculture soil texture triangle indicates sands and gravels for arc deltas (A), and finer silts and clays for bird's-foot deltas (B). Estuarine delta materials are complex and contain both fine and coarse materials.

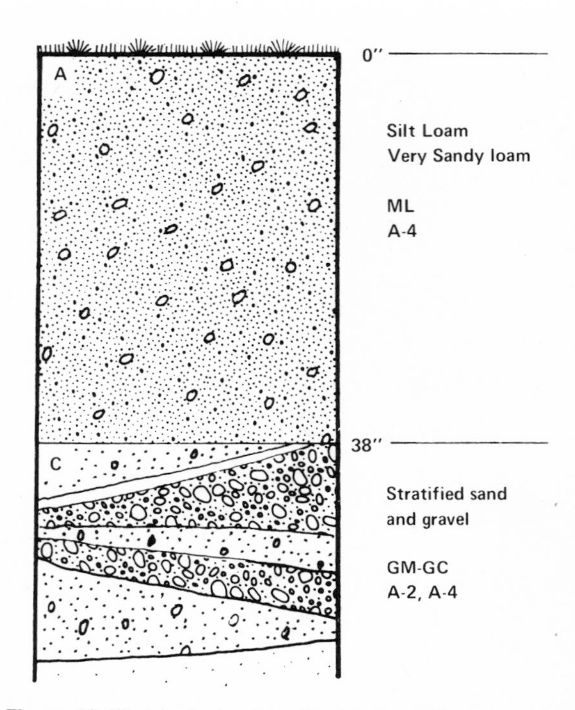

Figure 10.17. A typical soil profile developed in an arc delta in a humid climate.

Pond or Lake Construction

Ponds and lakes are not usually developed upon these landforms, since water is normally nearby, along the convex or outer edge. Small ponds and marinas are easily constructed back into the edge of the delta by excavating materials below the water elevation or the elevation of the groundwater table. Many of the small arc deltas in the Finger Lakes region of New York have marinas developed on their outer edges.

Foundations

Foundation operations in delta regions encounter wide variations in bearing capacities, owing to the underlying bedding of the soil materials. Arc deltas containing granular materials offer the most stable platforms, allowing the use of footings and rafts for small structures if elevations above the water table and flooding are sufficient. Heavier structures may use rafts or footings, but deltas containing silts and clay may require friction or point-bearing piles. Detailed engineering analysis is needed early in the planning process to determine the properties of the underlying materials for the proposed loads of the structure.

Table 10.5. Fluvial deltas: aerial and cartographic references

State	County or parish	Photo Agency	Photograph no.	Scale	Date	Corresponding quadrangles and geologic maps from USGS	Corresponding soil reports from SCS
Alaska	–	USGS*	Alaska 19-ABC	3250′/″	9/17/50	Kenai (A-2) (1:63,360), Prof. Paper No. 443	–
Alaska	–	USGS*	Alaska 22-AB	3333′/″	6/30/48	Icy Bay (1:250,000), I-271	–
Alaska	–	USGS*	Alaska 26-ABC	2083′/″	8/11/59	Mt. Fairweather (1:250,000), Prof. Paper No. 354-C	–
California	Ventura	ASCS	AXI-1FF-38-39	1667′/″	6/9/65	Ventura (7½′)	1970
California	Ventura	ASCS	AXI-1FF-46-49	1667′/″	6/9/65	Ventura (7½′)	1970
Louisiana	Plaquemines	USGS*	Louisiana 4-A	2500′/″	2/22/56	Pointe a La Hache (15′)	–
Louisiana	Plaquemines	USGS*	Louisiana 5-A	2500′/″	3/25/56	Empire, Venice (15′)	–
Louisiana	Plaquemines	USGS*	Louisiana 6-A	2500′/″	3/25/56	Venice (15′)	–
Louisiana	Plaquemines	USGS*	Louisiana 7-A	2500′/′	3/25/56	West Delta, East Delta (15′)	–
Louisiana	Plaquemines	USGS*	Louisiana 8-A	2500′/″	2/23/56	East Delta, Brenton Island (15′)	–
Louisiana	Plaquemines	USGS*	Louisiana 9-AB	2500′/″	3/25/56	Southwest Pass (15′)	–
Michigan	St. Clair	ASCS	XP-1EE-26-28	1667′/″	7/6/64	St. Clair Flats, Algonac (7½′)	1929
Michigan	St. Clair	ASCS	XP-1EE-19-21	1667′/″	7/6/64	St. Clair Flats, Algonac (7½′)	1929
New York	Tompkins	ASCS	ARU-1EE-53-54	1667′/″	7/6/64	Ludlowville (7½′)	1965
New York	Tompkins	ASCS	ARU-1EE-11-12	1667′/″	7/6/64	Ludlowville (7½′)	1965
New York	Tompkins	ASCS	ARU-1EE-74-75	1667′/″	7/6/64	Ludlowville (7½′)	1965

*From *Selected Aerial Photographs of Geologic Features in the United States,* U.S. Geological Survey Prof. Paper No. 590.

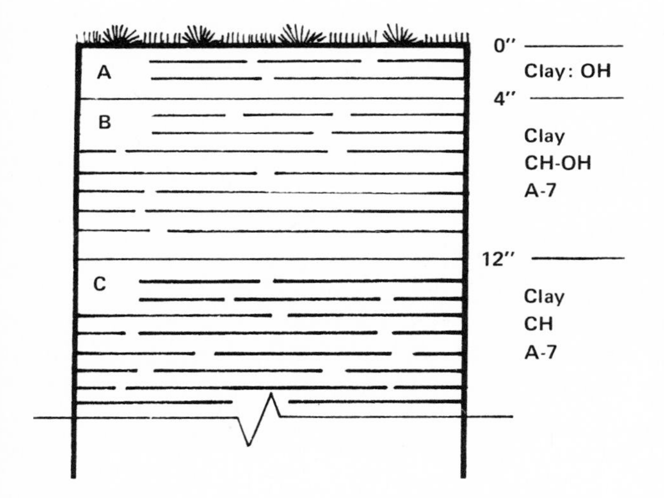

Figure 10.18. Soil profile in a bird's-foot delta. The soil composition in a bird's-foot delta is fine and may be of a plastic-organic nature.

Highway Construction

Highways are difficult to locate across delta formations because of the tendency of the channels to shift and the occurrence of organic deposits, lakes, poorly drained soils, plastic soils, and so on. Cold climates are susceptible to frost heave, requiring the use of granular subbase materials. Natural levees in bird's-foot delta formations provide for natural, elevated corridors.

Alluvium (Fans, Valley Fills, Continental Type)

Introduction

The term *alluvium* is used here to describe sediments deposited by streams that form alluvial fans, valley fills, or continental alluvium. Alluvial formations are composed of sediments of loaded streams which have changed gradient and velocity abruptly as they emerge from steep valleys onto relatively flat surfaces. Much of their bedloads and suspended loads are deposited at these change points. The alluvial formations discussed in this section are of common occurrence in arid and semiarid climates throughout the world.

Alluvial fans are the smallest of the three forms and are composed of sediments deposited when heavily loaded streams abruptly change gradient upon entering a gently sloping lowland area. The idealized planimetric shape of this landform is that of a fan that rises gently toward its apex or central point of origin at a 1- to 10-degree slope. The coarsest soil materials are found at the apex, where the initial reduction of velocity occurs; finer materials are deposited near the margin of the fan.

The initial cover of arid valleys is formed by alluvial fans and these, in coalescing, form *valley fills*. Valley fills may also contain other landforms, such as playa lake beds, which are sometimes found in the basins.

Continental alluvium also consists of sediments washed down from mountainous regions. These are deposited in lowlands to form broad, flat plains. As opposed to the other landforms described in this section, continental alluvium occurs on a regional scale and can cover thousands of square miles. Much of the vast plain from the Rocky Mountains across the southern United States to the Mississippi River is continental alluvium; it consists primarily of stratified, unconsolidated to semiconsolidated materials which may be partially cemented. The Great Plains are no longer under deposition processes, however, but are being eroded.

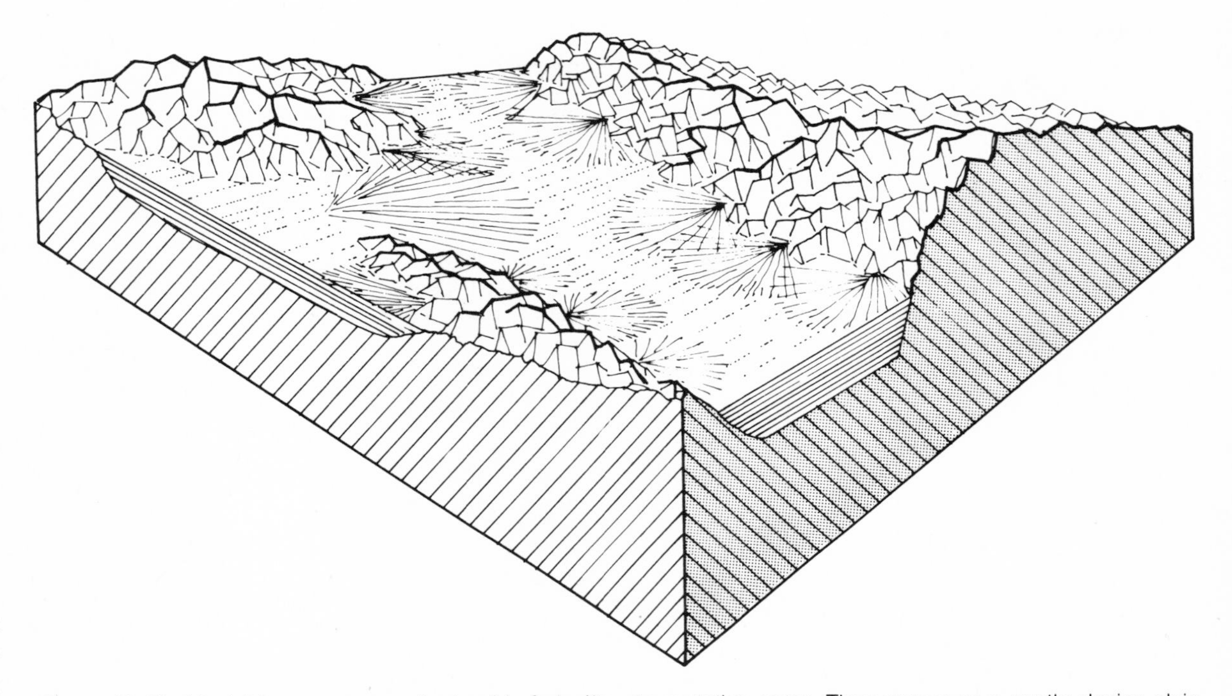

Figure 10.19. Alluvial fans are generally devoid of significant vegetative cover. They appear as a gently sloping plain adjacent to a valley edge. They are typically bounded on one side by rugged or mountainous topography which acts as a visual barrier. The other side gradually blends into the valley bottom. The fans offer a slightly elevated viewing position above their valley floors. They are visually sensitive to development because of their flat, elevated surface and lack of vegetation.

Interpretation of Pattern Elements

Alluvial deposits are easily identified by their distinct topographic-planimetric and drainage characteristics. Alluvial fans are identified by their fan outline and are distinguished from deltas by their perceptible slope toward the apex. Valley fills are characterized by the filled, flat valley bottom which contains many, somewhat-parallel, braided, inactive stream channels. Continental alluvium is identified by its flat topography, its coarse regional drainage which develops badland erosion features along its edges, and its circular depressions known as buffalo wallows.

The photographic key is divided into three sections, corresponding to each of these three forms.

Soil Characteristics

Alluvial deposits contain a wide variety of soil textures, but these can be described in general as stratified, unconsolidated to moderately consolidated, and partially cemented (as is common in deposits of continental alluvium). Alluvial fans consist of relatively coarse, unconsolidated, stratified soils of poorly graded, well-sorted sands and gravels, with occasional small, scattered pockets of silts and clays. These soils have little profile development; large boulders may be located near the apexes of the fans. In humid climates alluvial fans commonly contain a larger percentage of silts and clays.

Valley fills contain alluvial sediments of a wide textural range but are predominantly coarse, with scattered pockets of silts and clays. The soils are coarsest near the uplands and grade to finer materials in lowland depressions and playas. The soils are stratified and generally unconsolidated but may be partially cemented.

Continental alluvium develops several feet of silty loam which is underlaid by stratified, partially cemented materials of a generally sandy texture (although all textures can be found). Depressions contain a greater concentration of fines.

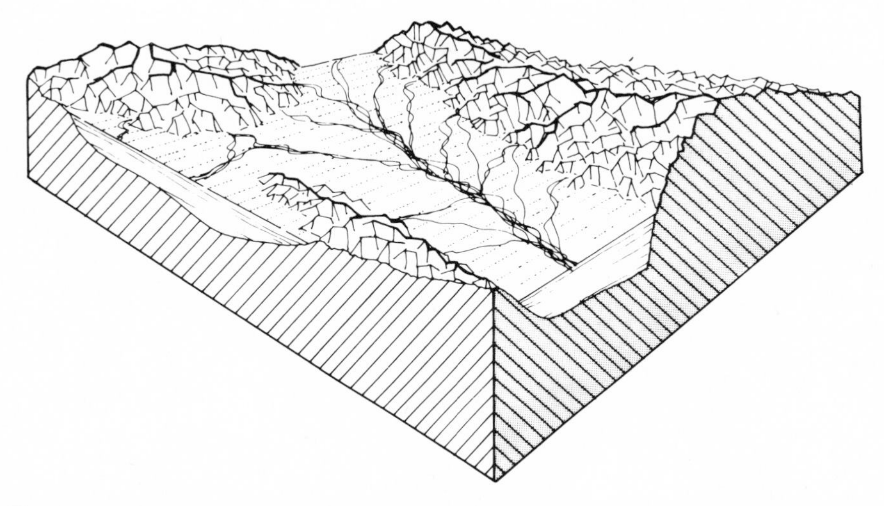

Figure 10.20. The valleys enclosing valley fills may be small, creating spatial closure, or they may be broad and large, thereby causing the fill to lose all definition of space and to appear as a broad plain. The flatness of the topography and the lack of vegetation make these areas visually sensitive for any projected development, especially in the form of high towers or vertical structures.

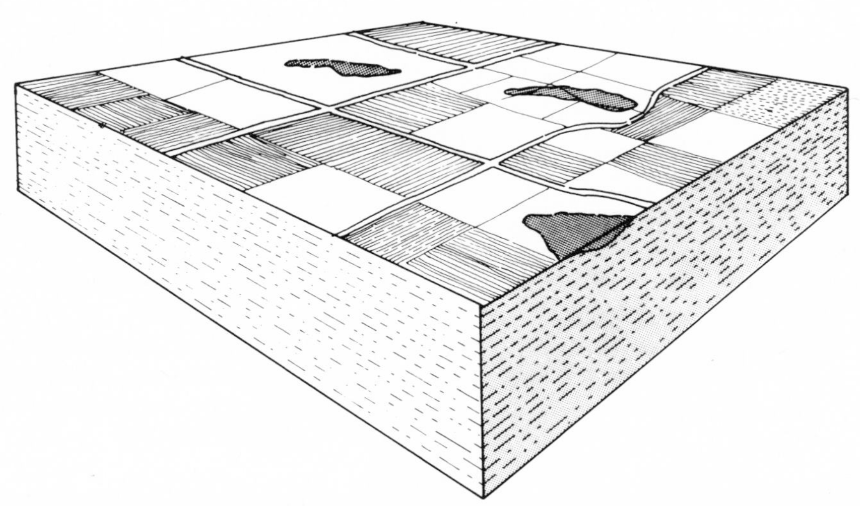

Figure 10.21. Continental alluvium is characterized by flat, monotonous plains interrupted by occasional dissection and associated badland topography. The flat surface and lack of vegetation on undissected plains make these landforms visually sensitive to structural development and vertical elements.

U.S. Department of Agriculture Classification

Alluvial fans are commonly classified by the U.S. Department of Agriculture as containing stratified sands and gravels. Very stony or rocky may be used to describe materials found near the apex of the fan.

Valley fills also have little profile development and are usually classified as stratified sands and gravels. Some stratified silty clays may be found in lowland depressions.

Continental alluvium typically develops a shallow profile of silt loam, sandy clay loam, or clay loam over stratified materials of all textures (but commonly indicated as stratified sand and gravel). Depressions contain a greater concentration of

Table 10.6. Fluvial landforms: Alluvium (fans, valley fills, continental)

Topography	*Drainage*	*Tone*	*Vegetation and land use*
Fan-shaped 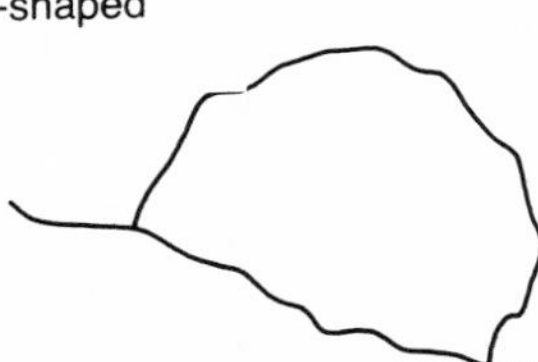 Alluvial fans have a fan-shaped outline with distinct boundaries. They slope 1 to 10 degrees toward the apex, the fan being convex in cross section.	Radial, braided 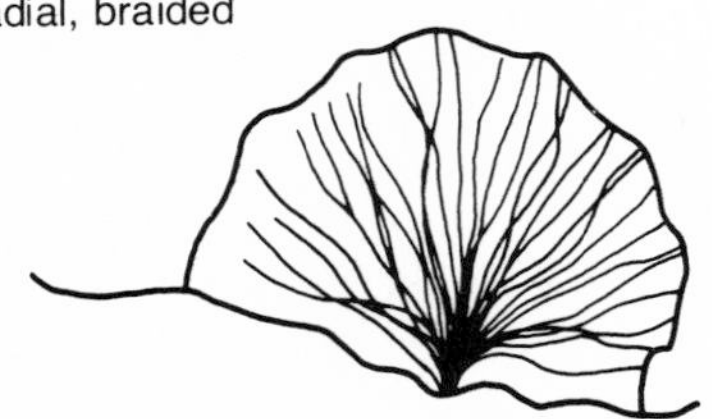 Generally, drainage is active only during severe storms, but the dry channels are readily observed and form, from the apex, a radial pattern of braided streams.	Light gray The exposed soils are generally light in tone, and the drainage channels are emphasized by vegetation. *Gullies* None Typically, gullies are not found outside the drainage channels.	Natural cover In arid climates fans have very sparse vegetation which is concentrated where greater moisture is available, as along drainage channels or near the fan fringe. Near the outer edge of fans, where water is readily available, development is common.
Filled valley bottoms 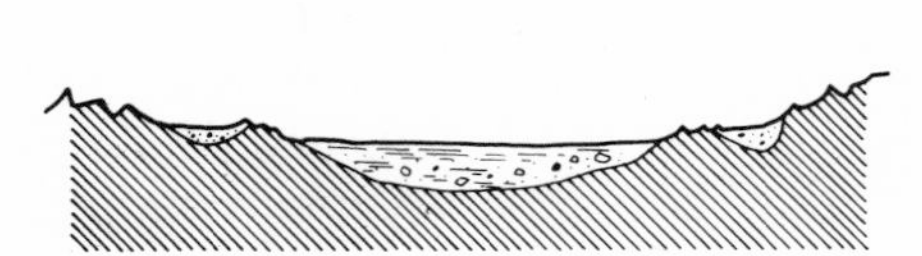 The topography may appear flat, but there is actually a gradual slope away from the highlands. The flat plain may be interrupted occasionally by resistant rock islands.	Parallel, braided 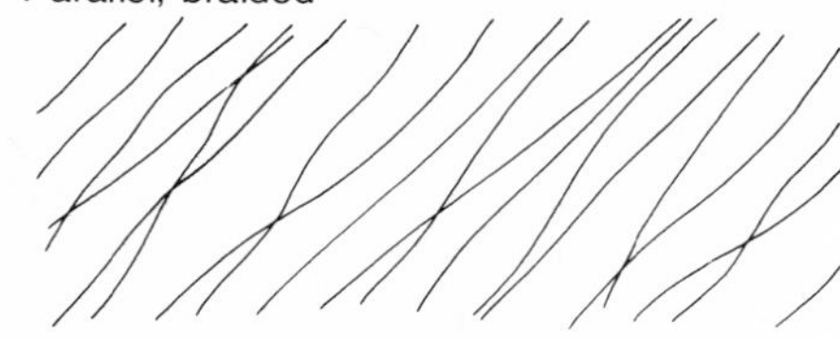 Valley fills are characterized by many dry, parallel, braided drainage channels. These are active during severe storms, when they shift channels and deposit vast amounts of alluvial sediments.	Uniform light gray The characteristic uniform light grays can be interrupted with irregularities, for example, alkaline deposits which have a scrabbled pattern. *Gullies* None Typically, gullies are not found outside the braided drainage channels.	Natural and cultivated The low rainfall and alkaline soil conditions are not favorable for the development of dense vegetative cover. The drainage channels contain denser concentrations of scattered scrub growth, reflecting the greater moisture availability. If irrigated, valley fills may be intensively cultivated with vegetable crops.
Continental			
Broad, flat plains 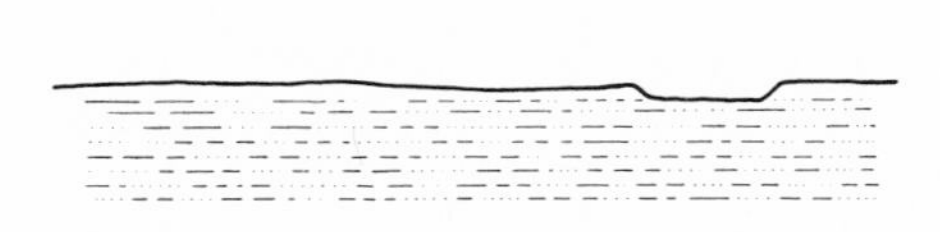 The characteristic broad, flat plains covering vast regions are occasionally broken by small, circular, flat-bottomed depressions caused by wind erosion which has enlarged ancient buffalo or cattle dust wallows.	Internal, dendritic  Drainage is mostly internal. Drainage features are neither apparent nor well-developed, owing to the low rainfall. Regional systems may dissect these forms, in which case the alluvial material easily erodes.	Uniform light gray Tones are light and uniform, with minor irregularities. Occasional deposits of windlaid sands, silts, and alkali appear light. *Gullies* Few, U-shaped U-shaped gullies are found where dissection occurs, such as along major drainage systems.	Natural and Cultivated Grain crop cultivation is common in semiarid regions, whereas more arid regions may be left in natural cover. Highly dissected regions are left in natural cover because of the severity of the topography.

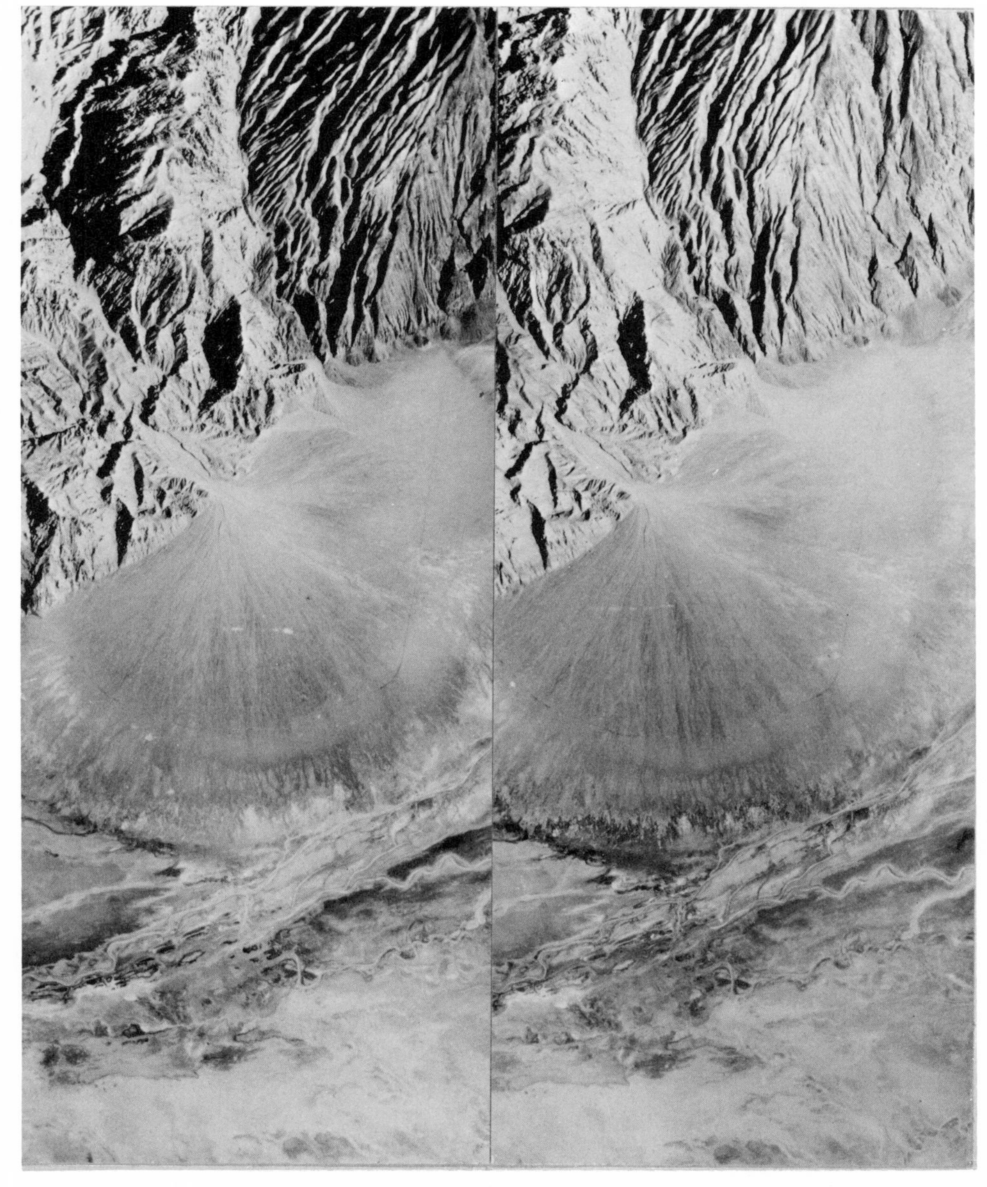

Figure 10.22. Coalescing alluvial fans adjacent to a triangular block fault in Inyo County, California. Photographs by the U.S. Geological Survey, California 26-A,B, 1:48,000 (4000′/″), November 27, 1948.

fine soils which are generally classified as plastic clays.

Unified and AASHO Classifications

Alluvial fans are typically described by symbols indicating coarse and poorly graded, such as GP and SP; other granular soils are designated SW, GW, GM, GC, SM, and SC. Localized deposits of ML or CL materials may be found. AASHO categories include mostly A-1 and A-2 materials with occasional pockets of A-4.

Valley fills are typically described by symbols indicating coarse and poorly graded, such as GP or SP; other granular soils are designated GW, SW, GM, GC, SM, and SC. Valleys have higher concentrations of CL, ML, or even CL-CH, CH, or other plastic categories. Descriptions by the AASHO system are characterized by A-1 and A-2 materials, with A-4 and some A-6 found in the lower depressions.

Continental alluvium generally contains materials finer in texture, since its sediments traveled a greater distance from their origin. Upper soil profiles are generally CL, and there are some SC or CL-CH variations. Depressions containing finer particles are classified as CL-CH, CH, or CL. AASHO symbols are typically A-6 or A-4, A-6 or A-7 being common in depressions.

Water Table

Significant water tables or groundwater resources may be found within these landforms, but they occur deep below the surface (except in valley fills where playas are found). Some seepage of water may be found near the fringes of alluvial fans where subsurface drainage encounters the surface. Circular depressions in continental alluvium may have ponded water, indicating a close proximity to the water table. After heavy

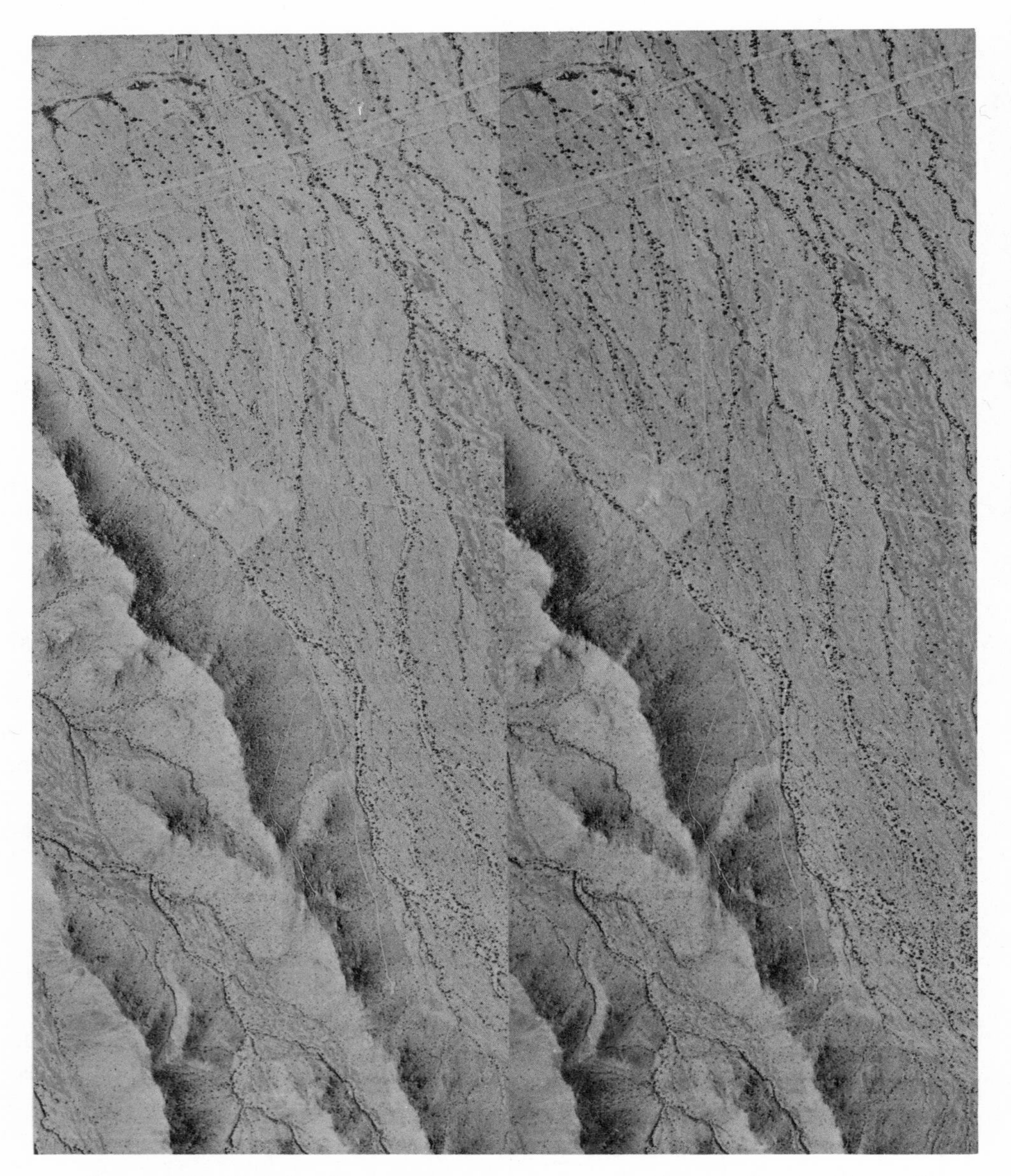

Figure 10.23. (A) Valley fill and coalescing alluvial fans in Maricopa County, Arizona. The concentration of vegetation emphasizes the drainage system. Photographs by the Agricultural Stabilization Conservation Service, DHP-9EE-33,34, 1:20,000 (1667'/"), January 31, 1964.

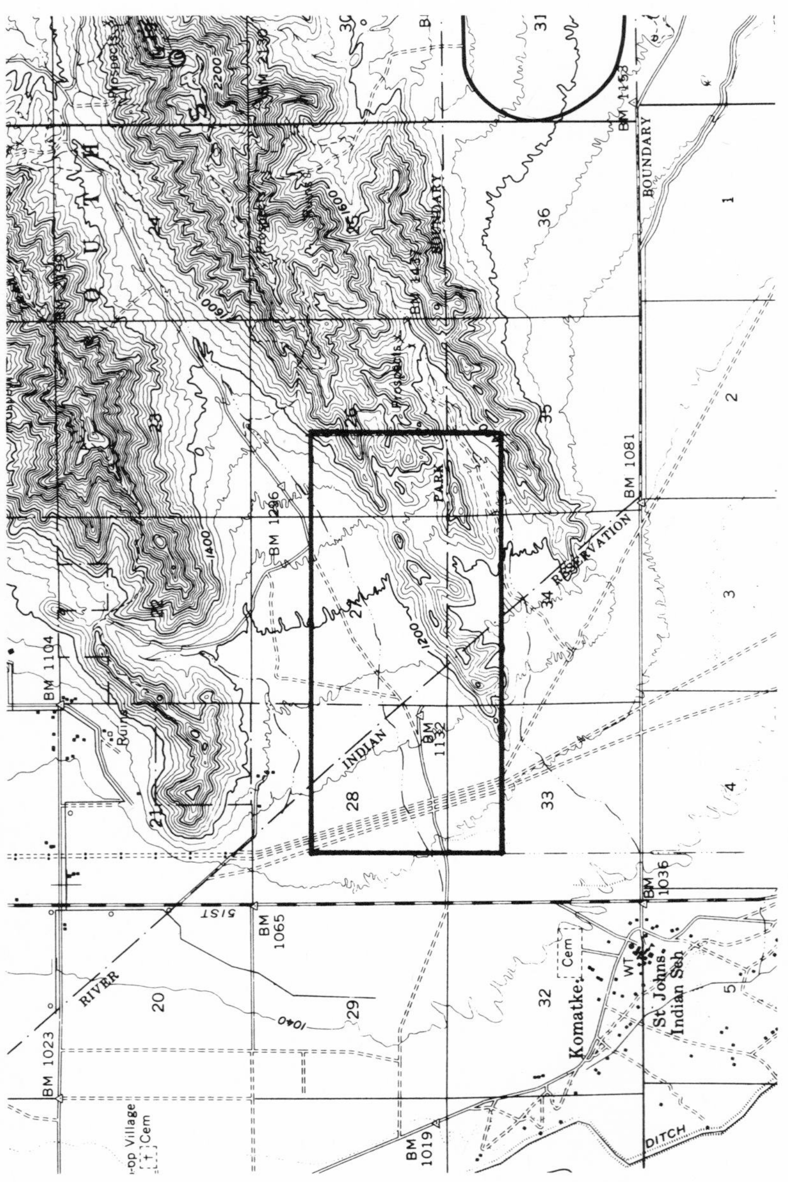

Figure 10.23. (B) U.S. Geological Survey Quadrangle: Phoenix (15').

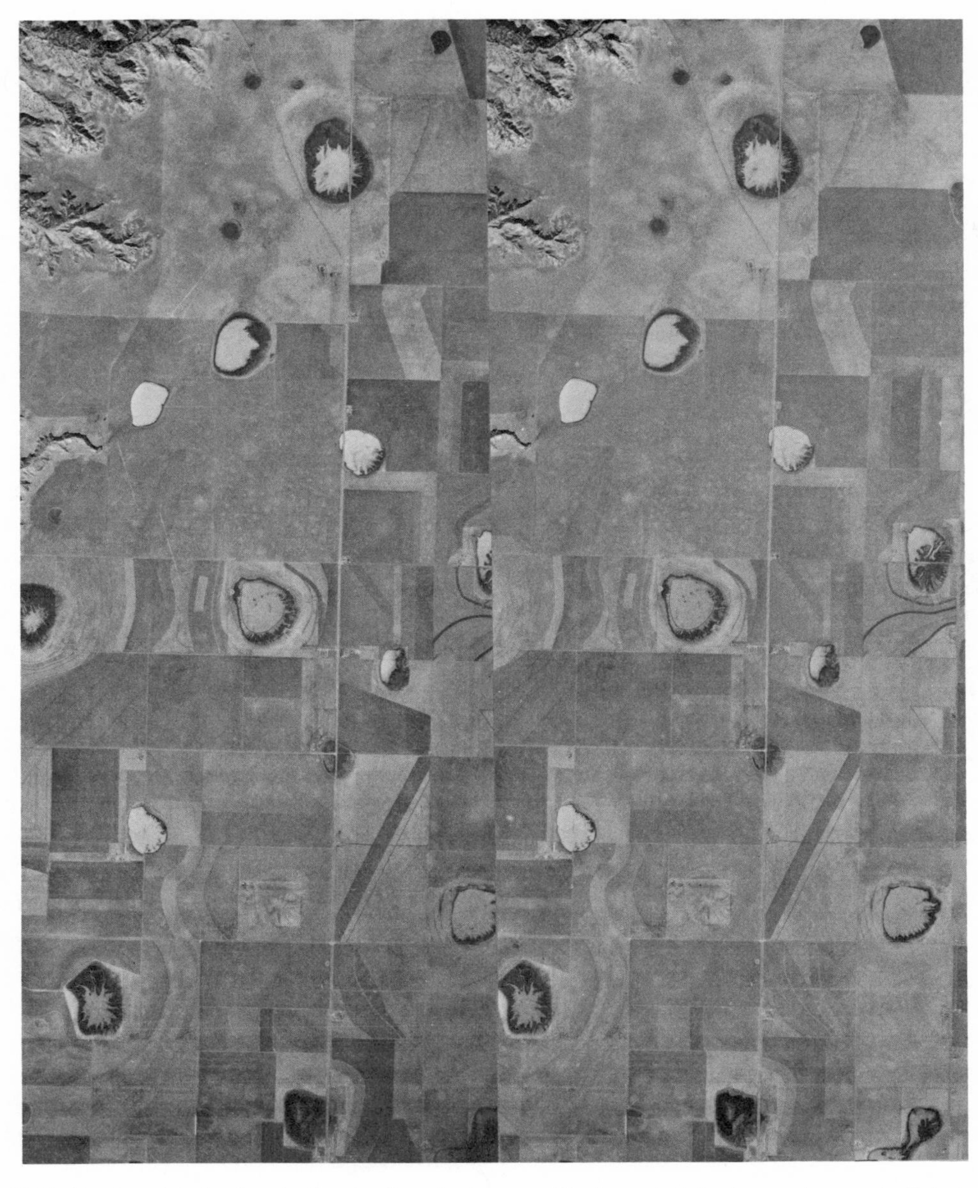

Figure 10.24. Continental alluvium in Crosby County, Texas, with characteristic depressions or buffalo basins. Photographs by the U.S. Geological Survey, Texas, 4-AB, 1:63,360 (5280′/″), January 23, 1954.

storms the water table in these forms may be found temporarily near the surface, especially in continental alluvium where surface soils are rather fine-textured. In the arid climates where these landforms are typically found, evaporation exceeds rainfall, and therefore the capillary soil moisture moves toward the ground surface where it is evaporated, leaving behind its dissolved salts which become alkaline deposits.

Drainage

Alluvial fans and the upper portions of valley fills are well drained, as a result of their stratified granular composition, the low rainfall, and the lack of subsurface drainage flow. The surface soils of continental alluvium are fine-textured and therefore prevent rapid mositure or water infiltration. The circular depressions associated with these landforms may have originated as solution pits which acted as points of internal drainage for the granular subsurface. It is believed that these depressions were later enlarged by the action of animals, who used them for dust wallows, and by wind erosion.

Soil Depth to Bedrock

Alluvial deposits are deep over bedrock except in young, newly formed deposits. Bedrock under alluvial fans and the upper portions of valley fills can be detected if it is within 20 feet of surface and if it disrupts the subsurface drainage flow. Patterns made by alkali or by the distribution of vegetation seeking differences in moisture content may be found on the surface, thus indicating the characteristics of the underlying rock type and structure. Rock outcrops are common in valley fill formations if the residual uplands have not been

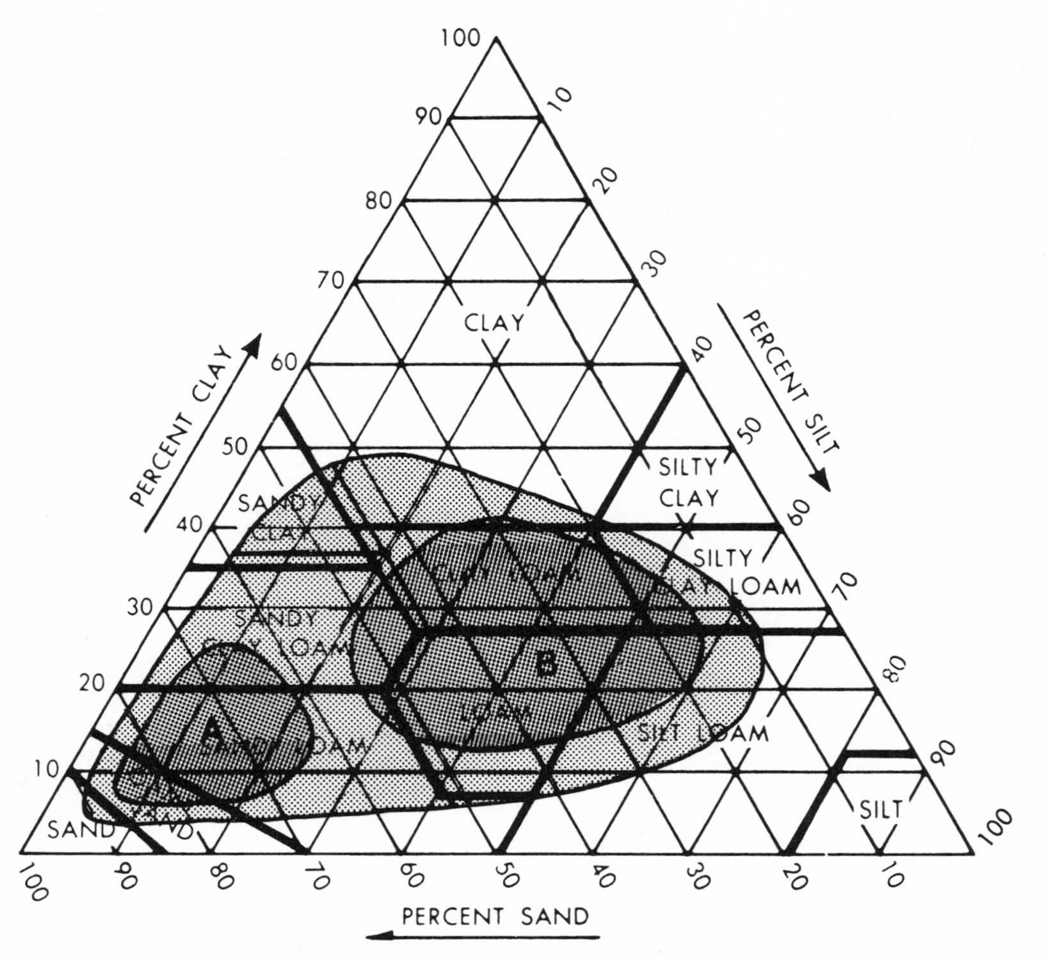

Figure 10.25. The U.S. Department of Agriculture soil texture triangle indicates coarse sands and gravels in alluvial fans and valley fill (A) and silt loam and sandy clay loam in continental alluvium (B).

completely weathered and eroded. Upper reaches of valley fills may have buried rock structures close beneath th. surface, and investigation of these requires seismic or other studies to determine accurate depths and profiles.

Issues of Site Development

Sewage Disposal

Septic tank leaching fields can be sited on alluvial fans or valley fills, but the rapid percolation rates in the granular materials may allow the effluent to contaminate groundwater resources. In arid valley fills the effluent tends to move toward the surface and evaporate. Continental alluvium contains silty surface soils which may have slow percolation rates and may be rather impervious. Intensive development operations should plan to construct treatment systems to guard against contamination of water resources in the subsurface sand and gravels.

Solid Waste Disposal

Alluvial fans and valley fills are not suitable sites for sanitary landfill operations, since the materials are rather pervious and allow percolation of leachate which will contaminate the underlying water resources. However, the bottoms of depressions may be sealed with clay to provide a site suitable for operation. Lower regions of valley fills may be more suitable if they contain fine materials and a water table at least 10 feet beneath the surface. Continental alluvium containing silts and clays provides more suitable sites for landfill operations, but care should be taken not to select a site whose base is in the more granular sand and gravel substratum. Some of the naturally occurring depressions found in these regions offer good site potential, since they contain thick, clay bottom layers which are relatively impervious. The silty surface soils are usually suitable as impervious cover materials.

Trenching

Trenching operations do not commonly encounter difficulties in these landforms, since the soil materials are fairly unconsolidated and deep over bedrock and the water table. Costs are approximately the same as those involved in removing deep, dry, moderately cohesive materi-

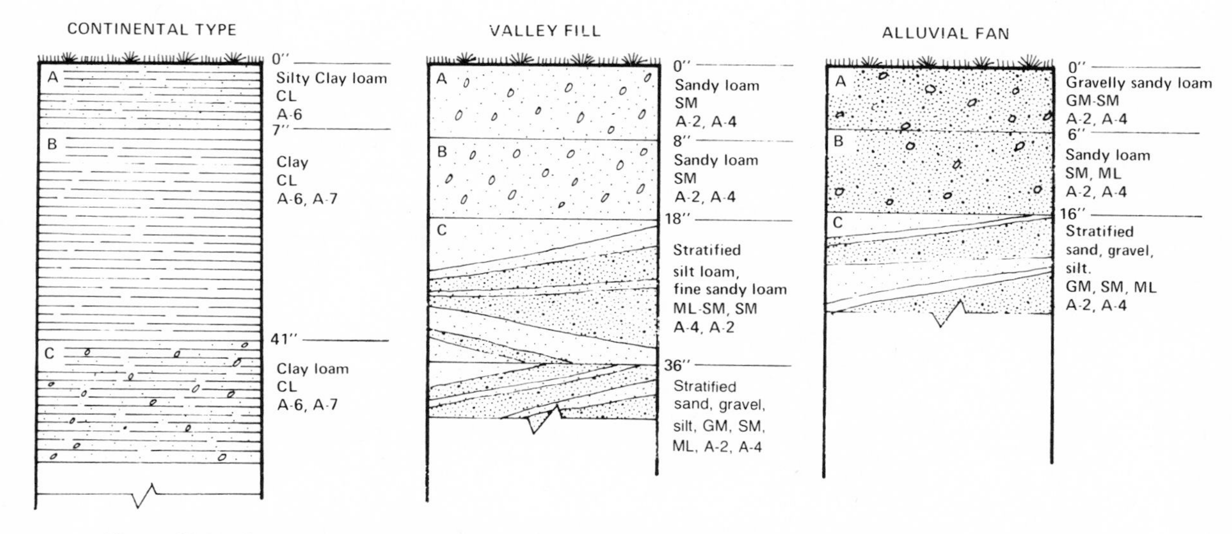

Figure 10.26. Soil profiles of continental alluvium, valley fill, and alluvial fan materials in a semiarid climate.

als. The finer soils found in surface horizons in continental alluvium may result in slightly higher costs and may be difficult to work under wet conditions.

Excavation and Grading

Significant grading and excavation operations are not required on these landforms since sites are flat. Pit-type excavations should not face any difficulties unless the water table is encountered; this requires either the use of special excavation equipment or dewatering. Typical grading and excavation costs are low, approximately the same as those associated with the removal of deep, dry, moderately cohesive materials. Bedrock may be encountered in the upland reaches of valley fills and present the additional expenses of rock removal.

Construction Materials (General Suitabilities)

Topsoil: Not Suitable to Poor. Alluvial fans and valley fills are not good potential sources of topsoil. Slightly better materials may be found on the surface of continental alluvium.

Sand: Excellent to Poor. Suitable sources of coarse, clean sands can be located in fans and valley fills; they are difficult to find in continental alluvium.

Gravel: Excellent to Poor. Alluvial fans and upland regions of valley fills provide excellent sources of gravels. Suitable sources are difficult to locate in continental alluvium, although some gravel materials may be found in stream bottoms where the plain has been partially dissected.

Aggregate: Excellent to Poor. Alluvial fans and upland valley fills offer excellent sources of aggregate. Suitable materials are difficult to locate in regions of continental alluvium.

Surfacing: Good. Sufficient local deposits of fines can usually be located to provide the right mixture and binder for road surfacing materials.

Borrow: Excellent to Fair. Well-graded materials do not naturally occur, but enough diversity exists across stratified beds to facilitate the mixing of desired grades while the material is being removed. Suitable mixtures may be more difficult to locate in continental alluvium. Compaction of noncohesive granular soils used for fill is accomplished with vibrators; of mixed-grained, moderately cohesive soils with pneumatic-tired rollers; and of cohesive soils with sheepsfoot rollers.

Building Stone: Not Suitable. Some large angular blocks can be found near the apex of alluvial fans and at the base of talus cones at the base of adjacent cliffs or steep slopes.

Mass Wasting and Landslide Susceptibility

Alluvial fans and valley fills, which are found in flat, lowland positions, are generally stable with respect to landslides and the processes of mass wasting. Excavation of vertical faces is constrained by the angles of repose of the unconsolidated materials, which average about 35 degrees. Where stream dissection has oversteepened slopes in dissected regions of continental alluvium, potential slide zones develop. The occurrence of any stratified layers of silt or clay facilitates sliding.

Groundwater Supply

Alluvial deposits, because of their coarse, stratified composition, are excellent sources of groundwater. However, in the arid climates where these forms occur, the water contains many dissolved salts and minerals and is therefore extremely hard, requiring minor treatment for use.

Table 10.7. Fluvial alluvium—aluvial fans: aerial and cartographic references

State	County	Photo Agency	Photograph no.	Scale	Date	Corresponding quadrangle and geologic maps from USGS	Soil reports and geologic maps
California	Inyo	USGS*	California 17-AB	3933'/"	9/10/55	Independence (15'), MF-254	—
California	Inyo	USGS*	California 23-ABC	5000'/"	10/11/52	Emigrant Canyon, Furnace Creek (15') Prof. Paper No. 494-A	—
California	Inyo	USGS*	California 24-AB	4000'/"	12/2/48	Bennetts Well, Furnace Creek (15') Prof. Paper No. 494-A	—
California	Inyo	USGS*	California 25-AB	4000'/"	11/27/48	Furnace Creek, Ryan, Bennetts Well (15')	—
California	Inyo	USGS*	California 26-AB	4000'/"	11/27/48	Funeral Peak, Bennetts Well (15') Prof. Paper No. 413	—
California	Inyo	USGS*	California 28-AB	4000'/"	11/25/48	Ash Meadows (15')	—
Idaho	Custer, Butte	USGS*	Idaho 6-AB	1667'/"	8/1/48	Idaho Falls (1:250,000)	—
Montana	Gallatin	USGS*	Montana 18-AB	833'/"	8/22/59	Hebgen Dam (15'), Prof. Paper No. 435-I	1931

*From *Selected Aerial Photographs of Geologic Features in the United States,* U.S. Geological Survey Prof. Paper No. 590.

Wells located along the center of valley fills encounter the highest alkali levels; those located in alluvial fans or upland reaches of valley fills provide water of better quality. Shallow wells can be located near the fringes of alluvial fans to tap a subsurface flow which may be under slight pressure. In some arid regions of the world, typically in the Near East, lines of wells are established perpendicular to the subsurface flow within valley fills and supply vast quantities of water.

Pond or Lake Construction

It is not feasible to construct ponds and large lakes in fans and upland regions of valley fills because of the high rate of evaporation and porosity of the soil materials. In some valley fills the water table is at the surface in the valley bottom and forms an arid lake or playa; this may be wet or dry as the water table fluctuates seasonally. These lakes are also very saline, because of their high concentrations of dissolved salts and other minerals. Continental alluvium generally contains fine surface soils which are sufficiently impervious for use as bottom materials. However, the lack of sufficient watershed development and the low rainfall do not provide sufficient potential for the creation of large reservoirs.

Foundations

Alluvial fans and the upper portions of valley fills consist of noncohesive, granular materials which provide suitable support for raft and footing foundations for low and moderate structural loads. Unconsolidated materials are predominate and, if heavy loads are to be considered, require compaction in place by vibrations caused by pile driving or vibroflotation. Friction piles are typically used for heavy loads since they help to compact the underlying soils. Both fans and upper portions of valley fills are subject to flooding and channel shifting, and this should be considered in site location. Local seams of fine-textured materials may occur, and these should be avoided because of their associated volume changes.

Central valley fill areas and continental alluvium provide suitable support for footings and rafts on compacted fill or gravel pads for low and moderate structural loads. Heavier structures may utilize friction piles or deep footings depending upon the specific conditions. Detailed subsurface investigations are needed early in the planning process to determine the characteristics of these highly variable underlying substrata.

Highway Construction

The many channels of alluvial fans and the upper regions of valley fills present drainage difficulties in the location and design of highways. For example, flooding and shifting channels are constraints; although many culverts may be needed to drain the multiple drainage channels, they are not recommended since they quickly fill with sediments and require continual maintenance. Some railroads and roads utilize saw-toothed earth dikes along the upslope edge of valley fills and alluvial fans in an attempt to channel floodwaters and sediments through fewer culverts, but these also require continual maintenance and cleaning. Present design practice in

Table 10.8. Fluvial alluvium — filled valleys: aerial and cartographic references

State or country	County	Photo Agency	Photograph no.	Scale	Date	Corresponding quadrangle and geologic maps from USGS	Corresponding soil reports from SCS
Arizona	Maricopa	ASCS	DHP-9EE-32-34	1667′/″	1/31/64	Phoenix (15′)	—
Arizona	Maricopa	ASCS	DHP-8EE-45-48	1667′/″	1/29/64	Phoenix (15′)	—
Arizona	Maricopa	ASCS	DHP-6EE-47-50	1667′/″	1/24/64	Lone Butte (7½′)	—
Arizona	Maricopa	ASCS	DHP-6EE-111-114	1667′/″	1/26/64	Lone Butte (7½′)	—
Arizona	Pinal	USGS*	Arizona 15-ABC	1667′/″	12/21/57	Sacaton Butte, Gila Butte SE (7½′)	—
Arizona	Pinal	USGS*	Arizona 16-ABC	5000′/″	4/29/53	Gila Butte, Casa Grande, Sacaton Butte, Signal Peak (15′)	—
California	Imperial	USGS*	California 39-AB	1667′/″	11/10/59	Kane Spring NE, Kane Spring NW (7½′)	1903‡
California	Inyo	USGS*	California 27-AB	4000′/″	12/2/48	Ash Meadows, Ryan (15′)	—
California	Inyo	USGS*	California 23-ABC	5000′/″	10/11/52	Emigrant Canyon, Furnace Creek (15′) Prof. Paper No. 494-A	—
California	Inyo	USGS*	California 28-AB	4000′/″	11/25/48	Ash Meadows (15′)	—
California	San Luis Obispo	USGS*	California 30-AB	1667′/″	7/22/57	McKittrick Summit, Painted Rock (7½′)	1928‡
California	San Luis Obispo	USGS*	California 31-ABC	1667′/″	8/2/57	Wells Ranch, Panorama Hills (7½′)	1928‡
Iran	—	USGS†	Iran 5-AB	4580′/″	7/6/56	USC & GS ONC H-7 (1:1,000,000)	—
New Mexico	Valencia, Socorro	USGS*	New Mexico 12-ABC	2375′/″	10/12/51	Mesa Aparejo (15′)	1929‡
Texas	Crosby	ASCS	CGP-2DD-190-191	1667′/″	1/8/63	Crosbyton (7½′)	1966
Texas	Crosby	USGS	Texas 4-AB	5280′/″	1/23/54	Lubbock (1:250,000)	1966

*From *Selected Aerial Photographs of Geologic Features in the United States,* U.S. Geological Survey Prof. Paper No. 590.
†From *Selected Aerial Photographs of Geologic Features outside the United States,* U.S. Geological Survey Prof. Paper No. 591.
‡Out of print.

these regions is to dip the road where major drainage channels occur, allowing flow across the road surface during floods; immediately after the water recedes, the sediments deposited within the road dip can be cleared with a bulldozer. Alignments along the lower edges of fans also encounter seepage zones and possible volume changes where underlying fine-textured materials occur. In any case the alignment of roads and highways across fans requires some type of active maintenance. Continental alluvium may have a surface plastic soil and therefore require the placement of a thin, granular fill to protect against the movements caused by volume-moisture changes.

Playas (Arid Lake Beds)

Introduction

Arid lake beds or playas (also called dry lakes, alkali flats, and salinas) are fluviolacustrine, stratified deposits of fine sands, silts, clays, and salts which occupy lowland basins or depressions in arid and semiarid climates. These landforms may be covered with water during wet seasons or may be permanently dry.

Playas are normally found in association with the landforms formed by the arid erosion cycle. As the upland rock areas are weathered and eroded, valleys are filled with sediments, or alluvium. Central basins in such valleys accumulate runoff or subsurface seepage waters, and form temporary lakes or salinas, whose waters hold in suspension silts, clays, and dissolved salts. As the waters evaporate during the dry season, the silts, clays, and salts are left on the ground surface in thin, stratified layers. In some areas the lake does not completely evaporate, but it nevertheless fluctuates considerably in its areal extent throughout the year. In rare instances ancient shorelines, which indicate a long-previous higher water elevation, can be distinguished.

Playas are generally devoid of vegetative cover, owing to the arid climate and the high concentration of salts in the soil. Their surfaces are rather flat and range from hard to soft or smooth to rough. These basins are subject to flooding after heavy rainfalls, but the flat topography dissipates the waters in shallow depths (1 or 2 feet) over a large area.

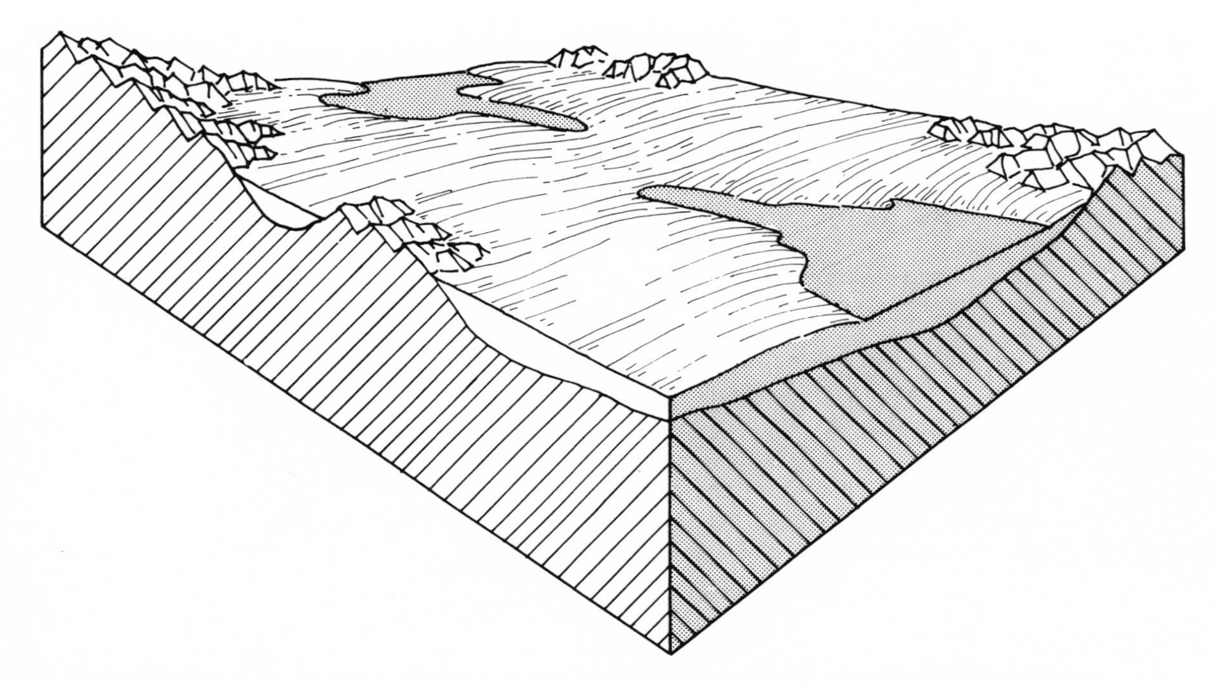

Figure 10.27. Playas have a flat, basinlike topography which is rather monotonous and devoid of vegetative cover. They do not provide any spatial variation, although beach ridges of old shorelines, when found, can offer limited, elevated viewing platforms. The basins are visually sensitive to land use development and to vertical structures, since they have no masking or screening capabilities.

Interpretation of Pattern Elements

Playas or arid lake beds are easily identified by their associated arid, filled-valley landforms and their characteristic flat basin position. There is no apparent drainage and little or no vegetative cover. Photographic tones are irregular and scrabbled. Irrigated lake bed deposits may be under intense cultivation.

Soil Characteristics

Playas consist of stratified, unconsolidated deep fine sands, silts, clays, and salts which grade into stratified alluvial sands and gravels in the substratum. There is little soil profile development because of the relative youth of the formations, the arid climate, and the continuing process of deposition. Variations in the alkali content influence the engineering characteristics and performance of the soils.

U.S. Department of Agriculture Classification

The U.S. Department of Agriculture commonly classifies playa deposits as alkali, stratified fine sands, silts, and clays. Subsurface alluvium is described as stratified sands and gravels.

Unified and AASHO Classifications

Playa soils are typically categorized by the Unified system as ML, CL, SM, SC, and plastic variations such as MH and CH. AASHO symbols include A-2, A-4, A-6, and occasionally A-3 or

Table 10.9. Fluvial landforms: Playas (arid lake beds)

Topography	*Drainage*	*Tone*	*Vegetation and land use*
Flat basin	None, evaporation	Light, scrabbled	Barren and cultivated
Playas are characterized by a very flat topography which occupies a large regional basin. Small-scale photographs or photographic indexes may be needed to determine the areal extent of the formation and to identify its transitional boundary with valley-fill alluvium. Old shorelines, which occur rarely, indicate previous water elevations (a).	The playa landform occupies a natural depression which does not develop any drainage features, owing to the flat basin topography and high water table. Ditches associated with intense agriculture indicate irrigation rather than drainage or attempts to lower the water table.	Large playas have uniform light tones over regional areas. Alkali and moisture differences are indicated by scrabbled, light and dark irregular tones. *Gullies* None	Most playas are either barren or have little vegetative growth, reflecting the arid conditions and the high alkaline level. Scattered scrub vegetation may occur, having a peppered appearance. Knowledge of local plant materials and their site preferences may aid in mapping moisture and alkalinity levels. Where irrigation has been instituted, intensive agriculture and truck farming cover the land surface with gridded patterns.

A-1. The substratum alluvial materials are defined in the section on alluvial-valley fill, "Fluvial: Alluvium."

Water Table

The water table in playas may fluctuate seasonally, but it is generally near the surface. Observing previous changes in water levels or floods may aid in establishing the seasonal profiles of the water table.

Drainage

Playas are naturally poorly drained, owing to their low basin topography, high water table, and fairly impervious soils. Flooding is common during wet seasons; also, after heavy storms in the surrounding mountains, the watersheds may drain into the adjacent valley fill.

Soil Depth to Bedrock

Playas are underlaid by alluvial valley fill materials which are deep and not directly associated with bedrock. Bedrock is not encountered in or considered a constraint on most construction practices.

Issues of Site Development

Sewage Disposal

Septic tank leaching fields are generally not suitable in these landforms because of the fluctuating water table, flooding hazard, and the fairly impervious soils. Intensive development requires construction of a suitable treatment system to maintain the quality of the water resources.

Solid Waste Disposal

Sites satisfying the requirements of a sanitary landfill operation in playas are difficult to locate because of periodic flooding.

Trenching

Trenching operations do not involve any significant difficulties in removing materials, since the soils are both unconsolidated and deep. Costs are approximately the same as those associated with removing deep, dry, moderately cohesive soil. However, if the water table is near the surface and encountered during trenching operations, costs may be increased by 35 to 40%. Pipelines (steel or concrete) placed within these formations are rapidly corroded by the high salt content and therefore need protection in order to prevent deterioration.

Excavation and Grading

Large, level sites can be created on these landforms with little grading or excavation. Pit-type excavations may encounter the water table, which makes the soils rather plastic, slippery, and difficult to handle and increases removal costs by 35 to 50%. Clam bucket or drag line devices can be used to excavate material beneath the water table, but the salt is highly corrosive to equipment. Minor surface grading is easily accomplished by light equipment, such as graders and scrapers.

Construction Materials (General Suitabilities)

Topsoil: Poor. Their high alkaline content generally makes these materials unattractive as commercial sources of topsoil.

Sand: Poor. Fine sands are occasionally found in playas, but they may be mixed or thinly interbedded with other fines. More suitable supplies can be found in the surrounding valley fill alluvium. Beach ridges, when they occur, provide suitable sources.

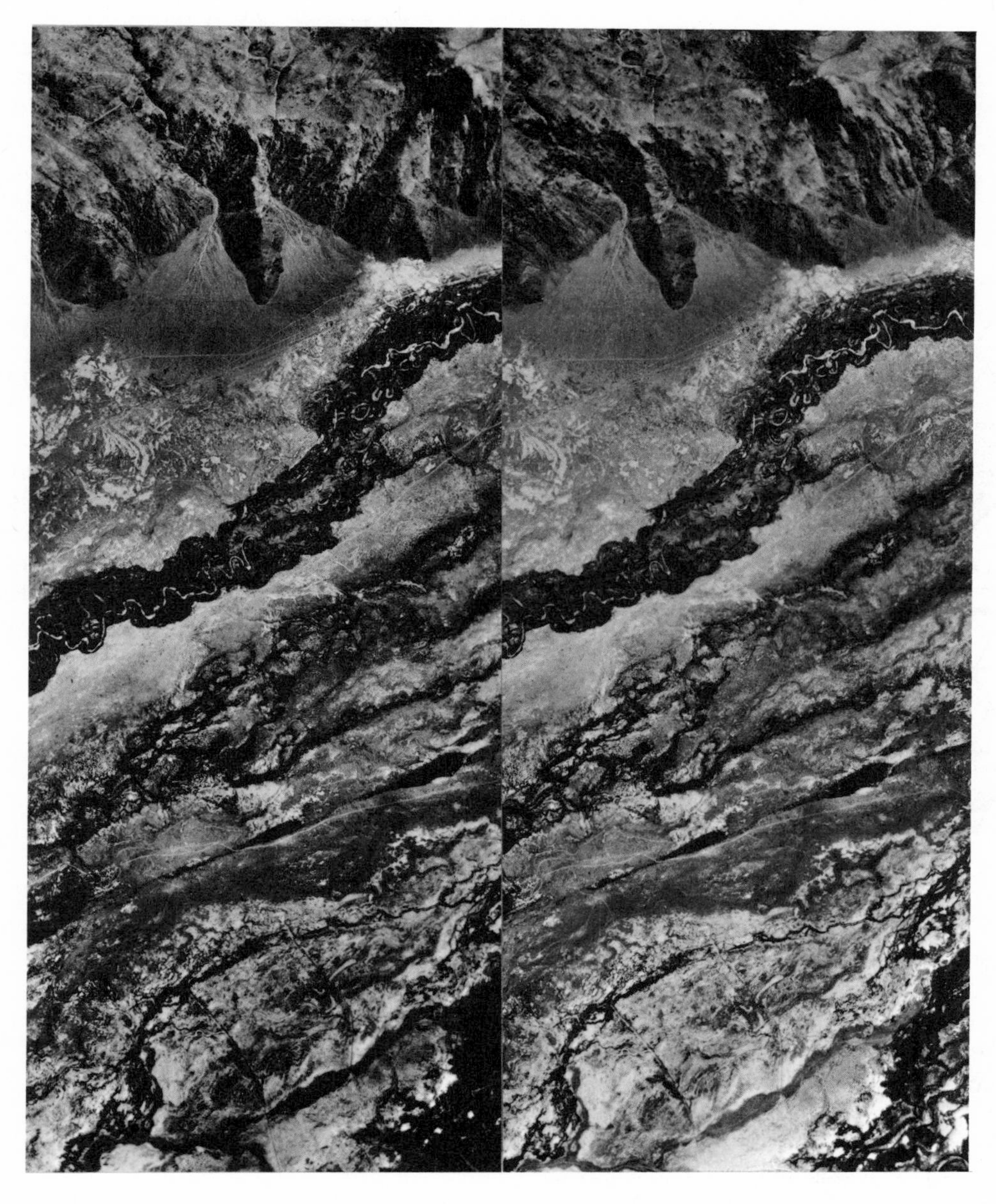

Figure 10.28. Playa or arid lake bed in Inyo County, California, well illustrates the scrabbled pattern associated with alkaline deposits. Photographs by the U.S. Geological Survey, California, 19-ABC, 1:37,000 (3933'/"), July 17, 1947.

Gravel: Not Suitable. Adjacent valley fill alluvial materials provide excellent sources.

Aggregate: Not Suitable. Adjacent valley fill alluvial materials or ancient beach ridges are excellent sources of aggregate materials.

Surfacing: Good. The fine, alkaline silts and clays are excellent sources of binder for surfacing materials for secondary roads. The best mixtures are obtained when these materials are mixed with granular materials found in adjacent valley fills or fans.

Borrow: Poor. The soils are generally fine-textured but are suitable for fills upon compaction by sheepsfoot rollers, under a relatively narrow range of moisture content (which is difficult to control in these regions).

Building Stone: Not Suitable

Mass Wasting and Landslide Susceptibility

Because of the depressed position of this landform, it is not susceptible to erosion or dissection and therefore does not develop a sliding potential.

Groundwater Supply

The groundwater table, as reflected by the water table, is found near the ground surface. The water elevation fluctuates according to climatic and seasonal changes, or if major storms in the upland watersheds feed the adjacent alluvial valley fill. The water contains high concentrations of dissolved minerals and salts and requires treatment before use.

Pond or Lake Construction

The playa represents a natural reservoir basin, but its ability to maintain a given water elevation is a function of the seasonal distribution and

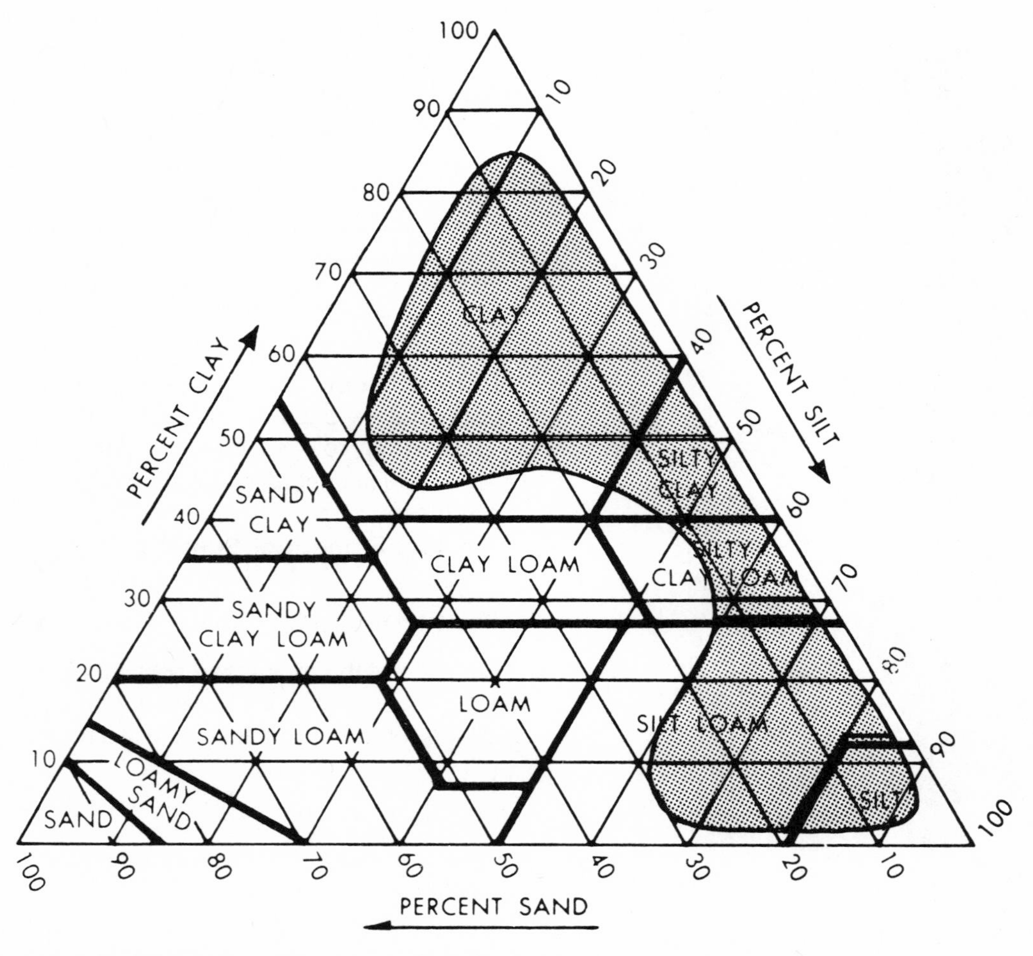

Figure 10.29. The U.S. Department of Agriculture soil texture triangle indicates clay and silty clay for playas.

amount of rainfall (both within the actual valley basin and in the surrounding upland watersheds), the storage capacity of the adjacent alluvial valley fill, and the rate of evaporation. Furthermore, if water within the watershed is used for other purposes than to feed the lake, such as irrigation, this will remove some water from the recharge system and effectively lower the lake bed or water table elevation.

Foundations

The plastic soils and high alkali content present difficulties in foundation construction in these landforms. Gravel pads can be placed over the plastic soils to separate a raft foundation from the corrosive salts, but large structures may require friction piles with protective coatings. The fine soils are subject to volume change during moisture fluctuations, and the entire area may be susceptible to shallow flooding, making elevated floor grades desirable. However, low dikes are easily constructed around developed areas to provide flood protection.

Highway Construction

Granular fills are necessary to elevate road surfaces above the plastic, fine-textures soils which are susceptible to volume change. Floods are generally shallow and of short duration and do not present severe constraints. Alignments are generally straight, but they may follow ancient beach ridges if these occur in the desired direction and location. Construction costs are relatively low, since there is no need for cuts or excavation, drainage structures, or culverts or bridging.

Table 10.10. Fluvial lake beds (playas): aerial and cartographic references

State	County	Photo Agency	Photograph no.	Scale	Date	Corresponding quadrangle and geologic maps from USGS	Corresponding soil reports from SCS
California	Inyo	USGS*	California 18-AB	3116′/″	9/27/47	Lone Pine, Mt. Whitney (15′), Prof. Paper No. 110	–
California	Inyo	USGS*	California 21-AB	3933′/″	10/10/55	Lone Pine, New York Butte (15′)	–
California	Inyo	USGS*	California 22-AB	3167′/″	7/16/47	Keeler, Olancha (15′)	–
California	Inyo	USGS*	California 25-AB	4000′/″	11/27/48	Furnace Creek, Bennetts Well, Ryan, Funeral Peak (15′)	–
California	Inyo	USGS*	California 26-AB	4000′/″	11/27/48	Funeral Peak, Bennetts Well (15′)	–
Montana	Phillips	USGS*	Montana 10-AB	1667′/″	7/10/53	Lewistown (1:250,000), I-327	–
Nebraska	Sheridan	USGS*	Nebraska 2-ABC	1667′/″	8/28/54	Antioch (15′)	1918†
Nevada	Churchill	USGS*	Nevada 1-AB	3600′/″	12/18/47	Carson Lake (15′), Prof. Paper No. 401	–
Nevada	Churchill	USGS*	Nevada 2-ABC	3600′/″	12/18/47	Fallon, Weber Reservoir (15′), Prof. Paper No. 401	–
Nevada	Churchill	USGS*	Nevada 3-AB	3600′/″	12/18/47	Carson Lake (15′), Prof. Paper No. 401	–

*From *Selected Aerial Photographs of Geologic Features in the United States,* U.S. Geological Survey Prof. Paper No. 590.
†Out of print.

Organic Deposits (Swamps, Bogs, Marshes)

Introduction

The organic deposits discussed in this section are the high concentrations of peat and organic soils that have accumulated largely from autogenic processes and which may be found on or between all other landforms. *Organic deposits* are primarily the result of the process of autogenic bog succession, in which certain freshwater bodies become filled over time with decayed vegetative debris. The environment is modified by the internal forces of growth and decay of the plants themselves, rather than by allogenic (outside) forces such as siltation. Autogenic succession may be aided by allogenic processes such as siltation, or by other geomorphic processes that deposit materials through sedimentation. Therefore, an organic depression filling itself largely by means of vegetative growth can be aided in the process by the sedimentation of soil materials.

The autogenic bog succession process begins when a freshwater body develops a thin zone of rushes and sedge-type vegetation along the shoreline shallows. As the rushes and sedges decompose, they form peat; this provides a stable platform for the growth of some varieties of trees. The trees in turn decompose to form woodly peat. As the process continues, the formerly open body of water literally grows shut, becoming filled with organic materials (Figure 10.30). Small, shallow ponds obviously complete this process more rapidly than larger, deeper ponds. It is not unusual to find an area of many ponds having a variety of shapes, sizes, and depths in which are represented the full range of the different stages of autogenic bog succession.

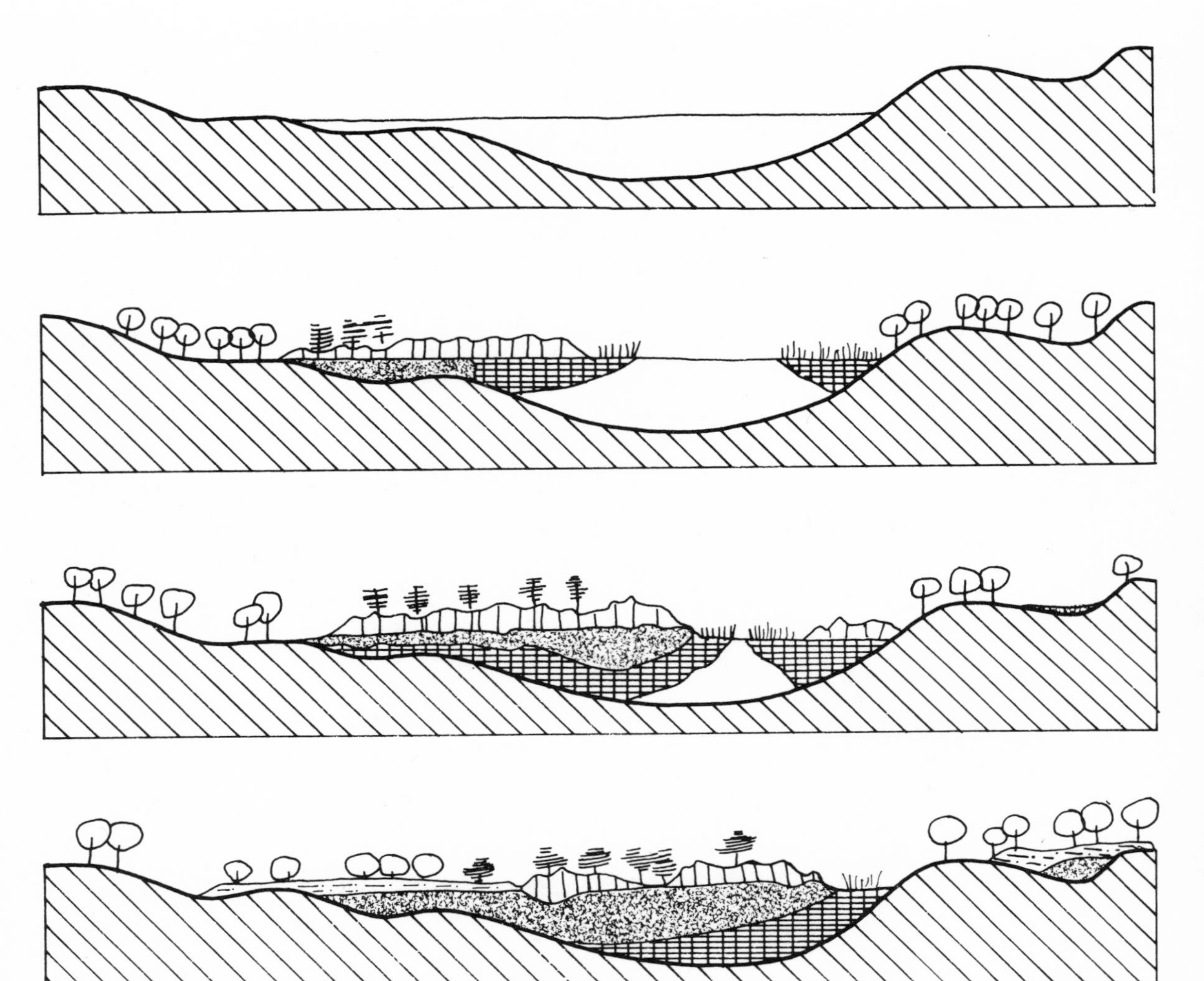

Figure 10.30. The successive stages of autogenic bog succession, whereby a pond is eventually completely filled with organic material.

Organic deposits are classified by their type of vegetative cover; marshes contain short marsh grasses, bogs contain both grasses and shrubby growth, and swamps are associated with tree cover. Organic deposits containing examples of all three types are often found. By carefully studying plant associations, ecologists can interpolate associated information about the composition of the peat, the degree of water level fluctuation, pond depth, projected successions and rates, water temperature and quality.

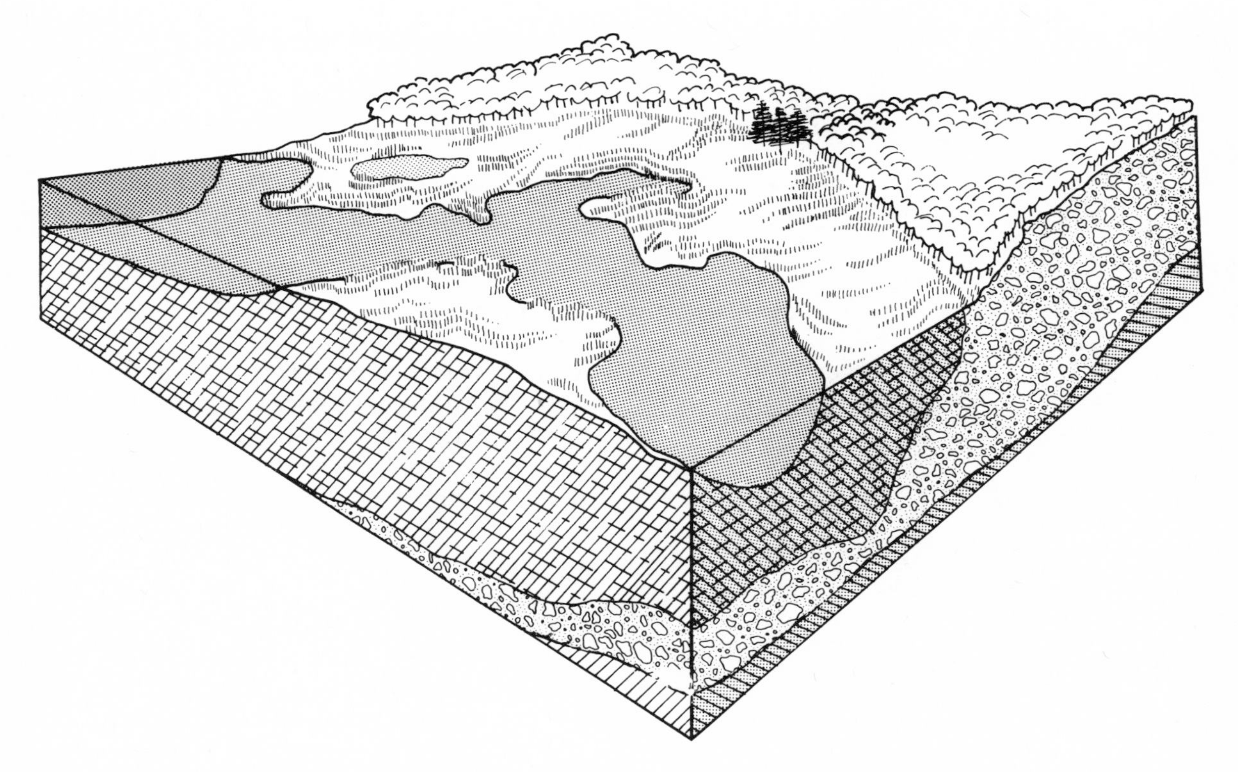

Figure 10.31. Organic deposits are typified by very intense vegetative growth which varies from low grasses to dense forests. The vegetative cover provides great diversity and spatial variation in these areas. From the surrounding, more elevated landforms, views may be had out over the wetland and, depending on the stage of succession, central areas of water may be visible. Compared with their adjacent and surrounding landforms, organic deposits generally have a higher degree of visual absorption because of their depressed positions and dense vegetative cover. Marshes comprised of low grasses are visually exposed; swamps and even bogs are visually highly absorptive. Visually,these landforms can offer points of needed diversity in regions of surrounding monotonous landforms such as till plains, glacial lake beds, or broad flood plains.

Interpretation of Pattern Elements

Organic deposits are characterized by a flat topography, heavy vegetative growth, and lack of surface drainage. The following topographic key can be utilized to distinguish organic deposits from other landforms, but it does not provide a classification of the different types of organic deposits.

Soil Characteristics

Organic deposits contain cumulose soils of organic origin which were formed in place from the decomposition of vegetative material. Shallow deposits consist primarily of peat, deeper deposits form organic silts and clays. The soils are generally very compressible and have low densities and high moisture, plasticity, and liquid limit levels.

U. S. Department of Agriculture Classification

The U. S. Department of Agriculture commonly classifies organic materials as muck or peat. Some underlying silts and clays may occur in deeper deposits.

Unified and AASHO Classifications

Both the Unified and AASHO classification systems refer to these materials as Pt or peat. Other Unified categories indicating highly plastic and/or organic materials include OH, CH, and MH. Organic materials of low plasticity are designated OL.

Water Table

The water table occurs directly at the landform surface or very close to it. It may fluctuate with seasonal conditions. Areas of standing water in the center of the deposit indicate the water elevation that can be expected throughout its areal extent.

Drainage

Organic deposits are saturated with water and are therefore poorly drained. Ditches are commonly used to lower the water table and facilitate drainage of the surface layer.

Soil Depth to Bedrock

Depth to bedrock is dependent on both the size and depth of the deposit and the nature of the underlying material. For example, depressions on exposed, impervious rock surfaces can develop organic deposits, and these are of common occurrence in regions where glaciation has removed most of the surface soil.

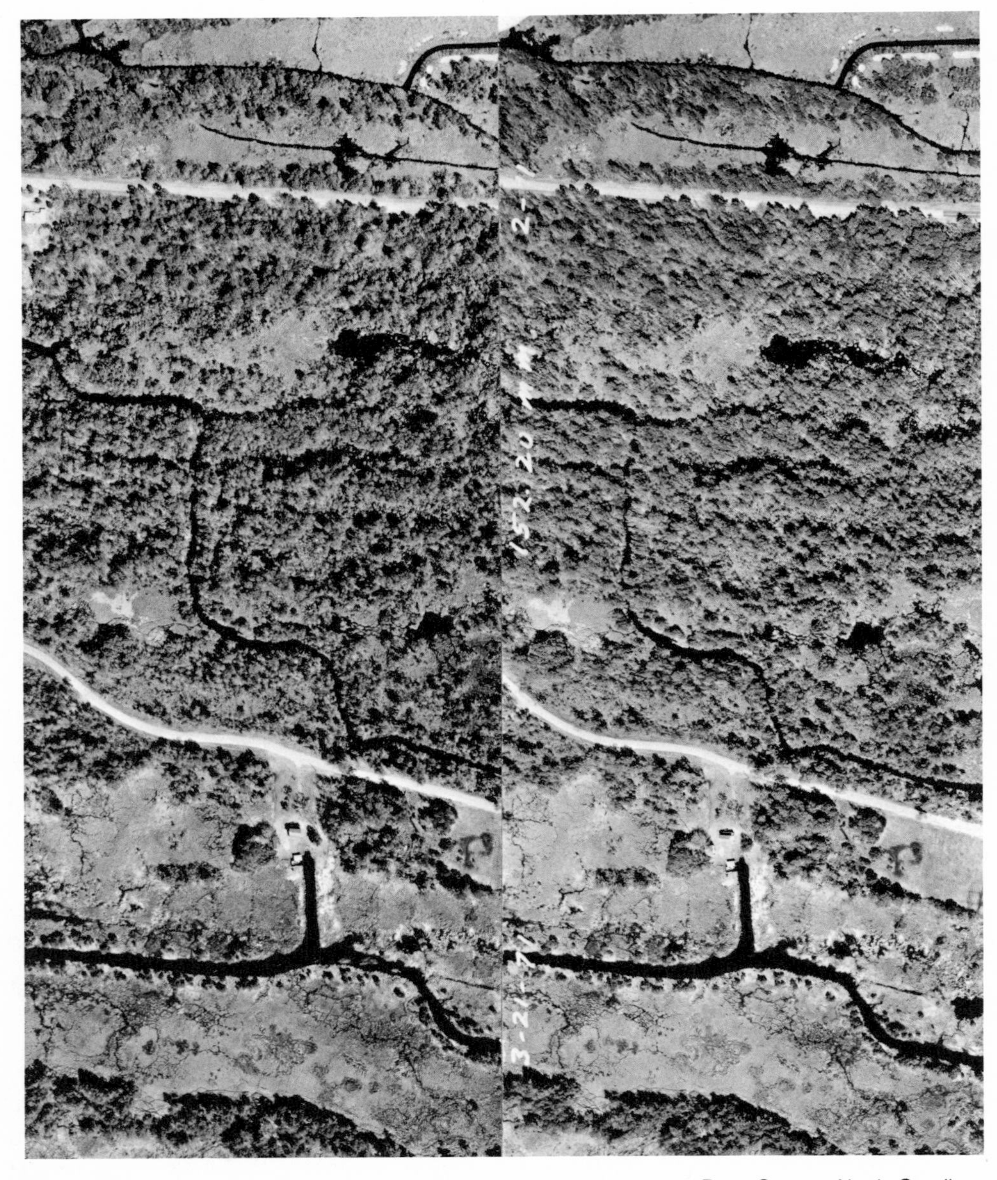

Figure 10.32. Organic materials in the form of marshes, bogs and swamps in Dare County, North Carolina, near Kitty Hawk. Photographs by Aerial Data Reproductions Associates, Riverside, New Jersey, 1:4,800 (400′/″), March 21, 1971.

Issues of Site Development

Sewage Disposal

Because of the associated high water table, septic tanks or other methods of on-site disposal are not suitable for use on or within organic materials.

Solid Waste Disposal

Sanitary landfill operations cannot be located within these deposits because of the high water table and the unsuitability of materials. Dumping in wetlands results in contamination of the water resources and is now illegal in many states.

Trenching

Trenching operations are difficult, owing to the high water table, the saturated conditions, and the flow tendencies of the materials. These areas are not easily drained, and the low supportive ability of peat makes it difficult to use any heavy equipment; dredges may be required. Costs are generally quite high, being three to almost four times those encountered in removing the same amount of deep, dry, moderately cohesive soil.

Excavation and Grading

Excavation of organic materials is difficult, owing to the saturated conditions and the tendency of the materials to flow. They are not easily drained, and their low supportive ability does not allow the use of heavy equipment directly on the land surface and may require the use of drag line, clam bucket, or dredging equipment. Costs are generally high, being three to four times those involved in removing the same amount of deep, dry, moderately cohesive soil.

Table 10.11. Fluvial landforms: Organic deposits

Topography	*Drainage*	*Tone*	*Vegetation and land use*
Flat depressions 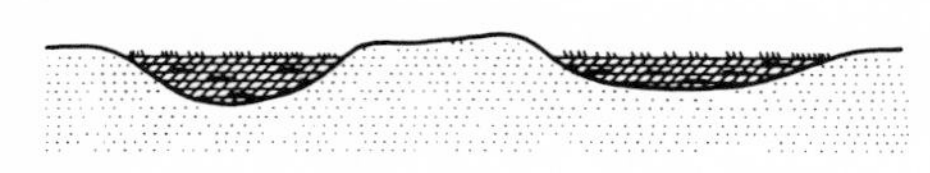 Organic deposits are characterized by very flat surfaces and by their occupying depressions in respect to the surrounding topography. In some instances there may be open water in the interior of the deposit.	None or artificial 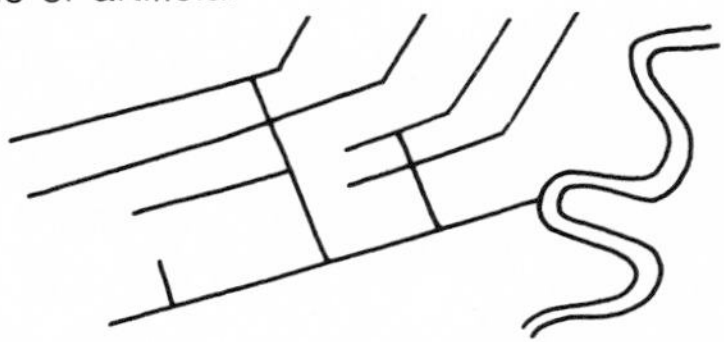 Generally a drainage pattern is not found across these deposits unless there is a surface outflow. In the latter case, a dendritic pattern develops, having a major trunk and short, branching tributaries. Artificial channels or ditches are common and indicate attempts to lower the water table, sometimes for mosquito control.	Light and dark The tones largely represent the different types of vegetative cover and are not to be considered indicators of soil materials. Exposed soils (peat) are very dark or black. *Gullies* None, ditched Because of the high water table and flat topography, gullies do not occur. Straight, dark lines are drainage ditches arising from attempts to lower the water table.	Vegetated and cultivated Most organic deposits are covered with a variety of vegetation and grass, which are indicators of the different stages of succession. In some regions partially drained organic deposits are utilized as intensive muck farms or truck farms specializing in vegetable crops. In New England many of the organic deposits of glacial origin are cultivated as cranberry bogs.

Construction Materials (General Suitabilities)

Topsoil: Good to Not Suitable. Excellent sources of organic peat soils can be found, but loamy soils are not common.

Sand: Not Suitable

Gravel: Not Suitable

Aggregate: Not Suitable

Surfacing: Not Suitable

Borrow: Not Suitable. Peat materials and highly organic soils cannot be compacted and are highly compressible and plastic.

Building Stone: Not Suitable

Mass Wasting and Landslide Susceptibility

Organic deposits are found in lowland depressions and have flat surfaces. Therefore, they do not provide the conditions for sliding or other processes of mass wasting.

Groundwater Supply

The presence of organic deposits indicates that the groundwater table is near the ground surface, but knowledge of the underlying material is needed when determining whether the water table is perched or is part of a larger soil aquifer system. Organic deposits in glacial outwash indicate zones where groundwater intercepts the ground surface. Lack of drainage channels through the organic deposit may indicate that it is underlaid by granular materials and that drainage therefore takes place primarily by subsurface flow. Wells drilled in organic deposits generally supply high yields, but the water may be stained with organic acids and sediments.

Pond or Lake Construction

All organic deposits formed by autogenic bog succession were initially clear bodies of water, forming ponds and lakes, and many organic deposits still contain a small pond or lake in the central area. Any attempt to dredge or otherwise remove the organic debris slows the natural process of succession but does not eliminate it. All lakes and ponds are destined to become weed-choked marshes and swamps.

Foundations

These zones are not normally considered for construction because of the very low supporting power of the material and the high water table. Construction is possible if the peat is filled and compacted by surcharging, using sand drains. Light structural loads can be supported by shallow foundations on surcharged fills, but heavier loads require the use of point-bearing piles if deposits are underlaid by materials of suitable bearing load. If uniform layers of silt are present, the operation of a pile driver will vibrate and settle the ground surface for hundreds of feet around the zone of operation.

Highway Construction

The low supporting ability, the high water table, and the plastic nature of the materials do not provide conditions favorable for highway location and construction. When highways must cross these areas, peat and organic material are typically removed and the area is backfilled with granular

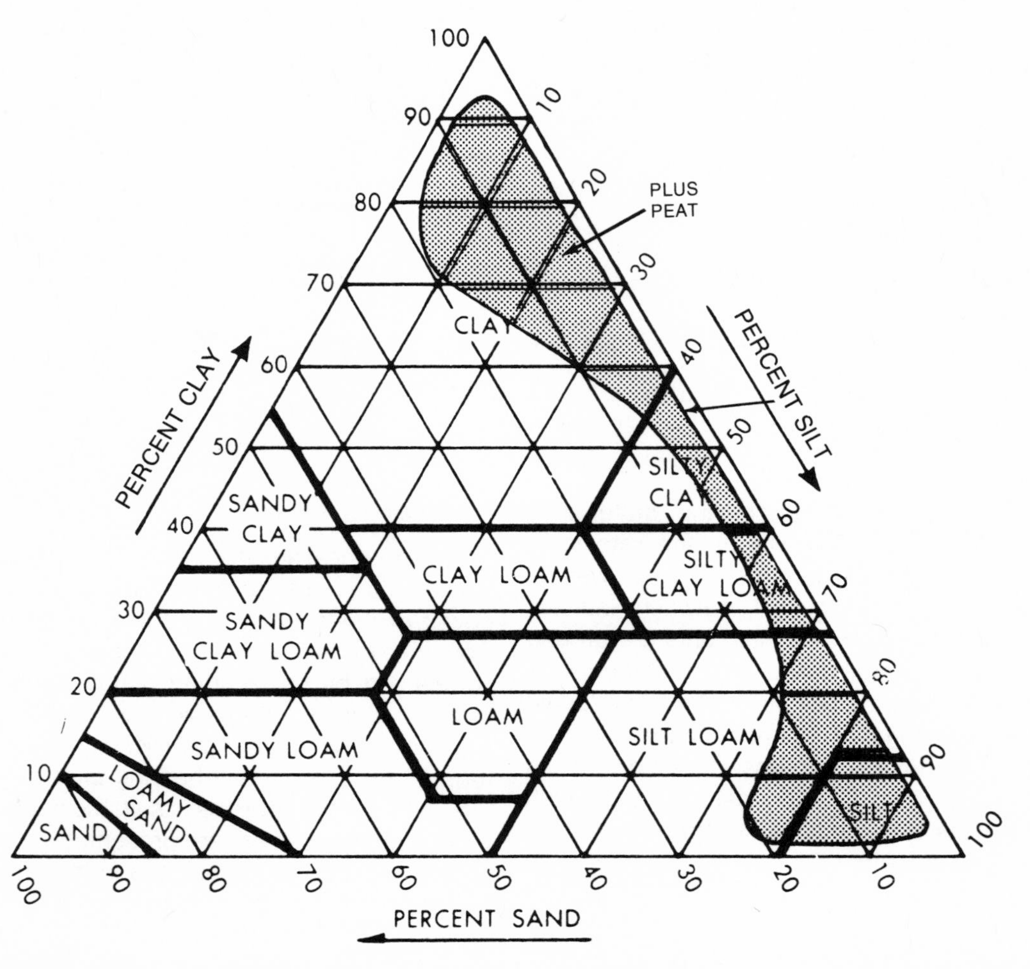

Figure 10.33. The U.S. Department of Agriculture soil texture triangle indicates silt and clay materials in organic deposits along with peat.

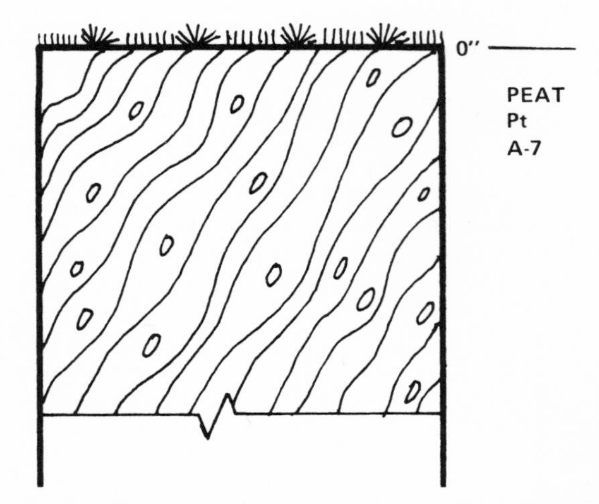

Figure 10.34. A typical soil profile developed in an organic deposit.

materials, thus distributing the applied load over a larger area. If compaction is accomplished by surcharge fills, settlement is to be expected, and flexible road surfaces are advised.

Table 10.12. Fluvial organic deposits: aerial and cartographic references

State	County	Photo Agency	Photograph no.	Scale	Date	Corresponding quadrangle and geologic maps from USGS	Corresponding soil reports from SCS
Alaska	–	USGS*	Alaska 10-AB	3333′/″	8/27/52	Mt. McKinley (1:250,000), Prof. Paper No. 373	–
Florida	Brevard	ASCS	CYS-5V-141-143	1667′/″	3/17/58	Deer Park NE (7½′)	–
Florida	Brevard	ASCS	CYS-3V-169-173	1667′/″	3/17/58	Eau Gallie, Melbourne West (7½′)	–
Florida	Sarasota	ASCS	DEW-2D-110-112	1667′/″	2/27/48	Laurel (7½′)	1959
Maine	Hancock	USGS*	Maine 2-ABC	2000′/″	5/1/56	Great Pond (15′)	–
Maine	Hancock	USGS*	Maine 3-ABC	2000′/″	5/7/56	Mt. Desert, Bar Harbor (15′)	–
Michigan	Cheboygan	USGS*	Michigan 5-AB	1416′/″	5/5/53	Tower (15′)	1939
Minnesota	Itasca	USGS*	Minnesota 2-AB	1416′/″	9/2/47	Calumet (7½′)	–
Montana	Teton	USGS*	Montana 3-ABC	3100′/″	7/13/55	Lake Theboe, Split Rock Lake (7½′)	–
North Carolina	Brunswick	ASCS	AOH-8F-43-46	1667′/″	12/1/49	Southport (7½′)	1937
North Carolina	Brunswick	ASCS	AOH-2GG-109-112	1667′/″	3/18/66	Lockwoods Folly (7½′)	1937
South Carolina	Horry	USGS*	South Carolina 1-ABC	1667′/″	2/23/58	Nixonville, Myrtle Beach (15′)	1918†
Virginia	Norfolk, Camden, Nansemond, Gates	USGS*	Virginia 7-ABC	5000′/″	12/8/59	Lake Drummond, Suffolk (15′)	1959, 1932
Wisconsin	Fond du Lac, Sheboygan	USGS*	Wisconsin 2-ABCD	1967′/″	4/30/52	Kewaskum (15′)	1911†, 1924†

*From *Selected Aerial Photographs of Geologic Features in the United States,* U.S. Geological Survey Prof. Paper No. 590.
†Out of print.

Coastal Plains

Introduction

The formation of coastal plains starts with sediments of gravel, sand, silt, and clay being carried by river systems to ocean shorelines. There waves and ocean currents sort the materials and deposit them on the ocean bottom in stratified layers which parallel the bottom and slope upward toward the land. If the land surface is uplifted or the ocean level goes down, the deposits are exposed and appear as a flat plain sloping gently toward the water. Coastal plains are actually physiographic provinces rather than simple landforms and contain within their areas other, associated landform features such as beach ridges, tidal flats, flood plains, and organic deposits. The term coastal plain is used here to classify regions of exposed, stratified, unconsolidated sediments of marine fluvial origin.

A newly uplifted or exposed coastal plain appears flat but actually slopes gently toward the ocean. It may exhibit linear shoreline features such as beach ridges or the original, submerged tidal channels with their broad dendritic form. These ancient tidal channels are filled with dense vegetation and become organic deposits. As the plain is exposed to the forces of erosion, an integrated, parallel drainage system is established; this removes materials from the less resistant deposits, which then form lowlands. More resistant layers (beds of sand or gravel, for instance) form linear uplands, or cuestas, whose strike generally parallels the present ocean coastline. Cuestas, because they are perpendicular to the initial drainage, modify the pattern to a trellis form (but this is apparent only on a regional scale).

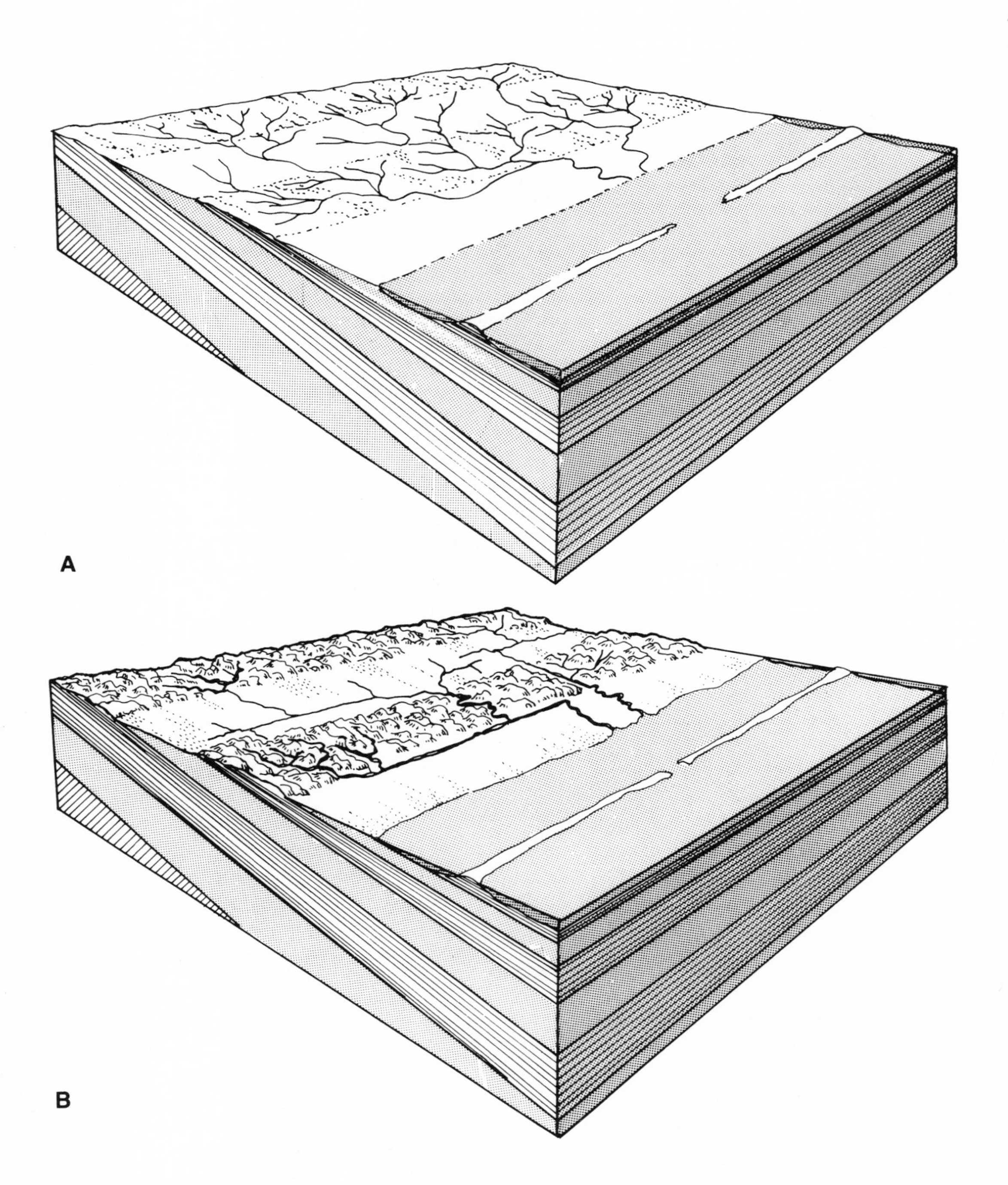

C

Figure 10.35. Development sequence of coastal plains. (A) Initial stage; land recently emerged. (B) Mature stage; erosion of lowlands and formation of cuestas. (C) Old stage; erosion has reduced relief to that of a low, flat plain.

As uplift and erosion continue, the coastal plain becomes highly dissected, developing a medium or fine dendritic drainage pattern and a very rugged topography. Eventually, erosional forces reduce even this rugged landscape to the peneplain stage, in which the drainage system again becomes predominately parallel and has broad meanders across a relatively flat plain (Figure 10.35).

In the states of North Carolina, South Carolina, and Georgia, young coastal plain deposits are characterized by strange elliptical depressions. These forms, or Carolina bays, are remarkably similar to one another in shape and have parallel northwest-southeast axes; they are narrower at their southeastern ends and may also have fringing, elevated sand ridges. It was originally postulated that the bays are remnants of a gigantic meteor shower, but this theory was quickly discounted by geologists for lack of supporting evidence. Their origin has since been more satisfactorily explained by the theory that they were formed from solution depressions accompanied by wind erosion. This theory holds that depressions, being wetter, would accumulate greater densities of vegetation. These would decay and form organic acids, and the acids would then dissolve the iron and aluminum from the sandy sediments, thus enlarging the depression and continuing its development through additional organic growth and decomposition. Since the water table of a coastal plain occurs near the surface, many of these depressions were at one time ponded; thus the influence of wind and waves could have helped form the parallel elliptical shapes, including the sandy ridges fringing their southeastern edges. The bays can be readily observed on aerial photographs. Typically, they have flat bottoms, may be ponded, and are covered with either dense swamp vegetation or marsh grass. They occur only in coastal plain physiographic regions.

Interpretation of Pattern Elements

The following photographic key is divided into two sections according to the characteristics associated with young and old coastal plains. Young coastal plains are characterized by a fairly level, undulating topography having regional parallel drainage and local drainage features such as Carolina bays or the broad dendritic swamps that indicate ancient tidal marshes. Young plains generally have forest cover in the wet lowlands and agriculture, forming irregular fields, in the drier uplands. Tones are varied but appear mottled in cleared fields.

Older, highly dissected coastal plains have a rugged topography and moderately steep slopes, indicating moderately cohesive soils. The drainage pattern is of a fine or medium dendritic form, having regional parallelism. Older plains tend to have a more uniform light tone. Cover is predominately forest, as a result of the steep slopes.

Soil Characteristics

A wide variety of soils is found in coastal plains. Typically, they include mixtures of sand, silts, and

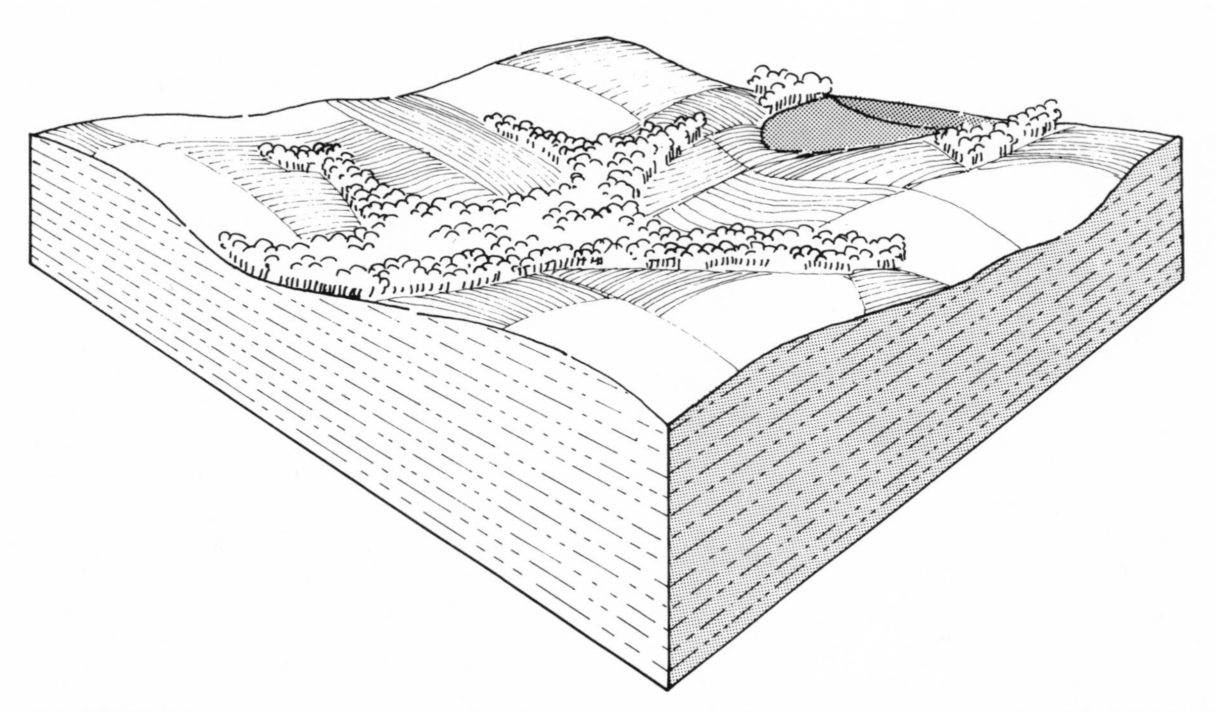

Figure 10.36. Young coastal plains characterized by a gently rolling undulating topography which offers little spatial variation and viewing capabilities. Dense vegetative cover generally occurs in lowland depressions; the drier uplands are cultivated. Localized views of the immediate surrounding area can be obtained on uplands where the vegetation has been cleared. The lowlands are visually the most absorptive because of the screening ability of the dense vegetative cover. If the vegetation is cleared, the uplands are sensitive to visual impact from development, and vertical structures will create the highest impact since they can be observed from surrounding uplands.

clay, and are unconsolidated and stratified in beds which slope gently toward the ocean. Gravels found in ancient beach ridges indicate a previous shoreline elevation. Uplands, having generally sandy soils, are well drained; lowlands, where soils are poorly drained, have the water table near or at the surface. These lowlands contain predominately silts, clays, and organic soils. Iron, cemented sands (bog iron) are common, formed by precipitation from iron-bearing swampwaters. Mapping of coastal plain soils can be fairly accurate if tonal distinctions and vegetative indicators are utilized to distinguish between the different moisture, textural, and drainage ranges.

U. S. Department of Agriculture Classification

The U. S. Department of Agriculture classifies most of the upland soils as sandy, loamy sand, or sandy loam. Subsurface soils are commonly sandy clay loam or clay loam. Soils in depressions are of finer texture and contain higher amounts of organic material; they are classified as loam, muck, peat, clay loam, or silt loam.

Unified and AASHO Classifications

The Unified classification system describes coastal plain upland soils as having surfaces designated SM or occasionally SP-SM, and subsurfaces designated SM, SC, or SP-SM. The AASHO system for the same soils indicates A-2 and A-4. Bedded layers of fine sands, or A-3 soils, can also be found. Soils in depressions are given Unified classifications OL, ML, CL, ML-CL, MH, CH, OH, and Pt, and AASHO classifications A-4, A-6, and A-7.

Water Table

The water table in young coastal plains is near the surface, even in the uplands, reflecting the small amount of regional uplift and the flat topography. Uplands average 6 to 10 feet to the water table; lowlands may have surface water table conditions. Swamps or other organic deposits occur where the water table is continuously at the surface. In highly dissected forms the water table is generally within 20 feet of the surface.

Drainage

The sandy upland soils are well-drained internally in both young and old deposits, having percolation rates greater than 2 inches per hour and as high as 10 inches per hour. The drainage rate in the subsurface decreases as the water table is approached, or where beds of more impervious silt or clay are encountered. Lowland areas are poorly drained because the water table occurs near the surface; ditching operations are effective in lowering the water table elevation.

Soil Depth to Bedrock

Soil materials are deep enough so that bedrock is not commonly encountered during normal construction operations.

Issues of Site Development

Sewage Disposal

Septic tank leaching fields are not difficult to locate if the percolation rates, slopes, and depth to water table fall above state minimum standards. The major constraint in young coastal plains is the water table; in older, dissected landforms, slopes are the constraint. It should be mentioned that, because most states do not have maximum percoltaion rate standards, contamination of the groundwater resources is possible even if the water table is below the minimum depth, for rapid percolation may allow effluent to reach the water table before it has been completely decomposed. Sanitary engineers should be consulted to determine the soil capabilities, and the risks involved in proposed sewage disposal systems. Intensive development requires the construction of suitable treatment systems if the ground and surface water resources are to be maintained.

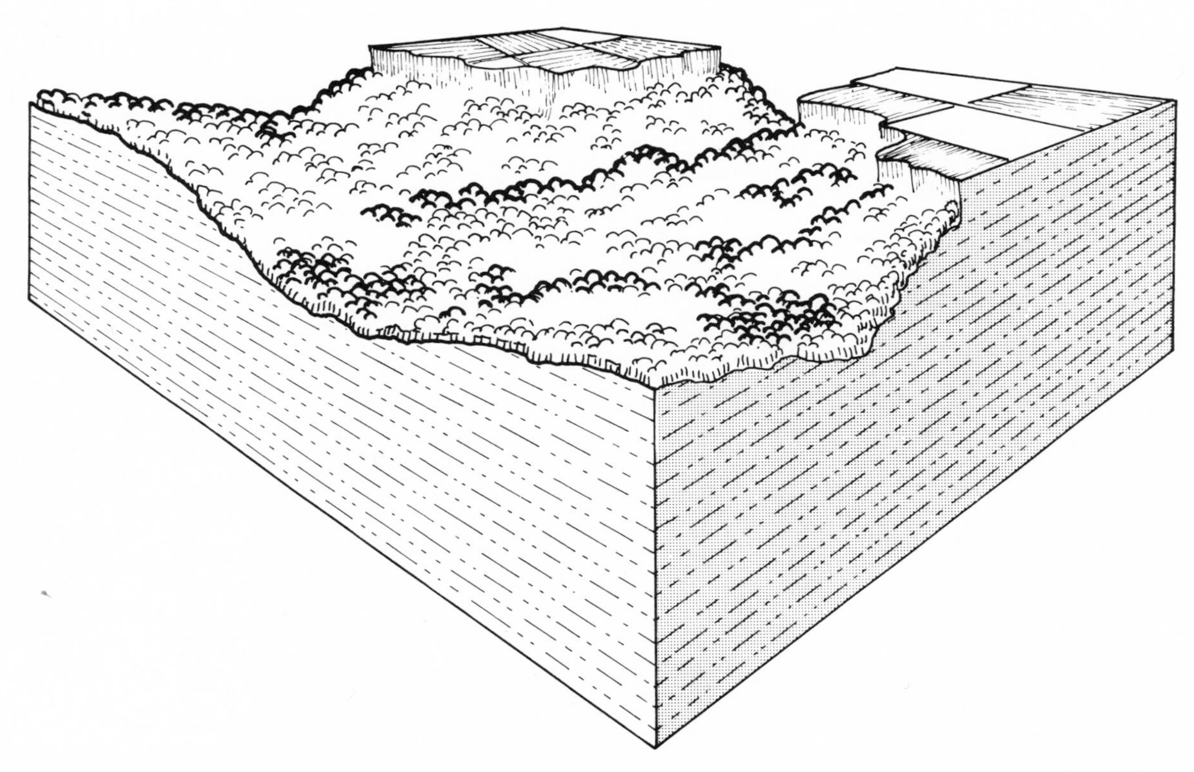

Figure 10.37. Highly dissected coastal plains are covered with forest vegetation which creates a tight, spatially closed landscape when traveling through the valleys. The uplands offer viewing potential of the surrounding region. The wooded hillsides and ridges are visually the most sensitive zones of the landform, and any cutting of vegetation and/or development in these areas will be very apparent.

Solid Waste Disposal

Because of the predominance of porous, granular subsoils, which are undesirable, suitable sites for sanitary landfill operations are difficult to locate. The porous soils and the close proximity of the water table allow leachate to percolate through the soil substratum and contaminate the groundwater resources.

Trenching

Trenching in young upland soils does not encounter significant difficulties, and costs are similar to those associated with the removal of dry, deep, moderately cohesive soils. On lower slopes and lowlands, the water table may be encountered, requiring pumping and drainage. Costs vary in accordance with the depth at which the water table is encountered and with the amount of organic material in the soil, but they are commonly 1.5 to 4 times as high as those for dry, moderately cohesive, deep soils. Older, highly dissected coastal plains create difficulties in trench alignments for liquid-carrying pipelines, and most routes follow the major valleys since crossing ridges requires the use of pumping stations.

Excavation and Grading

Little grading and excavation is needed in young coastal plains since the topography is flat or gently undulating. Light grading can be accomplished with light grader or scraper equipment. Pit-type excavations may encounter the water table and require some type of drainage or dewatering practice to prevent sand boils. Costs of removing drier, granular soils are low unless the water table is encountered. Lowlands containing deep peat and

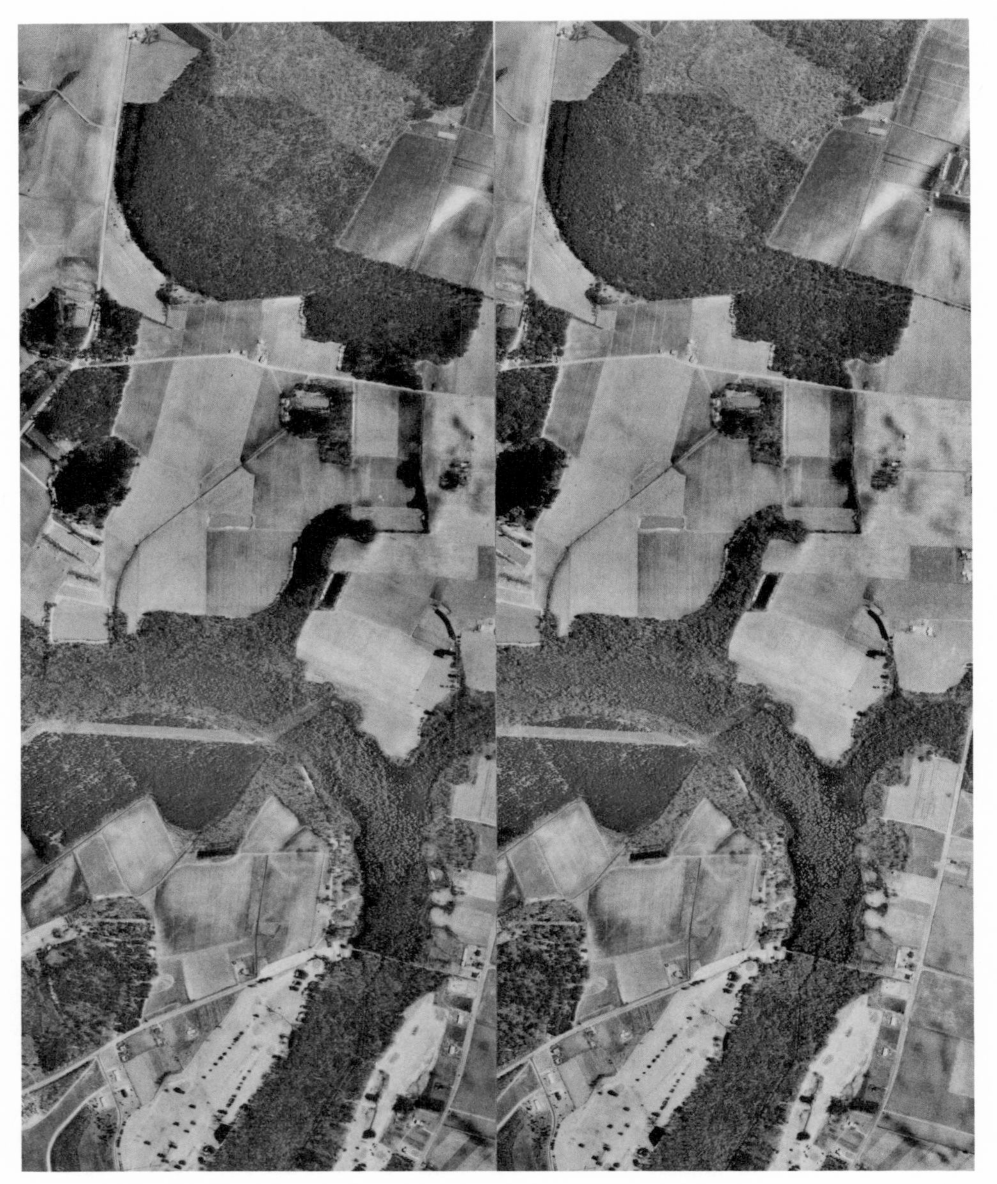

Figure 10.38. (A) Young coastal plain in Marion County, South Carolina, shows drainage courses of old tidal flats, Carolina bays, and mottled soil tones. Photographs by the Agricultural Stabilization Conservation Service, ASP-3JJ-161,162, 1:20,000 (1667′/″), February 26, 1968.

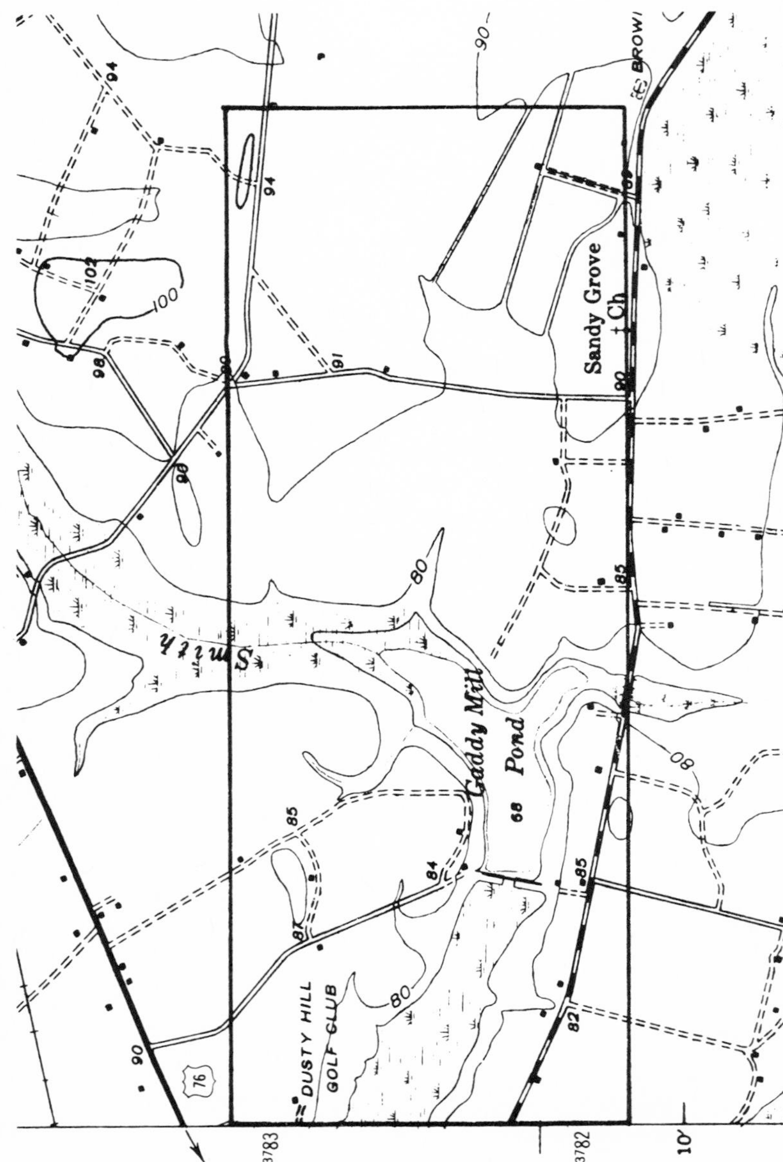

Figure 10.38. (B) U.S. Geological Survey Quadrangle: Mullins (7½′).

Table 10.13. Fluvial landforms: Coastal plains (young and old)
Young

Topography	*Drainage*	*Tone*	*Vegetation and land use*
Flat, undulating 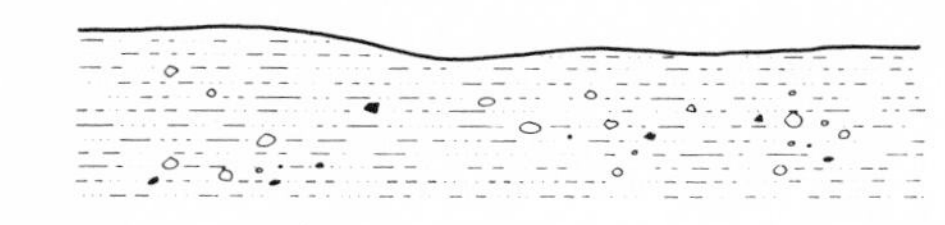 The topography is rather flat and softly undulating. Many swamps and organic deposits occupy depressions. The upland boundaries of these formations may be difficult to discern, but the lower edges are clearly defined by tidal flats or water.	Parallel, dendritic: Coarse 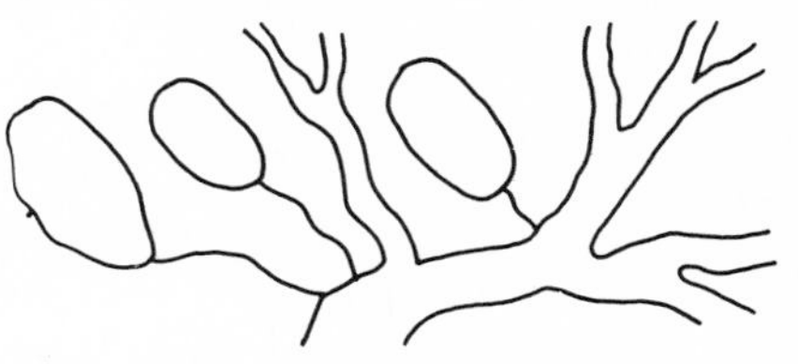 A coarse, parallel, regional pattern develops on coastal plains; however, a local, coarse, dendritic pattern with wide channels indicates abandoned tidal marshes. Many swamps or Carolina bays occupy depressions. Ancient shoreline landforms (beach ridges) may modify the regional drainage to a trellis form.	Mottled Light and dark tones are found throughout the deposit, with darks indicating higher levels of moisture and/or organic materials. Cleared fields have soft, mottled tones. *Gullies* Few, soft U-shaped The slight relief, porosity of materials, and youth of the formation are not conducive to gully development. Any gullies found have soft, U-shaped sections.	Cultivated and forested The lowlands are covered with dense forests. The uplands are cultivated, having many fields with irregular shapes. Contour plowing is practiced on steeper slopes to combat erosion. Lowlands may be ditched in attempts to lower the water table for reclamation or mosquito control.

Old

Topography	*Drainage*	*Tone*	*Vegetation and land use*
Dissected, cuestas 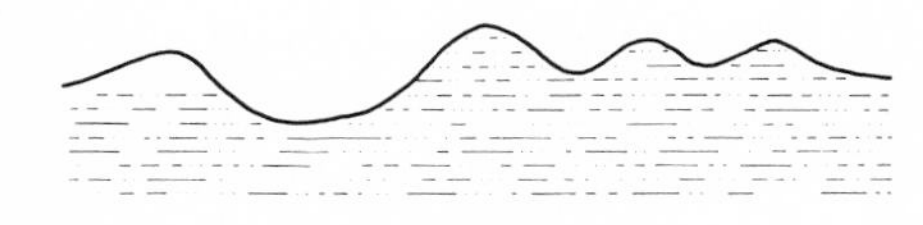 Old coastal plains are highly dissected but do not have the steep slopes associated with more consolidated sediments. The stratified materials are unconsolidated and readily erode, forming gentle, moderate slopes. Regionally, cuestas may be observed where more resistant beds are exposed.	Dendritic: Fine  The unconsolidated soils readily erode and develop a fine-textured dendritic drainage pattern with major channels following the broad valleys of abandoned tidal flats. Cuestas may develop some localized trellis patterns.	Uniform light Highly dissected coastal plains generally have rather uniform light tones in the uplands beneath the forest canopy and dark tones in valleys. *Gullies* U-shaped (V-shaped) Because of the high level of dissection, many gullies are found. U-shaped gullies indicate moderately cohesive mixtures of sand-clay; V-shaped gullies indicate coarser materials.	Forested Older, highly dissected coastal plains are generally covered with forest growth, reflecting the steep slopes. Little undergrowth may be observed, which indicates well-drained, sandy, rather droughty soils. More-level lowlands or uplands may have small areas in cultivation.

muck materials, however, escalate costs by as much as a factor of 4; drag lines, clam buckets, or dredges are needed to remove the saturated material.

Highly dissected coastal plains necessitate large amounts of cut and grading for almost all large construction projects. Since the material is unconsolidated, it can be removed with ordinary power equipment. Seepage and zones of potential sliding are common in these landforms and should be thoroughly investigated.

Construction Materials (General Suitabilities)

Topsoil: Excellent to Fair. The lowlands and lower slopes generally provide excellent topsoil materials which can be mixed for the desired composition.

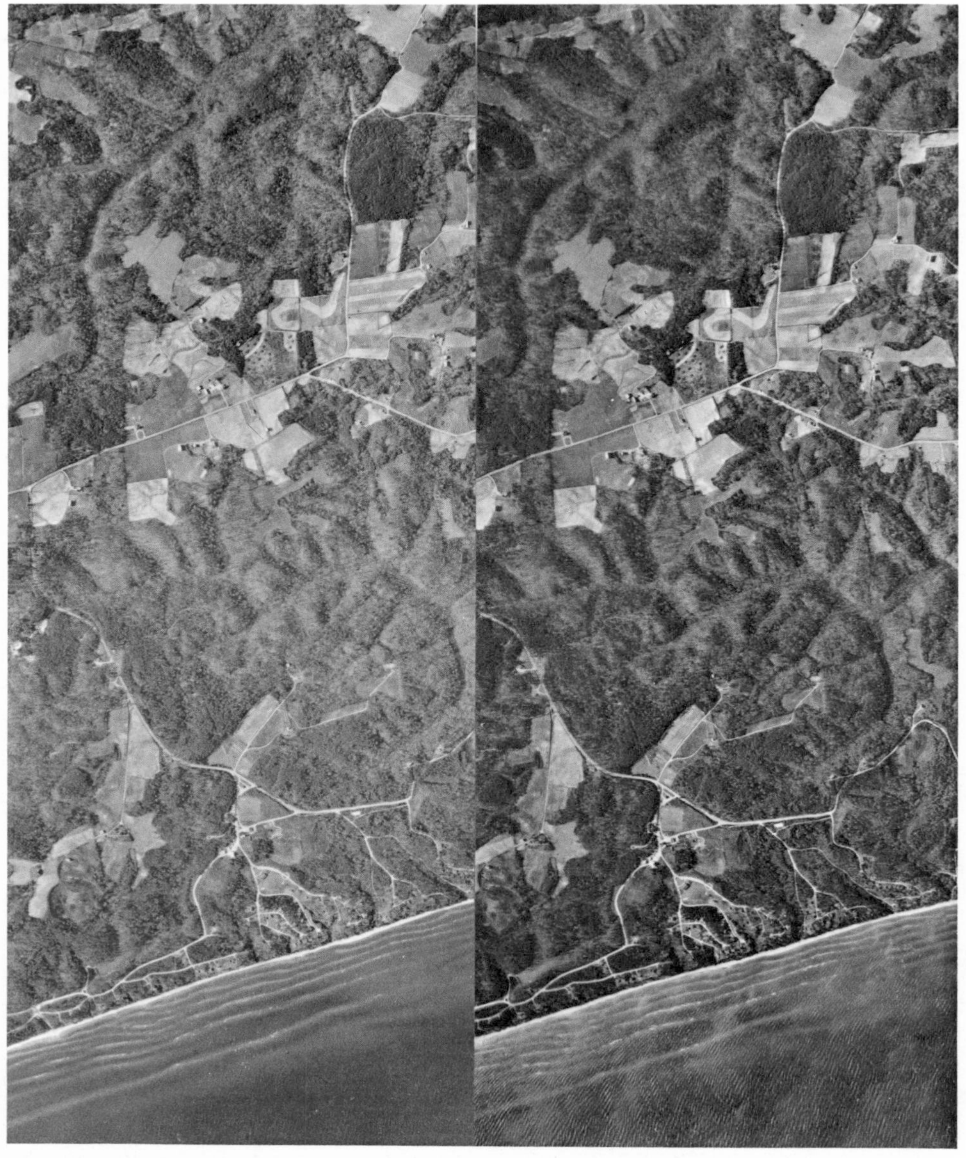

Figure 10.39. Dissected coastal plain in Calvert County, Maryland. Photographs by the U.S. Geological Survey, Maryland, 2-ABC, 1:28,400 (2358′/″), December 17, 1953.

Sand: Excellent to Fair. Poorly graded fine and coarse sands can be found in subsurface stratified layers. Ancient beach ridges have high potential as sources of these materials.

Gravel: Fair. Suitable sources of gravel cannot be found without difficulty; transport over considerable distances may be required. Beach ridges, where found, may offer excellent sources.

Aggregate: Fair to Poor. These materials are difficult to locate, and the sand and gravel mixtures suitable for aggregate may require transport from other landforms over considerable distances. Beach ridges offer the highest potential.

Surfacing: Excellent to Good. The common sandy-clay mixtures provide excellent materials for use in surfacing secondary roads. However, sufficient fines must be present to provide the necessary binder, or corrugated surfaces will quickly develop in the road surface.

Borrow: Excellent to Good. Sandy mixtures can be obtained that offer excellent materials for fill. These are typically compacted by pneumatic-tired rollers.

Building Stone: Not Suitable

Mass Wasting and Landslide Susceptibility

Young, flat or undulating coastal plains are relatively stable and do not have many of the conditions leading to sliding or other forms of mass wasting. Older, dissected plains, however, present many unstable conditions, owing to the interbedding of materials of different porosity and the internal drainage rate. The most prevalent forms of mass wasting are rotational slumps and slides caused by weight increase along hillslopes, created by a build-up of moisture and water over more impervious layers. Soil materials also move downslope by a combination of hillside creep, frost action, and erosion. Construction in highly dis-

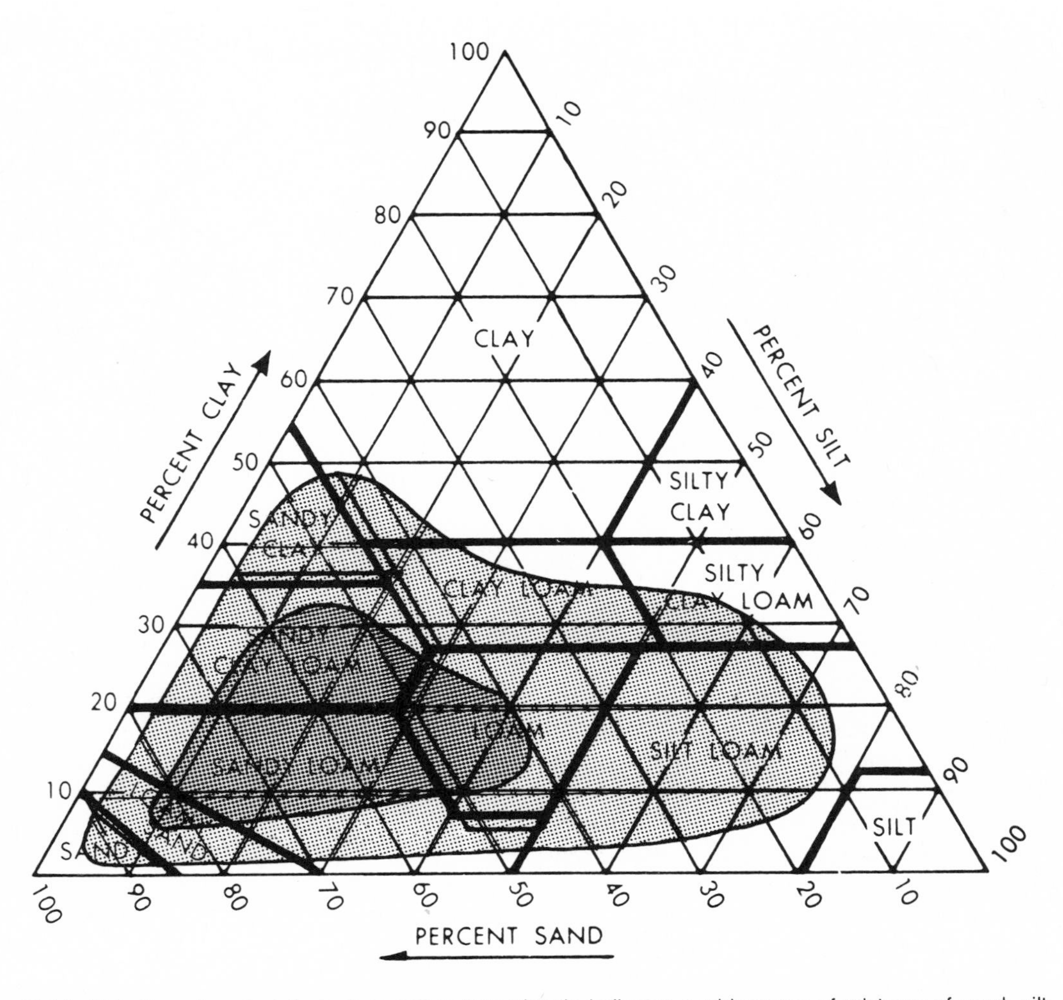

Figure 10.40. U.S. Department of Agriculture soil texture triangle indicates a wide range of mixtures of sand, silt, and clay. Most common are loamy sand, sandy loam, sandy clay loam, and clay loam.

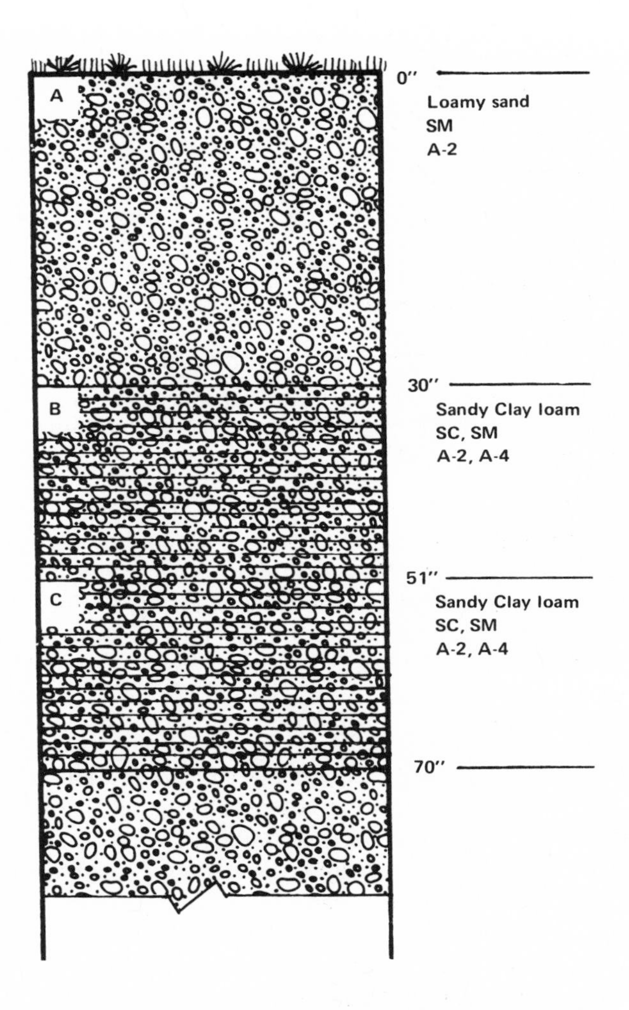

Figure 10.41. A typical soil profile developed in coastal plain materials in a humid climate.

sected regions should consider the potential for sliding and seepage before operations are initiated. Road cuts or other large cuts can utilize terracing techniques and drainage structures to minimize the sliding danger.

Groundwater Supply

The porous, granular, stratified underlying materials are excellent sources of groundwater and give high yields. In some areas artesian flows occur, since porous stratified beds contained between fairly impervious layers slope toward the ocean, thus creating favorable conditions for the development of a pressure head. In coastal

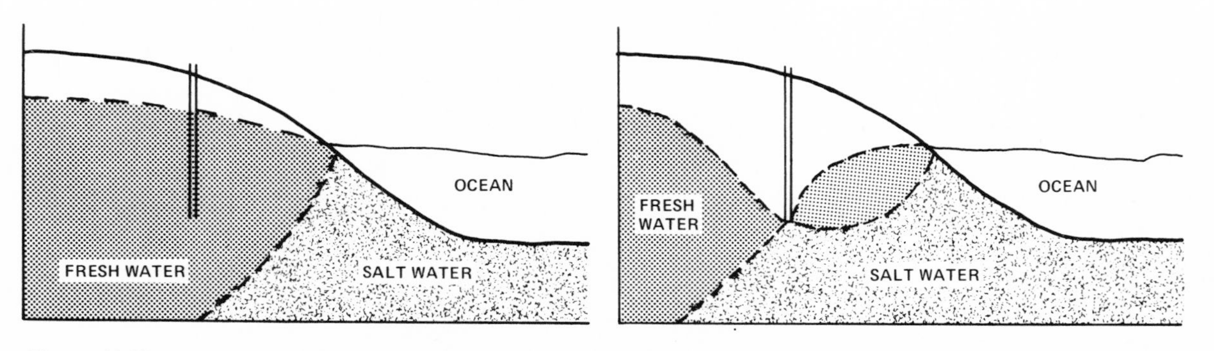

Figure 10.42. Saltwater intrusion. When a limited fresh groundwater supply is heavily used, the level of saltwater in the groundwater moves toward sea level. Wells previously supplying freshwater may become contaminated and no longer offer a suitable supply.

regions the freshwater table is underlaid by saltwater, and care must be taken not to overuse or otherwise diminish the freshwater layer since this causes saltwater to intrude into freshwater wells (Figure 10.42).

Pond or Lake Construction

Ponds and lakes can be sited in lowland areas of both young and highly dissected coastal plains. Lowland soils are fairly impervious when compacted and provide suitable, impervious reservoir bottoms. Small ponds can be created by dredging organic materials out of natural depressions and allowing the water surface to be maintained by the water table. It should be noted that this technique only provides a temporary pond, since over time vegetation will fill the depression with peat. However, if materials are removed to a sufficient depth, the process can be considerably slowed, since a deeper pond maintains a lower water temperature. Soil Conservation Service field agents can provide advice and capability estimates for small pond siting and construction.

Location of dam sites for major reservoirs in highly dissected coastal plains is extremely difficult because of leakage problems through the surrounding porous strata.

Foundations

Coastal plains contain rather sandy soils and are generally suitable for raft or footing foundations for small structures. Basements can be excavated in uplands where the water table is not near the surface, but foundation walls should be weatherproofed to protect against moisture penetration and seepage. Larger structures can utilize footings, piers, or a variety of piles, depending on the specific subsurface conditions and the structural building system being used. In areas with a high water table, clay layers near the surface may cause volume changes that are critical to surface-oriented foundations, such as rafts and footings. The natural soil materials of coastal plains are bedded horizontally and show many changes in content and texture through a vertical section; therefore, detailed subsurface analysis is critical in determining true load-bearing capabilities.

Highway Construction

Detailed soil investigations are needed to determine highway foundation conditions, since the soils show extreme vertical variations throughout these deposits. Lowlands, with organic, fine-textured materials and an associated high water table, require the use of fills or possibly bridging. Upland areas are better suited to highway construction, but clay layers near the surface may create volume change and frost heave problems. Alignments in young coastal plains are constrained primarily by the high water table and the wet, organic deposits in depressions.

Highly dissected coastal plains present major alignment difficultites and require heavy earthwork because of the rugged topography. Sliding potential requires special study and design of cut faces with terracing and/or drainage devices. Many culverts are needed in these areas to continue drainage of the many tributaries and gullies. Major highway alignments tend to follow the larger valleys where earthwork and drainage problems are fewer and costs are lower.

Table 10.14. Fluvial coastal plains: aerial and cartographic references

State	County	Photo Agency	Photograph no.	Scale	Date	Corresponding quadrangle and geologic maps from USGS	Corresponding soil reports from SCS
Florida	Brevard	ASCS	CYS-4V-10-13	1667′/″	4/23/58	Eau Gallie (7½′)	—
Florida	Sarasota	ASCS	DEW-3D-42-44	1667′/″	3/13/48	Englewood (7½)	1959
Maryland	Calvert	USGS*	Maryland 2-ABC	2358′/″	12/17/53	Prince Frederick (7½′)	1928
New Jersey	Burlington	USGS*	New Jersey 1-ABC	1667′/″	11/11/59	Columbus (7½′), GQ-160	—
North Carolina	Brunswick	ASCS	AOH-6F-115-119	1667′/″	11/19/49	Bolivia, Lockwoods Folly (7½′)	1937
North Carolina	Brunswick	ASCS	AOH-8F-43-46	1667′/″	12/1/49	Southport (7½′)	1937
North Carolina	Brunswick	ASCS	AOH-6F-67-69	1667′/″	12/1/49	Funston (7½′)	1937
North Carolina	Brunswick	ASCS	AOH-6AA-9-12	1667′/″	11/15/60	Snow Marsh (7½′)	1937
South Carolina	Dillon	ASCS	PB-1T-23-24	1667′/″	3/17/57	Florence (1:250,000)	1931
South Carolina	Dillon	ASCS	PB-1T-68-69	1667′/″	3/17/57	Florence (1:250,000)	1931
South Carolina	Dillon	ASCS	PB-1T-84-86	1667′/″	3/17/57	Florence (1:250,000)	1931
South Carolina	Dillon	ASCS	PB-1T-28-29	1667′/″	3/17/57	Florence (1:250,000)	1931
South Carolina	Dillon	ASCS	PB-1T-12-13	1667′/″	3/17/57	Florence (1:250,000)	1931
South Carolina	Horry	USGS*	South Carolina 1-ABC	1667′/″	2/23/58	Nixonville, Myrtle Beach (15′)	1918†
South Carolina	Marion	ASCS	ASP-4JJ-247-248	1667′/″	—	Mullins (7½′)	—
South Carolina	Marion	ASCS	ASP-3JJ-161-162	1667′/″	—	Mullins (7½′)	—
South Carolina	Marion	ASCS	ASP-2JJ-13-15	1667′/″	—	Mullins (7½′)	—

*From *Selected Aerial Photographs of Geologic Features in the United States,* U.S. Geological Survey Prof. Paper No. 590.
†Out of print.

Beach Ridges

Introduction

Beach ridges, or old shoreline deposits, are formed by the action of waves and represent previous shoreline elevations of oceans, glacial lakes, lakes, or large ponds. They are therefore associated with glacial lake beds, playas, active lake beds, or marine coastal plain landforms.

Beach ridges are composed of sandy or gravelly materials. They vary widely in spacing and size, depending upon the materials, the currents, and the waves involved in the deposition process. They are formed from sands and gravels carried to the oceans in the bedload of major rivers and transported and deposited by littoral currents and waves moving along the shore. Large volumes of these sediments are stored in offshore bars, whence they are carried toward the shore by waves. As a wave breaks, the particles of sand and gravel it carries are deposited as the velocity decreases and the wave recedes. This continuing process soon establishes a slightly elevated sand ridge whose elevation is remarkably constant, reflecting the highest waves that formed it. Such a formation can reach 50 feet in elevation and 150 feet in width. Beach ridges can also be quite small, tightly spaced, and less than 3 feet high. If the land is uplifted or the water level decreases, previously formed beach ridges are left high and dry, although probably still parallel to the coastline. Beach ridges are commonly found in parallel, curvilinear, or straight alignments, each of which represents a previous shoreline where the water elevation was constant for some period of time. The occurrence of these forms in coastal physiographic provinces indicates that the shoreline is one of emergence. Other landforms, such as tidal flats, sand dunes, and a variety of coastal bars and spits, are associated with beach ridges.

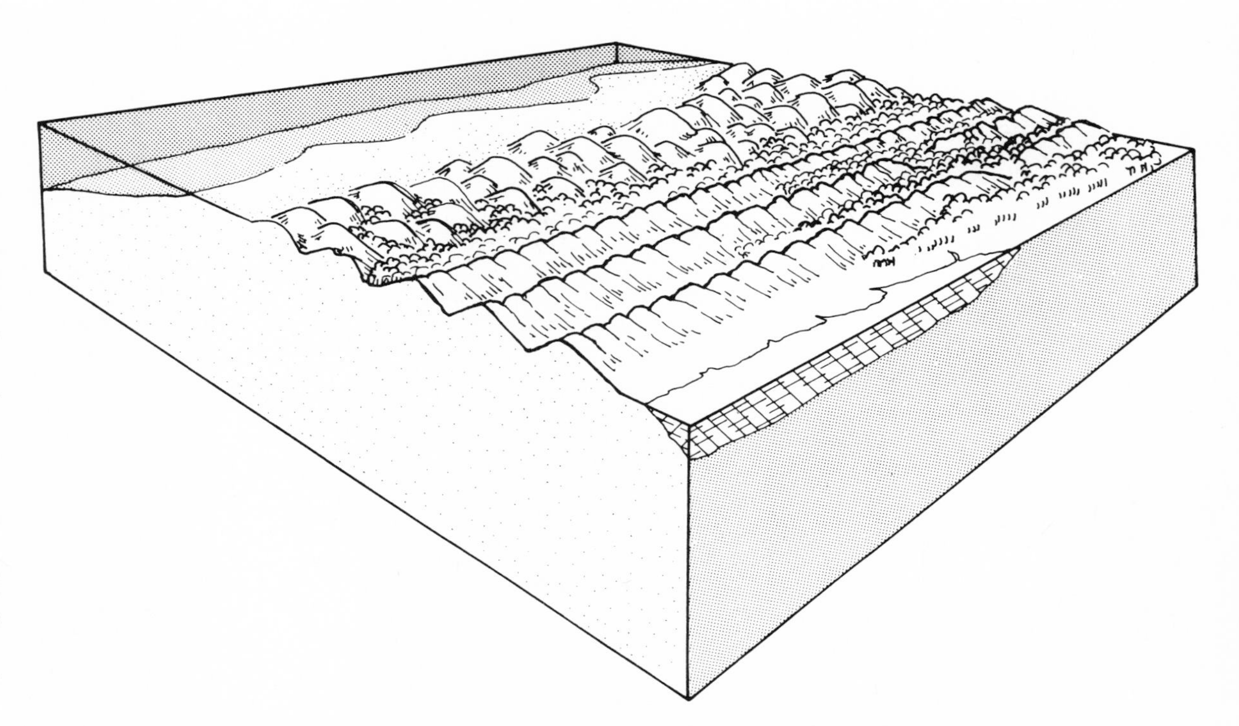

Figure 10.43. Beach ridges provide natural elevated linear pathways and viewing platforms across their contained environment. Their thinness allows viewing from both sides when traveling the ridge crest. These forms can add needed visual and spatial diversity to their commonly associated, flat, monotonous landscapes. In humid climates along coastal regions, the dense forest cover creates a visually absorptive landscape with good screening ability. Glacial beach ridges are often cleared to serve as highway and development corridors; because of their elevated position and lack of vegetation, they are visually sensitive.

If the beach ridge is composed mainly of fine sand, sand dunes and other forms indicating wind erosion and deposition will be found and will modify the idealized beach ridge form. Beach ridges composed of coarser sands and gravelly ridges are not as susceptible to wind erosion and appear with cleaner outlines and relatively sharper ridges. Detailed examination of the beach ridge section may show that the seaward side is more steeply sloping than the landward side. Also, the ocean side tends to have a cleaner edge, while the landward side is undulating or scalloped in appearance.

Interpretation of Pattern Elements

Beach ridges are easily identified from aerial photographs by their characteristic topographic form in which crests follow equal elevations along the length of each ridge. Several parallel ridges

Table 10.15. Fluvial landforms: Beach ridges

Topography	*Drainage*	*Tone*	*Vegetation and land use*
Parallel ridges	Internal	Light	Natural or cultivated
Beach ridges are easily identified by their parallel ridges, representing previous water elevations and paralleling an existing or ancient body of water, typically a lake or ocean. Sand ridges have soft, eroded outlines and segmented forms; coarser sands and/or gravels exhibit sharp crests and crisp outlines.	The porous soils and relatively small areal extent of these landforms do not facilitate the formation of runoff drainage systems. Swamps and other organic deposits are typical between ridges. Regional drainage systems, which generally flow perpendicular to the ridges, readily erode gaps.	The elevated position and the porous soils result in light photographic tones, usually emphasized by adjacent, dark organic deposits. Vegetative cover may mask or reverse tones. *Gullies* None, V-shaped The porous soils and the small areal extent do not facilitate gully formation. Some V-shaped gullies may be found on large, broad ridges.	Coastal beach ridges are usually left in their natural cover of either forest or scrub vegetation. Little undergrowth is found under forests, reflecting the droughty conditions. Broad, large ridges, or glacial beach ridges in humid or subhumid climates, are commonly cultivated and serve as natural, well-drained corridors for highways or development.

commonly occur together; these may either be straight or have a slight, gentle curve. The zone between ridges may contain a swamp or other organic deposit. Ridges that are predominately sand or gravel can be identified by the relative sharpness and crispness of outline of their crestlines. If wind erosion has masked or otherwise modified the original topographic form, sand ridges are indicated.

Soil Characteristics

Beach ridges are relatively young landforms and do not have developed soil profiles. Some surface soil development is found in glaciated regions, but the materials are predominately sandy or sands and gravels.

U.S. Department of Agriculture Classification

The U.S. Department of Agriculture commonly classifies these soil materials as beach forms or shoreline deposits consisting of stratified sands or stratified sands and gravels. Surface soils may be slightly developed on glacial ridges and are typically sandy loams, gravelly loams, or loamy sands.

Unified and AASHO Classifications

Surface soils may be indicated by SM or GM in the Unified categories, or A-1 or A-2 by the AASHO system. The underlying material is indicated by SP, GP, GP-GM, SP-SM, and occasionaly SW in older formations; AASHO categories include A-1, A-2, and occasionally A-3.

Water Table

The porous, granular, unconsolidated soil materials and their elevated position do not facilitate the formation of a perched or seasonal high water table. The occurrence of the water table is controlled by the underlying materials, but it is generally near the surface between ridges.

Drainage

The porous soils and their elevated position facilitate internal drainage. Capillary water does not normally occur, and therefore these forms are frost-free in cold climates.

Soil Depth to Bedrock

Bedrock is not associated with this landform and is not encountered during normal construction practices.

Issues of Site Development

Sewage Disposal

Septic tank leaching fields can usually be located on beach ridges if state minimum standards are followed, but the rapid percolation rate risks potential contamination of groundwater resources. Lower slopes, where the water table is nearer the surface, should be avoided. Sanitary engineers can provide soil capability studies for various on-site disposal systems and can define more clearly the potential hazards of groundwater contamination. Intensive development projects should utilize treatment systems if the quality of the groundwater resources is to be maintained.

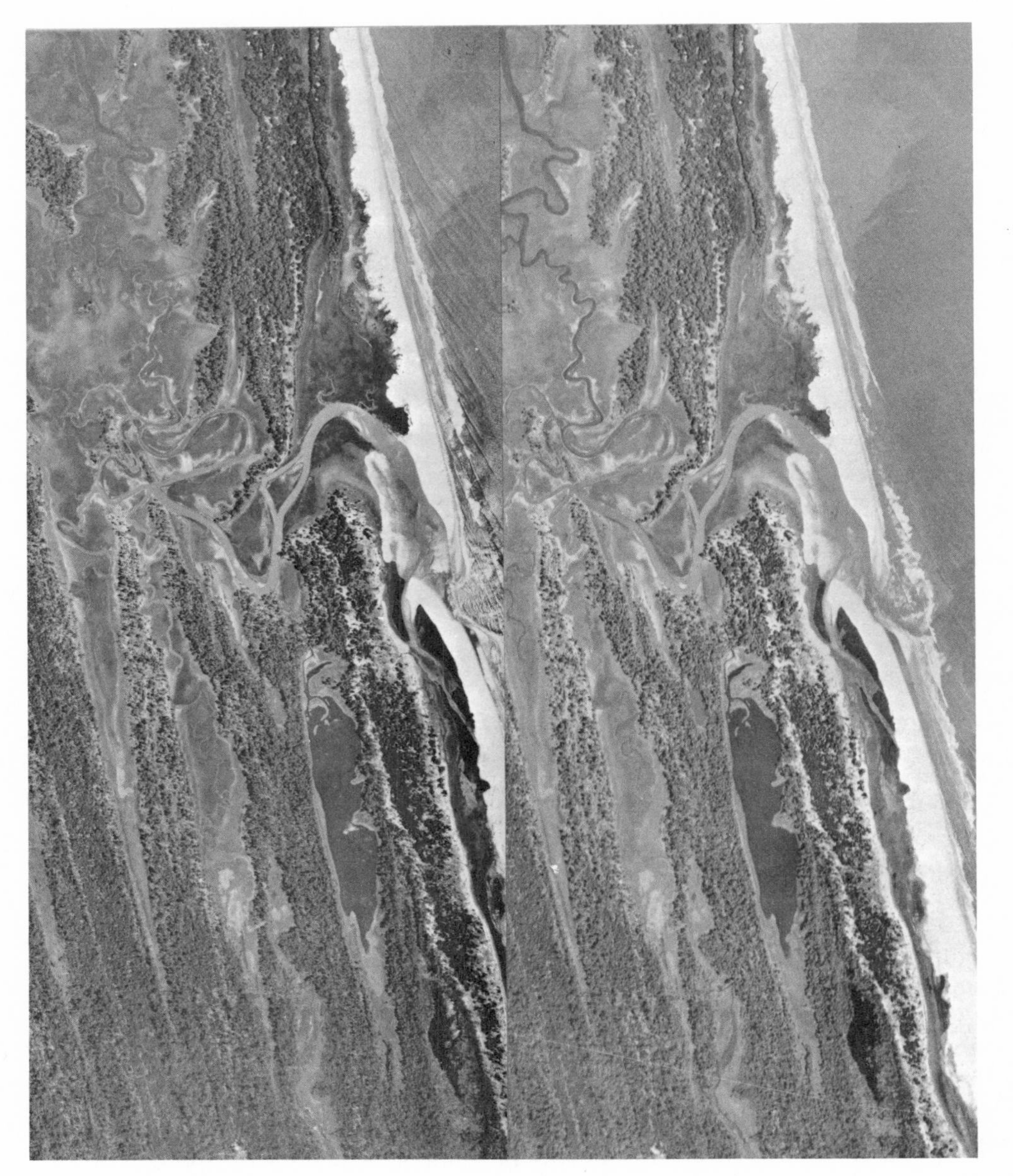

Figure 10.44. (A) Beach ridges adjacent to the coast of Charleston County, South Carolina, show their parallel elevated character above the surrounding tidal marshes. Photographs by the Agricultural Stabilization Conservation Service, CDV-7F-50,51, 1:20,000 (1667′/″), March 8, 1949.

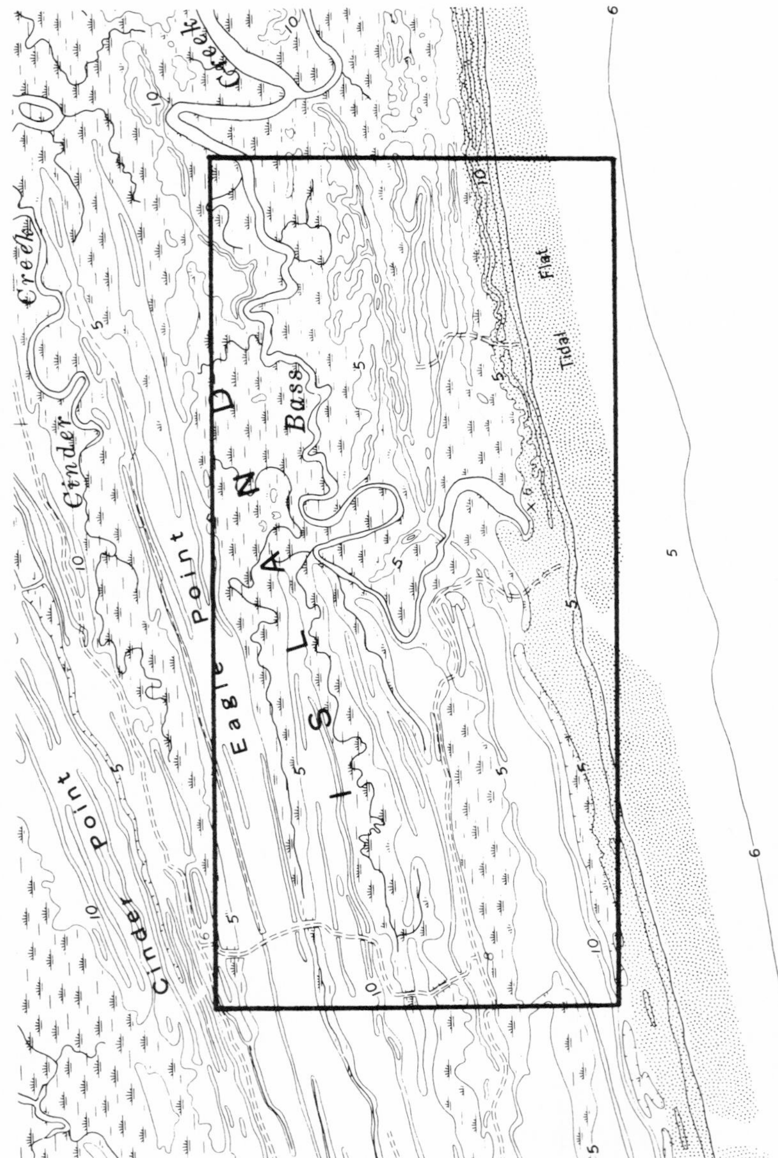

Figure 10.44. (B) U.S. Geological Survey Quadrangle: Kiawah Island (7½′).

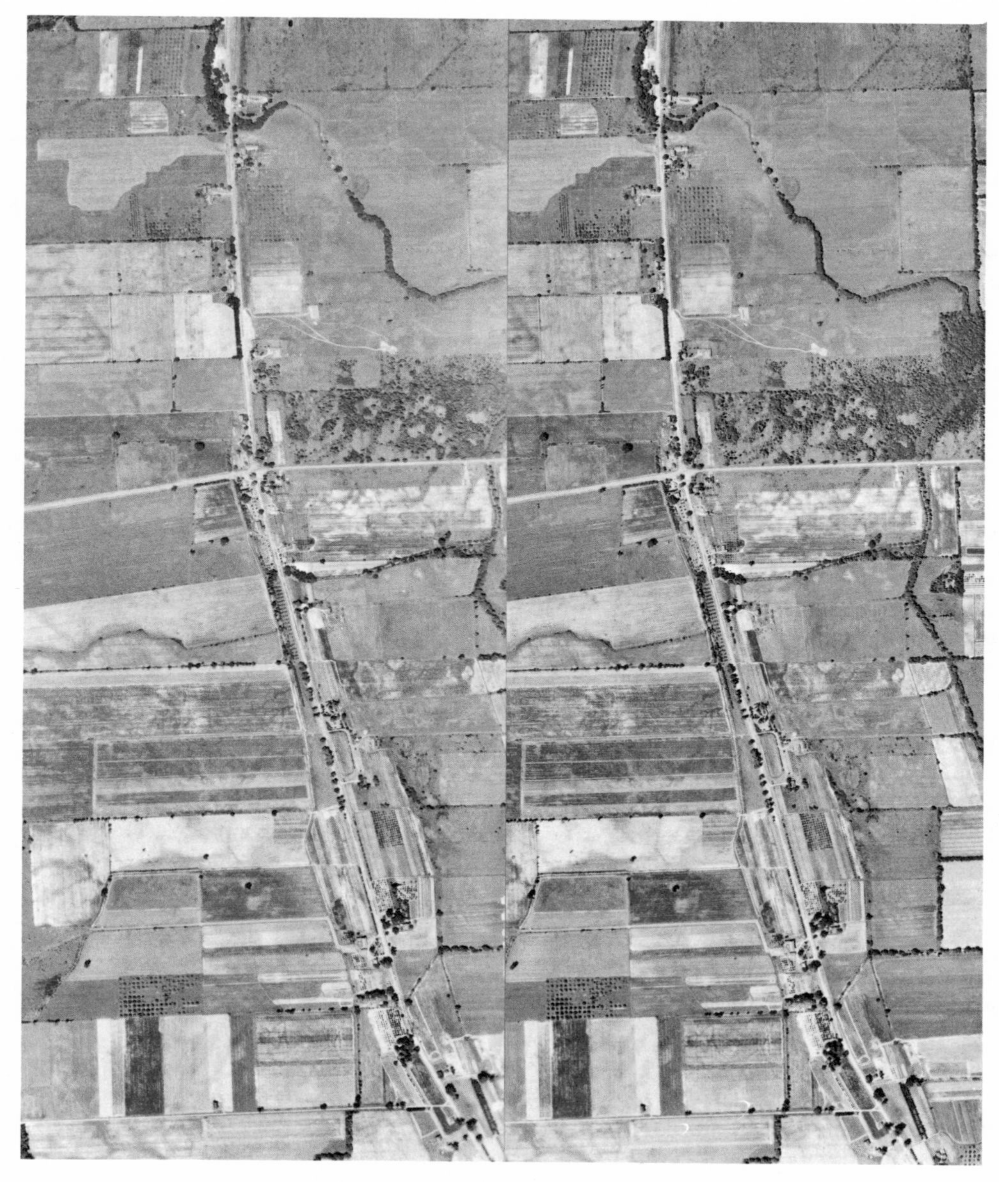

Figure 10.45. Glacial beach ridges in Niagara County, New York, show linear alignment and elevation above the surrounding glacial lake bed landform. Photographs by the Agricultural Stabilization Conservation Service, ARE-2V-83,84, 1:20,000 (1667′/″), August 9, 1958.

Solid Waste Disposal

Beach ridges do not provide suitable sites for the location of sanitary landfill operations since the porous nature of the contained soils facilitates percolation of leachate, and contamination of groundwater resources thus occurs.

Trenching

No significant problems are encountered during trenching operations in beach ridges, except that sheeting is necessary. The unconsolidated granular soils are easily removed at costs typical for dry, deep, moderately cohesive materials. Pipeline alignments can follow the natural direction of the formation, for it has little grade change along the crest unless it has been dissected previously by wind or water erosion.

Excavation and Grading

No significant problems are encountered when removing or grading beach ridge materials, except that sheeting and bracing are required for pit-type excavations. The soils are dry and unconsolidated and can be worked throughout the year; they neither develop frost in cold climates nor hold moisture during wet seasons. Costs are the same as those involved in removing deep, dry, moderately cohesive soils, but the sandy materials are abrasive to equipment. The amount of grading required for large construction operations depends upon the size and spacing of the ridges. Operations in swamps between ridges escalates costs, as described in the section on organic materials.

Construction Materials (General Suitabilities)

Topsoil: Not Suitable. Soils are poorly developed, droughty, and low in organic matter.

Sand: Excellent

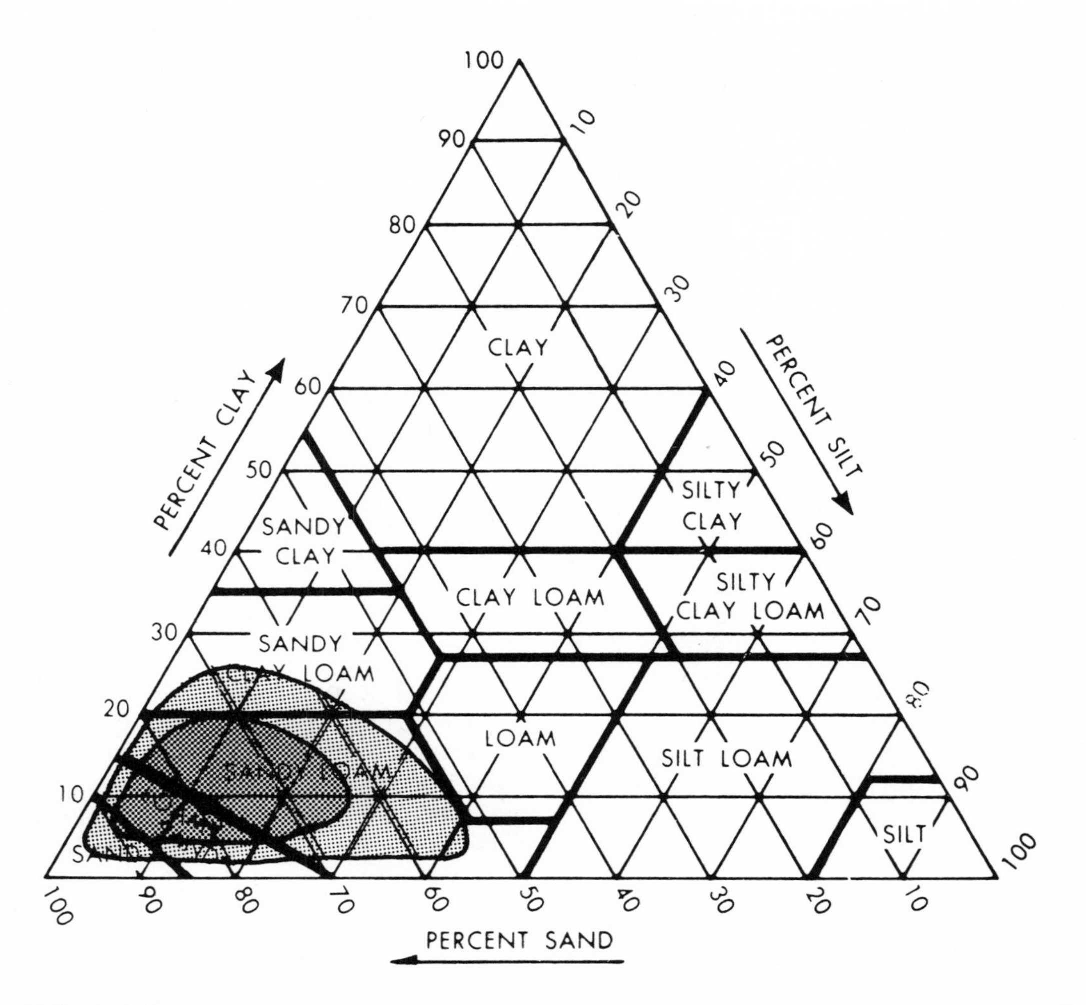

Figure 10.46. U.S. Department of Agriculture soil texture triangle indicates sandy loam or loamy sand with stratified sand and gravel found in the subsurface.

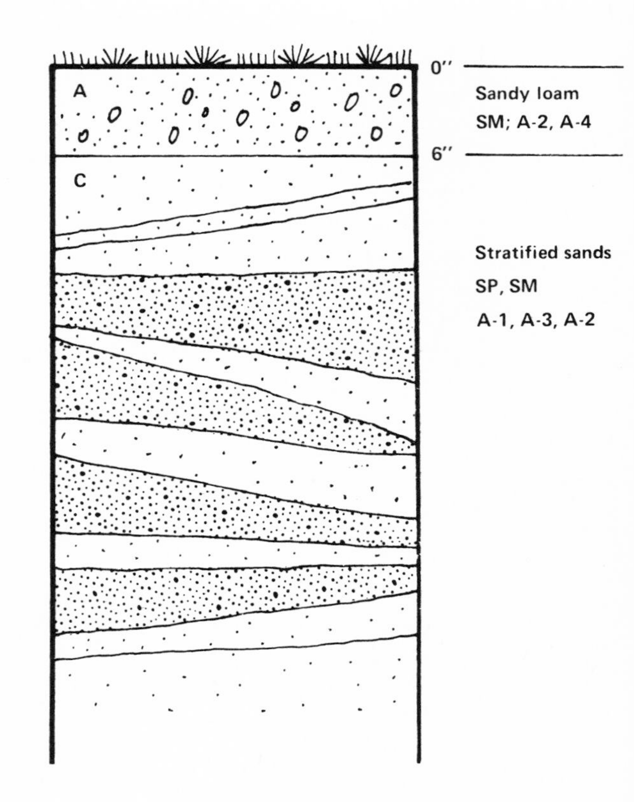

Figure 10.47. A typical soil profile developed in coastal beach ridges in a humid climate.

Gravel: Excellent to Not Suitable. Beach ridges may contain predominate mixtures of sand, or sands and gravels. When gravels occur, they provide an excellent source of these materials.

Aggregate: Excellent to Poor. Ridges containing mixtures of sand and gravel provide the best sources.

Surfacing: Excellent. Suitable surfacing materials for secondary roads can be obtained if granular materials from beach ridges are mixed with fines from adjacent landforms.

Borrow: Excellent. The granular soils of beach ridges are excellent sources of borrow materials and can be compacted for fills with vibrators or a combination of vibrators and rollers.

Building Stone: Not Suitable

Mass Wasting and Landslide Susceptibility

The uniform composition and low topographic relief of these landforms do not present conditions susceptible to mass wasting or sliding. The granular materials can hold vertical slopes of approximately 35 degrees.

Groundwater Supply

Beach ridges themselves do not contain water resources; the underlying and adjacent landforms determine the potential yields. In coastal regions beach ridges are underlaid by sandy coastal plains containing vast groundwater resources and high recharge capabilities. However, offshore islands or peninsulas of beach ridges contain a thin, freshwater lens floating over a saltwater table. Removal of the freshwater beyond its inherent recharge capability results in the intrusion of saltwater into the zone that was previously fresh, threatening both the stability of the vegetative associations and the quality of the water for human consumption.

Pond or Lake Construction

The lack of any developed watershed, the porous soils, and the elevated position of the landform do not facilitate development of major ponds or reservoirs. Small ponds can be created, but they need impervious bottom liners to prevent leakage and the water level can be maintained only by pumping.

Foundations

Beach ridges in their natural state contain unconsolidated, clean, granular soils. If an earthquake or similar type of shock occurs, the soils may consolidate, settle, and cause structural failure. Therefore, before any large structural loads are applied, these soils should be compacted in place by vibrations produced by pile drivers or vibroflotation techniques.

The granular soils, if compacted, have excellent bearing capacities and load-carrying abilities suitable for soil-associated foundation types for low and moderate loads. Small structures can utilize raft foundations or excavate basements and utilize continuous footings with a connecting slab. Heavier structures should utilize deep foundations, such as friction piles, which help to compress the soils during installation. Frost is not a problem in cold climates because of the inability of the sands to hold moisture.

Highway Construction

Beach ridges are excellent foundations for highway alignments because of their porous, frost-free, stable subsoils. Little grading and few drainage structures are needed if alignments follow the natural direction of the beach ridge. Because of its formation process, a beach ridge may have little elevation change over long distances. Thus, in an example mentioned by Belcher, a highway west of Nome, Alaska, was located on a beach ridge that had only a 3-foot change in elevation over 22 miles. Location of alignments either at a diagonal or perpendicular to the ridges requires slight grading, and encounters difficulties in crossing the depressed areas between the ridges, since they typically contain organic materials.

Table 10.16. Fluvial beach ridges: aerial and cartographic references

State	County	Photo Agency	Photograph no.	Scale	Date	Corresponding quadrangle and geologic maps from USGS	Corresponding soil reports from SCS
California	Inyo	USGS*	California 21-AB	3933′/″	10/10/55	Lone Pine, New York Butte (15′)	–
Florida	Franklin	USGS*	Florida 1-AB	3590′/″	11/1/42	Indian Pass, West Pass (7½′)	1915†
Indiana	Lake	ASCS	BFJ-1V-34-35	1667′/″	6/26/58	Gary (7½′)	1917†
Indiana	Lake	ASCS	BFJ-1V-96-97	1667′/″	6/26/58	Gary (7½′)	1917†
Michigan	Lenawee	ASCS	XV-2DD-35-37	1667′/″	9/9/63	Saline (15′)	1961
Michigan	Ontonagan	USGS*	Michigan 1-ABCDE	2275′/″	4/28/44	Bergland NE, Matchwood (7½′) Prof. Paper No. 504-B	–
Michigan	Lenawee	ASCS	XV-2DD-75-76	1667′/″	9/9/63	Saline (15′)	1961
Michigan	Presque Isle	USGS*	Michigan 6-AB	1415′/″	4/25/54	Grace, Onaway (15′)	–
New York	Niagara	ASCS	ARE-2V-83-84	1667′/″	8/9/58	Tonawanda (15′)	1947
New York	Niagara	ASCS	ARE-2V-108-109	1667′/″	8/9/58	Tonawanda (15′)	1947
New York	Niagara	ASCS	ARE-2V-116-117	1667′/″	8/9/58	Tonawanda (15′)	1947
North Dakota	Grand Forks	ASCS	ZZ-3CC-71-72	1667′/″	7/12/62	Grand Forks (1:250,000)	–
Ohio	Lorain	USGS*	Ohio 1-ABC	5000′/″	5/16/60	Avon, North Olmsted (7½′), I-316	–
South Carolina	Charleston	ASCS	CDV-7F-50-51	1667′/″	3/8/49	Kiawah Island (7½′)	–
Wisconsin	Door	ASCS	BHQ-3AA-244-245	1667′/″	9/19/61	Sister Bay (15′)	1916†
Wisconsin	Manitowoc	ASCS	BHY-2AA-145-147	1667′/″	10/6/61	Manitowoc (1:250,000)	1926

*From *Selected Aerial Photographs of Geologic Features of the United States,* U.S. Geological Survey Prof. Paper No. 590.
†Out of print.

Tidal Flats

Introduction

Tidal flats are found along ocean coasts where spits, bars, or other barriers provide protection from waves and allow the accumulation of organic debris and fine-textured sediments. These protected areas are subject to the fluctuations of tidal action which lays bare much of the area during low tide and floods it during high tide. The constant flow of water both in and out of the tidal flat, or estuary, develops and maintains the intricate system of drainage channels that characterizes this landform.

Tidal flats can be classified as tidal marshes, mud flats, and sand flats, according to their predominante materials. Tidal marshes, the most common, are typified by a peat surface layer underlaid by organic silt. Mud flats are also common and contain mixtures of silt and clay. Sand flats are mostly sandy and are found near tidal inlets where currents are relatively strong.

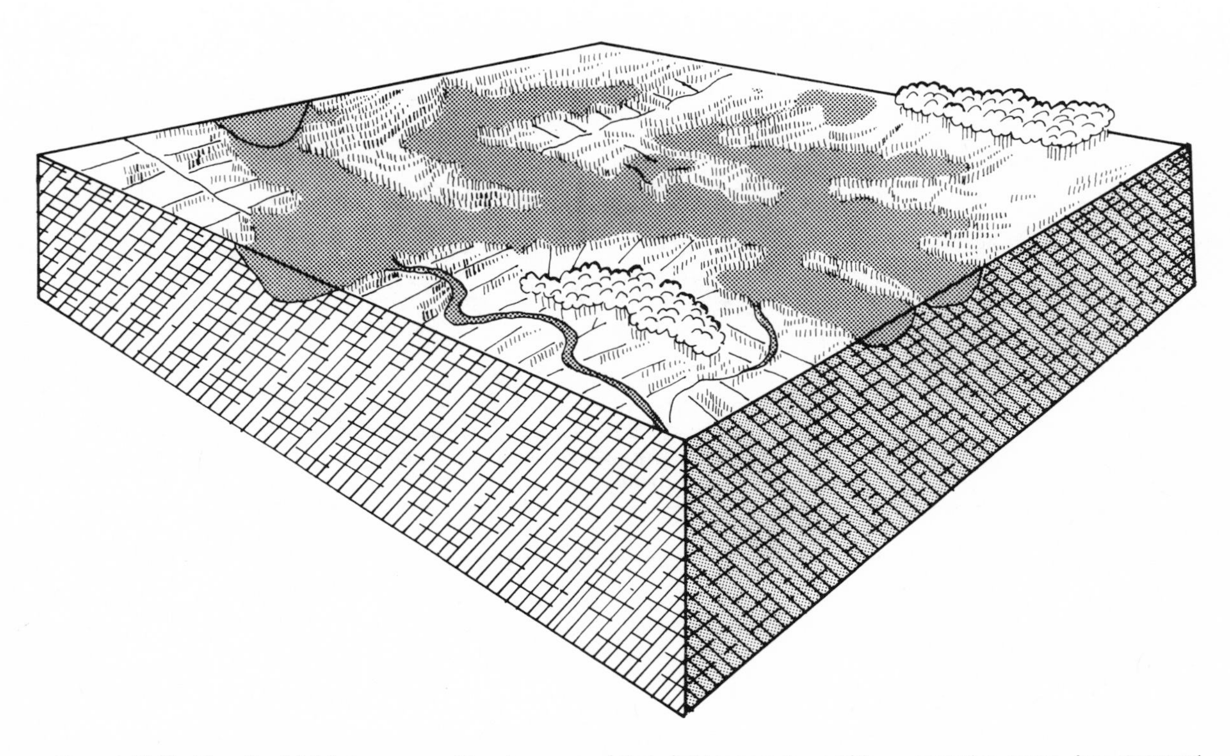

Figure 10.48. Visually, tidal flats are sensitive, because of their flat topography and low vegetative cover. Any structural development will be readily observed, owing to the lack of screening.

Interpretation of Pattern Elements

The photographic key for tidal flats is divided into three sections, according to type. Tidal marshes are readily identified by their dense vegetation and the unique drainage pattern of wide, wandering, dendritic channels. Mud flats are devoid of vegetative cover and have a similar drainage pattern, except that there are many small, detailed, hairlike appendages. Sand flats have light tones, no vegetative cover, and a well-developed surface drainage system.

Soil Characteristics

Tidal flats primarily contain organic materials mixed with fine silts and clays. Coarser sands may be found in main channels or within sand flats. Tidal marshes develop a thick organic peat or muck mat underlaid by organic silt. Mud flats consist primarily of silt and clay.

U.S. Department of Agriculture Classification

The U.S. Department of Agriculture classifies materials found in tidal marshes as peat, muck, or silt; in tidal mud flats as muck, silt, or clay; and in tidal sand flats as sands.

Unified and AASHO Classifications

The Unified classification system refers to tidal marsh materials as Pt underlaid by OL, OH, or MH; mud flats as CH or CH-MH; and sand flats as SP. AASHO categories include peat and A-7 for tidal marshes, A-7 for mud flats, and A-1 or A-3 for sand flats.

Water Table

As a result of tidal floodings, the water table, which is actually brackish, occurs above the ground surface during high tide.

Drainage

Drainage is controlled by tidal action, but these areas can be diked, drained, and reclaimed for agricultural lands, as is common in the Netherlands. Many tidal marshes along the coasts of the United States have been ditched to facilitate drainage for mosquito control.

Table 10.17. Fluvial landforms: Tidal flats (marsh, mud, sand)

Marsh

Topography	*Drainage*	*Tone*	*Vegetation and land use*
Flat	Tidal, dendritic	Varies	Marsh grass
The topography is flat and is bounded by land and water.	Wide, broad channels, forming a dendritic pattern, typify tidal marshes. Smaller channels occasionally close upon themselves, forming many irregular islands. Ditching is common for mosquito control.	The tone is determined by the vegetative cover but is typically light.	Dense cover of various marsh grasses is typical. In some areas the grass is cut and harvested.
		Gullies	
		None	

Mud

Topography	*Drainage*	*Tone*	*Vegetation and land use*
Flat	Tidal, dendritic	Uniform grays	None
The topography is flat and is bounded by land and water.	Wide, broad channels form a dendritic pattern which also has many hairlike channels.	Tidal mud flats typically have uniform gray tones. Changes in tone may indicate different water depths.	Little or no vegetation is found in mud flats because of the higher level of tidal flooding.
		Gullies	
		None	

Sand

Topography	*Drainage*	*Tone*	*Vegetation and land use*
Flat	Few, parallel channels	Light and dark	None
The topography is flat and is bounded by land and water.	Since the sand is internally well-drained, few drainage channels are developed. Some short, parallel channels without tributaries are found.	The exposed sand areas are very light or bright. Channels appear dark.	No vegetation or land use development is found because of the tidal flooding and the shifting of sandy materials.
		Gullies	
		None	

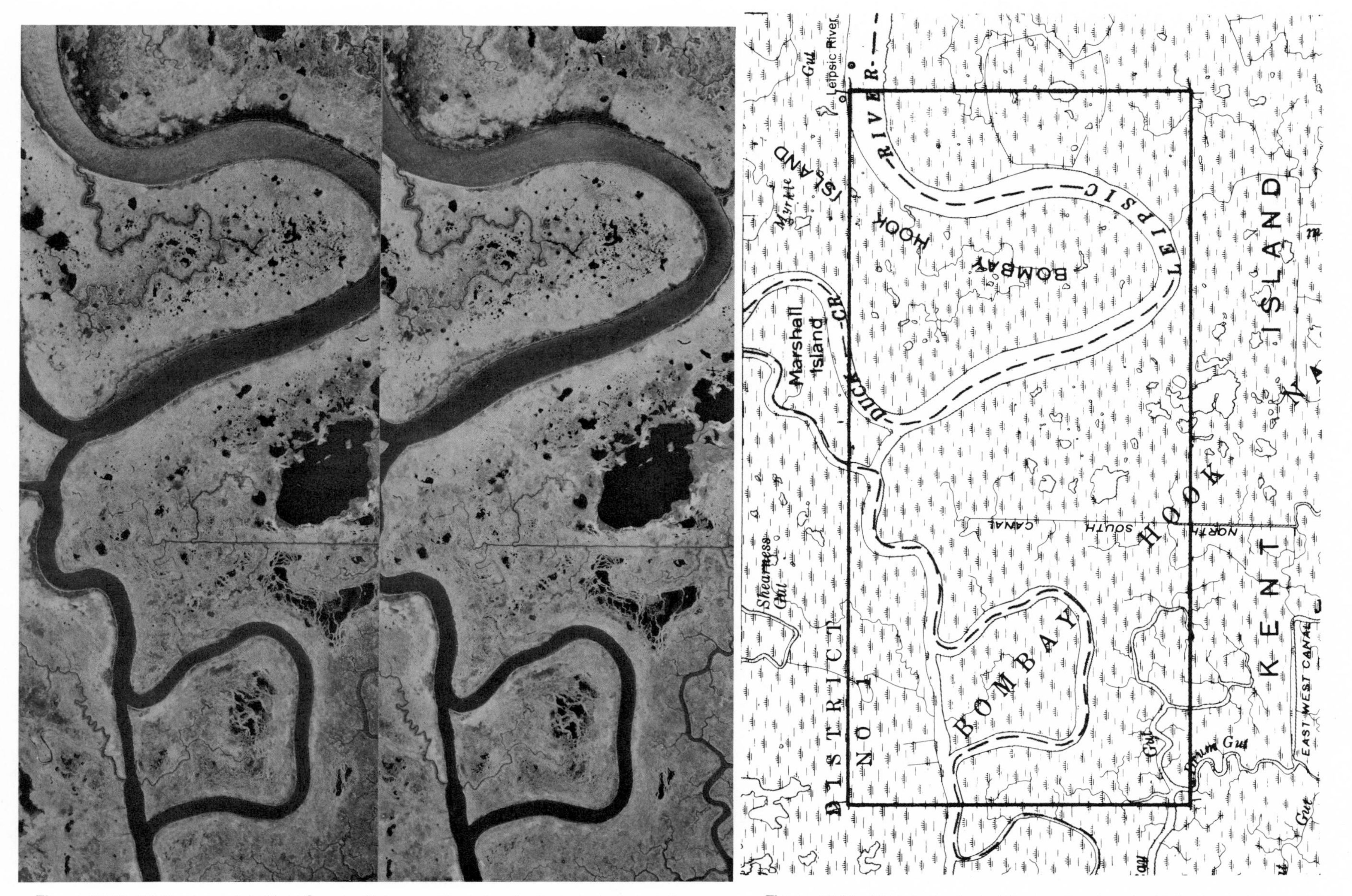

Figure 10.49. (A) Tidal marsh in Kent County, Delaware, illustrates the typical characteristics of these landforms. Photographs by the Agricultural Stabilization Conservation Service, AHP-2JJ-98,99, 1:20,000 (1667′/″), May 7, 1968.

Figure 10.49. (B) U.S. Geological Survey Quadrangle: Little Creek (7½′).

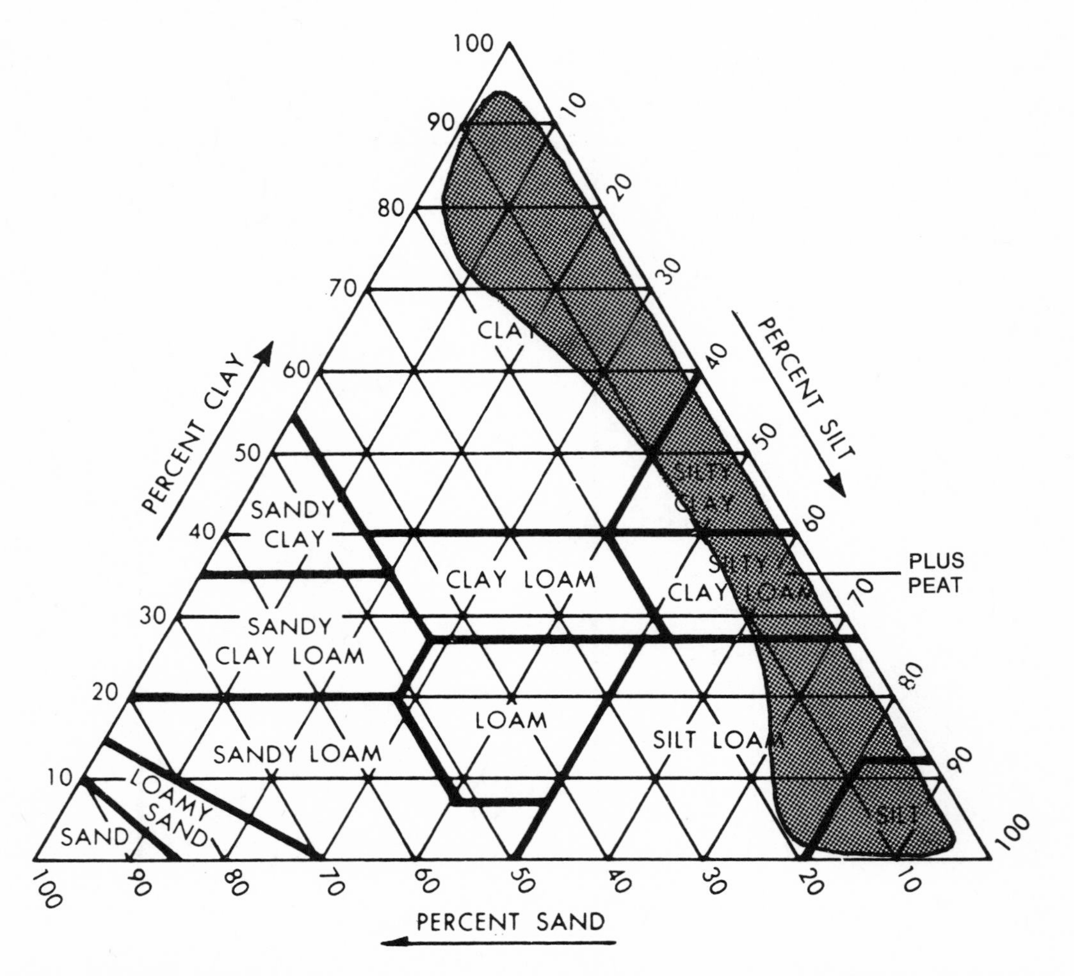

Figure 10.50. The U.S. Department of Agriculture soil texture triangle indicates sand for sand flats, and peat or organic silt and clay for marsh and mud flats.

Soil Depth to Bedrock

Soil depths to bedrock are very great unless the tidal flat has developed along and over a rocky shoreline, in which case depths are variable and independent of this landform. Most tidal marshes and mudflats are underlaid by sands.

Issues of Site Development

Sewage Disposal

Septic tank leaching fields cannot be operated within these landforms because of flooding and the unsuitability of the soil materials.

Solid Waste Disposal

Sanitary landfill operations cannot be sited within these landforms because of flooding and the unsuitability of the soil materials.

Trenching

Trenching is very difficult and expensive, owing to the low supportive ability of the materials, their tendency to flow, and drainage difficulties. Dredging equipment can be used for ditching operations. Costs are typically 2.5 to 4 times as high as those encountered in removing the same amount of deep, dry, moderately cohesive material.

Excavation and Grading

Excavation and grading of this landform rarely occur, since it is usually not considered for land use development. Excavation is expensive, reflecting the high water table and the compressibility of the materials and their tendency to flow. Dewatering is very difficult, but excavation walls can be stabilized by electro-osmosis in cohesionless or slightly cohesive silts. Costs are typicaly 2.5 to 6 times those associated with removal of deep, dry, moderately cohesive soils. The compressible surface does not support equipment unless stabilized by electro-osmosis or surcharging; Clam buckets, drag lines, or dredging equipment on barges are needed.

Construction Materials (General Suitabilities)

Topsoil: Not Suitable. Tidal flats are not suitable as sources of topsoil unless highly organic muck or peat is desired.

Sand: Not Suitable to Excellent. Sand flats provide excellent sources of sand materials with slight gradations.

Gravel: Not Suitable

Aggregate: Not Suitable

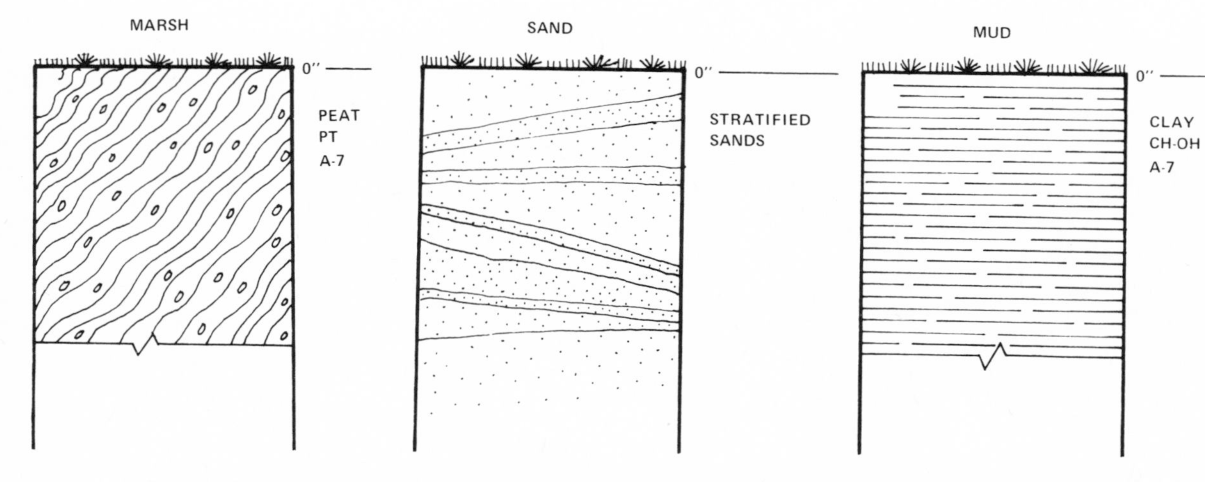

Figure 10.51. A typical soil profile developed in a marsh, sand, and mud tidal flat.

Surfacing: Not Suitable. The highly organic materials are generlly not suitable, but silt and clay deposits can be mixed with other granular materials to act as binder.

Borrow: Not Suitable to Good. Tidal marshes and mud flats do not offer suitable materials because of the fine-textured soils and high organic content. Sand flats, however, are good sources of sandy borrow materials.

Building Stone: Not Suitable

Mass Wasting and Landslide Susceptibility

The low, flat topography does not present conditions susceptible to mass wasting or sliding.

Groundwater Supply

The brackish water table occurs at and above the surface during high tide. Unless it is desalted, the water is not suitable for supplying fresh water needs or municipalities.

Pond or Lake Construction

Ponds and lakes are not constructed on these landforms because of the instability of the ground materials and the periodic salt water inundation.

Foundations

The ability of marsh and mud flats to support surface foundations is generally inadequate, even for low structural loads. Deep foundations are required, typically using point-bearing piles to transfer the applied loads to underlying strata of suitable bearing capacity. In most areas substrata of sand can be found beneath the organic silts that can provide suitable load support. Occasionally, piles can be driven that do not encounter any significant resistance by the time the pile is driven to its full extent. These conditions may require the use of floating pile foundations utilizing pile clusters, but these are rarely feasible and require a high safety ratio in their design.

Highway Construction

Highways crossing tidal marshes must compensate for the tidal flow of water by bridging over major channels. To carry and distribute the weight of the highway, organic materials are commonly removed and backfilled with granular light borrow. Deeper organic deposits may be compacted by surcharging, using sand drains as necessary. Settlement should be expected over time, as the organic mat is consolidated by the applied weight of the surcharge fill. Flexible road surfaces, such as asphalt, are recommended if settlement is expected to continue over a long period.

Table 10.18. Fluvial tidal flats: aerial and cartographic references

State	County	Photo Agency	Photograph no.	Scale	Date	Corresponding quadrangle and geologic maps from USGS	Corresponding soil reports from SCS
Florida	Dixie, Levy	USGS*	Florida 2-ABC	2733'/"	3/26/51	Suwannee, East Pass (7½')	–
Delaware	Kent	ASCS	AHP-2JJ-97-99	1667'/"	5/7/68	Little Creek (7½')	1918†
Delaware	Kent	ASCS	AHP-2JJ-6-7	1667'/"	5/7/68	Little Creek (7½')	1918†
Florida	Sarasota	ASCS	DEW-2D-68-70	1667'/"	2/26/48	Venice, Laurel (7½')	1959
Maryland	Baltimore	ASCS	AJO-4DD-30-33	1667'/"	5/16/64	Gunpowder Neck (7½')	1917†
Massachusetts	Barnstable	USGS*	Massachusetts 2-ABC	1667'/"	7/13/52	Hyannis (7½')	–
Massachusetts	Essex	ASCS	DPP-5LL-179-188	3333'/"	10/29/70	Newburyport East, Ipswich (7½')	1925
Massachusetts	Essex	ASCS	DPP-5LL-194-196	3333'/"	10/29/70	Ipswich, Gloucester (7½')	1925
North Carolina	Brunswick	ASCS	AOH-2GG-180-183	1667'/"	3/18/66	Snow Marsh (7½')	1937
North Carolina	Brunswick	ASCS	AOH-2GG-103-104	1667'/"	3/18/66	Southport (7½')	1937
North Carolina	Brunswick	ASCS	AOH-1AA-69-73	1667'/"	11/15/60	Snow Marsh (7½')	1937
South Carolina	Charleston	ASCS	CDV-4EE-240-242	1667'/"	11/12/63	Kiawah Island (7½')	1904†
South Carolina	Charleston	ASCS	CDV-5EE-58-60	1667'/"	11/12/63	Kiawah Island (7½')	1904†

*From *Selected Aerial Photographs of Geologic Features in the United States,* U.S. Geological Survey Prof. Paper No. 590.
†Out of print.

CHAPTER 11

MASS WASTING

INTRODUCTION

This chapter follows a format that differs from those used to describe landforms. The processes involved in mass wasting can be present in all landforms although some are more susceptible than others. For this reason the following discussion addresses conditions under which mass wasting is likely to occur. The past occurrence of these processes and a landform's future susceptibility to them should always be thoroughly considered in any terrain investigation; their effect on engineering projects and life and property can be devastating.

Mass wasting is a general term describing a variety of processes by which earth materials, some of them previously broken down by weathering processes, are moved by gravity. Unlike erosion it does not involve transporting agents, such as wind or ice; however, water is associated with it since an increase in pore water pressure can "trigger" an event. Also water, snow, or ice can increase the weight or mass of the materials and thereby increase their sliding potential. C. F. S. Sharpe (1938) has classified the various processes of mass wasting by their relative speed of movement and their ice or water content (see Table 11.1).

Sliding of earth and rock materials occurs when their shear strength can no longer support an increase in load. Shear strength varies for any earth material, depending upon the rate of loading, temperature, confining pressure, presence of pore fluids, and size and shape (slope) of the formation. When shear strength is exceeded, failures occur (1) because of a critical increase in the weight of a formation through absorption of moisture or loading with snow or ice, and/or (2) because the original slope angles have been increased by excavation, undercutting of streams, or previous actions of mass wasting, and/or (3) because of an increase in pore water pressure.

Generally, the materials and residual rocks of

Table 11.1. Classification of events of mass wasting by the speed of movement and the water or ice content*

Rate of movement			With increasing *ice* content	Rock or soil ⟷	With increasing *water* content	
Flow	Imperceptible	Glacial Transport	Solifluction (1 mm/day)	Creep (Rock creep) (Soil creep)	Solifluction (1 mm/day)	Fluvial Transport
Flow	Slow to rapid	Glacial Transport	Debris avalanche (30 m/hr$^+$) ↓		Earth flow (1 cm/hr) Mud flow (1 km/hr) Debris avalanche (1 ft/sec$^+$) ↓	Fluvial Transport
Slide	Slow to rapid	Glacial Transport	↓	Slump Debris slide Debris fall Rock slide Rock fall (30 m/hr$^+$)	↓	Fluvial Transport

* After C. F. S. Sharp, *Landslides and Related Phenomena*, 1938. (reprint) Pageant Books, Paterson, N.J., 1960.

mass wasting debris, or colluvium, do not develop the rounded shapes typically associated with water-transported materials. The flow process does not cause much rubbing together of particles, except for minor scrapes, gouges, or chips. The original shape of the rock is generally still apparent and exhibits many angularities. However, mass wasting debris may contain rounded rock fragments if the original material came from a waterlaid landform. In these situations the rock shapes are completely independent of the mass wasting process.

Mass wasting occur in all regions, in all landforms, and under many conditions, and a basic understanding of the forces and processes involved is helpful for identifying and defining their potential hazards. However, in order to determine site stability and to make design and construction recommendations for sites with problems resulting from mass wasting processes, specialists, such as geotechnical engineers or geologists, should be consulted early in the planning process.

PROCESSES OF MASS WASTING

The term *landslide,* as generally used, refers to most of the forms of rapid mass wasting, including debris avalanches, slumps, debris slides, falls, rock slides, and rock falls. The remaining types of mass wasting are labeled in accordance with their characteristic movement process, appearance, or materials, such as solifluction, mud flows, and creep. However this classification, following that of Sharpe (1938), is not very satisfactory in describing the actual processes that occur, for any event may have a set of sequences all of which have different movement rates. While the many terms that have been used to describe mass wasting events have created some confusion, the intent here is not to propose a new classification scheme, but rather to define and describe some of these terms now commonly used. Research currently underway by geomorphologists should result in a more comprehensive process classification.

Solifluction

Solifluction is a very slow process of mass wasting in which soil movement takes place at an imperceptible rate (less than 1 foot per year). When a high ice or water content exists, it increases the load, raises pore water pressure, reduces capillary tension, lowers shear strength, and thus allows gravity to move soil and rock debris downslope at a very slow rate. This process is very common in arctic and subarctic regions where permafrost exists. The permafrost forms a subsurface, impervious layer which inhibits drainage of the saturated surface materials, and since the surface tends to be more fluid than the subsurface, the resulting mixture of ice, soil, rock, and peat moves downslope slowly with a rolling motion, in a treadmill effect. It is very difficult to design linear functions, such as pipelines, highways, or railroads, across landscapes undergoing this process, and maintenance costs of such structures are high since alignments buckle and shift continuously and unpredictably.

Creep

Similar to solifluction, soil or rock creep is one of the slowest forms of soil-rock movement. Its downslope movement occurs in all landforms susceptible to volume change by frost

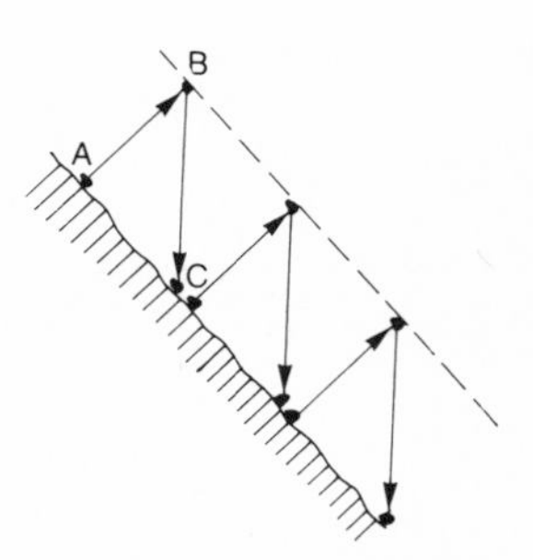

Figure 11.1. Downslope movement of soil or rock materials. The expansion of a soil or rock surface through frost action, wetting, or heating moves particles originally located at A to B. As the surface contracts by melting, drying, or cooling, particles move downward, under the influence of gravity, from B to C. The distance represented by AC is the net downslope movement of the particular soil or rock particle.

action or increase of moisture or pore water pressure (Figure 11.1). Other factors, such as expansion and contraction of the ground surface from heating and cooling, plant growth and decay, animal burrowing, solution weathering, and snow accumulation, can also promote soil and rock creep.

Creep activity can be observed in aerial photographs by noting tilted telephone poles, tilted or crooked fences parallel to the sideslope, separated fences perpendicular to the slope, tilted trees, and even tilted gravestones in cemeteries (Figure 11.3).

Mudflows and Earthflows

Mudflows or earthflows occur in conditions of high water saturation. The viscosity of the flow varies from an extremely thick, concrete-like flow to a thin sheet similar to muddy water. Flows move at rates up to 1 foot per second.

Mud flows are common in arid and semiarid climates and in zones of volcanic activity. In arid or semiarid mountainous areas, flows originate

Figure 11.2. Soil Creep in shale in Maryland. Photograph by G. W. Stose, U.S. Geological Survey.

during heavy rainstorms when large amounts of the soil debris that concentrates in valleys is eroded. In such cases a flowing front of thick soil materials may develop and be pushed by an accumulation of water behind it. The flow eventually encounters the lowlands, where the zone of thick, flowing material is breached by water, thereby allowing mud and water to flow and spread in a fanlike form over the valley floor. Such flows have tremendous densities enabling them to move very large objects; for example, huge residual boulders found in valleys in such regions have been transported and deposited by this process. In volcanic regions many valleys are filled with previous mudflows of volcanic ash. In these cases large deposits of unconsolidated ash, previously saturated by heavy rains, gain a flow consistency similar to that of concrete and move downslope, filling depressions, valleys, and basins.

During the 1964 earthquakes in Anchorage, Alaska, many large earthflows and mudslides were set off by vibrations and shock waves, and most of the damage from this event was attributable to the flows and slides rather than to the actual earthquake tremors. The earthquake vibrations caused clay layers underlying the city to liquify; large masses of material slid toward the ocean for approximately 1 minute after the earthquake shocks had ceased. The damage was limited by the presence of frost penetration and overlying layers of gravel, which helped to hold the soil masses together.

Slump

Slumps are the most common of the slide-type (Figure 11.4) processes of mass wasting; they occur where the hillside mass on a slope increases until the supporting power of the lower part of the slope is exceeded. Slumps tend to have a rotational movement wherein the materials move downward while rotating backward (Figure 11.5). If materials at the toe of the

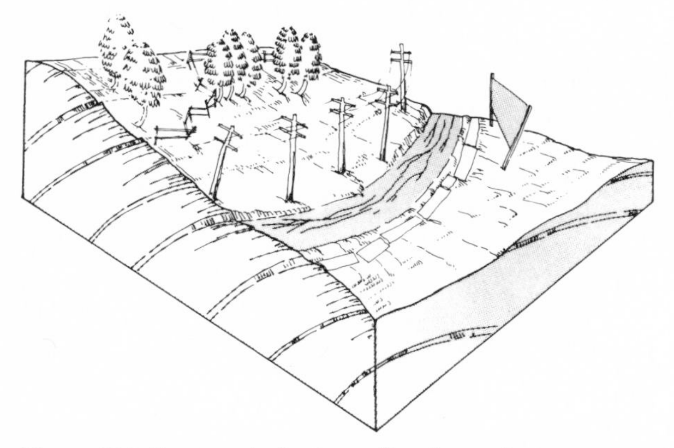

Figure 11.3. Features indicating soil and or rock creep.

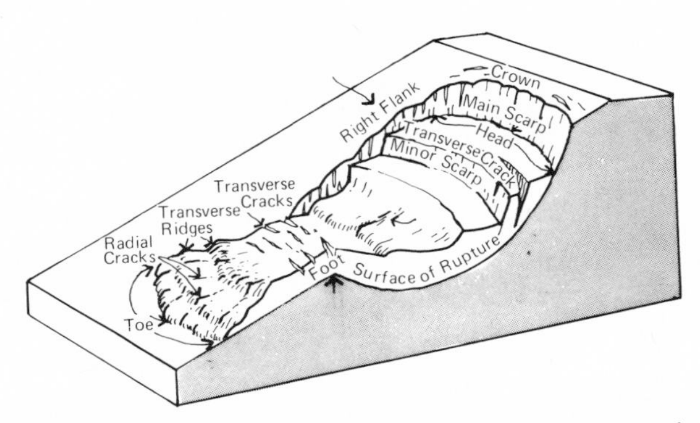

Figure 11.4. Typical landslide showing foot, toe, crown, and other features. (From *Landslides and Engineering Practice* by the Highway Research Board, 1958.)

slope begin to flow during slumps, they are classified as debris flows. Both slump and debris flows occur at all scales; they are common in landforms that contain interbedded materials of different permeability or resistance, and along stream banks oversteepened by erosion.

In 1965, a large slump and an associated debris flow occurred along the bank of the Nicolet River in Nicolet, Quebec, where construction activities had oversteepened a bank composed of interbedded silt and clay (Figure 11.6). Heavy rains had increased the soil weight and mass of the material of the cliff where the initial slump occurred, and the toe of the slope contained sufficient moisture to begin flowing. The debris flow from the first slump continued to remove the soil debris and thus created a secondary, unstable, vertical face that soon after also slumped and flowed. The subsequent slumps and flows continued for 7 minutes and carved a depression 500 feet long, 300 feet wide, and 10 to 20 feet in depth. Fortunately, only three lives were lost, but a school and several small structures were destroyed.

Avalanche

Avalanches are the most feared forms of mass wasting. They are the most rapid and can approach speeds of close to 100 miles per hour. Most avalanche movements are falls of large amounts of snow, ice, or rock from steep slopes where excessive materials have accumulated, and the movement is accelerated by entrapment of air and water which decrease the internal, frictional contacts.

Every winter snow avalanches that occur in the mountainous regions of the world kill many people and destroying whole towns or parts of towns. In 1962 a huge ice fall from a glacier near Ranrahirca, Peru, started an avalanche which killed about 3500 people. The avalanche contained an estimated 1 million tons of material as

Figure 11.5. Soil slump in Washington photographed 2½ hours after the event. Large rotational blocks typical of these slumps are apparent. Photograph by F. O. Jones, U.S. Geological Survey.

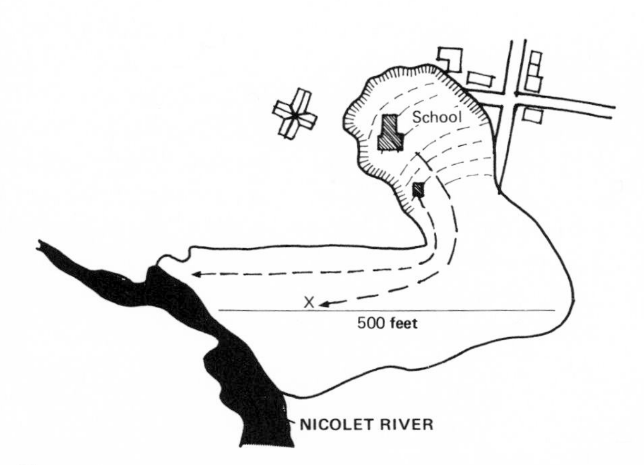

Figure 11.6. Nicolet River slump, Quebec, Canada, 1965. The slump and resultant flows along the bank of the Nicolet River moved over 200,000 cubic yards of materials in 7 minutes. Three people were killed, and a school and several other buildings were destroyed.

it moved from its origin at Huascaran peak (22,205 feet in elevation) 9 miles through the adjacent valleys and dropped approximately 13,000 feet in 7 minutes.

Rock avalanches are also very destructive. They contain primarily rock materials, and they move at high speeds with tremendous mass and energy. Rock avalanches are common where interbedded o: foliated rock structures are parallel to hillside slopes and where valley erosion has removed lateral support of the tilted materials.

In Frank, Alberta, Canada, in 1903, 70 people were killed in a rock avalanche when approximately 40 million cubic yards of rock broke loose from Turtle Mountain above the town and descended 3000 feet at about 60 miles per hour (Figure 11.7). The debris covered the valley floor in a path 2 miles wide and splashed up the farther valley wall to an elevation of 400 feet.

Another famous rock avalanche took place in 1963 at the Vaiont Reservoir in the Italian Alps, where a dam 870 feet high had been constructed across the Vaiont Canyon to create a deep reservoir (Figure 11.8). In the process of filling the reservoir, the limestone of the lower hillsides became saturated, filling solution cavities, expanding clay layers, and weakening the overlying rock structure. Then, just before the avalanche, there were 2 weeks of heavy rainfall. The increase in mass caused by the increased water content and the rise in pore water pressure caused a tremendous rock mass, 1 mile in width, 1.2 miles long, and over 500 feet high, to fall from the south valley wall into the Vaiont Reservoir. The debris created waves that extended 850 feet along the north side of the reservoir valley and overflowed the dam in 300-foot waves. The 875-foot deep reservoir was

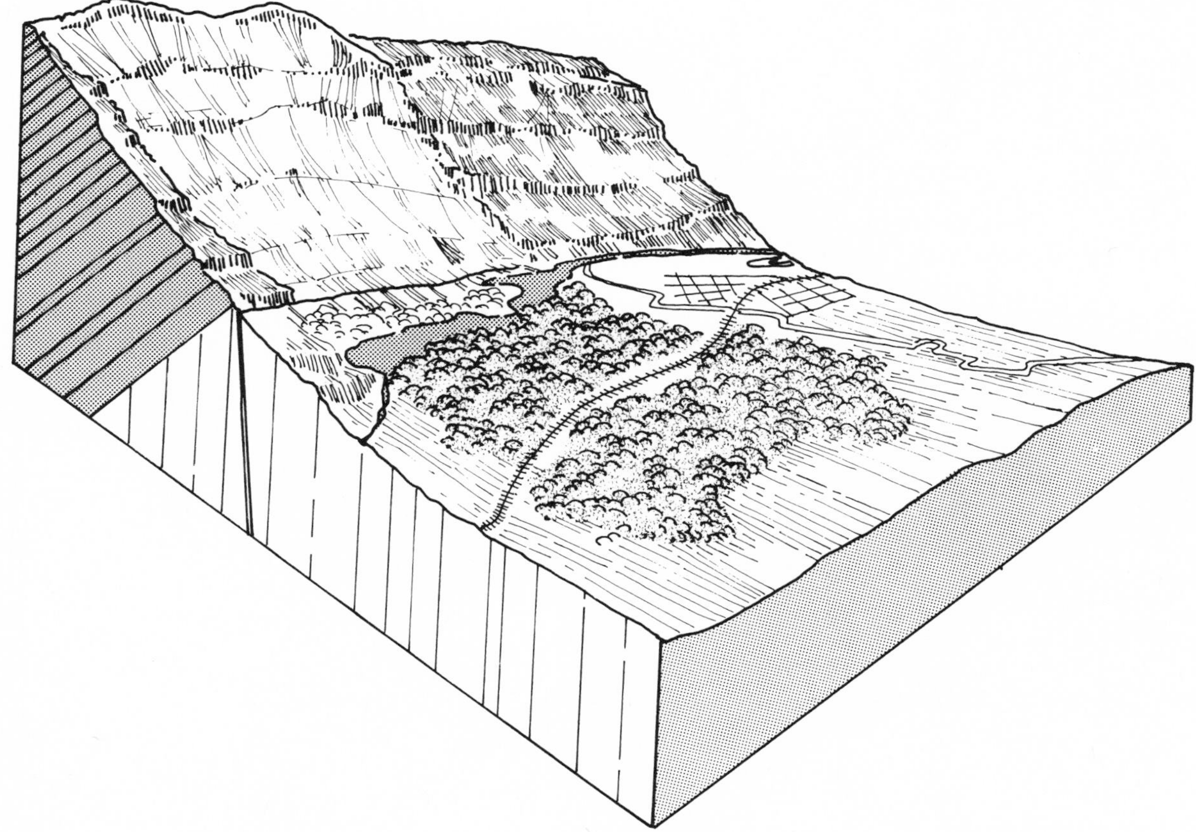

Figure 11.7. Turtle Mountain rock avalanche, Frank, Alberta, Canada, 1903. The rock avalanche killed 70 people in the town of Frank. Limestone with jointing patterns parallel to the hillside slopes provided planes of weakness; these were eroded by solution weathering, which removed much of the lateral support of the hillside.

filled with a mass of material which extended 1.2 miles upstream of the dam and rose to a height of 575 feet above the previous water level. The resulting floods and waves that swept through the downstream valley killed an estimated 3000 people, mostly around the town of Longarone. It is interesting to observe that for 2 weeks previous to the disaster no animal, wild or domestic, would venture onto the south valley face; they apparently sensed the instability of the rock mass.

Talus and Sliderock

Talus cones are formed at the bases of rock cliffs when mechanical weathering processes, supplying rock debris, are predominant. These cones of rock waste or sliderock face outward from the cliffs. Generally, the coarsest rock fragments are near the top of the cone and the smaller materials near the toe or base. Cones have very high angles of repose, typically greater than 45 degrees, reflecting their characteristic content of large, coarse, angular fragments of rock.

Colluvium

Colluvium is soil material that has been moved downslope by a process of mass wasting, such as creep, aided by frost action and erosion. Its materials are either not stratified or exhibit poor stratification, and they are either poorly sorted or well graded. Colluvium is easily distinguished from fluvial, lacustrine, or eolian landforms due to its lack of stratification, but it is hard to distinguish in residual soils and glacial landscapes. All regions of residual soil materials contain zones of colluvium fringing the bottoms of all significant slopes; it is important to identify these because of their unstable, unpredictable characteristics. Experience and local knowledge of a particular region help in identifying these areas but, in addition, field surveys and investigations should be made in suspected zones.

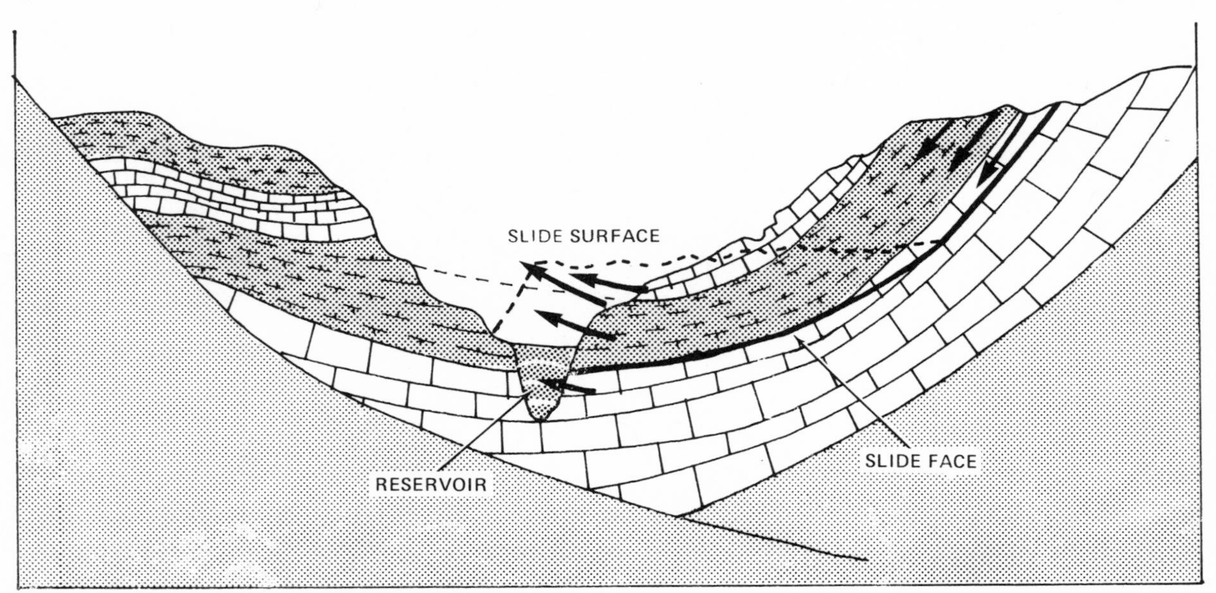

Figure 11.8. The Vaiont Reservoir Rock avalanche, Vaiont, Italy, 1963. Limestone beds riddled with solution voids and underlaid by impervious clay layers allowed hydrostatic pressures to develop in the limestone. There was no lateral support. Enough mass accumulated through several weeks of heavy rainfall to initiate failure. (After A. L. Bloom, *The Surface of the Earth*, Prentice-Hall, Englewood Cliffs, N.J., 1969, p. 47.)

CONDITIONS SUSCEPTIBLE TO SLIDING

Certain conditions correlate highly with activities of mass wasting, especially sliding. Any landform that exhibits one or more of the following conditions should be considered extremely hazardous. Figure 11.9 illustrates areas within the United States of relative landslide severity.

Saturated, Cohesive, Clayey Soils. Sloped areas of cohesive, clayey soils are susceptible to sliding because they increase in mass or weight with the addition of water. Removal of the natural, vegetative cover exposes these soils to greater water contact and may facilitate slumping and flow.

Loose, Granular Materials Having Low Shear Strength. Loose, granular materials gain readily in moisture content, thus increasing their mass until gravitational forces cause slide, slumping, and/or flows.

Interbedded Sedimentary Rock Parallel to Hillslopes. Interbedded sedimentary structures with bedding or joints parallel to valley walls are susceptible to sliding because they increase their mass through absorption of moisture or water, thus lubricating the bedding planes or joints, which in turn facilitates slippage.

Highly Foliated Metamorphic Rock. Foliations parallel to valley walls present conditions similar to those discussed for interbedded sedimentary rocks. Because of their slippery nature, the occurrence of such minerals as micas and serpentine increases the hazard.

Rotten or Decomposed Igneous or Metamorphic Rock. Rotten, loose, granular rock debris is susceptible to increases in moisture content and thus in its mass, which can create potential sliding conditions.

Fractures and Faults. When these linear features parallel or intercept slopes, they are susceptible to sliding, especially when there is no lateral support or when seepage helps to lubricate the rock surface.

Interbedded Materials of Differing Resistance or Permeability. In interbedded materials differences in resistance to weathering result in the undermining of the softer materials, and differences in permeability allow water to accumulate in elevated rock layers. Limestones over shale or clay layers, sandstone underlaid by an impervious stratum, and lava flows over tuff all represent conditions favorable for the creation of cliffs and overhangs susceptible to falls.

Seepage of Water along Hillslopes. The appearance of seepage water along hillslopes may indicate the development of high pore or hydrostatic pressures, especially if the water is found near the toe of the slope.

Colluvial Soils. These and other soils formed by earlier mass wasting events indicate previous unstable conditions which may still exist.

Water-Land Edges after Rapid Decreases in Water Level. Recession of flood waters or lowering of water levels because of reservoir drawdown create temporary unstable situations along the land edges. The high moisture content increases the mass of the land edge after the water level is lowered. Many small slumps can be observed along river banks after a rapid recession of flood waters.

The above conditions susceptible to mass wasting refer to those naturally occurring in undisturbed areas. Manmade fills are also susceptible to slumping or sliding if hydrostatic pressures build up, if seepage occurs under the fill, or if the fill is placed on a soil of high volume change. Fills in areas having these conditions should be sufficiently anchored and drained.

INTERPRETIVE CHARACTERISTICS OF SLIDES AND SLUMPS

Zones of previous sliding activity are easily identified on aerial photographs by characteristic crescent scarps and the hummocky topography exhibited by the debris flows. It is obviously more difficult to identify areas that have a potential for sliding or slumping, but the following characteristics may help to identify such zones. Because the key features in such cases

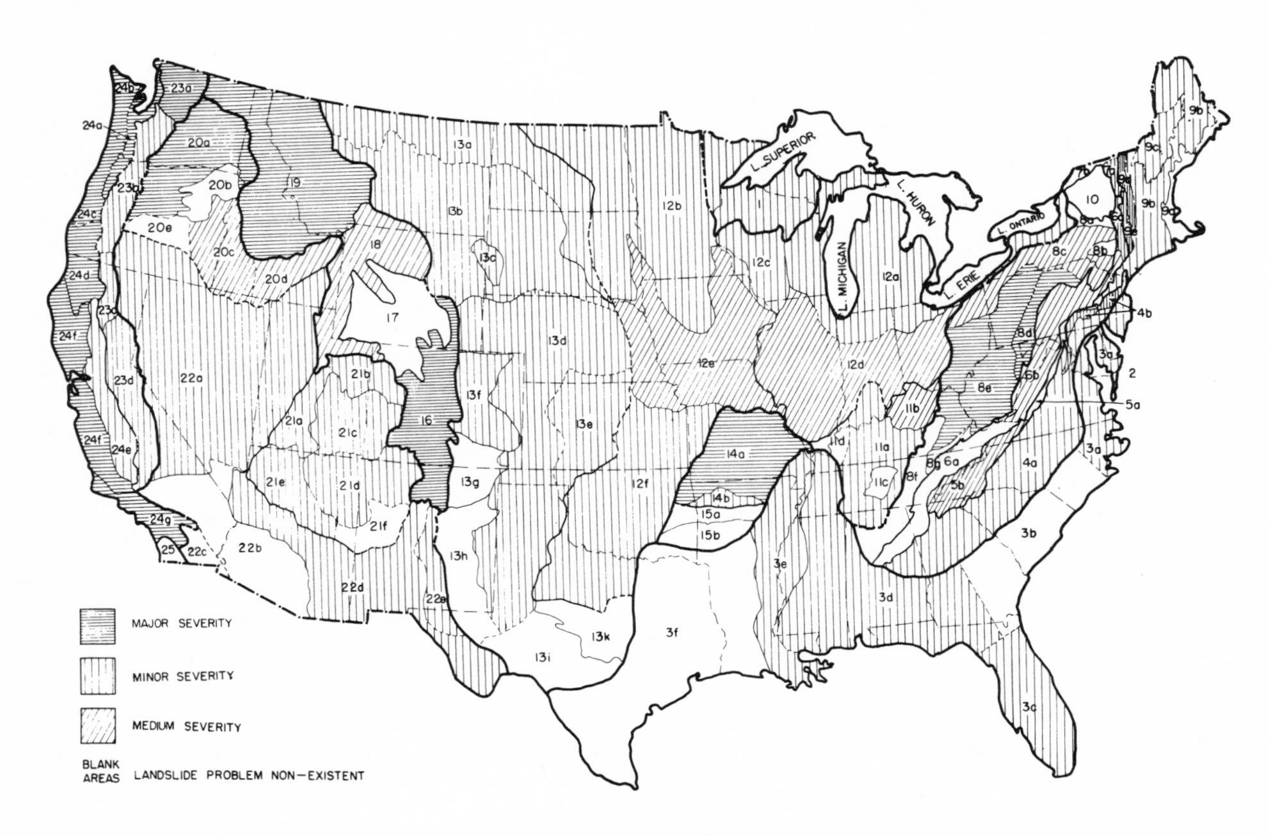

Figure 11.9. Landslide severity in the United States related to engineering projects. Map is based upon the "Physical Divisions" outlined by Fenneman. (Baker, R. F., and R. F. Chierzzi, "Regional Concept of Landslide Occurrence," Highway Research Board, Washington, D.C., Bulletin 216, 1959, p. 10.)

are rather small, large-scale photographs (greater than 1:9600) have been found to be the most useful.

Sharp Breaking Lines at the Scarp. The first sign of ground movement is the appearance of crescent-shaped, linear cracks in the surface soil. These may be difficult to see in their early stages since they may be obscured by vegetation.

Hummocky Ground Surfaces Below Cliffs. The appearance of hummocky materials indicates previous sliding activity and at least identifies materials highly susceptible to mass wasting. If this topography is graded or otherwise disturbed, sliding could start again.

Undrained Depressions along Hillside Toe or Crest. Undrained depressions indicate water seepage and accumulation and thus indicate a potential build-up of hydrostatic pressures.

Light Tones along Upper Edges of Hillsides or Cliffs. These tones, especially when linear, may indicate the formation of subsurface cracks; these facilitate drainage and cause the lighter tone. The appearance of these tones may precede the occurrence of actual breaks and scarps in the land surface.

Accumulation of Debris in Valleys and Stream Channels. Accumulations of soil materials in these areas indicate previous sliding and slumping, commonly associated with stream undermining of embankments.

Changes in Tone along Upper Areas of Cliffs or Embankments. Changes in tone near edges of embankments may indicate moisture differences in the subsoil, reflecting moisture accumulation and the development of hydrostatic water pressures.

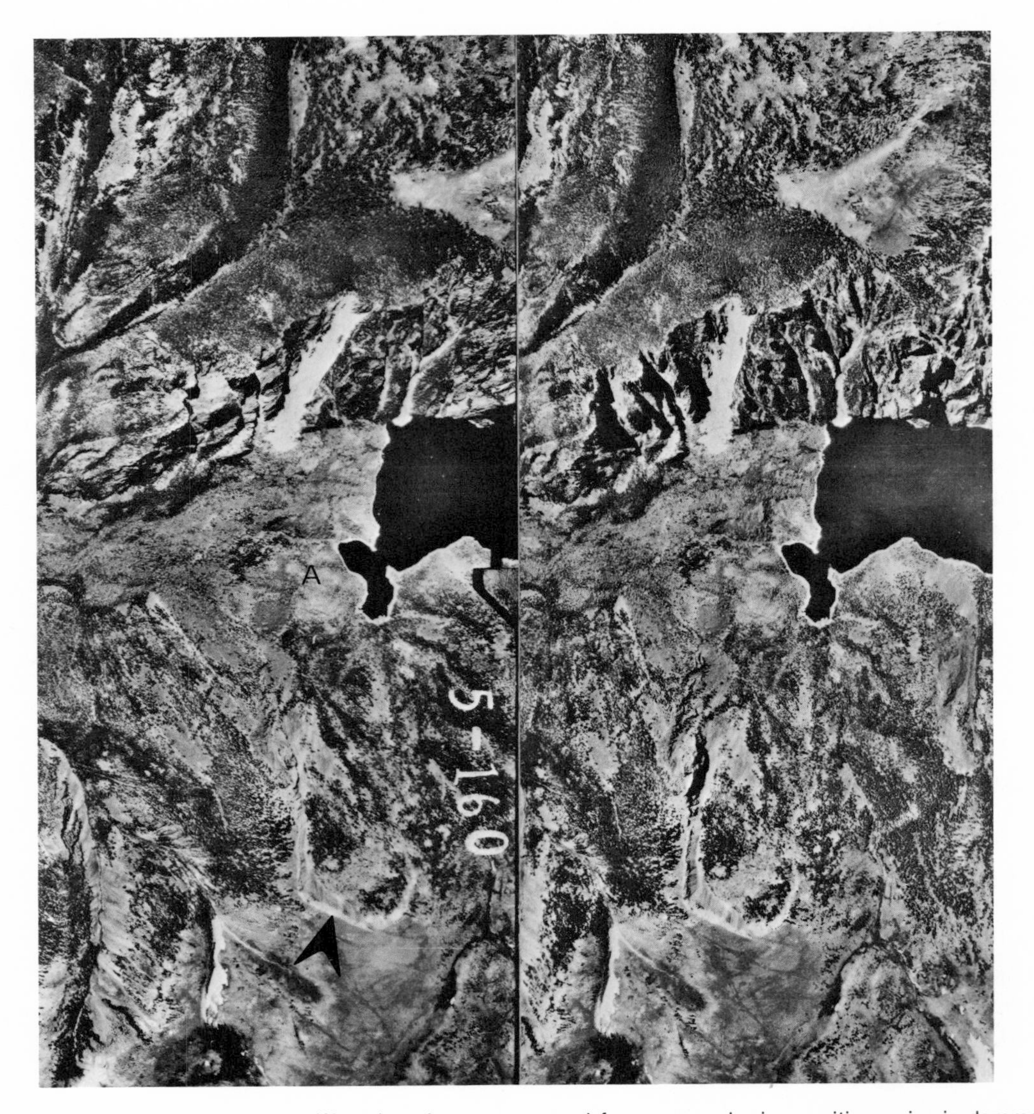

Figure 11.10. Landslide at Deep Lake, Wyoming. A canyon carved from precambrain granitic gneiss is dammed by a landslide more recent than the last glacial period. The scarp face of this slide, indicated by arrow, is still fresh in appearance. Note the hummocky topography in the canyon (A) typical of these deposits. The opposite valley shows other recent sliding events. Photographs from the U.S. Geological Survey, 1:37,400, July 1948.

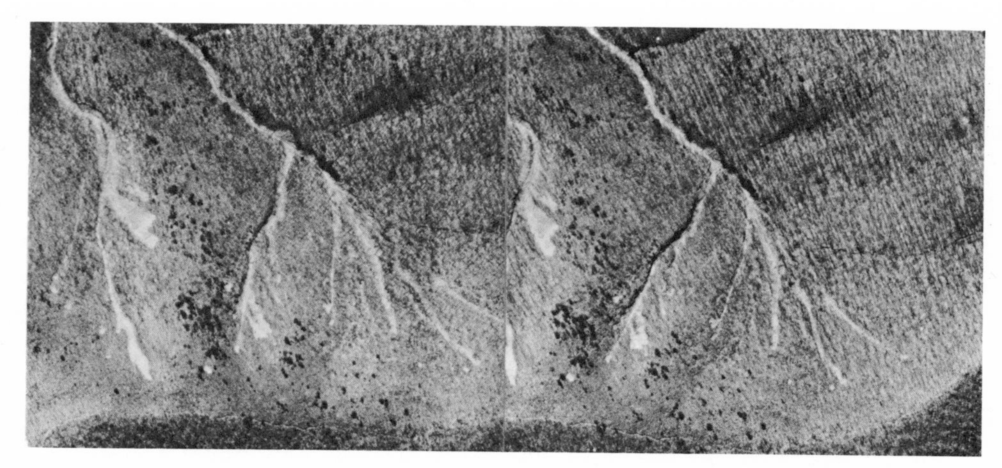

Figure 11.11. A metamorphic rock area in Tennessee showing avalanche scars resulting from a thunderstorm in 1951. Photographs from U.S. Geological Survey, 1:28,000, March 1963.

CONTROLLING POTENTIAL OR ACTIVE SLIDE AREAS

There are many methods of treating unstable areas to reduce the hazard of land movement. The most common are those that either reduce shearing stresses or increase the shearing resistance of the landmass. The reduction of shearing stress can be accomplished by excavation or drainage, or a combination of both. To increase shearing resistance, either restraining structures or methods of hardening the landmass can be employed. Table 11.2 summarizes some of the techniques used in past situations for the prevention and correction of landslides. The table is not all-inclusive; some unstable formations which will be encountered will represent unique conditions and require unique design recommendations.

Figure 11.12. A basalt plateau commonly has slump features along edge outcroppings as shown here near King Hill, Idaho. Note the circular slump scarp and the hummocky topography. Photographs from the U.S. Geological Survey, 1:20,000, October 1950.

Table 11.2. Methods for the prevention and correction of landslides*

Effect on stability of landslide	Method of treatment	Frequency of successful use†			Position of treatment of landslide	Best applications and limitations
		Fall	Slide	Flow		
Reduces shearing stresses	Excavation					
	Removal of head	N	1	N	Top and head	Deep masses of cohesive material
	Flattening of slopes	1	1	1	Above road or structure	Bedrock; also extensive masses of cohesive material where little material is removed at toe
	Benching of slopes	1	1	1	Above road or structure	Relatively small shallow masses of moving material
	Removal of all unstable material	2	2	2	Entire slide	
Reduces shearing stresses and increases shear resistance	Drainage					
	Surface:					
	Surface ditches	1	1	1	Above crown	Essential for all types
	Slope treatment	3	3	3	Surface of moving mass	Rock facing or pervious blanket to control seepage
	Regrading surface	1	1	1	Surface of moving mass	Beneficial for all types
	Sealing cracks	2	2	2	Entire, crown to toe	Beneficial for all types
	Sealing joint planes and fissures	3	3	N	Entire, crown to toe	Applicable to rock formations
	Subdrainage					
	Horizontal drains	N	2	2	Located to intercept and remove subsurface water	Deep extensive soil mass where ground water exists
	Drainage trenches	N	1	3	—	Relatively shallow soil mass with ground water present
	Tunnels	N	3	N	—	Deep extensive soil mass with some permeability
	Vertical drain wells	N	3	3	—	Deep slide mass, ground water in various strata or lenses
	Continuous siphon	N	2	3	—	Used principally as outlet for trenches or drain wells
Increases shearing resistance	Restraining structures					
	Buttresses at foot					
	Rock fill	N	1	1	Toe and foot	Bedrock or firm soil at reasonable depth
	Earth fill	N	1	1	Toe and foot	Counterweight at toe provides additional resistance
	Cribs or retaining walls	3	3	3	Foot	Relatively small moving mass or where removal of support is negligible
	Piling					
	Fixed at slip surface‡	N	3	N	Foot	Shearing resistance at slip surface increased by force required to shear or bend piles
	Not fixed at slip surface‡	N	3	N	Foot	Rock layers fixed together with dowels
	Dowels in rock	3	3	N	Above road or structure	Weak slope retained by barrier, which in turn is anchored to solid formation
	Tie-rodding slopes	3	3	N	Above road or structure	
Primarily increases shearing resistance	Miscellaneous methods					
	Hardening of slide mass					
	Cementation or chemical treatment					
	At foot	3	3	3	Toe and foot	Noncohesive soils
	Entire slide mass	N	3	N	Entire slide mass	Noncohesive soils
	Freezing	N	3	3	Entire	To prevent movement temporarily in relatively large moving mass

Table 11.2. (Continued)

Effect on stability of landslide	Method of treatment	Frequency of successful use†			Position of treatment of landslide	Best applications and limitations
		Fall	Slide	Flow		
	Electro-osmosis	N	3	3	Entire	Effects hardening of soil by reducing moisture content
	Blasting**	N	3	N	Lower half of landslide	Relatively shallow cohesive mass underlaid by bedrock. Slip surface disrupted; blasting may also permit water to drain out of slide mass
	Partial removal of slide at toe	N	N	N	Foot and toe	Temporary expedient only; usually decreases stability of slide

* Adapted from Highway Research Board, *Landslides and Engineering Practice,* Special Report No. 29, 1958, pp. 114-115.
† 1, Frequently; 2, occasionally; 3, rarely; N, not applicable.
‡ Used for correction only.
** Used for prevention only.

Table 11.3. Mass Wasting (miscellaneous landforms): Aerial photographic and cartographic references

State or country	County	Photo Agency	Photograph no.	Scale	Date	Corresponding Quadrangle and Geologic Maps from USGS	Corresponding soil reports from SCS
Bolivia	–	USGS*	Bolivia 1-ABCD	4166′/″	5/28/56	–	–
California	Los Angeles	USDA	AXJ-13K-101-102	1667′/″	6/4/53	–	1916†
Chile	Tarapaca	USGS*	Chile 1-AB	5000′/″	4/7/55	USAF Aero Chart, Point Angamos (1:1,000,000)	–
Colorado	Hinsdale	USGS‡	Colorado 9-AB	3333′/″	9/27/51	Durango (1:250,000), Prof. Paper No. 67	–
Montana	Gallatin	USGS‡	Montana 21-AB	833′/″	8/22/59	Tepee Creek (15′)	1931
Montana	Gallatin	USGS‡	Montana 22-AB	833′/″	8/22/59	Tepee Creek (15′)	1931
New Hampshire	Grafton	USDA	DXV-6N-2-3	1667′/″	7/11/55	Franconia (15′)	1939
New Mexico	Rio Arriba	USGS‡	New Mexico 2-AB	1750′/″	12/2/49	Lyden, Velarde (7½′)	–
New Mexico	Rio Arriba	USGS‡	New Mexico, 3-AB	3133′/″	7/23/63	Velarde, Taos Junction (7½′)	–
Tennessee	Tennessee, Swain, Sevier	USGS‡	Tennessee 2-AB	2333′/″	3/30/63	Great Smoky Mountain National Park, East ½ (15′)	1956
Virginia	Augusta	USGS‡	Virginia 2-AB	833′/″	3/13/55	Parnassus (15′)	1937
Washington	Snohomish	USGS‡	Washington 3-ABC	2308′/″	10/7/64	Glacier Peak (15′)	1947
Wyoming	Park	USGS‡	Wyoming 1-ABCD	3116′/″	7/7/48	Deep Lake (15′)	–

* From *Selected Aerial Photographs of Geologic Features outside the United States*, U.S. Geological Survey Prof. Paper No. 591.
† Out of print.
‡ From *Selected Aerial Photographs of Geologic Features in the United States,* U.S. Geological Survey Prof. Paper No. 590.

REFERENCES

Baker, R. F., "Determining the Corrective Action for Highway Landslide Problems," *Highway Research Board Bulletin,* No. 49, 1952.

Baker, R. F., and R. Chieruzzi, "Regional Concept of Landslide Occurrence," *Highway Research Board Report No. 216,* National Academy of Sciences, Washington, D.C., 1959.

Brunsden, D., *Slopes, Form and Process,* Institute of British Geographers, Oxford, 1971.

Cruden, D. M., and J. Krahn, "A Reexamination of the Geology of the Frank Slide," *Canadian Geotechnical Journal,* Vol. 10, No. 4, 1973.

Daly, R. A., *Igneous Rocks and the Depths of the Earth,* McGraw-Hill, New York, 1933.

Daly, R. A., W. G. Miller, and G. S. Rice, *Report of the Committee to Investigate Turtle Mountain, Frank, Alberta,* Geological Survey of Canada, Memorandum No. 27, 1912.

Eckel, E. B., ed., *Landslides in Engineering Practice,* Highway Research Board, Special Report No. 29, National Research Council, Washington, D.C., 1958.

Ernest, W. G., *Earth Materials,* Prentice-Hall, Englewood Cliffs, N.J., 1969.

Fairbridge, R. W., "Land Mass and Major Landform Classification," in R. W. Fairbridge, ed., *Encyclopedia of Geomorphology,* Reinhold, New York, 1968.

Froelich, A. J., "Geologic Setting of Landslides Along South Slope of Pine Mountain, Kentucky," *Highway Research Board Report No. 323,* National Academy of Sciences, Washington, D.C., 1970.

Highway Research Board, *Landslides and Engineering Practice,* Committee on Landslide Investigations, Highway Research Board, Special Report No. 29, Washington, D.C., 1958.

Holmsen, P., "Landslips in Norwegian Quick-Clays," *Geotechnique,* Vol. 3, 1953.

Kirkby, M. J., "Measurement and Theory of Soil Creep," *Journal of Geology,* Vol. 75, 1967.

LaChapelle, E., "Snow Avalanches," in R. W. Fairbridge, ed., *Encyclopedia of Geomorphology,* Reinhold, New York, 1968.

Ladd, G. E., "Landslides, Subsidences and Rockfalls as Problems for the Railroad Engineer," *Proceedings, American Railway Engineering Association,* Vol. 36, 1935.

Newland, D. H., "Landslides in Unconsolidated Sediments," *N.Y. State Museum Bulletin,* No. 187, Albany, N.Y., 1916.

Rapp, A., and R. W. Fairbridge, "Talus Fan or Cone; Scree and Cliff Debris," in R. W. Fairbridge, ed., *Encyclopedia of Geomorphology,* Reinhold, New York, 1968.

Salqueiro, P. R., "Landslide Investigation by Means of Photogrammetry," *Photogrammetria,* Vol. 20, No. 3, pp. 107–114, 1965.

Schumm, S. A., "The Role of Creep and Rainwash in the Retreat of Badland Slopes," *American Journal of Science,* Vol. 254, 1956.

Sharpe, C.F.S., *Landslides and Related Phenomena,* Columbia University Press, New York, 1938. (Reprinted in 1960 by Pageant Books, Paterson, N.J.)

Skempton, A. W., and D. J. Henkel, "A Landslide at Jackfield, Shropshire, in a Heavily Overconsolidated Clay," *Geotechnique,* Vol. 5, No. 2, 1955.

Ta Liang, and D. J. Belcher, "Airphoto Interpretation," in E. G. Eckel, ed., *Landslides and Engineering Practice,* Special Report No. 2, NAS-NRC Publication 544, Washington, D.C. (Highway Research Board), 1958.

Terzaghi, K., "Mechanism of Landslides," in *Application of Geology to Engineering Practice,* Geological Society of America (Berkey Volume), Wiley, New York, 1950.

Terzaghi, K., and R. B. Peck, *Soil Mechanics in Engineering Practice,* Wiley, New York, 1967.

Tompkin, J. M., and S. H. Britt, *Landslides, A Selected Annotated Bibliography,* Highway Research Board, Bibliography No. 10, Washington, D.C., 1951.

Zaruba, Q., and V. Mencl, *Landslides and Their Control,* American Elsevier, New York, 1969.

CHAPTER 12

CASE STUDIES

INTRODUCTION

This section presents selected projects illustrating a range of problem contexts, site locations, and scales where photointerpretation terrain analysis was applied. For all sites, little existing information was available pertinent to the required analysis. However, general soil surveys, statewide geologic maps, and topographic maps provided an initial overview. The landform terrain analysis approach was employed to provide more detailed geologic, surficial geologic, soil texture and depth, depth to water table, and vegetation information. The exact data categories used were a function of those desired for the defined analysis.

Site sizes ranged from 50 to over 20,000 acres and included studies for the location and design of housing, a resort community, retirement community, sand and gravel operation, and a ski area. The issues of development included: excavation and grading, sources of construction materials, potential zones of mass wasting, locations of septic tank leaching fields, and suitability analysis for the location of housing, highways, solid waste disposal sites, and major lakes. The cost of the required photointerpretation services including map preparation and field work varied from $.10 to $10.00 per acre depending upon total number of acres, amount of original existing information, degree of landscape unit complexity, and nature of data categories. In general the costs were less per unit area for larger sites, since the efficiency of using aerial photographs greatly increases in rugged and inaccessible terrain. However, in a site smaller than 10 acres, valuable information may still be gained from the study of its aerial photographs for a pattern of soil or rock not discernable on the ground may be perceived.

The following projects are briefly capsulized to illustrate a portion of the total analysis. They illustrate what can quickly (less than one week), accurately (90 percent), and efficiently accomplished with aerial photographs and proper terrain analysis procedures.

ANNAQUATUCKET

Located near Wickford, Rhode Island, this site of approximately 120 acres is bounded along its northern edge by the small, but picturesque, Annaquatucket River. The program under consideration included low-density housing (one-fourth acre lots) with a mixture of townhouses and a recreation area resulting in a gross density of one unit per acre. All housing had to rely upon septic tank leaching fields for sewage disposal, but water was available from the town.

The site, as partly delineated upon the 1:4,800 scaled aerial photographs, is partially wooded and contains spotted wetlands, ponds, and hummocky topography. Glaciation of the area during the last Wisconsin period resulted in a variety of glaciofluvial deposits, as suggested by the lack of apparent surface runoff (porous materials) and hummocky topography (associated with glacial eskers, kames, and kettleholes). The map (Figure 12.1) illustrates the surficial geology and major landforms of the site. More detailed soil interpretations were prepared including depth to seasonal high water table (Figure 12.2). As supported by this and other data, a cluster housing development with a mixture of housing types and densities, recreational areas, and trails was designed and constructed. The ability to have much of this data generated from aerial photographs resulted in a significant cost savings for the client and provided a more accurate overview of the site conditions for the client *and* the town planning board.

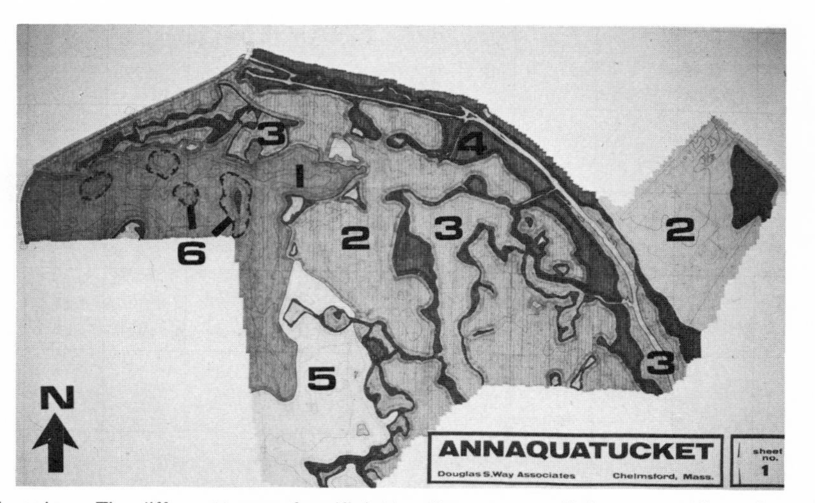

Figure 12.1. Surficial geology. The different types of surificial landforms are easily interpreted from the aerial photographs. (1) Esker crevasse fillings. (2) Ice-contact undifferentiated glaciofluvial sand and gravel. (3) Organic peat. (4) Ice contact glaciofluvial sand and gravel with 2 to 3 feet of surface organic matter. (5) Kame terrace consisting of sand and gravel. (6) Kettleholes created by melted ice blocks.

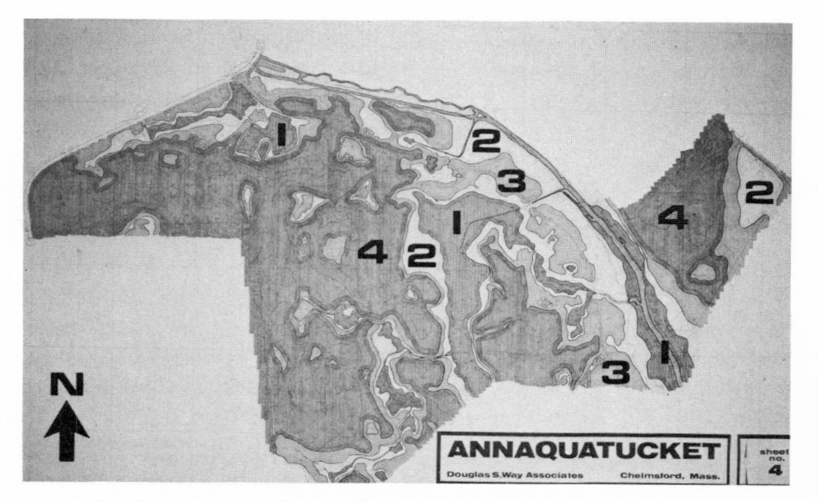

Figure 12.2. Depth to seasonal high water table. The depth to seasonal high water table is mapped, showing surface water (white). (1) Wetlands; (2) less than 1 to 3 feet to water table; (3) 3 to 5 feet; and (4) greater than 5 feet.

Figure 12.3. Annaquatucket. The stereo-triplet shows a portion of the site mapped in Figures 12.1 and 12.2. Portions of the esker, kettle holes, and kame terrace can be observed, but special attention should be given to comparing the aerial photographs to the depth to seasonal high water table map in Figure 12.2. Photographs by Aerial Data Reduction, Inc., Pennsauken, New Jersey, 1:4,800.

MOUNT WASHINGTON

A 26,000-acre site adjacent to Mount Washington, New Hampshire, was investigated to determine its potential for a year-round recreational community including a golf course, a ski area, possibly a lake, a variety of housing types and densities, and various other recreational facilities. The site is remote, large, and rugged. Little existing information made interpretation from aerial photographs attractive. In this instance all interpretation, mapping, and ground surveys were completed within one week.

Aerial photographs available at two different scales (1:20,000 and 1:62,500) were taken during the summer and late spring, respectively. The 1:62,500 spring photos provided excellent terrain condition indicators; leafless trees allowed the ground to be observed and distinctions to be made between conifers and deciduous types. The drainage patterns, gullies, and ravines were also apparent since the forest floor leaves had been removed from these areas by the spring thaw and rains.

Geology, landforms, soil texture and depth, stoniness, potential erosion zones, depth to water table, and vegetation types were interpreted by using both sets of photographs and field investigations. The stereo-triplet in Figure 12.5 indicates some of the major landforms present, and Figure 12.4 illustrates the soil textures and depths. This technique allowed for the completion of a general overview analysis within a short period of time at little expense. As priority areas were defined for further study, resources could be concentrated efficiently to derive the most feasible alternative development schemes.

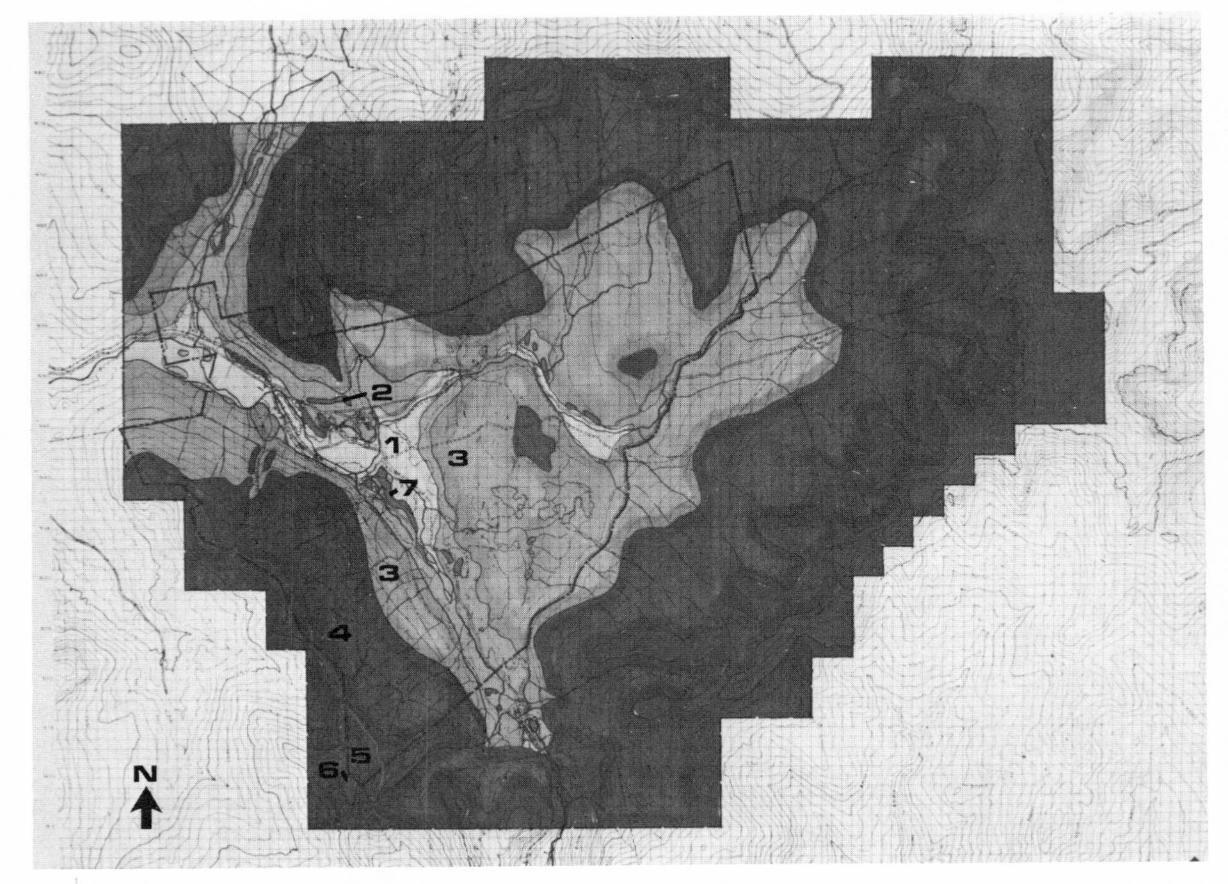

Figure 12.4. Mt. Washington. The soil texture and depth to bedrock is mapped. (1) Glacial outwash, sand and gravel, greater than 10 feet to bedrock. (2) Glacial kame terraces, sand and gravel, greater than 10 feet to bedrock. (3) Glacial till, stony sandy loam with large boulders, 5 to 10 feet to bedrock. (4) Glacial till, very stony bouldery sandy loam, 1 to 5 feet to bedrock. (5) Thin glacial till, 1 foot cover. (6) Exposed bedrock. (7) Glacial fluvial esker.

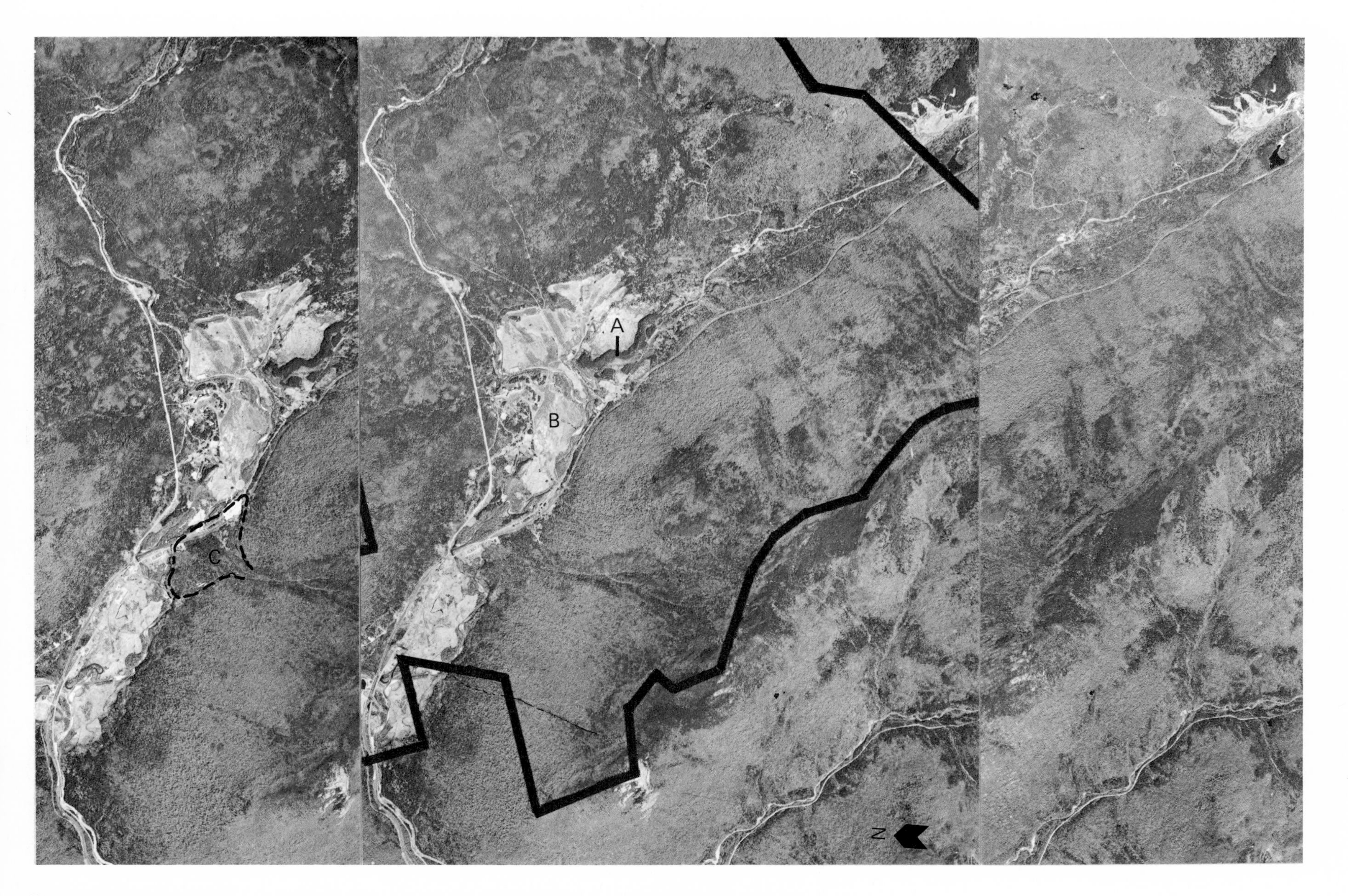

Figure 12.5. Mt. Washington. A portion of the site is shown in stereo. Note the glacial esker at A, outwash area B, and alluvial fan at the base of the ravine C. Photographs by U.S. Forest Service, 1:62,500, May 1958.

KIAWAH ISLAND

Kiawah Island, located off the coast of South Carolina, is composed of marine landforms: closely spaced beach ridges, tidal flats, and thin organic soils in interridge depressions. The original program concept was a resort recreation community consisting of housing, golf courses, and a major harbor facility designed to attract traffic from the adjacent intercoastal waterway. The island was subsequently sold to a group of Arab investors who are now developing a more intensive resort-hotel complex.

When observed on aerial photographs (see Figure 10.44) the island has a distinct set of linear ridges (10 to 12 feet above sea level) and depressions that generally parallel the shore. These had been created by previous ocean shoreline locations, each of which created a beach ridge. The lower areas between ridges encounter the seasonal high water table and have developed organic soils. The depth of these organic materials varies with the distance between ridges but averages less than five feet.

In addition to the typical soil interpretations, old aerial photographs and maps were obtained to determine long term erosion-deposition trends along the coast. Figure 12.6 shows the comparison of photos and maps in which major changes over the past 109 years can be observed. Development recommendations were made in consideration of these significant marine processes.

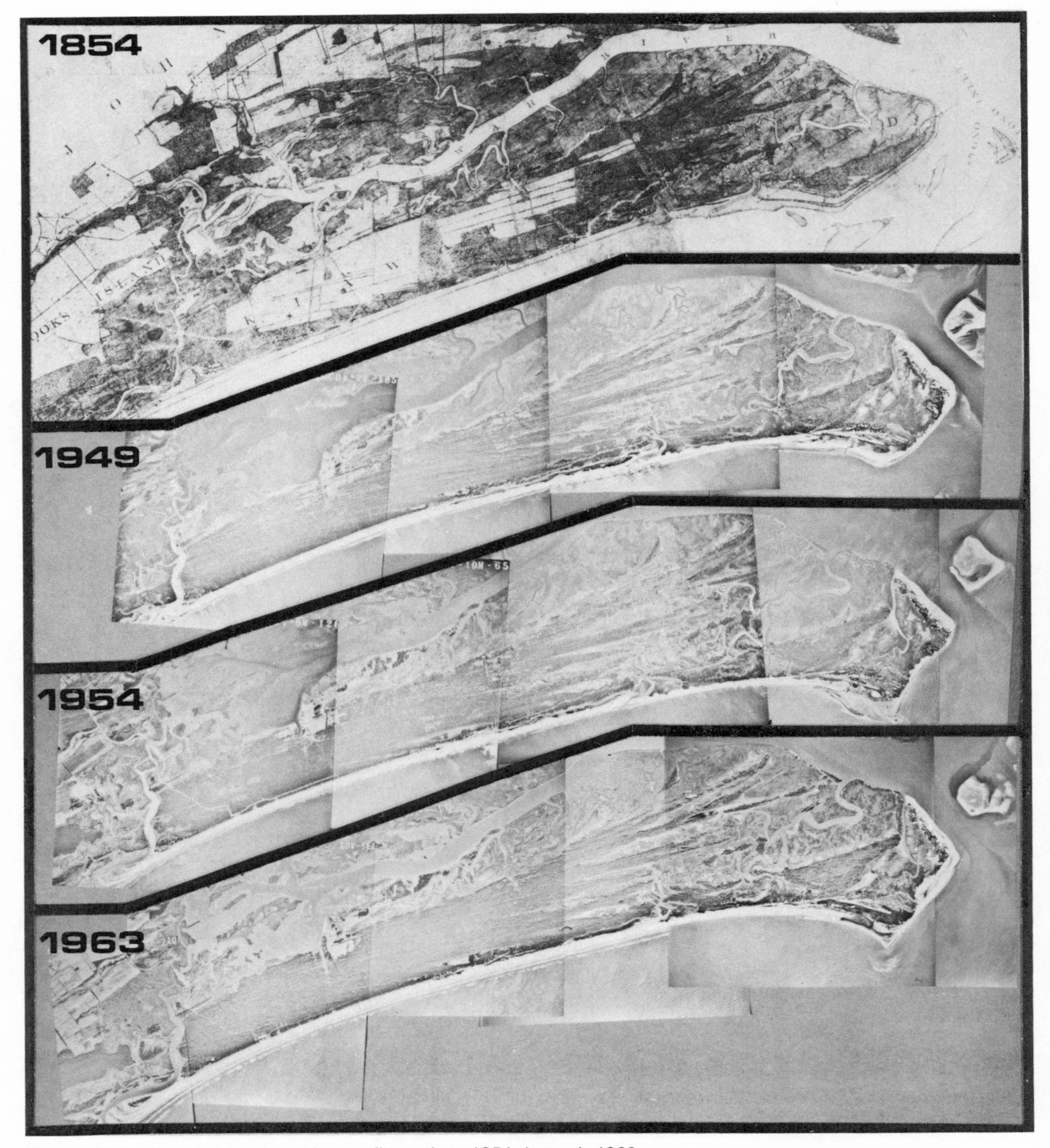

Figure 12.6. Kiawah Island and its configuration, 1854 through 1963.

Figure 12.7 A. Kiawah Island: Landforms. The major landform units are shown. (A) Active beach; (B) primary dunes; (C) sand flats; (D) shallow swamps; (E) well-defined beach ridges; (F) tidal flats.

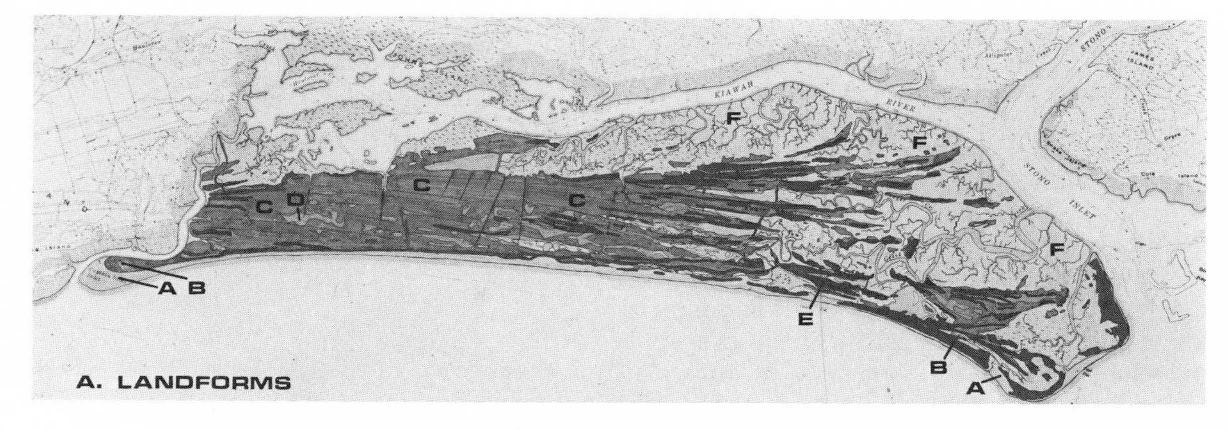

Figure 12.7 B. Kiawah Island: Development opportunity. The physical site conditions of soil, depth to water table, and percolation rate have been summarized to define general development opportunities. (A) Slight, few limitations; (B) moderate; (C) severe, wet soils but suitable if drained; (D) not suitable.

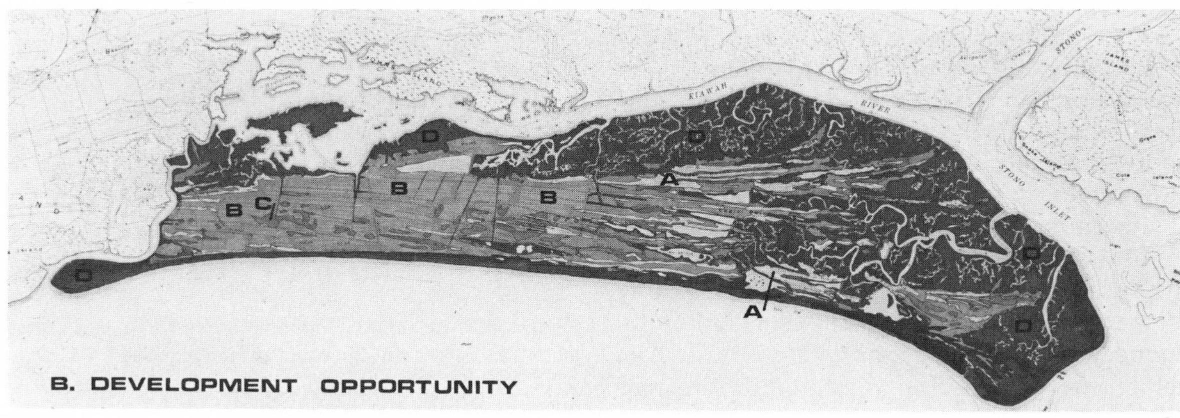

Figure 12.7 C. Kiawah Island: Visual analysis. The primary visual patterns and site amenities are mapped as they may influence site development. These features provide for the inherent site identity and should not be destroyed.

QUECHEE

For the past ten years a major resort-recreation development complex has been under construction and use near the town of Quechee, Vermont. Originally, over 4,000 acres of land were studied for capabilities concerning different housing types and densities, a championship golf course and elaborate club house facility, family ski area, hiking and riding trails, and major ponds and lakes. The initial concept and challenge was to create a "prototype" example of a successful four-season recreation development that would minimize environmental impacts, provide an exciting living community, be profitable, and maintain its quality and attractiveness during and long after construction. Could the qualities that attract people to the area be maintained and not destroyed during project implementation?

The aerial photograph presented in Figure 12.10 shows a portion of the Quechee site. Note the distinct linear trend in the microtopography of the uplands and the many rock outcrops suggesting foliated metamorphic rocks such as schist and gneiss. The major landform map (Figure 12.8) shows the many rock outcropings with overlying thin glacial till, glaciofluvial outwash, and postglacial flood plain deposits. The determination of the boundary location between lowland outwash and upland glacial till is difficult but important; the character of the soils changes significantly. The approximate boundary has been sketched on the aerial photograph, as defined by the slight change in slope and microtopographic conditions.

The soil depth to bedrock (Figure 12.9) was mapped; note the shallowness of the cover in most of the upland areas. This condition alone alerted the planning group that onsite septic tank leaching fields could not be satisfactorily located except for very low density sites such as five acre lots. Civil engineers were immediately consulted concerning alternative sewage disposal and water collection schemes.

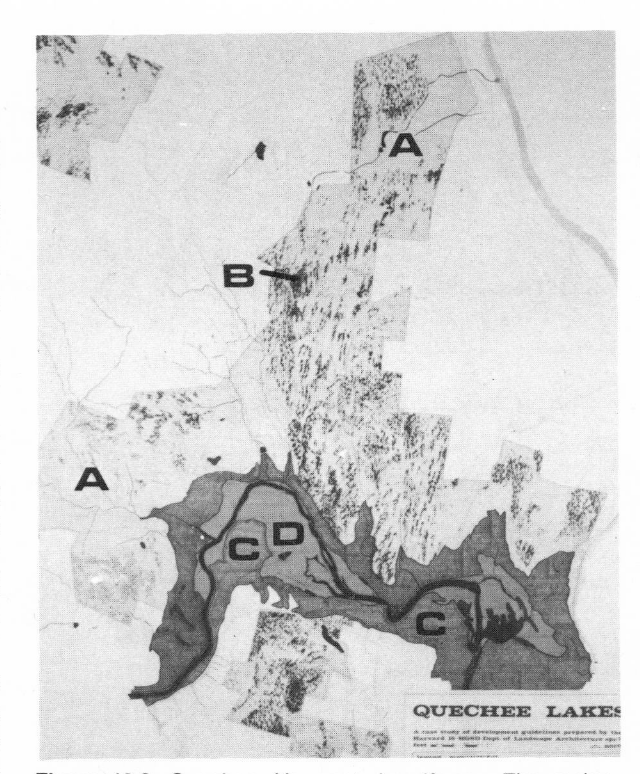

Figure 12.8. Quechee, Vermont: Landforms. The major landform units of the Quechee site are mapped. (A) Glacial till; (B) rock outcrops of schist and gneiss; (C) glaciofluvial outwash; (D) postglacial flood plain.

The deeper valley soils facilitated excavation for potential pond and lake construction but the granular composition, common with glacial outwash, implied that the water table would freely fluctuate following seasonal conditions. Any construction of a large body of water would therefore require an underlying impervious liner to minimize seepage losses during the drier seasons.

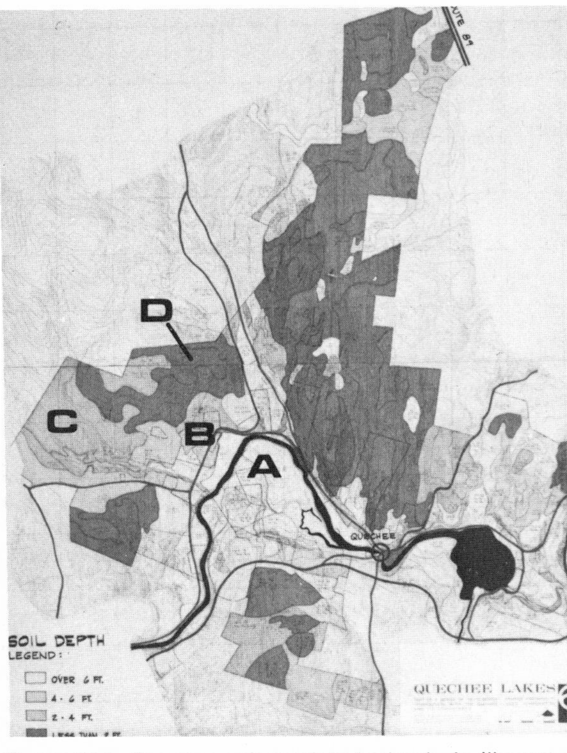

Figure 12.9. Quechee soil depth to bedrock. As illustrated by this map, the soils are very thin over much of the site. (A) Soils greater than 6 feet; (B) 4 to 6 feet; (C) 2 to 4 feet; (D) less than 2 feet.

Eventually, evaluation of all of the different utility infrastructure alternatives and program mixes resulted in a master plan for the project. Several lakes, a family ski area, championship golf course, and many other recreational zones were developed together with approximately 2,000 housing units in a mix from condominiums to five-acre farmsteads. The role of aerial photographic interpretation was important in quickly defining the range of conditions to be expected and thus allowing proper consultants to be contacted early in the planning process.

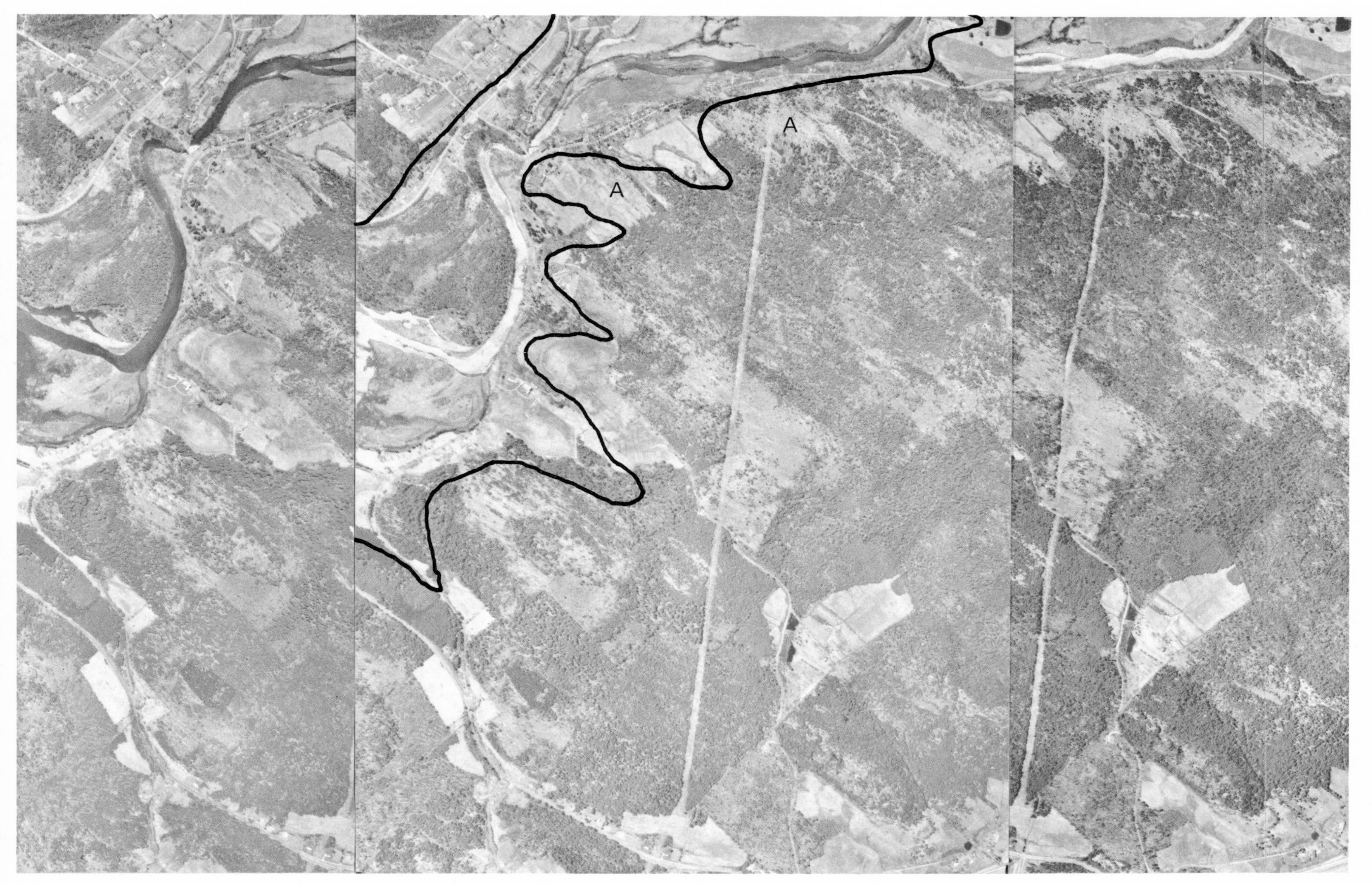

Figure 12.10. Stereo portrayal of the Quechee area. Note the boundary between glacial outwash and upland till and the rock-controlled topography at A. Photographs by Quinn Associates, Pennsylvania, 2000′/″, May 1969.

SAND AND GRAVEL EXPLORATION

A client in Rhode Island was searching for sand and gravel resources needed for a major construction project. The area shown in the aerial photographs (Figure 12.11) shows hummocky topography that may be associated with grannular glaciofluvial deposits. However, at issue is whether or not this topography is a result of underlying rock influence or unconsolidated sand and gravel. Note the subtle differences between the general areas A and B. Area A has a finer microtopography with many steep side slopes, some of which approach 20 to 30 degrees. The hills in area B are larger and seem more massive. Close examination may indicate rock outcroppings. Therefore, area A seems to offer potential sand and gravel resources; the topography expresses characteristics associated with these deposits, not underlying rock as in area B. The elevation of the groundwater table can be estimated to determine how many cubic yards of material could be removed above it. (In some areas excavation of sand and gravel below the water table require hard-to-obtain special permits.)

Figure 12.11. Area under consideration for potential sand and gravel excavation. Note the rock outcroppings at C. Photographs by Aerial Data Reduction, Inc., Pennsauken, New Jersey, 1:7,200.

GLOSSARY*

*Selected from *Glossary of Selected Geologic Terms,* by Stocks, W. L., and Varnes, D. J., Colorado Scientific Society, 1955; and the *Manual of Remote Sensing,* American Society of Photogrammetry, Falls Church, Va., 1975.

aa, n. The Hawaiian word for solidified lava characterized by an exceedingly rough, jagged, or spinose surface. Most of the surface is covered with a layer of loose fragmental clinkery material a few inches to several feet thick. Below the surface the clinker fragments may be stuck firmly together. Typically, the upper clinker layer is underlaid by, and grades into, a central massive layer (after Macdonald). Aa surfaces are extremely difficult to traverse on foot and are impassable to vehicles. Although a sequence of aa flows may contain considerable amounts of interbedded, relatively loose, clinkery material, the large open cavities characteristic of pahoehoe are uncommon.

ablation, n. Separation and removal of particular rock material by any process or processes acting either mechanically or by solution (Merriam-Webster). Usually the term is used to describe the wastage of glaciers by melting and evaporation so that debris released from the ice accumulates on the surface.

acutance, n. An objective measure of the ability of a photographic system to show a sharp edge between contiguous areas of low and high illuminance. This film property can now be measured, using microdensitometric techniques, instead of being qualitatively estimated by subjective visual perception.

additive color process, n. A method for creating essentially all colors through the addition of light of the 3 additive color primaries (blue, green, and red) in various proportions through the use of 3 separate projectors. In this type of process each primary filter absorbs the other 2 primary colors and transmits only about one-third of the luminous energy of the source. It also precludes the possibility of mixing colors with a single light source because the addition of a second primary color results in total absorption of the only light transmitted by the first color.

adobe, n. A general term widely used in the southwestern United States and in Mexico for clayey and silty material that can be made into sun-dried bricks. Adobe tends to fracture into cubical blocks on moderate drying. It is widely distributed both on bedrock and on alluvium, and may lie on slopes as well as on valley floors. It may or may not be calcareous. At some localities it has apparently been deposited by water, at other places by wind. No exact compositional limits have been set; the property of hardening so as to be useful for brick-making seems to depend on several factors among which is the presence of clloidal clay.

agglomerate, n. A mass of unsorted volcanic fragments which may be loose or consolidated into a solid mass by finer volcanic material which fills the interstices. It is usually localized within volcanic vents or at distances of not more than a mile therefrom.

aggregate, 1. **v.** To bring together; to collect or unite into a mass or sum. 2. **adj.** Composed of mineral crystals of one or more kinds, or of mineral or rock fragments (Merriam-Webster). 3. **n.** In the case of materials of construction, designating inert material which when bound together into a conglomerated mass by a matrix forms concrete, mastic, mortar, plaster, etc. (Am. Soc. Testing Materials, Standards on mineral aggregates, C 58-28T).

alkali flat, n. A plain or basin in an arid or semiarid region where alkali salts have become concentrated by evaporation and poor drainage. The so-called white alkalies consist mostly of sodium sulfate; the black alkalies are made up chiefly of sodium carbonate.

alluvial fan, n. A sloping, fan-shaped mass of loose rock material deposited by a stream at the place where it emerges from an upland into a broad valley or a plain. The highest point is at the apex of the fan, which is generally composed of boulders and cobbles that are dropped as soon as the stream emerges from its confining walls.

alluvium, n. A general term for all detrital material deposited permanently or in transit by streams. It includes gravel, sand, silt, and clay, and all variations and mixtures of these. It is usually applied to the deposits of streams in their channels and over their flood plains and deltas, but a few writers include material laid down in lakes and estuaries. Unless otherwise noted, alluvium is unconsolidated.

alpine glacier, n. The type of glacier that forms in the valleys and about the higher peaks of mountains. Such glaciers originate in rather broad amphitheatres at higher elevations and usually become concentrated in valleys at lower elevations.

angle of repose, n. The angle measured from a horizontal plane at which loose material will stand without sliding. It may be measured for either natural or artificial piles of material.

anthracite, n. Hard coal. This is a compact dense coal, iron black to velvet black in color. It is brittle; has a strong vitreous to submetallic luster, a more or less pronounced conchoidal fracture, a hardness of 2 to 2.5; and has specific gravity 1.4 to 1.8. Anthracite ignites less easily than other varieties of coal and burns with a short flame, but with great heat.

anticline, n. In its simplest form, an anticline is an elongate fold in which the sides or limbs slope downward away from the crest. This simple form may be greatly complicated during progressive stages of folding. The whole structure may be tilted so that one limb is horizontal, or bent over so that the surface dividing the fold symmetrically lies flat, or the fold may be so compressed that the limbs become parallel. However complexly modified, an anticlinal fold is assumed to have begun as a simple upward flexure which dies out in length by downward inclination of the axis and by lessening of dip and curvature of the strata around the ends of the fold. In a normal sequence of bedded rocks, the oldest beds are in the core of the anticline and the youngest lie on the flanks. The use of the term, however, is not confined to structures in sedimentary rocks, but is extended to folds in all rocks showing pronounced planar elements, such as the banding or foliation in metamorphic rocks. No size limits are implied, but the term is not generally used for small folds or crinkles having dimensions of inches or a few feet, and many of the very large structures of the order of 50-100 miles long have other characteristics that more properly class them as geanticlines, arches, anticlinoriums, etc. Although antilines and synclines (down folds) may greatly influence the development of erosional land forms, they are not in themselves a type of surface topography but are geologic structures.

aperture, n. The opening in a lens diaphragm through which light passes.

aquifer, n. A geologic formation or structure that transmits water in sufficient quantity to supply pumping wells or springs (Tolman). The terms water-bearing bed, water-bearing stratum, and water-bearing deposit are used synonymously with aquifer.

atoll, n. A roughly circular, elliptical, or horseshoe-shaped island or ring of islands of reef origin, composed of coral and algal rock and sand, around a lagoon in which there are no islands of noncoral origin.

attenuation, n. In physics, any process in which the flux density (or power, amplitude, intensity, illuminance,) of a "parallel beam" of energy decreases with increasing distance from the energy source.

azonal soils, adj. Any group of soils without well-developed profile characteristics owing to their youth or conditions of parent material, or relief that prevents the development of normal soil-profile characteristics (U.S. Dept. of Agriculture Yearbook, 1938).

badland, n. An area, large or small, characterized by extremely intricate and sharp erosional sculpture. Badlands usually develop in areas of soft sedimentary rocks such as shale, but may also occur in decomposed igneous rocks, loess, etc. The divides are sharp and the slopes are scored by intricate systems of ravines and furrows. Fantastic erosional forms are commonly developed through the unequal erosion of hard and soft layers. Vegetation is scanty or lacking and there is a notable lack of coarse detritus. Badlands occur chiefly in arid or semiarid climates where the rainfall is concentrated in sudden heavy showers. They may, however, occur in humid regions where vegetation has been destroyed, or where soil and coarse detritus are lacking.

bajada, n. A continuous alluvial apron extending along the base of a mountain range, formed by the coalescence of a series of alluvial fans. Some bajadas are separated from the actual mountain front by a carved-rock pediment into which they may gradually merge. Bajadas differ from pediments, which they resemble in surface form, in being surfaces of deposition rather than erosion. The term is usually restricted to semiarid and desert regions.

barrier reef, n. A type of coral reef that lies some distance from a continental coast or volcanic island. A lagoon of open water from a few hundred feet to several miles wide lies between the reef and the mainland. The reef itself is a long, narrow strip of coral rock and sand, broken in a few places by passages from the lagoon to the open ocean. It may be roughly circular if it encloses an island.

basalt, n. A word of ancient and unknown origin applied to a group of related igneous rocks that comprise the most common and widely distributed of all lavas. In the general usage, the term includes the majority of fine-grained, dark, heavy volcanic rocks.

base level, n. The theoretical limit toward which erosion constantly tends to reduce the land. Sea level is the general base-level, but in the reduction of the land there may be many temporary base levels below which, for the time being, the streams cannot reduce the land. These temporary base levels may be controlled by the level of a lake or a river into which the streams flow, or by a particularly resistant stratum or rock that the stream has difficulty in removing (Bryan).

bayou, n. A word of many local usages, especially in the lower Mississippi River basin, and Gulf Coast region of the United States. In general, a creek, secondary watercourse, or minor river, tributary to another river or body of water, may be called a bayou. Usually applied to a sluggish stream or slough in alluvial lowlands or swamps, or to an estuarial creek or inlet in the Gulf Coast region.

bed, n. Bed and layer refer to any tabular body or rock lying in a position essentially parallel to the surface or surfaces on or against which it was formed, whether these be a surface of weathering and erosion, planes of stratification, or inclined fractures. The body need not have been formed in a horizontal position, although its original position will have closely approached horizontality.

bench, n. A relatively flat, horizontal, or gently inclined surface, usually relatively long and narrow, which is bounded on one side by a steeper ascending slope and on the other side by a steeper descending slope. The name may be applied to ledges of rock shaped like steps or terraces, to forms cut by rivers, lakes, or oceans along their shores, and to steplike terraces artificially cut in mining and excavating. For a man-made bench the term **berm** is commonly used.

bentonite, n. A rock composed of any of the montmorillonite-beidellite group of clay minerals. It is derived from the alteration of volcanic tuff or ash. The color range of fresh material is from white to light green, or light blues. On exposure the color may darken to yellow, red, or brown. The term has been applied, sometimes erroneously, to various adsorptive clays used as drilling mud in oil-well drilling. Some bentonites swell on contact with water so that the size of a fragment may be increased many times, but this property is not entirely diagnostic and should be combined with other tests if an accurate determination is needed.

bituminous coal, n. This is a compact, brittle coal of a gray-black to velvet-black color. It has a lamellar, conchoidal, or splintery fracture, sometimes more or less cubical. The luster ranges from dull to pitchy; the specific gravity from 1.2 to 1.5. It gives a black to brownish-black streak. It burns with a yellow flame and gives a strong bituminous odor. It often shows distinct stratification through the varying luster of the different layers. Generally there are no traces of organic structures visible to the eye. Some varieties fuse or sinter together on heating, leaving a coherent residue or coke, and are thus called coking coals; others fail to do this and fall to powder.

blackbody, black body, n. (Symbol bb used as subscript) An ideal emitter which radiates energy at the maximum possible rate per unit area at each wave-length for any given temperature. A black body also absorbs all the radiant energy incident upon it. No actual substance behaves as a true blackbody although platinum black and other soots rather closely approximate this ideal. In accordance with Kirchhoff's law, a blackbody not only absorbs all wavelengths, but emits at all wavelengths and does so with maximum possible intensity for any given temperature.

block fault, n. A fault that aids in producing blocks in the earth's crust. A region that is divided by faults into differently elevated or depressed blocks is said to exhibit block faulting.

bog, n. A quagmire filled with decayed moss and other vegetable matter; wet spongy ground, where a heavy body is apt to sink; a small soggy marsh; a morass (Merriam-Webster).

bog ore, n. Iron-bearing material consisting of limonite and clay with iron silicates and siderite, humus, bog manganese, etc. It occurs in brittle flakes, thin beds, and irregular aggregates and is usually soft, porous, and of a brown color. It is thought to be formed by precipitation from iron-bearing swamp waters.

bottom-set beds, n. The layers of fine material carried out and deposited on the bottom of a body of water in front of a delta. These layers are progressively buried as the delta grows forward, and thus appear at the bottom of a complete cross section of a delta.

boulder train, n. A line or belt of glacial boulders that extends from the original rock outcrop, often for many miles, in the direction of glacial movement (Merriam-Webster).

butte, n. A conspicuous, isolated hill or small mountain with relatively steep sides. Most writers consider it to be smaller than a mesa, but there is great diversity of opinion as to whether the top should be flat, rounded, or pointed. It is usually considered to be an erosional remnant carved from flat-lying sedimentary rocks but some volcanic features such as cones and necks are locally called buttes. The terms butte and mesa are not used with complete distinction either in geologic literature or in popular usage.

caldera, n. Calderas are large volcanic depressions, more or less circular or cirquelike in form, the diameters of which are many times greater than those of the included vent or vents, no matter what the steepness of the walls or form of the floor. With rare exceptions all volcanic depressions of the characters defined are caused primarily by collapse.

caliche, n. In the arid and semiarid regions of the United States, a term applied broadly to secondary calcareous material occurring in a layer or layers at or near the surface. Caliche may be a soft horizon of lime accumulation in the soil, but more commonly the term refers to a cemented layer a few inches to many feet in thickness containing impurities of clay, sand, or gravel. The cementing material is essentially calcium carbonate, but may contain magnesium carbonate, silica, or gypsum. The upper surface is generally undulating, following the surface of the ground, and may show potholelike depressions. The lower surface may be extremely irregular or gradational with the underlying material. Most caliche deposits are of Pleistocene or Recent origin, and appear to form by a variety of processes whereby soil moisture evaporates or otherwise deposits its load of calcium carbonate. Caliche varies greatly in hardness and the resistance it offers to excavation. It has been used as road material in many parts of the southwestern United States.

canyon, n. A steep-walled valley or gorge in a plateau or mountainous area. The high and precipitous slopes impress the observer more than the flat land which may occur along the stream, and on this impression depends the distinction between canyons and other valleys (Bryan).

cathode-ray tube (CRT), n. A vacuum tube that generates a focused beam of electrons which can be deflected by electric and/or magnetic fields. The assembly contains an electron gun arranged to direct a beam upon a fluorescent screen. Scanning by the beam can produce light at all points in the scanned raster.

chalk, n. A soft, white, fine-grained variety of limestone that is especially characteristic of rocks of Cretaceous age in Britain and northwestern Europe. It is composed mainly of the calcareous shells of various marine microorganisms, but the matrix consists of fine particles of calcium carbonate some of which may have been chemically precipitated.

chernozem, n. A zonal group of soils having a deep, dark-colored to nearly black surface horizon, rich in organic matter, which grades below into lighter colored soils and finally into a layer of lime accumulation; developed under tall and mixed grasses in a temperate to cool subhumid climate (U.S. Dept. of Agriculture Yearbook, 1938).

chestnut soils, n. A zonal group of soils having a dark-brown surface horizon which grades below into lighter colored soil and finally into a horizon of lime accumulation, developed under mixed tall and short grasses in a temperate to cool and subhumid climate. They occur on the arid side of chernozem soils, into which they grade (U.S. Dept. of Agriculture Yearbook, 1938).

clastic, adj. Composed of broken fragments of minerals or rocks.

clay, n. 1. As a soil (or rock in the broad sense), clay is a fine-grained aggregate consisting wholly or dominantly of microscopic and submicroscopic mineral particles, derived from the chemical decomposition of rocks, which is plastic when wet and hard when dry. The distinctive physical properties are due to the presence of **clay minerals,** which are hydrous aluminum silicates that break down into colloidal, exceedingly minute shreds or flaky particles. Depending upon the percentage of coarse grains and the amount and type of clay minerals, clays are separated into: (1) **lean clays** of low to medium plasticity, cohesion, and compressibility that may be silty or sandy; and (2) **fat clays** that are greasy to the feel, highly plastic, cohesive, and compressible, tough and difficult to work when damp, impermeable, and very strong when dry. All gradations between these two types may be found. 2. As a particle-size term, clay has been used variously to denote all mineral particles of less than 2 to 5 microns diameter. For construction purposes, engineers generally class all fine-grained materials passing the 200-mesh sieve (less than .074 mm) as silt if it is relatively nonplastic and clay or clay soil if it is plastic, the division based upon certain values for liquid limit and plasticity index.

cohesion, n. The capacity of sticking or adhering together. In effect, the cohesion of soil or rock is that part of its shear strength which does not depend upon interparticle friction. In soils, true cohesion is attributed to the shearing strength of the cement or the adsorbed water films that separate the individual grains at their areas of contact. **Apparent cohesion** of moist soils is due to surface tension in capillary openings, and disappears completely on immersion.

colluvium, n. Earth material that has moved or been deposited mainly through the action of gravity. Talus piles, avalanches, and sheets of detritus moved by soil creep or frost action are examples.

compaction, n. The act or process of becoming compact. It is usually applied in geology to the changing of loose sediments to hard, firm rocks. With respect to construction work and soils engineering, compaction is any process by which the soil grains are rearranged to decrease void space and bring them into closer contact with one another, thereby increasing the weight of solid material per cubic foot.

consolidation, n. 1. In geology, any or all of the processes whereby loose, soft, or liquid earth materials become firm and coherent. Any action that increases the solidity, firmness, and hardness is important in consolidation. Geologically, cementation is probably the most important factor, followed by mechanical rearrangement of constituents through pressure, crystallization, and loss of water. The term also describes the change of lava or magma to firm rock. 2. In soil mechanics, consolidation is the adjustment of a saturated soil in response to increased load, involving the squeezing of water from the pores and decrease in void ratio. The rate of consolidation depends upon the rate at which the pore water escapes, and hence upon the permeability of the soil.

constructional land form, n. A land form created by accumulation of material; examples are volcanic cones, deltas, and flood plains. The term also applies to forms created by diastrophism, such as fault blocks or folds.

contact metamorphism, n. The process by which the texture, mineral assemblage, or chemical composition of rocks is changed owing to contact with or nearness to adjacent highly heated rocks.

coquina, n. A term applied to a conglomerate consisting predominantly of entire or broken calcareous shells with or without other hard parts of organisms. The constituent shells may be small, ranging down to 2 mm, or may be large fragments of shells, tridacna valves, or the like, one or two feet in diameter (Wentworth). Coquina is usually very porous but is strong enough for some structural purposes, and is also used for roadbeds in the southeastern United States.

coral, n. A general name for any of a large group of marine invertebrate organisms belonging to the phylum Coelenterata which are common in modern seas and have left an abundant fossil record in all periods later than the Cambrian. The term coral is commonly applied to the calcareous or hornlike skeletons left after death of the organism. As found in the fossil state, coral consists almost exclusively of calcium carbonate.

coulee, n. Commonly, in the northern plains of the western United States, any gully, dry wash, or intermittent stream valley of considerable size. More specifically, the term designates any of a number of steep-walled, trenchlike valleys cut in the Columbia River basalt in the state of Washington, and formerly occupied by glacial meltwater rivers. Also, rarely, a solidified stream or sheet of lava.

cross-bedding, n. A diagonal arrangement of bedding in sedimentary rocks, especially sandstone of wind or river origin, such that the layers are inclined at various angles to the more general planes of stratification or to the formation contacts.

crystalline rocks, n. The igneous and metamorphic rocks are commonly spoken of as crystalline rocks because they are mostly composed of more or less closely fitted mineral crystals that have grown in the rock itself.

cuesta, n. A hill or ridge with one steep face (escarpment slope) and an opposite gently inclined face (dip slope). The form is controlled by the erosion of an inclined resistant rock layer.

deflation, n. The sorting out and removal of relatively fine material by wind action.

deformation, n. 1. The process whereby rocks are folded, faulted, sheared, compressed, or the like by earth stresses, as in the growth of mountain ranges. 2. The result of the process (Merriam-Webster).

denudation, n. Broadly, the sum of the processes that result in the wearing down of the earth, including weathering, erosion, and transportation. A few writers restrict the term to the laying bare or uncovering of bedrock or a designated formation by the removal of overlying deposits.

desert, An arid region in which the vegetation is especially adapted to scanty rainfall, with long intervals of heat and drought, or, more rarely, is entirely lacking; a more or less barren tract incapable of supporting any considerable population without an artificial water supply. Rock disintegration in deserts predominates over rock decay, causing the slopes of stony detritus extending far up the mountain sides, the undrained basins, the salt pans or playas, and the areas of shifting sand, so common in deserts. Desert rainfall is usually less than ten inches annually (Merriam-Webster).

desiccation, n. The drying out of sediments, usually with shrinkage in volume.

destructional land forms, n. Forms resulting from the removal of substance. Examples are mesas, canyons, and cliffs caused by erosion.

differential weathering, n. The irregular and unequal effects of erosion and weathering due to the nonuniform resistance of rocks. The most obvious effect of differential erosion is that harder rocks come to stand higher than softer ones.

dike, n. A sheetlike body of igneous rock that fills a fissure in older rocks which it entered while in a molten condition.

dip, n. The angle which a stratum, sheet, vein, fissure, fault, or similar geological feature makes with a horizontal plane, as measured in a plane normal to the strike (Merriam-Webster).

doline, n. A funnel or bowl-shaped depression in limestone that has been formed chiefly by solution by groundwater that entered from the surface. The typical doline is more or less vertical and occurs at the intersection of two joint planes.

drift, n. Rock material of any sort deposited in one place after having been moved from another; as river drift. Specif., a deposit of earth, sand, gravel, and boulders transported by glaciers (glacial drift) or by running water emanating from glaciers (fluvioglacial drift) and distributed chiefly over large portions of North America and Europe, esp. in the higher latitudes (Merriam-Webster). It includes till, stratified drift, and scattered rock fragments (Flint).

drumlin, n. A smooth oval hill of glacial origin. Drumlins are composed almost exclusively of boulder clay but sometimes include lenslike masses of gravel and sand, and are thus mainly constructional in origin.

earth pillar, n. A high pillar of earth capped by a stone which has acted to protect the underlying material from the impact and erosion of rain. Earth pillars usually occur in groups and are carved out of such materials as boulder clay and landslide masses.

earthquake, n. A transient shock or series of vibrations of the earth caused by a disturbance of the elastic or gravitational equilibrium of the rocks at or beneath the surface.

effluent stream, n. A stream or stretch of a stream which, with respect to ground water, receives water from the zone of saturation. The upper surface of such a stream stands lower than the water table or other piezometric surface of the aquifer from which is it receives water (Meinzer, 1923).

electromagnetic radiation (EMR), n. Energy propagated through space or through material media in the form of an advancing interaction between electric and magnetic fields. The term radiation, alone, is used commonly for this type of energy, although it actually has a broader meaning.

eluviation, n. The movement of soil material from one place to another within the soil, in solution or in suspension, where there is an excess of rainfall over evaporation. Horizons that have lost material through eluviation are referred to as **eluvial** and those that have received material as **illuvial.** Eluviation may take place downward or sidewise according to the direction of water movement.

end moraine, n. An end moraine (also known as a **terminal moraine**) is a ridgelike accumulation of drift built chiefly along the terminal margin of a valley glacier or the margin of an ice sheet. It has a surface form of its own and is the result chiefly of deposition by the ice, or deformation by ice thurst, or both. The essential characteristic of end moraine is its close relationship to the terminus or margin of the glacier.

eolian, n., adj. Pertaining to wind. Designates rocks and soils whose constituents have been carried and laid down by atmospheric currents.

ephemeral stream, n. A stream or portion of a stream which flows only in direct response to precipitation. It recieves little or no water from springs and no long-continued supply from snow or other sources. Its channel is at all times above the water table. The term may be arbitrarily restricted to streams or portions of streams which do not flow continuously during periods of one month (Bryan).

erosion, n. The wearing away and removal of materials of the earth's crust by natural means. As usually employed, the term includes weathering, solution, corrasion, and transportation. The agents that accomplish the transportation and cause most of the wear are running water, waves, moving ice, and wind currents. Most writers include under the term all the mechanical and chemical agents of weathering that loosen rock fragments before they are acted upon by the transporting agents; a few authorities prefer to include only the destructive effects of the transporting agents.

erosion cycle, n. The succession of stages through which a newly uplifted land mass must pass before it is worn down to a peneplain or a surface near sea level. In the juvenile stages the surface is sharply cut by canyons; in the mature stage it may disappear and the topography be characterized by high steep hills and fairly open valleys; and in the old-age stages the land is so worn down that the streams meander sluggishly across a lowland (Merriam-Webster).

eskers, n. Relatively long, narrow, winding ridges of mixed sand and gravel. In longitudinal profile their crests are seen to be sinuous (Lahee, 1941). They are considered to have been deposited by streams of meltwater flowing through crevasses and tunnels in stagnant ice sheets.

estuary, n. A passage, as the mouth of a river or lake where the tide meets the river current; more commonly an arm of the sea at the lower end of a river; a firth. In physical geography, a drowned river mouth, caused by the sinking of the land near the coast (Merriam-Webster).

evaporites, n. A group of sedimentary deposits whose origin is largely due to evaporation. A few are included that develop through metamorphism of other evaporites. Only gypsum, anhydrite, and rock salt are of great quantitative importance.

exfoliation, n. The breaking or spalling off of thin concentric shells, scales, or lamellae from rock surfaces. The action is due to changes in temperature, the action of frost, and, in the opinion of some observers, to obscure chemical effects.

eye base, n. The distance and orientation of

the line between centers of rotation of the eyeballs of an individual. It differs from interocular distance in that it is oriented.

far infrared, n. A term for the longer wavelengths of the infrared region, from 25 μm to 1 mm, the generally accepted shorter wavelength limit of the microwave part of the EM spectrum. This is severely limited in terrestrial use, as the atmosphere transmits very little radiation between 25 μm and the millimeter regions.

fault, n. A break in materials of the earth's crust on which there has been movement parallel with the surface along which the break occurs. A fault occurs when rocks are strained past the breaking point and yield along a crack or series of cracks so that corresponding points on the two sides are distinctly offset. One side may rise, or sink, or move laterally with respect to the other side. Faults may be well exposed and easily located by direct observation or they may be located by indirect evidence such as (1) lines of springs, (2) repetition of the same strata, (3) omission of strata, (4) displacement of topographic features, (5) differential weathering of shattered rocks along the fault, and (6) peculiar drainage patterns. Depending on the nature of movement, and the relation to the affected rocks, a great number of different types of faults have been named.

fault block, n. A mass of rock bounded laterally by faults.

fiducial marks, n. Index marks (usually 4), rigidly connected with the camera lens through the camera body, which form images on the negative. The marks are adjusted so that the intersection of lines drawn between opposite fiducial marks defines the principal point.

finger lake, n. A long, narrow lake, usually one of a group, occupying a glacial trough in a mountainous region. A group of such lakes occurs in central New York State.

fiord, fjord, n. A deep, narrow, and steep-walled inlet of the sea formed in most instances by intense glacial erosion of a valley. Some fiords may have been invaded by ocean water after glaciation and owing to general subsidence of the land, but most are the direct result of erosion by tongues of ice that actually entered the ocean and moved along the bottom.

fissility, n. The tendency possessed by some rocks of splitting into thin sheets either along the bedding planes or along cleavage planes induced by fracture or flowage. The term is not applied to minerals.

flatirons, n. Triangular hills or rock layers carved by streams crossing a hogback ridge. Usually a flatiron is a resistant plate of dipping rock adhering to the dip-slope side of a hogback ridge and having a narrow apex at the top and a broad base where it dips below the surface.

foliation, n. The banding or lamination of metamorphic rocks as contrasted with the stratification of sediments. Foliation implies the ability to split along approximately parallel surfaces due to the parallel distribution of layers or lines of one or more conspicuous minerals in the rock. The layers may be smooth and flat or they may be undulating or strongly crumpled.

footwall, n. The lower or underlying wall of an inclined fault, vein, ore deposit, coal bed, etc. It is the surface upon which the miner obtains a footing, hence the name. Some writers apply the term only to the actual surface, others include the entire mass of rock underneath the fault or mineral deposit.

fracture, n. 1. The form or kind of surface obtained by breaking in a direction other than that of cleavage in crystallized minerals, and in any direction in massive minerals. 2. A crack in a rock large enough to be visible to the unaided eye. It may be a joint, fault, or fissure, but use of the term usually implies that the surfaces of the break are not in absolute contact.

gap, n. A short notchlike depression across a ridge, connecting lowlands on each side. A stream may or may not be present.

geyser, n. A special type of hot spring that throws forth intermittent jets of hot water and steam. The action results from the contact of ground water and rock or vapor hot enough to generate steam under conditions that prevent continuous circulation.

glacier, n. A body of ice (usually with some névé) consisting of recrystallized snow, lying wholly or largely on land, and showing evidence of present or former flow (Flint, 1947).

gneiss, n. A term originally applied to a more or less banded metamorphic rock with the mineral composition of granite. As now employed it designates a foliated metamorphic rock with no specific composition implied, but having layers that are mineralogically unlike and consisting of interlocking mineral particles that are mostly large enough to be visible to the eye.

graben, n. A long narrow block of the earth's crust that has been relatively depressed by normal faults along the sides. Although most grabens appear as topographic depressions or trenches the term refers to the fundamental structure and not the topography.

granite, n. A true granite is a visibly granular, crystalline rock of predominantly interlocking texture, composed essentially of alkalic feldspars and quartz. Feldspar is generally present in excess of quartz, and accessory minerals (chiefly micas, hornblende, or more rarely pyroxene) are commonly present.

granularity, n. The graininess of a developed photographic image, evident particularly on enlargement, that is due either to agglomerations of developed grains or to an overlapping pattern of grains.

gray body, n. A radiating surface whose radiation has essentially the same spectral energy distribution as that of a blackbody at the same temperature, but whose emissive power is less. Its absorptivity is nonselective. Also spelled grey body.

gray scale, n. A monochrome strip of shades ranging from white to black with intermediate shades of gray. The scale is placed in a setup for a color photograph and serves as a means of balancing the separation negatives and positive dye images.

groundwater, n. Subsurface water in a zone of saturation; phreatic water. The term is not meant to include a temporary saturated zone at or near the ground surface that is produced immediately after precipitation or by thawing. If a **water table** exists, ground water is the water below the water table. Subsurface water above the zone of saturation is vadose water.

gumbotil, n. A sticky, gray, plastic clay consisting chiefly of the mineral beidellite and formed under conditions of poor drainage from certain types of till. It may contain fragments of more or less altered rock that were originally mixed with the finer particles.

hardpan, n. A loosely used term designating a relatively hard or impervious layer beneath the soil or in the subsoil that offers exceptionally great resistance to digging or drilling. As the term is widely used in an essentially popular nontechnical manner for a variety of materials, it should always be accompanied by specific information as to occurrence and composition.

hard water, n. Water characterized by the presence of dissolved mineral salts, especially those of magnesium and calcium.

hogback, n. A long sharp ridge carved by differential erosion from a steeply dipping resistant layer or series of layers of igneous or sedimentary rock. The term is usually restricted to ridges carved from beds dipping at angles greater than 20 degrees; beds dipping at angles of less than 20 degrees give rise to cuestas. In a hogback the two slopes are approximately equal in steepness; in a cuesta the escarpment slope is notably steeper than the dip slope.

hoodoo, n. A natural rock pile or column of fantastic shape. Hoodoos usually occur in groups in regions of arid or semiarid climate. They are usually cut in horizontal rocks but their development is facilitated by joints and vertical fissures and by the presence of layers of varying hardness.

hornito, n. A small steep-sided spatter cone built by the accretion of clots of molten rock blown by hot gases from a vent in the solidified crust of a lava flow.

hummock, n. A rounded or conical knoll or hillock, or a rise of ground of no great extent above a level surface (Merriam-Webster).

humus, n. Dark-colored, organic, well-decomposed soil material consisting of the residues of plant and animal materials together with synthesized cell substances of soil organisms and various inorganic elements. The composition of humus varies widely according to the conditions of formation and with the stage of decomposition, but it generally consists largely of lignin and protein. It is highly colloidal, has high cation-exchange capacity, will absorb much water with swelling, shrinks on drying, but does not have the pronounced cohesion of clay-mineral colloids.

hydrostatic pressure, n. A state of stress in which the pressure exerted perpendicular to a reference surface does not vary with the orientation of the surface; a uniform pressure from all directions, as beneath the surface of a homogeneous fluid.

igneous rocks, n. Rocks formed by solidification of hot mobile rock material (**magma**) including those formed and cooled at great depths (**plutonic** rocks), which are crystalline throughout, and those which have poured out on the earth's surface in the liquid state or have been blown as fragments into the air (**volcanic** rocks). Igneous rocks comprise the bulk of the earth's crust. They occur in bodies with a variety of shapes such as flows, dikes, sills, and batholiths, and may be recognized in the field by their form, geologic relations, and usually by the texture and mineral content of the hand specimen. In a general way the igneous rocks are classified by the degree to which they have crystallized and by the kinds of minerals of which they are composed.

image-motion compensator, n. A device installed with certain aerial cameras to compensate for the forward motion of an aircraft while photographing ground objects. True image-motion compensation must be introduced after the camera is oriented to the flight track of the aircraft and the camera is fully stabilized.

impermeable, adj. Not permeable; not permitting passage, as of a fluid, through it substance; impassable; impervious (Merriam-Webster).

infiltration, n. 1. The permeation or percolation of liquid or gas among the grains or pores of a rock. 2. Material that infiltrates or is deposited by infiltration.

influent stream, n. A stream or stretch of a stream, not separated from the water table by an impervious layer, the surface of which stands higher than the water table in the locality through which it flows. Such a stream contributes water to the zone of saturation.

infrared (IR), n. Pertaining to or designating the portion of the EM spectrum with wavelengths just beyond the red end of the visible spectrum, such as radiation emitted by a hot body.

infrared, photographic, n. Pertaining to or designating the portion of the EM spectrum with wavelengths just beyond the red end of the visible spectrum, such as radiation emitted by a hot body (over 500° C); generally defined as from 0.7 to about 1.0 μm, or the useful limits of film sensitivities.

inselberg, n. A small mountain or hill standing above a desert pediment or peneplain. It is surrounded by more or less level rock surfaces or by debris derived from and overlapping its slopes.

interstitial water, n. Water above the zone of rock flowage that occurs in the interstices of the rocks.

intrazonal soils, n. Any of the great groups of soils with more or less well-developed soil characteristics that reflect the dominating influence of some local factor of relief, parent material, or age over the normal effect of the climate and vegetation (U. S. Dept. of Agriculture Yearbook, 1938).

joint, n. A fracture or parting plane along which there has been little if any movement parallel with the walls.

kame, n. A fluvioglacial deposit occurring as a mound, knob, or hillock in which one or more sides were in contact with the glacier ice. Kames are diverse in size, shape, and composition and generally, but not universally, consist of poorly sorted, poorly stratified material.

karst topography, n. A type of topography developed in a region of easily soluble limestone bedrock. It is characterized by vast numbers of depressions of all sizes, sometimes by great outcrops of fluted limestone ledges, sinks and other solution passages, almost total lack of surface streams, and large springs in the deeper valleys.

landform, n. A terrain feature formed by natural processes, which has a definable composition and range of characteristics that occur wherever that landform is found.

lava, n. A general name for molten rock poured out upon the surface of the earth by volcanoes and for the same material that has cooled and solidified as solid rock.

lava tunnels, n. Long caverns beneath the surface of a lava flow; in exceptional cases they may be 12 miles long. They are due to the withdrawal of magma from an otherwise solidified flow (Billings). Also called **lava tubes**.

levee, n. An embankment along the shore of a river or arm of the sea to prevent overflow. A natural levee is one built by a river in times of flood by deposition of material upon the banks. Natural levees are relatively low and wide.

liquid limit, n. (Soils) The liquid limit of a soil is the water content, expressed as a percentage of the weight of the oven-dried soil, at the boundary between the liquid and plastic states. The water content at this boundary is arbitrarily defined as the water content at which two halves of a soil cake will flow together for a distance of 0.5 in. along the bottom of the groove separating the two halves, when the cup is dropped 25 times for a distance of 1 cm. (0.3937 in.) at the rate of two drops per second (Am. Soc. Testing Mat., D423-54T).

lithification, n. The consolidation of liquid or loose materials into solid rock. The term thus includes the solidification of molten lava and the compaction or cementation of loose sediments into rock. See **consolidation**.

littoral currents, n. Currents that move generally parallel to and adjacent to the shore line.

loam, n. A soil or earth composed of a mixture of clay, silt, and sand. Clay loam and sandy loam are terms applied when clay or sand particles occur in excess of the usual amounts. Although organic matter is not always present in large amounts, the term loam usually implies a fertile soil.

loess, n. An unconsolidated or weakly consolidated sedimentary deposit composed dominantly of silt-sized rock and mineral particles deposited by wind. Loess may contain appreciable amounts of fine sand or clay, or both, but most of the particles are generally within the size range 0.01 to 0.05 mm. The mineral composition of loess is variable, depending on the source of material. The particles are mostly fresh and angular and are generally held together by calcareous cement or binder.

magma, n. Hot mobile rock material generated within the earth, from which igneous rock results by cooling and crystallization. It is usually conceived of as a pasty or liquid material, or a mush of crystals together with a noteworthy amount of liquid phase having the composition of silicate melt.

magnetometer, n. An instrument for measuring changes in the earth's magnetic field and used extensively in airborne geophysical surveying. Three broad categories of this device are used; namely, fluxgates, nuclear magnetic resonance detectors, and optically pumped systems. The former two generally operate at sensitivities of 1 gamma (10^{-5} gauss) and the latter type is used for high sensitivity surveys with a sensitivity approaching 1/100th of a gamma.

marble, n. In the strict geologic sense marble is the metamorphic form of limestone, including the dolomites and magnesian limestones, in which the calcite and the mineral dolomite are recrystallized. The term "marble" is also a trade name applied to any carbonate rock of good color and texture and hard enough to take a polish. True marble is a granular crystalline rock made up of calcite or dolomite grains cemented or intergrown and interlocking by additional calcite. It is white; the mottling, banding, and colors of ornamental varieties are due to impurities such as oxides of iron and organic matter. As the ordinary varieties are composed of calcite they will effervesce readily with dilute hydrochloric acid; the dolomitic varieties will effervesce only if the acid is applied to a fresh scratch.

marl, n. An old term loosely applied to a variety of materials most of which occur in loose, earthy, or friable deposits and contain a relatively high proportion of calcium carbonate or dolomite. In the coastal plain area of the United States the term has been applied to fine-grained calcareous sands, to deposits of unconsolidated shell, to more or less calcareous clays and silts, and to sediments containing glauconite. The term is also applied to the deposits of glacial lakes in which the percentage of clacium carbonate may be as low as 30 percent. Usually marl is gray, but yellow, green, blue, and black varieties are not uncommon. The calcareous nature of marl may be demonstrated by the effervescence in acid.

marsh, n. Marsh consists of wet, periodically flooded areas covered dominantly with grasses, cattails, rushes or other herbaceous plants. **Marsh** is mainly covered with grasses and grasslike plants, while **Swamp** is covered with trees (U. S. Dept. of Agriculture, Soil Survey Manual, 1951). Various types may be distinguished according to the influence of tides and presence of salt or fresh water.

massif, n. I. A principal mountain mass. 2. A block of the earth's crust bounded by faults or flexures and displaced as a unit without internal change; a large fault block of mountainous topography (Merriam-Webster). An intrusive body of moderate size; a diameter of between 10 and 20 miles may be regarded as typical. The term does not imply any opinion on the mode of formation or emplacement of that mass (Balk). The term "massive" is used by some writers in preference to "massif"

mass movement, n. A general term for a variety of processes by which large masses of earth material are moved by gravity either slowly or quickly from one place to another. The rapid translocation of material in avalanches, landslides, and related events is one phase of mass movement but the slower, less noticeable actions of earth flowage, soil-creep, and solifluction probably accomplish greater effects. Same as **mass wasting.**

mesa, n. A Spanish word meaning table. Most writers use the term to describe level or nearly level masses of land of relatively small extent that stand distinctly above the surrounding country and are bounded on all sides by steep erosion scarps

metamorphic rocks, n. One of the three great groups of rocks. Metamorphic rocks are formed from original igneous or sedimentary rocks through alterations produced by pressure, heat, or the infiltration of other materials at depths below the surface zones of weathering and cementation. Rocks that have undergone only slight changes are not usually considered to be metamorphic; for practical purposes the term is best applied to rocks in which transformation has been almost complete or at least has produced characteristics that are more prominent than those of the original rock. Metamorphic rocks are more or less reconstructed in place while remaining essentially solid. New minerals and textures come into being which are stable under the conditions that produce the change. Through metamorphism marble is produced from limestone and dolomite, quartzite from sandstone, and phyllites, gneisses, and schists from other types of rocks. The latter types are mostly streaked and banded, and slates tend to split in well-defined thin layers.

monadnock, n. An isolated high point, which, because of greater resistance to erosion or some other cause, remains projecting above the general surface of an almost level plain.

mosaic, n. An assemblage of overlapping aerial or space photographs or images whose edges have been matched to form a continuous pictorial representation of a portion of the earth's surface.

mosaic, controlled, n. A mosaic that is laid to ground control and in which prints are used that have been ratioed and rectified as shown to be necessary by the control.

mosaic, semi-controlled, n. A mosaic composed of corrected or uncorrected prints laid to a common basis of orientation other than ground control.

mosaic, strip, n. A mosaic consisting of one strip of photographs or images taken on a single flight.

mosaic, uncontrolled, n. A mosaic composed of uncorrected prints, the detail of which has been matched from print to print without ground control or other orientation.

mosaicking, v. Assembling of photographs or other images whose edges are cut and matched to form a continuous photographic representation of a portion of the earth's surface.

mottled, adj. Covered with irregular spots; said of negatives, prints, or image texture.

muck, n. 1. Fairly well decomposed organic material, relatively high in mineral content, dark in color, and accumulated under conditions of imperfect drainage (U.S. Dept. Agriculture Yearbook, 1938). 2. Broken rock or ore that results from blasting during mining operations.

multiband system, n. A system for simultaneously observing the same (small) target with several filtered bands, through which data can be recorded. The term is usually applied to cameras, may be used for scanning radiometers which utilize dispersant optics to split wavelength bands apart for viewing by several filtered detectors.

muskeg, n. In northern United States and Canada: a swamp or bog in an undrained or poorly drained area of alluvium or glacial till, or, more especially, in a rocky basin filled with water-saturated muck, decayed vegetal matter, and sphagnum moss incapable of sustaining much weight. The surface is commonly hummocky. In Alaska, more widely applied to any mossy and swampy ground regardless of topographic environment.

nadir, n. 1. That point on the celestial sphere vertically below the observer, or 180° from the zenith. 2. That point on the ground vertically beneath the perspective center of the camera lens.

near infrared, n. The preferred term for the shorter wavelengths in the infrared region extending from about 0.7 micrometers (visible red), to around 2 or 3 micrometers (varying with the author). The longer wavelength end grades into the middle infrared. The term really emphasizes the radiation reflected from plant materials, which peaks around 0.85 micrometers. It is also called solar infrared, as it is only available for use during the daylight hours.

névé, n. (French) Consolidated, granular snow not yet changed to glacier ice—a more or less dense and settled, although permeable, aggregate of medium to large individual grains and welded together by frequent alternations of melting and freezing on original snow crystals, and in which one finds numerous layers of ice. More generally—the overall snow cover which exists during the melting period and sometimes from one year to another (Miller). More or less synonymous with the German term **firn,** which is old snow that has outlasted one summer at least and been transformed into a dense heavy material as a result of frequent melting and freezing. There is some tendency to use **firn** for the material itself and **névé** for the area of snow accumulation above a glacier.

nunatak, n. An island of bedrock in a glacial field, a hill projecting through, and entirely surrounded by, the ice (Lahee, 1941).

ooze, n. Ooze in a sedimentary sense is any soupy deposit covering the bottom of any water body. Specifically, the term relates to more or less calcareous or siliceous deposits that cover extensive areas of the deep ocean bottom.

ore, n. Scientifically, the word "ore" comprehends all metal-bearing minerals which are commercial sources of the metals, percentages not being considered. Technically, an ore is a metal-bearing mineral, or aggregate of such minerals, mixed with barren matter, called "gangue," and capable of being mined at a profit. By contrast, where the element of profit is uncertain or impossible, the term "mineral deposit" may be used instead of "ore deposit."

organic, adj. In the biologic sense organic means pertaining to, or of the nature of organisms; plants and animals. In the chemical sense it means containing carbon as an essential ingredient.

organic deposits, n. This group of sediments embraces those deposits made up principally of material which originally formed part of the skeleton or tissues of living organisms.

outcrop, 1. **n.** A part of a body of rock that appears, bare and exposed, at the surface of the ground. In a more general sense the term applies also to areas where the rock formation occurs next beneath the soil, even though it is not exposed. 2. **v.** To appear exposed and visible at the surface of the earth. The noun outcrop is often used in this way as a verb in place of **crop out.**

outlier, n. An isolated mass of rock surrounded on all sides by outcrops of older rocks. The older beds need not rise to the level of the outlier and enclose it bodily, but as viewed on a geologic map that does not show the third dimension the older rocks appear to surround the outlier. Outliers may be isolated remnants or cappings of hills separated from a main cliff by erosion; they may be remnants of younger rocks on the centers of synclines; or they may be remnants on the downthrown sides of faults.

outwash, n. Detrital material removed from a glacier by meltwater and laid down by streams beyond the glacier itself.

overburden, n. A term used by geologists and engineers in several different senses. By some it is used to designate material of any nature, consolidated or unconsolidated, that overlies a deposit of useful materials, ores, or coal, especially those deposits that are mined from the surface by open cuts. As employed by others **overburden** designates only loose soil, sand, gravel, etc., that lies above the bedrock. The term should not be used without specific definition.

pahoehoe, n. The Hawaiian word for solidified lava that is characterized by a smooth, billowy, or ropy surface having a skin of glass a fraction of an inch to several inches thick. Pahoehoe is distinguished from the **aa** type by its smooth surface, and probably also by such internal characteristics as higher content of glass, and by the gas bubbles being more numerous, more spheroidal, and having smoother walls. Pahoehoe flows often contain large open or partially filled lava tubes which served as under surface conduits for the advancing lava. Lava flows often change from the pahoehoe type to aa as they advance, particularly if stirred up by passing over rough terrain. The change is apparently related to increased viscosity through cooling, loss of volatile constituents, and crystallization (After Macdonald).

panorama, n. A photograph of a wide expanse of terrain, normally taken on or near the earth's surface; more often a series of adjoining or overlapping photographs.

parallax, n. The apparent change in the position of one object, or point, with respect to another, when viewed from different angles. As applied to aerial photos, the term refers to the apparent displacement of two points along the same vertical line when viewed from a point (the exposure station) not on the same vertical line.

parallax wedge, n. A simplified stereometer for measuring object heights on stereoscopic pairs of photographs. It consists of two slightly converging rows of dots or graduated lines printed on a transparent templet which can be stereoscopically fused into a single row or line for making parallax measurements to the nearest 0:002 inch.

parent material, n. The disintegrated rock material, usually unconsolidated and unchanged or only slightly changed, that underlies and generally gives rise to the solum, or true soil, by the natural process of soil development from such material; also called **source material** (Merriam-Webster, addenda).

peat, n. A dark-brown to yellowish matted mass of semicarbonized plant material in which remains of leaves, twigs, stems, and roots are discernible. The plants entering into the composition of peat are those that grow in marshes and swamps, especially mosses of the genus *Sphagnum*. Although there may be a large proportion of inorganic material, true peat, because of its high carbon content, will ignite and burn freely when dry. It is much used in open-grate fires. The places where peat forms are known as **peat bogs** or **peat moors.**

pedalfer, n. A soil in which there has been a shifting of alumina and iron oxide downward in the soil profile but with no horizon of carbonate accumulation (U. S. Dept. of Agriculture Yearbook, 1938).

pediment, n. A gently inclined erosion surface of low relief typically developed in arid or semiarid regions at the foot of a receding mountain slope. The surface is usually underlain by rocks of the upland, and thus may cut across rocks of various compositions and structures. The pediment may be bare or mantled by a thin layer of alluvium which is in transit to the adjoining basin. Occasionally, the term is used for erosion surfaces developed upon earlier basin fill as well as upland rock. Some pediments resemble alluvial fans and aprons in outward form.

pellicular water, n. Water adhering as films to the surfaces of openings and occurring as wedge-shaped bodies at junctions of interstices in the zone of aeration above the capillary fringe (Tolman).

peneplain, peneplane, n. A peneplain is an extensive land area of very low relief produced in the ultimate stage of a normal cycle of subaerial erosion. The surface may be nearly level and generally bevels underlying rocks without regard to their hardness and structure, although isolated hills of resistant rock, called **monadnocks,** may rise somewhat above the surface of the plain. The altitude of the surface as formed is close to ultimate base level, sea level, but most of those seen today have been uplifted and dissected.

perched ground water, n. Ground water occurring in a saturated zone separated from the main body of ground water by unsaturated rock (Tolman).

percolation, n. The movement of water or other liquid, under hydrostatic pressure developed naturally underground, through the interstices of the rock or soil. The flow of water through large openings such as caves is not included.

permafrost, n. Permanently frozen ground or, more correctly, ground that remains below freezing temperatures for two or more years. Permafrost underlies much of the arctic and subarctic regions where it may form a continuous or discontinuous layer, or streaks and lenses within nonfreezing (**talik**) material. The top of the permafrost may lie beneath a few inches to several feet of ground that is frozen in winter but thawed in summer, the **active layer.** The bottom of the permafrost lies at depths ranging from a few feet to over a thousand feet. Also known in Europe as **tjale** (Scan.) and **Dauerfrostboden** (Ger.). **Pergelisol** has been proposed as a more exact term for perennially frozen ground.

permeable, adj. Pertaining to a rock or soil having a texture that permits passage of liquids or gases under the pressure ordinarily found in earth materials. Same as **pervious.**

petrifaction, n. The process of petrifying, or changing into stone; conversion of organic matter, including shells, bones, etc., into stone, or a substance of stony hardness. Petrifaction is produced by the infiltration of water containing dissolved mineral matter, as calcium carbonate, silica, etc., which replaces the organic material particle by particle, sometimes with original structure retained (Merriam-Webster).

physiographic province, n. A region that has unit structures, any specified kind, and unit geomorphic history (von Engeln).

piedmont, adj. Lying or formed at the foot of mountains.

placer, n. A deposit of sand or gravel, usually of fluvial origin, containing particles of gold or other valuable mineral.

plasticity, n. The property of a material that enables it to undergo permanent deformation without appreciable volume change or elastic rebound, and without rupture. Solids such as metals, minerals, and rocks possess plasticity under the proper conditions of rate of deformation, heat, and/or pressure. Clay and clay soils also exhibit plasticity. Plasticity of soils is a colloidal property, as mineral grains do not possess plasticity unless they are reduced to colloidal size. In soils, plasticity is a function of mineralogy, grain shape, and surface area and absorption of water films, as well as grain size. The term plastic flow or plastic deformation is used in geology to describe the stretch and flow of materials along cleavage and gliding planes and through the modification of space lattices under powerful confining pressures. There is no notable loss of cohesion.

plasticity index, n. (Soils) The range of water content, expressed as a percentage of the oven-dried soil, through which the soil is plastic. It is defined as the liquid limit minus the plastic Clayey soils, for example, have higher plasticity indexes than non-clay soils, because they remain plastic over a wider range of water content.

plastic limit, n. (Soils) The plastic limit of a soil is the lowest moisture content, expressed as a percentage of the weight of the oven-dried soil, at which the soil can be rolled into threads 1/8 inch in diameter without the thread breaking into pieces (Am. Soc. Testing Mat., Procedures for Testing Soils, D 424-39).

plateau, n. A tableland or flat-topped area of considerable extent elevated above surrounding country on at least one side.

playa, n. A Spanish word meaning literally shore or strand; a level or nearly level area that occupies the lowest part of a completely closed basin and that is covered with water at irregular intervals and for longer or shorter periods of time, forming a temporary lake. It is generally composed of evenly stratified beds of clay or silt that have been deposited in a temporary lake and that may contain large amounts of soluble salts. The surface is usually devoid of vegetation and may be either hard or soft and smooth or rough. Playas are frequently called "dry lakes" and other terms such as "alkali flats" or "salinas" are used either in a general sense or to designate playas of special types (Bryan).

pluton, n. A body of intrusive igneous rock of any shape or size. At first, such a general term may seem to be of little value, but when the data are insufficient it is far better to call a body a pluton than to use a term with a definite meaning (Billings).

podzol soil, podsol soil, n. A zonal group of soils having an organic mat and a very thin organic-mineral layer above a gray leached layer which rests upon an illuvial dark-brown horizon, developed under the coniferous or mixed forest, or underneath vegetation in a temperate to cold moist climate. Iron oxide and alumina, and sometimes organic matter, have been removed from the A and deposited in the B horizon. From the Russian for like, or near, ashes (U. S. Dept. of Agriculture Yearbook, 1938).

porosity, n. The property of a rock of containing interstices without regard to size, shape, interconnection, or arrangement of openings. It is expressed as a percentage of total volume occupied by interstices (Tolman).

quicksand, n. Saturated sand into which a heavy object easily sinks. The lack of bearing power may be due to seepage pressure of water percolating through the sand in an upward direction or it may be due to inherent instability of the structure of the sand, unaided by seepage pressure (Terzaghi and Peck).

radar, brute force, n. A radar imaging system employing a long physical antenna to achieve a narrow beamwidth for improved resolution.

radar, synthetic aperture (SAR), n. A radar in which a synthetically long apparent or effective aperture is constructed by integrating multiple returns from the same ground cell, taking advantage of the Doppler effect to produce a phase history film or tape that may be optically or digitally processed to reproduce an image.

radiometer, n. An instrument for quantitively measuring the intensity of EMR in some band of wavelengths in any part of the EM spectrum; usually used with a modifier, such as IR radiometer or microwave radiometer. Most radiometers measure the difference between the source radiation incident on the detector and a radiant energy (blackbody) reference. Comparison between the two is often achieved by mechanically interposing a reflective chopper, so that both sources can be viewed consecutively by the same detector, or by electrically switching, as in a microwave radiometer.

razorback, n. A sharp narrow ridge. There is little or no implication as to geologic structure, hence the term is not quite so specific as **hogback.**

"Red Beds", n. A name used in older reports for sedimentary rocks of the western United States, largely of Permian and Triassic age, that are prevailingly red in color. They are mostly shale and sandstone; gypsum is common.

regolith, n. The superficial mantle of unconsolidated debris that nearly everywhere covers the solid or "bed" rock and forms the surface of the land. It may consist of soil in the agronomic sense, broken rock, volcanic ashes, alluvium, aeolian sand, glacial material, or any other residual or transported product of rock decay.

residual soil, n. Soil formed in place from the disintegration products of the underlying rock.

roche moutonnee, n. The roch moutonnee is a knob of rock which regularly exhibits a gently inclined, striated and grooved, smoothed, or even polished slope on the end against which the ice impinged, the onset or stoss side. The long axis of the roche moutonnee, regardless of the structural characteristics of the rock, is oriented in the direction of ice movement. In sharp contrast with the onset end, the lee end is steep, even precipitous, and regularly has a rough, hackly surface. The irregularities of this surface are intimately related to the structure, chiefly the joint or other fracture pattern of the rock (von Engeln).

rock cleavage, n. As originally defined, rock cleavage is any structure by virtue of which a rock has the capacity to part along certain well-defined surfaces more easily than along others. Geologists usually employ the terms for secondary structures produced by metamorphism or deformation rather than for original structures such as bedding or flow structure.

rock flour, n. Very finely ground rock material of silt and clay size formed by the abrasive action of glaciers. It consists predominantly of angular, unweathered mineral fragments and thus does not possess the cohesion characteristic of fine-grained materials composed of clay minerals. Rock flour is common in glacial outwash stream and lake deposits.

rock glacier, n. An accumulation of broken rock fragments and finer-grained material that moves slowly but perceptibly downhill under its own weight and with the aid of interstitial ice. Rock glaciers usually occur in high mountains and somewhat resemble true ice glaciers in location, shape, and movement.

runoff, n. The water that flows off the surface of the land in visible surface streams; also the quantity of water that thus runs off. It may be expressed in gallons, cubic feet, or acre feet. Unless otherwise specified the term always implies surface drainage; the term groundwater runoff applies to water removed below the surface.

sag and swell, n. The undulating topography characteristic of sheets of till. The till usually is thick enough to completely obliterate all traces of former topography, and the postglacial drainage is then controlled by the surface configuration of the till.

salina, n. A salt marsh, pond, or lake, enclosed from the sea (Merriam-Webster). A place where deposits of crystalline salt are found.

salt dome, n. A more or less circular uplift of sedimentary rocks caused by the pushing up of a body of salt or gypsum. The salt usually occurs as a central core or plug and the surrounding sediments are pushed up at sharp angles. The movement of the salt is thought to result from its lower specific gravity in relation to the overlying rocks.

salt pan, n. Any undrained natural depression, as an extinct crater, tectonic basin, or the like, in which water gathers and leaves a deposit of salt on evaporation (Merriam-Webster).

scabland, n. The terms "scabland" and "scabrock" are used in the Pacific northwest to describe areas where denudation has removed or prevented the accumulation of a mantle of soil, and the underlying bare rock is exposed or is covered largely with its own coarse angular debris (Worcester).

scanner, n. 1. A device that scans, and by this means produces an image. 2. A radar set incorporating a rotatable antenna, or radiator element, motor drives, mounting, etc. for directing a searching radar beam through space and imparting target information to an indicator.

scree, n. A heap of rock waste at the base of a cliff or a sheet of coarse debris mantling a mountain slope. By most writers "scree" is considered to be a synonym of "talus" but it is a more inclusive term. Whereas talus is an accumulation of material at the base of a cliff, scree also includes loose material lying on slopes without cliffs.

sedimentary rock, n. Sedimentary rocks are those composed of sediment: mechanical, chemical, or organic. They are formed through the agency of water, wind, glacial ice, or organisms and are deposited at the surface of the earth at ordinary temperatures. The materials from which they are made must originally have come from the disintegration and decomposition of older rocks, chiefly igneous. The distinguishing feature of many sedimentary rocks is their arrangement in parallel layers. Along with this are certain so-called initial or original structures such as ripple marks, cross-bedding, mud cracks, fossils, and raindrop impressions. Two broad categories may be recognized, the classic or fragmental rocks such as sandstone, conglomerate, and shale, formed from other rocks, and the chemical and biochemical sediments; formed by precipitation from solutions; as, limestone, gypsum, and salt.

seepage, n. Water or other liquid that has oozed or seeped through porous material. Also, the amount of such material expressed in terms of volume.

shear, n. An action or stress, resulting from applied forces, which causes or tends to cause two contiguous parts of a body to slide relative to each other in a direction parallel to their plane of contact; also called **shearing stress**

(Merriam-Webster). In geology shear is often used synonymously with couple, and may be employed to describe the sliding of rocks past each other along fractures (Billings).

shear strength, n. The internal resistance offered to shear stress, as measured by the maximum shear stress, based on original area of cross section, that can be sustained without failure. The criteria of failure, such as fracture, percent of strain, or continuous flow should be defined in designating shear strength. The shear strength of a paticular rock or soil varies greatly, depending upon the rate of loading, temperature, confining pressure, presence of pore fluids, size, and shape of the specimen. In soil mechanics, the shear strength of a soil is measured by the shear stress that may be sustained without excessive deformation under specified conditions of rate of loading, confining pressure, and pore water pressure.

shrinkage limit, n. (Soils) The shrinkage limit of a soil is that moisture content, expressed as a percentage of the weight of the oven-dried soil, at which a reduction in moisture content will not cause a decrease in the volume of the soil mass, but at which an increase in moisture content will cause increase in the volume of the soil mass (Am. Soc. Testing Mat., Procedures for Testing Soils, D 427-39).

slaking, n. Loosely, the crumbling and disintegration of earth materials when exposed to air or moisture. More specifically, the breaking up of dried clay when saturated with water, due either to compression of entrapped air by inwardly migrating capillary water or to the progressive swelling and sloughing off of the outer layers. Also, the disintegration of tunnel walls in swelling clay due to inward movement and circumferential compression.

soil, n. The term "soil" has various shades of meaning depending upon the general field in which it is being considered. To the agronomist and pedologist the soil is the superficial stratum of the regolith consisting of: (1) mineral matter that has originated from rocks by the action of weathering processes; (2) organic matter both living and dead that has usually accumulated over a period of time through the biologic activities and death of plants and animals; (3) soil moisture, containing mineral and organic matter in colloidal state or true solution; and (4) soil air. As all these factors are essential to the growth of vegetation, soil is, in the opinion of the pedologist, that earth material which has been so modified and acted upon by physical, chemical, and biological agents that it will support rooted plants. Most geologists adhere to the basic concept of the pedological definition of soil, that it should be capable of supporting plant life, but many consider the term to include all loose unconsolidated material above the bedrock that has in any way been altered or weathered from its original condition. In this sense the term is nearly synonymous with **regolith**. Others consider a rather indefinite "upper few feet" of the regolith as the soil. To the engineer the term soil is synonymous with the word regolith.

soil moisture, n. Pellicular water of the soil zone. It is divided by the soil scientist into available and unavailable moisture. Available moisture is water easily abstracted by root action and is limited by field capacity and the wilting coefficient. Unavailable moisture is water held so firmly by adhesion or other forces that it cannot usually be absorbed by plants rapidly enough to produce growth. It is commonly limited by the wilting coefficient (Tolman).

solifluction, n. The slow downhill flowage or creep of soil and other loose material that is saturated with water. It is especially active in subarctic regions and some high mountains during melting periods.

solum, n. The upper part of the soil profile, above the parent material, in which the processes of soil formation are taking place. In mature soils this includes the A and B horizons, and the character of the material may be, and usually is, greatly unlike that of the parent material beneath.

spectrometer, n. A device to measure the spectral distribution of EMR. This may be achieved by a dispersive prism, grating, circular interference filter with a detector placed behind a slit. If one detector is used, the dispersive element is moved as to sequentially pass all dispersed wavelengths across the slit. In an interferometer-spectrometer, on the other hand, all wavelengths are examined all the time, the scanning effect being achieved by rapidly oscillating two, partly reflective (usually parallel) plates so that interference fringes are produced. A Fourier transform is required to reconstruct the spectrum.

steptoe, n. A bedrock island protruding through a lava flow.

stereoscope, n. A binocular optical instrument for assisting the observer to view two properly oriented photographs or diagrams to obtain the mental impression of a three-dimensional model.

stress, n. The intensity at a point in a body of the internal forces or components of force which act on a given plane through the point. Stress is expressed in force per unit area (pounds per square inch, kilograms per square millimeter, etc.). The stress or components of stress acting perpendicular to a given plane is called the normal stress. A normal stress may be either a tensile stress or a compressive stress depending upon the nature of the force. The stress or component of stress acting tangential to the plane is called the shearing stress.

subtractive color process, n. A method of

creating essentially all colors through the subtraction of light of the 3 subtractive color primaries (cyan, magenta, and yellow) in various proportions through use of a single white light source.

sun synchronous, n. An earth satellite orbit in which the orbital plane is near polar and the altitude such that the satellite passes over all places on earth having the same latitude twice daily at the same local sun time.

swamp, n. Swamp consists of naturally wooded areas, all or most of which are covered with water much of the time. **Tidal swamp** is influenced by salty tidal water, and **freshwater swamp** is influenced by nontidal fresh water (U.S. Dept. of Agriculture, 1951).

swell-and-swale, n. The type of topography characteristic of the ground moraine of a continental glacier. Gentle, well-rounded hills alternate with corresponding subdued depression.

table mountain, n. A mountain with a flat, tablelike top.

tectonic, adj. Pertaining to the rock structures and external forms resulting from the deformation of the earth's crust.

thermal infrared, n. The preferred term for the middle wavelength ranges of the IR region, extending roughly from 3 micrometers at the end of the near infrared, to about 15 or 20 micrometers where the far infrared commences. In practice the limits represent the envelope of energy emitted by the earth behaving as a grey body with a surface temperature around 290° K (27° C). Seen from any appreciable distance, the radiance envelope has several brighter bands corresponding to windows in the atmospheric absorption bands. The thermal band most used in remote sensing extends from 8-13 micrometers.

tidal marsh, n. A type of marsh that occurs only along coasts, in shallow bays and lagoons, and is alternately submerged and laid bare by the tides.

till, n. That part of glacial drift deposited directly by ice, without transportation or sorting by water, consisting generally of an unstratified, unsorted, unconsolidated to moderately consolidated, heterogeneous mixture of clay, sand, gravel, and boulders. Also called **boulder-clay**. The wide range in grain size is typical, although the dominant size and mineralogic and lithologic composition is determined in large part by the rocks from which the till was derived.

top-set beds, n. The nearly horizontal layers of sediment deposited on top of a growing delta.

topsoil, n. A general term applied to the surface portion of the soil including the average plow depth (surface soil) or the A horizon, where this is deeper than plow depth. It cannot be precisely defined as to depth or productivity except in reference to a particular soil type (U.S. Dept. of Agriculture Yearbook, 1938).

toughness, n. Toughness is best described as resistance to impact. This property depends on the interlocking condition of the mineral constituents and on the individual tenacity of the minerals.

trap, n. A term originally applied to igneous rocks that are neither coarsely crystalline nor cellular. It is still used in a general and noncommittal sense by engineers and geologists for dark-colored, heavy, igneous rocks composed essentially of ferromagnesian minerals, basic feldspars, and little or no quartz. Among the specific rock-types included under the term are: basalt, peridotite, diabase, and fine-grained gabbro. The term is commonly employed for such rocks used in road making.

tsunami, tunami, n. A sea wave generated by submarine earthquakes, landslides, or volcanic action. Characteristics are great speed of propagation, long wave length, and low observable amplitude on the open sea, though tsunamis may pile up to great heights on entering shallow water.

unconformity, n. A surface that separates one set of rocks from another younger and bedded set, and that represents a period of nondeposition, weathering, or erosion, either subaerial or subaqueous, prior to the deposition of the younger set. Although most commonly applied to an interruption in the continuity of a sequence of sedimentary rocks, the term is also used for breaks in a sequence of layered volcanic or pyroclastic rocks, if erosion or weathering is evident, and for the break between eroded igneous or metamorphic rocks and younger bedded rocks. Two general types of unconformity are recognized. Where beds above and below the break are parallel the term **disconformity** is applied. If the beds below the break are not parallel to those above, or if the rocks below have lost their beddning by metamorphism or are plutonic, the break is a **nonconformity.** Unconformities are local or regional according to their areal extent. The lapse in time represented by the break may be long or short, though there is some tendency to call disconformities of very small time value **diastems,** and to apply **disconformity** to breaks represented elsewhere by a rock unit or units of formation value.

valley train, n. Sand and gravel deposited in a valley by drainage from a valley glacier.

varves, n. The regular layers or alternations of material, in sedimentary deposits, that are due to annual seasonal influences. Each varve represents the deposition during a year and consists ordinarily of a lower part deposited in summer and an upper finer-grained part deposited in the winter. Varves of silt- and clay-size material occur abundantly in glacial lake sediments, and varves are believed to have been recognized in certain marine shales and in slates.

volcano, n. An opening in the earth from which hot rocks and other material are expelled. If the material accumulates around the vent it will build up a cone which may reach the proportions of a large mountain. This cone may also be called a volcano. Volcanoes may be classified according to the material ejected, the internal structure, the stage of erosion of the cone, or the predominant mode of eruption.

volcanism, n. The phenomena related to or resulting from volcanic action.

wash, n. A piece of ground washed by the action of a sea or river; the dry bed of an intermittent stream, sometimes at the bottom of a canyon. Also, loose or eroded surface material of the earth transported and deposited by running water.

watershed, n. The whole area that contributes water to a particular river, lake or basin; also, the divide or height of land from which the natural drainage flows in opposite directions.

water table, n. A water table is the upper surface of a zone of saturation except where that surface is formed by an impermeable body. No water table exists where the upper surface of a zone of saturation is formed by an impermeable body (Meinzer, 1923).

weathering, n. The various chemical and mechanical processes acting at or near the surface of the earth to bring about the disintegration, decomposition, and comminution of rocks. Weathering is accomplished in place; no transportation of the loosened or altered particles is involved.

yazoo, n. The phenomenon of a deferred tributary junction (sometimes referred to as a yazoo) is another common feature where trunk streams have extensive flood plains. As a meander enlarges and moves progressively downstream by meander sweep it carries with it the mouth of a tributary. The tributary is thus compelled to a prolonged course over the back lands parallel to the main stream (von Engeln).

zonal soil, n. Any of the great groups of soils having well-developed soil characteristics that reflect the influence of the active factors of soil genesis—climate and living organisms, chiefly vegetation (U.S. Dept. of Agriculture Yearbook, 1938).

APPENDIX A

SOIL CLASSIFICATION SYSTEMS

Soil classification systems may be confusing to the novice; published descriptions are often outdated as a result of their continuous updating or minor revisions. The most widely used systems are described in this section, but professionals wishing to use them should obtain current descriptions from the parent organization. The Unified soil classification system is emphasized since it categorizes soils in relation to engineering purposes.

THE WORLD SEVENTH APPROXIMATION

The most recent soil classification system is the World Seventh Approximation. The National Cooperative Soil Survey of the U.S. Department of Agriculture has adopted this classification system, and the U.S. Soil Conservation Service supplies any information available on it and tries to answer questions about changes in the classification.

The Seventh Approximation system lists 10 basic soil orders. The orders represent broad climatic groupings of soils (except for entisols and histosols, which occur in many different climates). Soil suborders (nearly 40) give further definition in relation to the presence or absence of waterlogging, or differences resulting from climate or vegetation. The suborders are further divided into great groups according to the presence or absence of soil horizons and their general arrangement. Each great group is then divided into subgroups which represent either the central (typic) segment of one great group or integrades, having properties of one major great group and minor properties of other great groups. Soil families are derived from the subgroups and based upon properties important to the propagation of vegetation; these include soil texture, temperature, mineralogy, permeability, consistency, and horizon thickness. Lastly, families are divided into series that have genetic horizons formed from a parent material and are similar in soil profile arrangement except for surface texture. Soil series names are determined by the name of the town nearest which the soil was originally identified. Names such as Gloucester, Merrimac, Miami, Essex, Genesee, and Sudbury are typical. The soil series are usually further divided and classified into soil types by making distinctions of surficial texture; that is, a soil type is described by the soil series name followed by a textural description, for example, Merrimac fine sandy loam. Planners and engineers are usually concerned with the soil series and soil type levels, because the detailed information and profile description they provide are needed to determine capabilities for various land users.

U.S. DEPARTMENT OF AGRICULTURE

Since 1899 approximately one-third of the United States has had soil surveys prepared by the Soil Survey Division of the U.S. Department of Agriculture. The published maps and reports provide soils information primarily for farmers and ranchers, on management and growth yield characteristics of crops, but the information may also be useful for engineers and planners for site selection and evaluation and for prospective buyers and developers of land. These surveys, however, are not intended to supply the detailed soils information needed for specific engineering applications.

The published soil surveys of the department vary considerably in accuracy and mapping technique. The older surveys are more general and have many inaccuracies, but they may be of some value in verifying large-scale photointerpretation projects or indicating broad soil groupings. More recent surveys, especially those completed since the late 1950s, contain detailed pedological descriptions and interpreted capability measures which may be of value to the planner-engineer. Capability measures vary from county to county but generally include crop yield estimates for defined management techniques, potential woodland production, rangeland suitabilities, engineering characteristics, community planning interpretations, soil capabilities for drainage and irrigation, and capabilities for recreation and wildlife habitat. The most recent surveys are mapped and published on an aerial photograph base to allow rapid and accurate location of a particular soil type in the field.

The U.S. Soil Conservation Service has recently revised its mapping format to include more detailed soil capability interpretations of the kind that might be useful to community planners. However, it is not clear how their interpretations are derived, and therefore their conclusions are sometimes inaccurate. For example, soil capabilities are given for single-unit houses, but the types of foundations are never defined; yet the capability of soils to support houses relates specifically to the type and depth of foundation used. There are also many equally important factors other than soils that should be considered for the location of land uses—potential environmental impact on water surface and subsurface systems, wildlife habitat, quality stands of vegetation and sensitive visual zones, just to mention a few—and these are not considered in the judgment. It is therefore important to use these capability interpretations only as general indicators, since technological developments and the consideration of other factors can quickly outdate and limit the capability interpretations.

Published county or area soil surveys for a specific locale are available free from local Soil Conservation Service offices to land users or professional workers in the areas surveyed. All surveys in print can be obtained free from the Information Division, Soil Conservation Service, Washington, D.C. 20250, or purchased from the Superintendent of Documents, U.S. Government Printing Office, Washington, D.C. 20402. Also, many libraries, colleges, and universities have copies of the reports and surveys covering their own county and state. Appendix C lists the available soils surveys.

The classification system used by the Department of Agriculture is the World Seventh Approximation, and the county soil maps are presented at the series and types levels. The textural descriptions are defined by a textural triangle (Figure A.1). This diagram is used throughout the landform sections of this book to illustrate the ranges of soil texture associated with the parent material of each landform.

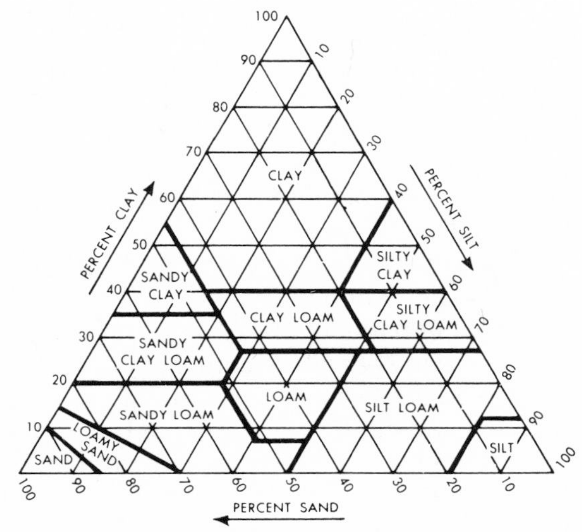

Figure A.1. U.S. Department of Agriculture soil textural triangle.

THE AMERICAN ASSOCIATION OF STATE HIGHWAY OFFICIALS (AASHO)

The American Association of State Highway Officials classification system is widely used by highway engineers. The system was developed by the U.S. Bureau of Public Roads in 1928 and has been revised many times, with the latest revision being in 1945 by the Highway Research Board.

The classifications are based upon observed field performances of soils under highway pavements, and the soils are grouped according to their load-carrying capacities. There are seven major soil groups, ranging from A-1, which has the best subgrade bearing capacity, to A-7, which offers the least stable foundation. The following is a summary of the various groupings and subgroupings of the AASHO classification system.

Granular Materials (35% or Less Passing through a No. 200 Sieve).

A-1 Well graded mixtures of gravel, ranging from coarse to fine, with or without a non-plastic or slightly plastic soil binder
- *A-1-a* Predominantly stone fragments or gravel
- *A-1-b* Predominantly coarse sands

A-2 Mainly granular materials between A-1 and A-3
- *A-2-4* Predominantly A-2 but having binder
- *A-2-5* Characteristics similar to soil types A-4 and A-5

A-2-6 Predominantly A-2 but having binders

A-2-7 Similar to the A-6 and A-7 groups

A-3 Predominantly sandy but deficient in coarse material and soil binder

Silt-Clay Materials (More Than 35% Passing through a No. 200 Sieve).

A-4 Predominantly silt content with moderate to small amounts of coarse material and small amounts of sticky colloidal clay

A-5 Similar to A-4 soils except that very poorly graded soils containing materials such as mica and diatoms are included

A-6 Predominantly clay content containing moderate to negligible amounts of coarse material

A-7 Predominantly clay soils but more elastic than A-6 because of the presence of one-sized silt particles, organic matter, mica flakes, or lime carbonate

A-7-5 Similar to A-7 soils but having moderate plasticity indexes in relation to the liquid limit; may be highly elastic as well as subject to considerable volume change

A-7-6 Similar to A-7 soils but having high plasticity indexes in relation to the liquid limit and subject to extremely high volume change

THE UNIFIED SYSTEM

During World War II, Arthur Casagrande of Harvard University developed the Unified soil classification system for the U.S. Army Corps of Engineers. In 1952 the U.S. Bureau of Reclamation, in cooperation with the Corps of Engineers and Dr. Casagrande, revised the original classification system to include capabilities for embankments and foundations and for roads and airfields. The result is the current Unified soil classification system (see Table 3.3).

The Unified system places soils in three divisions: coarse-grained, fine-grained, and organic. Coarse-grained soils contain less than 50% fines that pass through a no. 200 sieve; fine-grained soils contain greater than 50% fines that pass through a no. 200 sieve. The organic soils are identified by visual examination. The 15 soil groups in the three divisions are designated by easily remembered letter symbols derived from descriptive soil unit terms relating to the liquid limit (high or low), major soil textural fraction, and relative gradation

Table A.1. AASHO classification of soils and general ratings

General classification	Granular materials (35% or less passing no. 200 sieve)							Silt-clay materials (more than 35% passing no. 200 sieve)			
	A-1		A-3	A-2				A-4	A-5	A-6	A-7
Group classification	A-1-a	A-1-b		A-2-4	A-2-5	A-2-6	A-2-7				A-7-5[*], A-7-6[†]
Sieve analysis											
Percent passing											
No. 10	50 max.	—	—	—	—	—	—	—	—	—	—
No. 40	30 max.	50 max.	51 min.	—	—	—	—	—	—	—	—
No. 200	15 max.	25 max.	10 max.	35 max.	35 max.	35 max.	35 max.	36 min.	36 min.	36 min.	36 min.
Characteristics of fraction passing no. 40 sieve											
Liquid limit	—	—	—	40 max.	41 min.	40 max.	41 min.	40 max.	41 min.	40 max.	41 max.
Plasticity index	6 max.	6 max.	N.P.[‡]	10 max.	10 max.	11 min.	11 min.	10 max.	10 max.	11 min.	11 min.
Group index	0	0	0	0	0	4 max.	4 max.	8 max.	12 max.	16 max.	20 max.
Usual types of significant constituent materials	Stone fragments gravel and sand		Fine sand	Silty or clayey gravel and sand				Silty soils		Clayey soils	
General rating as subgrade	← Excellent to good →							← Fair to poor →		← Very poor →	

*Plasticity index of A-7-5 subgroup is equal to or less than liquid limit minus 30.

†Plasticity index of A-7-6 subgroup is greater than liquid limit minus 30.

‡Nonplastic.

(well-graded or poorly graded) of the soil in question.

Coarse-Grained Soils. Coarse-grained soils are divided into gravels and gravelly soils (G) and sands or sandy soils (S). Gravelly soils have particles that are mostly too large to pass through a no. 4 sieve; most sandy soils can pass through a no. 4 sieve. The eight major groups and several subgroups of the Unified classification of coarse soils are listed below, with the letter symbol used in making graphic sections and in mapping.

GW (Well-graded gravels). These soils consist of clean, well-graded gravels containing few or no fines. Any fines that may be present do not interfere with internal drainage characteristics or strength properties.

GP (Poorly graded gravels). These soils contain clean, poorly graded gravels or sand-gravel mixtures referred to as gap-graded soils. These soils do not have any appreciable content of fines that would affect drainage or strength characteristics.

GM (Silty gravels). These predominanty gravel mixtures contain fines of which more than 12% pass through a no. 200 sieve. The fines have little or no plasticity. Well-graded mixtures of gravel-sand-silt, along with poorly graded mixtures of silty gravel, can be found in this group. There are two subdivisions (*d* and *u*) of GM soils; these allow one to make finer distinctions between the liquid limit and the plasticity index and are primarily for use in road and airfield capability ratings. The suffix *d* is used when the liquid limit is 28 or less and the plasticity index is 6 or less; the suffix *u* is used when the liquid limit is greater than 28.

GC (Clayey gravels). These predominantly gravel mixtures contain fines of which more than 12% pass through a no. 200 sieve. The fines are claylike and have low to high levels of plasticity. Well-graded mixtures of gravel-sand-clay and poorly graded materials of clayey gravels are classified in this group.

SW (Well-graded sands). This group contains soils consisting primarily of clean, well-graded sands or gravelly sands with few or no fines. Any fines that are present do not affect the internal drainage or strength properties.

SP (Poorly graded sands). This group contains predominantly sandy soils consisting of clean sands or gravelly sands with few or no fines. The fines do not affect the internal drainage or strength properties.

SM (Silty sands). This group contains fines of which more than 12% pass through a no. 200 sieve and which have little or no plasticity. Well-graded and poorly graded mixtures of sand may be present. As with the GM group, the SM soils can be subdivided into two minor categories related to their liquid limit and plasticity index. The suffix *d* is used when the liquid limit is 28 or less and the plasticity index is 6 or less; the suffix *u* is used when the liquid limit is greater than 28.

SC (Clayey sands). These sands contain a minimum of 12% fines that can pass through a no. 200 sieve, are claylike, and have low to high levels of plasticity. This group includes both well-graded sands and poorly graded mixtures that fulfill the other characteristics.

Fine-Grained Soils. The Unified system distinguishes seven categories of fine-grained soils by their textural composition, organic content, and liquid limit. Silts are defined as fine-grained soils that plot below the A line on the plasticity chart; clays are those soils that plot above the A line (Figure A.2). An exception to this rule are organic clays, which plot below the A line. The liquid limit of 50 divides these soils into groups of high (H) or low (L) liquid limit and related plasticity.

ML (Inorganic silts and very fine sands). Low plasticity. Soils in this group have liquid limits below 50 and lie below the A line on the plasticity chart. Included in this group are inorganic silts, very fine sands, rock flour, silty or clayey fine sands, or clayey silts, all having slight plasticity. Some kaolin clays and loess soils are included in this group.

CL (Inorganic clays). Low to medium plasticity. The soils of this group have liquid limits below 50 and lie above the A line on the plasticity chart, having low to medium plasticity. Inorganic clays, gravelly clays, sandy clays, silty clays, and lean clays are included. Many of the glacial soils of the north central United States are classified in this group.

OL (Organic silts). Low plasticity. These fine-grained soils have liquid limits below 50 and are plotted below the A line on the plasticity chart. Organic silts and organic silty clays of low plasticity are found in this group.

MH (Inorganic silts). Low plasticity. Soils in

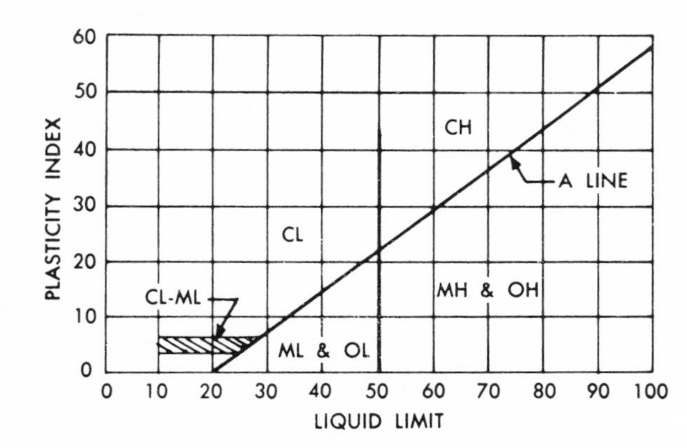

Figure A.2. A Line on the plasticity index-liquid limit chart, and associated units of the Unified classification system.

Table A.2. Characteristics and ratings of the Unified categories pertinent to roads and airfields*

Major Divisions (1)	(2)	Symbol: Letter (3)	Symbol: Hatching (4)	Symbol: Color (5)	Name (6)	Value as Subgrade When Not Subject to Frost Action (7)	Value as Subbase When Not Subject to Frost Action (8)	Value as Base When Not Subject to Frost Action (9)	Potential Frost Action (10)
COARSE-GRAINED SOILS	GRAVEL AND GRAVELLY SOILS	GW		Red	Well-graded gravels or gravel-sand mixtures, little or no fines	Excellent	Excellent	Good	None to very slight
		GP		Red	Poorly graded gravels or gravel-sand mixtures, little or no fines	Good to excellent	Good	Fair to good	None to very slight
		GM d		Yellow	Silty gravels, gravel-sand-silt mixtures	Good to excellent	Good	Fair to good	Slight to medium
		GM u		Yellow	Silty gravels, gravel-sand-silt mixtures	Good	Fair	Poor to not suitable	Slight to medium
		GC		Yellow	Clayey gravels, gravel-sand-clay mixtures	Good	Fair	Poor to not suitable	Slight to medium
	SAND AND SANDY SOILS	SW		Red	Well-graded sands or gravelly sands, little or no fines	Good	Fair to good	Poor	None to very slight
		SP		Red	Poorly graded sands or gravelly sands, little or no fines	Fair to good	Fair	Poor to not suitable	None to very slight
		SM d		Yellow	Silty sands, sand-silt mixtures	Fair to good	Fair to good	Poor	Slight to high
		SM u		Yellow	Silty sands, sand-silt mixtures	Fair	Poor to fair	Not suitable	Slight to high
		SC		Yellow	Clayey sands, sand-clay mixtures	Poor to fair	Poor	Not suitable	Slight to high
FINE-GRAINED SOILS	SILTS AND CLAYS LL IS LESS THAN 50	ML		Green	Inorganic silts and very fine sands, rock flour, silty or clayey fine sands or clayey silts with slight plasticity	Poor to fair	Not suitable	Not suitable	Medium to very high
		CL		Green	Inorganic clays of low to medium plasticity, gravelly clays, sandy clays, silty clays, lean clays	Poor to fair	Not suitable	Not suitable	Medium to high
		OL		Green	Organic silts and organic silt-clays of low plasticity	Poor	Not suitable	Not suitable	Medium to high
	SILTS AND CLAYS LL IS GREATER THAN 50	MH		Blue	Inorganic silts, micaceous or diatomaceous fine sandy or silty soils, elastic silts	Poor	Not suitable	Not suitable	Medium to very high
		CH		Blue	Inorganic clays of high plasticity, fat clays	Poor to fair	Not suitable	Not suitable	Medium
		OH		Blue	Organic clays of medium to high plasticity, organic silts	Poor to very poor	Not suitable	Not suitable	Medium
HIGHLY ORGANIC SOILS		Pt		Orange	Peat and other highly organic soils	Not suitable	Not suitable	Not suitable	Slight

*After The Asphalt Institute, *Soils Manual,* College Park, Maryland, Table VI-2, 1969.

†LL, Liquid limit.

Compressibility and Expansion (11)	Drainage Characteristics (12)	Compaction Equipment (13)	Unit Dry Weight lb. per cu. ft. (14)	Typical Design Values	
				CBR (15)	Subgrade Modulus k lb. per cu. in. (16)
Almost none	Excellent	Crawler-type tractor, rubber-tired roller, steel-wheeled roller	125-140	40-80	300-500
Almost none	Excellent	Crawler-type tractor, rubber-tired roller, steel-wheeled roller	110-140	30-60	300-500
Very slight	Fair to poor	Rubber-tired roller, sheepsfoot roller; close control of moisture	125-145	40-60	300-500
Slight	Poor to practically impervious	Rubber-tired roller, sheepsfoot roller	115-135	20-30	200-500
Slight	Poor to practically impervious	Rubber-tired roller, sheepsfoot roller	130-145	20-40	200-500
Almost none	Excellent	Crawler-type tractor, rubber-tired roller	110-130	20-40	200-400
Almost none	Excellent	Crawler-type tractor, rubber-tired roller	105-135	10-40	150-400
Very slight	Fair to poor	Rubber-tired roller, sheepsfoot roller; close control of moisture	120-135	15-40	150-400
Slight to medium	Poor to practically impervious	Rubber-tired roller, sheepsfoot roller	100-130	10-20	100-300
Slight to medium	Poor to practically impervious	Rubber-tired roller, sheepsfoot roller	100-135	5-20	100-300
Slight to medium	Fair to poor	Rubber-tired roller, sheepsfoot roller; close control of moisture	90-130	15 or less	100-200
Medium	Practically impervious	Rubber-tired roller, sheepsfoot roller	90-130	15 or less	50-150
Medium to high	Poor	Rubber-tired roller, sheepsfoot roller	90-105	5 or less	50-100
High	Fair to poor	Sheepsfoot roller, rubber-tired roller	80-105	10 or less	50-100
High	Practically impervious	Sheepsfoot roller, rubber-tired roller	90-115	15 or less	50-150
High	Practically impervious	Sheepsfoot roller, rubber-tired roller	80-110	5 or less	25-100
Very high	Fair to poor	Compaction not practical			

Note:

1. Column 3, division of GM and SM groups into subdivisions of d and u are for roads and airfields only. Subdivision is on basis of Atterberg limits; suffix d (e.g., GMd) will be used when the liquid limit is 25 or less and the plasticity index is 5 or less; the suffix u will be used otherwise.
2. In column 13, the equipment listed will usually produce the required densities with a reasonable number of passes when moisture conditions and thickness of lift are properly controlled. In some instances, several types of equipment are listed because variable soil characteristics within a given soil group may require different equipment. In some instances, a combination of two types may be necessary.
   a. Processed base materials and other angular materials. Steel-wheeled and rubber-tired rollers are recommended for hard, angular materials with limited fines or screenings. Rubber-tired equipment is recommended for softer materials subject to degradation.
   b. Finishing. Rubber-tired equipment is recommended for rolling during final shaping operations for most soils and processed materials.
   c. Equipment size. The following sizes of equipment are necessary to assure the high densities required for airfield construction:

      Crawler-type tractor—total weight in excess of 30,000 lb.

      Rubber-tired equipment—wheel load in excess of 15,000 lb., wheel loads as high as 40,000 lb. may be necessary to obtain the required densities for some materials (based on contact pressure of approximately 65 to 150 psi).

      Sheepsfoot roller—unit pressure (on 6- to 12-sq.-in. foot) to be in excess of 250 psi and unit pressures as high as 650 psi may be necessary to obtain the required densities for some materials. The area of the feet should be at least 5 per cent of the total peripheral area of the drum, using the diameter measured to the faces of the feet.
3. Column 14, unit dry weights are for compacted soil at optimum moisture content for modified AASHO compaction effort.
4. In column 15, the maximum value that can be used in design of airfields is, in some cases, limited by gradation and plasticity requirements.

this category have liquid limits above 50 and are plotted below the A line on the plasticity chart. Included are inorganic silts; micaceous or diatomaceous, fine sandy or silty soils; and elastic silts.

CH (Inorganic clays). High plasticity. These soils have liquid limits above 50 and are plotted above the A line on the plasticity chart. Fat clays, gumbo clays, bentonite, and certain volcanic clays are found in this group.

OH (Organic clays). Medium to high plasticity. Organic clays and organic silts of high plasticity are found in this category. They have liquid limits above 50 and are plotted below the A line on the plasticity chart.

Organic Soils. Organic soils, or peat, are characterized by having greater than 50% organic debris. They include peat, humus, grass, leaves, branches, and other decomposed or partially decomposed organic matter.

Pt (Peat and highly organic soils). Highly organic soils are very compressible and have undesirable construction characteristics.

Field Classification. All soils mapping, even when done from aerial photographs, must rely upon field verification sampling. Field classification procedures, however, involve subjective judgments and interpretations made in relation to standards and measures such as those outlined in this section. Therefore, it is important to understand and be familiar with the standards being used in order to insure the highest accuracy in field sampling.

Geotechnical engineers record the types of soils encountered during field borings in relation to color, stiffness, and other attributes. The following terms are commonly used in field classification.

Sand and gravel. Cohesionless particles of rounded, subangular, or angular fragments, including sands having diameters up to 1/8 inch and gravel 1/8 to 8 inches.

Hardpan. Any subsurface soil layer that offers significant resistance to the penetration of drilling tools.

Inorganic silt. Silt types ranging from less plastic varieties containing angular grains of quartz (rock flour) to more plastic varieties consisting of flake-shaped particles. This material is classified as fine-grained and consists of silt-sized particles.

Organic silt. Fine-grained materials of silt size, containing a mixture of small particles of organic matter.

Clay. Fine-grained, clay-sized particles derived from the chemical decomposition of rock. They are generally plastic within a broad range of water content. Gumbo applies to clays, common in the western United States, that are very sticky, appear waxy, and are very tough when in the plastic state.

Peat. Material consisting entirely of organic matter derived from decayed and decaying vegetative debris.

Certain other soil materials are characterized by their associated, unique geological origins. They are easily identified in the field and are generally noted in boring logs when encountered. They include till, tuff, loess, modified loess, diatomaceous earth, bog lime, marl, adobe, caliche, varved clay, and bentonite. (Definitions of these terms can be found in the glossary.)

Field classification procedures are not difficult, but some practice is necessary to attain confidence in making identifications. The following equipment is desirable for field testing: a small shovel or spade, a 3-foot soil auger, a supply of clean water, standard sieves no. 4 and no. 200, small sample bags, and labels.

The following field procedure should be used for accurately identifying and classifying soils according to the Unified system soil groups. The process is in key form; that is, it employs a process of elimination beginning with coarse soils and ending with fine soils. All observations and pertinent information should be recorded as various tests are carried out.

(1) Obtain a representative sample of the soil.

(2) Estimate the size of the largest particle.

(3) Remove materials larger than 3 inches and estimate their weight by percentage composition.

(4) Spread the remaining sample on a flat surface or in the hand and classify as predominantly coarse-grained or predominantly fine-grained. If a no. 200 sieve is available, a distinction between fine and coarse can be made by observing the amount of dry or washed sample that does or does not pass through the sieve. (Grains that pass through a no. 200 sieve are barely visible to the naked eye.)

(5) If the material is coarse, classify as sand or gravel by estimate or sieve analysis. (Gravels are retained on a no. 4 sieve; predominantly sandy soils, if dry or washed, pass through.)

(6) If the material is a gravel or sand, classify as clean or as mixed with appreciable fines. Appreciable fines are present if more than 12% by weight of the sample material passes through a no. 200 sieve.

(7) If the sand or gravel mixture is clean, decide whether it is well-graded (W) or poorly graded (P) and classify as GW, GP, SW, or SP.

(8) If the gravel or sand contains more than 12% fines, decide whether the fines are silty (M) or clayey (C), and classify as GM, GC, SM, or SC. The "teeth test" can be used to distinguish silts from clays; if the sample is bitten, clay tends to stick to the teeth. Also, sands, which feel gritty,

Table A.3. Engineering use chart*

TYPICAL NAMES OF SOIL GROUPS	GROUP SYMBOLS	IMPORTANT PROPERTIES				RELATIVE DESIRABILITY FOR VARIOUS USES *									
						ROLLED EARTH DAMS			CANAL SECTIONS		FOUNDATIONS		ROADWAYS		
													FILLS		
		PERMEABILITY WHEN COMPACTED	SHEARING STRENGTH WHEN COMPACTED AND SATURATED	COMPRESSIBILITY WHEN COMPACTED AND SATURATED	WORKABILITY AS A CONSTRUCTION MATERIAL	HOMOGENEOUS EMBANKMENT	CORE	SHELL	EROSION RESISTANCE	COMPACTED EARTH LINING	SEEPAGE IMPORTANT	SEEPAGE NOT IMPORTANT	FROST HEAVE NOT POSSIBLE	FROST HEAVE POSSIBLE	SURFACING
WELL-GRADED GRAVELS, GRAVEL-SAND MIXTURES, LITTLE OR NO FINES	GW	PERVIOUS	EXCELLENT	NEGLIGIBLE	EXCELLENT	—	—	1	1	—	—	1	1	1	3
POORLY GRADED GRAVELS, GRAVEL-SAND MIXTURES, LITTLE OR NO FINES	GP	VERY PERVIOUS	GOOD	NEGLIGIBLE	GOOD	—	—	2	2	—	—	3	3	3	—
SILTY GRAVELS, POORLY GRADED GRAVEL-SAND-SILT MIXTURES	GM	SEMIPERVIOUS TO IMPERVIOUS	GOOD	NEGLIGIBLE	GOOD	2	4	—	4	4	1	4	4	9	5
CLAYEY GRAVELS, POORLY GRADED GRAVEL-SAND-CLAY MIXTURES	GC	IMPERVIOUS	GOOD TO FAIR	VERY LOW	GOOD	1	1	—	3	1	2	6	5	5	1
WELL-GRADED SANDS, GRAVELLY SANDS, LITTLE OR NO FINES	SW	PERVIOUS	EXCELLENT	NEGLIGIBLE	EXCELLENT	—	—	3 IF GRAVELLY	6	—	—	2	2	2	4
POORLY GRADED SANDS, GRAVELLY SANDS, LITTLE OR NO FINES	SP	PERVIOUS	GOOD	VERY LOW	FAIR	—	—	4 IF GRAVELLY	7 IF GRAVELLY	—	—	5	6	4	—
SILTY SANDS, POORLY GRADED SAND-SILT MIXTURES	SM	SEMIPERVIOUS TO IMPERVIOUS	GOOD	LOW	FAIR	4	5	—	8 IF GRAVELLY	5 EROSION CRITICAL	3	7	8	10	6
CLAYEY SANDS, POORLY GRADED SAND-CLAY MIXTURES	SC	IMPERVIOUS	GOOD TO FAIR	LOW	GOOD	3	2	—	5	2	4	8	7	6	2
INORGANIC SILTS AND VERY FINE SANDS, ROCK FLOUR, SILTY OR CLAYEY FINE SANDS WITH SLIGHT PLASTICITY	ML	SEMIPERVIOUS TO IMPERVIOUS	FAIR	MEDIUM	FAIR	6	6	—	—	6 EROSION CRITICAL	6	9	10	11	—
INORGANIC CLAYS OF LOW TO MEDIUM PLASTICITY, GRAVELLY CLAYS, SANDY CLAYS, SILTY CLAYS, LEAN CLAYS	CL	IMPERVIOUS	FAIR	MEDIUM	GOOD TO FAIR	5	3	—	9	3	5	10	9	7	7
ORGANIC SILTS AND ORGANIC SILT-CLAYS OF LOW PLASTICITY	OL	SEMIPERVIOUS TO IMPERVIOUS	POOR	MEDIUM	FAIR	8	8	—	—	7 EROSION CRITICAL	7	11	11	12	—
INORGANIC SILTS, MICACEOUS OR DIATOMACEOUS FINE SANDY OR SILTY SOILS, ELASTIC SILTS	MH	SEMIPERVIOUS TO IMPERVIOUS	FAIR TO POOR	HIGH	POOR	9	9	—	—	—	8	12	12	13	—
INORGANIC CLAYS OF HIGH PLASTICITY, FAT CLAYS	CH	IMPERVIOUS	POOR	HIGH	POOR	7	7	—	10	8 VOLUME CHANGE CRITICAL	9	13	13	8	—
ORGANIC CLAYS OF MEDIUM TO HIGH PLASTICITY	OH	IMPERVIOUS	POOR	HIGH	POOR	10	10	—	—	—	10	14	14	14	—
PEAT AND OTHER HIGHLY ORGANIC SOILS	PT	—	—	—	—	—	—	—	—	—	—	—	—	—	—

*Courtesy of Bureau of Reclamation, U.S. Department of the Interior.
Note: No. 1 is the best.

can be distinguished from silts and clays, which feel smooth and leave a stain when rubbed between the fingers.

(9) Fine-grained soils are identified by tests of dilatancy, dry strength, and toughness (see Table A.2). The following procedures apply to samples from which particles greater than 1/64 inch have been removed.

(a) Dilatancy (reaction to shaking). Take a soil sample equivalent to ½ cubic inch and add water to make the soil soft but not sticky. Place the sample in the open palm of one hand and, holding the hand horizontal, strike it sharply against the other hand. A positive reaction is obtained if water appears on the surface of the sample and gives it a glossy or livery appearance. If the sample is then squeezed, the water will disappear from the surface; upon continued disturbance the sample eventually cracks or crumbles. Silt (glacial rock flour) has the quickest dilatancy reaction, while plastic clay has no dilatancy reaction. Slow to moderate reactions are obtained from inorganic silts and clays (MH and CL).

(b) Dry strength (crushing characteristics). Take another portion of the sample, approximately ½ cubic inch, and add water until a consistency similar to putty is obtained. Allow the sample to dry completely in the air or sun and then crush it between the fingers, thus testing its dry strength. The results of this test fall on a relative scale from none to high strength, and experience is required in order to obtain accurate results. Crushing strength of soil is determined by the character and quantity of the colloidal moisture within the soil; dry strength increases proportionately with plasticity.

Inorganic clays (CH) have high dry strengths, while inorganic silts (MH) have only very slight dry strength. Silty fine sands (ML) and silts (OL) have similar dry strengths but are distinguished from each other by feeling the powder from a dried specimen. Fine sands feel somewhat gritty, while silts feel like smooth flour.

Table A.4. Identification of fine-grained and organic soils utilizing field analysis techniques

Soil type	Group symbol	Dilatancy	Dry strength	Toughness consistency
Inorganic silts, very fine sands, silty or clayey sands	ML	Quick to slow	None to slight	None
Inorganic clays of low to medium plasticity, gravelly clays, sandy clays, silty clays	CL	None to very slow	Medium to high	Medium
Organic silts and organic silty clays of low plasticity	OL	Slow	Slight to medium	Slight
Inorganic silts, micaceous or diatomaceous, fine, sandy or silty soils, elastic silts	MH	Slow to none	Slight to medium	Slight to medium
Inorganic clays of high plasticity	CH	None	High to very high	High
Organic clays of medium to high plasticity, organic silts	OH	None to very slow	Medium to high	Slight to medium
Peat and other high organic soils	Pt	Readily identified by color, odor, spongy feel, and fibrous texture		

(c) Toughness (consistency near the plasticity limit). A moist sample of the soil, approximately ½ cubic inch, is molded to the consistency of putty. A ribbon ⅛ inch in diameter is formed by rolling the sample between the hands, folding, and rerolling repeatedly. As this process continues, the moisture content in the ribbon is gradually reduced. The sample becomes stiffer and eventually, when the plastic limit is reached, crumbles. After the ribbon crumbles the pieces are molded back together and kneaded until the lump crumbles. The tougher the ribbon as it nears the plastic limit and the stiffer the lump when it crumbles, the more potent is the colloidal clay fraction in the soil. This is typically true for inorganic clays (CH). Weakness of the ribbon at the plastic limit and quick loss of coherence of the lump below the plastic limit indicate either inorganic clays of low plasticity (CL) or kaolin-type clays and organic clays that occur below the A line on the plasticity chart. Highly organic clays (OH, OL) feel very weak and spongy at the plastic limit and may have an "organic" odor.

A catalog of equipment necessary for field or laboratory soil testing can be obtained from: Soiltest, Inc., 2205 Lee Street, Evanston, Illinois, 60202 (312-869-5500).

APPENDIX B

DATA AND AERIAL PHOTOGRAPHIC SOURCES

At the start of any planning process, a primary concern is the definition, acquisition, and categorization of data. Data inventory is a costly process and one for which stringent organization is needed in order to provide, fully and economically, the complex range of data to be manipulated later within the planning process. This appendix identifies and describes the data sources presently or potentially available to the land planning professions that can be used for accurate, current, and relevant data inventories. Each source is described in relation to its particular efficiencies. Sources of aerial photographs are given particular attention because of their potential for saving time and money in data acquisition. The discussion does not attempt to evaluate data sources or data-collecting techniques such as remote sensors because of the complicated cost-benefit issues involved.

Although the specific data needed in any new project obviously vary with the planning issues, site size and location, budget, schedule, and level of requirements for accuracy and analytic flexibility, most resource analysis methods currently used consistently rely upon the same limited number of sources for necessary information about climate, topography, surface water, vegetation, and distribution of land uses. In most cases these data are contained in documents publically available and provided by the government, most notably the U.S. Geological Survey quadrangle maps and the Soil Conservation Service reports.

Thus most existing methods of inventory and analysis are largely governed by the data available. However, analysis does not and should not have to be biased by the availability of data; that is, the fact that data exist and are available does not alone justify their use, nor does the absence of data mean that the analysis cannot be undertaken.

Recently, many analysts, notably some working in large, undeveloped areas, have recognized that it may be more efficient to gather new data rather than to collect, collate, and reorganize existing, sometimes scanty or erroneous data. Thus Angus G. Hills in Canada, C. S. Christian in Australia, and Donald Belcher of Cornell University have developed analysis techniques that derive a substantial portion of their resource material from the interpretation of aerial photographs.

ISSUES OF DATA COLLECTION

Many problems must be taken into account when considering data collection and sources. The individual requirements of each job dictate to a large extent the cost limitations on data acquisition, and if little money is available for an inventory, limits on the accuracy of acquired data must be expected. Furthermore, accurate data sources do not provide relevant and accurate information by themselves: in the case of aerial photographic analysis, they must be interpreted and recorded on a professional level of competence; that is, a professional who has never used aerial photographic interpretation procedures will not be able to map soil and rock conditions accurately from aerial photographs.

In gathering any data, a system of cross-checking must be established in order to provide a measure of accuracy. One of the best techniques is to use several data sources rather than to rely upon just one. And even if budget and schedule severely limit the number of sources used, field checking of existing and interpreted data should always be incorporated as a minimum control.

Every data source has inherent problems involving accuracy and resolution. For example, as reported by the Rand Corporation, panchromatic black-and-white photographs can delineate 6-inch parking strips when taken at 100,000 feet. On the other hand, results of a study made at Raytheon show that airborne radar (SLAR) can distinguish only 50-foot objects and is suitable for 1:24,000 scale mapping. Again, long-wavelength electromagnetic sensors struggle for any recognizable resolution. Such problems of cost-accuracy-resolution will be solved to some degree over time as technology develops and provides more feasible applications for data acquisition.

Categories of Data

One of the most critical stages in project planning is the identification of the data categories relevant to the problems and issues of a project. It is not enough merely to identify broad data categories; the scale and accuracy of each must also be defined, as must the type of description necessary for the subcomponents. For example, if topographic slope information is required, how should it be described for the problem under consideration? For example, is a slope gradient listing such as 0 to 10%, greater than 10 to 20%, and greater than 20% satisfactory? Or is more detail required, such as 0 to 3%, greater than 3 to 8%, greater than 8 to 12%, greater than 12 to 15%, greater than 15 to 25%, greater than 25 to 35%, greater than 35 to 60%, and greater than 60%? The scale and accuracy of the data must also be considered in relation to the topographic map from which the slope data is interpreted. In some cases the level of detail of the classification may far exceed the accuracy limits of a topographic base map, and the classification will therefore be useless. Table B.1 lists many of the data categories, primarily those relating to physical resources, typically used by resource planners.

SOURCES OF DATA

After the desired data categories, subvariable listings, and related scale and accuracy are defined, a search is initiated to determine whether the data already exists in usable form or whether it must be generated from aerial photographic interpretation, field surveys, or the use of special survey consultants. Many local, state, and federal agencies are continuously collecting data, which is often published and available upon request. The following list describes the major existing sources of information relating to physical site resources.

Local Agencies

Data often exists and is available from local, state, federal, or private agencies. In most instances this data is available free as a public service. Local sources of data should always be investigated before new surveys are initiated, for the amount and type of existing data determine the scope of new surveys and photography. A check with town halls, county offices, and local commercial firms will identify any useful data, such as well records, soil borings, or soil surveys, that may have been previously contracted for and documented.

Field Surveys

Field surveys are generally more expensive per square mile, but they usually provide the most accurate information. They should always be a part of even the most sophisticated data-

Table B.1. Data categories previously used for the inventory and analysis of physical site resources by selected planners and planning agencies*

Climate	Water
Precipitation	Watersheds
Temperature	Drainage patterns
Radiation	Ground water
Fog	Aquifers
Wind	Lakes
Frost	Rivers, streams
Storms	Wetlands
Air drainage	Springs
Geology	Waterfalls
Bedrock outcrops	Flood (potential)
Bedrock type	Water quantity
Unique formations	Water quality
Stable formations	Water temperature
Depth to bedrock	Dissolved and suspended solids
Building materials	
Structural type	Biological productivity
Structural age	Shoreline type (use)
Weathering type	Shoreline quality
Volcanic activity	
Economic minerals	Vegetation
Landform type	Vegetation type
Soils	Vegetation quality
Soil type	Specimen stands
Soil texture	Shore and bank community
Soil depth	
Internal drainage	Fields
Erodability	Forest areas
Bearing capacity	Natural associations
Permeability	Understory
Instability	Overstory
Stoniness	
Productivity	Wildlife
Topography	Wildlife type
Elevation	Wildlife habitat quality
Slopes	Prime habitats
Slope rim	Major ecotones
Relief variation	Uniqueness
Orientation	Wilderness

*After C. S. Steinitz et al., *A Comparative Evaluation of Resource Analysis Methods*, U.S. Army Corps of Engineers, Research Contract DACW33-68-CO152, 1969.

gathering system, so that accuracy limits may be checked. Since some resource patterns cannot be easily identified on the ground and thus require aerial overview observations, field surveys should be used in conjunction with other data sources to verify the accuracy of interpretations made from remote sensors. (In some cases, however, field surveys are not possible, for example, when it is inadvisable to show interest in a particular property before the land is purchased or development plans are final.)

Deep soil borings and seismic surveys in conjunction with laboratory analysis provide detailed data from which structural-engineering designs can evolve. Field studies of this nature for selected areas must be contracted from soils engineering firms to obtain the necessary information relating to engineering performance standards. Because of the high cost of these detailed surveys, it is not feasible to use them to collect data for large land areas; they are generally used after specific sites have been identified. The purpose of many terrain analysis projects is to identify areas of prime development potential where detailed surveys should be initiated.

Topographic Surveys

Topographic contour information and base map compilation are necessary components of data mapping and analysis. A record of data patterns and coordinate location is only as accurate as the map upon which it is presented. For very generalized data surveys, existing topographic information from U.S. Geological Survey quadrangle maps may be sufficient. However, as a project proceeds and detailed planning starts, more accurate planimetric and topographic information is necessary in order to derive and map more details, such as cut and fill grading estimates or drainage and property line location, and to locate points in the field accurately. This detailed mapping can be provided by a competent photogrammetric firm. *Photogrammetric Engineering,* the professional magazine of the American Society of Photogrammetry, is an informative guide to these organizations.

U.S. Geological Survey

Since 1882 the U.S. Geological Survey has provided general-purpose maps of varying scales and degrees of accuracy for all of the United States. Many different series and scales are available to satisfy a wide range of user demands. The maps commonly used by planners are the 1:24,000 scale (200 feet/inch) quadrangle maps, the 1:62,500 scale (5280 feet/inch) quadrangle maps, and the 1:250,000 (approximately 4 miles/inch) small-scaled maps. The most useful maps have proved to be the 1:24,000 scale ones because of the greater amount of detail shown. Their contour interval is typically 10, 20, and occasionally 5 feet, and they include a wide range of information such as: road types, bridges, tunnels, railroads, power transmission lines and towers, dams, large buildings, urbanized areas, storage tanks, quarries, mines, borrow pits, campsites, shoreline rock or reef features, horizontal-vertical control stations, spot elevations, municipal boundaries, fence or field lines, section corners, boundary monuments, levees, washes, sand areas, gravel beaches, glaciers, water wells, springs, rapids, channels, depth soundings, dry lake beds, intermittent streams, aqueduct tunnels, waterfalls, intermittent lakes, small washes, marshes, inundated areas, submerged marshes, woodlands, orchards, vineyards, mangrove swamps, scrub vegetation, and wooded marshes. Other special maps available from the U.S. Geological Survey include:

Metropolitan Area maps (1:24,000), covering major metropolitan areas;

National Park Series (scale varies), covering national parks, monuments, and historic sites;

State base maps (mostly 1:500,000 and 1:1,000,000);

United States maps (scale varies).

Additional information concerning the different types of maps available and their symbolism can be obtained from *Topographic Maps,* a publication available from:

National Cartographic Information Center (NCIC)
U.S. Geological Survey
507 National Center
Reston, Virginia 22092
703-860-6045

Mapping of geological features has been completed for the entire United States at a scale of 1:5,000,000, but mapping for individual states leaves much to be desired. The geological and surficial geological maps available are listed in Appendix D. Certain areas of the United States have had bedrock and/or surficial geological mapping completed at the same scales as the other U.S. Geological Survey quadrangle maps, and the status and availability of these maps can be obtained from the NCIC. Appendix E lists maps of the United States that are useful sources of general information and that are available from the Geological Survey as well as from other mapping organizations.

Another publication of the U.S. Geological Survey is *The National Atlas.* The atlas is available from the U.S. Government Printing Office, Washington, D.C.; it provides a wide range of physical, historical, economic, sociocultural, and administrative unit maps within a single volume containing 765 maps in 335 pages. Included in this atlas are maps describing coverage of the United States by aerial photographs, detailed topographic maps, and navigation charts. The atlas, while too general in content for direct application by planners, nonetheless can provide an important contribution to a planning project by identifying the general setting and character of a site.

U.S. National Ocean Survey

The U.S. National Ocean Survey of the U.S. Department of Commerce is responsible for mapping and charting the coastal areas of the United States. Copies of old maps drawn as early as the late 1700s in the old pictorial cartographic style can be ordered from the U.S. National Ocean Survey. These maps are extremely useful for documenting past influences of hurricanes and storms in coastal areas. Sequences of maps can be used to illustrate long-term erosion or deposition of sand materials and can thus help to identify the nature—static or dynamic—of a particular coastal site. For further information regarding these old topographic surveys, contact the U.S. Department of Commerce, National Ocean Survey, Washington, D.C. 20235.

Soil Conservation Service

The Soil Conservation Service of the U.S. Department of Agriculture has been conducting and publishing soil surveys since 1899. These surveys are intended to provide information on soils to the general public and are available free from the Information Division, Soil Conservation Service, Washington, D.C., to land users or professionals who require the information. Only one-third of the United States has been mapped in detail, and the maps most useful to planners are those completed after 1957. The Soil Conservation Service does not project the completion of its mapping of the United States within the foreseeable future. The current mapping status of soils information can be obtained from the Soil Conservation Service in Washington, D.C. (see Appendix C).

The Soil Conservation Service has many field districts and offices scattered throughout the United States. Their function is to assist those who need information on soils and to compile soil maps; they can make valuable contributions to a data inventory by supplying unpublished, first-hand information. This is especially true in relation to complex categories such as ecology and wildlife. For further information contact the Soil Conservation Service, U.S. Department of Agriculture, Washington, D.C.

Other Government Agencies

Depending upon site location, other United States government agencies with responsibilities over land areas may be of assistance. These include (but are not limited to) the Department of Commerce, the Forest Service, the Bureau of Outdoor Recreation, the National Park Service, the Department of Wildlife and Fisheries, the Army Corps of Engineers, the Bureau of Reclamation, and the National Aeronautics Space Administration. Usually, the central office of each agency in Washington, D.C., knows the specific data available for a particular region or will direct inquiries to the relevant field office.

SOURCES OF AERIAL PHOTOGRAPHIC COVERAGE

Many projects do not have the time or financing to initiate their own aerial survey to obtain current photographs of a site. However, inexpensive stereophotographic coverage of many sites in the United States can be purchased from several government agencies or private sources. It should be noted that many government photographs available may be less than desirable for photointerpretation purposes; the scale may be too small, the photographs may have been taken at the wrong time of year, have resolution too poor for project application, or be very old and outdated (for some projects). However, older photographs are desirable for comparing changes in land use and development. Many of the agencies listed below have been regularly taking aerial photographs of much of the United States since the 1940s or even the late 1930s.

A series of catalogs is now available from the National Cartographic Information Center (NCIC) that illustrates the general availability of aerial photographs for all sections of the United States. This Aerial Photographic Summary Record System (APSRS) has compiled the photographic coverage of the Agriculture Stabilization Conservation Service (ASCS), the Geological Survey (GS), National Aeronautics and Space Administration (NASA), Soil Conservation Service (SCS), National Ocean Survey (NOS), Forest Service (FS), the Defense Mapping Agency (DMA), state highway departments, and various commercial firms. Five catalogs, showing the most recent coverage, are available. In addition, microfiche computer printouts can be purchased; these specify for each area percentage of cloud cover, type of camera, scale, date, and agency project code. Catalogs are $1.00, and microfiche printouts range between $1.50 and $3.00. Catalog 1 covers Alaska; catalog 2 the United States from longitude 125° to 110°; catalog 3 from 110° to 95°; catalog 4 from 95° to 80°; and catalog 5 from 80° to 65°. These materials can be purchased from the NCIC, User Services Section, U.S. Geological Survey, 507 National Center, Reston, Virginia 22092.

When ordering photographs from government agencies or private companies cited in this section, for which the exact photograph number is not known, it is advisable to enclose a copy of a U.S. Geological Survey quadrangle map with the area of interest outlined in red pencil. To insure that your order will be processed efficiently, stereo- or single-photograph coverage, size, finish, and weight of paper should also be specified.

Agricultural Stabilization Conservation Service

The Agricultural Stabilization Conservation Service (ASCS) holds the single largest collection of aerial photographs of the United States. Most of the photographs are black-and-white and were taken at scales of 1:20,000 to 1:40,000.

Photographs should be ordered on the standard forms available from the Aerial Photographic Division of the ASCS and should include information on the desired scale, quantity of each photograph, county symbol, roll number, exposure number, and the size, weight, and finish of paper desired. The standard paper sizes and associated scales available are listed in Table 3.1 for both 1:20,000 and 1:40,000 scale photographs. Additional information is available from:

Agricultural Stabilization Conservation Service (ASCS)
Aerial Photography Division
ASCS-USDA
2505 Parley's Way
Salt Lake City, Utah 84109
801-524-5856 FTS-588-5856

Coverage: Most of the United States (status book available upon request).
Film Type: Black and white (Spring, summer, or fall).
Scale: Most 1:20,000; some 1:40,000.

Table B.2. Scale and paper size relationships for contact prints and enlargements from 1:20,000 and 1:40,000 scale photography

Paper size	Approximate scale from 1:20,000 photography	Approximate scale from 1:40,000 photography
9½″ x 9½″	1″ = 1667′	1″ = 3334′
13″ x 13″	1″ = 1320′	1″ = 2640′
17″ x 17″	1″ = 1000′	1″ = 2000′
24″ x 24″	1″ = 660′	1″ = 1320′
24″ x 24″	1″ = 330′*	1″ = 660′*
38″ x 38″	1″ = 400′	1″ = 800′

*This size and scale relationship exists if one-quarter of the original photograph is ordered. The numbering system begins with A in the upper right-hand quarter and continues clockwise through quarters B and C, ending with D in the upper left-hand quarter. If quadrants are to be ordered, A, B, C, or D should be specified (for example, ABC-2JJ-168C).

U.S. Forest Service

The Forest Service of the U.S. Department of Agriculture photographs primarily forest lands under its jurisdiction. Most of the photographs have been taken at a scale of 1:15,840 with an 8¼-inch lens, providing a contact scale of 1320 feet/inch. The U.S. Forest Service also photographs much of its holdings with black-and-white infrared to aid in timber volume estimations. (Photographic coverage of this nature is indicated in their indexing booklet by the suffix R.) A publication entitled *Aerial Photography Status Maps* is available from the U.S. Forest Service and illustrates their photographic coverage within the United States. This booklet, as well as information concerning photographs of all states not covered by regional offices (see Figure B.1) is available from:

U.S. Department of Agriculture
Forest Service
P. O. Box 2417
Washington, D.C. 20013
ATTN: Ms. Audrey B. MacNamara
703-235-8638, FTS-235-8638

Coverage: National forest areas (status book available upon request).
Film Types: Black and white, some black and white infrared and color.
Scales: Most 1:15,840.

Inquiries and orders concerning photographic coverage of the western United States must be addressed to each region shown in Figure B.1.

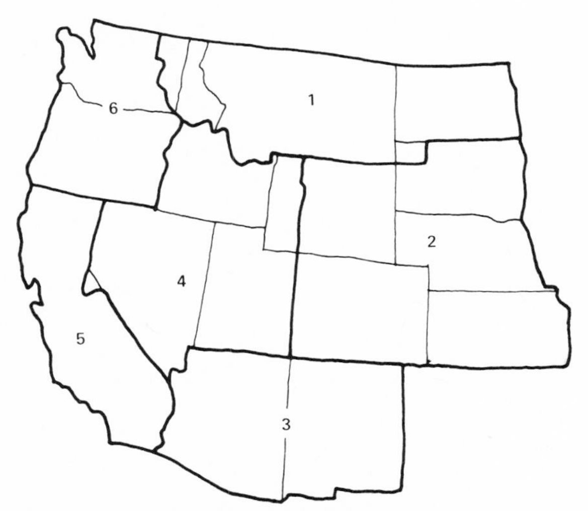

Figure B.1. District regions of the U.S. Forest Service west of the Mississippi River and addresses of their photographic laboratories. (1) Federal Building, Missoula, Montana 59801. (2) Federal Center Building, No. 85, Denver, Colorado 80225. (3) Federal Building, 517 Gold Avenue, S.W., Albuquerque, New Mexico 87101. (4) Forest Service Building, Ogden, Utah 84403. (5) 630 Sansome Street, San Francisco, California 94111. (6) Post Office Box 3623, Portland, Oregon 97208. For Alaska, contact Regional Forester, Forest Service, Box 1628, Juneau, Alaska 99801.

Soil Conservation Service

The Soil Conservation Service has aerial photographs of much of the United States, primarily at a scale of 1:20,000, or 1,667 feet/inch. The SCS also has scattered, larger-scaled photographs (1:3,000 to 1:12,000) suitable for stereoplotter use. Listings of flight strips of this photography and of other coverage appear in the Soil Conservation Service publication *Status of Aerial Photography.* Copies of this booklet and photograph orders can be directed to:

U.S. Department of Agriculture
Soil Conservation Service
Cartographic Division
6505 Belcrest Rd.
Hyattsville, Maryland 20782
301-436-8756, FTS-436-8756

Coverage: Scattered United States (status book available upon request).
Film Type: Black and White (usually during nongrowing seasons).
Scales: Most 1:20,000; some 1:45,000–1:55,000.

U.S. Geological Survey

The U.S. Geological Survey has extensive photographic coverage of the United States and its territories at a wide variety of scales. Requests for general information should be sent to any of the following:

NCIC-Reston (Headquarters)
USGS, 507 National Center
Reston, Virginia 22092
703-860-6045, FTS-928-6045

NCIC-Mid-Continent
USGS, 1400 Independence Rd.
Rolla, Missouri 65401
314-364-3680, ext. 107
FTS-276-9107

NCIC-Rocky Mountain
USGS, Topographic Div.
Stop 510, Box 25046
Denver Federal Center
Denver, Colorado 80225
303-234-2326, FTS-234-2326

NCIC-Western
USGS, 345 Middlefield Rd.
Menlo Park, California 94025
415-323-2427, FTS-467-2427

EROS Data Center
USGS, Sioux Falls
South Dakota 57198
605-594-6511, ext. 151
FTS-784-7151

Coverage: Most of the United States (status maps available upon request). The Geologic Survey also has high altitude photography for most of the United States at a scale of 1:60,000 or 1:120,000.
Film Type: Black and white; some color and color infrared.
Scales: Varies, but commonly 1:24,000.

Persons interested in obtaining a wide variety of landform photographs should consider purchasing Geological Survey Professional Papers, No. 590 and No. 591. Professional Paper No. 590, *Selected Aerial Photographs of Geologic Features in the United States,* by C. S. Denny, C. R. Warren, D. H. Dow, and W. J. Dale, lists 317 sets of aerial photographs with corresponding topographic and geological maps of numerous features. Each example has a written description and a listing of photographs which can be ordered separately. Professional Paper No. 591, *Selected Aerial Photographs of Geologic Features in Areas outside the United States,* by C. R. Warren, D. L. Schmidt, C. S.

Denny, and W. J. Dale, describes 67 sets of photographs and maps of unique geological features scattered throughout the world and illustrates a small sample of each set. The full-sized contact prints and their corresponding maps must be ordered separately from the U.S. Geological Survey. Both these publications are extremely useful as guides for obtaining material to be used in teaching aerial photographic interpretation.

National Ocean Survey

The National Ocean Survey, formerly the Coast and Geodetic Survey, has photographic coverage of the coastal regions of the United States. The scales vary. Color, color infrared, and black-and-white infrared photographs may be available, in addition to the usual panchromatic black-and-white. A general index is not available, but inquiries about coverage and orders can be directed to:

NOAA, National Ocean Survey
6001 Executive Blvd.
Rockville, Maryland 20852
ATTN: Coastal Mapping Division
C-3415, Mr. Howard Wolfe
301-443-8601

Other Governmental Agencies

Many other agencies hold photographic coverage of land areas under their direct jurisdiction. The NCIC Aerial Photographic Summary Record System has attempted to catalog this diverse coverage into one summary record. They should be contacted directly for any coverage related to: U.S. Air Force, U.S. Navy, NASA, Bureau of Land Management, Bureau of Reclamation, Bureau of Indian Affairs, National Park Service, Tennessee Valley Authority, the Army Corps of Engineers, Defense Mapping Agency, and so on.

Old Aerial Photographs

The National Archives maintains older aerial photographic films (pre-1950) for many governmental agencies. The coverage for some areas of the United States extends back to the late 1930s and early 1940s and can be quite useful in the study of the progression of land use growth or in the investigation of terrain conditions. Requests for further information should be directed to:

National Archives (GSA)
Cartographic Records Division
Washington, D.C. 20408
202-962-0173

State Highway Departments

Many states maintain through their highway or public works departments aerial photographs for highway planning and design. These photographs, usually black-and-white, taken at relatively large scales, 1:12,000 or 1:7,200, are very useful for detail landform and soils mapping. If taken by a commercial firm under contract to the state, they can be obtained in a short period of time.

Private Companies

Many reputable private companies in the United States have photographed sites, both large and small, on a contract basis. Copies of these photographs are more expensive than those from government sources, and permission in writing from the original contractor is usually necessary.

In many cases a client for a planning project has had prepared, during earlier contracts, a base map showing the extent, boundaries, and topography of the properties being analyzed, and often aerial photographs have been made to facilitate stereoplotting; these may be available from the firm that prepared the base map. A request directed to the client will often locate such photographs, which may be more suitable for interpretation than those available from private or federal agencies.

Some companies periodically fly large areas speculatively for later sales and mapping opportunities. Large companies within a project area should be contacted for possible coverage. The names and addresses of these firms can be found in professional journals such as *Photogrammetric Engineering and Remote Sensing.*

Photographic Coverage of Areas outside the United States

Photographic coverage of areas outside the United States is usually available, but it is rather difficult to obtain. Most western nations have programs of aerial photography and mapping for their own countries, and they can be contacted directly regarding the availability of this coverage. Most of these same nations are also covered by the U.S. Air Force aerial photographs; these photographs are available through the Photographic Records and Services Division of the Aeronautical Chart and Information Center, Washington, D.C., or from:

Defense Mapping Agency

Building 56
U.S. Naval Observatory
Washington, D.C. 20305
202-254-4406

Permission for release must be granted by diplomatic representatives of the country photographed and necessary security clearances must be obtained. In general, photographs of Canada and Mexico are obtained with the least difficulty.

Information concerning photographic coverage of Canada can be obtained from:

National Air Photo Library
Room 180, Surveys and Mapping Building
615 Booth Street
Ottawa 4, Ontario, Canada

Satellite and Other Space Programs

Photographs and images acquired by any of the United States space programs are available from:

EROS Data Center
U.S. Geological Survey
Sious Falls, South Dakota 57198
605-594-6511 (ask for User Services)
FTS-784-7151

Several foreign receiving stations have been constructed that are now responsible for collecting and disseminating LANDSAT imagery for areas roughly 2,000 km in every direction from their locations. The following stations must be contacted directly for coverage available within their area of jurisdiction. The EROS Data Center does not maintain a central library listing coverage available from these foreign stations. However, images taken before these centers were in operation can be obtained from the EROS Data Center.

Canada (Prince Albert, Saskatchewan)
Integrated Satellite Information
Box 1630
Prince Albert, Saskatchewan
Canada S6U 5T2

Canada (St. John's, Newfoundland)
Department of Forestry & Agriculture
Building 810
Pleasantville
St. John's, Newfoundland A1A 1P9

Brazil (Cuiaba)
Instituto De Pesquisas Espaciais (INPE)
Attn: Divisao de Banco de Dados
Av. Dos Astronautas, 1758 - Caixa Postal 515
12.200 - Sao Jose dos Campos, S.P. - BRASIL

Italy (Fucino)
Telespazio - SES
Corso D'Italia
L-3 Roma, ITALY

APPENDIX C

SOIL CONSERVATION SERVICE: PUBLISHED SOIL SURVEYS

ALABAMA

1964 Baldwin
1914 Barbour
1961 Calhoun
1959 Chambers
1924 Cherokee
1939 Colbert
1929 Coosa
1921 Crenshaw
1962 Cullman
1960 Dale
1938 Dallas
1958 De Kalb
1955 Elmore
1965 Fayette
1965 Franklin
1971 Greene
1939 Hale
1968 Houston
1954 Jackson
1931 Lauderdale
1959 Lawrence
1950 Lee
1953 Limestone
1944 Macon
1958 Madison
1920 Marengo
1959 Marshall
1930 Mobile
1916 Monroe
1960 Montgomery
1918 Morgan
1944 Morgan
1930 Perry
1917 Pickens
1967 Randolph
1941 Sumter
1938 Wilcox
1937 Winston

ALASKA

1963 Fairbanks Area
1971 Homer-Ninilchik Area
1962 Kenai-Kasilof Area
1916 Kenai Peninsula
1968 Matanuska Valley Area
1960 Northeastern Kodiak Island Area
1973 Salcha-Big Delta Area
1974 Susitna Valley

ARIZONA

1967 Beaver Creek Area
1921 Benson Area
1927 Buckeye-Beardsley Area
1941 Casa Grande Area
1974 East Maricopa and Northern Pinad
1928 Gila Bend Area
1964 Holbrook Show-Low Area
1974 Long Valley Area
1930 Nogales Area
1973 Organ Pipe Catus National Mt.
1970 Safford Area
1926 Salt River Valley Area
1921 San Simon Area
1954 Sulphur Springs Valley Area
1941 The Duncan Area
1972 Tucson-Avra Valley Area
1938 Upper Gila Valley Area
1942 Virgin River Valley Area
1941 Yuma Desert Area

ARKANSAS

1972 Arkansas
1961 Bradley
1967 Chicot
1968 Cleveland
1974 Crittenden
1916 Craighead
1968 Cross
1972 Desha
1971 Franklin
1969 Greene
1971 Mississippi
1973 Ouachita
1974 Phillips
1966 St. Francis
1969 Washington
1968 Woodruff

CALIFORNIA

1966 Alameda Area
1931 Alturas Area
1965 Amador Area
1969 Antelope Valley Area
1945 Bakersfield Area
1937 Barstow Area
1920 Brawley Area
1971 Carson Valley Area
1923 Coachella Valley Area
1952 Coalinga Area
1939 Contra Costa
1971 Eastern Fresno Area
1974 Eastern Santa Clara Area
1964 Eastern Stanislaus Area
1930 El Cajon Area
1918 El Centro Area
1974 El Dorado Area
1968 Glenn
1918 Grass Valley Area
1946 Kings
1937 Lodi Area
1952 Los Banos Area
1962 Madera Area
1974 Mariposa
1956 Mendota Area
1962 Merced Area
1938 Napa Area
1948 Newman Area
1974 Palo Verde Area
1942 Pixley Area
1927 Placerville Area
1954 Sacramento Area
1941 Sacramento San Joaquin Delta Area
1925 Salinas Area
1969 San Benito
1973 San Diego
1961 San Mateo Area
1958 Santa Barbara Area
1972 Santa Barbara Area (Northern)
1958 Santa Clara Area
1944 Santa Cruz Area
1927 Santa Ynez Area
1974 Shasta Area
1919 Shasta Valley Area
1972 Sonoma
1930 Suisan Area
1974 Surprise Valley Home Camp

1974 Tahoe Basin
1967 Tehama
1943 Tracy Area
1970 Ventura Area
1940 Visalia Area
1942 Wasco Area
1971 Western Riverside Area
1972 Yolo

COLORADO

1974 Adams
1947 Akron Area
1973 Alamosa Area
1971 Arapahoc
1926 Arkansas Valley Area
1974 Baca
1932 Brighton Area
1968 Crowley
1967 Delta-Montrose Area
1966 Elbert (eastern part)
1927 Fort Collins Area
1962 Fraser-Alpine Area
1955 Grand Junction Area
1929 Greeley Area
1930 Longmont
1968 Morgan
1971 Phillips
1966 Prowers
1970 Sedgwich
1961 Trout Creek Watershed

CONNECTICUT

1962 Hartford
1966 Tolland

DELAWARE

1971 Kent
1970 New Castle
1974 Sussex

FLORIDA

1954 Alachua
1974 Brevard
1954 Collier
1958 Dade
1960 Escambia
1962 Gadsden
1958 Hillsborough
1923 Lake
1958 Manatee
1971 Okeechobee
1960 Orange
1972 Pinellas
1927 Polk
1959 Sarasota
1966 Seminole
1965 Suwannee
1965 Washington

GEORGIA

1971 Banks and Stephans
1941 Bartow
1912 Ben Hill
1969 Ben Hill-Irwin
1973 Berrien and Lanier
1922 Bibb
1916 Brooks
1974 Bryan and Chatham
1968 Bulloch
1925 Calhoun
1948 Candler
1971 Carroll and Haralson
1941 Catoosa
1973 Cherokee, Gilmer and Pickens
1927 Clarke
1968 Clarke-Oconee
1974 Cobb
1972 Dawson, Lumpkin and White
1942 Dade
1939 Decatur
1923 Dooly
1968 Dougherty
1961 Douglas
1928 Elbert
1917 Floyd
1960 Forsyth
1958 Fulton
1965 Gordon
1967 Gwinnett
1963 Habersham
1941 Hall
1929 Hart
1963 Hart
1967 Houston and Peach
1914 Jackson
1916 Jasper
1930 Jefferson
1968 Jenkins
1972 Lamar, Pike, and Upson
1927 Lee
1918 Madison
1937 McDuffie
1961 McIntosh
1916 Meriwether
1965 Meriwether
1974 Montgomery, Toombs, and Wheeler
1965 Morgan
1968 Pierce
1926 Quitman
1974 Schley and Sumter
1964 Spalding
1914 Tattnall
1914 Terrell
1959 Tift
1939 Toombs
1954 Towns
1964 Treutlen
1963 Twiggs
1950 Union
1964 Walton
1965 Wayne
1929 Worth

HAWAII

1974 Island of Hawaii
1972 Kauai, Oaku, Mavi, Molokai, and Lanai
1955 The Territory of Hawaii

IDAHO

1930 Benewah
1973 Bingham
1943 Blackfoot-Aberdeen Area
1939 Bonner
1972 Canyon
1965 Gem County Area
1929 Gooding Area
1950 Idaho Falls Area
1971 Kooskia
1902 Lewiston
1969 Teton Area Idaho-Wyoming
1921 Twin Falls Area

ILLINOIS

1971 Douglas
1972 Edwards and Richland
1969 Gallatin
1966 Jersey (State)
1964 Johnson (State)
1970 Lake
1974 Logan
1969 Montgomery
1968 Pulaski and Alexander
1964 Wabash

INDIANA

1969 Allen
1947 Bartholomew
1916 Benton
1928 Blackford
1946 Brown
1958 Carroll
1974 Clark and Floyd
1955 Cass
1922 Clay
1974 Daviess
1937 Dubois
1974 Elkhart
1960 Fayette and Union
1966 Fountain
1950 Franklin
1946 Fulton
1922 Gibson
1925 Hancock
1974 Hendricks
1971 Howard
1940 Jennings
1948 Johnson
1943 Knox
1922 Kosciusko
1972 Lake
1944 La Porte
1967 Madison
1946 Martin
1927 Miami
1950 Morgan
1955 Newton
1953 Noble
1930 Ohio and Switzerland
1964 Owen
1967 Parke
1969 Perry
1938 Pike
1968 Pulaski
1925 Putnam
1931 Randolph
1937 Rush
1962 Scott
1974 Shelby
1973 Spencer
1950 St. Joseph
1940 Steuben
1971 Sullivan
1959 Tippecanoe
1944 Vanderburgh
1930 Vermillion
1914 Warren
1939 Washington
1925 Wayne

IOWA

1963 Adams
1958 Allamakee
1940 Audubon
1967 Bremer
1930 Calhoun
1969 Cass
1919 Cedar
1940 Cerro Gordo
1923 Clarke
1916 Clay
1969 Clay
1925 Clayton
1928 Crawford
1973 Crawford
1920 Dallas
1940 Davis
1939 Decatur
1922 Delaware
1921 Des Moines
1938 Franklin
1921 Greene
1929 Guthrie
1974 Guthrie
1930 Hancock

1935 Howard
1961 Humboldt
1939 Ida
1967 Iowa
1941 Jackson
1921 Jasper
1922 Jefferson
1960 Jefferson
1914 Lee
1917 Linn
1960 Lucas
1927 Lyon
1939 Marion
1920 Mills
1916 Mitchell
1959 Monona
1931 Monroe
1914 Muscatine
1940 Osceola
1923 Plymouth
1928 Pochahontas
1960 Polk
1914 Pottawattamie
1929 Poweshick
1916 Ringgold
1924 Sac
1915 Scott
1961 Shelby
1941 Story
1950 Tama
1954 Taylor
1962 Van Buren
1930 Washington
1918 Wayne
1971 Wayne
1914 Webster
1922 Winneshiek
1968 Winneshiek
1972 Woodbury
1922 Worth

KANSAS

1938 Allen
1931 Bourbon
1960 Brown
1974 Chase
1926 Clay
1915 Cowley
1974 Crawford
1973 Edwards
1965 Finney
1965 Ford
1960 Geary
1969 Grant
1968 Gray
1961 Greeley
1961 Hamilton
1971 Harper
1968 Haskell
1973 Hodgeman
1928 Johnson
1963 Kearny
1938 Kingman
1972 Lane
1964 Logan
1930 Marion
1913 Montgomery
1974 Morris
1963 Morton
1930 Neosho
1968 Pratt
1966 Reno
1967 Republic
1959 Saline
1965 Scott
1965 Seward
1970 Shawnee
1973 Sherman
1961 Stanton
1961 Stevens
1965 Wichita
1931 Woodson

KENTUCKY

1964 Adair
1969 Barren
1963 Bath
1973 Boone, Campbell and Kenton
1966 Caldwell
1945 Calloway
1974 Calbway and Marshall
1964 Clark
1965 Elliott
1974 Estill and Lee
1968 Fayette
1965 Fourteen Co. Eastern Ky. Reconnaissance
1964 Fulton
1921 Garrard
1953 Graves
1972 Grayson
1968 Harrison
1967 Henderson
1966 Jefferson
1915 Jassamine
1919 Logan
1950 Marshall
1930 Mercer
1967 Metcalfe
1920 Muhlenberg
1971 Nelson
1916 Shelby

LOUISIANA

1962 Acadia
1928 Beauregard
1962 Bossier
1968 East Baton Rouge
1918 La Salle
1931 Livingston
1921 Natchitoches
1974 Ovachita
1919 Sabine
1973 St. James and St. John The Baptist
1959 St. Mary
1968 Tensas
1960 Terrebonne
1922 Washington

MAINE

1964 Aroostook Northeastern Part
1964 Aroostook Southern Part
1974 Cumberlandy
1963 Penobscot
1973 Somerset (Southern Part)
1955 Waldo
1952 York

MARYLAND

1973 Anne Arundel
1928 Calvert
1971 Calvert
1929 Caroline
1964 Caroline
1969 Carroll
1974 Cecil
1974 Charles
1963 Dorchester
1960 Frederick
1974 Garrett
1968 Howard
1930 Kent
1961 Montgomery
1967 Prince Georges
1931 Queen Annes
1966 Queen Annes
1966 Somerset
1970 Talbot
1962 Washington
1970 Wicomico
1973 Worcester

MASSACHUSETTS

1929 Franklin
1967 Franklin
1969 Plymouth

MICHIGAN

1929 Alger
1924 Alpena
1967 Arenac
1931 Bay
1928 Branch
1974 Charlevoix
1939 Cheboygan
1927 Chippewa
1942 Clinton
1927 Crawford
1974 Emmet
1972 Genesee
1972 Gladwin
1966 Grand Traverse
1941 Ingham
1967 Ionia
1937 Iron
1926 Jackson
1972 Lapeer
1973 Leelanau
1961 Lenawee
1929 Luce
1971 Macomb
1922 Manistee
1939 Mason
1925 Menominee
1950 Midland
1960 Montcalm
1930 Montmorency
1968 Muckegon
1951 Newaygo
1938 Oceana
1969 Osceola
1931 Oscoda
1973 Ottawa
1921 Reconnaissance Ontonagon
1924 Roscommon
1938 Saginaw
1961 Sanilac
1939 Schoolcraft
1929 St. Clair
1974 St. Clair
1926 Tuscola
1930 Washtenaw

MINNESOTA

1916 Anoka
1968 Carver
1965 Crow Wing
1960 Dakota
1961 Dodge
1957 Fairbault
1958 Fillmore
1974 Hennepin
1929 Houston
1930 Hubbard
1958 Isanti
1939 Kanabec
1926 Lake of the Woods Reconnaissance
1954 Le Sueur
1970 Lincoln
1955 McLeod

1927 Mille Lacs
1958 Nicollet
1974 Norman
1941 Pine
1972 Pope
1949 Rock
1942 Roseau
1959 Scott
1968 Sherburne
1973 Steele
1971 Stevens
1965 Wabasha
1965 Waseca
1968 Wright

MISSISSIPPI

1921 Alcorn
1971 Alcorn
1958 Bolivar
1965 Calhoun
1974 Chickasaw
1926 Claiborne
1963 Claiborne
1965 Clarke
1959 Coahoma
1965 Covington
1959 De Soto
1971 George
1932 Greene
1915 Grenada
1967 Grenada
1930 Hancock
1924 Harrison
1916 Hinds
1959 Humphreys
1961 Issaquena
1927 Jackson
1964 Jackson
1973 Lee
1959 Leflore
1963 Lincoln
1938 Marion
1972 Marshall
1966 Monroe
1916 Newton
1960 Newton
1963 Panola
1922 Perry
1968 Pike
1974 Pontotoc
1957 Prentiss
1958 Quitman
1962 Sharkey
1959 Sunflower
1970 Tallahatchie
1967 Tate
1966 Tippah
1944 Tishomingo
1956 Tunica
1968 Walthall
1964 Warren
1961 Washington

MISSOURI

1974 Barton
1962 Boone
1974 Caldwell
1964 Daviess
1914 DeKalb
1971 Dent
1914 Grundy
1914 Harrison
1953 Holt
1954 Jasper
1914 Johnson
1911 Leclede
1945 Linn
1956 Livingston
1964 Moniteau
1971 Pemiscot
1922 Ray
1956 St. Charles
1917 Texas
1968 Worth

MONTANA

1944 Big Horn Valley Area
1959 Bitterroot Valley Area
1953 Central Montana (Reconnaissance)
1931 Gallatin Valley Area
1967 Judith Basin
1929 Lower Flathead Valley Area
1939 Lower Yellowstone Valley Area
1940 Middle Yellowstone Valley Area
1929 Northern Plains of Montana (Reconnaissance)
1971 Powder River Area
1967 Treasure
1960 Upper Flathead Valley Area
1943 Upper Musselshell Valley Area
1958 Wibaux
1972 Yellowstone

NEBRASKA

1954 Blaine
1921 Boone
1972 Boone
1937 Boyd
1938 Brown
1941 Cass
1956 Cherry
1915 Dawes
1921 Deuel
1965 Deuel
1929 Dixon
1916 Dodge
1931 Dundy
1963 Dundy
1939 Frontier
1930 Furnas
1964 Gage
1924 Garden
1938 Garfield
1938 Gosper
1937 Greeley
1962 Hall
1974 Harlan
1938 Hayes
1970 Hitchcock
1938 Holt
1964 Hooker
1974 Howard
1937 Keya Paha
1962 Kimball
1930 Knox
1948 Lancaster
1937 Loup
1969 McPherson
1917 Morrill
1922 Nance
1960 Nance
1950 Otoe
1921 Perkins
1973 Phelps
1974 Polk
1967 Red Willow
1937 Rock
1939 Sarpy
1965 Saunders
1968 Scotts Bluff
1974 Seward
1931 Sherman
1929 Stanton
1968 Thayer
1965 Thomas
1972 Thurston
1932 Valley
1964 Washington
1923 Webster
1974 Webster
1937 Wheeler

NEVADA

1967 Las Vegas and Eldorado Valleys Area
1965 Lovelock Area
1968 Pahranagat-Penoyer Area
1974 Surprise Valley-Home Camp
1974 Tahoc Basin

NEW HAMPSHIRE

1968 Belknap
1949 Cheshire and Sullivan
1943 Coos
1939 Grafton
1953 Hillsboro
1965 Merrimack
1959 Rockingham
1949 Strafford
1973 Strafford

NEW JERSEY

1925 Bergen Area
1971 Burlington
1926 Camden Area
1966 Camden
1927 Freehold Area
1962 Gloucester
1972 Mercer
1969 Salem

NEW MEXICO

1958 Bluewater Area
1968 Cabezon Area
1958 Curry
1928 Deming Area
1971 Eddy Area
1930 Fort Summer Area
1974 Harding
1974 Hidalgo
1974 Lea
1932 Lovington Area
1959 Portales Area
1930 Rincon Area
1967 Roosevelt
1933 Roswell Area
1960 Southwest Quay Area
1970 Torrance Area
1967 Zuni Mountain Area

NEW YORK

1942 Albany and Schenectady
1956 Allegany
1932 Broome
1971 Broome
1971 Cayuga
1940 Cattaraugus
1932 Chemung
1933 Chemung
1961 Cortland
1955 Dutchess
1929 Erie
1958 Franklin
1969 Genesee
1960 Lewis
1956 Livingston
1938 Monroe
1973 Monroe
1947 Niagara
1972 Niagara
1938 Onondaga

1958 Ontario and Yates
1939 Orleans
1940 Otsego
1937 Rensselaer
1969 Schoharie
1942 Seneca
1972 Seneca
1931 Steuben
1925 St. Lawrence
1946 Sullivan
1953 Tioga
1965 Tompkins
1940 Ulster
1938 Wyoming
1974 Wyoming
1916 Yates

NORTH CAROLINA

1960 Alamance
1955 Avery
1917 Beaufort
1937 Brunswick
1920 Buncombe
1954 Buncombe
1938 Carteret
1937 Chatham
1951 Cherokee
1941 Clay
1929 Craven
1922 Cumberland
1927 Davie
1959 Duplin
1931 Franklin
1953 Graham
1954 Haywood
1943 Henderson
1964 Iredell
1948 Jackson
1938 Jones
1933 Lee
1927 Lenoir
1914 Lincoln
1929 Macon
1956 Macon
1942 Madison
1952 Mitchell
1930 Montgomery
1921 Onslow
1937 Pamlico
1957 Pasquotank
1926 Rockingham
1967 Scotland
1940 Stokes
1937 Surry
1947 Swain
1948 Transylvania
1974 Transylvania
1970 Wake
1942 Warren
1932 Washington
1958 Watauga
1974 Wayne
1925 Wilson
1962 Yadkin
1952 Yancey

NORTH DAKOTA

1944 Billings
1974 Burleigh
1971 La Moure and Parts of James River Valley
1914 Lamoure
1942 McKenzie
1951 Morton
1964 Sargent
1968 Stark
1918 Traill
1966 Tri-County Area
1972 Walsh
1970 Wells

OHIO

1938 Adams
1965 Allen
1973 Astabula Area
1938 Athens
1930 Brown
1927 Butler
1971 Champaign
1958 Clark
1923 Clermont
1962 Clinton
1968 Columbiana
1969 Delaware
1971 Erie
1960 Fairfield
1974 Fayette
1955 Huron
1925 Lake
1938 Licking
1939 Logan
1943 Lucas
1971 Mahoning
1916 Marion
1916 Miami
1974 Monroe
1925 Muskingum
1928 Ottawa
1960 Paulding
1914 Portage
1969 Preble
1930 Putnam
1967 Ross
1940 Scioto
1913 Stark
1974 Summit
1954 Tuscarawas
1972 Van Wert
1973 Warren
1966 Wood

OKLAHOMA

1965 Adair
1939 Alfalfa
1962 Beaver
1968 Blaine
1973 Caddo
1938 Carter
1970 Cherokee and Delaware
1943 Choctaw
1960 Cimarron
1954 Cleveland
1974 Coal
1967 Comanche
1963 Cotton
1931 Craig
1973 Craig
1959 Creek
1963 Dewey
1966 Ellis
1939 Garfield
1967 Garfield
1931 Grant
1937 Greer
1967 Greer
1960 Harper
1968 Hughes
1961 Jackson
1973 Jefferson
1967 Kay
1962 Kingfisher
1931 Kiowa
1931 LeFlore
1970 Lincoln
1960 Logan
1966 Love
1940 Major
1969 Major
1937 Mayes
1974 McCurtain
1938 McIntosh
1939 Murray
1956 Noble
1952 Okfuskee
1969 Oklahoma
1968 Okmulgee
1964 Ottawa
1959 Pawnee
1937 Pittsburg
1971 Pittsberg
1941 Pontotoc
1973 Pontotoc
1963 Roger Mills
1966 Rogers
1970 Sequoyah
1964 Stephens
1930 Texas
1961 Texas
1930 Tillman
1974 Tillman
1942 Tulsa
1968 Washington
1941 Washita
1950 Woods
1938 Woodward
1963 Woodward

OREGON

1973 Alsea Area
1949 Astoria Area
1954 Baker Area
1920 Benton
1921 Clackamas
1929 Columbia
1970 Curry Area
1958 Deschutes Area
1972 Marion
1966 Prineville Area
1964 Sherman
1973 South Umpqua Area
1964 Tillamook Area
1948 Umatilla Area
1974 Yamhill Area

PENNSYLVANIA

1967 Adams
1939 Armstrong
1970 Berks
1946 Bucks
1962 Carbon
1963 Chester and Delaware
1958 Clarion
1916 Clearfield
1966 Clinton
1967 Columbia
1954 Crawford
1972 Dauphin
1960 Erie
1973 Fayette
1938 Franklin
1969 Fulton
1944 Huntingdon
1931 Indiana
1968 Indiana
1964 Jefferson
1959 Lancaster
1969 Lehigh
1971 Mercer
1967 Montgomery
1955 Montour and Northumberland
1974 Northampton
1969 Pike
1958 Potter
1973 Susquehanna
1929 Tioga
1946 Union
1938 Wayne
1968 Westmoreland
1929 Wyoming
1963 York

PUERTO RICO

1965 Lajas Valley Area
1942 Puerto Rico

RHODE ISLAND

1939 Kent and Washington
1942 Newport and Bristol
1943 Providence

SOUTH CAROLINA

1937 Abbeville
1966 Bamberg
1963 Calhoun
1971 Charleston
1962 Cherokee
1960 Darlington
1931 Dillon
1938 Edgefield
1974 Florence and Sumter
1929 Greenwood
1915 Hampton
1919 Kershaw
1973 Lancaster
1963 Lee
1922 Lexington
1965 Marlboro
1960 Newberry
1963 Oconee
1943 Pickens
1972 Pickens
1916 Richland
1962 Saluda
1968 Spartanburg
1943 Sumter
1928 Williamsburg
1965 York

SOUTH DAKOTA

1971 Bennett
1959 Brookings
1966 Codington
1974 Davison
1922 Grant
1963 Hand
1974 Lake
1921 McCook
1964 Minnehaha
1971 Shannon
1974 Todd
1921 Union
1969 Washabaugh

TENNESSEE

1947 Bedford
1953 Benton
1959 Blount
1958 Bradley
1948 Claiborne
1955 Cocke
1959 Coffee
1950 Cumberland
1955 Decatur
1972 Dekalb
1965 Dyer
1964 Fayette
1958 Franklin
1968 Giles
1948 Grainger
1958 Greene
1904 Greeneville Area
1946 Hamblen
1947 Hamilton
1926 Hardin
1963 Hardin
1960 Henderson
1922 Henry
1958 Henry
1958 Houston
1946 Humphreys
1941 Jefferson
1956 Johnson
1955 Knox
1969 Lake
1959 Lawrence
1946 Lincoln
1961 Loudon
1958 Marion
1959 Maury
1957 McMinn
1974 Meigs
1953 Norris Area
1973 Obion
1953 Perry
1963 Putnam
1948 Rhea
1942 Roane
1968 Robertson
1956 Sevier
1970 Shelby
1953 Sullivan
1967 Warren
1958 Washington
1964 Williamson

TEXAS

1974 Andrews
1965 Armstrong
1963 Bailey
1938 Bee
1966 Bexar
1958 Brazos
1948 Brown
1941 Cameron
1962 Carson
1937 Cass
1974 Castro
1959 Cherokee
1963 Childress
1964 Cochran
1974 Coke
1974 Coleman
1930 Collin
1969 Collin
1973 Collingsworth
1966 Crosby
1960 Dawson
1968 Deaf Smith
1922 Dickens
1970 Dickens
1943 Dimmit
1971 El Paso
1916 Eastland
1964 Ellis
1973 Erath
1932 Falls
1946 Fannin
1966 Fisher
1964 Foard
1960 Fort Bend
1929 Frio
1965 Gaines
1930 Galveston
1966 Gray
1974 Hale
1967 Hall
1960 Hansford
1932 Hardeman
1972 Hardeman
1922 Harris
1961 Haskell
1974 Hemphill
1925 Hidalgo
1965 Hockley
1969 Howard
1939 Hunt
1965 Jefferson
1974 Jim Hogg
1940 Kaufman
1974 Kent
1967 Kinney
1962 Lamb
1959 Lynn
1942 Maverick
1974 McCulloch
1958 McLennan
1967 Menard
1928 Midland
1973 Midland
1969 Mitchell
1972 Montgomery
1965 Nuecess
1973 Ochiltree
1930 Polk
1930 Randall
1970 Randall
1919 Reconnaissance Northwest
1922 Reconnaissance West Central
1919 Red River
1970 Runnels
1916 San Saba
1931 Scurry
1973 Scurry
1972 Starr
1968 Sutton
1974 Terrell
1962 Terry
1974 Travis
1927 Victoria
1974 Wharton
1932 Wheeler
1962 Wilbarger
1926 Willacy
1938 Williamson
1964 Yoakum
1940 Zavala

UTAH

1920 Ashley Valley Area
1960 Beryl-Enterprise Area
1970 Garbon-Emery Area
1968 Davis-Weber
1919 Delta Area
1959 East Millard Area
1939 Price Area
1958 Richfield Area
1959 Roosevelt-Duchesne Area
1946 Salt Lake Area
1974 Salt Lake Area
1962 San Juan Utah Area
1921 Uinta River Valley Area
1972 Utah (Central Part)
1942 Virgin River Valley Area

VERMONT

1971 Addison
1974 Chittenden
1959 Grand Isle
1937 Reconnaissance of Entire State

VIRGIN ISLANDS

1932 Reconnaissance of St. Croix Island
1970 Virgin Islands of the U.S.

VIRGINIA

1917 Accomack and Northampton
1940 Albemarie
1937 Augusta
1954 Bland
1967 Carroll
1974 Charlotte
1952 Culpeper
1963 Fairfax

1956 Fauquier
1958 Fluvanna
1930 Grayson
1938 Halifax
1914 Henrico
1941 Isle of Wight
1953 Lee
1960 Loudoun
1962 Mathews
1956 Mecklenburg
1932 Nansemond
1959 Norfolk
1963 Northumberland and Lancaster
1960 Nottoway
1971 Orange
1958 Prince Edward
1945 Princess Anne
1961 Rappahannock
1931 Rockbridge
1945 Russell
1951 Scott
1948 Smyth
1937 Southampton
1974 Stafford and King George
1948 Tazewell
1945 Washington
1954 Wise

WASHINGTON

1967 Adams
1971 Benton
1951 Clallam
1972 Clark
1974 Columbia
1974 Cowlitz
1958 Island
1952 King
1974 King
1945 Kittitas
1939 Kitsap
1954 Lewis
1960 Mason
1955 Pierce
1962 San Juan
1960 Skagit
1956 Skamania
1947 Snohomish
1917 Spokane
1968 Spokane
1958 Thurston
1964 Walla Walla
1953 Whatcom
1958 Yakima

WEST VIRGINIA

1968 Barbour
1966 Berkeley
1918 Braxton and Clay
1922 Grant and Mineral
1941 Greenbrier
1972 Greenbrier
1927 Hampshire
1930 Hardy and Pendleton
1961 Jackson and Mason
1973 Jefferson
1960 Marshall
1923 Mercer
1925 Monroe
1965 Monroe
1938 Pocahontas
1959 Preston
1914 Raleigh
1931 Randolph
1921 Tucker
1967 Tucker-Randolph
1970 Wood and Wirt

WISCONSIN

1958 Barron
1961 Bayfield
1974 Brown
1962 Buffalo
1925 Calumet
1930 Crawford
1961 Crawford
1916 Door
1973 Fond duLac
1961 Grant
1922 Green
1974 Green
1922 Green Lake
1962 Iowa
1970 Kenosha and Racine
1918 Jackson
1960 La Crosse
1966 Lafayette
1926 Manitowoc
1916 Milwaukee
1971 Milwaukee and Waukesha
1970 Ozaukee
1964 Pepin
1923 Pierce
1968 Pierce
1915 Portage
1915 Reconnaissance South Part of North Central
1959 Richland
1974 Rock
1925 Sauk
1927 Trempealeau
1928 Vernon
1969 Vernon
1971 Walworth
1971 Washington
1921 Washington and Ozaukee
1917 Waupaca
1927 Winnebago
1915 Wood

WYOMING

1928 Basin Area
1955 Campbell
1971 Goshen (Southern Part)
1939 Johnson
1939 Sheridan
1969 Teton Area Idaho-Wyoming
1940 Uinta

APPENDIX D

STATE GEOLOGIC MAPS*

*Courtesy of Geologic Inquiries Group, U.S. Geological Survey, Dept. of the Interior, Washington, D.C., 1971

State geologic maps are published by a variety of organizations. Most are issued by the state geologic agencies, some by the U.S. Geological Survey, and a few by professional societies or state universities.

This list brings together information about the state maps, including important features of the maps, their availability, and instructions for ordering.

Free or inexpensive small-scale maps have been issued for many states. These are cited following the references to larger-scale maps for each state.

TO ORDER MAPS:

Map prices and the addresses of publishers other than the U.S. Geological Survey are given in the text. Sales tax applies to orders sent to addresses within the States.

Maps published by the U.S. Geological Survey can be ordered from:

Distribution Center
U.S. Geological Survey
1200 South Eads Street
Arlington, Virginia 22202

or

Distribution Center
U.S. Geological Survey
Federal Center
Denver, Colorado 80225

GEOLOGIC MAPS

Alabama

Geology, 1926. Scale, 1:500,000. Color. Cross sections. Order from Geological Survey of Alabama, P.O. Drawer O, University, Ala. 35486. Out Of Print. $2.50.

Alaska

Geology, 1957 (compiled in 1954). Scale, 1:2,500,000. Color. Index map of sources. Order from U.S. Geological Survey. $2.00.

Surficial geology, 1964 (compiled in 1960). Scale, 1:1,584,000. Publ. as U.S. Geol. Survey Map I-357. Color. List of sources. Order from U.S. Geological Survey. $2.50.

Arizona

Geology, 1969. Scale, 1:500,000. Color. Photo index map of sources. Highways. Topography. Order from U.S. Geological Survey. $2.50.

Geology, 1958–1961. Scale, 1:375,000. Color. Topography. Index map of sources.

County maps:

Coconino, 1960. $1.00
Cochise, 1959. $.75
Gila, 1959. $.75
Graham and
  Greenlee, 1958. $.75
Maricopa, 1957. $.75
Mohave, 1959. $.75
Navajo and
  Apache, 1960. $1.00
Pima and
  Santa Cruz, 1960. $.75
Pinal, 1959. $.75
Yavapai, 1958. $.75
Yuma, 1960. $.75

Order from Arizona Bureau of Mines, University of Arizona, Tucson, Ariz. 85721. Sold separately;

as a set—$8.75; or as part of folio including text, cross sections, mineral resources maps, outcrop map, and index to mining districts—$24 to State residents, $25 to others.

Geology and highways, 1967. Scale, 1:1,000,000. Color. Text includes physiographic and geologic history maps. Order from Arizona Geological Society, P.O. Box 4489, University Station, Tucson, Ariz. 85717. $1.75.

Arkansas

Geology, 1929. Scale, 1:500,000. Publ. by Arkansas Geol. Survey. Color. Cross sections. 1949 highways. Out of Print.

California

Geology, 1958–1969. Scale, 1:250,000. Color. Topography. Highways. Index map of sources for each sheet. The complete set consists of 27 map sheets and a legend sheet. Most sheets cover 1° latitude and 2° longitude.

Atlas sheets:

Alturas, 1958
Bakersfield, 1965
Chico, 1962
Death Valley, 1958
Fresno, 1966
Kingman, 1961
Long Beach, 1962
Los Angeles, 1969
Mariposa, 1967
Needles, 1964
Redding, 1962
Sacramento, 1965
Salton Sea, 1967
San Bernardino, 1969
San Diego-El Centro, 1962
San Francisco, 1961
San Jose, 1966
San Luis Obispo, 1959
Santa Ana, 1966
Santa Cruz, 1959
Santa Maria, 1959
Santa Rosa, 1963
Trona, 1963
Ukiah, 1960
Walker Lake, 1963
Weed, 1964
Westwood, 1960

Order from Division of Mines and Geology, Dept. of Conservation, 1416 9th St., Room 1323, Sacramento, Calif. 95814. $1.50 per sheet plus 55% sales tax.

Generalized geology, 1966. Scale, 1:2,500,000. 17 × 21½ inches. U.S. Geol. Survey Map I-512. Color. Order from U.S. Geological Survey. $0.25.

Colorado

Geology, 1935. Scale, 1:500,000. Publ. by U.S. Geol. Survey in cooperation with Colorado State Geol. Survey and Colorado Metal Mining Fund. Color. Index map of sources. Order from U.S. Geological Survey. $2.50.

Connecticut

Geology, 1956. Scale, about 1:250,000. Publ. by Connecticut Geol. and Natural History Survey. Black and white; geologic contacts outlined in red. Highways. Index map of sources. Order from State Librarian, Distribution and Exchange Agent, State Library, Hartford, Conn. 06115. $0.25 plus 3½% sales tax. Map also included *in* Bull. 84, "Explanatory Text to the Geologic Map of Connecticut,"—$1.00 plus tax.

Glacial geology, 1930. Scale, 1:125,000. Publ. *in* Connecticut Geol. and Natural History Survey Bull. 47, "The Glacial Geology of Connecticut." Color. Topography. Out of Print.

Delaware

Generalized geology, 1966. Scale, 1:300,000. Color. Cross section. Text. Main highways. Order from Delaware Geological Survey, University of Delaware, Newark, Del. 19711. Free. Restricted distribution.

District of Columbia

Geology, 1958. Scale, 1:62,500. Publ. *in* U.S. Geol. Survey Water-Supply Paper 1776, "Geology and Groundwater Resources of Washington, D.C., and Vicinity," 1964. Color. Highways. Order from Superintendent of Documents, Government Printing Office, Washington, D.C. 20402. $2.25.

Sedimentary formations, 1947. Scale, 1:31,680. Color. Topography. Highways. Order from U.S. Geological Survey. $1.00.

Geology, 1901. Scale, 1:62,500. Publ. as U.S. Geol. Survey Geol. Folio 70. Color. Topography. Cross sections. Text. Out of Print.

Florida

Geology, 1965. Scale, about 1:2,000,000. Publ. as Florida Geol. Survey Map Series No. 18. Color. Cross section. Text. Map also included in State Special Pub. No. 5, "Summary of the geology of Florida and a guidebook to the classic exposures," 1964 (rev. ed.). Order from Florida Bureau of Geology, P.O. Drawer 631, Tallahassee, Fla. 32302. Map, single copies free; Special Pub. 5, $4.00.

Georgia

Geology, 1939. Scale, 1:500,000. Color. Cross sections. Out of Print. Blue-line print available for $2.00 from Georgia Dept. of Mines, Mining and Geology, 18 Hunter St., S.W., Atlanta, Ga. 30334.

Generalized geology, 1968. Scale, about 1:1,600,000. Folder, 8½ × 11 inches. Includes brief description of the rocks and life of the geologic ages shown. Color. Order from address given above. Single copies free.

Hawaii

Geology and ground water. 1939–1960. Scales, 1:62,500 and 1:125,000. Publ. by Hawaii Dept. of Public Lands, Div. of Hydrography. Color. Topography. Cross sections. Maps of one or more islands were published each in a separate bulletin; most bulletins are out of print. Due to limited supply, distribution of in-print bulletins restricted to libraries and educational institutions.

Out of print

Oahu (1:62,500) Bull. 2, 1939, "Geologic Map and Guide to the Island of Oahu, Hawaii."

Lanai (1:62,500) and Kahoolawe (1:125,000). Bull. 6, 1940, "Geology and Ground-water Resources of the Islands of Lanai and Kahoolawe, Hawaii."

Maui (1:62,500) Bull. 7, 1942, "Geology and Ground-water Resources of the Island of Maui, Hawaii."

Hawaii (1:125,000) Bull. 9, 1946, "Geology and Ground-water Resources of the Island of Hawaii."

Free. Restricted distribution

Molokai (1:62,500) Bull. 11, 1947, "Geology and Ground-water Resources of the Island of Molokai, Hawaii."

Niihau (1:62,500) Bull. 12, 1947, "Geology and Ground-water Resources of the Island of Niihau, Hawaii."

Kauai (1:62,500) Bull. 13, 1960, "Geology and Ground-water Resources of the Island of Kauai, Hawaii."

Order from U.S. Geological Survey, Rm. 332, First Insurance Bldg., 1100 Ward Ave., Honolulu, Hawaii 96814.

Black and white geologic maps of all eight major islands, at scales ranging from about 1:200,000 to 1:1,000,000, are published in "Geology of the State of Hawaii," by Harold F. Stearns, 1966. Available from Pacific Books, Publishers, Palo Alto, Calif. 94302, or through bookstores. $8.50. (Maps taken from Hawaii Div. of Hydrography Bull. 8, "Geology of the Hawaiian Islands," by the same author, 1946.)

Idaho

Geology, 1947. Scale, 1:500,000. Publ. by U.S. Geol. Survey in cooperation with Idaho Bur. of Mines and Geology. Color. Cross sections. Index map of sources. Order from U.S. Geological Survey or Idaho Bureau of Mines and Geology, University of Idaho, Moscow, Idaho 83843. $2.50 plus 3% sales tax.

Illinois

Geology, 1967. Scale, 1:500,000. Color. Columnar and cross sections. Inset maps show structure, subsurface, and surficial geology. Order from Illinois State Geological Survey, 121 Natural Resources Bldg., University of Illinois, Urbana, Ill. 61801. Single copies free.

Surface geology, 1960. Scale, about 1:1,000,000. Publ. by Univ. of Illinois *in* Bull., v. 61, no. 16, and Engineering Experiment Sta. Circ. No. 80, "Surface Deposits of Illinois," 1963. Black and white. Order from University of Illinois, Engineering Experiment Station, Urbana, Ill. 61801. $2.00 plus 5% sales tax.

Generalized geology, 1961. Scale, 1:1,000,000. 17 × 29 inches. Color. Physiographic base. Available separately or *in* State Geol. Survey Educational Series 7. Black and white edition, 8 × 11 inches, also available. Order from State address given in first reference. Single copies free.

Glacial geology, 1970. Scale, 1:2,500,000. 8½ × 11 inches. Color. Available separately or *in* State Geol. Survey Bulletin 94, "Pleistocene Stratigraphy of Illinois," 1970. Order from State address given in first reference. Single copies free.

Indiana

Geology (bedrock and surficial), 1961. Scale, 1:250,000. Publ. by Indiana Geol. Survey in cooperation with neighboring States. Surficial deposits in color; bedrock shown by gray-line patterns. Topography. Highways. When complete will be 8 sheets, each covering 1° latitude and 2° longitude.

Completed sheets:

Indianapolis, 1961
Danville, 1967
Vincennes, 1971
Chicago, 1971

Order from Publications Section, Geological Survey, Dept. of Natural Resources, 611 N. Walnut Grove, Bloomington, Ind. 47401. $1.25 per sheet, plus 10% for packaging and mailing.

Generalized bedrock geology, 1956, and *Glacial geology,* 1958. Scale, 1:1,000,000. 12½ × 26 inches. Publ. as Indiana Geol. Survey Atlas of Mineral Resources of Indiana, Map Nos. 9 and 10, respectively. Color. Cross sections on bedrock map. Order from above address. $0.50 each, plus $0.10 postage.

Generalized bedrock, 1970. Scale, 1:1,800,000. 9 × 14 inches. Publ. as Indiana Geol. Survey Misc. Map No. 16. Color. Cross section. Order from State address given in first reference. 25 cents.

Iowa

Geology, 1969. Scale, 1:500,000. Publ. by Iowa Geol. Survey. Color. Highways. Order from Iowa Geological Survey, Geological Survey Bldg., Iowa City, Iowa 52240. $1.20 folded; $1.50 rolled.

Glacial drift, 1914. Scale, 1:1,000,000. Publ. by Iowa Geol. Survey. Color. Shows distribution of drift sheets throughout State. Out of Print.

Generalized bedrock geology, 1969 (color), and *Glacial geology,* 1955 (black and white). Scale, about 1:2,400,000. 8½ × 11 inches. Order from State address given in first reference. Bedrock map, $0.15; glacial map free.

Kansas

Geology, 1964. Scale, 1:500,000. Publ. as Kansas Geol. Survey Map M-1. Color. Highways. Order from State Geological Survey, University of Kansas, Lawrence, Kansas 66044. $2.50.

Kentucky

Geology, 1954. Scale, 1:1,000,000. Rock units shown by red pattern; black and white cross sections. Order from Kentucky Geological Survey, 307 Mineral Industries Bldg., 120 Graham Ave., Lexington, Ky. 40506. $0.50 plus 5% sales tax. 10% charge for 4th class mail; 30% charge for 1st class.

Louisiana

Geology, 1959. Scale, about 1:1,267,000. Publ. by Louisiana Geol. Survey. Color. Out of Print. (Black and white ed., $0.50.) Color map included with State Water Resources Bull. 1, "Groundwater in Louisiana." Order from Louisiana Geological Survey, P.O. Box G, University Station, Baton Rouge, La. 70803. $1.50.

Maine

Geology, 1967. Scale, 1:500,000. Color. Topography. Highways. Order from State Geologist, Department of Economic Development, State House, Augusta, Maine 04330. $3.00 plus 4½% sales tax.

Maryland

Geology, 1968. Scale, 1:250,000. Color. Cross section. Highways. Source map. Order from Maryland Geological Survey, Latrobe Hall, Johns Hopkins University, Baltimore, Md. 21218. $3.00 plus 4% sales tax.

Generalized geology, 1967. Scale, about 1:1,600,000. 8½ × 11 inches. Color. Order from Board of Natural Resources, State Office Bldg., Annapolis, Maryland 21404. $0.05.

Massachusetts

Geology, 1916. Scale, 1:250,000. Publ. *in* U.S. Geol. Survey Bull. 597, "Geology of Massachusetts and Rhode Island," 1917. Color. Topography. Covers Mass. and R.I. Out of Print.

Michigan

Bedrock geology, 1936. Scale, 1:500,000. Publ. as Michigan Geol. Survey Division Publ. 39. Color. Cross sections. Two sheets. Order Southern Peninsula sheet from Publications Room, Dept. of Natural Resources, Stevens T. Mason Bldg., Lansing, Mich. 48926. $1.00 plus 4% sales tax. Northern Peninsula sheet. Out of Print.

Glacial deposits, 1955 and 1957. Scale, 1:500,000. Publ. as Michigan Geol. Survey Publ. 49. Color. Topography. Index map of principal sources. Two sheets. Order from address given above. Southern Peninsula sheet (1955), $1.00; Northern Peninsula sheet (1957), $1.00 plus 4% sales tax.

Generalized bedrock geology, 1968. Scale, 1:2,500,000. 11 × 15 inches. Publ. as Michigan Geol. Survey Division Small Scale Map 2. Color. Cross sections. Order from address given above. Single copies free.

Minnesota

Bedrock geology, 1966. Scale, 1:250,000. Publ. by Minnesota Geol. Survey. Color. Topography. Highways. When complete will have 11 sheets.

Completed:
St. Paul sheet, 1966
Order from Minnesota Geological Survey, University of Minnesota, Minneapolis, Minn. 55455. $2.00 plus 3% sales tax.

Geology, 1932. Scale, 1:500,000. Publ. by Minnesota Geol. Survey. Color. Cross sections. Two sheets. Order from University of Minnesota Press, Minneapolis, Minn. 55455. $1.75 plus $0.15 postage and 3% sales tax.

Bedrock geology, 1970. Scale, 1:1,000,000. Publ. as Minnesota Geol. Survey Map M-14. Color. Highways. Order from Minnesota Geological Survey, University of Minnesota, Minneapolis, Minn. 55455. $3.00 plus 3% sales tax.

Mississippi

Geology, 1969. Scale, 1:500,000. Color. Order from Mississippi Geological Survey, P.O. Box 4915, Jackson, Miss. 39216. $4.00 plus $0.50 postage and handling.

Missouri

Geology, 1961. Scale, 1:500,000. Color. Highways. Order from Div. of Geological Survey and Water Resources, P.O. Box 250, Rolla, Mo. 65401. $1.00.

Montana

Geology, 1955. Scale, 1:500,000. Publ. by U.S. Geol. Survey in cooperation with Montana Bur. Mines and Geology. Color. Index map of sources. Order from U.S. Geological Survey. $3.50.

Glacial deposits, 1961. Scale, 1:500,000. Publ. as U.S. Geol. Survey Map I-327. Color. List of sources. Shows glacial deposits of Montana east of Rocky Mountains (approx. NE half of State). Order from U.S. Geological Survey. $1.50.

Nebraska

Geology, 1969. Scale, 1:1,000,000. Color. Order from Conservation and Survey Div., 113 Nebraska Hall, University of Nebraska, Lincoln, Nebr. 68508. $0.75 plus 2% sales tax.

Nevada

Geology, 1957- Scale, 1:200,000 and 1:250,000. Publ. by U.S. Geol. Survey and Nevada Bur. of Mines. When complete will be 14 county geologic maps. Mineral Co. in color, others in black and white except as noted. All sheets except Elko Co. show topography. List of sources.

U.S. Geol. Survey sheets (scale, 1:200,000). Order from U.S. Geol. Survey.

Douglas, Lyon, Ormsby, and Washoe Cos. (Superseded, except for Washoe Co., by Bull. 75 listed below)	MF-80	1961	$0.50
Clark Co. (Superseded by Bull. 62 listed below)	MF-138	1958	Out of print
Eureka Co. (Superseded by Bull. 64 listed below)	MF-178	1961	0.50
Lincoln Co.—Pink patterns used for some formations.	MF-206	1961	1.00
Humboldt Co.	MF-236	1961	1.00
Esmeralda Co.—Red patterns used for some formations	MF-298	1965	0.50

Nevada Bur. of Mines (scale, 1:250,000). Order from Nevada Bureau of Mines, University of Nevada, Reno, Nev. 89507.

Elko Co., *in* Bull. 54, "Geology and Mineral Resources of Elko Co., Nevada." (Geologic map of Elko County available separately for $ 1.00.)	1957	Out of print
Mineral Co., *in* Bull. 58, "Geology and Mineral Deposits of Mineral Co. Nevada." (Geologic map and mineral resources map set sold separately for $1.00.)	1961	$ 2.50
Clark Co., *in* Bull. 62, "Geology and mineral resources of Clark Co., Nevada."	1965	$ 5.00
Eureka Co., *in* Bull. 64, "Geology and mineral resources of Eureka Co., Nevada."	1967	$ 5.00
Lyon, Douglas, and Ormsby Cos., *in* Bull. 75, "Geology and mineral deposits of Lyon, Douglas, and Ormsby Cos., Nevada." (Geologic and mineral resources map set sold separately for $2.00.)	1969	$ 3.00
Washoe and Storey Cos., *in* Bull. 70, "Geology and mineral deposits of Washoe and Storey Cos., Nevada." (Geologic and mineral resources map set sold separately for $3.00)	1970	$ 6.00

Lincoln Co., *in* Bull. 73, "Geology and mineral deposits of Lincoln Co., Nevada."	1970 $10.00

Note: Mineral Co., *in* Bull. 58, reprinted 1970. New price is $4.00 (Geologic map and mineral resources map no longer available separately.)

New Hampshire

Bedrock geology, 1955. Scale, 1:250,000. Publ. by U.S. Geol. Survey in cooperation with New Hampshire State Planning and Develop. Comm. and Div. of Geol. Sciences, Harvard Univ. Color. Cross sections. Index map of sources. Topography. Structural map inset. Order from U.S.

Geological Survey or Div. of Economic Development, State House Annex, Concord, N.H. 03301. $2.00.
Map also included with "The Geology of New Hampshire," Part II, 1956, Publ. by New Hampshire State Planning and Develop. Comm. Order from State address given above. $3.50.

Surficial geology, 1950. Scale, 1:250,000. Publ. by New Hampshire State Planning and Develop. Comm. Color. Topography. Order from State address given above. $1.00.
Map also included with "The Geology of New Hampshire" Part I, 1951. Order from State address given above. $1.50.

New Jersey

Geology, 1950. Scale, 1:250,000. Publ. as New Jersey Bur. of Geology Atlas Sheet No. 40. Color. Cross sections. Highways. Out of Print.

Generalized geology, 1959. Scale, about 1:1,000,000. Folder, 8 × 12 inches. Includes brief description of the rock units and their major resources. Color. Order from New Jersey Bureau of Geology and Topography, P.O. Box 1889, Trenton, N.J. 08625. $0.25.

New Mexico

Geology, 1965. Scale, 1:500,000. Color. Index map of sources. Highways. 1 map sheet; 1 legend sheet. Order from U.S. Geological Survey. $2.50.

Geology and highways, 1961. Scale, about 1:1,635,000. Publ. by New Mexico Geol. Soc. Color. Order from New Mexico Geological Society, Campus Station, Socorro, N. Mex. 87801. $1.25 rolled; $1.00 folded, plus $0.25 mailing charge.

Generalized geology, 1969. On postcard, 6 × 8 inches. Color. Cross sections along south and east borders. Order from State Bureau of Mines and Mineral Resources, New Mexico Institute of Mining and Technology, Campus Station, Socorro, N. Mex. 87801. $0.10.

New York

Geology, 1961. Scale, 1:250,000. Publ. as New York State Museum and Science Service Map and Chart Series No. 5, 1962. Color. Highways. Topography. 5 sheets—Niagara, Finger Lakes, Hudson-Mohawk, Lower Hudson (N.Y. City), and Adirondack; text contains cross sections and physiographic diagram. Out of Print.

Generalized geology, 1966. Scale, 1:1,000,000. 22½ × 35 inches. Publ. *in* New York State Museum Educational Leaflet 20, "Geology of New York—A Short Account." Color. Physiographic base. Order from New York State Museum and Science Service, State Education Dept., Albany, N.Y. 12224. $1.00; map only, $0.50; payable to New York State Education Dept. Black and white map, 8½ × 11 inches, available free.

North Carolina

Geology, 1958. Scale, 1:500,000. Publ. *in* North Carolina Div. of Mineral Res. Bull. 71, "Explanatory Text to the Geologic Map of North Carolina." Color. Highways. Order from Div. of Mineral Resources, Dept. of Conservation and Development, P.O. Box 2719, Raleigh, N.C. 27602. Map and text., $3.00 plus 3% sales tax.

Generalized geology, 1966. Scale, 1:2,500,000. 8½ × 14 inches. Color. Order from address given above. $0.25 plus tax.

North Dakota

Bedrock geology, 1969. Scale, 1:1,000,000. Publ. by North Dakota Geological Survey as Misc. Map 10. Color. Inset map showing generalized glacial geology. Order from North Dakota Geological Survey, University Station, Grand Forks, N.D. 58202. $0.50 over the counter; $1.00 by mail.

Glacial deposits, 1963. Scale, 1:500,000. Publ. by U.S. Geol. Survey in cooperation with the North Dakota Geol. Survey, as Map I-331. List of sources. Order from U.S. Geological Survey. $1.50.

Generalized bedrock geology, 1965, and *Glacial geology,* 1965. Scale, about 1:2,000,000.

8½ × 11 inches. Publ. as North Dakota Geol. Survey Misc. Maps 8 and 9, respectively. Order from State address given in first reference. $0.15 each.

Ohio

Geology, 1920 (Glacial boundary added, 1947). Scale, 1:500,000. Color. Highways. Cross sections. Bedrock map with line showing limits of glaciation. Order from Ohio Div. of Geological Survey, 1207 Grandview Ave., Columbus, Ohio 43212. $2.00 plus 4% sales tax and 10% postage charge.

Glacial deposits, 1961. Scale, 1:500,000. Publ. as U.S. Geol. Survey Map I-316. Color. Index map of sources. Order from U.S. Geological Survey. $0.75.

Generalized bedrock geology, 1966(?), and *Glacial geology,* 1966. Scale, 1:2,000,000. 8½ × 11 inches. Color. Order from State address given in first reference. $0.02 each.

Oklahoma

Geology, 1954. Scale, 1:500,000. Publ. by U.S. Geol. Survey in cooperation with the Oklahoma Geol. Survey. Color. Index map of sources. Order from U.S. Geological Survey or Oklahoma Geological Survey, University of Oklahoma, Norman, Okla. 73069. $2.50.

Geology, 1969. Scale, 1:250,000. Publ. by Oklahoma Geol. Survey in its Hydrologic Atlas series, sheet 1 of 4. Color. Topography. Highways. A total of nine Atlases are planned, each covering 1° latitude by 2° longitude (all the State except the Panhandle).
Completed:

Map HA-1, Fort Smith quadrangle, 1969
Map HA-2, Tulsa quadrangle, 1971.

Order from Oklahoma Geological Survey, University of Oklahoma, Norman, Okla. 73069. $3.00 per set, folded in envelope.

Oregon

Geology, 1961. Scale, 1:500,000. Publ. as U.S. Geol. Survey Map I-325. Oregon west of 121st meridian (approx. west half of State). Color. Cross sections. Tectonic map inset. Index map of sources. Order from U.S. Geological Survey. $2.00.

Generalized geology, 1969. Scale, 1:2,000,000. 14 × 26 inches. Publ. as U.S. Geol. Survey Map I-595. Color. Order from U.S. Geological Survey. $0.25.

Pennsylvania

Geology, 1960. Scale, 1:250,000. Publ. by Pennsylvania Geol. Survey. Color. Highways. Cross sections. Index map of sources. Order from Bureau of Publications, P.O. Box 1365, Harrisburg, Pa. 17125. $3.75 plus 6% sales tax.

Generalized geology, 1962. Scale, about 1:1,900,000. 9 × 12 inches. Publ. by Pennsylvania Geol. Survey as Map No. 7 and *in* Educational Series No. 4, "Pennsylvania Geology Summarized." Color. Order from Pennsylvania Geological Survey, Main Capitol Annex, Harrisburg, Pa. 17120. Single copies free.

Puerto Rico

Geology, 1964. Scale, 1:240,000. Publ. as U.S. Geol. Survey Map I-392. Black and white. Highways. List of sources. Order from U.S. Geological Survey. $0.50.

Rhode Island

Bedrock geology, 1971. Scale, 1:125,000. Publ. *in* U.S. Geol. Survey Bull. 1295, "Bedrock geology of Rhode Island." Color. Topography. Highways. Index of sources. Order from Superintendent of Documents, Government Printing Office, Washington, D.C. 20402. $1.25.
Map available separately. From U.S. Geological Survey. $1.00. (Order: Plate I of Bull. 1295, "Geologic Map of Rhode Island," 1971.)

South Carolina

Geology (Cretaceous and Tertiary), 1936. Scale, 1:500,000. Publ. *in* U.S. Geol. Survey Bull. 867, "Geology of the Coastal Plain of South Carolina." Color. Includes most of Coastal Plain. Out of Print.

Geology (Crystalline rocks), 1965. Scale, 1:250,000. Publ. by U.S. Geol. Survey as Map I-413 and *in* Bull. 1183, "The crystalline rocks of South Carolina," 1965. Color. Topography. Highways. List of sources. Shows mainly Piedmont Province (approx. NW half of State). Order from U.S. Geological Survey. Map, $1.25; Bull. 1183, Out of Print.

Generalized geology, 1961(?). Scale, about 1:1,800,000. 12 × 17 inches. Publ. as South Carolina Div. Geology Map MS-6. Black and white. Order from Division of Geology, State Development Board, P.O. Box 927, Columbia, S.C. 29202. $0.05.

South Dakota

Geology, 1953. Scale, about 1:570,240. Publ. by South Dakota Geol. Survey. Color. Out of Print.
Geology, 1951. Scale, 1:500,000. Publ. by U.S. Geol. Survey. Color. Order from U.S. Geological Survey. $1.50.

Generalized geology, 1964. Scale, about 1:2,500,000. 8½ × 11 inches. Publ. as South Dakota Geol. Survey Educational Series No. 2. Color. Order from South Dakota Geological Survey, Science Center, University, Vermillion, S.D. 57069. $0.25 plus $0.10 for mailing.

Tennessee

Geology, 1966. Scale, 1:250,000. Color or black and white. Index map of sources. Diagrammatic sections. Main Highways. 4 sheets. Order from Div. of Geology, Dept. of Conservation, G-5 State Office Bldg., Nashville, Tenn. 37219. $2.00 a sheet or $8.00 a set, plus 4½% sales tax.

Generalized geology, 1970. Scale, about 1:2,900,000. 8½ × 11 inches. 5 × 7 postcard map also available. Publ. by Tennessee Div. of Geology. Color. Order from Div. of Geology, Dept. of Conservation, G-5 State Office Bldg., Nashville, Tenn. 37219. Single copies free; additional copies 15 cents each (10 cents for postcard).

Texas

Geology, 1965. Scale, 1:250,000. Publ. by Texas Bur. of Econ. Geology. Color. Topography. Highways. When complete, will be 37-sheet atlas, each sheet covering 1° latitude and 2° longitude.

Completed sheets:
- Amarillo, 1969
- Beaumont, 1968
- Houston, 1968
- Lubbock, 1967
- Palestine, 1968
- Perryton, 1970
- Plainview, 1968
- Sherman, 1967
- Texarkana, 1966
- Tyler, 1965
- Van Horn-El Paso, 1968
- Waco, 1970

Order from Bureau of Economic Geology, University of Texas, University Station, Box X, Austin, Texas 78712. $2.50 per sheet, plus 4% sales tax.

Geology and highways, 1963. Scale, 1:1,800,000. Color. Geology taken from 1933 Texas map; highways from 1958 Texas Highway Comm. map. Order from Dallas Geological Society, P.O. Box 2867, Dallas, Texas 75221. $1.00 folded; $1.25 rolled, plus 4% sales tax.
Map also included in 1966 printing of Univ. of Texas Bull. 3232, v. 1, "The Geology of Texas," 1933. Order from Bureau of Economic Geology, University Station, Box X, Austin, Tex. 78712. Bull. with map, $3.00 plus 4% sales tax.

Geology, 1937. Scale, 1:500,000. Publ. by U.S. Geol. Survey. Color. Cross sections. Explanatory notes. Physiographic and tectonic map insets. Out of Print.

Generalized geology, 1966(?). On postcard, 5 × 7 inches. Color. Order from Bureau of Economic Geology, address given above. $0.20.

Utah

Geology, 1961–1964. Scale, 1:250,000. Color. Topography. Index map of sources. Highways. In 4 sheets: NE ¼, 1961; NW ¼, 1962; SW ¼, 1963; SE ¼, 1964. Order from Utah Geological and Mineralogical Survey, 103 Utah Geological Survey Bldg., University of Utah, Salt Lake City, Utah 84112. $4.00 per sheet plus 10% mailing charge.

Generalized geology, 1966. Scale, about 1:2,500,000. 8½ × 11 inches. Publ. as Utah Geol. and Mineralog. Survey Map 23. Color. Correlation chart. Order from address given above. $0.15 plus $0.10 minimum mailing charge.

Vermont

Bedrock geology, 1961. Scale, 1:250,000. Publ. by Vermont Geol. Survey. Color. Topography. Highways. Cross sections. Index map of sources. Tectonic and metamorphic map insets. Order from Vermont Department of Libraries, Montpelier, Vt. 05602. $4.00 plus 3% sales tax.

Surficial geology, 1970. Scale, 1:250,000. Publ. by Vermont Geol. Survey. Color. Topography. Highways. Cross sections. Index map of sources. Inset maps showing drift sheets and ice directions, and maximum extent of glacial and post-glacial lakes. Order from address given above. $4.00 plus 3% sales tax.

Generalized bedrock geology, 1970. Scale,

1:1,000,000. 8½ × 11 inches. Color. Major structural features. Order from address given above. $0.15 (In lots of 100 for schools, $0.10 each).

Virginia

Geology, 1963. Scale, 1:500,000. Publ. by Virginia Div. of Mineral Res. Color. List of sources. Order from Virginia Div. of Mineral Resources, Box 3667, Charlottesville, Va. 22903. Rolled, $5.00; folded, $4.00 plus 4% sales tax.

Generalized geology, 1964. Scale, about 1:2,850,000. 8½ × 11 inches. Color. Economic minerals listed for each geologic period. Order from address given above. $0.15.

Washington

Geology, 1961. Scale, 1:500,000. Color. Highways. Index map of sources. Order from Div. of Mines and Geology, Dept. of Natural Resources, P.O. Box 168, Olympia, Wash. 98501. $3.00 (Specify rolled or folded).

Generalized geology, 1969. Scale, 1:2,000,000. 11 × 20 inches. Publ. as U.S. Geol. Survey Map I-583. Color. Order from U.S. Geological Survey. $0.25.

West Virginia

Geology, 1968. Scale, 1:250,000. Color. Columnar and cross sections. Tectonic, basement, and systemic map insets. Two sheets, each 42 × 58 inches. Order from West Virginia Geological and Economic Survey, P.O. Box 879, Morgantown, W. Va. 26505. Folded, $6.50 plus $0.20 sales tax; rolled, $7.50 plus $0.23 sales tax.

Generalized geology, 1968. Scale, 1:2,000,000. 8½ × 11 inches. Color. Order from address given above. $0.15 plus $0.01 sales tax.

Wisconsin

Geology, 1949. Scale, 1:1,000,000. Color. Order from Wisconsin Geological and Natural History Survey, 1815 University Ave., University of Wisconsin, Madison, Wis. 53706. $0.50.

Generalized bedrock geology, 1963 (color), and *Glacial geology,* 1956 (two colors). Bedrock map (scale, about 1:3,300,000) includes cross section and text. Glacial map (scale, about 1:2,750,000) includes text. Both 8½ × 11 inches. Order from address given above. Single copies free.

Wyoming

Geology, 1955. Scale, 1:500,000. Publ. by U.S. Geol. Survey in cooperation with Geol. Survey of Wyoming. Color. Highways. Index map of sources. Out of Print.

APPENDIX E

ADDITIONAL MAPS

The maps listed are those the author believes to be of good reference as supporting material. They are listed under the agencies from which they are available. Prices vary but are generally $5.00 per set unless otherwise indicated.

American Geographical Society, New York, New York. "Potential Natural Vegetation of the Conterminous United States," by Kuchler, A. W., 1:3,168,000. 1964.

Donald J. Belcher and Associates, Inc. Ithaca, New York. "Origin and Distribution of United States Soils," by Belcher, D. J., Gregg, L. E., Jenkins, D. S., and Woods, K. B., 1946.

Geological Society of America, New York, N.Y. "Glacial Map of the United States, East of the Rocky Mountains," 1st. Edition, 1:1,750,000, 1960.

United States Geological Survey, Washington, D.C.; Denver, Colorado. "National Atlas," 765 maps and 335 pages of the physical and cultural resources of the United States, 1971, $100.00.

"One Hundred Topographic Maps Illustrating Specified Physiographic Features," Assembled by Upton, W. B., Jr., U.S. Geological Survey, Scale varies, 1955, $40.00.

"Geology Map of North America" (2 sections), by the U.S. Geological Survey, 1:5,000,000, 1965.

"Tectonic Map of North America" (2 sections), by the U.S. Geological Survey, 1:5,000,000, 1965.

"Geologic Map of the United States" (4 sections), by the U.S. Geological Survey, 1960.

"Basement Rock Map of the United States," by Bayley, R. W., Muehlberger, W. R., U.S. Geological Survey, 1:2,500,000, 1968.

"Map Showing Relation of Land and Submarine Topography: Nova Scotia to Florida" (3 sections), by Vchupi, E., U.S. Geological Survey, 1960.

"Water Resources Development in the United States," by the U.S. Geological Survey.

Index

About the Author

Douglas S. Way is Associate Professor of Landscape Architecture at Harvard University where he specializes in land planning, applied geomorphology, and techniques of terrain analysis and remote sensing. He studied civil engineering at Iowa State, received his Bachelor's degree in landscape architecture from the University of Wisconsin, a Master's degree in landscape architecture from Harvard University, and is a Ph.D. candidate at Clark University. He has been a visiting lecturer at over thirty universities and has been a Distinguished Visiting Professor at Utah State University.

The author's firm, Douglas Way Associates, specializes in air photo terrain analysis and land planning and has worked on many national and international development projects. He has also been involved in the preparation of a number of environmental impact statements, serves as a consultant to prominant public and private groups, and has been called as an expert witness in several environmental litigations.

Professor Way is an Associate of the American Society of Landscape Architects; a Member of American Society for Testing of Materials and the Committee of Soil and Rock for Engineering Purposes; and is a Member of the American Society of Photogrammetry and Chairman of the Engineering Applications Committee.